ENVIRONMENTAL GEOLOGY

An Earth System Science Approach

▸ ABOUT THE COVER

The landscape in the Palouse area of Washington state is an example of the complex interactions among Earth's systems. Sediments in the Palouse Hills in the foreground were deposited by the wind during the Pleistocene epoch and have been evolving under atmospheric and other Earth processes ever since. The green fields of cultivated winter wheat and barley thrive on rain from occasional thunderstorms like the one foreshadowed by the clouds gathering above. Water seeping down through the vegetation and soil adds to or replenishes the groundwater that permeates the basalts hidden underneath the fields. The mountains in the background, over the Idaho state line, are made up of metamorphic rocks and granite intrusions, and owe their existence to Earth's interior heat, which melts rock, rearranges Earth's continents, and creates mountains. They mark the edge of the North American craton—the ancient, stable core of the continent, where continental drift attached the younger landmasses to the west. In all, the cover image reminds us how Earth's major systems—atmosphere, hydrosphere, biosphere, pedosphere, lithosphere, and deep interior—interact within the whole Earth system in ways that constantly shape and reshape our environment.

The Palouse area of Washington was named for a tribe of American Indians. The tribe is extinct, but the name lives on in the hills, rivers, and town of Palouse, and in the Appaloosa breed of horses. (Kunio Owaki/ The Stock Market)

An Earth System Science Approach

DOROTHY J. MERRITTS
Franklin and Marshall College

ANDREW DE WET
Franklin and Marshall College

KIRSTEN MENKING
Vassar College

W. H. FREEMAN AND COMPANY
NEW YORK

We dedicate this book to those who are curious about Earth, to those who marvel at its processes, and to those who seek to understand the interconnectedness of its many environments. The complexity of the whole Earth system, from its empyrean reaches to clayey depths, reminds us how much we still have to understand.

ACQUISITIONS EDITOR: Holly Hodder
DEVELOPMENT EDITORS: Elizabeth Zayatz and Susan Seuling
PROJECT EDITOR: Georgia Lee Hadler
COVER AND TEXT DESIGNER: Vicki Tomaselli
ILLUSTRATIONS: Network Graphics and Tomo Narashima
PHOTO RESEARCHERS: Kathy Bendo and Debra P. Hershkowitz
PRODUCTION COORDINATOR: Julia De Rosa
COMPOSITION: Electronic Publishing Center and Progressive Information Technologies
MANUFACTURING: Von Hoffman Press
MARKETING MANAGER: Nicole Folchetti

Library of Congress Cataloguing-in-Publication Data
Merritts, Dorothy
Environmental geology: an earth system science approach
Dorothy Merritts, Andrew de Wet, Kirsten Menking.
p. cm.
Includes bibliographical references and index.
ISBN 0-7167-2834-6
1. Environmental geology. I. De Wet, Andrew. II. Menking, Kirsten.
III. Title.
QE38.M47 1997
550—dc21
97-21337
CIP

Printed in the United States of America

First printing, 1997

Brief Contents

Contents

8 The Groundwater System 232

9 The Atmospheric System 260

Boxed Features

Case Study

Geologist's Toolbox

Global and Environmental Change

Cycling Examples

Preface

Many profound changes in the Earth sciences during the 20th century take geoscientists into the 21st century with a new perspective on how Earth works and how geology should be taught. We are among a growing number of geologists who are convinced that an integrated, Earth system, interdisciplinary approach to environmental geosciences provides essential insights into the workings of the whole Earth and is crucial to the development of scientific literacy. Our primary goal is to guide students toward a personal understanding of Earth's varied environments, Earth as a system, and the local and global ramifications of natural events and human actions.

Our own introductory course in environmental geosciences from an Earth system perspective is original but not unique. Increasingly, geology and geosciences departments are offering courses with titles such as Environmental Geosystems, Environmental Earth Sciences, and Environment and the Earth System. An integrated Earth system perspective is advocated by the American Geophysical Union, the Keck Undergraduate Geology Consortium, the National Science Foundation, and the National Research Council. The involvement of lead author Dorothy Merritts in incorporating an Earth system perspective into the college curriculum for nonscience as well as science students led to her appointments to committees supported by these organizations to help develop publications recommending this approach and demonstrating its use in the undergraduate science curriculum. In developing our course, however, we found no textbook that consistently weaves together basic concepts of Earth system science and environmental geosciences. To compensate, we created, tested in class, and refined a set of course materials so comprehensive that they evolved, with the urging and advice of many colleagues, into this book.

We discovered that our colleagues in the environmental geosciences wanted in a textbook many of the same things we wanted: chapters on scientific and systems thinking; on individual systems such as soils, surface water, groundwater, atmosphere, oceans, and energy; and on interpreting and predicting environmental change. These topics are treated more fully in this environmental geology textbook than in others at the introductory level. At the same time, we were careful to select and cover the fundamentals of physical geology that are essential to understanding the Earth system and contemporary environmental and resource issues.

Content and Organization

This book moves from general concepts to individual systems to global environmental change. Part I, "Fundamental Concepts," is a set of three chapters: The first grounds the reader in scientific thinking; the second presents the concept of systems in general and the Earth system in particular, with emphasis on budgets, cycling, and residence time; and the third considers the meaning of environmental change over geologic time.

In our own teaching, we find that focusing on the evolution, processes, resources, hazards, waste problems, and policy issues for one environmental system at a time is preferable to saving all discussion of environmental issues for the end of the course. Therefore, Parts II and III concentrate on individual systems and their relationships to one another. Part II, "Solid Earth Systems," consists of three chapters covering lithosphere and pedosphere systems and their effect on the environment. These chapters examine plate tectonics and rock-forming processes, solid Earth materials, the resources and hazards of the lithosphere, and soils and soil-forming processes. Part III, "Fluid Earth Systems," contains four chapters, on surface water, groundwater, atmosphere, and ocean.

The content of the chapters in Parts II and III is organized into sections on processes, resources, and environmental hazards, wastes, and policy. Chapter 7, "The Surface Water System," for example, begins with the processes that distribute surface water and shape streams and rivers, lakes and wetlands. Next comes an examination of the many types of resources provided by surface water, including riverine, coastal, and glacial wetland ecosystems. The chapter ends with an examination of the hazards and waste problems associated with surface water, including flooding, drought, and pollution, and a discussion of relevant environmental laws and regulations.

Part IV, "Energy and Change in Earth Systems," considers the driving mechanisms of all processes and change on Earth, with three chapters: energy as a system; causes and geological evidence of environmental change; and significant events in the evolution and probable future of Earth's environment.

The relationship of the biosphere to the other Earth systems is invoked in every chapter. The chapter on the pedosphere (Chapter 6), for example, describes the biological processes that contribute to soil formation and illustrates the relationship of the biosphere to the nitrogen cycle. The chapter on causes of environmental change (Chapter 12) discusses the role of biological processes in shaping Earth's atmosphere and lithosphere over geologic time. The chapter on lithosphere hazards and resources (Chapter 5) considers the biospheric costs and benefits of volcanism and the devastating ecological effects of environmentally irresponsible mining practices and the role of the biosphere in the copper cycle.

Pedagogical Approach

We are keenly aware of the need to present scientific information in a way that captures the interest of students who are taking environmental geology as their first and perhaps their only college science course. Throughout all chapters, we emphasize issues relevant to everyday experience and use frequent examples and case studies. Explanations are written clearly and with a minimum of technical jargon; they are enhanced by vividly rendered and carefully labeled diagrams of structures and processes, flow charts, and maps; selected tables and graphs; and impressive photographs.

Chapter Openers

Each chapter begins with a case study and photograph relating the subject of the chapter to Earth systems and environmental issues. For example, Chapter 2, "Dynamic Earth Systems," begins with a photograph of Mount Pinatubo erupting and an introductory paragraph about the worldwide effects of such an event on Earth's interconnected systems, and within the chapter an opening case links subsequent "volcano weather" events with the Pinatubo eruption. A chapter outline gives the student both a preview of the content and a review device, and a list of goals establishes a context for the chapter.

CHAPTER 2

Dynamic Earth Systems

After 600 years of quiet, Mount Pinatubo—a volcanic mountain in the Philippines—erupted in June 1991. Its plume of ash and gas affected global climate for several years.

The Concept of Systems
- Types of Systems
- Dynamic Systems That Tend Toward a Steady State

The Planetary Evolution of Earth
- Origin of the Universe and Our Solar System
- Differentiation of Earth

Earth's Environmental Systems
- The Lithosphere
- The Pedosphere
- The Hydrosphere
- The Atmosphere
- The Biosphere

Earth's Energy System
- States of Energy
- Sources of Energy
- Energy Budget of Earth
- Human Consumption of Energy

Feedback Links Among Earth Systems
- Positive and Negative Feedback
- Evidence of Global Feedback

Change, Cycles, and Earth System Dynamics
- The Rock Cycle
- The Hydrologic Cycle

On our amazing Earth a natural event at one location can have complex and far-reaching effects. When a large volcano such as Mount Pinatubo erupts, the effects are felt around the world for years, from the movements of Earth's crust that cause a volcanic eruption, to the debris that spreads through the atmosphere and circles the globe, and to the ensuing changes in climates in different parts of the world. Such interconnectedness is the result of linked entities of matter and energy known as Earth systems. To understand Earth, its processes, and the effects of those processes on us, we need to understand these smaller systems that make up the whole Earth system. Consequently, in this chapter we will:

- Examine the concept of systems and why it is a powerful tool for understanding how Earth works.
- Briefly review Earth's planetary evolution and the major systems that make up the planet.
- Identify the forces that drive Earth processes and examine how feedback mechanisms either amplify or regulate them.
- Look at ways in which the rock cycle and the hydrologic cycle circulate matter and energy through the whole Earth system over time.

29

Art Program

Diagrams are vividly rendered and thoroughly labeled, making geologic structures and processes clear and easily remembered.

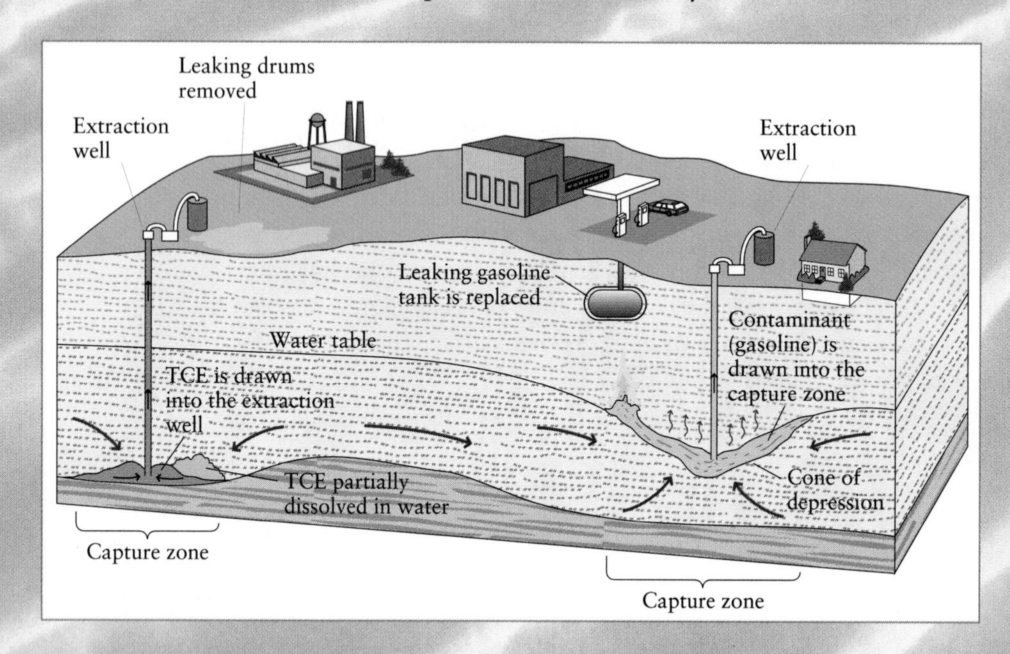

Flow charts are constructed for clarity and visual interest, and they make concepts such as budgets, reservoirs, and fluxes easy to understand.

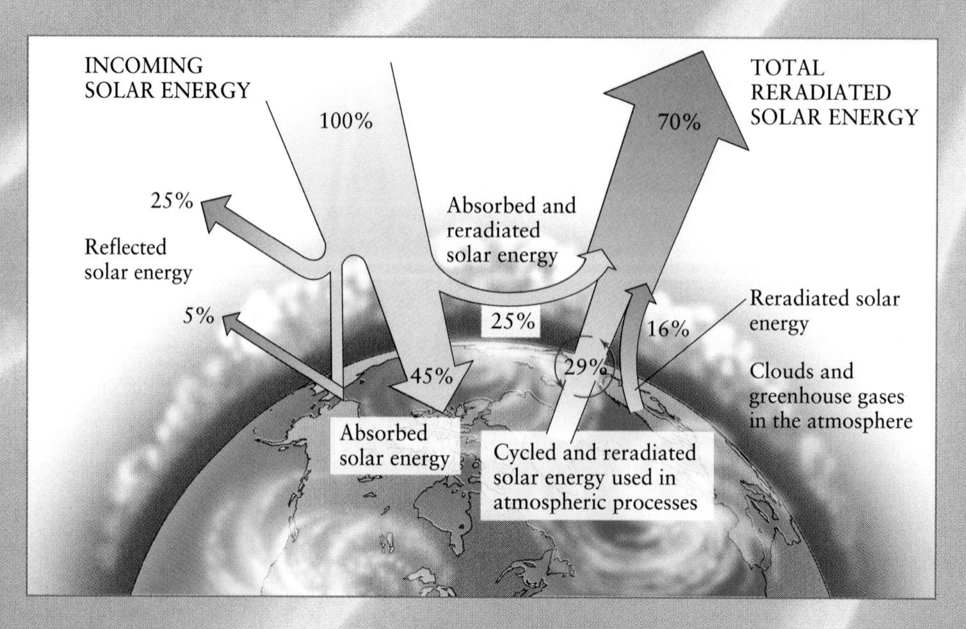

Color photographs and digital images provide striking illustrations, from the minuscule to the majestic, of the material presented in the text discussion.

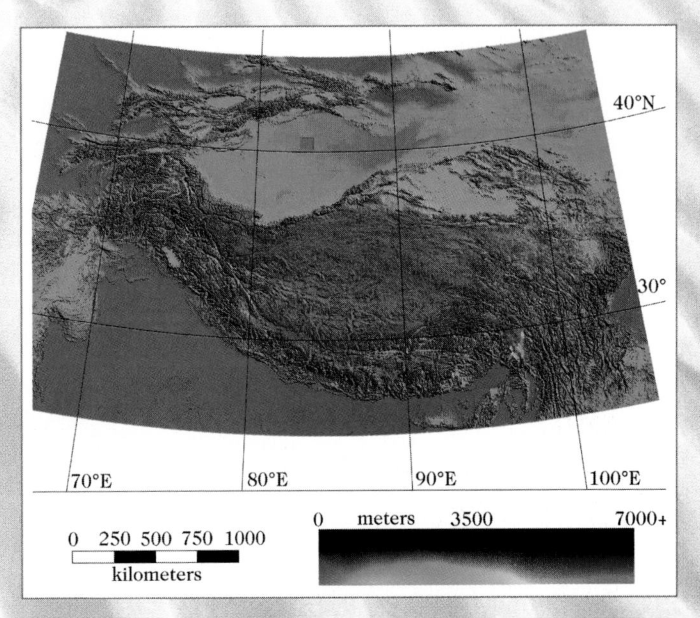

Real-World Examples

Chapters include frequent examples of how geologists are solving environmental problems. Chapter 8, "The Groundwater System," for example, discusses the recent development of in situ bioremediation as a practical means of cleaning gasoline and other organic pollutants from aquifers.

Cycling Examples

Unique to this textbook are illustrated discussions in many chapters of the cycling of matter and energy through different Earth systems.

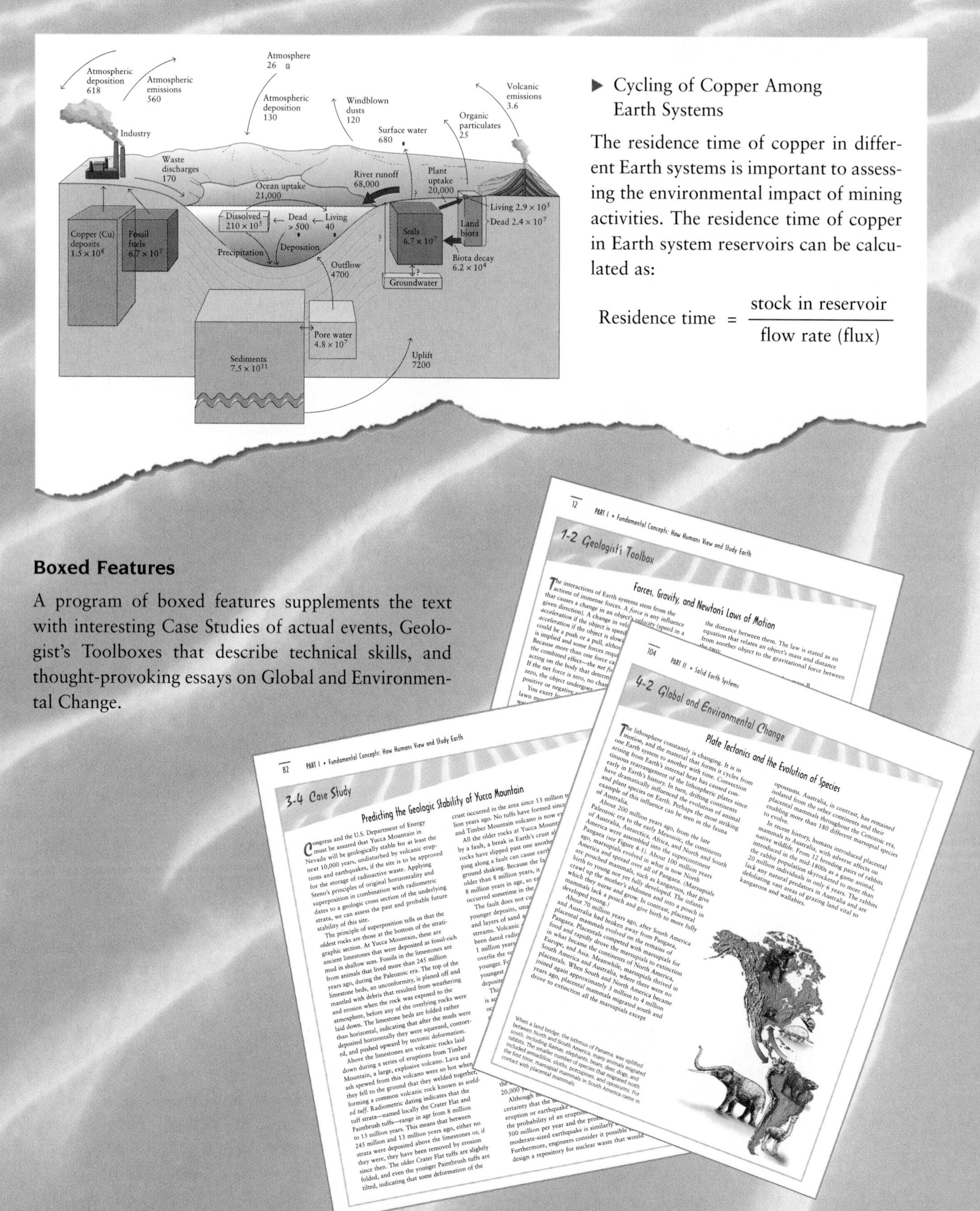

$$\text{Residence time} = \frac{\text{stock in reservoir}}{\text{flow rate (flux)}}$$

Boxed Features

A program of boxed features supplements the text with interesting Case Studies of actual events, Geologist's Toolboxes that describe technical skills, and thought-provoking essays on Global and Environmental Change.

Section Summaries

A list of the important ideas is provided at the end of each major section within a chapter to encourage students to pause, think about what they have read, and ensure that they understand it before going on to the next section.

- New lithosphere forms along rifts at mid-ocean spreading centers on divergent boundaries, and old lithosphere is destroyed at subduction zones along convergent boundaries. The amounts of lithosphere created and destroyed are equal, keeping the lithosphere system in a steady state.
- Plate boundaries are zones of active rock deformation, faulting, and folding. Faulting occurs at shallow depths (generally less than 10 to 30 km), where rocks are cool and brittle, whereas folding generally occurs at greater depths, where rocks are warmer and more ductile.
- The three basic types of faults are normal, reverse, and strike-slip. Although each occurs at all types of plate boundaries, normal faults are dominant at divergent boundaries, where crust is stretched and pulled apart; reverse faults are dominant at convergent boundaries, where crust is compressed and shortened; and strike-slip faults are dominant at transform plate boundaries, where crustal blocks slip horizontally past each other.

Closing Thoughts

This unique feature is a short essay in which the authors reflect on the chapter and identify significant links to human affairs. Closing Thoughts gives the authors the opportunity to share with students their thoughts about the philosophical and economic implications of an environmental problem and to contrast them with the scientific aspects.

Closing Thoughts

Rocks contain clues to their past: the depths to which they were buried, the temperatures to which they were heated, and the places to which they have traveled. The first step in discovering these clues is identifying the rocks and the minerals within them. An experienced geoscientist can tell much from even small rock exposures. By mapping different rock types and determining their ages, geoscientists have reconstructed much of Earth's long and eventful history. Since the recognition of plate tectonics, a revolutionary concept that forever changed our understanding of Earth, the stories geoscientists discover from rocks are even more fantastic: Rocks from the south pole once were close to those now located in the northern hemisphere, near modern-day Las Vegas; fragments of plates and island arcs once were sutured to the coasts of North America, forming modern-day New Jersey and Alaska. Rocks also contain information about past climates, atmospheres, environments, oceans, and life-forms that existed when the rocks formed. The clues in rocks and the tales they tell also help us to find mineral resources, to understand earthquakes, and to predict volcanic eruptions, as discussed in the next chapter.

Chapter Summary, Key Terms, Questions, Exercises, and Readings

Chapters end with a traditional Summary that complements the earlier end-of-section summaries, a list (with page numbers) of the Key Terms highlighted in boldface type within the chapter, a set of Review Questions that helps students understand and remember important ideas, a set of Thought Questions that encourages students to apply what they have learned to questions not specifically addressed in the chapter, a set of Exercises that encourages students to solve quantitative problems and explore their own environment, and a list of Suggested Readings, accessible to the interested nonscientist, that enhances the major topics of the chapter.

Appendix and Glossary

At the end of the book are five appendixes and a Glossary. Appendixes include the Linnaean classification system, a periodic table of the elements, a conversion table of metric units and their English equivalents, a list of the properties of common minerals, and an annotated list of accessible recommended journals and Web sites. The comprehensive Glossary includes definitions of environmental geosciences terminology that may be unfamiliar to the student.

Supplements

Instructor's Resource Manual

Teaching from a new textbook—especially one that takes a fresh approach—always takes some extra time at the start. This Resource Manual helps the instructor cut the necessary start-up time to a minimum. Instructors who want to design or further develop their own environmental geology courses will find plenty of class-tested materials and tips from the authors' own experience. Each *Resource Manual* contains a special *Instructor's Resource CD-ROM* and many other useful features:

- **Sample Syllabi** suggest how the order of chapters can be varied and how the textbook can be used in courses of differing lengths.
- **Lecture Notes** for each chapter offer ideas for approaching specific topics based on the experiences of our textbook authors.
- **Supplements Ideas** suggest how instructors can coordinate all the items in the supplements package—including the *Web Site* and the *Earth Matters: Environmental Geology CD-ROM*—to get the most out of them.
- **Original Case Studies** provide excellent supplementary material for lectures. Each is directly tied to a chapter in the book and consists of about five paragraphs of text. Some come with original diagrams on transparency masters and each is keyed to a slide in the *Slide Set*.
- **Questions** for each chapter are offered in the *Resource Manual* for short quizzes and to spark class discussions.

Instructor's Resource CD-ROM

Packaged with the *Instructor's Resource Manual,* this CD-ROM offers hundreds of photographs and diagrams from the textbook and other sources in presentation manager software that makes it easy to display images during a lecture. The CD-ROM also offers a selection of articles focusing on environmental geology from *Scientific American* magazine and a collection of Web addresses relating to environmental geology.

Slide Set With Lecture Notes

Geological slides that specifically depict environmental topics are often difficult to find. The *Slide Set* offers instructors 120 images of this type, covering every chapter. Each slide is sharp and crisp, for best projection quality. An accompanying booklet of lecture notes describes each slide in vivid detail and gives a black and white thumbnail representation of the images.

Overhead Transparency Set

A selection of full-color transparencies contains key images from the textbook.

Test Banks

Fifty multiple-choice questions for each chapter are offered in the *Test Bank*. It is available in printed, Windows, and Macintosh formats.

Web Site

Instructors and students alike will find useful materials on the Web site for this book at: www.whfreeman.com/environmentalgeology

Students will profit from an interactive multiple-choice practice quiz for each chapter called Q&A. They will receive valuable feedback on every answer as well as directions, based on their answers, to areas they need to study further. The site also contains a continually updated variety of original content offerings and links to sites on the Web relating to environmental geology.

Earth Matters: Environmental Geology CD-ROM

Jeremy Dunning and Lawrence Onesti, Indiana University, Bloomington

Anyone using *Environmental Geology* will want this new CD-ROM. *Earth Matters* takes an interactive multimedia approach to the topic, with a focus on problem solving and learning by doing. The CD-ROM is comprised of seven modules that cover the broad range of enviromental geology. They are:

- Materials and Processes
- Earthquakes and Volcanoes
- Surface Processes
- Groundwater
- Natural Hazards
- Waste Disposal
- Resources and Sustainability

Accompanying the CD-ROM is a print booklet of lessons drawn from the disc material, which can be used as an enviromental geology lab component.

To receive a copy of *Earth Matters* and all the supplements to *Environmental Geology,* instructors should contact their local W. H. Freeman and Company sales representative.

Acknowledgments

We have enjoyed preparing this book and thinking of those who will read it. But we have come to realize that writing a textbook requires the effort of far more than one or a few authors. This book has benefited from the energy, enthusiasm, and creativity of dozens of people, including some of our own students, our families and colleagues, and many other professors and scientists who helped make this book as accurate, useful, and interesting as possible for its student readers. We are deeply grateful to all these people. Colleagues who provided us with frequent and thorough expert advice include astronomers Dana Bachman and Mike Seeds, chemists Richard Moog and Scott van Arman, and geologists Rus O'Connell, Carol de Wet, Roger Anderson, Fred Graybeal, Jane Woodward, and John Stamatakos. The geology and environmental studies faculty at Franklin and Marshall College contributed generously of their time and expertise. They not only provided substantial amounts of expert advice but also supported our efforts even as work on the book took us away from the activities of our own departments. To these fellow faculty, whom we regard as dear friends and admire as gifted teachers and scientists, we owe special gratitude: Edward Beutner, Leslie Burlingame, David Hawkins, Stan Mertzman, Rob Sternberg, Steve Sylvester, Roger Thomas, and Robert Weibe. Students and research assistants who helped us gather information include Oona O'Connell, Kate Seeds, Venantas Miskinis, Adam Love, Kathy Anspach, Andrew Greene, Jordan Mueller, and Allison Schill.

Tremendous and capable support came from a team of editors, artists, and assistants at W. H. Freeman and Company. We are truly astonished at the level of scientific understanding and appreciation that they brought to the project, helping us to refine our book as each chapter went through many stages to make it as clear as possible. Mike Perman and Jerry Lyons were the first to approach us, and they quickly convinced us that we would have ample and experienced support with their team. We were fortunate to work with Sonia DiVittorio during early phases of the text's preparation, and then enjoyed the chance to work extensively with Holly Hodder, Elizabeth Zayatz, Susan Seuling, Georgia Lee Hadler, Patrick Shriner, Kathy Bendo, and Meredith Nightingale. Through every stage, we were firmly and enthusiastically guided by Mary Shuford, a woman with an amazing ability to manage authors with dreams that go beyond what might actually be possible in a single textbook.

Our editors also helped us garner the intellectual input of reviewers, and this part of the textbook-writing process has been especially rewarding to us. We were very pleased to have the responses of other scientists with the common goal of sharing our enthusiasm and learning with our students. We are most grateful to:

David Alexander
University of Massachusetts, Amherst

Ramon Arrowsmith
Arizona State University

Vic Baker
University of Arizona

Edward C. Beutner
Franklin and Marshall College

Paul R. Bierman
University of Vermont

Arthur L. Bloom
Cornell University

Lawrence W. Braile
Purdue University

Oliver Chadwick
University of California, Santa Barbara

Fred E. Clark
University of Alberta

Jack Donahue
University of Pittsburgh

Jim B. Finley
Shepherd Miller, Inc.

Peter G. Flood
University of New England Australia

Sharon L. Gabel
State University of New York, Oswego

Tom Gardner
Trinity University

Alexander E. Gates
Rutgers University

Ron Harris
West Virginia University

Bruce E. Herbert
Texas A&M University

Peter L. Kresan
University of Arizona

Lawrence Lundgren
University of Rochester

Cathryn A. Manduca
Carleton College

David R. Montgomery
University of Washington

B. Moorman
University of Calgary

Stephen O. Moshier
Wheaton College

Lawrence J. Onesti
Indiana University

Frank J. Pazzaglia
University of New Mexico

Jill S. Schneiderman
Vassar College

Steven C. Semken
Diné College

Dorothy L. Stout
Cypress College

Lisa Esquivel Wells
Vanderbilt University

We also wish to thank our colleagues who took time out of their busy schedules at the October 1996 Geological Society of America meeting in Denver to share their vision for teaching the environmental geoscience course with us and for their reactions to our text. Their input helped us in countless ways. The participants were:

Ramon Arrowsmith
Arizona State University

Cinzia Spencer-Cervato
University of Maine

Mark J. Johnsson
Bryn Mawr College

Peter L. Kresan
University of Arizona

Sharon Locke
Bates College

David T. Long
Michigan State University

Cathryn A. Manduca
Carleton College

Steven C. Semken
Diné College

Jill S. Schneiderman
Vassar College

About the Authors

▶ **Dorothy Merritts** is an environmental geologist at Franklin and Marshall College. She holds an undergraduate degree in geology from Indiana University of Pennsylvania, a master of science degree in engineering geology from Stanford University, and a doctorate in surface processes, soils, and tectonics from the University of Arizona. She has worked on mine reclamation, done research with the U.S. Geological Survey on earthquake hazards in California, and studied groundwater resources, streams, and soil processes. As an expert on flooding, she served from 1994 to 1996 on the National Research Council's Committee on Alluvial Fan Flooding, a committee selected to advise the Federal Emergency Management Agency on flood-hazard issues.

At Franklin and Marshall College, Professor Merritts helped to found the school's first Environmental Studies Program, of which she is chair. She teaches courses in Earth systems and environmental geosciences, groundwater hydrology, geomorphology, and environmental studies. Using field experiences as a primary teaching tool, she takes students on tubing expeditions, rents small airplanes to fly students over sinkholes, leads students into underground mines, and requires every student to keep a field notebook. Her efforts to improve the teaching of science to nonscientists led to her appointment to the Committee on Undergraduate Science Education at the National Research Council in 1993, and she was instrumental in helping to write and develop the committee's influential handbook for science teaching.

She remains involved with efforts to incorporate Earth system science into the college curriculum, and in 1996–1997 served on the Spheres of Influence panel convened by the American Geophysical Union and the National Science Foundation.

▶ **Andrew de Wet** is an environmental geologist at Franklin and Marshall College. He holds an honors degree in geology from the University of Durban, South Africa, and a doctorate from Cambridge University, England. Trained as a petrologist, he has done field work in Greece, the United Kingdom, South Africa, and the United States, and he approaches environmental issues from both a theoretical and a practical perspective. His interest in Earth system science has focused on land use and dynamic-systems modeling. His research integrates remote sensing, geophysical techniques, and geographic information systems with field work and laboratory analysis.

Professor de Wet's capacity for visualizing complex systems has led him to innovations in research and in teaching. He emphasizes the complexity inherent in natural systems, but clarifies the issues using graphic representations

and demonstrations, including the use of Stella computer modeling in discovering links and feedbacks between system components. Professor de Wet was instrumental in creating and refining the illustration program for this textbook. He has published articles on the teaching of geology in the *Journal of Geological Education* and the *Journal of Women and Minorities in Science and Engineering* and is active in promoting the participation of women in science. He is a member of the Geological Society of America, the American Geophysical Union, and the American Society of Photogrammetry and Remote Sensing; he has been affiliated with NASA's programs in remote sensing and geographic information systems.

▸ **Kirsten Menking** is a professor of geology at Vassar College. She earned a bachelor's degree in geology in 1990 from Occidental College, where she conducted research in paleomagnetism, and a Ph.D. in Earth sciences in 1995 at the University of California, Santa Cruz, where she participated in a U.S. Geological Survey Project to study lake sediments and reconstruct cycles of glaciation in the Sierra Nevada. As a visiting professor at Franklin and Marshall College, she taught courses in introductory environmental geosciences, climate change, and landscape evolution.

Professor Menking's postdoctoral research, at the University of New Mexico, focused on the climate record contained within Pleistocene epoch Lake Estancia. Her research interests include using sediments to unravel Earth's history of climatic change, linking paleoclimatic data to atmospheric and hydrologic processes, and analyzing the evolution of landforms in response to climatic and tectonic processes. In teaching, she enjoys relating the material to students' everyday experiences and using simple, hands-on experiments to illustrate geologic principles. She works actively to increase the participation of women and ethnic minorities in the Earth sciences and is a member of the Geological Society of America, American Geophysical Union, and the Association of Women Geoscientists.

Environmental Geology

An Earth System Science Approach

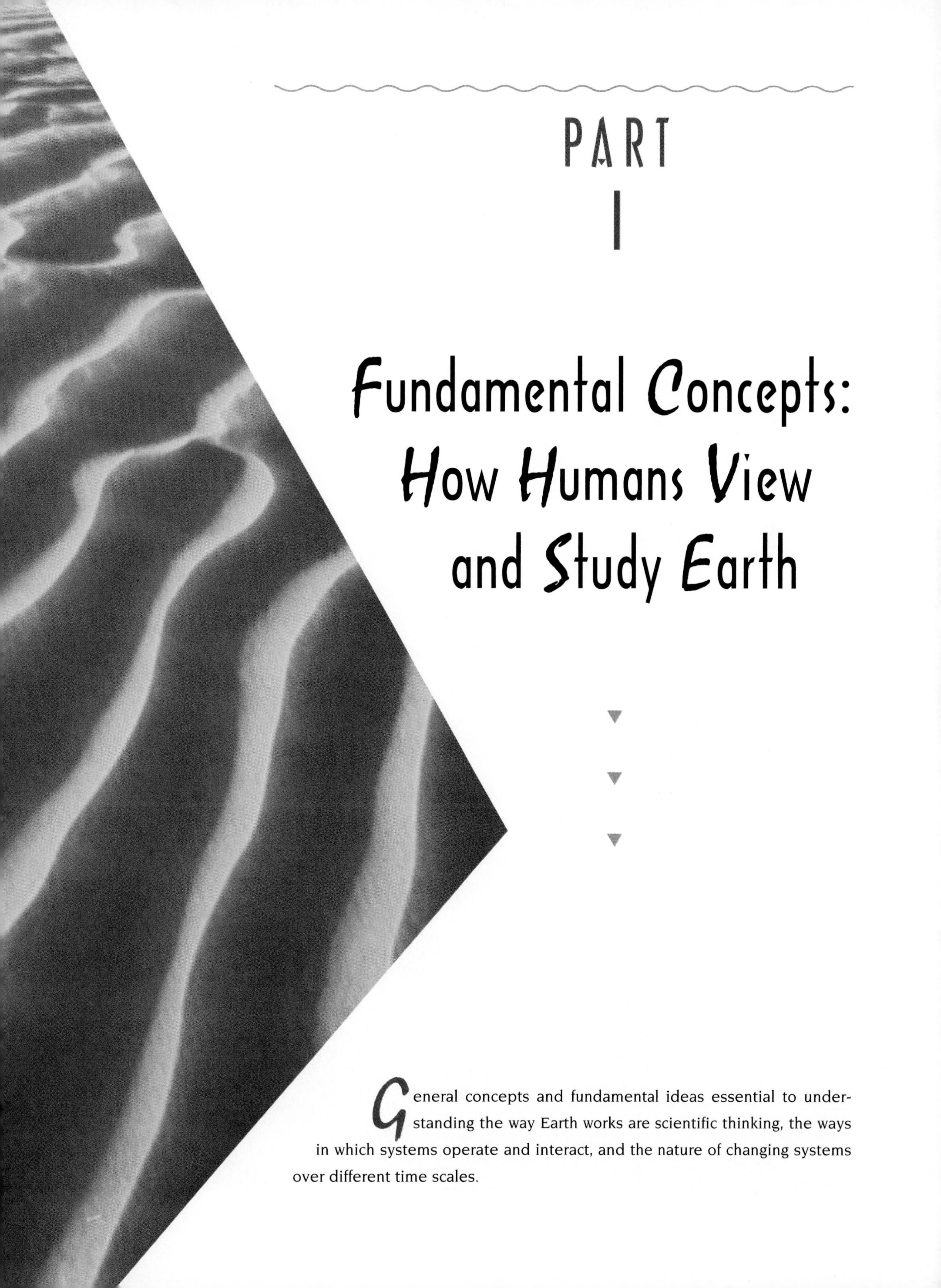

PART I

Fundamental Concepts: How Humans View and Study Earth

General concepts and fundamental ideas essential to understanding the way Earth works are scientific thinking, the ways in which systems operate and interact, and the nature of changing systems over different time scales.

CHAPTER 1

A scientist wearing a protective mask samples the amount of sulfurous gas seeping from Galeras volcano in the Andean mountains, Colombia. Making such measurements requires the scientist to risk his or her life for the sake of understanding volcanism and predicting future eruptions.

Introduction to Environmental Geosciences

What is science, and why is it important to our lives? At one time, natural catastrophes such as earthquakes and volcanoes were regarded as the punishment of angry gods. Voltaire reported after the Lisbon earthquake of 1755 that Portuguese authorities ordered public burnings of heretics to keep "the earth from quaking" again (*Candide*, 1759). As recently as 1963, when the Agung volcano in Bali, Indonesia, erupted during a rare religious ceremony and killed more than 1500 people, some survivors refused international aid because they believed that their suffering was necessary to soothe displeased gods. Scientific thinking, in contrast, is based on the idea that natural phenomena—including earthquakes, volcanoes, climate changes, and floods—have physical causes that can be discovered and used to make predictions about subsequent events.

A scientific understanding of the way our planet works is essential to understanding Earth's varied environments and the conditions that can threaten them. In this chapter, we will examine some fundamental questions about science and environmental geosciences:

- What distinguishes scientific thinking from other types of thought?
- Why is the scientific method the most effective strategy yet devised for learning about physical events?
- What are the limitations of scientific thinking?
- How is science used in the interest of conserving Earth's resources and avoiding environmental risks?
- What is the advantage of Earth system science for understanding the global environment?

In the Earth sciences, the scientists' laboratory is the environment, and the experiments that scientists monitor are the processes of nature itself. Natural phenomena such as volcanic eruptions or global climate change provide the scientist with observations that can be used to determine the causes of those events and, perhaps, to predict subsequent events. Although scientists do much more than make observations—they also perform the important task of developing hypotheses to explain the causes of natural phenomena—the observations form the underpinnings of scientific investigation.

Making the observations to study Earth systems presents scientists with difficult, but exciting, challenges. Observing phenomena such as climate change can require years of effort and sophisticated tools to measure slight variations in the composition of the atmosphere, but it's a relatively safe challenge. Observing the behavior of volcanoes, in contrast, requires that scientists take samples of hot, poisonous gases and newly erupted molten rock (lava), and the investigation has cost some scientists their lives. Stanley Williams is a volcanologist who survived an eruption in 1993 that claimed the lives of all six colleagues who were with him. Team members were sampling gases seeping from the crater of the Galeras volcano in the Andes Mountains in 1993. During this procedure, the volcano began to shake and, within moments, debris was blasted up to several kilometers above the crater, raining down on the scientists and killing all but Williams. Williams was badly injured, and it was several hours before others arrived to drag him to safety, but he lived to help discover a way to predict such sudden eruptions.

Williams and other colleagues found that the composition of the gas seeping from the volcano changed just before the eruption. In particular, the amount of sulfurous gas dropped as cracks in the "plumbing system" that moves heat and fluids to the surface became clogged. Gas pressure built up until the summit of the volcano exploded violently, unfortunately at the time of the scientists' visit.

Williams's discovery is a breakthrough but still requires scientists to sample gases from an active volcano. Other scientists have been working to develop new equipment that can monitor volcanic gases from stations located up to 2 km (kilometers) from a volcano. A spectrometer can detect sulfurous gas from afar by measuring the absorption of light in the atmosphere (Figure 1-1). These scientists found that if the amount of sulfurous gas in the air above a volcano changes, the absorption of light measured by the spectrometer varies. The different groups of scientists have solved two problems from their observations: how to predict explosive volcanic eruptions and how to monitor active volcanoes without risking one's life.

FIGURE 1-1 Researchers can monitor the amount of sulfurous gases seeping from a volcano, as here at Mt. Etna in Italy, from as far away as 2 km with a remote spectrometer (the instrument in the center of the photograph).

Science As a Way of Knowing

Science is both a body of knowledge and a *way of knowing*, that is, a way of understanding the world and its processes. The word is derived from the Latin *scientia*, which means knowledge, but science is much more than a fixed body of facts and information. It is also a dynamic process of intellectual activity driven by human curiosity. The scientific approach to understanding our environment is based on the assumption that natural events have physical, and therefore ultimately knowable, causes. Scientific knowledge is acquired through the systematic and disciplined testing of ideas by careful experimental design, methodical data collection, and objective reasoning and analysis.

Science has been described as "incomparably the most successful" way of knowing, because its means of reasoning and results are objective and testable, and it has been able to predict many events. The same cannot be said for nonscientific ways of knowing. An astrologer could predict your future based on the pattern of stars and planets in the sky at the time of your birth, but there is no observable connection between the stars and a person's life.

Throughout the world water "witches" are hired to help locate underground water supplies with magical sticks and other items that supposedly twitch because of supernatural or otherwise unknown forces when the witch stands above the buried water. Again, there is no observable connection between the underground water and the twitching of the various items.

Astrology, water witching, and similar ways of knowing are labeled *pseudoscience* (false science) because they often are represented as science when in fact they are not. The way to identify pseudosciences is to see how they contradict scientific thinking:

- Their fundamental assumption is that natural events can be controlled by supernatural forces.
- Their findings are biased rather than objective; they are accepted when they support a particular idea but rejected when they do not.
- Their ideas cannot be tested.
- Their results cannot always be duplicated.
- Their predictions are correct no more often than one would expect from chance occurrences.

The Scientific Method

Science is distinguished from nonscientific ways of knowing by its reliance on the **scientific method.** This is an objective way of exploring the natural world, reasoning, and testing ideas that results in an ability to predict the outcome of certain conditions. The particular procedure of investigation may vary, but the scientist must always:

- Observe phenomena (and review relevant scientific work on similar phenomena).
- Develop a **hypothesis,** an explanation of the observations that is based on physical principles.
- Predict the likely outcomes of the hypothesis.
- Design an experiment or field investigation to test predictions based on the hypothesis.
- Compare the real and predicted outcomes of the tests.
- Accept, reject, or modify the hypothesis.

If testing results in acceptance of a hypothesis, scientists will continue to test the idea under different conditions. A hypothesis that has survived repeated testing and is able to accurately explain a wide variety of phenomena becomes viewed as a **theory.** If such a theory explains many types of phenomena and has survived all challenges, it is considered a universal law or a unifying theory (Figure 1-2). The universal law of gravitation helps to explain why rocks fall and planets orbit their suns. The unifying theory of plate tectonics explains why volcanoes erupt and ocean basins are formed and destroyed.

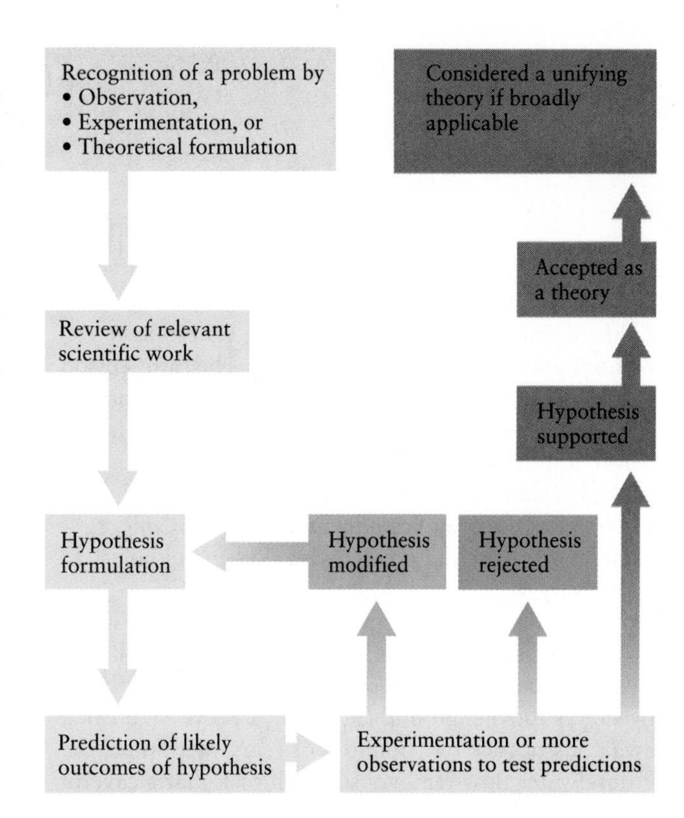

FIGURE 1-2 The scientific method is based on systematic testing of ideas, or hypotheses, developed by scientists to explain observations of natural phenomena.

Approaches to Scientific Reasoning

Scientists tend to approach their research in distinctive ways. With the *empirical approach,* the scientist compares results of tests to predictions and looks for positive or negative correlations. The case study in Box 1-1 describes one scientist's empirical approach to predicting where sinkholes would appear on a farmer's land. Her observation that sinkholes always occur along fractures is an example of a positive correlation.

The *experimental approach* is based on creating a model of whatever is to be studied, subjecting it only to the desired conditions, and analyzing the results. A geologist, for example, might build a physical model of a fracture system with rock and soil to determine if flowing water could enlarge fractures and cause sinkholes. Although the experimental approach is used in geology, it is more common as a laboratory method in the biological, chemical, and physical sciences; proteins, chemicals, and pulleys are far less unwieldy to experiment with than sinkholes and groundwater.

A third approach to scientific reasoning is the *theoretical approach,* in which the scientist uses logic and mathematical proofs to develop general principles for making

1-1 Case Study

Using the Scientific Method to Locate Sinkholes

A farmer in central Pennsylvania was worried about three sinkholes that had opened on his property without warning in a two year period. The third episode occurred while his elderly father was plowing, causing driver and tractor to drop 4 m (meters; about 13 ft) before hitting solid ground; fortunately, the man only broke his shoulder. The farmer urgently requested that a geologist at nearby Franklin and Marshall College examine the land and predict where other sinkholes were likely to occur.

The geologist knew from research published in scientific journals that sinkholes occur along fractures in the solid rock underlying the soil. These breaks are enlarged by slowly moving groundwater that dissolves and carries away bits of rock over millions of years. With time, the groundwater flow removes so much rock that a cavern forms, and its roof becomes thinner and thinner until, in some cases, only a veneer of soil remains. A sinkhole appears when the roof becomes so thin that the soil above it collapses into the cavern. A man on a 4-ton tractor can speed up such collapse by riding onto the thin cavern roof. Sinkholes do not occur above all fractures or all caverns—only those where the roof is very thin or unstable. The bigger the cavern, the bigger the sinkhole.

The job for the scientist was to locate all underground fractures and caverns, determine their size, and, finally, determine the thickness of soil and rock remaining above each. In a sense, she had to be able to see through the ground to what is beyond our normal view. The scientist contacted another researcher, who had spent much of his career studying sinkholes, and learned that measurements of the flow of electricity through the ground can be used to locate subterranean caverns. The resistance of the ground to the flow of electricity depends on the type of rock in the subsurface and whether or not pockets of air and water, such as those in caverns, exist. Electricity can flow more easily through water than through rock, because rock is a poor conductor. The scientist set up electrodes in a gridlike array across the farmer's field and measured electrical resistivity at dozens of sites.

From this experiment emerged a map of the field that identified a prominent band along which electrical resistivity was very low. The scientist deduced that this

Sinkholes can open without apparent warning when the ceilings of underground caverns thin sufficiently to collapse.

and testing predictions. Theoretical scientists tend to rely on mathematical models, often supplemented by computer simulations. This approach is generally most common in physics and chemistry, but it has an important place in biology and Earth sciences as well. For example, scientists have created computer simulations of past global climates that work with projected climatological data to predict global change far into the future (Figure 1-3).

The Link Between Science and Technology

The words *science* and *technology* often are used interchangeably, but they are not at all the same. Technology is the application of scientific knowledge for practical purposes. Computers, satellites, deep-sea submersibles, telescopes, heart monitors, hearing aids, and plumbing all would have been impossible if scientific analysis had not

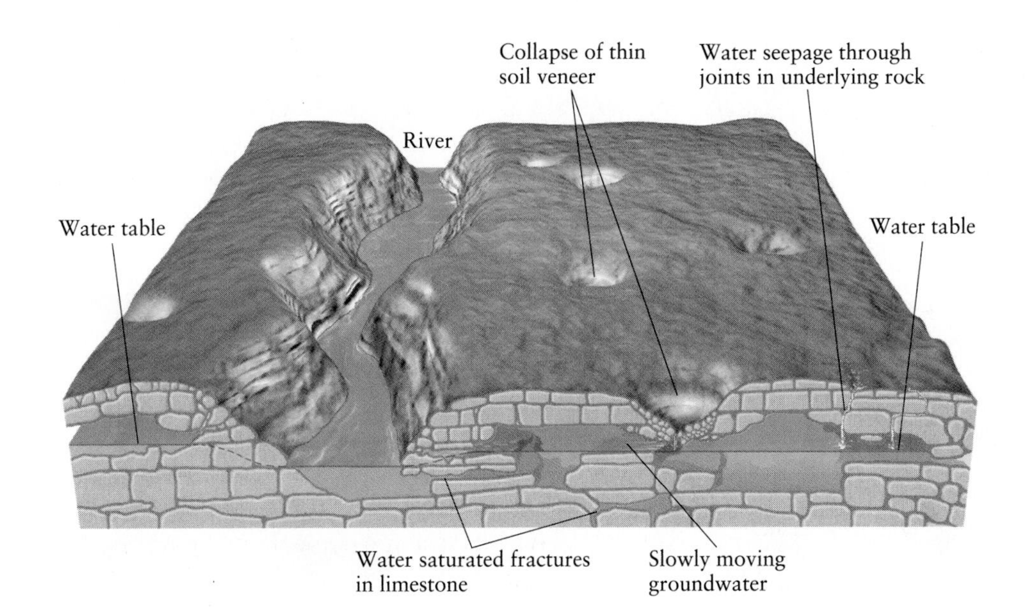

Sinkholes form when underground limestone caverns become hollowed out by water erosion. Limestone dissolves in water, and the dissolved matter is carried away to nearby streams. The ceilings of caverns become thin and ultimately collapse.

was the surface trace of a vertical fracture that had filled with air and moist, weathered bits of rock. Using this map to guide her ground observations, she noted that all three sinkholes occurred roughly along the band, and she found water-loving plants such as willow trees and a small spring from which groundwater emanated to the surface. She then flew over the field in a small airplane to take aerial photos. From these, she identified a dark zone of soil along the same trend. The darkness was due to the moist, clay-rich soil that accumulated along the enlarging fracture. The clay was left behind when other minerals from the rock were dissolved and carried away by the water.

The scientist now knew where underground caverns were, but not their size or which ones had the thinnest or most unstable roofs. To test her hypothesis that the area of low electrical resistivity was a fracture along which sinkholes occur, she drilled into the ground and, indeed, hit a subsurface cavern. The scientist was able to predict where future sinkholes would occur and with further work could predict how large the caverns were and which roofs were most likely to collapse. If other sinkholes were to appear along this fracture, they would further confirm the scientist's hypothesis. If no other sinkholes appear as predicted, the hypothesis may have to be rejected or modified.

been applied in practical ways to develop new technologies. Thomas Edison's (1847–1931) invention of the light bulb, for example, is based on the principles of electricity.

The nonscientist sometimes is skeptical of the value of scientific research that does not seem to have practical merit. Scientists typically respond to this criticism in two ways. First, they argue that the primary aim of science is to achieve greater understanding of natural phenomena and that this goal alone has merit. If research were driven only by the desire to increase human comfort or to win wars, many discoveries that have had an astounding impact on our lives might not have been made. Second, scientists argue that most technological advances result from cumulative advances in science, even though individual bits of scientific work might not be recognized as practical. While Thomas Edison was

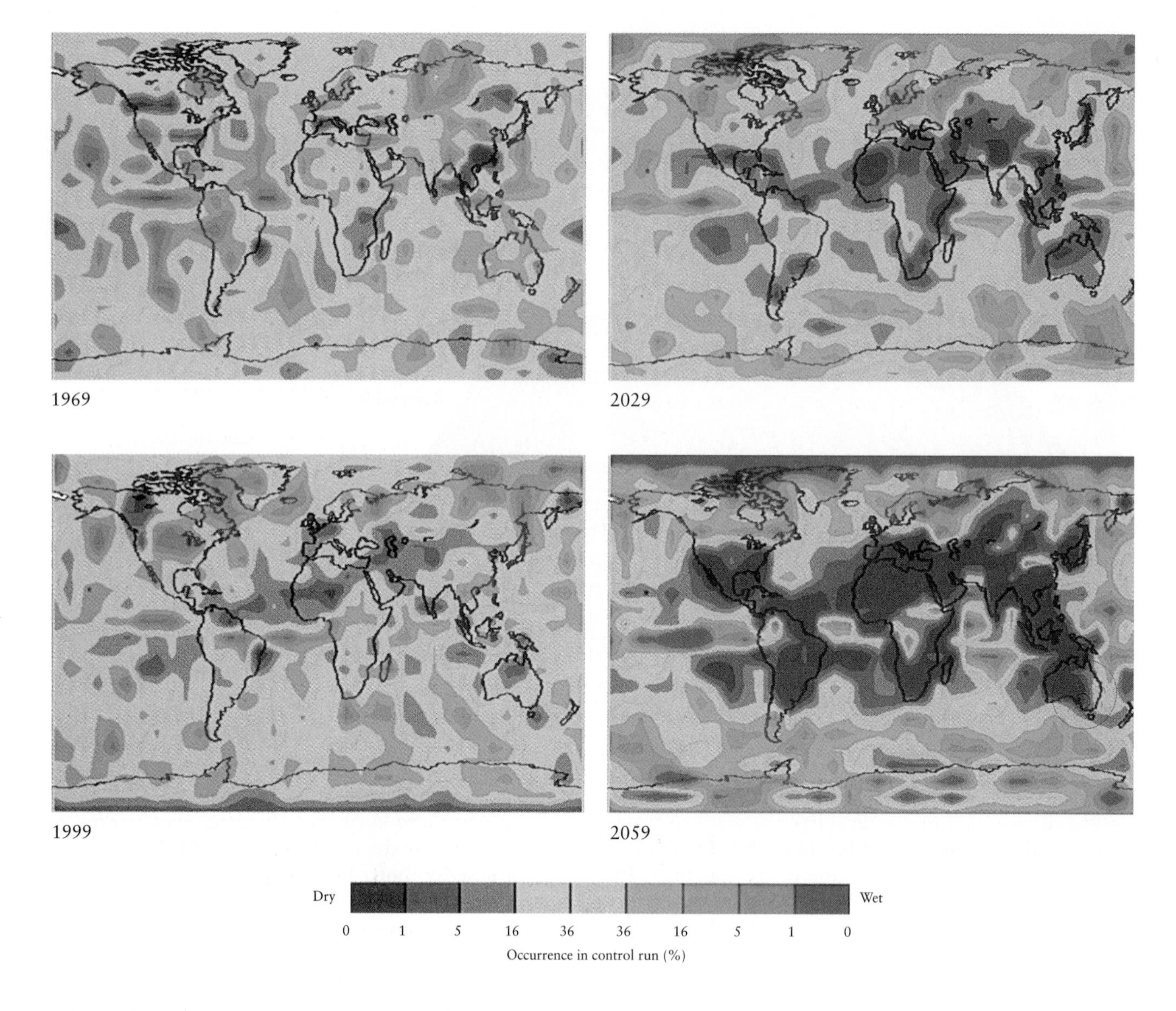

FIGURE 1-3 Because computers can make rapid calculations involving huge amounts of mathematical data, computer models are especially useful for simulating complex natural processes under any number of conditions. Here, a computer climate model predicts future occurrences of summer drought.

exploring the nature of electricity in his New Jersey laboratory in 1879, many critics said his efforts were a waste of time, because they could see little of practical value in electricity. Legend has it that Edison replied: "Well, I don't know what we can do with it, but I bet it will be taxed some day." Within a few years, electric utility companies had appeared. Today, light produced by electricity is so prevalent that all the major cities of the world can be seen from space at night, and sale of electricity now produces billions of dollars in tax revenue every year.

The Limitations of Science

Science has been very successful in its ability to predict some geologic events. Geologists predicted the volcanic eruption of Mount Pinatubo in the Philippines weeks before it happened in 1991, saving thousands of lives (see Chapter 2). Scientists also have predicted successfully the warming of Earth because of atmospheric pollution, the locations of rich mineral and groundwater resources, and the depletion of some resources as a result of over-exploitation.

Although science is not yet able to predict the timing of every natural disaster, it has been successful in identifying where certain types of disasters are likely to occur and in suggesting ways to cope with them. The earthquakes that damaged the northern Los Angeles region in 1994 and Kobe, Japan, in 1995 struck with no apparent warning, causing thousands of deaths and billions of dollars in damages. Still, on the recommendation of scientists newer buildings had been constructed for earthquake resistance and the public had been educated about how to respond during an earthquake. Without these precautions, the devastation might have been even more severe.

Not all predictions by scientists have been correct. In the 1960s a few scientists warned of an impending doomsday before the end of the 20th century due to rapid human population growth, which they referred to as a *population bomb*. Nearly 30 years and 2 billion people later, however, population continues to grow, and a few scientists even argue that Earth can support several billion more people before any major calamity occurs. (We will address the issue of population growth again, later in the chapter.)

Because science has led to many great technological achievements, some people see it as not only the most successful but also the most powerful way of knowing about our world. Yet, science is a way of finding out about nature, and nothing more. It is limited to making sense of the physical world. Powerful in one arena, it is limited in others. For example, science can provide us with the knowledge necessary to predict how much global warming might occur as a result of burning fossil fuels in power plants. But the scientific approach cannot help us make the decision to change our life-styles and habits so that we burn less fossil fuel and are responsible for less impact on global climate. That decision is a matter of human values and ethics. If burning fossil fuels also causes acid rain that can damage—and even destroy—forests, the decision to regulate power plants so as to reduce their emissions is political, not scientific, for it is based on human judgment of how much value trees have to us.

Some fundamental questions in science might be answerable, yet still are unsolved despite the many advances of the past few centuries. An example is the following: "Why, in our experience, do objects fall?" You might automatically think that this question has been answered, and that in fact you know the answer to it—gravity. But what is gravity? Gravity is not a reason for why objects fall, only the scientific name for the phenomenon. Science has many names for things, but a name is not the same as knowledge and understanding. Throughout this book, you should question each new term. Ask yourself what its meaning is, and whether the term is just a name, or if it also provides understanding.

- The scientific way of learning about our world is relatively new in human history, but it has proved more successful than any other in helping us to understand natural phenomena and their causes.
- Of the different approaches in scientific methodology, the observational is used most often in the Earth sciences, because it is the most practical for studying the way Earth works. With the advent of computer technology, however, many Earth scientists now use computers to simulate Earth processes and geologic time spans (millions to billions of years).
- Science provides us with answers to many questions, but it cannot help us to make ethical or moral decisions that depend on human values.

History of Science

Scientific discoveries and knowledge have affected how humans view themselves and their surroundings. The modern view of environmental systems was shaped by many preceding world views, some of which have been discarded and others of which have been modified. A brief review of this history helps to put our modern world view into perspective (Figure 1-4).

Scientists are restrained by a system of checks and balances that support the search for objective truth. The two most important checks on scientists' findings are the continuous testing of ideas and assumptions under varying sets of conditions and the constant dialogue among scientists to verify the repeatability of results. The most important balance is a persistent skepticism—the attitude that a hypothesis or theory, no matter how well-favored, can and must be rejected if it is consistently contradicted by new evidence.

Many major scientific discoveries and revelations are associated with periods when dominant scientific views are challenged and discarded. In the 16th and 17th centuries, for example, the prevailing view was that Earth is at the center of the solar system, and that all other planets and the Sun revolve around Earth. An Italian scientist, Galileo Galilei (1564–1642), agreed with an earlier scientist, Copernicus (1473–1543), that the Sun is at the center of the solar system. The Catholic Church was so invested in the Earth-centered view that the Pope threatened Galileo with torture. Although Galileo recanted his theory of the solar system in writing just before his death, it became accepted by a majority of scientists and, ultimately, by the Catholic Church. Galileo had touched off a scientific revolution and a new view of the universe.

FIGURE 1-4 From the ancient philosophers to the pioneers of plate tectonics theory and Earth system science, prominent scientists have made major contributions to our understanding of geoscientific processes.

Era	Scientists
EARTH SYSTEM SCIENCE	**Sherwood Rowland** (b. 1927) American chemist who sparked the worldwide effort to eliminate the use of chlorofluorocarbons (CFCs) by proposing, in 1974 with Mario Molina, that CFCs destroy the protective layer of ozone in the atmosphere; the hypothesis was affirmed in 1985 when a large hole in the ozone layer was discovered **James Lovelock** (b. 1919) British-American biochemist who proposed the Gaia hypothesis, that Earth's living things affect the environment in ways that keep it balanced in their favor
THE PLATE TECTONICS REVOLUTION	**Dan McKenzie** (b. 1942) British geophysicist who proposed the hypothesis that Earth's outermost layer is broken into large, moving plates **W. Jason Morgan** (b. 1926) American geologist who developed the geometric techniques for explaining how crustal plates move **Harry Hess** (1906–1969) American geologist who developed the idea that new crust is created at spreading centers along the sea floor **Alfred Wegener** (1880–1930) German meteorologist whose investigation of matching coastlines, rock formations, and fossil types on opposite sides of the Atlantic led to his proposal, in 1912, of the idea of continental drift
THE AGE OF EARTH AND EVOLUTION	**Marie Curie** (1867–1934) Polish-French chemist who was the first to separate radium from uranium ore **Charles Darwin** (1809–1882) British natural historian who developed the idea of environmental influence on the evolution of species by natural selection **Charles Lyell** (1797–1875) British geologist who wrote an influential book, *The Principles of Geology* **James Hutton** (1796–1797) Scottish geologist, regarded as the father of geology and shaper of the principle of uniformitarianism
THE SCIENTIFIC REVOLUTION	**Isaac Newton** (1642–1727) English mathematician and physicist who invented calculus and reflecting telescopes and formulated the laws of motion and of universal gravitation **Nicolaus Steno** (1638–1686) Birth name Niels Stensen. Danish physician, anatomist, and geologist who was among the first to recognize that rocks contain clues to past events and their timing; he developed the principles of deposition of sedimentary rock: superposition (younger layers are deposited above older layers) and original horizontality (layers are deposited horizontally) **Johannes Kepler** (1571–1630) German mathematician who discovered three of the fundamental laws of planetary motion **Galileo Galilei** (1564–1642) Italian mathematician and physicist who discovered the laws of dynamics and built telescopes to study the Moon and Venus **Nicholaus Copernicus** (1473–1543) Polish astronomer who set out to test the hypothesis that the Sun is the center of the solar system
THE DAWN OF SCIENCE	**Ptolemy** (AD 100–170) Egyptian mathematician who believed he had proved that Earth is the center of the solar system **Aristotle** (384–322 BC) Greek philosopher who studied astrology, biology, and astronomy and invented logic; he believed fossils were created by a plastic-forming force and that a Prime Mover (God) was the major force setting all objects, including planets, in motion

Galileo, Newton, and the Changing World View

The work of Galileo changed much more than humanity's view of the position of Earth and the Sun in the solar system. Galileo was one of the first to test hypotheses in the way that modern science is practiced. By combining experimentation and mathematics, Galileo identified laws that are uniform everywhere on Earth. Thanks to his work, scientists who followed him produced a cascade of discoveries that marked major advances in science.

Galileo made significant contributions to *mechanics*, the study and understanding of motion, that were critical to subsequent inventions of many types of machines, including the steam engine, during the Industrial Revolution. For example, using objects such as lead spheres in his laboratory, he established the nature of free fall and projectile motion. Galileo used mathematics to develop theoretical interpretations of nature and was able, therefore, to explain and predict the motion of different objects. At the time, this was astonishing. The general principles and equations he developed enabled him to devise instruments such as the thermometer, barometer, pendulum clock, and telescope.

The year Galileo died, 1642, a man was born who changed the fabric of science completely, with astounding consequences to society. Isaac Newton (1642–1727), a mathematics professor at Cambridge University in England, subjected "the phenomena of nature to the laws of mathematics." Newton focused specifically on the phenomena of gravitational attraction and the motion of different objects in the universe.

Newton proposed that even if one were to throw something with a large force, it would not fly away from Earth indefinitely in a straight line. Instead, it would follow a typical curved path, because the gravitational attraction of Earth would pull it toward Earth's center. That is why a ball shot from a cannon flies in an arc that ultimately returns it to Earth. If one could throw something with sufficient force to balance Earth's gravitational pull, the object would go all the way around Earth at a constant distance from the surface and continue indefinitely to orbit Earth. The object would not move in a straight line away from Earth because another force is acting to deflect its course into an orbit.

Newton identified this force as *gravity*, the attraction between bodies based on their mass. Newton's universal law of gravitation states that the amount of attraction between any two objects depends on the mass of each object and the distance between them (Box 1-2). The larger the body, the larger its gravitational attraction. Newton verified this idea by demonstrating that the orbital path of the Moon is exactly what one would predict based on the gravitational attraction of Earth. The force moving the Moon away from Earth is counterbalanced by gravitational force, which pulls the Moon's path into an orbit.

Newton, unquestionably one of the world's greatest geniuses, developed most of these ideas at the age of 23, in 1665 while visiting his mother's country manor to escape the plague in London. True to the legend, he was sitting under an apple tree, watching the Moon in the sky, when an apple fell, and the thought that the motion of the two might be related by the same physical laws flashed into his mind. Within a year, he had developed the mathematical reasoning to test this idea.

Newton did not publish his theoretical results until 1687 in a book titled *Principia mathematica philosophiae naturalis* (Mathematical Principles of Natural Philosophy). This book caught the imagination of the intellectual world of western Europe—it was published in edition after edition, and scholars had it strategically placed on their desks when their portraits were made. Alexander Pope (1688–1744), the English master of poetic and political wit, paid homage to Newton in a poem that reworded the opening to the book of Genesis:

> Nature and Nature's laws lay hid in night;
> God said, *Let* Newton *be*! and all was Light.

The educated world became fascinated with science as a means of understanding, and perhaps mastering, nature. People came to regard the universe as a perpetual-motion machine and all things in the universe as parts of that machine. This mind-set had distinct advantages. In the 18th and 19th centuries, Newtonian physics was applied with success to hydraulics, light, electricity, and magnetism. If the laws of science are so rational and - future events so predictable, then humans had within their grasp the possibility of great power and control. Science held an exalted position as the crowning achievement of the human species. This attitude led to unprecedented technological changes that spurred the Industrial Revolution and ultimately gave people such capabilities as splitting atoms to release nuclear energy, exploring the ocean floors, altering climates, and traveling into outer space.

Darwin, Wegener, and Changing Views of Time and Place

In the late 1800s another scientific revolution was to occur, this time in biology and geology rather than physics and astronomy, and it was to have even greater impact on the human self-image. The naturalist Charles Darwin (1809–1882) proposed that organisms evolve through a process of natural selection, and that humans evolved from other primates (see Appendix 1). At the same time, geologists using evidence in rocks to unravel clues of past environments and events in Earth history began to suspect that Earth must be quite old for so much to have occurred (see Chapter 3). In fact, numerous life-forms had come and gone on Earth long before even the mammals appeared that in turn led to primates and one specific primate type, *Homo sapiens*. A new sense of time, history, and evolution began to pervade the human consciousness.

In the middle of the 20th century, a revolution occurred in the Earth sciences. The observations of many geologists working worldwide led to development of the **theory of plate tectonics**, which holds that Earth's rigid outer layer is broken into about a dozen large plates that constantly are moving toward, away from, and past

1-2 Geologist's Toolbox

Forces, Gravity, and Newton's Laws of Motion

The interactions of Earth systems stem from the actions of immense forces. A *force* is any influence that causes a change in an object's *velocity* (speed in a given direction). A change in velocity is called *positive acceleration* if the object is speeding up and *negative acceleration* if the object is slowing down. A force could be a push or a pull, although no direct contact is implied and some forces require no such contact. Because more than one force can act on a body, it is the combined effect—the *net force*—of all the forces acting on the body that determines what happens to it. If the net force is zero, no change occurs. If it is not zero, the object undergoes a change in velocity—either positive or negative acceleration.

You exert forces when you throw a baseball, push a lawn mower, or lift your body to do a chin-up. The weight of your own body is a force, because your mass accelerates toward Earth in response to Earth's gravitational attraction. Gravity is the force that drives many environmental processes, including debris flows, landslides, river water flowing in channels, and coastal erosion. For example, water flows in rivers because, essentially, it is falling downhill in response to Earth's gravitational force.

Newton developed the universal law of gravitation in the early 1600s from studying why the planets revolve about the Sun. According to Newton's law of gravity, every particle of matter in the universe attracts every other particle with an attractive force that is proportional to the product of the masses of the objects, and inversely proportional to the square of the distance between them. The law is stated as an equation that relates an object's mass and distance from another object to the gravitational force between the two:

$$\text{Force} = \frac{G \times \text{mass A} \times \text{mass B}}{(\text{distance between A and B})^2}$$

To accurately relate mass and distance to force, a proportionality constant, *G*, is required. *G* is called the gravitational constant because it has the same value everywhere in the universe. The value of *G*, determined experimentally in 1798, is 6.67×10^{-11} m^3/sec^2 kg (cubic meters per second squared-kilogram).

Try calculating the force of gravitational attraction between a classmate sitting next to you and yourself. You are attracting each other with a force that is dependent only on four things: your mass (A), your classmate's mass (B), the distance between the two of you, and the constant *G*. If you are both the same mass, 50 kg, and the distance between you is 2 m, then the force between you is

$$\begin{aligned}\text{Force} &= \frac{G \times \text{mass A} \times \text{mass B}}{(\text{distance between A and B})^2} \\ &= (6.67 \times 10^{-11}\ \text{m}^3/\text{sec}^2\ \text{kg} \\ &\quad \times\ 50\ \text{kg} \times 50\ \text{kg})/(2\ \text{m} \times 2\ \text{m})^2 \\ &= 4.2 \times 10^{-8}\ \text{m/sec}^2\ \text{kg} \\ &= 4.2 \times 10^{-8}\ \text{N}\end{aligned}$$

each other (Figure 1-5). These plate motions are driven by forces that originate deep in Earth's interior.

Before the acceptance of plate tectonics theory, geologists held that the continents and ocean basins are fixed and stable. In 1912 German meteorologist Alfred Wegener (1880–1930) proposed that the present-day continents look like pieces of a jigsaw puzzle because they once were assembled (about 200 million years ago) into a supercontinent that he called Pangaea. Wegener claimed that the continents of today are the drifting fragments of Pangaea. As evidence, Wegener demonstrated that rock formations and fossils on the coasts of South America and Africa match, as do their coastlines, even though they are separated by the Atlantic Ocean. Wegener's observational approach seemed sound, but no scientists could explain the physical cause of *continental drift*, nor

Newton (N) is the name given to the force required to accelerate a mass of 1 kg by 1 m/sec^2. One newton has about the same weight (or force) as an apple, which has a mass of about 100 g (grams). The gravitational attraction between you and your classmate is very tiny, about one-billionth of a newton. In comparison, the force of gravity pulling you toward Earth is about 490 N, if your mass is 50 kg. In other words, your weight is 490 N. If gravitational attraction exists between all objects with mass, why are they not all pulled uncontrollably together? Earth has vastly more mass than any of the objects on it, so its gravitational attraction is far greater. This means that the net force on water and rocks, people and plants, dust and air, is downward toward Earth's center.

The net force on debris at Earth's surface is downward, toward the Earth's center, so loose rocks on a slope move downhill during an avalanche, as here in Zion National Park, Utah. Loose debris from canyon walls has fallen onto the canyon floor, which itself has a downhill slope. Debris continues to slide and roll down the canyon, forming a churning mass that raises dust as pieces of rock are broken and shattered.

did anyone understand the mechanisms by which continents could move. It was not until the 1960s, with the discovery that Earth's outer layer is broken into plates and that interior forces are capable of moving them across Earth's surface, that the concept of continental drift was taken seriously (see Chapter 4).

The theory of plate tectonics explains not only continental drift but also many Earth processes that previously were thought to be unrelated. The collision, separation, and sliding of separate plates helps to explain earthquakes, volcanoes, the growth of mountains, the changing shape of oceans, and other phenomena that ultimately affect all of Earth's systems. Because such diverse events as climate change, coal formation, and species migration all have their roots in this unifying theory, plate tectonics will be discussed throughout this book.

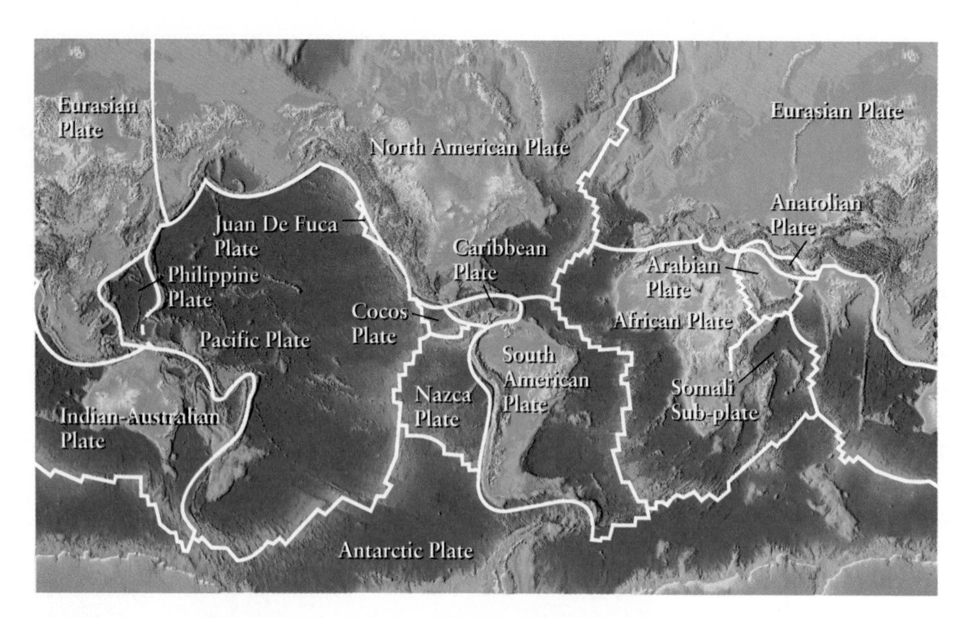

FIGURE 1-5 The theory of plate tectonics holds that Earth's outermost layer is broken into large, moving plates. Their motion contributes to volcanism, earthquakes, and climate change, as well as many other Earth processes.

- Galileo's work in the 1600s heralded the scientific revolution, largely because it confirmed a Sun-centered—rather than Earth-centered—solar system.
- Building on Galileo's contributions, Newton discovered fundamental laws of motion and gravity that led to the technological achievements of the Industrial Revolution and a mechanistic view of the universe.
- Darwin revolutionized biology and geology in the 1800s with his theory of evolution through natural selection.
- In the early 20th century Wegener amassed observational evidence that supported the movement of continents over Earth's surface, and unwittingly inspired the plate tectonics revolution that occurred several decades after his death.

Evolution and Environment

To understand our place in nature and our relationship to the environment, we need to know something about the changes in Earth's environments with time and the evolution of our species. Studies over the past century indicate that Earth formed about 4.6 billion years ago. About a billion years later, after Earth's crust, atmosphere, and oceans had formed (see Chapter 2), single-celled organisms (bacteria) emerged in the oceans. Over time, the developing plant life in the oceans injected enough oxygen as waste into the atmosphere to make it breathable, and about 3 billion years ago the first multicelled organisms crawled out of the oceans and onto the continents. Meanwhile, most anaerobic organisms, which cannot survive in an oxygen-rich environment, died out.

The oxygenated environment supported the evolution of millions of marine and terrestrial species, but many that once flourished have since died out. The number of marine life-forms, for example, has grown over the past 600 million years, but not without significant periods of extinction (Figure 1-6). The sharpest drop in the number of marine life-forms occurred 245 million years ago, when scientists estimate that nearly 96 percent of all marine species disappeared, possibly in part because of lethal climate changes resulting from massive volcanism. About 65 million years ago, almost half of all species were wiped out, including all species of dinosaurs. Most scientists attribute this mass extinction to a giant asteroid impact that affected Earth's climate and ecosystems.

With the extinction of some species came the evolution of others, so that the number of species on Earth has increased over time. This **biodiversity** (diversity of life) is attributable in large part to plate tectonics. The continents on Earth's moving plates are carried along as the plates collide, slip past each other, or split apart so that oceans form and widen between them. The separation of continents from Pangaea in the past 200 million years has created diverse new environments, including extensive nearshore areas that provide rich habitats. It also has isolated some species from others, which evolved into new life-forms over time. Our own species appeared on the African continent about 100,000 years ago.

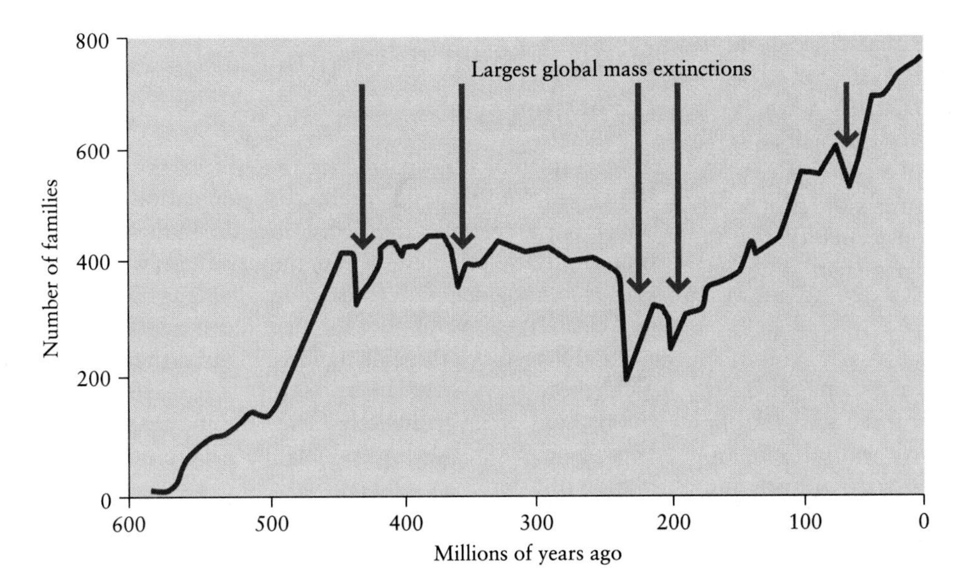

FIGURE 1-6 The number of plant and animal species has increased over the past 600 million years, although many species have become extinct. This graph tracks the fluctuation in the estimated number of *families* (groups of related species; see Appendix 1) of marine organisms. Arrows point to the five largest global mass extinctions. The general trend has been toward greater biological diversity until the present, but a sixth major decrease in number of species is occurring now, as a result of human activity.

The Evolution of Humans

After the extinction of the dinosaurs 65 million years ago, mammals flourished and evolved on Earth. Primates appeared in African forests during this time. Primates are an *order* of mammals that includes apes and humans (see Appendix 1). Between 2 million and 4 million years ago the climate of tropical Africa became drier and warmer and forests shrank, possibly encouraging some early primates to walk upright into the encroaching savannas (grasslands). A large-brained, upright-walking, talkative primate evolved in this grassland environment some 50,000 to 100,000 years ago. It was our own species, which we now call *Homo sapiens,* for "thinking man." For most of the time since *Homo sapiens* appeared, its members traveled about in small groups of hunters and gatherers, often living in caves and along the banks of rivers. Some groups remained in Africa while others moved northward and eastward throughout Europe and Asia, possibly in search of more food or in response to changing climates and environment.

From about 25,000 to 12,000 years ago, human beings endured ice-age conditions. Ice sheets covered large parts of North America and Europe (as well as Antarctica and Greenland), and smaller glaciers throughout the world covered much more area than they do today (Figure 1-7). With significant amounts of water bound in extensive ice sheets, sea level dropped, exposing land bridges between several continents. Using these newly exposed bridges, people migrated to previously unreachable lands. People had to cope with climates ranging from hot and dry to cold and ice-covered. Those who were best able to adapt to their new environment lived longest and passed their genes to the greatest number of offspring. Gradually, humans in a particular environment developed particular characteristics. Those living in hot

FIGURE 1-7 From about 25,000 to 12,000 years ago large ice sheets existed on the North American and Eurasian continents, as well as on Antarctica and Greenland, which are still buried by ice. Enough seawater was locked up in the ice to lower the sea level and expose land bridges, which allowed human migration from Europe to North and South America. Global warming since then has melted much of the ice and raised sea level, submerging the land bridges once again.

deserts evolved with tall, slim bodies that dissipate heat quickly, while those living in cold regions evolved with short, stocky bodies that conserve heat. As a result of evolution and adaptation, human beings diversified into different races, all with a degree of ability unrivaled among other animals to create and use tools from raw Earth materials—in short, to develop technology—and to communicate their thoughts with language.

Human Effects on Environment

About 10,000 years ago global climate became generally warmer, and ice sheets in North America and Europe melted. Early in this period of relative warmth our ancestors discovered how to grow their own food crops and to domesticate animals. As the climate grew even warmer and drier in many parts of the world, humans learned to divert stream water into irrigation ditches and to control the flow of water in reservoirs behind dams.

With this Agricultural Revolution, more people could be supported in smaller areas than hunters and gatherers required, and the first cities arose about 7000 years ago. Accumulation of the world's people in increasingly larger cities continues to this day. In turn, the closer proximity of humans in urban areas spurred the transfer of ideas and new technologies. People became increasingly reliant on metals for toolmaking and on fuels such as wood for energy. The Industrial Revolution, which began in the 1700s, marks a time of significantly increased consumption of energy and production of manufactured goods. People discovered a means of converting ancient solar energy stored in fossil fuels (coal and, later, oil and natural gas) into mechanical energy, a large part of which is used now to grow and transport agricultural food products around the world.

As a result of our growing ability and willingness to manipulate our environment, human population has expanded in numbers unprecedented for a species of such large creatures. In the last few hundred years, the number of humans has grown from less than half a billion to 1 billion in the early 1800s, 2 billion in 1930, 4 billion around 1975, 5 billion in 1987, and 5.8 billion in 1997. The population is projected to increase to as much as 8 or 9 billion by the year 2010.

Human impact on Earth's resources and environment is great, and it is proportional to the number of people. Like all other species, humans compete for and use essential resources—air, water, food, and space. The negative environmental impacts of the Industrial Revolution include air and water pollution, soil erosion, the concentration of human wastes, and the spread of disease. As a result, environmental awareness and legislation have increased in many nations, especially in the last few decades of the 20th century.

Critical Issues in Environmental Geosciences

Several related issues have contributed to a growing environmental awareness. Foremost is population growth, which has accelerated markedly since the Industrial Revolution. With population and industrial growth come ever-increasing demands for such resources as fuels, land, and clean, fresh water. Environmental degradation is the inevitable result of these demands. Finally, although people have always lived in areas prone to earthquakes, floods, and other geologic hazards, the increasing numbers of people in these areas heightens the potential for human disaster.

Human Population Growth

As global population approaches 6 billion, concern is mounting over how many more people will be added in the future. Human population growth has been nearly impossible to predict accurately because it depends on a bewildering array of variables, from advances in agriculture, sanitation, and medicine to the influences of culture, religion and medical practices. The rate of human population increase has varied throughout human history, but its most striking feature is an accelerated increase in the last few hundred years. This pattern can be explained in part by a model known as exponential growth (Figure 1-8).

Exponential growth occurs when the amount of something, such as the number of people on Earth, increases by a fixed percentage so that both the base amount and the added quantity become larger and larger, even though the

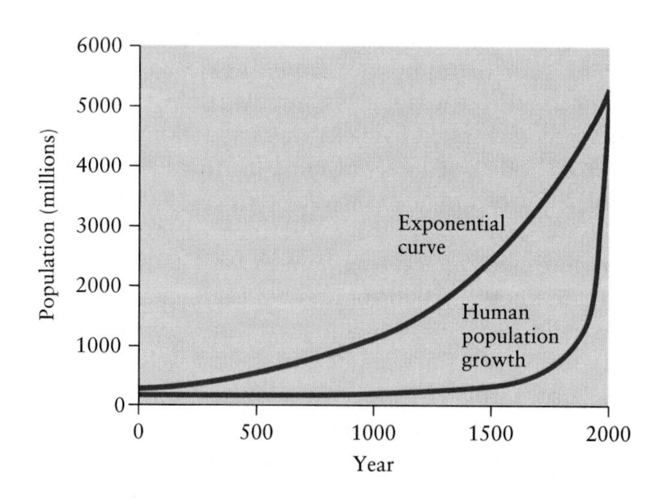

FIGURE 1-8 The actual rate of human population growth over the past 2000 years has been less gradual than the rate that would have resulted if human population growth had been truly exponential since AD 1.

rate of increase does not. Consider a population of 1 billion people with an annual growth rate of 2 percent. In one year the population will increase by 2 percent of 1 billion, or 20 million. In the second year the amount of growth will be 2 percent of 1 billion *plus* 20 million (or 20.4 million people). Each year the number of people added increases although the rate of growth does not.

The Implications of Exponential Growth One of the most important aspects of exponential growth is the suddenness with which a growing population can fill all available space. In *The Limits to Growth*, Donella Meadows and her co-authors illustrate this aspect of exponential growth with a riddle:

> A water lily growing on a pond doubles in size each day. It is twice its original size on the second day and four times its original size on the third day. This is an example of exponential growth, because growth occurs at a constant rate (100 percent) each day. In 30 days, the water lily will have grown so much that it will completely cover the pond, stifling other plant and animal life. In order to save the pond, you must cut back the lily plant once it covers half the pond. When will that be?

Think about this. If the lily plant doubles each day, and will cover the whole pond in 30 days, it will cover half the pond on the 29th day. It grows as much on the last day as it did on all previous days added together. You have only the 29th day to save the pond.

Human population is growing in a manner somewhat similar to exponential growth, doubling once every 40 to 46 years at present growth rates. As our numbers grow at this astonishing pace, it takes less and less time to add another billion people. It took 2 million years of human history to add the first billion people, 130 years to add the second billion, 30 years to add the third, 15 years to add the fourth, and only 12 years to add the fifth billion. The sixth and seventh billion people will be added just before and after the year 2000, unless fertility and mortality rates change drastically. In other words, population growth is gathering considerable momentum. At present growth rates, it takes less than 2 years to replace the number of people who died in all wars fought in the world in the past 200 years, about 165 million. One of the greatest shortcomings of the human species is its unwillingness to face the consequences of accelerating population growth.

The Consequences of Rapid Population Growth It may be true that the more people who are born, the greater the chance that someone with creativity and intelligence will come along and develop new technologies for alternative energy, or new ways to minimize waste production during industrial processes. The more realistic scenario, however, is that our present environment will not survive the strain. Increasingly, plant and animal species will become extinct, habitat will be destroyed, air and water will become fouled, and people will die of famine during droughts or in wars fought over diminishing resources. If human population were to double unchecked until one person exists for every square meter of land on Earth, we might find life to be difficult and unpleasant (Figure 1-9). All humans could lead better lives with higher standards of living if the number of humans is kept below some undetermined threshold.

FIGURE 1-9 With industrialization, cities have grown even larger; some megacities have populations of more than 10 million people. Poverty and crowding are juxtaposed against signs of affluence, as here in Rio de Janeiro, Brazil.

Resources and Sustainable Development

Does Earth have some finite "carrying capacity," some threshold number of people beyond which it cannot sustain the human population with clean air, clean water, and adequate nourishment? To answer this question, we need to consider not only the number of people but also the quantity of resources necessary for their survival.

A *resource* is anything we get from our environment that meets our needs and wants. Some essential resources, such as air, water, and edible biomass (plant and animal matter), are available directly from the environment. Other resources are available only because we have developed technologies for exploiting them. These include oil, iron, and groundwater. In general, people in affluent and highly industrialized countries use far more resources than are required for basic survival. The United States, for example, has only 4.8 percent of the world's population, yet it consumes about 33 percent of the world's processed nonrenewable energy and mineral resources.

Resources are classified according to their degree of renewability into three major types: potentially renewable, nonrenewable, and perpetual (Figure 1-10). **Potentially renewable resources** can be depleted in the short term by rapid consumption and pollution, but in the long term

(a)

(b)

(c)

FIGURE 1-10 (a) Soil, a potentially renewable resource, is put at risk by the demands for food associated with worldwide population growth. (b) New technologies and ever-larger machines are used to mine coal, a nonrenewable resource, at increasing rates. (c) Solar energy is a perpetually available resource. Reflective panels can focus this energy on a "power tower" where it is converted into electricity.

they usually can be replaced by natural processes. The highest rate at which a potentially renewable resource can be used, without decreasing its potential for renewal, is its *sustainable yield.* If the sustainable yield is exceeded, the base supply of the resource can shrink so much that the resource can become exhausted—used up. Soil formation, for example, occurs at rates of about 2 to 3 cm (centimeters) per thousand years, making it a potentially renewable resource. Unwise farming practices, however, can cause soil loss of 6 to 8 cm per decade. Soil cannot be renewed at this rate because its formation is dependent on plants and soil moisture, both of which are gone once the soil itself is lost. In some parts of the world, all soil has been removed and so is essentially nonrenewable.

Nonrenewable resources, such as fossil fuels and metals, are finite and exhaustible. Because they are produced only after millions of years under specific geological conditions, they cannot be replenished on the scale of human lifetimes. The formation of oil, for example, requires that buried plant and animal matter be subjected to tens of millions of years of squeezing and heating by geologic processes, a rate of formation so slow that the resource cannot be considered renewable.

Some nonrenewable resources, such as copper and iron, can be recycled or reused to conserve the supplies. *Recycling* involves collecting, melting, and reprocessing manufactured goods. Resources commonly recycled include aluminum, paper, and iron. *Reuse* involves repeated use of manufactured goods in the same form. Glass bottles commonly are reused. Other resources, such as fossil fuels, cannot be recycled or reused, because their combustion during energy production converts them to ash that falls to Earth, exhaust fumes that go into the atmosphere, and heat that ultimately leaves Earth as low-temperature radiation. Nonrenewable resources become economically depleted when so much—typically about 80 percent of the resource—is exploited that the remainder is too expensive to find, extract, and process. Oil buried in rocks deep beneath the Antarctic ice sheet, for example, would be far more expensive to find, extract, and refine than is oil trapped beneath the shallow salt beds of Texas.

Perpetual resources are those that are inexhaustible on a human time scale of decades to centuries. Examples include solar energy, which has fueled numerous reactions and processes on Earth for its entire 4.6 billion year history, heat energy from interior of the Earth, and energy generated by Earth's surface phenomena such as wind and water flowing downhill.

One of the most significant changes in human history is the increased demand for, and use of, energy. Before the Agricultural Revolution 10,000 years ago, people's energy demands were primarily for food. On average, a person uses about 10 megajoules (MJ) of energy (2000 to 3000 food calories) per day. By 1991 daily per capita energy consumption in the United States and Canada was more

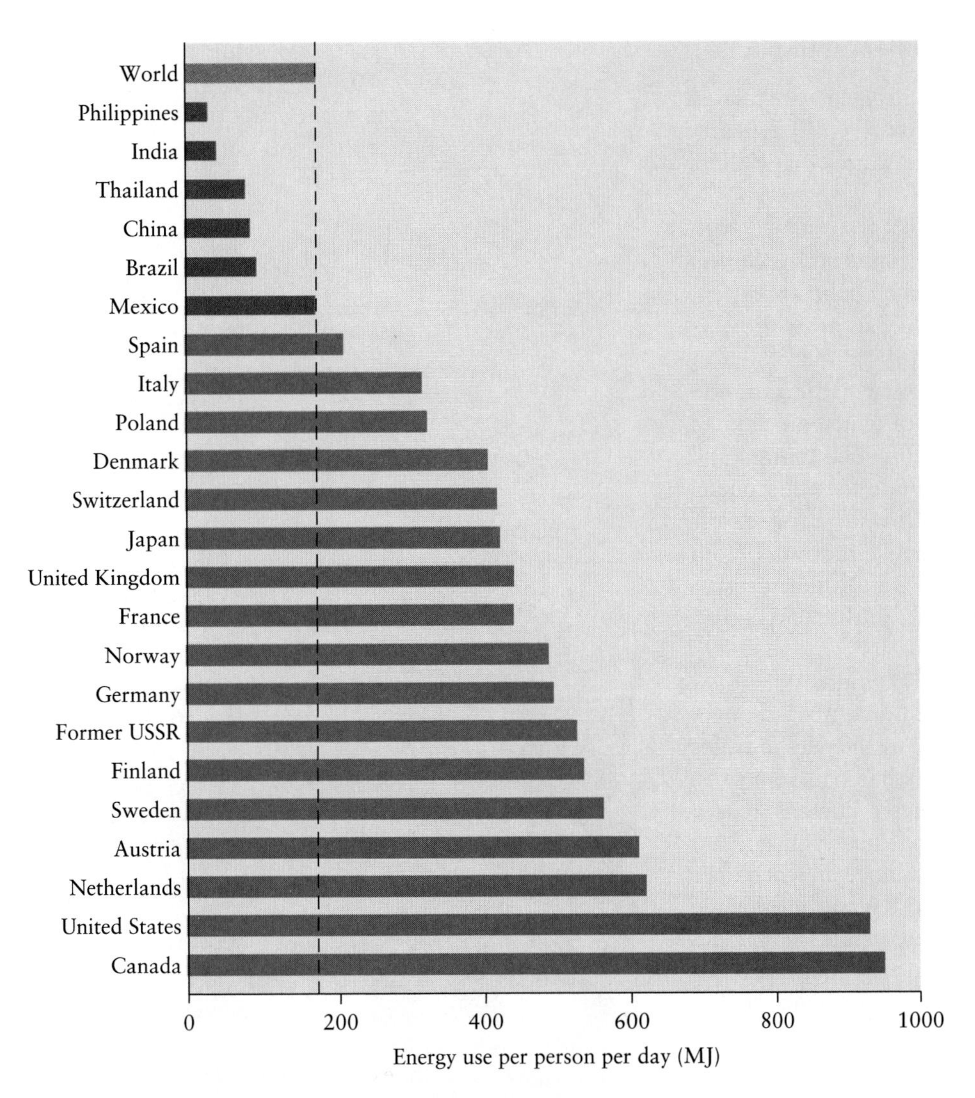

FIGURE 1-11 Canada and the United States use far more energy per person per day than does any other nation. The world average per capita daily energy use is slightly less than 200 MJ (dashed line). A joule (J) is a measure of energy; 1 J is the amount of energy used when a force of 1 N (newton) moves something a distance of 1 m in the direction of the force. A megajoule (MJ) is equal to 1 million joules. Not only do amounts of energy used vary from one country to another, but so also do types of energy used (see Chapter 11). For example, India uses more plant and animal matter as a fuel source than most other nations shown here, whereas China uses large amounts of wood and coal, and the United States uses large amounts of coal and petroleum.

than 900 MJ, spent mostly in industry, transportation, and temperature control (Figure 1-11). Much of this energy consumption goes toward the production and transport of food. Between 1950 and 1979, world population increased from 2.5 billion to about 4 billion people and world food production doubled, but the amount of oil used by farmers to produce this food increased fivefold.

Production and consumption of renewable, nonrenewable, and perpetual resources follow different growth curves (Figure 1-12). For potentially renewable resources, such as wood, the curve is similar to that of the growth of a biological population: Production rises exponentially for some time, then levels off at a state of no growth that is equal to the rate of replenishment. For nonrenewable resources, such as fossil fuels and minerals, production rates rise exponentially, reach a maximum value, then decline exponentially as supplies approach zero. For perpetual resources, such as solar energy, production can rise exponentially for essentially unlimited time periods. It is clear that humans can rely on nonrenewable energy resources only for a limited time, before turning to a perpetual source of energy.

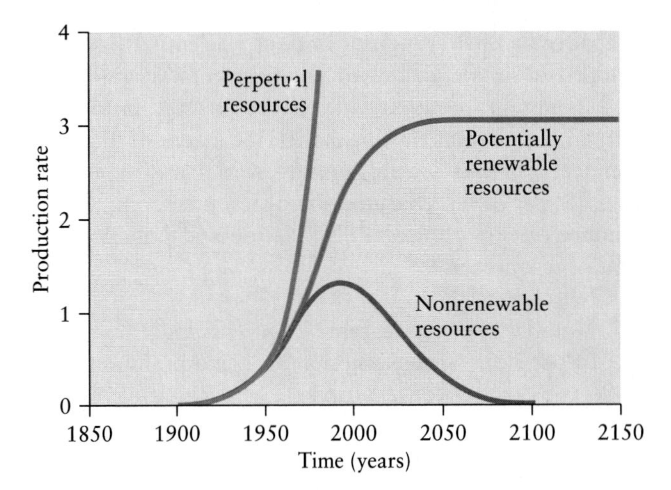

FIGURE 1-12 As demand for a resource increases, more of it is produced. For nonrenewable resources, production cannot increase forever, because ultimately the resource will be exhausted. Renewable resources can provide lasting supplies if the rate of production is similar to the rate of renewal. Only for perpetual resources is unlimited increase in production possible.

Pollution, Wastes, and Environmental Impact

Many of the world's environmental problems are related to the wastes produced by human activities. To a greater or lesser extent, all humans produce wastes that pollute and degrade air, water, and land.

Pollution is the contamination of a substance with another, undesirable, material. Environmental pollution results when a pollutant degrades the quality of an environment. Common pollutants in urban areas are carbon dioxide (CO_2), sulfur dioxide (SO_2), and oxides of nitrogen (NO_x molecules with one or two atoms of oxygen and one atom of nitrogen), all of which are created by combustion of fossil fuels in automobile engines and industrial power plants and released into the atmosphere. These molecules react with gases and water vapor in the atmosphere to produce smog and acidic rainwater, both of which make the air and water unhealthful, causing damage to plants, fish, land animals (including people), and even stone.

Wastes are unwanted by-products and residues left from the use or production of a resource. Early human habitation sites are often identifiable by the piles of waste left behind, including shells, bones, charcoal from fires, and broken pottery. Sometimes, early humans reused wastes in innovative ways, building huts from animal bones, for example. People tended to move on when their waste piles became substantial. Most waste piles made by small bands of people were biodegradable, so they did little long-term environmental damage. In contrast, modern humans in industrialized societies have much higher population densities and technologies that enable them to concentrate large amounts of waste, much of which is toxic and infectious, in small areas (Figure 1-13).

A Model of Environmental Impact Scientists have developed a simple model of environmental degradation and pollution to assess the environmental impact of human populations. In this model, the extent of the environmental impact depends on three variables: (1) population, (2) per capita consumption of resources, and (3) the amount of environmental degradation and pollution per unit of resource used:

$$\text{Number of people} \times \text{per capita consumption of resources} \times \text{degradation and pollution per unit of resource used} = \text{environmental impact}$$

According to this model, three types of environmental degradation can occur. The first, **people overpopulation,** occurs when a large group of people has insufficient resources. Although the group's per capita consumption of resources is low, the per capita amount of environmental degradation is high because this group must scavenge all available resources in order to survive. Environmental protection means little in the face of starvation. In parts of northern and western Africa, for example, large areas of land have been denuded of most vegetation and famine has occurred repeatedly in the 20th century.

(a)

FIGURE 1-13 (a) At early human habitation sites, people sometimes recycled their wastes, as shown by this hut built of mammoth bones and animal hides. (b) This 1880 cartoon protests the dumping of garbage off Coney Island and other New York City beaches.

The second type of environmental degradation, **consumption overpopulation,** occurs when a small number of people use resources at such a high rate per person that they cause very high levels of resource depletion and pollution. Ranking first among modern examples is the United States, which has less than 5 percent of the world's population but consumes more than 33 percent of nonrenewable energy and mineral resources and produces more than 33 percent of the world's pollution. As a result, despite its relatively small population, it has a tremendous environmental impact. In contrast, China has about 21 percent of the world's people, four times as many as the United States, but causes less than one-tenth the environmental damage of the United States. For this reason, it sometimes is noted that the environmental impact of one child born in the United States is equivalent to 20 children born in many developing nations, such as India, where resource use and pollution per person are much smaller.

As nations develop, the extent of their environmental impact due to consumption changes rapidly. In the 19th century, for example, during a rush to settle all parts of the United States, forests were destroyed for farming, and many sites were severely degraded by destructive mining practices. Since then, increasing environmental awareness in the United States has resulted in much effort to prevent such degradation in future development efforts. Likewise, as China enters the global market of free trade and as other nations become more developed, it is likely that the nature of their environmental impact will change.

The third type of environmental degradation, **pollution overpopulation,** occurs when a small or a large number of people use technologies that are grossly polluting. In this case, the amount of pollution produced per unit of resource used is so high that extreme environmental degradation occurs. Industrialized countries in eastern Europe and the former Soviet Union have amassed a legacy of environmental pollution that will haunt their citizens for decades to come (Figure 1-14).

Minimizing Environmental Impacts Solving the problem of environmental degradation caused by rapid population growth is essential, but the solution is not simple. Since the Industrial Revolution, the increasing access to medicine and sanitary supplies of water has reduced death rates and thus contributed to the rapid population growth of the past few centuries. As a result, reducing population will require a change in the birth rate. Solutions for reducing birth rates are related to education and standard of living, among other things. In most countries of the world, the higher the level of education and standard of living of women, the more likely they are to use birth control methods and have fewer children. Reduced birth rates in most developed nations, as well as in China and India, caused worldwide population growth rates to drop from 2.06 percent in the late 1960s to 1.73 percent in the late 1970s, and to 1.6 percent in the 1990s. This consistent drop in population growth rates is encouraging, but rates have to drop in many more nations to level off global population growth.

These facts argue for striving to educate and raise the standard of living of more women in the world. Such efforts are supported by different international groups, including the United Nations, but much more progress could be made, for in many nations women are largely

FIGURE 1-14 In eastern Europe, environmental pollution and degradation are extreme, and it will take decades to clean up the damage.

uneducated and illiterate. An important goal for the 1990s and the third millennium is to continue the reduction in population growth rates.

Minimizing environmental degradation caused by consumption can be achieved through reduced resource consumption per person and coordinated efforts to recycle or reuse products over and over again. Attention to the problems associated with increased consumption of resources and sustainable development has reached international levels. In June 1992 representatives of 150 nations and 1400 nongovernmental organizations, nearly 8000 journalists, and many others met in Rio de Janeiro, Brazil, for the United Nations Conference on Environment and Development, informally called the Earth Summit. The most positive results of the Earth Summit are the nations' commitments to:

- Establish new international institutions, such as the Sustainable Development Commission, a United Nations body.
- Develop strategies for national reporting measures.
- Develop financial mechanisms to support international environmental efforts.
- Increase public and nongovernmental participation in resolving environmental conflicts.

It will be some years before the long-term outcome of these efforts will be noticeable, but certainly the Earth Summit has heightened the world's awareness of environmental problems and focused attention on reasonable solutions. Recycling programs, for example—especially for paper—are becoming more common in many developed nations.

Minimizing environmental degradation caused by pollution can be achieved through use of more efficient technologies to minimize the amount of waste (much of which produces pollution) generated during industrial and resource extraction processes. Efforts to develop waste minimization procedures for different industries are now the focus of research and testing at many universities and industrial labs. Some new factories in Denmark, for example, produce nearly zero waste, designing their activities so that waste products from one stage are used as input to other stages of the manufacturing process.

As environmental regulations become more stringent, many companies are developing new technologies to extract and process resources that produce fewer wastes than ever before. New methods to mine copper that require no excavation of Earth are being tested in Arizona (see Chapter 5). The copper is removed from Earth by solution techniques, so that heating and melting of rocks to separate precious metals are not required. The result is a substantial (or nearly complete) reduction in the production of fumes, noxious gases, and ash that once was typical of the exhaust material belched from smokestacks at mine sites.

Natural Disasters, Hazards, and Risks

The noted American historian Will Durant once quipped that "civilization exists by geological consent, subject to change without notice." **Natural disasters** are sudden and destructive environmental changes that happen as a result of long-term geologic processes but appear to occur without notice. Earthquakes, for example, are the result of hundreds or thousands of years of geologic strains, but the breaking point occurs relatively suddenly. A *geologic hazard* is a natural phenomenon or process with the potential for disaster. Rocks perched at the top of a steep hill constitute a hazard. A destructive rock avalanche is a disaster.

Like human-made disasters such as oil spills and air pollution from factory smoke, natural disasters can pollute the environment. Radon, a radioactive gas that occurs naturally in some rocks and soils, can seep into the basements of houses over a period of years, gradually increasing human exposure to this carcinogen and leading to the sudden appearance of cancers in relatively large numbers of people in a small geographic area. Similarly, carbon dioxide gas seeping from a volcano in Cameroon, in Africa, accumulated at the bottom of a lake in a crater atop the volcano and suddenly belched from the lake in 1981, suffocating thousands of sleeping residents within minutes.

Volcanic eruptions and earthquakes are natural disasters caused almost entirely by the interior processes that move Earth's tectonic plates. Floods and landslides, in contrast, are natural disasters that can be worsened markedly by human activities. Removing natural vegetation from hillslopes, for example, can increase the runoff of rainwater and thereby increase flooding in streams. Removing vegetation also destroys roots, which anchor the soil on the slopes. Landslides occur when the loosened soil and other debris slip suddenly downhill.

Risk refers to the magnitude of potential death, injury, or loss of property due to a particular hazard. The risk of death in the United States each year by an earthquake or volcano (<0.1 death per million people) is far less than the risk of death by automobile accidents (300 deaths per million people) or fires (0.5 death per million people). In California, the risk of death by earthquakes and volcanoes is greater than it is in Ohio or Kansas. In fact, much of the United States west of the Rocky Mountains is considered to be a high-risk area for earthquakes, and all active volcanoes are in western states. Nevertheless, some parts of the central and eastern United States have been devastated by large earthquakes in the past, and in many ways are even more vulnerable to damage than western areas (see Chapter 5).

- Environments on Earth are ever-changing, as they have been since Earth formed about 4.6 billion years ago. In response to changing environments, different life-forms—including our own species—have evolved by processes of natural selection.
- As Earth entered a period of relative global warmth about 10,000 years ago, people began to develop agriculture (the Agricultural Revolution), and since then they have altered dramatically the composition of ecosystems on Earth. The environmental ramifications of agriculture have been substantial.
- Since the Agricultural Revolution, the rate of human population growth has been rapid, accelerating in the years since the Industrial Revolution. Most of the recent growth is attributed to a decrease in death rates because of better sanitation and medical care.
- With industrialization has come a marked increase in consumption of energy, and in particular of the use of nonrenewable sources of energy (fossil fuels).
- Decreasing environmental degradation would require that population growth stabilize; that we develop perpetual sources of energy, such as solar energy; and that we decrease the use of nonrenewable resources that require extensive mining and that produce pollutants during combustion.
- As human population increases, so does the hazard associated with natural disasters. Mitigating natural disasters requires increased effort from geoscientists to understand natural phenomena and their causes.

Environmental Geosciences and Earth System Science

Worldwide recognition of geologic and human-induced environmental issues has sparked the development of new fields of study, especially combined disciplines such as geochemistry, biogeochemistry, and environmental geoscience. Because environmental changes and geologic hazards are the result of the interactions of many Earth systems with one another, they need to be addressed by broadly trained environmental scientists who can cross traditional disciplinary boundaries. As scientists learn more about the interaction of Earth systems, they can make increasingly effective recommendations for conserving resources, minimizing environmental degradation, and reducing the potential for death and destruction associated with geologic hazards.

This book is about the environmental geosciences, which deal with Earth, its origin, its composition, its many interacting systems, and the way it works. Environmental geoscientists study issues related to cleaning up the environment, finding and using resources such as petroleum and water, predicting and mitigating natural disasters such as earthquakes and volcanic eruptions, and studying global change, particularly the dual threat of global warming and rising sea level. Cleanup problems include how to prevent clogging of rivers by soil erosion, how to select optimal sites for solid waste (landfills) and hazardous waste repositories, and how to purify groundwater contaminated by millions of improperly located and poorly maintained waste dumps (Figure 1-15).

FIGURE 1-15 A hydrogeologist at a military base traces the flow of contaminants in groundwater on Cape Cod, Massachusetts. Each pipe indicates a spot where scientists can measure the contaminants. The stream of lights defines the shape and flow direction (to the left) of a plume of contaminated groundwater.

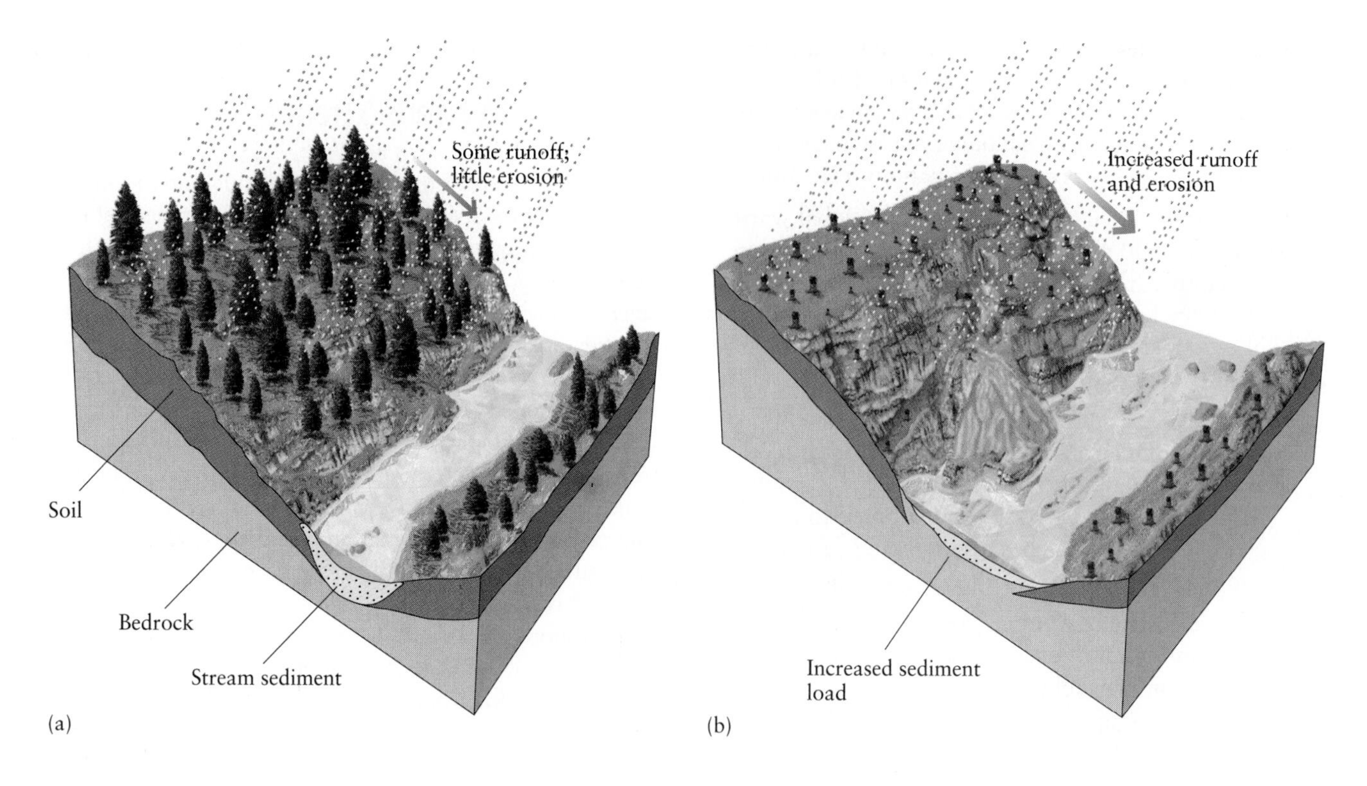

FIGURE 1-16 Effects of deforestation on hillslopes and streams exemplify the integration of Earth systems. (a) In a heavily forested area, trees slow the movement of rainwater so that it seeps slowly through the leaf litter on the forest floor and into the soil. The tree roots hold the soil, and little erosion occurs. (b) Deforestation exposes the lightly vegetated soil. Rain erodes the soil and sends it flowing down the hillside and into the stream. The water level rises, causing flooding and more erosion.

Linked Environmental Systems

The environmental impacts of deforestation provide an example of the interrelated nature of Earth systems and of the need for multidisciplinary studies of such impacts. If a forest is cleared for farming, the soil system is unprotected from rainfall. The loosened soil can easily wash away, and landslides and mudflows can occur. All this material washes into streams and rivers, increasing their sediment loads, clogging their channels, and raising their water levels so that flooding becomes more likely. The quantity of rainwater flowing rapidly off the barren, exposed hillslopes also exacerbates flooding (Figure 1-16). In contrast, an intact forest would minimize soil erosion due to rainfall by:

- Catching some of the rainwater on leaves so that it drips slowly to the forest floor.
- Absorbing rainwater through the organic mat on the forest floor.
- Absorbing rainwater into the roots of trees and plants, which ultimately return water to the atmosphere.
- Causing rainwater to percolate down through the soil to the underlying groundwater table.

Thus, high quality streams and rivers depend upon healthy ecosystems, which in turn depend upon healthy soils.

Earth System Science: An Emerging Environmental World View

Because scientists now understand Earth as a complex group of interacting systems, they study it from a new perspective called **systems thinking.** A **system** is a group of interrelated and interacting objects and phenomena. Systems thinking enables us to construct models of systems that then can be explored to determine the possible impact of a change, such as the emission of gases from a volcano into the atmosphere, on the different parts of the system.

Scientists James Lovelock and Lynn Margulis have suggested that, contrary to the mechanistic view of Earth, Earth systems interact in ways that make the planet function more like an organism than a machine. They named this idea the **Gaia hypothesis,** for the Greek goddess who personifies Earth. In the myth, Gaia (Earth) sprang from Chaos and gave birth to Uranus (Heaven) and Pontus (Ocean). Her name is in the word root *geo-*, as in geology and geography. The Gaia hypothesis states broadly that life on Earth has had an influence on the evolution of the physical Earth, determining, for example, the composition of the atmosphere and ocean sediments. Furthermore, life serves as a control system with feedback mechanisms for stabilizing global temperature and other attributes of the planet. Together, the living and the inanimate components of Earth have co-evolved as coupled systems. The Gaia hypothesis now is receiving much serious attention from scientists as they test its predictions and statements.

Earth system science focuses on the interconnections among different Earth systems and the changes that occur in them with time. Earth system science views the whole Earth as a single system in which matter and energy are cycled through numerous subsystems, including the lithosphere (outermost solid Earth), pedosphere (disintegrated and decomposed rock and soil), hydrosphere (water), biosphere (life), and atmosphere. Humans are a part of the whole Earth system, depending upon it for resources, affecting its environments, and responding to its changes.

In this book we use Earth system science as a pattern of thought that helps us to view and understand Earth and our environment. The next chapter is devoted to an exploration of Earth system science and how it can be used to model environmental processes. Thereafter, it is used throughout each chapter to discuss different Earth systems and global change.

- Since the scientific revolution that began during the Renaissance, many separate scientific disciplines have developed, including physics, chemistry, and geology.
- Viewing environmental systems as interconnected, rather than isolated, requires a multidisciplinary approach, such as that provided by Earth system science.

Closing Thoughts

Before the development of science, nature seemed unpredictable and frightening. People felt able to predict and control natural disasters such as floods, droughts, and volcanic eruptions only through appeals to supernatural forces. At the same time, they learned how to control their immediate environment with tools, weapons, and hunting and farming skills. As agricultural practices spread to all continents, humans cleared forests, marked property lines, and planted fields. The consequences to the larger environment were rarely considered. The historic destruction of vast temperate-zone forests, including the forests of Europe, the Great Forest of eastern North America, and the rain forest of the Pacific northwest, is seldom acknowledged, but it was similar to the modern destruction of tropical rain forests. In all cases, the impact on indigenous plant life, animal life, and human communities was ignored, to catastrophic effect.

With the scientific revolution of the 17th century came a new wave of technological advances, leading to the Industrial Revolution in the 19th century, a mechanistic world view, and a heightened potential to alter the environment. The 19th century also saw Darwin's theory of the evolution of species and the beginnings of geological science. By the end of the 20th century plate tectonics was recognized as geology's unifying theory and Earth system science had become the predominant approach to studying the environment. In general, humans have become increasingly conscious of their own impact and dependence on the environment, and of the interdependence of different parts of Earth's system.

Together, science and technology have enabled virtually exponential population growth. Increasing understanding of phenomena such as climate change has alerted us to the ill effects of overpopulation and overconsumption on the global environment, but our attitudes will determine whether or not we adapt our life-styles to these realities. Controlling population growth, encouraging conservation, and learning to rely on renewable and perpetual resources requires efforts beyond science—in politics, philosophy, and societal values and priorities.

Summary

- Science is both a body of knowledge and a way of understanding the world and its processes. It is based on the assumption that natural events have physical, discoverable causes. The fundamental tool of science is the scientific method, which entails systematic, disciplined testing of hypotheses by careful experimental design, methodical data collection, and objective reasoning and analysis.
- Technology refers to the practical applications of scientific knowledge such as the production of objects necessary for human sustenance and comfort.
- Major scientific discoveries and revelations often are associated with scientific revolutions, when dominant views of nature are challenged and new views developed. The latest revolution in Earth sciences was the recognition of plate tectonics as a unifying theory. Plate tectonics is based on the observation that Earth's outermost physical layer is broken into at least a dozen plates, which are constantly in motion.
- Biological species have evolved and become extinct throughout much of Earth history, but more species have existed in the past few million years than ever before. However, recent human activities—in particular the destruction of habitat—are causing global mass extinctions.
- Four issues are especially critical in the environmental geosciences: human population growth, use of Earth resources, pollution and environmental degradation, and geologic hazards.
- Earth can be viewed as a system in which matter and energy are cycled through its many interconnected parts. The study of how the whole Earth is affected by a change in any of its subsystems is called Earth system science.

Key Terms

scientific method (p. 5)
hypothesis (p. 5)
theory (p. 5)
theory of plate tectonics (p. 11)
biodiversity (p. 14)
exponential growth (p. 16)
potentially renewable resource (p. 17)
nonrenewable resource (p. 18)
perpetual resource (p. 18)
pollution (p. 20)
people overpopulation (p. 20)
consumption overpopulation (p. 21)
pollution overpopulation (p. 21)
natural disaster (p. 22)
systems thinking (p. 24)
system (p. 24)
Gaia hypothesis (p. 25)
Earth system science (p. 25)

Review Questions

1. How does science differ from pseudoscience? Explain and give an example of each.
2. How are science and technology different from, and similar to, each other?
3. What are the basic steps in the scientific method?
4. What are scientific models, and why are they useful for making predictions about the environment?
5. How did the scientific findings of Galileo, Newton, Darwin, and Wegener change the perceived relationship between humans and nature?
6. What is gravitation and how does it affect Earth's processes?
7. Why has the number of species on Earth varied over geologic time?
8. What is the danger of exponential growth in human populations?
9. What types of overpopulation have been identified and what are some useful approaches to the environmental problems they present?
10. How do potentially renewable, nonrenewable, and perpetual resources differ from one another? Give examples of each.
11. What is a natural disaster? Give an example. What are some similarities between geologic and human-induced disasters?

Exercises

1. If the human population doubles in the next 40 years, what will it be? Do you think it will reach this number? Why or why not?
2. Why does a highly industrialized nation, such as the United States, have as much environmental impact, or more, as a developing country with many more people, such as China? (Use the basic form of the equation for environmental impact to explain your answer.)

Suggested Readings

Allen, John (ed.), published annually. *Annual Editions: Environment (year)*. Guilford, CN: The Dushkin Publishing Group.

Cohen, Joel E., 1995. *How Many People Can the Earth Support?* New York: W. W. Norton.

Horiuchi, Shiro, 1992. "Stagnation in the Decline of the World Population Growth Rate During the 1980s." *Science* 257:761–765.

WorldWatch Institute, published annually. *State of the World*. New York: W. W. Norton.

Meadows, Donella, Behrens III, Willia, Meadows, Dennis, Randers, Jorgen, 1972. *The Limits to Growth*. New York: A Potomac Associates Book, Universe Books.

National Aeronautics and Space Administration, Earth System Sciences Committee, 1988. *Earth System Science: A Closer View*. Boulder, CO: University Corporation for Atmospheric Research.

Population Reference Bureau, published annually. *World Population Data Sheet*. Washington, DC: Population Reference Bureau.

Wilson, E. O., 1995. *The Diversity of Life*. Cambridge, MA: Belknap Press of Harvard University Press.

World Resources Institute and International Institute for Environment and Development, published annually. *World Resources*. New York: Basic Books.

CHAPTER 2

After 600 years of quiet, Mount Pinatubo—a volcanic mountain in the Philippines—erupted in June 1991. Its plume of ash and gas affected global climate for several years.

Dynamic Earth Systems

On our amazing Earth a natural event at one location can have complex and far-reaching effects. When a large volcano such as Mount Pinatubo erupts, the effects are felt around the world for years, from the movements of Earth's crust that cause a volcanic eruption, to the debris that spreads through the atmosphere and circles the globe, and to the ensuing changes in climates in different parts of the world. Such interconnectedness is the result of linked entities of matter and energy known as Earth systems. To understand Earth, its processes, and the effects of those processes on us, we need to understand these smaller systems that make up the whole Earth system. Consequently, in this chapter we will:

- Examine the concept of systems and why it is a powerful tool for understanding how Earth works.
- Briefly review Earth's planetary evolution and the major systems that make up the planet.
- Identify the forces that drive Earth processes and examine how feedback mechanisms either amplify or regulate them.
- Look at ways in which the rock cycle and the hydrologic cycle circulate matter and energy through the whole Earth system over time.

Public concern over the threat of global warming has been mounting. The 1980s and 1990s brought alarming evidence that industrial, automobile, and other sources of emissions from the combustion of coal, oil, and gas were releasing enough carbon dioxide to trap increasing amounts of heat in the atmosphere. At the same time, forests all over the world were being cleared for lumber and grazing land, destroying millions of acres of trees and other green plants that absorb carbon dioxide. Global temperature had been rising and was expected to rise even higher. Yet the early 1990s brought several years of global cooling.

This drop in global temperature had nothing to do with human activity. It was due, instead, to the massive release of Sun-blocking gases and ash from the catastrophic 1991 eruption of Mount Pinatubo in the Philippines. The years 1992 and 1993 were marked by "volcano weather," a pattern of weather anomalies that seem to be linked to the effects of a volcanic eruption. Some of these weather oddities caused considerable damage:

- August 1992—Hurricane Andrew in the Bahamas and the southeastern United States killed 62 people and injured hundreds, destroyed 25,000 homes and wrecked 100,000 more. Total damage was estimated at $20 billion to $25 billion.
- September 1992—Hurricane Iniki in Hawaii killed 3 people, left 100 people injured, and damaged or destroyed thousands of homes. Total damage was estimated at $2 billion.
- Summer 1993—Prolonged rainfall and flooding in the western United States caused more than 50 deaths and did $12 billion worth of damage.
- August 1992 to August 1993—Average global temperature decreased by 0.5°C, as compared to a 30-year average, a temporary reversal of the trend of increased global warming since the late 1800s.

The eruption of Mount Pinatubo did more than cool the planet with particles that blocked sunlight. It also spread shattered rocks and ash across the landscape, clogging streams, causing floods, and altering storm systems all over the world. These global ramifications of a single geologic event illustrate why an Earth system approach is necessary for understanding complex, interacting Earth systems and chains of events.

The sequence of linked atmospheric and hydrologic events that might have led to the hurricanes, storms, and global cooling of 1992 and 1993 began beneath Earth's rocky crust. Mount Pinatubo is located where buoyant molten rock, called magma, rises, collects in magma chambers beneath the surface, and eventually erupts onto the surface.

After more than 600 years of calm, the magma collecting beneath Mount Pinatubo began its final ascent in the spring of 1991. Carefully monitoring the mountain's unrest with equipment to detect changes in location of the rising magma, scientists predicted the occurrence of the cataclysm to within hours, saving thousands of lives by enabling timely evacuation. On June 15, magma reached the surface in a catastrophic eruption that killed 900 people who had remained in the area and destroyed the homes of 200,000 others. Nearly 7 km^3 of debris was hurled 30 km into the atmosphere (7 cubic kilometers is equivalent to a cube of about 1.2 miles in each dimension). The airborne debris included almost 4 km^3 of tiny solidified particles called volcanic ash and 30 billion kg (kilograms) of sulfur dioxide gas. This gas had been dissolved in the magma, but once in the atmosphere it reacted with water vapor to form small particles of sulfuric acid. The volcanic ash and sulfuric acid particles had circled the globe within 22 days and persisted in the atmosphere for over 2.5 years (Figure 2-1). Because these materials reflect solar radiation, less sunlight than usual reached Earth's surface and global temperature fell by nearly 0.5°C within a year.

The debris from the eruption also affected soil and water systems over a large area. Ashfall deposits more

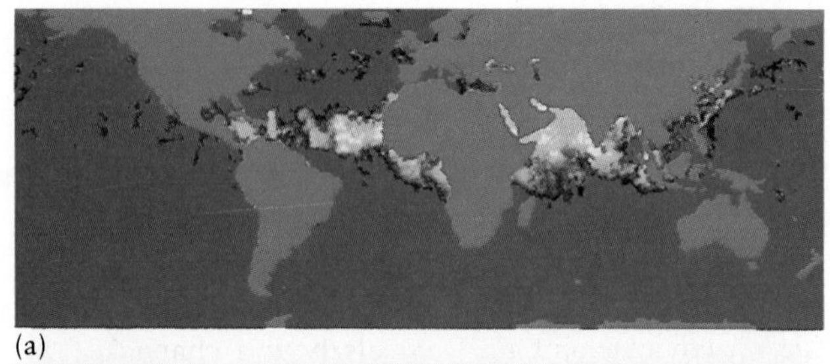
(a)

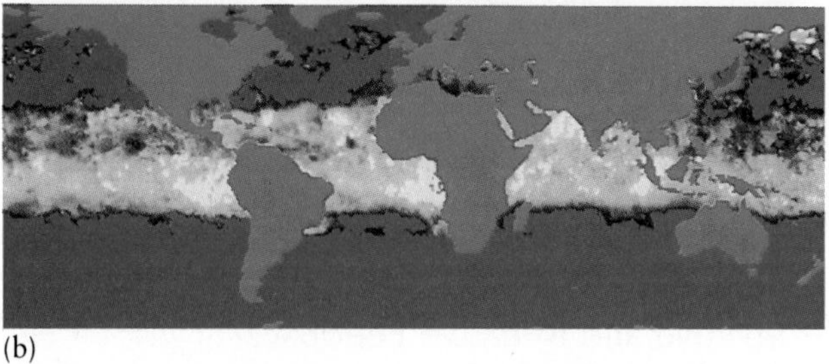
(b)

FIGURE 2-1 Maps compiled from satellite data document the distribution of volcanic ash and acid particles from the June 15, 1991, eruption of Mount Pinatubo. Concentrations are shown as shades of yellow—brown indicates the lowest concentration, white the highest. (a) Data recorded June 19–27 maps a nearly normal distribution of these particles in the atmosphere, with a concentration only slightly heavier than normal over the Indian Ocean. (b) Data recorded August 8–14 maps a vast belt of particles circling the globe around the equator.

than 5 cm thick covered a 4000-km^2 area around Mount Pinatubo. With time, the volcanic ash deposits will enrich the soil and increase its agricultural productivity, just as previous eruptions have done for millennia, but their immediate effect was to destroy plant life. Loose ash combined with rains from a coincidental typhoon and formed mudflows—thick slurries of mud and water—that spread out as far as 40 km from the volcano's flanks, choking streams and flooding valleys.

The eruption of Mount Pinatubo was exceptionally large and destructive, but otherwise it was just one of many volcanic eruptions that occur each year. Throughout most of its 4.6 billion year history, volcanic eruptions have modified the surface of Earth by bringing molten rock to the surface and releasing gases, including water vapor, into the atmosphere. In fact, the gases released from volcanic eruptions formed Earth's early atmosphere, and the water vapor that returned to Earth's surface as rain filled its ocean basins. Where two moving plates collide and one is subducted—that is, it sinks beneath the other—sediment, rock, and water are dragged back into Earth's hot interior. Partly as a result of this cyclical process of destruction and renewal of rock, environments at the surface are ever-changing, remaining stable for only a few million years at most.

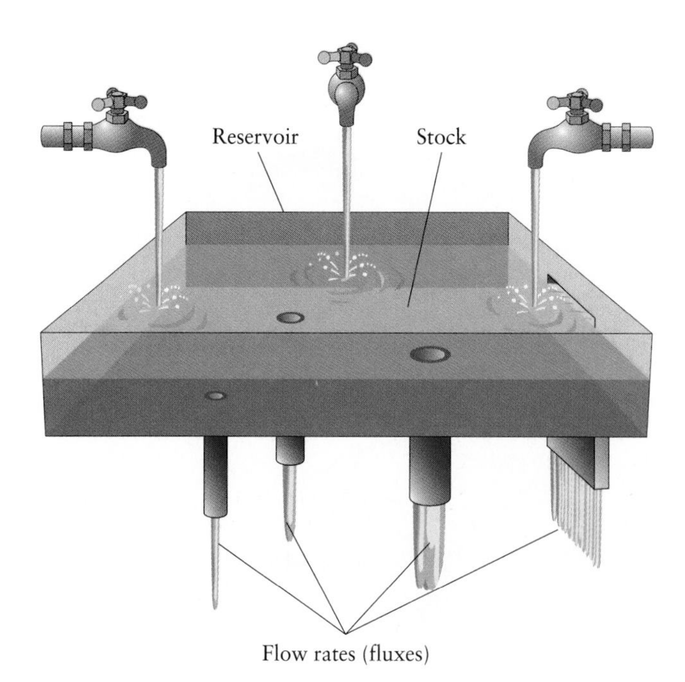

FIGURE 2-2 A system consists of reservoirs, which contain a particular material, such as water in this example of a sink. The content of the reservoir is the reservoir's stock, and the flow of that stock from one reservoir to another is called the flow rate, or flux. The stock is maintained in a steady state if the inflows and the outflows are in balance.

The Concept of Systems

Understanding chains of events such as those linked with the eruption of Mount Pinatubo is made simpler by viewing Earth as a whole system composed of interconnected systems. A system, you recall from Chapter 1, is a group of interrelated and interacting objects and phenomena. The Earth system approach is a way of modeling an environment that reveals how a change in one system affects the whole Earth system.

Systems consist of reservoirs and fluxes. A **reservoir** is a container in which a store or supply of a particular material resides. A sink, for example, may hold in reserve a given volume of water, maintained by running into the sink as much water as drains out (Figure 2-2). The content of a reservoir, whether matter or energy, is known as its **stock. Fluxes** are the movements of material and energy from one reservoir to another. Fluxes, also called flow rates, are measured in amounts per unit of time. The sink is a reservoir, the amount of water that it holds in reserve is its stock, and the rates at which water flows into and out of the system are its fluxes. One focus of the systems approach is to learn, by measuring the fluxes of energy and matter into and out of a system and from one reservoir to another, how systems stay in balance. In nature, we would look at a lake, for example, much as we do the sink. The quantity of water in the lake depends on the flux of water into the lake with streams and precipitation and the flux of water out of the lake with streams and evaporation.

Types of Systems

A system can be described as open, closed, or isolated, depending on how freely matter and energy can cross the system's boundaries (Figure 2-3). An **open system** allows both matter and energy to flow in and out. A lake is an open system because matter (gases, water molecules, sediment, organic materials) and radiant energy (energy from the Sun) both can enter and leave the lake. Open systems are the most common in nature. **Closed systems** are characterized by the ability to exchange energy but not matter across their boundaries. Finally, an **isolated system** does not allow either energy or matter to *cross* its boundaries; in other words, an isolated system has no interactions with its surroundings. Although closed systems rarely occur naturally on Earth and no system on Earth can truly be isolated, scientists sometimes will assume that a system is completely closed or isolated in order to simplify their model and calculations. For example, scientists sometimes treat the whole Earth as a closed system because very little matter enters or leaves

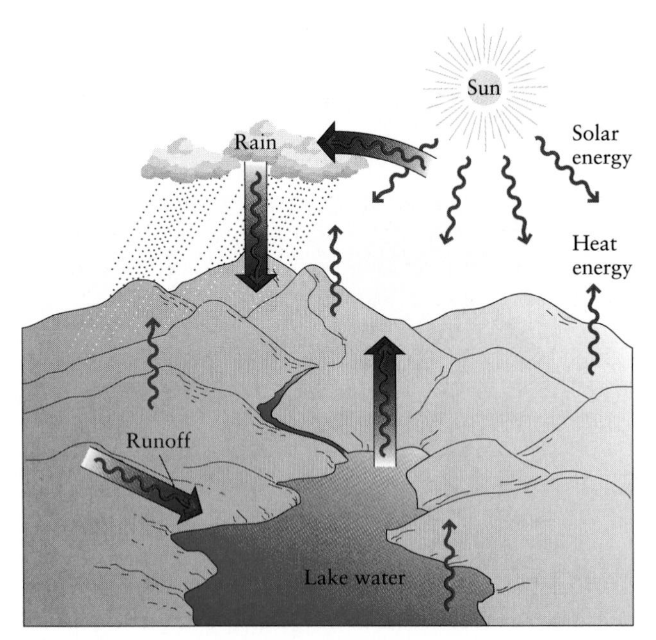

(a) Open system

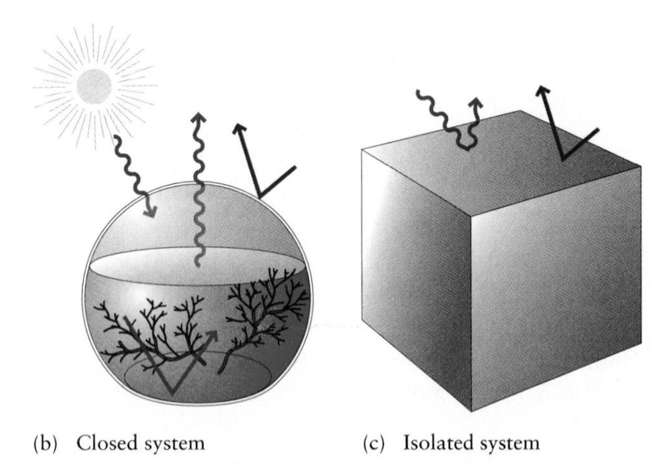
(b) Closed system (c) Isolated system

FIGURE 2-3 Systems are classified as open, closed, or isolated depending on the nature of their boundaries and interactions with their surroundings. (a) An open system allows matter and energy in and out. (b) A closed system allows only energy in and out. (c) An isolated system blocks matter and energy from entering or leaving.

it over time periods of millions of years. Aside from the occasional meteorite and small amount of cosmic dust to reach Earth's surface, and the minor amounts of light elements that escape from the outer atmosphere, Earth's mass has been relatively constant for more than 4 billion years.

Systems can further be described as dynamic or static. A **dynamic system** is one in which energy is used to do work that causes the condition, or state, of the system to change with time. A **static system** is one in which no work is done and no change in state occurs. **Energy** is the ability to do work, which is defined as change effected when a force is applied (see Box 1-2). Energy is so important to the operation of environmental systems that it sometimes is treated as a system in itself. The manner in which changes to a system occur is known as a *process;* common processes that enact change in environmental systems are volcanism, mountain building, erosion, and flooding.

Most Earth systems are dynamic. Life on Earth, for example, is a dynamic system in which processes such as growth and reproduction are powered by energy from the Sun. The state of this system has changed with time in that the numbers and types of species on Earth have increased for several billion years and occasionally have plummeted during periods of major extinctions. In contrast, the Moon is a relatively static system in which little change occurs with time, and its appearance does not alter.

Dynamic Systems That Tend Toward a Steady State

In most dynamic systems, the flow of matter or energy or both into a reservoir is equal to the flow out. As a result, the stock of the material or energy in the reservoir remains constant with time. Systems in which inflows tend to balance outflows remain in a **steady state.** The ocean, for example, is considered to be in a steady state because its stock of water and dissolved substances has changed very little over the past few thousand years

Further in the past and over a longer time period, however, the dynamic ocean system was not always in a steady state. About 20,000 years ago, sea level was nearly 120 m lower than it is today because ice sheets were much more extensive during the full-glacial conditions prevalent at that time and much of the world's water was frozen in large ice sheets (Figure 2-4). Between 18,000 and 6000 years ago, global warming melted all but the Greenland and Antarctic ice sheets. The ocean reservoir could not maintain its former steady state because its fluxes were not in balance; the inflow of water from melting ice was greater than the outflow of water through evaporation. As a result, the stock of ocean water increased with time and caused a rise in sea level.

- A system is a group of interconnected and interacting objects and phenomena that can be separated into reservoirs (or stocks) of matter and energy, connected by fluxes of matter and energy from one reservoir to another.
- In an open system, both matter and energy can flow back and forth across the system's boundaries. In a closed system, only energy can flow back and forth across the system's boundaries.
- Dynamic systems change with time. Most Earth systems are both open and dynamic.

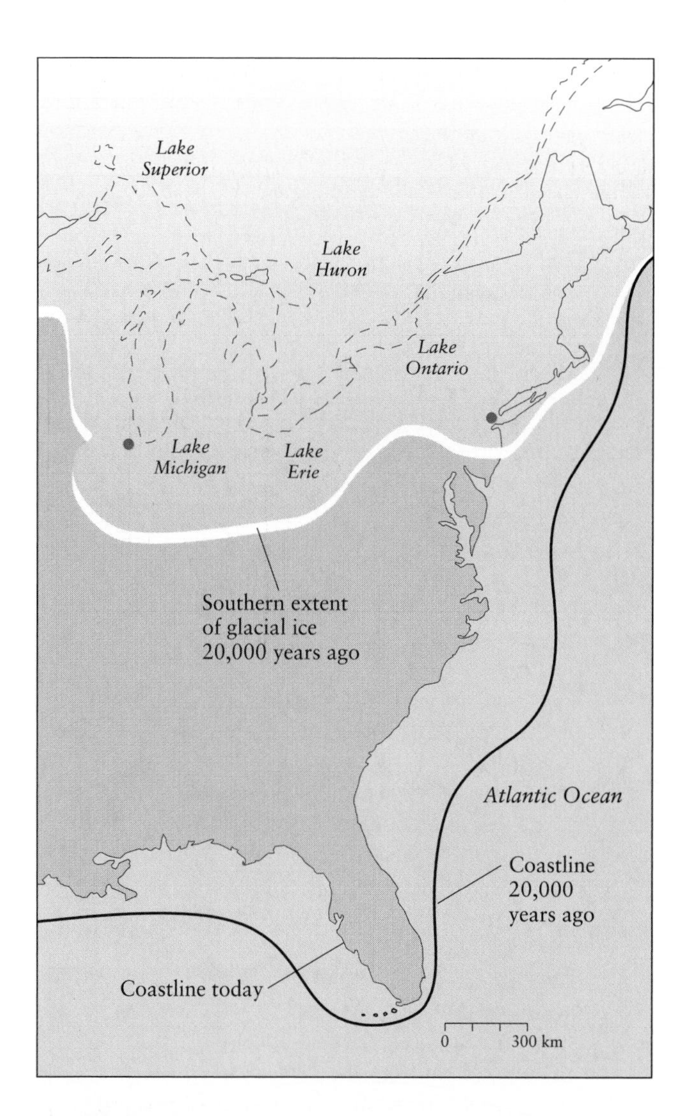

FIGURE 2-4 Twenty thousand years ago, ice sheets existed in Greenland, Antarctica, North America, and northern Eurasia, so sea level was some 120 m lower than it is today. Since then, the North American and Eurasian ice masses melted and sea level rose to its present position, remaining relatively stable for about the past 6000 years.

The Planetary Evolution of Earth

The whole dynamic Earth system is powered partly by the planet's own internal energy but mostly by energy from a star—the Sun—around which it revolves. Earth's internal and environmental systems are rooted in the way the planet evolved within our solar system, which itself evolved within the universe. Years of research by astronomers and physicists have produced a widely accepted theory of the origin of the universe, stars, and planets. Properly called the cosmic singularity theory, it is more commonly referred to as the theory of the "Big Bang."

Origin of the Universe and Our Solar System

According to the **Big Bang theory,** some 10 billion to 20 billion years ago an explosive cosmic event occurred from which all energy and matter are derived. Unimaginably hot, pure energy in the form of gamma rays filled the universe, cooling as it expanded and decaying to form subatomic particles of matter. [The physicist Albert Einstein (1879–1955) explained the basic equivalence of energy and matter as $E = mc^2$, in which E is the energy released by conversion of mass (m), and c is the speed of light, about 298,000 km per second.] Within a few minutes, these particles combined to form protons and neutrons, the building blocks of atomic nuclei (Box 2-1). The number of protons in an atom determines its chemical identity, and neutrons add to the atom's mass.

The hydrogen nucleus, with only one proton and one neutron, was the first type of atomic nucleus to form. Because temperatures and densities were still extremely high, nuclear reactions occurred that fused pairs of hydrogen nuclei into single helium nuclei. By the time the universe was about three minutes old, it had expanded and cooled to the point at which nuclear fusion stopped. Of the atomic nuclei that already had formed, about 75 percent were hydrogen and 25 percent were helium. Gravity drew the hydrogen and helium together, condensing them and thereby heating them into glowing gas clouds from which, hundreds of millions of years later, galaxies of stars were born.

Within galaxies, dust and gas left over from initial star formation continuously gravitate into clouds called *nebulae* (the Latin word for "clouds"). One star-forming nebula active today is shown in Figure 2-5. Scientists have evidence that our Sun and solar system evolved from such a nebula. Every star, including our Sun, is composed mostly of hydrogen and helium held together by its own gravity and generating energy through nuclear fusion reactions at its center. Continued nuclear fusion and other reactions in the bodies of stars have created all the heavier elements in the universe.

The *nebular hypothesis* of the origin of solar systems holds that nebulae condense to become stars, and planets are created as a by-product of star formation. Gravitational attraction draws the matter in a slowly rotating nebula into a flattened whirling disk, with the bulk of the matter at the center, where it develops into a star. Around this central star, the remaining materials in the nebula gravitate into larger bits of rock and metal that coalesce into planets orbiting the new star. Our solar system is believed to have formed this way about 4.6 billion years ago (Figure 2-6).

Ours is not the only planetary solar system in the universe. Our Sun is one of the 500 billion stars in the Milky Way galaxy, and astronomers estimate that there are at least 100 billion more galaxies in the universe. Star formation continues, and it seems that the universe still is

FIGURE 2-5 Astronomers propose that our solar system developed about 5 billion years ago from a spherical, rotating cloud of stars, dust, and glowing gas called a nebula. The Trifid Nebula, observed with a powerful telescope, is an example of a stellar "nursery," where stars are born.

making planets as well, because astronomers have identified dust, debris, and planets around other stars. Somewhere in the universe, stars and solar systems similar to our own probably exist.

Differentiation of Earth

During the first half-billion years of Earth history, so much debris was present in the solar system that Earth was bombarded by meteors, which are bits of rocky matter left over from the period of planet formation. As the Sun became luminous, its radiation created solar winds, which forced much of the remaining debris out of the solar system. Since then, Earth's mass has remained essentially static, with only relatively minor additions from cosmic dust and occasional meteors. Earth's chemical and physical characteristics, however, had just begun to take shape.

Soon after it formed from accreting debris, our planet began to heat up. Heating resulted from three

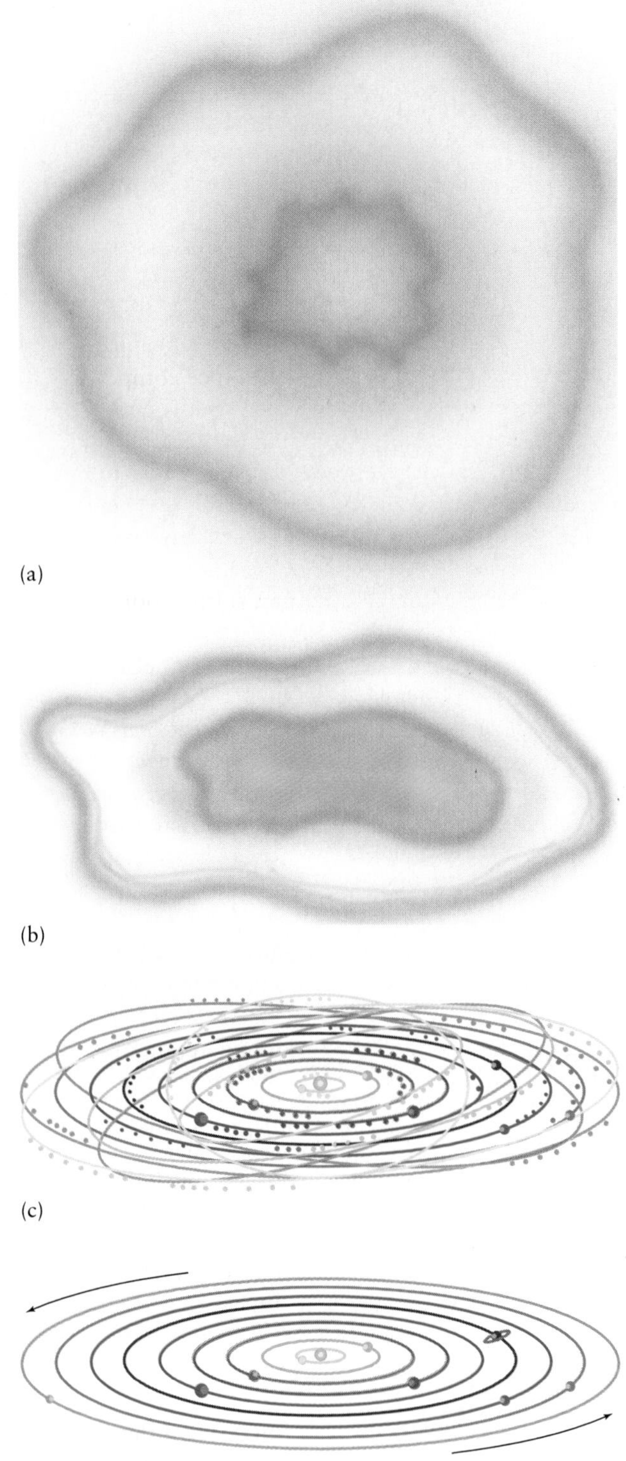

FIGURE 2-6 Evolution of the solar system. (a) Matter is drawn together in a nebula because of the force of gravity. (b) The roughly spherical, slowly rotating nebula contracts, forming a flattened rapidly rotating disk with matter concentrated at the center. (c) The center of the disk becomes a star, about which revolve rings of debris. In the case of our solar system, this star is the Sun. (d) Material in the rings condenses into planets.

2-1 Geologist's Toolbox

Atoms, Elements, and Bonds

Early in the 19th century, chemists realized that matter is made of submicroscopic units. Some of these tiny units could not be further subdivided; these were named *atoms*, from a Greek word meaning "uncuttable." An **element** is a substance composed of only a single type of atom, and an **atom** is defined as the smallest particle of an element that retains the characteristics of the element. Your body is made up of about 5.3×10^{27} atoms, of which 63.1 percent are the element hydrogen, 25.6 percent are oxygen, 9.5 percent are carbon, 1.3 percent are nitrogen, and the remainder are mainly calcium, sulfur, sodium, magnesium, and iron. Most of the atoms in your body occur in chemical combinations called compounds.

When atoms combine chemically, they lose their individual characteristics and form distinctive particles called molecules. A **compound** is a substance composed of a single type of molecule, and a **molecule** is defined as the smallest particle of a compound that retains all the characteristics of the compound. The most abundant compound on Earth's surface and in your body is water.

Each element is assigned a symbol, often but not always the first two letters of its name (e.g., calcium is Ca), on a chart called the periodic table of elements. The chemical formula for a compound contains the chemical symbol for each type of element it contains, followed by a subscript number if more than one atom of a particular element is present. Water, for example, is a compound that consists of two atoms of hydrogen and one atom of oxygen, so the chemical formula for a molecule of water is H_2O. The characteristics of water are very different from the characteristics of hydrogen and oxygen gases. A single molecule of water retains all the characteristics of water, but if water is broken down into its component elements it ceases to be water.

Observing that different elements have different masses per unit of volume (a cubic centimeter of hydrogen weighs far less than 1 cm^3 of lead, for example), scientists next began to determine the masses of atoms with respect to each other. This relative mass is called *atomic mass*. The symbol for atomic mass is μ, the Greek letter *mu*. If atoms have mass and their masses differ, atoms must be made of even smaller particles that themselves have mass. In the beginning of the 20th century, scientists discovered that atoms were indeed made up of subatomic particles.

The first subatomic particles to be identified were **protons**, which have a positive electrical charge, and **electrons**, which have a negative electrical charge. In 1932, scientists identified **neutrons**, which have no electrical charge. The simplified model of the atom that we now use consists of a central core or *nucleus* (containing protons and neutrons) with electrons moving in orbits around the nucleus. Protons and neutrons have very similar masses (1.0072764 and 1.0086649 μ, respectively), while electrons are much smaller and less massive (0.00005486 μ).

Scientists now refer to elements by their *atomic number*, the number of protons in the nucleus, because the number of protons determines an element's physical and chemical nature. It was found that each element is unique because its atoms have a specific and unique number of protons. The modern periodic table reflects this by being based on the number of protons (see Appendix 2). The arrangement of elements in the periodic table reflects increasing atomic number and complexity of atomic structure.

In an uncharged atom, the number of protons and electrons must balance, so the more protons in an atom, the more electrons it contains. Electrons are thought to occupy spheres, called *shells*, in which the electrons spin about a central nucleus containing the relatively more massive protons and neutrons. The simplest configuration of subatomic particles is one proton (+1) in the nucleus and one electron (–1) orbiting it. This arrangement forms the hydrogen atom, with an atomic number of one and a net charge of zero. If an atom has two protons (+2), 2 neutrons (0), and 2 electrons (–2), it is the element helium, with an atomic number of 2 and a net charge of zero. Each successive element as arranged in the periodic table has a greater number of protons, neutrons and electrons.

In all atoms, the protons and neutrons are packed closely together, while the electrons orbit relatively far away. Even so, the atoms are very tiny, with an average diameter of 10^{-10} m. If the size of the nucleus is

(continued)

2-1 Geologist's Toolbox

(continued)

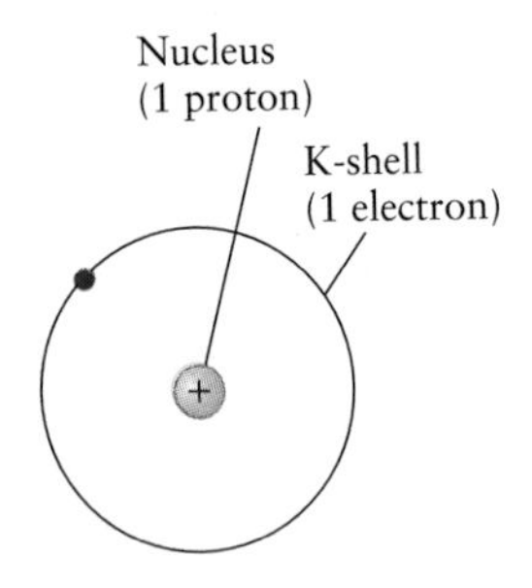

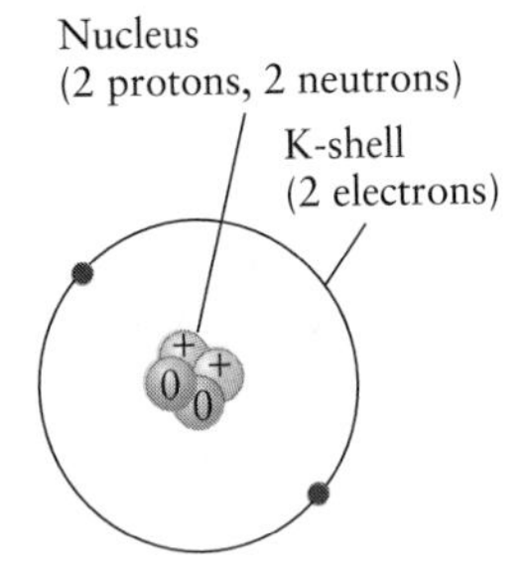

The atomic structure of hydrogen and helium. The drawings are not to scale because the nuclei are much, much smaller than is possible to show.

imagined to be as large as the head of a pin, the electrons would be about 100 m away. From the nucleus outward, the electron shells are labeled K, L, M, N, O, P, and Q. Each shell going out from the nucleus can hold more electrons. The K shell can accommodate only two electrons, while the Q shell can accommodate 98.

Shell	Maximum number of electrons possible in shell	Example
K	2	Hydrogen K:1
L	8	Carbon K:2; L:4
M	18	Silicon K:2; L:8; M:4
N	32	Iron K:2; L:8; M:14; N:2
O	50	Silver K:2; L:8; M:18; N:18; O:1
P	72	Gold K:2; L:8; M:18; N:32; O:18; P:1
Q	98	Uranium K:2; L:8; M:18; N:32; O:21; P:9; Q:2

The columns in which elements are located in the periodic table are related to their chemical properties, which in turn are linked to the number of transferable electrons in their outer shells (*valence electrons*) and hence to their tendency to gain, lose, or share electrons. As atomic number increases, electrons fill up the innermost, or K, shell, first, and then start to fill the L shell. For example, the hydrogen atom has one electron in the K shell, whereas the helium atom has two. The K shell is full in the case of the helium atom, but hydrogen has a strong tendency to lose its only electron.

Ions, charged atoms, are formed when atoms lose or gain electrons from partially filled shells. An atom that loses an electron has a net positive charge of (+1) and is called a **cation.** Atoms that gain electrons and have a net negative charge are called **anions.** Those elements in the first and second columns of the periodic table, which include hydrogen, have a strong tendency to lose electrons from their outer shells. Those in the last column, which includes helium and neon, have completely filled outer shells and have no tendency to gain or lose electrons. Ions of opposite electrical charge are attracted to one another and can form *ionic bonds,* as in the example of sodium and chlorine. When atoms share electrons in their outer shells with other atoms, they form *covalent bonds.*

All the elements known to exist in nature can be constructed from atoms formed by combinations of different numbers of protons, neutrons, and electrons. Two elements that are environmentally important are carbon and nitrogen, both of which cycle through Earth's outermost systems and form important nutrients essential to life. Other elements of environmental importance include sulfur, which is a main component of acid rain and acid mine drainage, and chlorine, which is associated with ozone depletion. Radon, a radioactive gas, and its decay products pose an environmental threat to human health, as they are carcinogens.

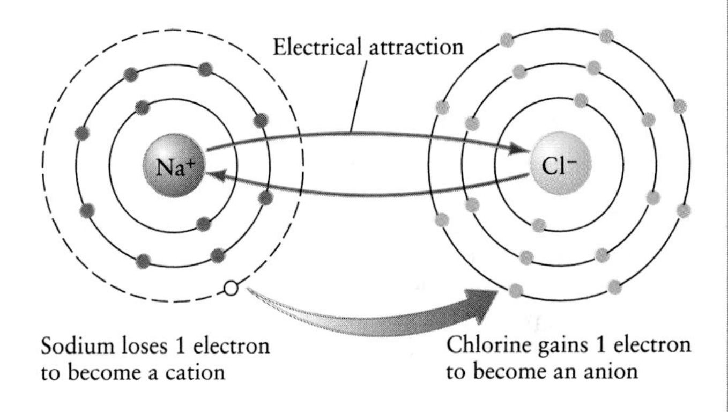

Sodium and chlorine atoms become ions and bond with each other to form a chemical compound, sodium chloride (same as table salt).

(Opposite) Electron shells of atoms. In some elements, such as those shown here, the outermost electron shell is only partly filled. For example, the L shell can hold 8 electrons, but the carbon atom, with 6 electrons—2 of which are in the K shell—has only 4 electrons in the L shell. The element with atomic number 10, neon, has just enough electrons to fill completely the K and L shells.

processes: collision of debris that coalesced to form Earth, compression of debris in the growing planet, and radioactive decay of chemical elements. The temperature of the interior of Earth rose to about 2000°C, high enough to melt iron. Between 4.5 billion and 4.3 billion years ago, the intense heat of the primordial Earth melted much, or perhaps all, of its mass, initiating a process that formed concentric layers with different chemical compositions and physical characteristics.

Chemical Differentiation With Earth largely molten, most of its materials moved freely. Nickel and iron, two of the heaviest elements, sank toward Earth's center to form the dense **core** of the hot, liquid Earth. Lighter elements, particularly oxygen, silicon, aluminum, magnesium, calcium, potassium, and sodium, floated toward Earth's surface. When Earth's surface began to cool and solidify, these elements formed a rigid **crust,** which is Earth's thinnest and least dense rock layer. Between the core and the crust lies the **mantle,** a vast layer of rock rich in oxygen, silicon, iron, and magnesium and therefore of a composition and density intermediate between the crust's and the core's (Figure 2-7). Most of the rocky matter in Earth's crust and mantle consists of **silicate minerals,** which are solid compounds formed largely from silicon and oxygen (Box 2-2).

Physical Differentiation There are physical as well as chemical distinctions between Earth's layers, the result of differences in the structure of minerals and in the way they behave under different conditions of temperature and pressure. Heating a solid makes it less dense and more malleable. Compressing a solid makes it denser and more rigid.

From Earth's surface to its inner core, both pressure and temperature increase (Figure 2-8). Within Earth's crust alone, the geothermal temperature gradient is about 1°C per 28 m: If you descend vertically 28 m into a mine shaft, you will feel a temperature increase of 1°C. At the bottom of the deepest mine on Earth, about 2 km down, the temperature is about 71°C, or 128°F! Below 70 km the temperature is high enough relative to the pressure of the overlying rock to cause the solid mantle to become somewhat ductile, or plastic. At about 70 km, therefore, a boundary separates a rigid layer called the **lithosphere** from the underlying, ductile layer, which is called the **asthenosphere,** even though no compositional change or melting occurs. The asthenosphere extends beneath the lithosphere from 70 to 200 km. Lower than 200 km, increasing pressure results in a layer of dense, more rigid rock known as the **mesosphere** (from the Greek word *mesos* for "in the middle"). Between 2891 and 5150 km, the residual temperature from when Earth melted is so high that its **outer core** is still molten, despite the great pressure from the overlying rock. At depths greater than 5150 km, however, the pressure is so immense that the **inner core** is solid, even though it is compositionally similar to the outer core and at least as hot.

Additional layers exist above the lithosphere and outside Earth's surface, forming the soils, oceans, atmosphere, and other environmental systems. As we learned from the volcanic eruption of Mount Pinatubo, our planet's interior systems have considerable influence over its environmental systems.

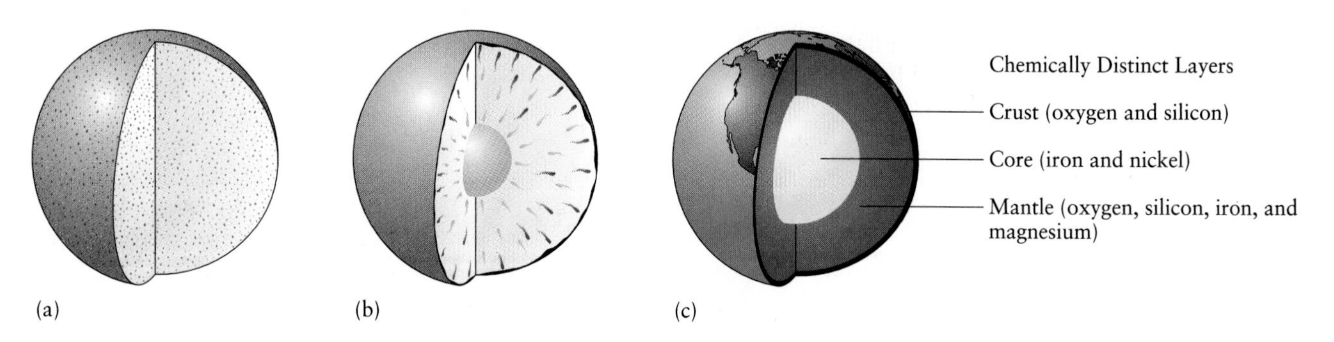

FIGURE 2-7 (a) Early Earth was a homogenous, rocky mixture of mostly iron, oxygen, silicon, and magnesium; there were no continents, oceans, or atmosphere. (b) During the process of differentiation that resulted from catastrophic melting of most or all of Earth's mass, dense elements sank to the center and light elements floated upward. (c) As a result, the Earth is compositionally layered in concentric spheres from a dense core rich in iron and nickel, to a thick mantle of mostly oxygen, silicon, iron, and magnesium, and a light crust of mostly oxygen and silicon.

- Astronomers have developed a theory for the origin of the universe, the Big Bang, in which all matter and energy originated from a cosmic explosion 10 billion to 20 billion years ago.
- The universe consists of at least 100 billion galaxies; our own, the Milky Way galaxy, contains some 500 billion stars.
- Solar systems, consisting of stars and orbiting planets, develop from coalescing gases and particles in a cloud of matter known as a nebula.
- Our solar system developed from a nebula about 4.6 billion years ago, and Earth formed from bits of rocky and metallic matter that coalesced into a single planet.
- Early Earth heated up to about 2000°C as it formed, initiating a major period of complete—or nearly complete—melting and resultant segregation of its elements into chemically distinct layers: core, mantle, and crust.
- Because of changes in pressure and temperature with depth, Earth consists of layers that have distinctive physical properties: solid inner core, liquid outer core, solid mesosphere, semisoft asthenosphere, and solid lithosphere.

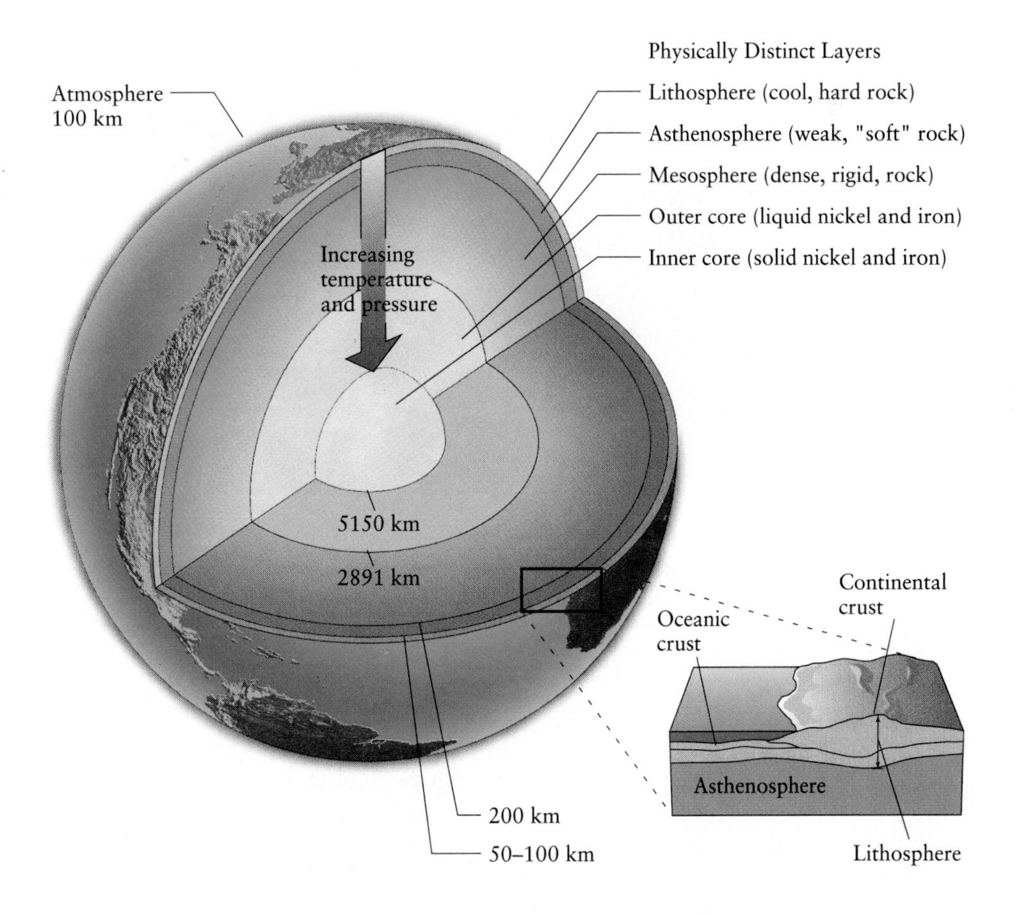

FIGURE 2-8 Earth is zoned physically as well as chemically. The outermost 50 to 100 km, including the crust and top of the mantle, is very hard and rigid, so it can break; this is the lithosphere. Below it lies the asthenosphere, a weak zone with the consistency of stiff taffy. Under the asthenosphere lies the mesosphere, which is solid and hard, and below the mesosphere lies the liquid outer core. The inner core, although chemically similar to the outer core, is solid and rigid because of the immense pressures in the center of Earth.

2-2 Geologist's Toolbox

Minerals

A **mineral** is any naturally occurring element or compound that has a specific chemical composition or range of compositions. Most geologists restrict the definition of a mineral to nonorganic crystalline solids, such as quartz, diamond, and halite (table salt). Each mineral has different properties, which depend on (1) the types of elements of which it is made, (2) the way the atoms are arranged, and (3) the kinds of chemical bonds that hold the atoms together.

Minerals can consist of only one element, as in the minerals gold (Au), diamond (C), and copper (Cu), or two or more elements, as in the minerals quartz (SiO_2), calcite ($CaCO_3$), and the talc used in talcum powder ($Mg_3Si_4O_{10}(OH)_2$). As will be discussed in Chapter 4, the type of element in a mineral provides the basis of a mineral classification scheme, with those minerals containing only one element called native elements, those containing oxygen anions called oxides, those containing sulfur anions called sulfides, and so forth. Each of these groups of minerals has different characteristics derived from the elements that form it.

Many of the major mineral resources contain the metals iron, copper, manganese, and nickel, the so-called transition elements that have valence electrons in more than one outer shell. The ability of these valence electrons to move from one outer shell to another

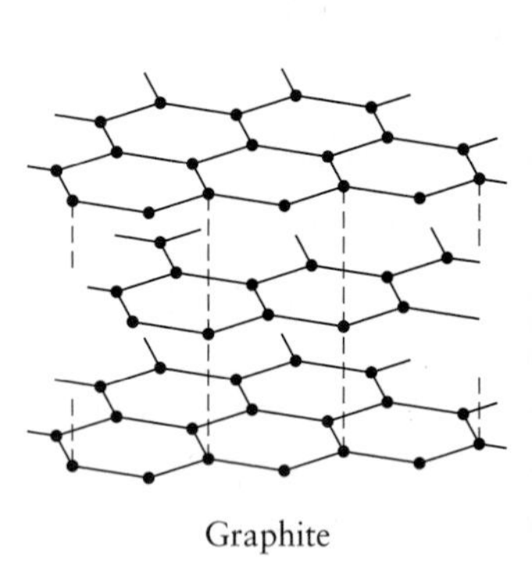

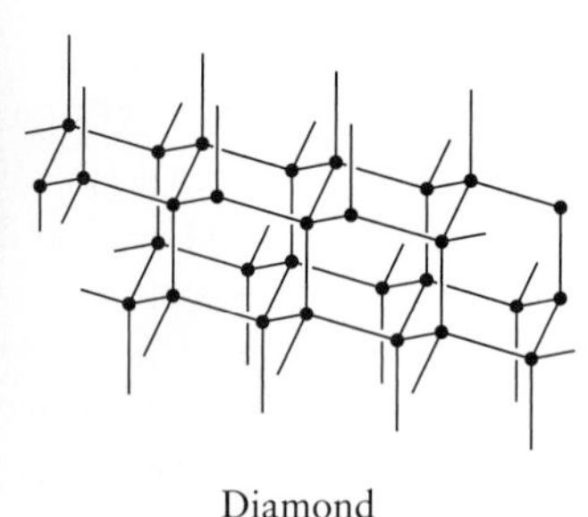

Two different arrangements of carbon atoms result in two different minerals: graphite, one of the softest minerals, and diamond, the hardest known mineral at Earth's surface.

Earth's Environmental Systems

The outer Earth can be organized into five major systems that continuously interact with one another, with solar energy, and with the internal heat energy of the planet to produce climates and environments at Earth's surface. Roughly from the bottom up, these environmental systems are the lithosphere (rock), the pedosphere (weathered, broken particles of rock capped with soil), the hydrosphere (water bodies and frozen water), the biosphere (plant and animal life), and the atmosphere (air). Solar energy, internal heat energy, and gravitational energy also can be treated as a single energy system for the whole Earth. These systems tend to permeate each other. Water, for example, resides not only in large pools such as oceans and rivers but also in the soil, in the air, and in living things. Similarly, living things reside on and in rocks, soils, bodies of water, and even the atmosphere. Nevertheless, each "sphere" is an identifiable reservoir that can be viewed as an open, dynamic system into and out of which flow energy and matter.

Earth's five surface systems differ from each other in composition and physical properties (Figure 2-9). The lithosphere is composed mostly of silicates, solid com-

results in many of the important physical properties of metals, including their ductility and high electrical and heat conductivity relative to other elements.

If a compound has more than one possible arrangement of atoms, each form has a separate mineral name. The minerals diamond and graphite both consist of pure carbon, so one might expect them to have similar properties. Yet diamond is the hardest mineral known on Earth and graphite is so soft that it is used to make pencils, for it leaves a mark on paper when the mineral is broken apart. The difference is due partly to the arrangement of the carbon atoms. Diamond forms deep in the lithosphere, under conditions of very high pressure, so its atoms are densely packed, whereas graphite forms at shallower depths and its atoms are not as densely packed.

Of all the elements on Earth, only a small number are found in abundance in the crust and mantle. Therefore, only a few dozen minerals are common at Earth's surface. Of these rock-forming minerals, the most abundant are the *silicates,* a group of minerals in which silicon and oxygen combine to form the silicate anion $(SiO_4)^{4-}$. In the silicate anion, four oxygen ions surround one silicon ion. Silicon has an atomic radius a little less than one-third that of oxygen. The tight packing of four oxygen atoms about the relatively small silicon atom creates a tetrahedral structure, which has four faces and resembles a pyramid.

Silica tetrahedra, found in most magmas, link together to form a variety of mineral structures, ranging from simple chains of tetrahedra to complex sheets and three-dimensional frameworks. Because these structures and the cations that often join them form most minerals found in Earth's crust, including quartz and mica, we will consider them again in more detail in Chapter 4.

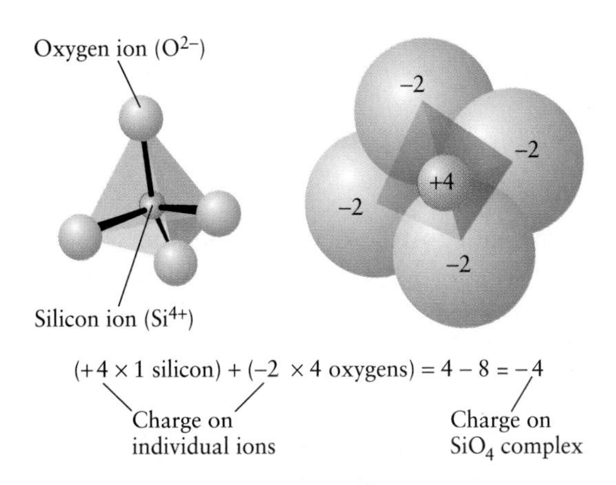

The silica tetrahedron. The drawing on the left shows the tetrahedral structure, a polygon with four faces. On the right, atoms are drawn to scale, showing their sizes relative to one another. The silicon ion is about one-third the size of the oxygen ion.

pounds primarily of oxygen and silicon with a few other elements including aluminum and sodium. The pedosphere is a store of elements derived from the breakdown of rock and carbon-rich organic matter that have recombined into liquid, solid, and gaseous compounds. The hydrosphere is mostly hydrogen and oxygen combined to form water molecules in the gaseous (vapor), liquid (oceans and fresh water), and solid (ice) states. The biosphere is composed mostly of the elements hydrogen, carbon, and oxygen in the solid, liquid, and gaseous states. The atmosphere consists chiefly of free nitrogen and oxygen in their gaseous states.

The Lithosphere: Earth's Rocky Outer Skin

Chemically, the outermost layer of Earth is its largely silicate crust. Physically, the outermost layer is the rigid lithosphere, or rock system, which begins near the top of Earth's mantle and includes the crust. The crust varies markedly both physically and chemically beneath continents and oceans. Both types of crust are relatively light because of their high silicon and oxygen content, but **oceanic crust** contains more iron and magnesium and therefore is denser than **continental crust.** Oceanic crust also is much thinner, averaging about 5 km in thickness

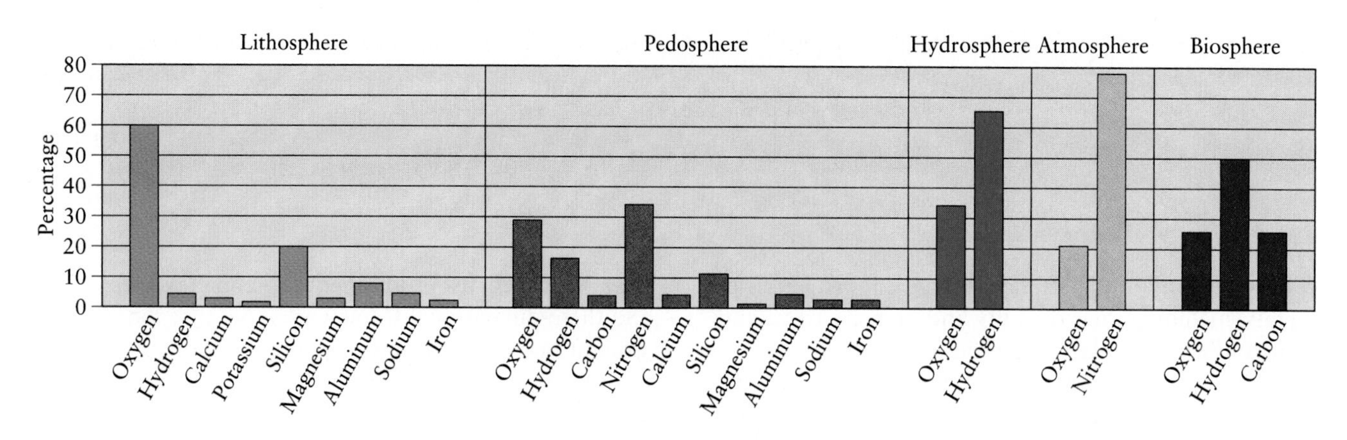

FIGURE 2-9 Composition of each Earth sphere in numbers of atoms relative to a total of 100. (Because some atoms, such as hydrogen, are very small and light, a comparison based on *weight* of atoms rather than *number* of atoms would yield different results.) Both the atmosphere and the pedosphere contain substantial amounts of nitrogen—vital to plant and animal life. In the hydrosphere, oxygen and hydrogen atoms combine to form water molecules, while the biosphere contains relatively large quantities of hydrogen, oxygen, and carbon, the elements from which living organisms are formed.

compared with 35 to 65 km under continents (see Figure 2-8). The rocks and minerals we use in building and manufacturing are all extracted from the lithosphere.

The lithosphere resembles a giant spherical jigsaw puzzle with more than a dozen separate plates drifting about on Earth's surface (see Figure 1-5). The creation, jostling, and destruction of these lithospheric plates generate earthquakes and volcanoes, as we will see in Chapters 4 and 5. The plates move because currents of thermal energy rise from the hot interior of Earth, causing the weak, plastic asthenosphere underlying the plates to flow. A tiny amount of the asthenosphere, about 2 or 3 percent, is liquid. This liquid is the source of most molten rock, or magma, that rises through the lithosphere to Earth's surface, as at Mount Pinatubo in 1991.

Dynamic processes related to the motion, creation, and destruction of lithospheric plates are referred to as tectonism, from the Greek word *tekton,* which means "builder." Tectonic processes have been continuously building, demolishing, and rebuilding Earth's lithospheric plates and surface features for at least the past 2.5 billion years, ever since Earth's crust solidified from cooling magma. Since then, the plates have been moving across Earth's surface, breaking apart, colliding with each other, and continuously changing the locations of continents, ocean basins, mountain belts, and island chains. The collision of lithospheric plates typically causes one to dive down beneath the other into the mantle, a process called *subduction.* Catastrophic events such as earthquakes and volcanic eruptions occur in seconds or minutes, but the tectonic processes that give rise to these events in the lithosphere operate on time scales of hundreds of thousands to millions of years.

The Pedosphere: Where Soils Form

Atop the lithosphere lies the **pedosphere,** the entire layer of disaggregated and decomposed (weathered) rock debris and organic matter at the surface of exposed landmasses (Figure 2-10). Soil is the topmost part of the pedosphere. The prefix *ped,* derived from the Latin word for "foot," is used to refer to this layer because it lies underfoot. The oxygen, water, acids, and organic matter abundant at Earth's surface contribute to the transformation of rock and organic matter into pedospheric materials. With depth and distance from the surface, the pedosphere gradually becomes indistinguishable from the lithosphere.

Soil is the unconsolidated material that forms from rock particles and organic matter at the boundary zone between the pedosphere, hydrosphere, atmosphere and biosphere. Soil is arranged in layers or "horizons," with the horizon poorest in organic matter at the bottom and the horizon richest in organic matter at the top. The thickness of a soil depends on the amount and type of organic matter it contains, and is rarely more than a few meters in any environment. Even though most soils are only 1 to 2 m thick, all terrestrial agricultural activities and food production depend on them for nutrients and support. The total thickness of the pedosphere might be as much as 100 to 200 m in humid, tropical areas where rainfall and temperature are high and weathering processes extend deeply into the crust, or as little as zero in a cold, dry desert. In contrast, the lithosphere is 1000 times thicker.

The pedosphere is an open system into and out of which move charged atoms (ions), solid particles, and gases. Minerals weathered from rocks and sediments are mixed with organic matter produced by plants and animals, and the mixture becomes further decomposed to produce elements and nutrients vital to life. The composition of solid mineral matter throughout the pedosphere is similar to that of Earth's crust, consisting almost completely of only eight elements: oxygen, silicon, aluminum, iron, magnesium, calcium, sodium, and potassium. The organic matter in soil, in contrast, contains other essential elements that are vital to plant life, such as hydrogen, carbon, nitrogen, and phosphorus. As a resource, soils containing both mineral and organic nutrients are vital to food production in agriculture.

Rates of soil-forming processes are affected by climate, rock type, organic matter, topography, and time. Most soil-forming processes operate on time scales of thousands to millions of years. Natural processes of soil erosion on hillslopes generally are similar to those of soil formation. Unfortunately, more rapid processes of soil destruction, such as erosion due to deforestation or poor farming practices, operate on much shorter time scales—years to hundreds of years. If the rate of destruction of soil exceeds that of formation, the soil stock, or reservoir,

FIGURE 2-10 The walls of this deep mine pit in Utah illustrate the zone of disintegration and decomposition of the lithosphere that characterizes the pedosphere, of which soil forms only the topmost meter or so. In this example, the upper zone (several hundred meters thick) has been leached of many elements, including copper, and now is mostly iron and oxygen. However, the copper was redeposited below the leached layer, in the lighter-colored (whitish) zone, forming a rich mineral deposit that now is being mined. The steps, or benches, carved along the walls of the pit enable machinery and trains to enter and leave the mine. The Bingham Mine is 5 km wide, 8 km long, and nearly 2 km deep, making it the largest open copper mine in North America.

will become depleted with time, just as the amount of water flowing into or out of the oceans affects the level of the sea surface.

The Hydrosphere: Earth's Distinguishing Planetary Characteristic

Rain, river flow, and waterfalls are processes that we take largely for granted, but they occur solely on Earth, and not on any other planet in our solar system. Earth is the only planet in our solar system with abundant water and the unique combination of temperature and pressure at its surface that allows water to exist in all three states of matter: solid, liquid, and gas (Figure 2-11). Earth's abundance of liquid water gives the planet a shimmering, fluid, blue appearance when viewed from space (Figure 2-12).

Venus and Mars, like Earth, have atmospheres resulting from outgassing, the escape of gases from their interiors, but temperatures on Venus were higher than those on Earth during early planetary evolution. On Venus, therefore, outgassing was more intense, releasing more water vapor, which the heat broke down into its component atoms, hydrogen and oxygen. The light hydrogen escaped to space, so it could not bond again with the oxygen to form water. On Mars, abandoned river channels indicate that water did flow at the surface in the planet's geologic past, but the distance of the planet from the Sun—50 percent greater than that of Earth—makes it too chilly for water to exist unfrozen today. The planet's surface temperature at present (–60°C near the equator and –123°C near the poles) is far below the freezing temperature of water.

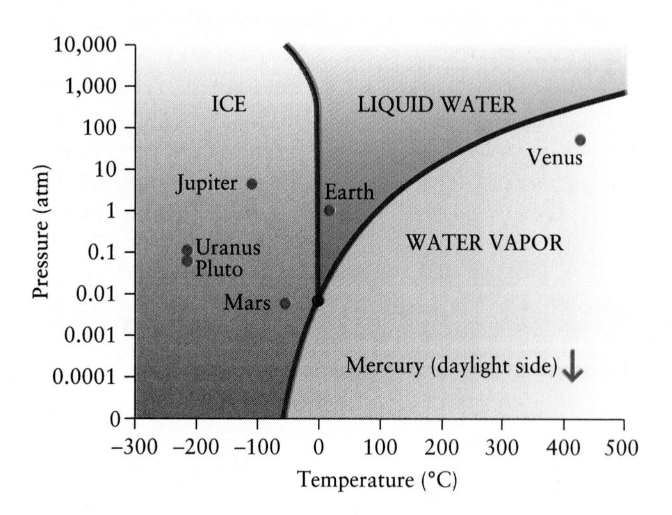

FIGURE 2-11 Temperature and pressure conditions under which water is a solid (ice), a liquid (water), or a gas (water vapor). Of all the planets in the solar system, only Earth has a suitable environment for liquid water at its surface. (Mercury's extremely low pressure, 10^{-9} atmospheres, is indicated by a downward pointing arrow. Data are not available for Saturn and Neptune).

FIGURE 2-12 In this view of Earth from space, the shimmering blue envelope of water is the hydrosphere. The white clouds in the atmosphere consist of water vapor.

Water permeates all of Earth's surface systems, yet the movement of water on Earth is confined within the **hydrosphere,** a zone about 10 to 20 km thick, extending from a depth of several kilometers in Earth's crust to an upper limit of about 12 km in the atmosphere. If all water on Earth (1.4 billion km^3) were evenly distributed, it would form a layer nearly 3 km thick. Most of this water—97 percent—occurs in the oceans and is salty because the runoff of dissolved rock matter from continents into the oceans contains ions. Of the 3 percent of water not in oceans, 2.7 percent occurs in solid form, as ice. Ice covers about 10 percent of the land surface today, although it often has covered as much as 30 percent during cooler, glacial climatic conditions. The remaining 0.3 percent of water occurs as fresh water on continents (rivers, lakes, and groundwater) and as water vapor in the atmosphere.

The hydrosphere started to form about 4 billion years ago during the period of global melting, when some oxygen and other light elements, such as hydrogen and nitrogen, floated up from the molten Earth as gases. Because of the gravitational attraction of Earth's mass, some of these gaseous elements—especially the heavier ones—were unable to escape to space. Rather, they remained, some of them combining with one another, to form an envelope of gaseous matter around the Earth. Although hydrogen, containing only one proton, is the lightest element that exists, it so readily combines with oxygen to form water (H_2O) that much of it remained in this early atmosphere. As Earth continued to cool, the water vapor in the atmosphere began to condense and rain down, filling the ocean basins.

Little water has been added to or lost from the surface of Earth since the first rains that fell from the newly formed atmosphere. Although volcanic eruptions still contribute some water vapor and other substances to the atmosphere and hydrosphere, outgassing is much less pronounced since much of Earth cooled to a solid state.

Water in its three states migrates over different time scales at the surface of Earth. As water vapor in the atmosphere, water moves rapidly about the globe, on time scales of hours to days and months. The movement of water in its liquid state in streams and oceans occurs more slowly, over periods of days to thousands of years. In its frozen form, water in ice caps, ice sheets, and glaciers migrates over periods of thousands to tens of thousands of years. Relative to other liquids, water can store large amounts of heat energy. In areas that receive large amounts of solar radiation, water stores solar energy as heat and becomes warmer, whereas in areas that receive less solar radiation water releases its stored heat energy and becomes cooler. As a result, the movement of water from one reservoir to another affects local to global climatic conditions and helps to regulate Earth's surface temperature.

The Atmosphere: Earth's Envelope of Gases

The Greek word *atmos* means "vapor," and provides the name for the atmosphere, an envelope of gases that surrounds Earth's surface. The **atmosphere** extends from just below the planet's surface, where gases penetrate openings such as caves in the lithosphere and animal burrows in the pedosphere, to more than 10,000 km beyond Earth's surface, where gases gradually thin and become indistinguishable from the solar atmosphere.

Almost all the volume of gas in the atmosphere is composed of nitrogen (78 percent), oxygen (21 percent), argon (0.9 percent), and carbon dioxide (0.03 percent). Other gases, including neon, helium, nitrous oxide, methane, and ozone, occur in trace amounts. The percentage of water vapor at the base of the atmosphere varies considerably, from 0.3 percent on a cold, dry day to as much as 4 percent on a hot, wet day, but the water content decreases rapidly to nearly zero just a few kilometers above Earth's surface.

Venus and Mars are the most Earthlike planets in the solar system, yet they have atmospheres that consist almost entirely of carbon dioxide. The difference between their atmospheres and Earth's seems to be related to the existence of life on Earth, because biological activity produces oxygen. Some of Earth's oxygen combines high up in the atmosphere to form a protective layer of ozone, a gas that absorbs some of the Sun's ultraviolet rays, which can damage the delicate molecules that form living tissue. Other gases in the atmosphere store and transfer thermal energy at Earth's surface, modulating the planet's temperature. If Earth had no atmosphere, its surface temperature would be at least 33°C cooler.

The Biosphere: Where Life Exists

The **biosphere** is the aggregate of life-forms living in, on, and above Earth. It is a thin layer of living organisms, most of which use solar energy to build cells and to function. Between 2 million and 4 million different species of organisms are thought to exist on Earth: The majority depend on *photosynthesis,* the process by which green plants convert solar energy, carbon dioxide, and water into food energy. These organisms thus are confined to a zone in which they can receive solar radiation. This zone overlaps part of the atmosphere, the land surface (the top of the lithosphere), the uppermost part of the pedosphere, and the illuminated (euphotic) zone of water bodies (Figure 2-13). In other words, the biosphere is distributed among all the other environmental systems.

The composition of the biosphere is similar to that of Earth's other environmental systems, but much closer to that of the hydrosphere than the lithosphere, as living cells generally are 60 to 90 percent water. Like the atmosphere and hydrosphere, the biosphere is dominated by lighter elements. In fact, no elements with atomic numbers higher than 53 (iodine) are found in living cells except, rarely, in trace amounts. By number of atoms, the biosphere consists of 99 percent hydrogen, oxygen, carbon, and nitrogen. These four elements are found in all organisms on Earth.

A major difference between the biosphere and all the other major Earth systems is its high percentage of carbon, an element that occurs in very small amounts elsewhere (see Figure 2-9). Carbon atoms form strong bonds with one another, enabling them to develop elaborate chains of unlimited length, as well as rings and branching structures, that can attach to other atoms and molecules. These complex organic (i.e., carbon-containing) macromolecules sometimes consist of thousands of atoms and form more compounds than any of the other 102 elements on Earth, which is one reason most chemistry students consider organic chemistry to be among their hardest courses. The fundamental organic molecules are carbohydrates, fats, proteins, and nucleic acids. All fossil fuels (oil, coal, and natural gas) were formed from organic molecules that originated in the biosphere, but were buried by sediments to become part of the lithosphere after the death of organisms that produced them.

In the early history of the planet, oxygen in the atmosphere was insufficient for animal life to exist. Earth's primitive atmosphere contained large amounts of methane (CH_4), but these molecules were split apart by solar radiation, and the carbon combined with oxygen to form carbon dioxide (CO_2). Early life-forms, including bacteria,

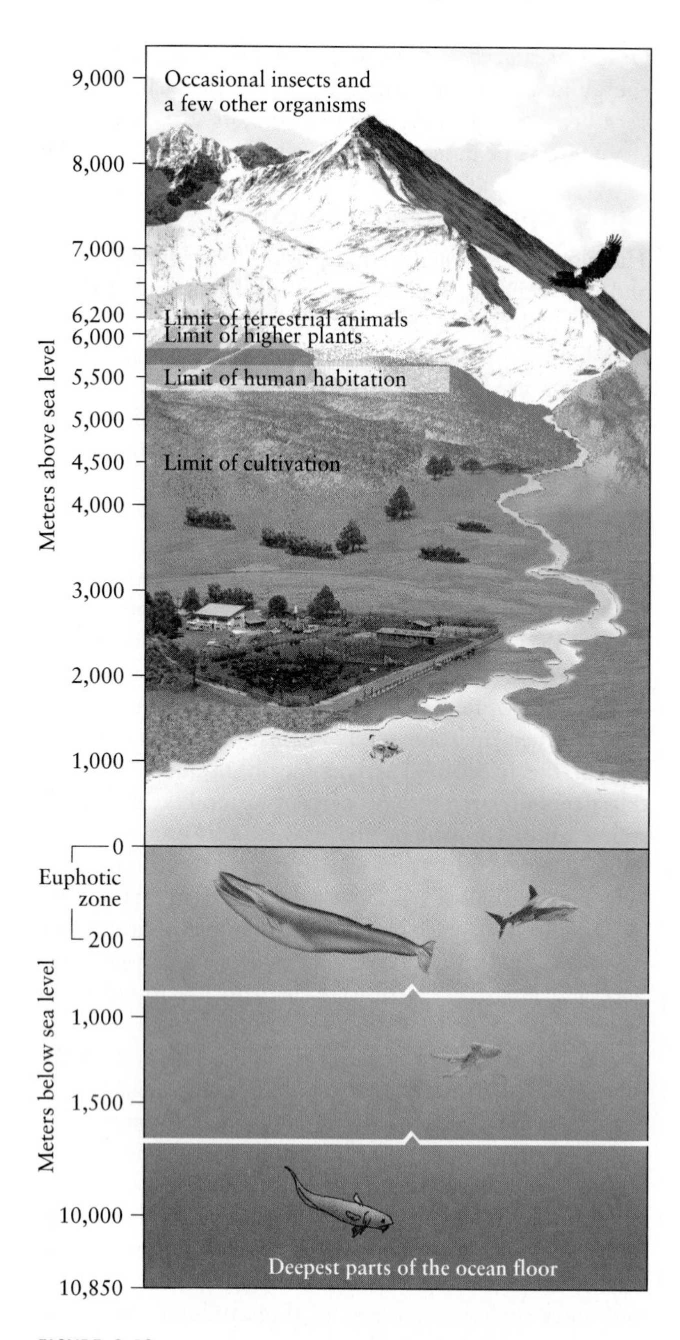

FIGURE 2-13 The vertical extent of life on Earth is referred to as the biosphere. The area below sea level is not drawn to scale.

blue-green algae, and phytoplankton, used the CO_2 for metabolism, producing oxygen as a by-product. With time, the amount of CH_4 in the atmosphere decreased, while the amounts of CO_2 and free oxygen increased. The continued evolution and spread of plants—which increased amounts of oxygen in the atmosphere—enabled oxygen-breathing animals to evolve; some of these animals (herbivores) relied on consuming plant matter for energy, while others (carnivores) preyed on animals to fuel their own metabolism.

Earth is the only planet in the solar system known to have life. The role of water in the evolution of life is essential, in part because of the many special properties of the water molecule, such as its high heat capacity. Life began in the oceans (see Chapter 3), and those life-forms that moved onto the continents carried their ocean environments with them, in their cells. It should come as no surprise, then, that the human body is nearly two-thirds water, and the water flowing through the bloodstream and bathing the cells carries on the task of the ocean by supplying the body with nutrients and removing waste products. To stay alive, terrestrial creatures require a supply of fresh water and, consequently, are greatly dependent on the processes that involve water on land. In this sense, the biosphere and hydrosphere are intimately interconnected.

- The lithosphere is a rigid, rocky layer of Earth about 70 km thick that overlies a more ductile layer, the asthenosphere. The lithosphere is broken into plates that float above the asthenosphere and move when heat from Earth's interior induces flow in the asthenosphere. The outermost part of the lithosphere is a chemically distinct layer of continental and oceanic crust.
- The pedosphere, generally less than 100 to 200 m thick, is the entire layer of weathered rock debris and organic matter at the surface of exposed landmasses. The relatively organic-rich upper meter or so of the pedosphere, called soil, provides the nutrients essential to all agricultural production on land.
- Earth's gaseous atmosphere and largely liquid hydrosphere formed since early Earth differentiated into physically and chemically distinct layers as the result of outgassing of volatile elements from Earth in its early, molten state and during volcanic eruptions. Both atmosphere and hydrosphere serve to modulate Earth's surface temperatures, keeping them within a fairly limited range.
- Because of biological activity, Earth's atmosphere is rich in oxygen compared with the atmospheres of nearby planets. In turn, Earth's atmosphere shields the biosphere from harmful ultraviolet rays from the Sun by efficiently absorbing them.
- The biosphere of Earth is unique in our solar system and formed soon after the early atmosphere and hydrosphere developed. It is an exceptionally thin layer, but contains millions of different species that have evolved since the earliest, simple forms of life.

Earth's Energy System

Energy can be considered a sixth environmental system. The *energy system* extends deeper into Earth than the five surface systems as well as farther into space, and it powers the whole Earth. Earth generates energy in its own interior and receives energy from the Sun. Energy from these two sources, plus small amounts from gravitational attraction between the Moon, Sun, and Earth, flows through all the other Earth systems. Since energy is the ability to do work, all processes and changes would cease if Earth received no energy (Figure 2-14). Earth's access to energy and ability to use it makes the planet a dynamic system.

Our Moon, in contrast, is a static system, considered geologically and environmentally dead because its interior

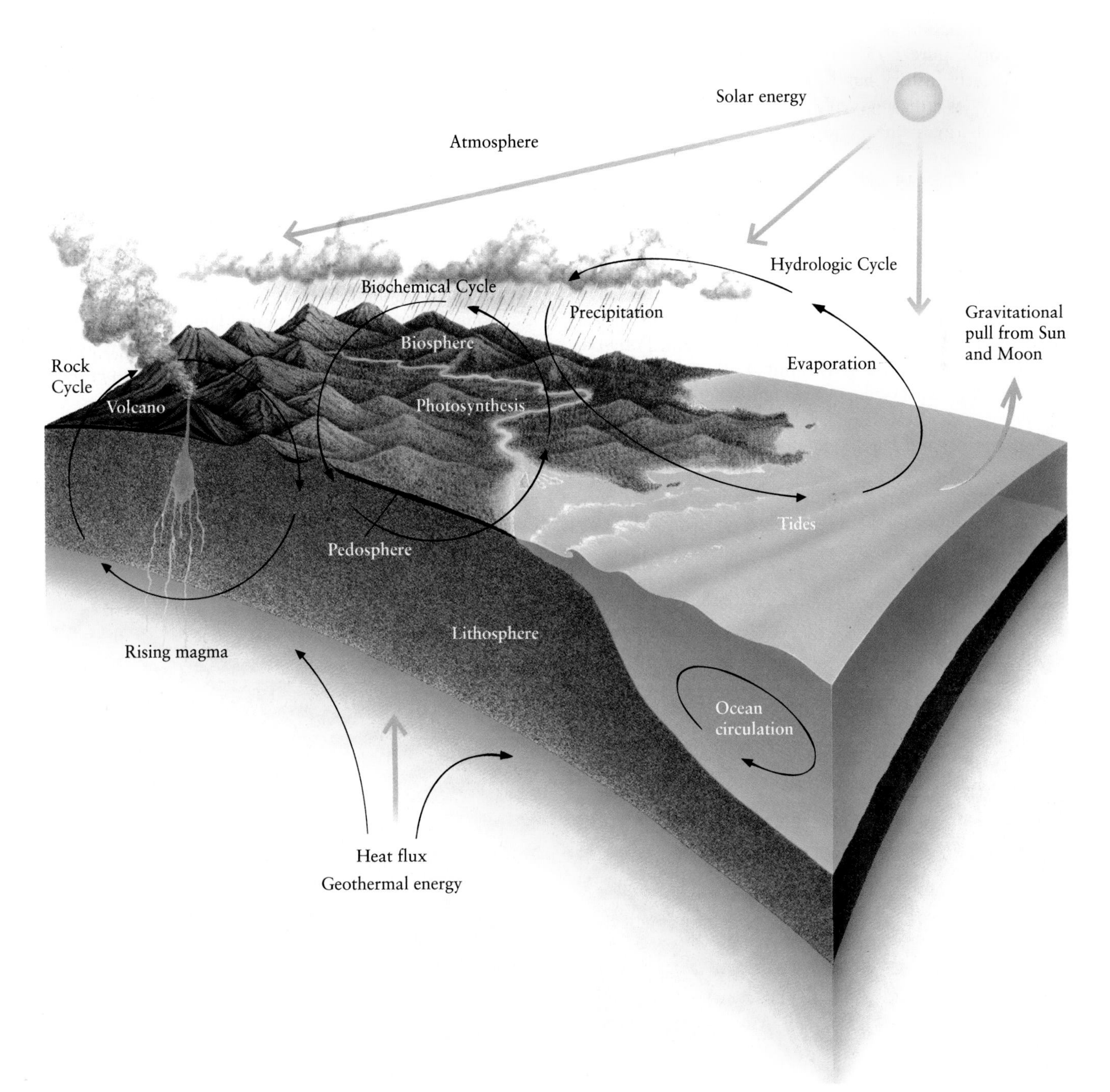

FIGURE 2-14 Energy is the driving force for all processes on Earth. Matter and energy flow cyclically through the hydrosphere (hydrologic cycle), biosphere (biochemical cycle), atmosphere, pedosphere, and lithosphere (rock cycle) at the surface of the Earth.

is cold and lacking in energy and because it has no atmosphere to capture and transfer solar energy. As a consequence, there is no volcanism on the Moon, no mountain building, no subduction, no soil formation, and no flowing water in rivers or storms. The Moon's surface is still, so craters and pits formed during the period of meteor bombardment more than 4 billion years ago appear fresh, as if formed yesterday. Earth was not spared this heavy bombardment, yet few craters appear on its surface. Most of those that do are subtle remnants of their original form, and are much younger than the craters on the Moon (Figure 2-15). On Earth, most evidence of the planet's early history has been nearly or completely destroyed by weathering, volcanism, and other processes that use energy to enact change.

States of Energy

Energy exists in several states, including kinetic and potential. **Kinetic energy** (KE) is the energy of a body in motion and is defined as $KE = 0.5\ mv^2$, where m is the mass of the body and v is the velocity of the body. **Potential energy** (PE), in contrast, is the energy of a body that results from its position within a system. Potential energy is related to the gravitational attraction between bodies and is expressed as $PE = mgh$, where m is mass of the body, g is acceleration of the body due to the gravitational force, and h is the relative height of the object.

Boulders perched on a hillside contain potential energy as long as they are not moving. If a process—perhaps an earthquake—causes the ground to give way, the boulders will roll downhill under the influence of gravity. Before the rockfall begins, the rocks contain potential energy proportional to their distance from Earth's center. As they roll and bounce down the slope, their potential energy is converted into kinetic energy and **thermal** (heat-related) **energy** (due to the motion of atoms). By the time the boulders reach the bottom of the hill, all their potential energy has been converted into thermal energy as a result of frictional resistance. This example demonstrates a fundamental principle of energy: Energy is completely conserved when work is done. While the energy has undergone a change in state (from potential to kinetic and thermal), the amount of potential energy before the work was done equals the amount of kinetic and thermal energy expended to do the work.

Energy can be transferred between bodies in the universe, as in the transfer of thermal energy. All bodies contain internal thermal energy, which arises from the random motions of their atoms. The term *heat* describes the transfer of thermal energy from one body to another. In a body receiving thermal energy, the added energy causes the atoms in the body to speed up. The temperature of a body is a measure of the average speed at which its atoms move. When heat is transferred to a body, its temperature rises because its atoms are moving faster.

Energy is measured in many different units, including foot-pounds, British thermal units, and electron volts, but the most common convention among geoscientists is to measure energy in either joules or calories. One *joule* (J) is defined as 1 $kg\text{-}m^2/sec^2$ (see Appendix 3). The *calorie*

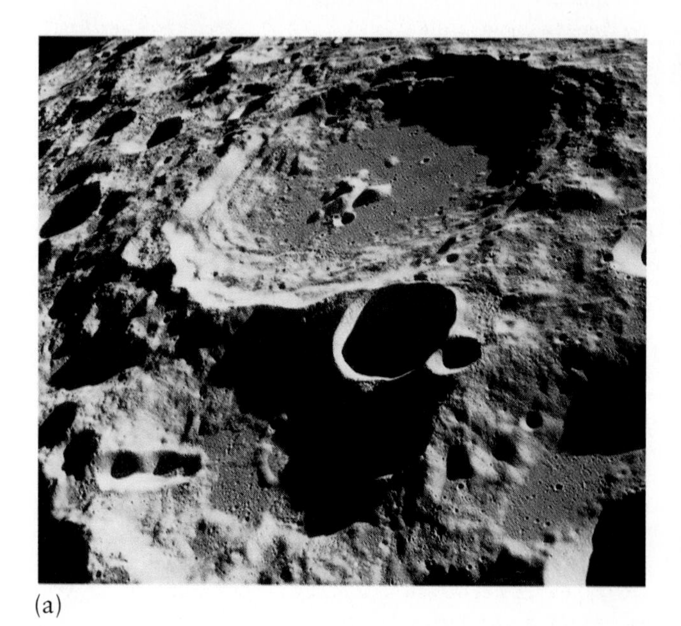

(a)

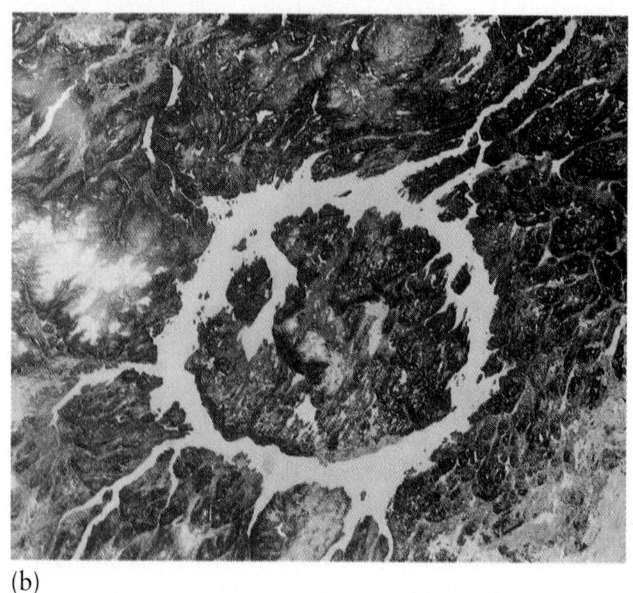

(b)

FIGURE 2-15 Impact craters on (a) the Moon and (b) Earth look very different. Those on the Moon's surface appear fresh and recent, while the one on Earth's surface is barely discernible except as a flattened outline.

(cal) is defined as the amount of energy needed to raise the temperature of 1 g of liquid water by 1°C. One calorie is equivalent to 4.184 J. When contemplating how many calories our food contains, we refer to Calories (upper case C), each of which is equivalent to 1000 cal, or 4184 J. A candy bar, for example, contains about 400 Calories, or 400,000 calories, enough to raise the temperature of 400 kg of water by 1°C. The human body (typically 50 to 80 kg) needs to maintain a temperature of 37°C, so it uses more calories when the surrounding temperature is lower than the body's temperature. Because about 70 percent of the human body is water, much of its caloric energy is used to maintain the temperature of water at 37°C.

Sources of Energy

Three sources of energy, two internal and one external, power all the processes on Earth. The foremost source, supplying 99.98 percent of Earth's energy, is the Sun. Smaller amounts of energy are created within Earth through spontaneous radioactive decay of some elements in rocks, and some residual heat from early Earth formation still exists. An even smaller amount of energy is supplied by the gravitational attraction within the Earth-Sun-Moon system and drives the movement of ocean water as tides. The inflow and outflow of energy at Earth's surface can be quantified for the three sources of energy to the whole Earth system: solar radiation, inter-nal thermal energy, and tidal energy. Earth receives 1.73×10^{17} watts (W; 1 W = 1 J/sec) of power from the Sun. In comparison, the heat energy produced within Earth itself yields only 32×10^{12} W of power, and the energy related to tides, a mere 3×10^{12} W. (Power is energy per unit time.)

Solar Energy In the Sun, hydrogen atoms at extremely high temperatures and pressures combine to form helium atoms in a process called fusion. The fusion reaction releases radiant energy that is transmitted through space in all directions, and Earth intercepts some of this energy.

Some incoming solar energy (less than 1 percent) is converted by plants to a form of chemical potential energy. During photosynthesis, plants combine solar energy with water and carbon dioxide to produce the carbohydrates on which the whole biosphere depends for food and fuel. When we burn wood in a campfire, we release its stored chemical energy, which we can use then to warm ourselves or to cook food. Under the proper geological conditions, some biomass may be converted to oil, coal, or natural gas. These are called fossil fuels because they come from plants and animals that lived many tens to hundreds of millions of years ago. When we drive cars, we are using solar energy that is many tens to hundreds of millions of years old!

The amount of energy stored within fossil fuels is extremely small compared with the amount of solar energy received by Earth. In any 20 day period, Earth receives as much energy from the Sun as is stored in all the existing fossil fuel reserves. For this reason, solar energy could become an important energy source in the 21st century.

Earth's Internal Energy Earth makes its own supply of internal energy through the spontaneous decay of certain elements in minerals. In this process, known as *radioactive decay,* parent atoms convert to other atoms by losing or gaining subatomic particles in their nuclei (discussed more fully in Chapter 3). Radioactive decay also releases relatively small amounts of thermal energy that are transferred through the surrounding rock.

Although much smaller than the influx of solar energy, it is Earth's internal heat energy, along with some residual heat from early Earth formation, that drives lithospheric plate motions, which in turn result in earthquakes and volcanism. As plates move, energy can be stored in rocks, much as you store energy in an elastic band by stretching it. When such rocks finally snap, as a band stretched too far would, the amount of energy released can be quite large. One of the largest recent earthquakes in California, the Northridge event just north of Los Angeles in 1994, released about 2×10^{15} J of energy. This amount is equivalent to about 1000 nuclear bombs the size of those dropped on Japan during World War II.

Gravitational Attraction Earth derives energy from gravitational attractions between itself and the Moon, the Sun, and the other bodies in the solar system. Although minor compared even with Earth's internal energy, the energy derived from gravitation causes both the crust and ocean water to change shape with time in a regular fashion, resulting in rock tides as well as ocean tides. The kinetic energy of winds and waves—like all kinetic energy—ultimately is converted to heat by friction and lost to space.

Energy Budget of Earth

The energy system of Earth is in an approximately steady-state condition; the amount of energy received by the whole Earth system is approximately equal to the amount of energy flowing out of the system. Because the Sun supplies all but 0.02 percent of Earth's energy, the inflow and outflow of solar energy is a close approximation of Earth's energy budget. An energy budget, like a monetary budget, is an accounting of inputs and outputs. A steady-state condition is like a balanced financial budget.

As shown in Figure 2-16, more than half of all incoming solar radiation is returned to space before doing any work at Earth's surface. About 25 percent is reflected to space by particles or clouds in the atmosphere and another 5 percent is reflected by Earth's surface, while 25 percent is absorbed by clouds and the atmosphere. This leaves about 45 percent of incoming solar radiation to be absorbed by Earth's surface. About two-thirds of

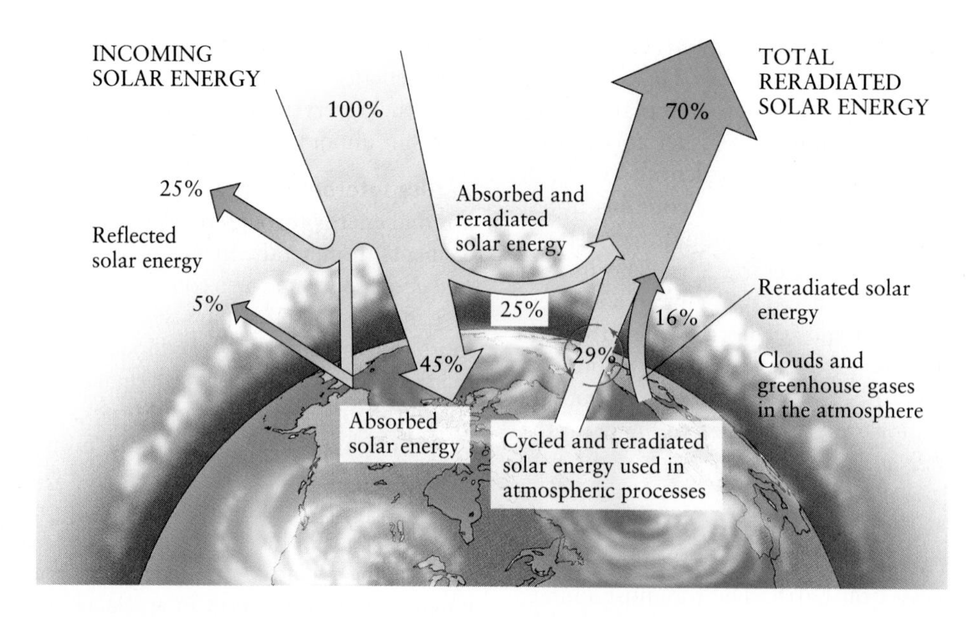

FIGURE 2-16 Some solar energy is reflected and some is absorbed and reradiated to outer space. Inflow of energy is equal to outflow, keeping Earth's energy budget balanced.

this absorbed energy (29 percent of the original incoming radiation) is cycled through Earth's environmental systems and reradiated, while one-third (16 percent of the original incoming radiation) is converted to heat (thermal energy). The wavelengths of reradiated energy are longer than those of the original (or reflected) solar radiation. Because clouds and greenhouse gases such as carbon dioxide trap energy at longer wavelengths relatively easily, reradiated energy has a strong influence on Earth's climate. If you add the percentages in Figure 2-16, you can see that inflow of solar energy is equal to outflow, with 30 percent of the inflow leaving Earth as reflected energy and 70 percent leaving as reradiated energy. Earth's energy budget is balanced.

Human Consumption of Energy

Humans have devised many methods to harness the different energies discussed above. For millennia, people have used wind and moving water to grind grains in mills and, more recently, to generate electricity. We have learned how to enhance the natural process of radioactive decay using fission—a process that splits atoms apart to obtain nuclear energy. Solar collectors are used to capture and utilize solar energy. In the United States, the amount of incoming solar energy is 500 times that of the nation's current energy consumption, but very little solar energy has been collected and used so far. We are more familiar with the use of fossil fuels since it is oil, coal, and natural gas that power most cars, heat most homes and water, and generate most of the electrical energy needed to run various appliances and utilities.

Of all the different sources of energy available to the world, fossil fuels satisfy 91 percent of total energy demand. However, the type of fuel used and the per capita use of energy vary dramatically from country to country, as well as from region to region within countries. The United States uses more energy than any other country in the world. With only 5 percent of the world's population, the United States consumes about 25 percent of the world's annual energy demand. Despite its vast energy consumption, however, the United States does not lead the world in consumption of energy per person. This distinction belongs to Canada, where low temperatures, vast distances between cities, and industries that depend on extraction of natural resources, such as logging, combine to make per capita energy consumption higher than anywhere else in the world (see Figure 1-11).

- Three sources of energy are responsible for all processes on Earth: solar energy, internal thermal energy, and gravitational energy.
- Solar energy contributes 99.98 percent of the energy that drives processes on Earth, but Earth's internal thermal energy drives plate tectonics, while the gravitational attraction among Earth, Moon, Sun, and other bodies in the solar system drives the tides.
- Earth's energy system is in a steady state; the amount of energy received by the whole Earth system is approximately equal to the amount of energy flowing out of the system.

Feedback Links Among Earth Systems

All Earth systems are linked, so changes in one can affect processes operating in another, as the aftermath of the Mount Pinatubo eruption illustrates. The outcome of an event such as a volcanic eruption is governed by feedback mechanisms by which the event triggers subsequent events that either promote or reverse the effects of the initial event.

Positive and Negative Feedback

Positive feedback is the process by which a change promotes continued change in the same direction. An example of a positive feedback process occurs as a result of the interaction between Earth's ice sheets and atmosphere, as shown in Figure 2-17(a). Unlike dark surfaces, which absorb solar energy and convert it into thermal energy, the bright white surface of clean glacial ice reflects 40 to 90 percent of the sunlight that strikes it. As a result, the atmosphere immediately above a growing ice sheet cools and promotes the formation of more ice, which reflects more light and leads to yet more cooling. The process of glacial ice formation is thus a positive feedback process because all the changes it leads to occur in the same direction. Positive feedback processes are destabilizing in that they promote a cascade of events that propels the system toward accelerating change.

Negative feedback is the process by which change in one direction leads to events that reverse the direction of

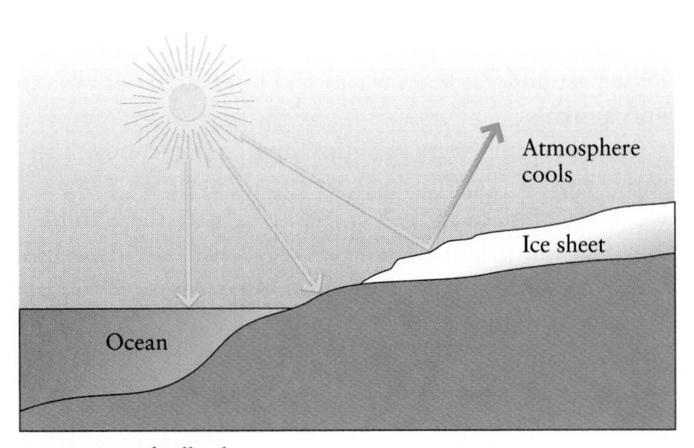

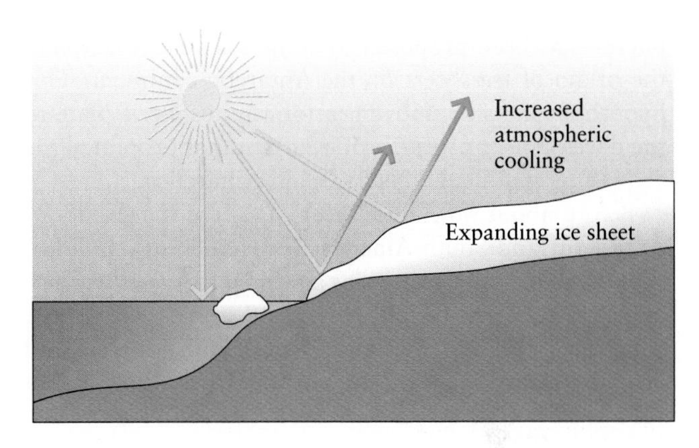

(a) Positive feedback

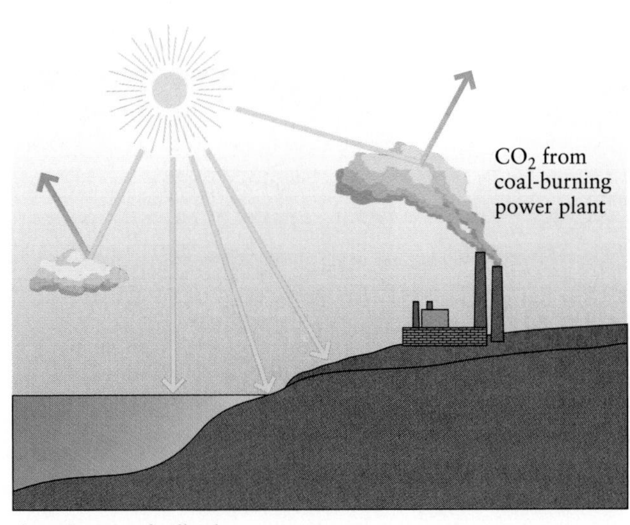

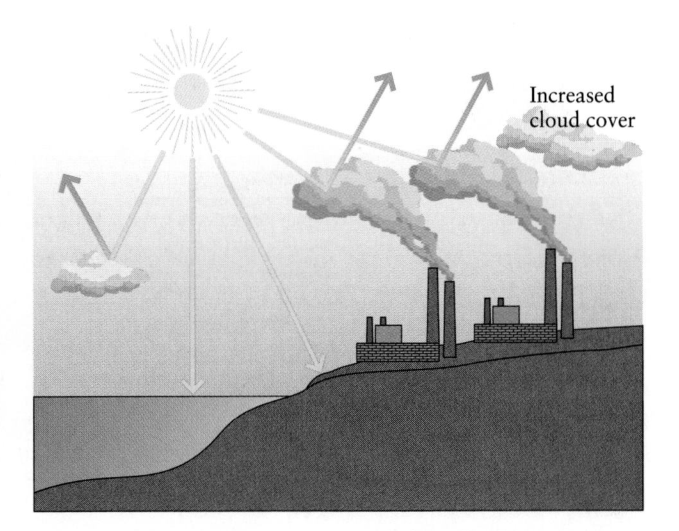

(b) Negative feedback

FIGURE 2-17 (a) An expanding ice sheet shows how positive feedback is a destabilizing process. (b) A simplified model of the greenhouse effect shows how negative feedback is a self-regulating, stabilizing process.

change. Increased emissions of water vapor, carbon dioxide, and other heat-trapping gases from power plants and automobiles would logically lead to global warming in a simple positive feedback process. The picture is much more complex, however, because a warmer atmosphere also can result in more cloud cover. Clouds reflect solar energy back into space before it can reach Earth's surface and be transformed into heat. The result is global cooling due to the increased density of clouds, which themselves resulted from global warming, as shown in Figure 2-17(b). Because an initial event that causes warming results in a process that causes cooling, this chain of events is a negative feedback process. Negative feedback processes are stabilizing because they counteract the effect of the initial event and help to regulate the system so that it maintains a steady state.

Evidence of Global Feedback: Growth of the Antarctic Ice Sheet

An intricate example of the interconnections among three systems—the lithosphere, hydrosphere, and atmosphere—has been proposed by some scientists to explain the origin of ice sheets on the Antarctic continent. This hypothesis links the movement of a lithospheric plate to the development of an oceanic current that promoted the growth of continental glacial ice in Antarctica.

Before about 40 million years ago, the continents of Antarctica and South America were joined in one lithospheric plate, and neither was glaciated (covered with ice masses). Between 40 million and 30 million years ago, Antarctica began to split apart from South America and drift southward (Figure 2-18). As the two continents separated, the ocean filled the growing basin between them, forming the Drake Passage. With time, the Drake Passage enlarged and an ocean current developed that completely encircles Antarctica and the south pole. This circular current deflected warm ocean currents moving south from the equator and deprived Antarctica of a significant source of heat. At the same time, parts of the Antarctic continent were rising as a result of tectonism, forming mountains and a plateau. These elevated land masses cooled further because atmospheric temperature decreases about 7°C for each 1-km rise in elevation.

With the continent so chilled, ice deposits began to accumulate. The solar radiation reflected to the atmosphere from the bright white surface of the ice mass was so high that a positive feedback process developed and enhanced the cooling trend. If all ice could be removed from Antarctica, its temperature would rise by 15° to 25°C just because of the difference in reflectivity between ice and rock. Although it seems reasonable to assume that the position of Antarctica over the south pole should alone be sufficient to cause it to be ice covered, the continent has been at that location in the geologic past and yet been warm and populated with plants and animals.

It is evident from just this one example how intricately linked are the Earth's primary systems. Ongoing work by climate modelers indicates that the complete story of the origin of ice on Antarctica might include more than the opening of the Drake Passage. The ice growth probably also involved a change in the atmosphere's composition that stemmed from other system changes and feedbacks.

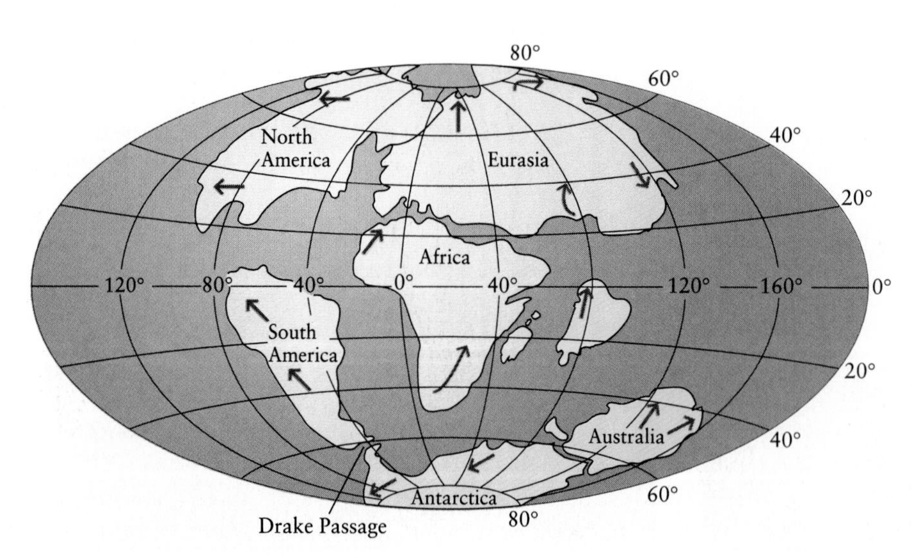

(a) Earth 30 million years ago

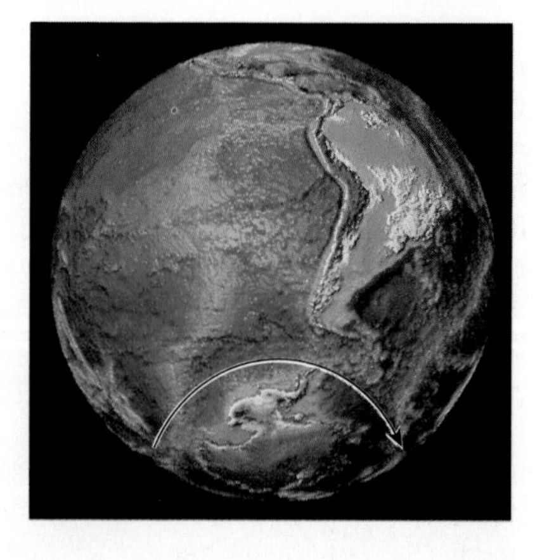

(b) Earth today

FIGURE 2-18 (a) About 30 million years ago, the continents of Antarctica and South America were separating, as Antarctica drifted southward and South America drifted northward. (b) Today, the Drake Passage allows circulation of a current (arrow) that completely encircles Antarctica and the south pole.

Change, Cycles, and Earth System Dynamics

Interdependent environmental systems are characterized not only by feedback mechanisms but also by the circulation of matter and energy in endless loops or cycles. Nearly all matter on Earth is recycled over and over again. For example, when fossil fuel is burned in an automobile, its carbon is released to the atmosphere. From there, the carbon might be extracted by trees during photosynthesis, and from them it might be returned to the atmosphere during a forest fire, or buried in sediment to become a fossil fuel yet again. We will discuss this important cycle—the carbon cycle—in Chapters 6 and 10. Here we explore two other cycles that link all the environmental systems and the energy system—the rock cycle and the hydrologic cycle.

The Rock Cycle

In terms of a human life span, rock seems eternal. For example, the owner of land underlain by many connected caves recently built a house and gift shop directly above one of the largest caverns, now a popular tourist site in Pennsylvania. To the owner, the rock beneath his home seems permanent and unchanging, despite nearby sinkholes and the cave's evidence that over millennia rock disintegrates and changes form. Likewise, mountains seem fixed and eternal to many people, yet to the geologist they are evidence that young rocks have been pushed upward by plate tectonics processes. The continual creation, destruction, and recycling of rock into different forms is known as the **rock cycle,** and it is driven by plate tectonics, solar energy, and gravity.

The rock cycle begins when tectonic forces drive molten rock from Earth's mantle toward the surface. The hot melt is less dense than the surrounding, cooler mantle rock and rises buoyantly through it, just as hot smoke rises through the cooler air around it. As the hot melt rises, it cools and freezes into **igneous rock,** so called from the Latin *ignis,* "fire." Water, wind, biological activity, and other environmental stresses may weather the rock, dissolving and breaking it up into particles that eventually move downward and accumulate in layers of sediment (Latin *sedere,* "to sink down"). Under sufficient pressure, sediments harden into **sedimentary rock.** Environmental forces may again disintegrate the rock. Alternatively, tectonic forces either may drive sedimentary and other types of rock back into the mantle and remelt it into magma, or subject the rock to enough heat and pressure to transform it, without melting, into **metamorphic rock** (from a Greek word meaning "to change form"). All rocks on Earth can be classified as igneous, sedimentary, or metamorphic.

Igneous Rock The earliest solid rocks on Earth were igneous rocks, formed between 4.5 billion and 4 billion years ago as the planet differentiated from a homogenous melt and its surface cooled and froze into a rigid silicon- and oxygen-rich crust. More igneous rock was created as additional magma floated toward the surface and either poured out volcanically onto the surface or lost heat and solidified slowly within the crust. As magma cools, its atoms slow down and ultimately freeze into orderly arrays of interlocking crystals. *Lava,* which is magma that has erupted onto the surface, spews out glowing hot but already has lost some of the original magma's heat. During Earth's early formation, some lava solidified into islands of igneous rock, much like the volcanic islands of Hawaii and Iceland today. Many of these early landmasses probably were destroyed by remelting in the turbulent Earth, but others survived, grew with time, and developed into precursors of the continents we know today.

Most magmas are derived from partial melting of the upper mantle and form *basalt,* a dark, relatively dense igneous rock that makes up most of the world's oceanic crust. In addition to the lightweight elements silicon and oxygen, basalt contains enough heavier elements, particularly magnesium and iron, to give the rock its color and density. It also contains a substantial amount of water. This water is important because it lowers the melting point of the basalt. Wet basalt recycled into the mantle melts easily, but the new magma contains less magnesium and iron, and becomes enriched in the lightweight silicon and oxygen.

The rock of the early continents had to be light enough to rise above sea level, so some of the wet basalt that formed first must have been refined into less dense rocks by a second cycle of melting. Remelting occurred as bits of early crust were dragged down with subducting plates to depths where temperatures are higher. Once again, as the wet basalt was reheated, partial melting resulted in magma progressively more enriched in silicon, oxygen, and other elements from minerals with low melting temperatures. A magma of this composition produced even less dense rocks when it rose to the surface. These igneous rocks, reacting with gases and liquids at Earth's surface, were transformed into even lighter sedimentary rocks.

Sediments and Sedimentary Rock Igneous rocks exposed at Earth's surface react with surface gases and liquids and become weathered. Clues in ancient rocks indicate that the processes of weathering observed worldwide today occurred as well in the geologic past. *Weathering* involves both chemical and physical processes that alter and break the rock into particles at the site where it is exposed. Loose debris then may be transported away from the site of weathering by flowing water, wind, gravity, and glaciers. The processes of sediment transport are called *erosion.* Gravitational attraction causes loose rock particles to sift down through water bodies and the atmosphere to settle in depressions on the ocean floor and on landmasses—a process called *sedimentation.* The accumulation of sediment in low areas is especially common along coastlines, where rivers drop their loads of gravel, sand, and mud as they enter the ocean. The pressure of layer upon layer of accumulating sediments squeezes and cements the loose particles into solid, sedimentary rocks.

If sedimentation since the formation of the early continents had progressed unimpeded at today's rate, Earth now would be covered with a layer of sediment more than 10 km thick. Because this is not the case, other processes must destroy sediments and sedimentary rocks at about the same rate that they are formed. Much sedimentary rock is eroded to form new sediments and sedimentary rocks, but two other processes completely destroy or change the rocks. One is remelting, which occurs when sediments or sedimentary rocks are returned to the higher temperatures and pressures of the mantle. The only way to get sediments, which are light, into the mantle is to drive them down with a subducting lithospheric plate. This process of subduction and remelting recycles the sedimentary materials into new igneous rocks.

Metamorphic Rock Sedimentary rocks can be recycled not only into igneous rock through subduction into the mantle but also into metamorphic rock through stresses that occur within the lithosphere. Metamorphism refers to the physical and chemical changes that occur when a rock is exposed to crustal temperatures or pressures high enough to change its crystal structure but too low to melt its minerals. The rock is recrystallized into a new, much harder metamorphic rock while still in the solid state. Both sedimentary and igneous rocks can be metamorphosed to become metamorphic rocks, and even metamorphic rocks can be remetamorphosed.

▸ Cycling of Rock Among Rock Reservoirs

Four major reservoirs contain the three basic rock types—igneous, sedimentary, and metamorphic—in the crust and mantle (Figure 2-19). The largest reservoir is the mantle itself, with 4000×10^{18} metric tons of solid rock and magma. The second largest rock reservoir is the continental crust, with 18×10^{18} metric tons of low-density, crystalline, igneous and metamorphic rocks. The ocean floor, with 5×10^{18} metric tons of crystalline igneous rocks (basalt) is the third largest reservoir. Sediments and sedimentary rocks form the smallest reservoir (2.5×10^{18}) metric tons, but cover 75 percent of Earth's surface.

Rock cycle processes such as weathering and melting create fluxes of matter from one reservoir to another. As in most Earth systems, the fluxes into and out of a reservoir tend to balance each other, keeping the reservoir in a steady state. On the ocean floor, 65×10^9 metric tons

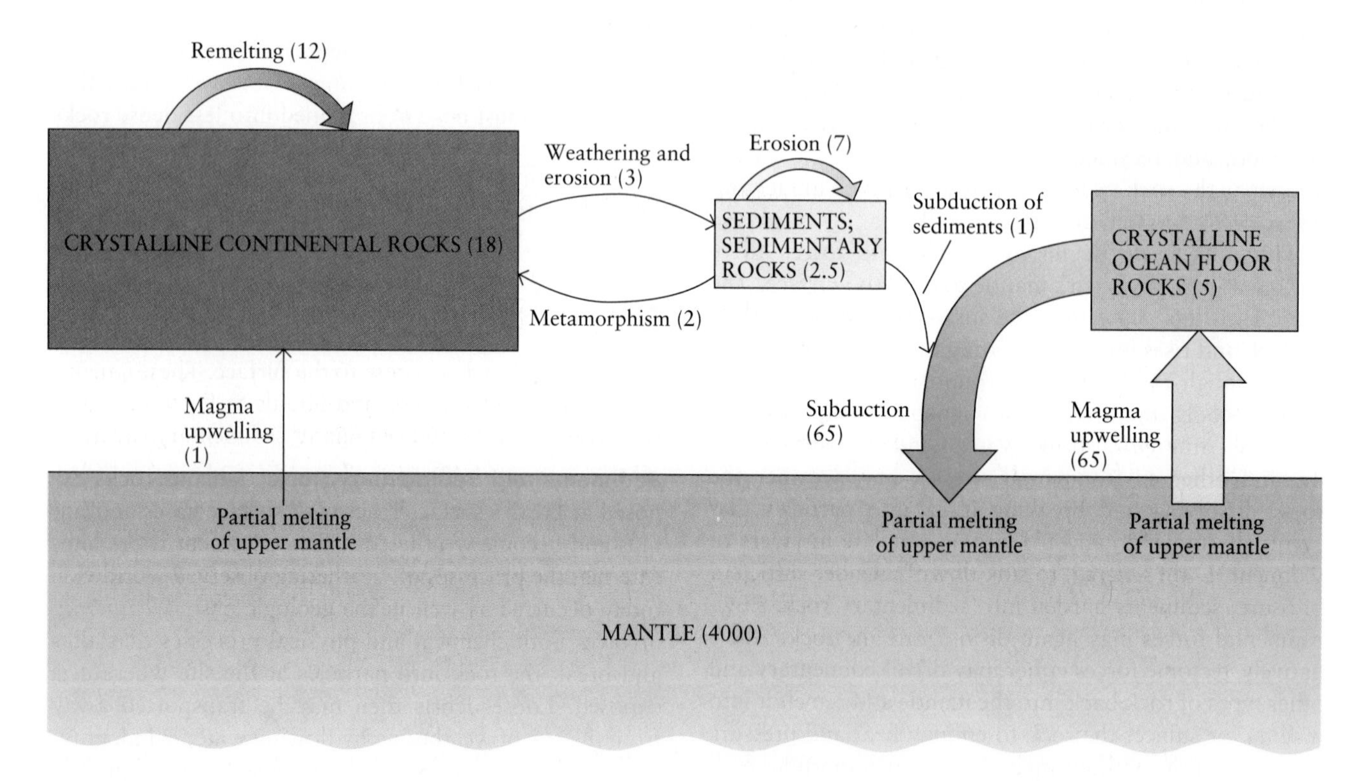

FIGURE 2-19 The rock cycle portrayed as a system containing four major reservoirs (rectangles) and the flows, or fluxes (arrows), of matter among them. The masses of the reservoirs are expressed in units of 10^{18} metric tons—a metric ton is 1000 kg, which is about 1.1 tons. Flow rates are expressed in units of 10^9 metric tons per year. Boxes and arrows are drawn in relative scale. Note that the mantle reservoir is so large that it cannot be shown in its entirety.

of magma flow out to form new oceanic crust each year, while the same amount of oceanic crust is returned to the mantle when plates collide and one is forced down under the other. Heating and partial melting of the subducted plate replenishes the supply of magma in the mantle. Evidence of this balance is twofold. First, the total mass of oceanic crust has not increased despite frequent ocean-floor volcanism. Second, although the oldest continental rocks on Earth have been dated at about 4 billion years, the oldest rocks in oceanic crust have been dated at only 180 million years.

Similarly, the volume of continental crust is maintained in a steady state, although the flux of rock through the continental crust is slow compared with the flux of rock through the oceanic crust. The relatively small amount of igneous rock added to continental crust from the mantle through igneous processes (10^9 metric tons per year) is balanced by the return of sedimentary rocks to the mantle through subduction. Although continental crust has accumulated over a time scale of billions of years, over millions of years the amount of crust added is similar to the amount destroyed.

The Hydrologic Cycle

Unlike rock, which changes physical and chemical state and location over long periods of geologic time, large volumes of water change physical state and location over relatively short periods. From dark clouds to raindrops on a lake, from snowflakes to ice in a glacier, or from a cold meltwater stream to a chilly sea with floating icebergs, water moves easily among atmospheric, continental, and oceanic reservoirs (Figure 2-20). The movement of water from one reservoir to another is the **hydrologic cycle**, and it is the fundamental, unifying concept in the study of water on Earth. Before the hydrologic cycle was understood, there was no consistent, natural explanation for why rivers hold water long after the most recent rain has fallen or why water bubbles from springs high in the mountains. Finding the right place to sink a well for groundwater was an unreliable business presided over by people thought to have supernatural powers. Even today, in parts of the United States, "dowsers" still are engaged to find well sites, although their success rates are no better than chance.

The simplest model of the hydrologic cycle shows the flux of water through *evaporation* from oceans to precipitation on continents and back again to oceans through *runoff*. Evaporation however, can occur anywhere on Earth, and much water is returned to the atmosphere from evaporation over landmasses, lakes, and rivers. Other processes involved in the movement of water are *transpiration* (the release of water vapor by plant and animal cells) and *infiltration* of water below the ground surface along openings such as pore spaces in sediments or fractures in rocks. Once in the ground, water seeps downward until it reaches a zone in which all openings are saturated with water. In this zone, water flows as *groundwater* toward areas of low elevation and pressure, such as streams and springs. Eventually, this water, too, will return to the atmosphere or ocean and continue to cycle (Figure 2-21).

FIGURE 2-20 Water is continuously changing state and location at Earth's surface, transforming from water vapor (gas) in clouds to liquid raindrops or solid snowflakes. Snow might become ice in a glacier. Solid ice can be transformed to water vapor in the atmosphere on a warm, dry day, or it can be transformed to meltwater and flow into nearby streams, lakes, or oceans.

Solar Energy and the Hydrologic Cycle The driving force for motion in the hydrologic cycle is the unequal distribution of solar energy from Earth's poles to its equator. The equator gets more than twice as much solar energy as the poles. Water molecules on the Sun-warmed ocean surface absorb enough thermal energy to escape into the atmosphere as water vapor. Water moving in ocean currents, clouds, and air masses serves as a global heat exchanger, transferring thermal energy from areas of excess in the tropics to areas of deficit near the poles, and causing climatic variations around the planet.

Because the amount and distribution of solar radiation reaching Earth's surface vary over time, both global climate and the hydrologic cycle change constantly. Many factors cause this variation in solar radiation intensity, including the changing inclination of Earth with respect to the Sun and the changing amount of cloud cover in the atmosphere (discussed in detail in Chapters 12 and 13). During the most recent period of maximum glaciation, about 18,000 years ago, more water was stored on continents and less in the oceans than today. In addition,

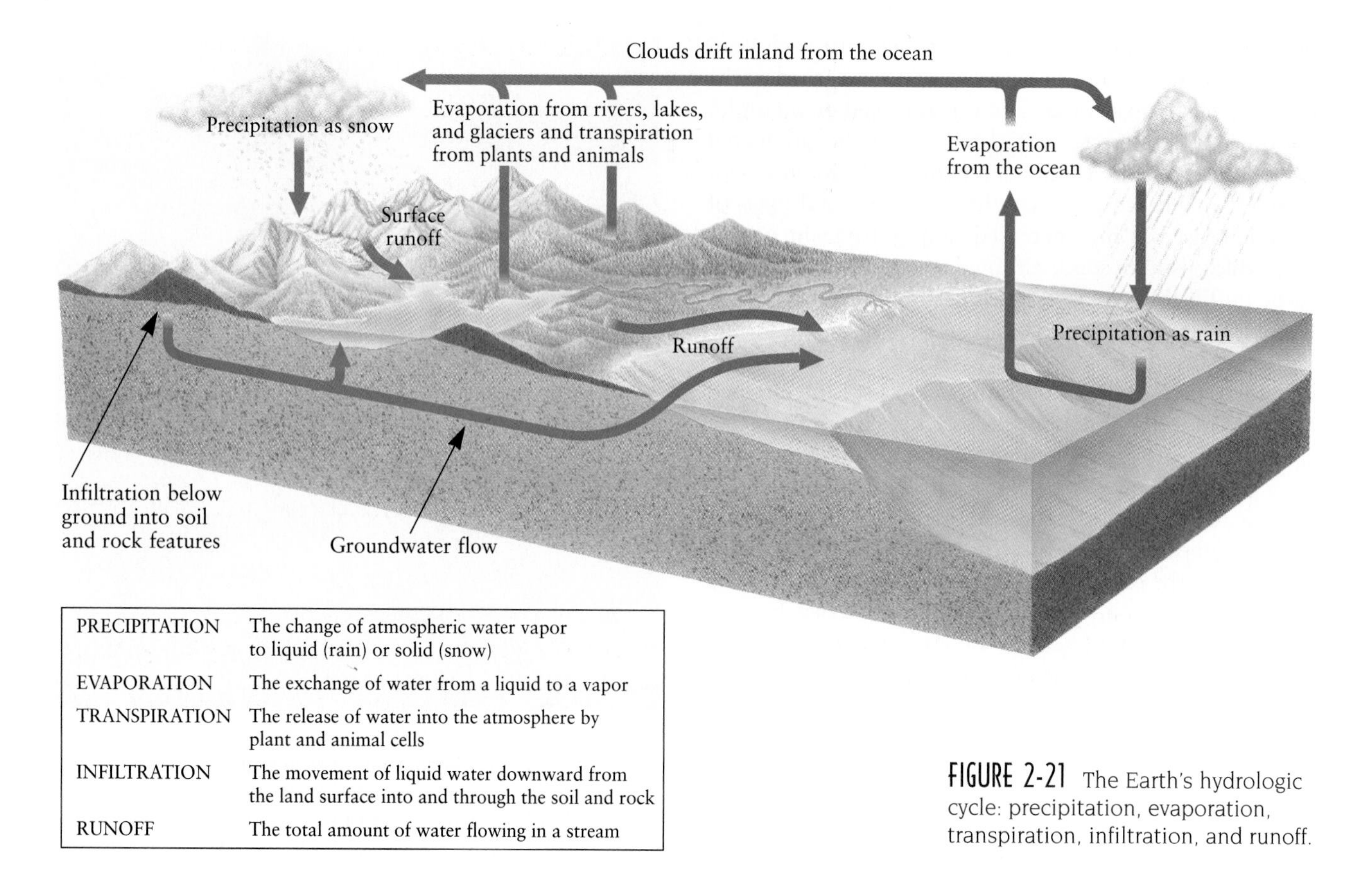

PRECIPITATION	The change of atmospheric water vapor to liquid (rain) or solid (snow)
EVAPORATION	The exchange of water from a liquid to a vapor
TRANSPIRATION	The release of water into the atmosphere by plant and animal cells
INFILTRATION	The movement of liquid water downward from the land surface into and through the soil and rock
RUNOFF	The total amount of water flowing in a stream

FIGURE 2-21 The Earth's hydrologic cycle: precipitation, evaporation, transpiration, infiltration, and runoff.

FIGURE 2-22 This space shuttle view of Rub Al Kali desert of Yemen shows an intricate network of streams that carries little water today. These dry streambeds indicate a major change in the local hydrologic budget of the region, from wetter to drier conditions, between 18,000 and 10,000 years ago.

the potential for precipitation to evaporate from continents decreased with the drop in thermal energy. Consequently, the amount of surface water available for runoff and infiltration was greater. Extremely dry regions with surface water deficits today, such as the Rub Al Kali desert of Yemen, were rich in surface water between 18,000 and 10,000 years ago, during the end of the last full glacial period (Figure 2-22). The runoff of this surface water carved channels that today only rarely contain water.

The Rub Al Kali desert is only a small part of the evidence indicating that past climatic changes altered hydrologic conditions around the world. Consequently, future changes in climate—such as global warming—are likely to affect the hydrologic cycle and water availability. Some Earth system scientists foresee greater ecological and socioeconomic dangers from alterations in rainfall patterns than from a few degrees of increase in annual mean temperature.

▸ Cycling of Water Among Oceans, Atmosphere, and Continents

On Earth, three main reservoirs store water mass: the oceans, the atmosphere, and the continents. Figure 2-23 shows the annual fluxes of water among these reservoirs and Table 2-1 shows the stocks of water in Earth's subreservoirs, such as glaciers and biota (living things). Each reservoir is both the recipient of inflows and the source of outflows to other reservoirs. The amount of time that a water molecule typically resides in a given reservoir is called the **residence time.** Since the hydrologic cycle has maintained an approximately steady state over hundreds of years, the residence time of water in each reservoir is equal to the amount of stock in that reservoir divided by its flux. The following calculations of residence time are based on the stocks and fluxes shown in Figure 2-23:

Residence time of water in oceans

$$= \frac{\text{stock}}{\text{flux}}$$

$$= \frac{\text{stock of water in oceans}}{\text{inflow (or outflow) of water into oceans}}$$

$$= \frac{1{,}338{,}000 \text{ thousand km}^3}{(47 + 458) \text{ thousand km}^3\text{/year}}$$

$$= 2650 \text{ years}$$

Residence time of water in the atmosphere

$$= \frac{\text{stock of water in the atmosphere}}{\text{inflow (or outflow) of water into the atmosphere}}$$

$$= \frac{12.9 \text{ thousand km}^3}{(72 + 505) \text{ thousand km}^3\text{/year}}$$

$$= 0.02 \text{ years } (\sim 8 \text{ days})$$

Residence time of water on continents

$$= \frac{\text{stock of water on continents}}{\text{inflow (or outflow) or water to continents}}$$

$$= \frac{47{,}971.71 \text{ thousand km}^3}{119 \text{ thousand km}^3\text{/year}}$$

$$= 403 \text{ years}$$

These values represent the *average* residence time of a molecule of water in each reservoir. Actual values for an individual water molecule might range from seconds to tens of thousands of years.

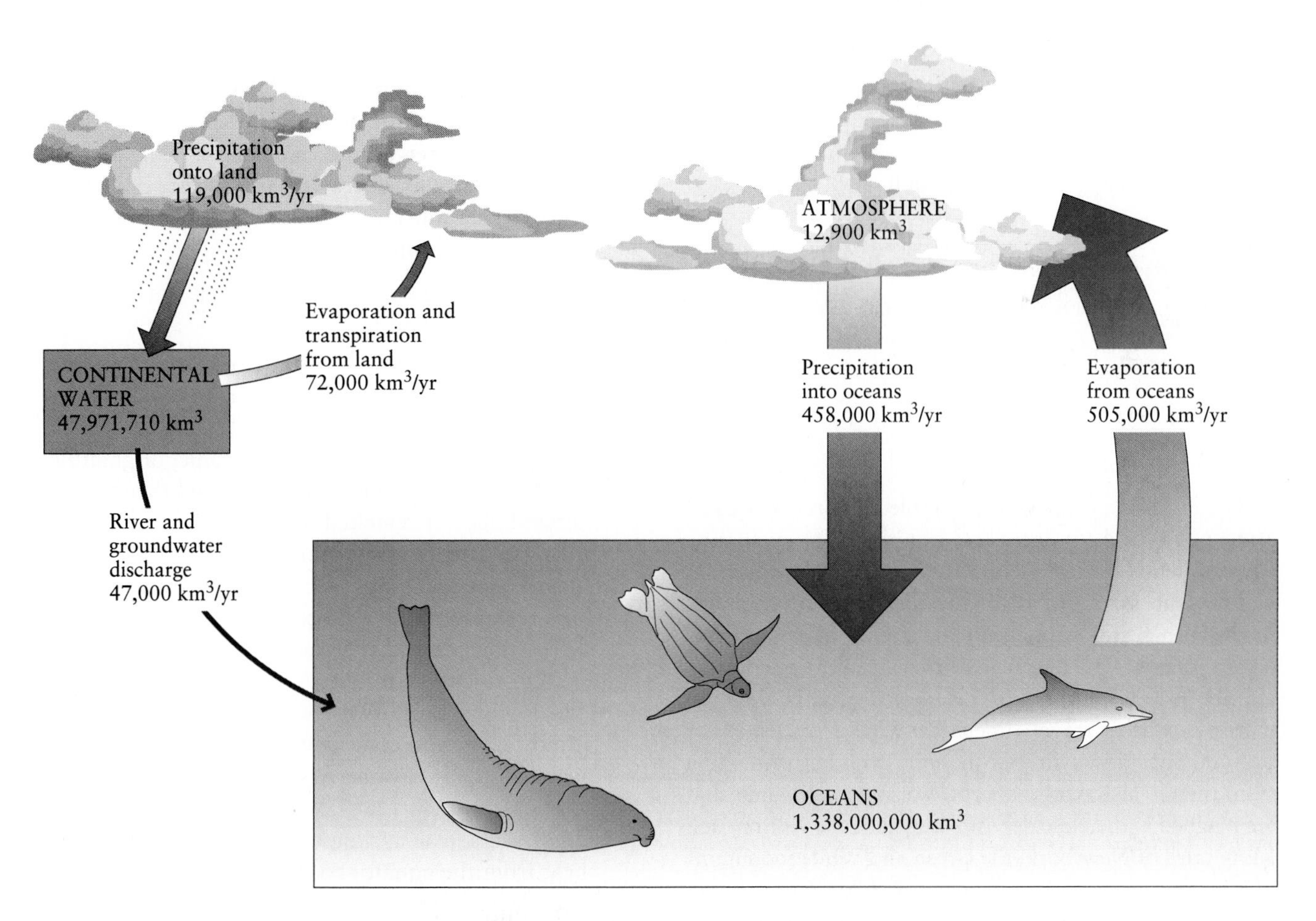

FIGURE 2-23 The hydrologic cycle, shown as a system with three major reservoirs, and the fluxes of matter among them. Data are given in Table 2-1. Boxes representing the ocean and continental reservoirs are drawn to scale relative to one another; the atmosphere would be so small that it is not shown as a box but rather just as clouds. Arrows representing fluxes are drawn to scale relative to one another.

TABLE 2-1 Stocks of Water on Earth

Form	Volume (km^3)	Total water (%)
Oceans	1,338,000,000	96.54
Glaciers, permanent snow	24,064,000	1.74
Groundwater*	23,400,000	1.69
Fresh water	10,530,000	0.76
Saline water	12,870,000	0.93
Permanent ground ice	300,000	0.022
Lakes	176,400	0.013
Soil moisture	16,500	0.001
Atmospheric water	12,900	0.001
Rivers	2,120	0.0002
Biota	1,120	0.0001
Wetlands and other sources	10,960	0.008
Total amount of water	1,385,984,000	100†
Total amount of fresh water	35,029,000	2.53

*Less than half this groundwater, or 10,530,000 km^3, is fresh water. The remainder is saline.

† Percentages total slightly more than 100 because of rounding.

Source: *World Water Balance and Water Resources of the Earth*, UNESCO, 1978.

The residence times of water molecules in the oceans, atmosphere, and continents are significant for evaluating environmental issues. Acidic emissions from volcanoes and the tall stacks of fossil fuel–burning power plants move with water vapor into the atmosphere and thus have residence times similar to that of atmospheric water, about 8 days. For this reason, acid rain is common within a week's travel time downwind from emission sources. Sites downwind of the industrial belt extending from the Great Lakes to southwestern Pennsylvania that are plagued with acidic deposition include the Adirondack Mountains of New York, the Green and White mountains of Vermont, and southern Canada. Acidic rainwater and snowfall damage forests and ultimately can lead to their destruction.

The Hydrologic Budget Just as energy in a system can be accounted for in a budget analysis, so can water in the hydrologic cycle be evaluated. The inflow and outflow of water to a system are compared, and from them the excess or deficit water is determined. In the case of water in the ocean, for example, inflow occurs through precipitation and runoff from continents, while outflow occurs through evaporation. If inflow is greater than outflow, a surplus of water will occur in the oceans and sea level will rise. If, on the other hand, outflow is greater than inflow, a deficit of water will occur and sea level will fall.

A hydrologic budget can be developed for the oceans, the whole Earth, or smaller areas such as continents and even cities. An example of the hydrologic budget for the U.S. mainland indicates that flow rates have been altered by human activity. Before extensive irrigation and industrial water use (primarily for cooling), 71 percent of the precipitation was returned to the atmosphere as evaporation and transpiration, while the remaining 29 percent made its way into groundwater and streams and, ultimately, to the oceans. Today, however, 7 percent of streamflow is withdrawn for human use, and only 5 percent makes it way back to the oceans. The remaining 2 percent returns to the atmosphere by evaporation from irrigation water on agricultural fields, from reservoirs, and so forth. As a result, 73 percent of the precipitation now makes its way back to the atmosphere as evaporation and transpiration. Although 2 percent may seem like a small amount, it is enough to have a noticeable local impact, as in areas where decreased infiltration has reduced replenishment to underground water supplies.

- Rocks can be classified into three basic types—igneous, originating in the hot mantle; sedimentary, originating as broken and dissolved particles of preexisting rock; and metamorphic, originating as sedimentary and igneous rocks and then altered but not remelted. In order of decreasing size, the reservoirs that hold these rocks are the mantle (igneous), the continental crust (igneous and metamorphic), the oceanic crust (igneous), and sediments and sedimentary rocks.

- The major reservoirs for water are the oceans, atmosphere, and continents. Residence time of water in the atmosphere is only 8 days, whereas that on continents is 403 years and in oceans 2650 years. The cycling of water redistributes heat from the equator to the poles. Soils, streams, and glaciers are among the continental subreservoirs for water. The greatest stock of water on Earth is contained in the ocean reservoir. The greatest stock of fresh water is frozen in thick ice sheets on Greenland and Antarctica.

Closing Thoughts

This chapter began by discussing the global effects of a single volcanic event—the 1991 eruption of Mount Pinatubo in the Philippines. The resulting injection of sulfurous gases and volcanic ash into the atmosphere was followed by "volcano weather" and a measurable lowering of global temperature over the next few years. In many ways, Earth behaved as a group of interconnected and interactive open systems, with matter and energy flowing from one to another. From this example, it is clear that matter erupted from the interior of Earth can alter the global environment at its surface, and that effects of single events can be complex and multiple.

Matter and energy have been recycled among Earth's systems throughout its history. At least 1000 volcanoes on Earth are active or potentially active, and hundreds of eruptions occur each year, although most are smaller than that of Mount Pinatubo in 1991. Some scientists refer to the emission of gases and debris from volcanoes as Earth's "breathing."Ashfalls, mudflows, hurricanes, and lowered temperatures can cause enormous devastation to human lives and livelihoods—but not, over the long term, to the environment. These events are by-products of processes that are essential to the way Earth works.

Although the scale of human impact on Earth is still small in comparison with that of geologic processes such as volcanism, continental drift, waxing and waning ice ages, or mountain building, humans are by no means an insignificant force. We certainly are affecting processes that occur over time spans of decades to hundreds of years. Our impact on the hydrologic and rock cycles is measurable. Whether or not the ensuing changes in the cycling of matter and energy result in an environment that is less well suited to our species' particular needs than is that of the past few thousand years remains to be seen.

Summary

- Earth is best understood from an Earth system perspective because the activities of our planet are interrelated by the cycling of matter and energy through Earth's various systems.
- The environment can be divided into six major systems: the lithosphere (rock), pedosphere (weathered rock and soil), hydrosphere (water), atmosphere (air), biosphere (life), and energy system.
- Cycling of matter and changes in the Earth system are the result of processes that are driven by energy—solar energy, the energy from radioactive decay of minerals in Earth, and gravitational energy.
- Geologic processes, such as continental drift or volcanism, can affect multiple Earth systems over a period of time, each change resulting in another. Earth systems are characterized by both positive and negative feedback processes when they respond to changes in conditions.
- The impact of humans on the rock cycle is very small compared to the impact of geologic processes, but for the hydrologic cycle human impact is affecting the quality of our environment over a period of years to centuries.

Key Terms

reservoir (p. 31)
stock (p. 31)
flux (p. 31)
open system (p. 31)
closed system (p. 31)
isolated system (p. 31)
dynamic system (p. 32)
static system (p. 32)
energy (p. 32)
steady state (p. 32)
Big Bang theory (p. 33)
element (p. 35)
atom (p. 35)
compound (p. 35)
molecule (p. 35)
proton (p. 35)
electron (p. 35)
neutron (p. 35)
cation (p. 37)
anion (p. 37)
core (p. 38)
crust (p. 38)
mantle (p. 38)
silicate mineral (p. 38)
lithosphere (p. 38)
asthenosphere (p. 38)
mesosphere (p. 38)
outer core (p. 38)
inner core (p. 38)
mineral (p. 40)
oceanic crust (p. 41)
continental crust (p. 41)
pedosphere (p. 42)
soil (p. 42)
hydrosphere (p. 44)
atmosphere (p. 45)

biosphere (p. 45)
kinetic energy (p. 48)
potential energy (p. 48)
thermal energy (p. 48)
positive feedback (p. 51)
negative feedback (p. 51)
rock cycle (p. 53)
igneous rock (p. 53)
sedimentary rock (p. 53)
metamorphic rock (p. 53)
hydrologic cycle (p. 55)
residence time (p. 57)

Review Questions

1. How did the eruption of Mount Pinatubo in 1991 affect global climatic conditions?
2. Why are Earth's systems considered to be open? Why is the whole Earth system considered to be closed?
3. Is the amount of annual solar radiation (energy) reaching Earth's surface an example of a flux or a stock?
4. How and when did the lithosphere, hydrosphere, pedosphere, atmosphere, and biosphere form on Earth?
5. How are the lithosphere and crust different from each other?
6. What are the major processes that operate in the lithosphere and rock cycle?
7. What are the major processes that operate in the hydrosphere and hydrologic cycle?
8. What are the two most abundant elements in Earth's modern atmosphere?
9. What are the three main sources of energy on Earth?
10. What are the residence times of water in the oceans, atmosphere, and continents? How is the concept of residence time important to environmental concerns?

Thought Questions

1. What do you predict would happen if the flux of rock from Earth's mantle were to increase? For example, there is some evidence that periods of intensified volcanic activity have occurred during Earth's history. How would this affect the rock cycle?
2. Is the following an example of a positive or negative feedback? Explain your answer. Development of an area that was tropical forest results in removal of all vegetation and exposure of bare soil. Rainfall runs off the devegetated surface and increases the flow of water in streams. Increased runoff also has greater power to remove soil, thus causing gullying and erosion. As the amount of soil decreases, the amount of infiltration of rainfall into the soil is lessened, resulting in even more runoff, and consequently even more soil erosion and gullying.
3. Why would scientists try to model Earth systems? What would be the benefits and limitations?
4. Why can't plants (and hence food or trees) be grown easily in the pedosphere if the upper meter or so of soil is eroded away?
5. Can the oceans ever contain more water than they do at present, and hence result in a rise in sea level? Can they ever contain less water than at present, and hence result in a drop in sea level? Has either of these types of change occurred in Earth's history?
6. If the residence time of water in the atmosphere were much longer, say 20 days rather than 8, how would environmental pollution resulting from industrial emissions be affected?

Exercises

1. Using Figure 2-19, calculate the residence time of rock in the continental crust, in sedimentary rock, in oceanic crust, and in the mantle. For which of these are the fluxes the greatest and the least?
2. Make a flow chart that illustrates the connections among different Earth systems and cycles for a single event, such as a storm that causes coastal erosion or overgrazing that results in soil erosion.

Suggested Readings

Berner, R. A., and Berner, E. K., 1987. *The Global Water Cycle.* Englewood Cliffs, NJ: Prentice-Hall.

Broecker, Wallace S., 1986. *How to Build a Habitable Planet.* Palisades, NY: Eldigio Press.

Butcher, Samuel S., Charlson, Robert J., Orians, Gordon H., and Wolfe, Gordon V. (eds.), 1992. *Global Biogeochemical Cycles.* London: Academic Press, Harcourt Brace Jovanovich.

Graedel, Thomas E., and Crutzen, Paul J., 1993. *Atmospheric Change: An Earth System Perspective.* New York: W. H. Freeman and Company.

Graedel, Thomas E., and Crutzen, Paul J., 1995. *Atmosphere, Climate, and Change.* New York: W. H. Freeman and Company.

Gregor, C. Bryan, Garrels, Robert M., Mackenzie, Fred T., and Maynard, J. Barry, 1988. *Chemical Cycles in the Evolution of the Earth.* New York: John Wiley and Sons.

Kaufmann III, William J., and Comins, Neil F., 1996. *Discovering the Universe.* 4th ed. New York: W. H. Freeman and Company.

McFarland, Ernest L., Hunt, James L., and Campbell, John L., 1994. *Energy, Physics, and the Environment.* Winnipeg, Canada: Wuerz Publishing.

National Aeronautics and Space Administration, Earth System Sciences Committee, 1988. *Earth System Science: A Closer View.* Boulder, CO: University Corporation for Atmospheric Research.

Piel, Gerard, 1992. *Only One World.* New York: W. H. Freeman and Company.

Readings from Scientific American, 1990. *Managing Planet Earth.* New York: W. H. Freeman and Company.

Schlesinger, William H., 1991. *Biogeochemistry: An Analysis of Global Change.* San Diego: Academic Press.

Scientific American, September 1990. *Special Issue: Energy for Planet Earth.*

Smil, Vaclav, 1997. *Cycles of Life: Civilization and the Biosphere.* New York: W. H. Freeman and Company.

Wyman, R. L., 1991. *Global Climate Change and Life on Earth.* London: Routledge, Chapman, and Hall.

CHAPTER 3

The Monterey Formation, a sequence of banded sedimentary rocks named for the town of Monterey, occurs throughout much of western California and is an important source of fossil fuels. Geoscientists have calculated that 1 m of the banded rock represents nearly 5000 years of sedimentation. The average thickness of the formation is 300 m. Layers in this photograph represent approximately 30,000 years of geologic time.

Geologic Time and Earth History

Earth's environments have changed dramatically and repeatedly over a history that reaches back nearly 5 billion years. To understand how geological events and human activities can affect our modern environment, we need to understand the environmental changes of the past.

Geoscientists learn about past environments primarily from clues embedded in sedimentary rocks. The Monterey Formation, for example, contains light-colored layers with evidence of marine animals and plants alternating with dark-colored layers bearing mud evidently washed from land to sea during winter storms. Over time, pressure and heat eventually compacted and hardened these horizontal layers of loose sediment into solid, sedimentary rocks. Subsequently, movements of Earth's lithospheric plates compressed, raised, and tilted the layers, deforming them and generating enough heat to turn their organic compounds into petroleum and gas.

The Monterey Formation has a distinct bottom and top that represent the beginning and end of a specific environment that existed in a certain place during a particular span of geologic time. To illuminate the principles of geologic time, in this chapter we will:

- Learn how geologists developed a relative time scale from fossil evidence and rock formations.
- Distinguish between relative and absolute geologic time scales.
- Examine the principles of radiometric dating that led to an absolute geologic time scale.

The ability to date geological events can be crucial to life on Earth. Consider the problem of radioactive waste disposal. Exposure to radioactive emissions damages living cells and causes cancer. The United States has 109 nuclear power plants to generate electricity but has no fully approved repository to dispose permanently of their spent fuel rods, which are highly radioactive. Until a site is approved and a storage facility is built, all high-level nuclear waste is stored at or near the power plants, often in pools of cool water. Most of these temporary storage sites will be full by the year 2010, and the U.S. Department of Energy is working to authorize and construct a permanent disposal site before then.

Yucca Mountain in Nevada has been proposed as a site for storing this nuclear waste. The mountain is miles from human habitation, and the waste would be buried in solid rock hundreds of meters beneath it (Figure 3-1). Because the waste will remain highly radioactive for thousands of years, scientists must determine whether or not the facility will remain intact and safe from disturbance and migration of wastes for at least the next 10,000 years. The major technological risk is the challenge of time: No known human engineering works have survived the rigors of rain, disintegration, and erosion this long. The Egyptian pyramids, for example, are only about 5000 years old. The major geological threat is the presence of volcanic rocks in the area. Are the rocks young enough that a source of magma still could exist beneath the surface and erupt again before most of the radioactivity of the material to be stored has dissipated?

The age of a volcanic rock reveals how long ago the magma from which it formed erupted onto Earth's surface as lava, then cooled and froze to its present solid state. Scientists determine this date by measuring the relative amounts of radioactive elements in the rocks in a process called radiometric dating.

Yucca Mountain's rocks formed between 8 million and 13 million years ago, when a large, highly explosive volcano spewed rocks and ash over great distances. This volcano is long gone, but much smaller volcanoes erupted and grew atop the older volcanic rocks within the past few million years. The youngest volcanic cones formed between about 10,000 and 1 million years ago. Statistical analysis of this information led scientists to estimate that the chance of a future eruption in the area is about 1 in 500 million per year. Thus, the volcanoes pose an insignificant risk. Although an active volcano could form and erupt in the next 10,000 years, the possibility is remote.

Time, Space, and Earth Processes

The meaning of a 10,000 year time span is hard to grasp when the human lifetime is only 70 to 80 years, and geologic time, which is measured in spans of hundreds

FIGURE 3-1 The possibility of using Yucca Mountain, in the desert of southern Nevada, as a storage site for thousands of tons of nuclear waste is being intensively investigated.

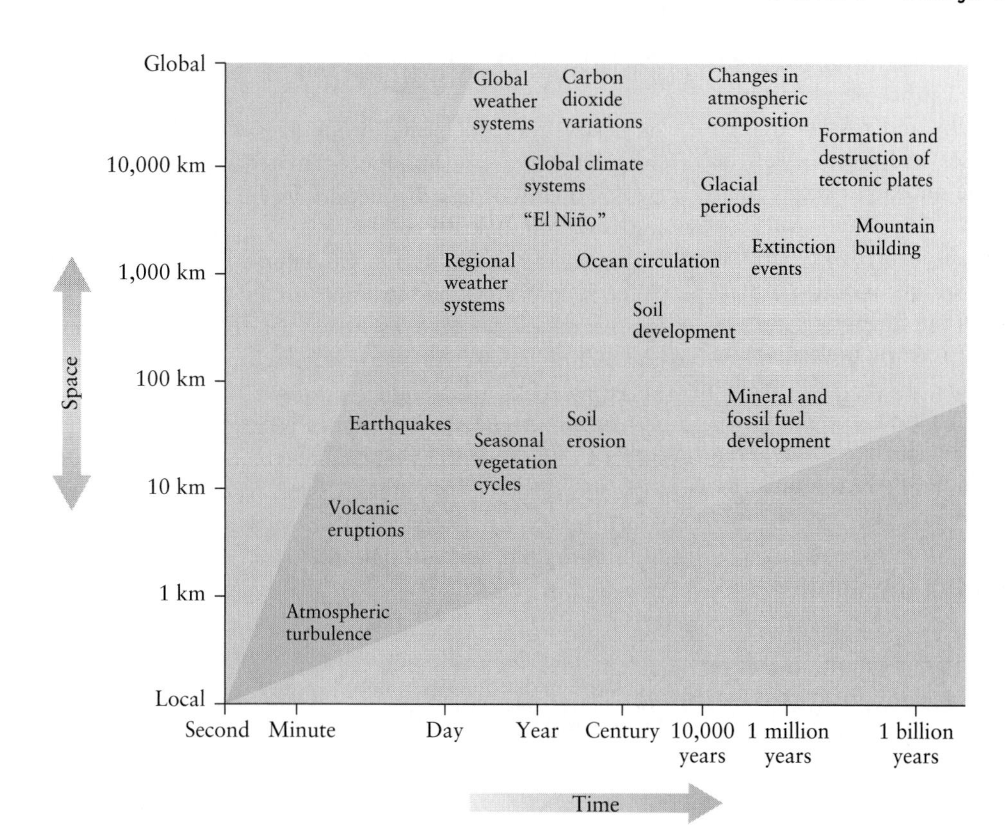

FIGURE 3-2 This time-space diagram illustrates temporal and spatial scales of different Earth processes. The scales on both axes are not linear, but logarithmic. In other words, each successive increment on the axes represents a tenfold increase in time or distance (an order of magnitude).

of thousands to billions of years, may seem beyond comprehension. Nevertheless, an understanding of the time scales and the space scales of Earth's processes is vital to predicting geologic hazards and using resources wisely. Consider our dependence on a finite store of fossil fuels such as petroleum. The rich stores of oil in Earth's crust—such as those in the Monterey Formation—took millions of years to form and are distributed in rock formations that extend over distances of tens to hundreds of kilometers. They could be depleted, however, in a matter of centuries at our current consumption rates.

Soil takes less time to form than oil does. Still, a layer of organic-rich soil 1 to 2 m thick takes tens to hundreds of thousands of years to form over distances of hundreds of kilometers. Although soil usually is considered a renewable resource, it takes long enough to form relative to rates of soil erosion that there is cause for concern about depletion in some areas.

Some geologic processes operate at relatively short time scales but affect large areas. In the southern hemisphere, the cold-weather arrival of a warm ocean current known as El Niño, so named because it occurs at Christmastime, is an event that recurs in 3 to 10 year cycles, with extreme events recurring every few decades. El Niño causes droughts on some landmasses and floods on others, affecting regional climates for thousands of kilometers. Other hazards, such as earthquakes and volcanoes, operate over short time scales and have greatest impact over short distances, less than 100 kilometers. Emissions from some large volcanic eruptions however, can affect climate on a global scale, as did the 1991 eruption of Mount Pinatubo discussed in Chapter 2.

Figure 3-2 shows the relationship between time and space for fossil fuel development, soil development, earthquakes, and other geologic processes.

Time's Arrow

Time is the fourth dimension in a four-dimensional universe. The other three dimensions—length, width, and depth—are spatial. Unlike the spatial dimensions, time has direction and is irreversible. An object in space can move in any direction from a given point, but an object existing at some point in time can move only forward. You cannot retrace your steps in history, as in a video playing in reverse. This unidirectional, linear nature of time is referred to as "time's arrow." The cycles of Earth's systems, such as the rock cycle, are repetitive, but they occur within the overall direction of time's irreversible arrow. Earth therefore has a history in which events can be placed in order relative to one another. Past occurrences, such as the formation of Earth or the appearance of the human species, can be arranged in sequence from earliest to latest. In a procedure known as **relative age dating**, layers of rocks and fossils—which contain clues about past processes and events—can be arranged in order relative to one another, even if their absolute ages are unknown.

In the 20th century, scientists discovered ways to use the radioactive decay of unstable atoms in rocks, minerals, and organic matter to determine the actual ages of Earth's materials. This technique is known as **absolute age dating**. Relative and absolute dating usually are combined, so that different events first are placed in sequence relative to one another and then are bracketed in time by those events for which absolute dates are available.

Using Earth's History to Predict the Future

Geoscientists assume that the causes of past events can be used to predict future events. This assumption comes ultimately from the fundamental premise that processes occurring on Earth today are the same as those that occurred in the past. Known as the **principle of uniformitarianism**, it often is paraphrased as "the present is the key to the past." Uniformitarianism does not imply that the rates of all processes remain exactly the same with time. Rather, it means that processes observed today can be used to explain events that occurred earlier in Earth's history because the physical laws (such as the law of gravitation) that control these processes remain constant. For example, by emitting large volumes of particulate matter and gases into the atmosphere, Mount Pinatubo's eruption in the Philippines in 1991 affected global climate for several years thereafter. Similar but older volcanic deposits occur on surrounding mountain slopes and plains. Using the principle of uniformitarianism, Earth scientists infer that the volcano has erupted repeatedly over the past few thousands of years and, furthermore, that past large eruptions might also have affected global climate.

The idea that the present is the key to the past can be extended to predict what might happen in the future. By interpreting the geologic record, scientists predicted the 1980 eruption of Mount St. Helens in the state of Washington and closed the area to the public in time to save many lives (Box 3-1). Similar reasoning is the basis for the intense scrutiny of rocks and sediments at the proposed nuclear waste site at Yucca Mountain.

- Geologic processes typically take place over spans of hundreds of thousands to billions of years.
- An understanding of contemporary geologic processes can be used to interpret the rock record and predict future geologic events.

Relative Geologic Time

As new ways to determine rock ages were developed in the past few centuries, the estimated age of Earth increased many orders of magnitude. In the 17th century, Earth was thought to be about 6000 years old. Today, it is considered to be about 4.6 billion years old, about 700,000 times the age estimate of the mid-1600s. The age estimates changed as new scientific understanding and technologies enabled greater accuracy and precision in dating rocks and geologic events.

In 1664, Archbishop Ussher of Ireland announced that he had determined the age of Earth. By tracing genealogical evidence in the Bible, Ussher reckoned that Earth was created in the year 4004 BC and so was only about 6000 years old. His estimate was later refined by another religious scholar who fixed the exact time of Earth's birth as 9 AM on October 26, 4004 BC.

Early scientists, in contrast, observed Earth's rocks for clues to Earth's age. Two people stand out as having ideas far ahead of those of their contemporaries, one of them centuries before Ussher's work. The first was Leonardo da Vinci (1452–1519), the Italian artist, scientist, and engineer, and the second was Niels Stensen (1638–1686), a Danish physician.

Leonardo's Interpretation of Fossils

Leonardo da Vinci studied layers of sand, gravel, and fossil debris in the mountainsides of the Apennines, in central Italy (Figure 3-3). (A fossil is the naturally preserved remains of a plant or animal.) He observed that the fossilized shells and other skeletal debris resembled creatures such as clams, crabs, and oysters found along modern beaches. Since the time of Aristotle (384–322 BC), fossils were interpreted variously as shapes that developed in the rock itself and just happened to resemble living creatures; Creation's mistakes, which never became living creatures; or as the remains of victims of the biblical Flood. In Leonardo's time, Earth's surface was thought to be stable and unchanging, having been formed during the Creation and modified only once thereafter, during the Flood.

Leonardo's contemporaries could not deny that the fossils looked like the seashells along the coast, so some suggested that the shells had been transported to the mountainsides during the Flood. Leonardo, however, thought otherwise. He noted that the assemblages of fossils were often fully intact and that the shells of clams and other immobile organisms appeared to be in the same positions in which, in life, the organisms grew—their "growth positions." He concluded that they had not been moved. Furthermore, relying on his observations of live clams, he concluded that the fossilized ones could not

3-1 Case Study

Predicting the Eruption of Mount St. Helens

On May 18, 1980, the Mount St. Helens volcano in Washington erupted in a spectacular explosion, but no one was surprised and few people were hurt. The public had been alerted months earlier to the likelihood of an eruption, thanks to a geologic investigation that began in the 1970s, when the U.S. Geological Survey assigned geologists Dwight Crandell and Donal Mullineaux to evaluate the hazards of dormant, but potentially active, volcanoes in the Cascade Range of the Pacific Northwest.

Crandell and Mullineaux recorded evidence of past eruptions to reconstruct the history of each volcano. While hiking and driving over many thousands of square kilometers, they identified and mapped rocks and sediments stemming from lava flows, pyroclastic flows (flows of fluidized masses of hot, shattered fragments erupted from a volcano), ashfalls, and mud- and debris flows. By comparing the thickness of each deposit, how much area it covers, and its position in sequence with other deposits, they were able to determine the relative age of each deposit. The oldest layers, logic dictated, lay at the bottom, and each successive upper layer was younger. Using absolute age-dating methods to date trees killed during ancient eruptions and rocks emitted from the volcanoes, they also determined the absolute ages of some deposits. In all, they identified hundreds of deposits, each from a different eruption that had occurred in the past 5000 years.

Of all the Cascades volcanoes, Mount St. Helens had been demonstrably the most active and explosive during the past 4500 years, with eruptions occurring on average every 225 years. With good reason, the Native American name for Mount St. Helens is Loowit, the Lady of Fire. The most recent eruptive period was between 1831 and 1857, in the early days of the Oregon Trail, the California gold rush, and the arrival of settlers in the Pacific Northwest.

In 1978, Crandell and Mullineaux warned, "In the future, Mount St. Helens probably will erupt violently and intermittently, just as it has in the recent geologic past, and these future eruptions will affect human life and health, property, agriculture, and general economic welfare over a broad area . . . the current quiet interval will not last as long as 1000 years; instead, an eruption is more likely to occur within the next 100 years, and perhaps even before the end of this century."

Only two years later, in March 1980, earthquakes below the volcano's north slope provided a clear sign of unrest inside its magma chamber. An oval crater developed on the volcano's summit, and the north flank of the volcano was rapidly swelling upward and outward. Intensified earthquake activity (10,000 earthquakes by mid-May) centered under the growing bulge warned that magma was moving upward. A formal forecast that an eruption probably would occur within weeks was issued to the public, and the area was closed with roadblocks in April 1980.

Shortly after dawn on May 18, two geologists, Dorothy and Keith Stoffel, began circling the summit crater of Mount St. Helens in a small airplane, taking

(continued)

Along the bank of a river near Mount St. Helens, a geologist examines numerous deposits associated with the volcano's eruptions. Near the bottom of the stratigraphic section, a poorly sorted mixture of ash, mud, and large boulders was deposited about 13,000 years ago after an eruption led to a catastrophic mudflow. Above this mixture is an oxidized, reddish layer of finer airborne particles ejected from the volcano about 11,000 years ago, and above it is a thin blue-gray layer deposited by an ash cloud that swept down the flanks of the volcano about 3300 years ago. A bouldery layer from the 1980 eruption lies atop these older deposits.

3-1 Case Study

(continued)

The eruption of Mount St. Helens on May 18, 1980.

photographs from an altitude of 400 m. At 8:32 AM, they were directly above the crater's steep sides when the whole north side of the summit "began to ripple and churn . . . as one gigantic mass," as they later recalled. Geologist David Johnston also was watching from his station about 10 km northeast of the volcano. While the Stoffels circled the crater, Johnston radioed headquarters, shouting, "Vancouver! Vancouver! This is it!"

Those were Johnston's last words. A 500°C blast of gas and ash rocketed both upward and to the northeast, blowing the top off the mountain and incinerating everything in its path. Heavier material surged down the north slope of the volcano, hugging the ground, moving down valleys and up ridges, engulfing the landscape in a fiery flow of rock. The Stoffels watched in awe. In their small plane, they heard and felt nothing. The pilot took a steep nosedive to gain speed, then headed south to flee the deadly cloud. Johnston's post was destroyed and he was never found.

From a geologic point of view, the 1980 eruption of Mount St. Helens is just another event in the continuous growth and evolution of a continent on a dynamic and constantly changing Earth. Volcanism has continued at various places on Earth ever since a crust solidified over its hot, molten interior 4 billion years ago. This particular eruption was a sudden catastrophe that completely changed the landscape in a matter of minutes, but geologists analyzing ash layers from previous eruptions in the Cascades and monitoring the volcano's day-to-day activity were able to predict this one in time to minimize its human toll. Like the other volcanoes in the Cascades, Mount St. Helens has erupted repeatedly over tens of thousands of years and surely will erupt again.

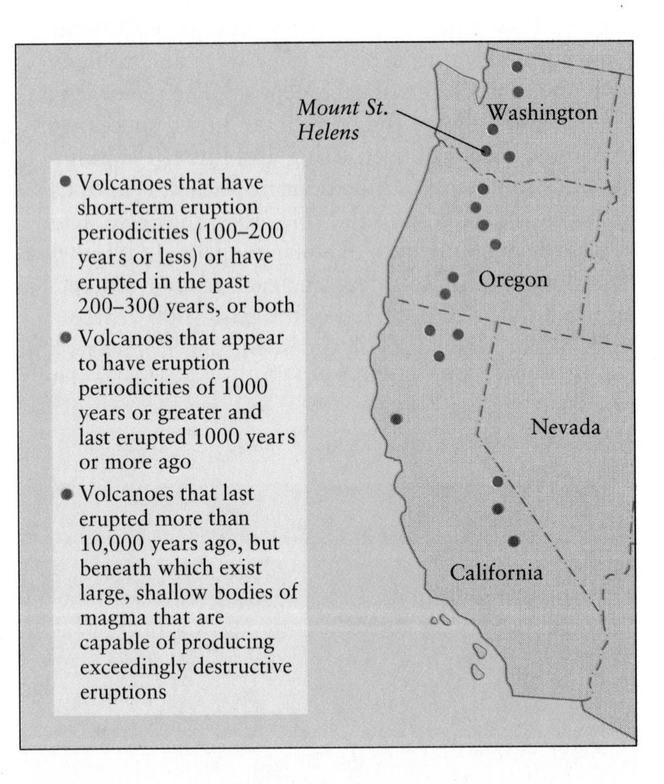

The Cascade Range, a chain of lofty volcanoes, extends from British Columbia to northern California in the Pacific Northwest and includes Mount St. Helens. Volcanoes in mid-California also are indicated.

FIGURE 3-3 Leonardo da Vinci examined fossils similar to the clams, snails, and other shallow, nearshore marine organisms exposed in this outcrop near Ancona, Italy. These fossils and the sedimentary rocks that enclose them are less than several million years old.

have moved to the mountainsides on their own during the Flood. Clams dig narrow furrows along which they lift and drop themselves, moving only a meter or so per day. They would have needed far longer than the 40 day duration of the biblical Flood to move over 300 km (185 miles), especially going up a steep mountainside. Leonardo suggested that a sea once covered the land and that the land might have been raised since then. He also noted the occurrence of earthquakes and ground movement in the area and suggested that they might be related to uplift of the rock. In all his deductions, Leonardo was correct, but few people knew of his ideas until his journals were made public many years later.

Steno's Laws of Rock Layering

Niels Stensen—his name often latinized to Nicolaus Steno—was the second person with prescient ideas regarding time and Earth history. Born almost 200 years after Leonardo, Steno moved to Italy from his native Denmark. Independently, he came to the same conclusion as had Leonardo on the origin of fossils. Steno did something else as well—he recognized in the layers of rocks, or **strata**, of Tuscany, a record of time, a sequence in which lies recorded a history of events. The word *strata* (singular *stratum*), Latin for "bedcovering," or "blanket," is used by geologists to describe the sheetlike rock layers. The study of strata is called **stratigraphy**. Steno inferred time from a spatial arrangement of rocks. This is the essence of geology, a science that did not yet exist as such.

Using geometry, Steno was the first to formulate laws for the formation of rock layers. One of his laws is known as the **principle of superposition**. As explained by Steno, the sequence of events that leads to layers of rock begins with the settling of particles of rock at the bottom of a column of water. Subsequent deposits, forming one layer after another, later become compressed into a series of strata, as shown in Figure 3-4. The column of water might be within a river, sea, or lake. Sediment also can be carried and deposited by wind and ice. Steno realized that in a sequence of strata, a given stratum is younger than those below it and older than those above it, hence the term "superposition."

A second law, related to the first, is the **principle of original horizontality**: Strata are originally deposited in uniform, horizontal sheets. Because of plate tectonics and mountain-building processes that deform rocks, strata often become tilted, as shown in Figure 3-4. The Monterey Formation shown at the beginning of this chapter was tilted by plate movements. Even if strata are completely overturned, the original direction of deposition often can be determined from fossil evidence and other clues. As long as the direction of deposition is known, the relative age of each layer can be determined using Steno's laws.

Fossils and the Geologic Column

As powerful as Steno's principles were, they did not enable scientists, miners, and engineers to correlate rock layers from different locations. At a given site, one knows the relative age of each layer from the principle of superposition, but cannot tell the relative ages of layers between sites. For example, during the 1700s mining for coal and minerals increased throughout Europe, and miners began to recognize general sequences of rocks locally. In order to communicate with one another about the strata, they came up with names that referred to properties of the strata, such as their smell ("Stinking Vein") or thickness ("Ten-Inch Vein"). In coal-rich Somerset, England, coal miners knew that a Ten-Inch Vein was usually found below a "Shelly Vein." They had no way of knowing, however, if a Stinking Vein in southeast Somerset was the same as a Stinking Vein in northwest Somerset.

The solution to the problem of correlating rocks existing at different sites was in the fossils within the rocks. An English surveyor named William Smith (1769–1839, nicknamed Strata Smith) studied the fossils he found while surveying canals in the area of Somerset. He realized that each layer of rock contained a particular and

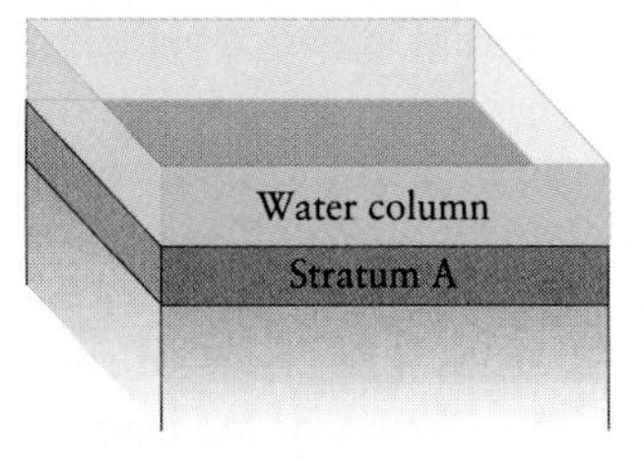

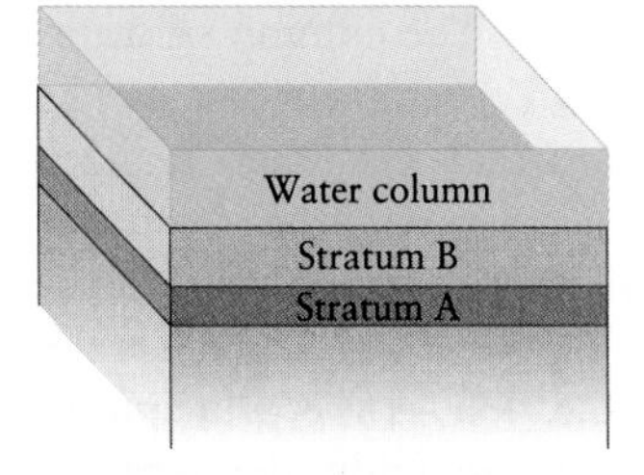

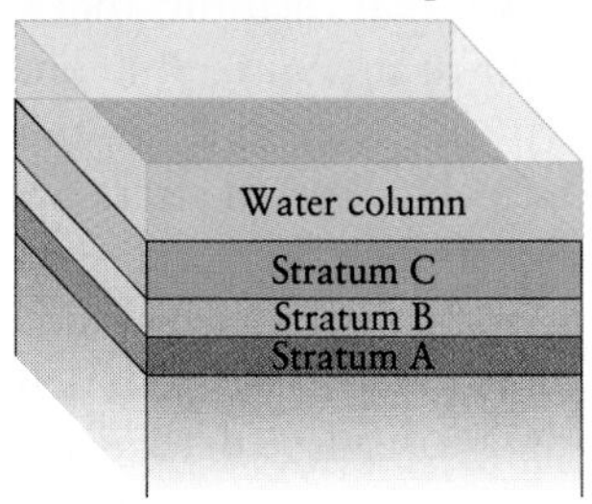

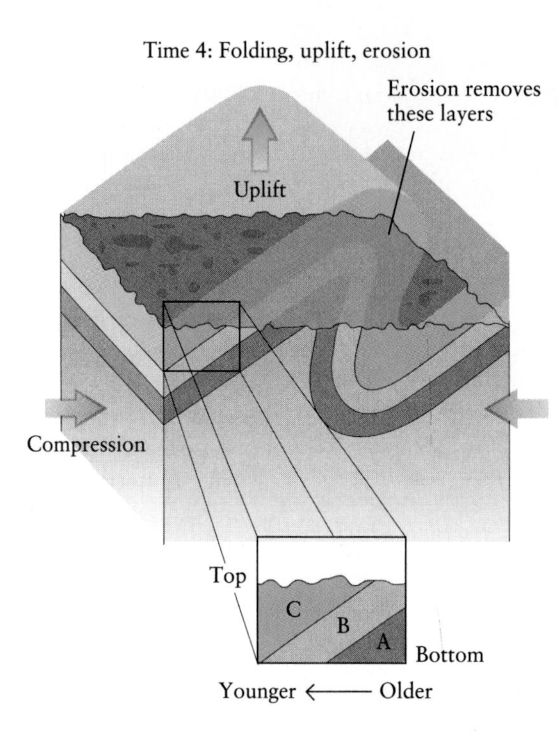

FIGURE 3-4 Steno's laws, based on the geometry of sedimentary layers, can be used to deduce geologic history. In the sequence of rocks illustrated here, for example, three layers (A, B, C) originally deposited in a lake environment were later raised and tilted. During uplift and mountain building, sedimentary layers sometimes are overturned. However, it usually is possible to determine which layer is the original stratigraphic top from clues such as the shapes of burrows made by worms in lake-bottom muds.

characteristic group of fossils. When friends showed him their fossil collections, he could identify—to their amazement—exactly the stratum from which the fossils came.

Smith found that some fossils exist only in certain strata, and never in younger ones, as if the life-forms had become extinct. New forms appeared in younger strata, then they, too, disappeared from the record. Each fossil assemblage is unique in time. Smith used this knowledge to formulate the **principle of faunal and floral succession**, which states that assemblages of species of plants and animals succeed each other in time. Smith and his contemporaries in the early 1800s had no idea why such changes occurred. Nevertheless, the fact that succession exists resulted in a powerful tool for correlating and mapping strata (Box 3-2).

Smith and geologists after him applied the principles of superposition and faunal and floral succession to map fossil-containing rocks worldwide. For convenience in mapping, rocks were grouped into units called **formations**, which are layers that have similar properties and appearance and the same stratigraphic age based on their fossil content. In essence, each represents a specific environment, or conditions of origin. When all the known strata were combined in their relative chronological order, geologists had created a **geologic column**, a composite stratigraphic section for the world. With time, the geologic column grew longer and more complex, leading many to conclude in the 19th century that Earth must surely be more than 6000 years old.

The Geologic Time Scale

As the geologic column developed, geologists gave names to the specific intervals of geologic time during which rocks were deposited. These time units were used to construct a **geologic time scale** based entirely on stratigraphy and fossils, with no notion as yet of absolute ages (Figure 3-5). As the year is divided into hierarchical units of months, days, minutes, and seconds, geologists divided the geologic past into hierarchical units, called eons, eras, periods, and epochs. An **eon** is the largest unit of geologic time, and only three eons are recognized. The oldest is the Archean eon (from the Greek for "ancient"). Sedimentary rocks of the Archean are sparsely sprinkled with microscopic fossils of primitive unicellular organisms. The next eon is the Proterozoic (from the Greek for "early life"). Sedimentary rocks of the Proterozoic eon also contain unicellular organisms, but these are more plentiful and more complex than those of the Archean eon. The latest eon, beginning with multicellular organisms large enough to be seen without a microscope and stretching to the present day, is the Phanerozoic (from the Greek for "visible life"). From age-dating techniques developed in the 20th century, it is now known that each of these eons represents hundreds of millions to billions of years.

The Phanerozoic is the only eon divided into eras with commonly used names. **Eras** represent time spans from many tens to hundreds of millions of years. Early geologists identified eras in the Phanerozoic eon by major boundaries in the rock and fossil record, boundaries at which as many as 70 to 80 percent of all previously existing species disappeared, marking times of major extinctions. Eras are divided into **periods,** which in turn are divided into smaller time units, **epochs,** which sometimes are as short as 10,000 years. As with eras, the boundaries of periods and epochs also correspond with changes in marine and terrestrial fossils, but the changes are not as prominent as those that mark the boundaries of eras.

The Phanerozoic eon is divided into three eras based on its fossil record: Paleozoic ("old life"), Mesozoic ("middle life"), and Cenozoic ("recent life"). During the Paleozoic era, trilobites flourished in the oceans, fishes evolved, amphibians took to the land, and reptiles wandered early forests. This era ended fairly abruptly, and about 70 percent of these species disappeared, to be replaced on land by dinosaurs, conifer trees, and flowering plants during the Mesozoic era. The Mesozoic also ended fairly abruptly, when dinosaurs and many other species became extinct, making way for mammals, including humans, to become prominent in the Cenozoic era.

The Cenozoic era is divided into two periods, the Paleogene (65 million to 23 million years ago) and the Neogene (23 million years ago to the present). The Neogene period is the time in which *Homo*, the genus, or group of closely related species, to which humans belong, began to spread across Africa, Asia, and Europe (see Appendix 1). It also marks a time of global cooling which culminated in Earth's fifth ice age in the past few million years. During this time, ice sheets repeatedly have advanced and retreated across North America and Europe. For these reasons, the two most recent epochs of the Neogene period, the Pleistocene and Holocene, sometimes are referred to as the "Age of Humans" or the "Age of Ice."

Evolution: A Biological Timekeeper

While piecing together the geologic time scale, geologists came to realize that Earth must be quite old in order for so many different life-forms to have come, flourished, and vanished. One such scientist was the British geologist Charles Lyell (1797–1875), a contemporary of the famed naturalist Charles Darwin (1809–1882). When Darwin left for his five year round-the-world expedition on HMS *Beagle* in 1831, he took along a copy of Lyell's textbook, *Principles of Geology.* During those years, Darwin pondered the nature of different species and wondered if they might have changed from one form to another over time.

While visiting the Galápagos Islands off Ecuador, for example, Darwin recognized that different species of finches live on each island. After returning home and studying his notes and samples, he developed the idea that the finches had adapted themselves to their local environment by some process (or processes) that resulted in evolutionary change from one species to another with time. He likened the process to that used by breeders when preferentially breeding animals with certain traits, such as cows without horns but with high milk yield. He reasoned that nature might act as a breeder by a process of natural selection, a process that winnows populations of those members least fit for a given environment, resulting in greater reproduction of those best adapted.

As natural variation in traits within a species is quite large, Darwin suspected that with time, processes of selection could result in the evolution of new species as well as the extinction of old ones. After returning home, Darwin refined this hypothesis of natural selection and the origin of species and later published his ideas in his classic book, *On the Origin of Species.* Darwin realized that the processes of natural selection he envisioned would require immense amounts of time to occur, but how immense he was unable to determine.

Darwin's ideas explain how the principle of faunal and floral succession works. As new species evolve, they sometimes migrate from one continent to another, depending on the arrangement of continents. If something causes extinction—perhaps rapid environmental change or a catastrophic asteroid impact—some species are terminated. Evolution, migration, and extinction of species are the reasons that distinct fossil assemblages occur in specific strata, that strata can be correlated from one site to another, and that geologic history can be reconstructed. Even though it was to be another half century before radioactivity was discovered and applied to age dating, evolution made it possible to devise a relative geologic time scale from the rock record of life on Earth.

- In sedimentary rock formations, fossils and stratigraphic sequences are clues to the relative ages of the rock layers.
- The geologic time scale is marked by mass extinctions and the appearance of new life-forms.

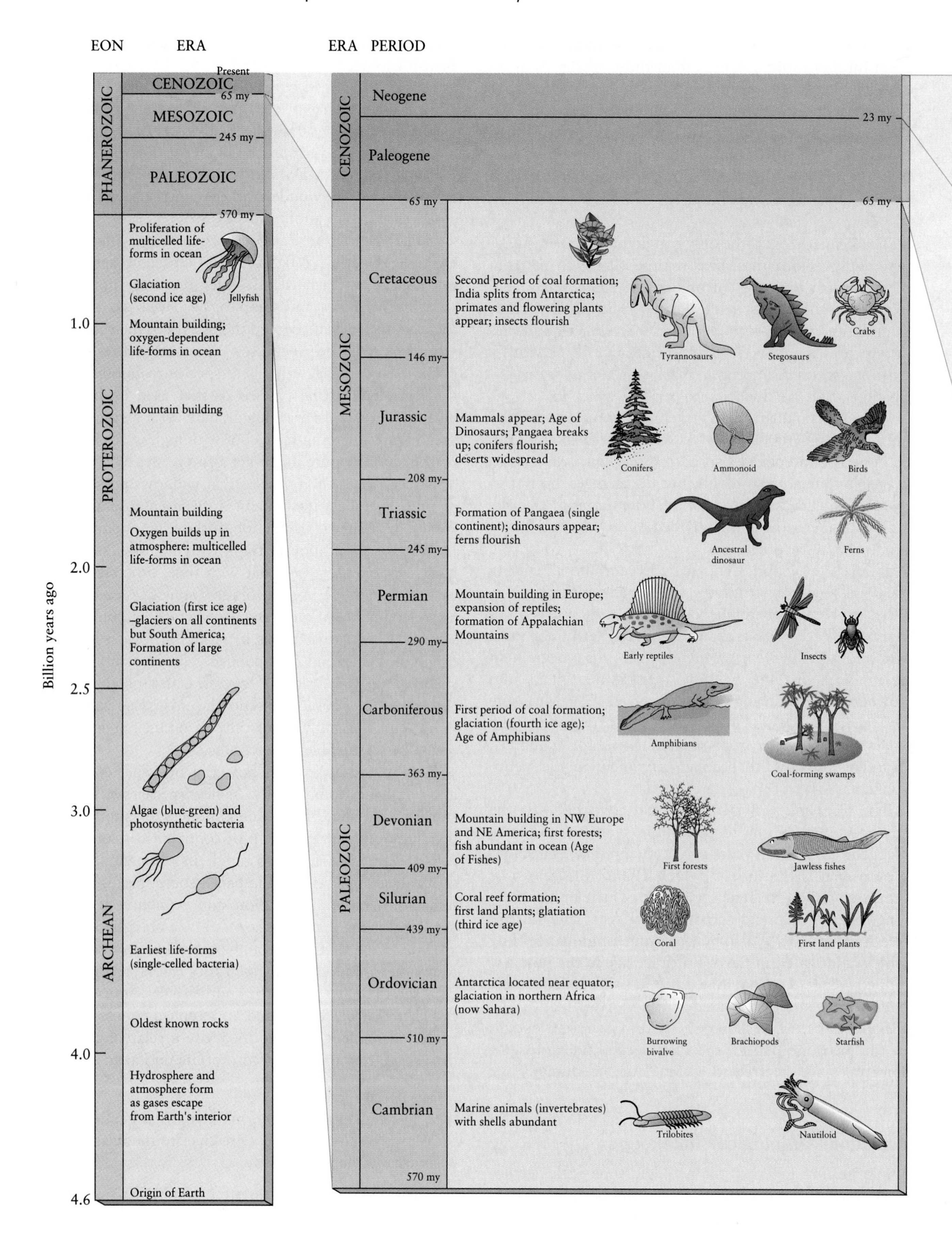

EON
ERA
ERA
PERIOD
Present
CENOZOIC
65 my
MESOZOIC
245 my
PALEOZOIC
570 my
PHANEROZOIC
PROTEROZOIC
ARCHEAN
Billion years ago
1.0
2.0
2.5
3.0
4.0
4.6
Proliferation of multicelled life-forms in ocean
Jellyfish
Glaciation (second ice age)
Mountain building; oxygen-dependent life-forms in ocean
Mountain building
Mountain building
Oxygen builds up in atmosphere: multicelled life-forms in ocean
Glaciation (first ice age) –glaciers on all continents but South America; Formation of large continents
Algae (blue-green) and photosynthetic bacteria
Earliest life-forms (single-celled bacteria)
Oldest known rocks
Hydrosphere and atmosphere form as gases escape from Earth's interior
Origin of Earth
CENOZOIC
Neogene
23 my
Paleogene
65 my
MESOZOIC
Cretaceous
Second period of coal formation; India splits from Antarctica; primates and flowering plants appear; insects flourish
Tyrannosaurs
Stegosaurs
Crabs
146 my
Jurassic
Mammals appear; Age of Dinosaurs; Pangaea breaks up; conifers flourish; deserts widespread
Conifers
Ammonoid
Birds
208 my
Triassic
Formation of Pangaea (single continent); dinosaurs appear; ferns flourish
Ancestral dinosaur
Ferns
245 my
PALEOZOIC
Permian
Mountain building in Europe; expansion of reptiles; formation of Appalachian Mountains
Early reptiles
Insects
290 my
Carboniferous
First period of coal formation; glaciation (fourth ice age); Age of Amphibians
Amphibians
Coal-forming swamps
363 my
Devonian
Mountain building in NW Europe and NE America; first forests; fish abundant in ocean (Age of Fishes)
First forests
Jawless fishes
409 my
Silurian
Coral reef formation; first land plants; glatiation (third ice age)
Coral
First land plants
439 my
Ordovician
Antarctica located near equator; glaciation in northern Africa (now Sahara)
Burrowing bivalve
Brachiopods
Starfish
510 my
Cambrian
Marine animals (invertebrates) with shells abundant
Trilobites
Nautiloid
570 my

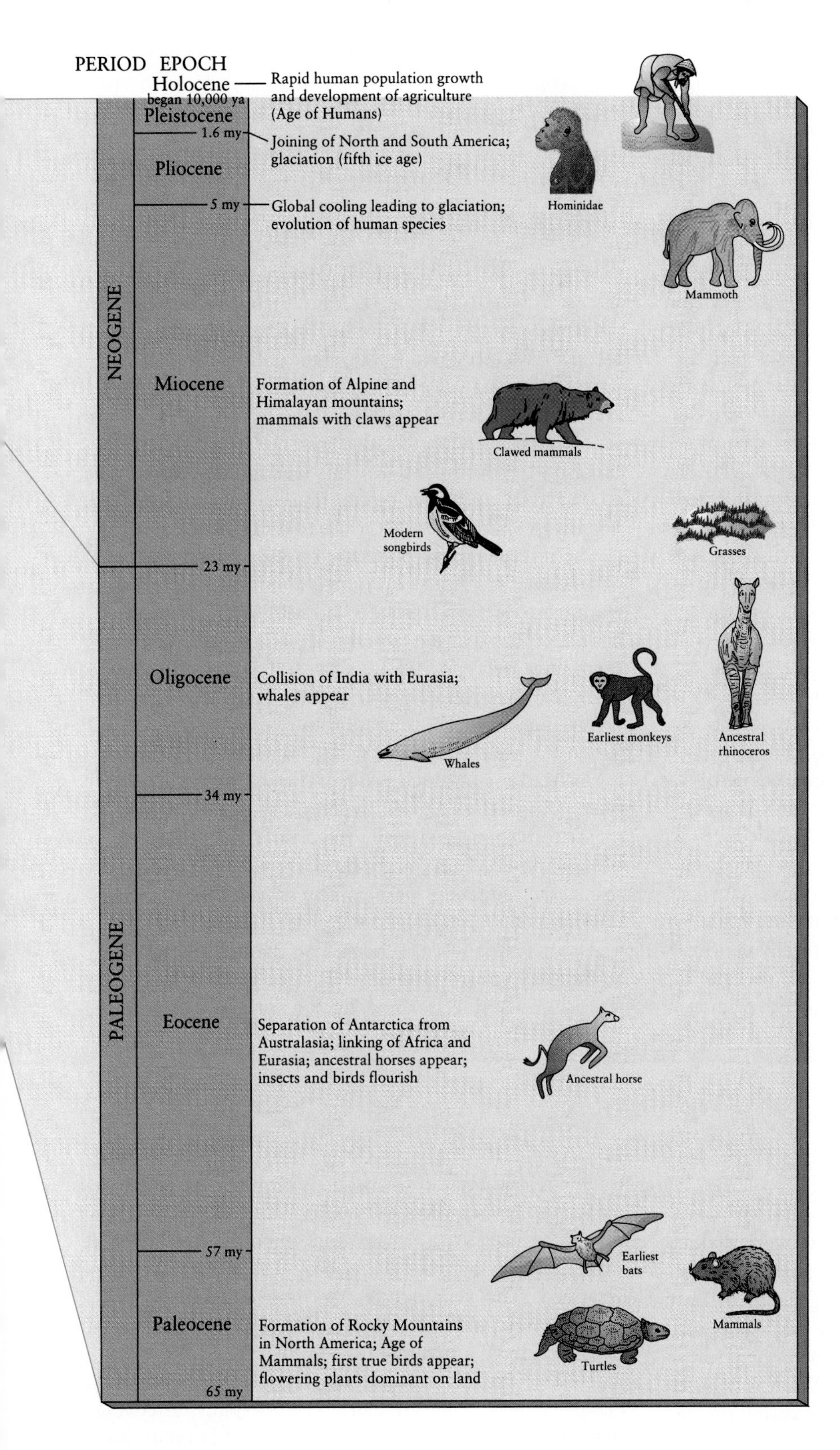

FIGURE 3-5 The geologic time scale for Earth is constructed from stratigraphic and fossil records. Age estimates are based on radiometric dates of igneous rocks obtained in the 20th century. Examples of common life-forms are shown at the approximate times at which they appeared or became abundant.

3-2 Geologist's Toolbox

How to Interpret a Stratigraphic Section

As early geologists, miners, and surveyors developed an understanding of rock strata and the fossils within them, they were able to apply William Smith's principle of faunal and floral succession to correlate rock formations from one locale to another. In 1725, a complete vertical section of strata was published for the county of Somerset in southwestern England. This section shows that coal layers on the southeast and northwest sides of a ridge are the same, even though they occur at different depths below the surface. Miners had recognized distinctive traits of different coal seams, such as their thickness and hardness, but it was the characteristic assemblages of shells, ferns, and other fossils within them and the rocks around them that enabled the miners to make definite correlations between strata at different sites.

Geologists continue to rely on stratigraphic sections and on the principles of original horizontality, superposition, and faunal and floral succession to infer the geologic history recorded by the rocks. Steno's law of superposition, for example, indicates that the youngest rocks in the Somerset section are the yellowish clay, marl, and limestone at the top of the section. (Marl is a loose, earthy mixture of clay and calcium carbonate typically deposited in freshwater environments, while limestone is mostly calcium carbonate typically deposited in marine-water environments.) The older rocks in the section are the interbedded coal and other sedimentary layers, with the Ten-Inch Vein the oldest of the coal seams. The coal-bearing rocks are inclined nearly 45°, while the younger rocks are horizontal. According to Steno's law of original horizontality, the older rocks have been tilted since deposition. The older rocks also are broken by a fault, a plane along which rocks are offset, so that layers on the southeast are raised relative to those on the northwest. For this reason, the coal-bearing strata would no longer be continuous from southeast to northwest if they were returned to a horizontal position.

The following interpretation of the geologic history of Somerset takes into account the information gleaned from using Steno's laws. The bottom layer of coal-bearing rocks was deposited first. The fossils within them indicate deposition in a coastal area with a fluctuating sea level and coastline. When the area was under water, marine shells, sand, and mud were deposited. When the area was above water, swamps formed, leaving behind organic-rich peats and mud. Subsequently, tectonic processes caused the entire sequence of sediments to be squeezed and altered into rock, tilted, broken along a fault, and raised above water. As a result, the sediments were compressed and heated and the organic compounds in the peats became coal.

An ancient landscape formed on the uplifted rocks. Scoured by streams and other surface processes, a nearly flat erosion surface formed. Such a surface, called an

Absolute Geologic Time

At the end of the 19th century, scientists were divided in controversy over Earth's age. Some prominent physicists and chemists argued for a young Earth of only 20 million to 100 million years, while geologists and those who supported ideas about evolution argued for an old Earth, perhaps billions of years old. What was needed to end this dispute was a means of measuring the absolute, or numerical, age of Earth.

Radioactivity: Nature's Atomic Clock

Meanwhile, physicists and chemists in France, Germany, and England were rapidly making discoveries about a strange new energy source that would soon lead to a reliable method for measuring the absolute ages of Earth's materials. French physicist Henri Bequerel (1852–1908) discovered in 1896 that uranium minerals emit invisible rays that leave tracks on photographic plates, even if wrapped. The year before, German physicist Wilhelm Roentgen (1845–1923) had called the mysterious rays produced from a cathode tube in his lab X rays.

Two scientists working together in France, Marie (1867–1934) and Pierre Curie (1859–1906), discovered that these mystery rays emanated not only from uranium but also from the elements thorium, polonium, and radium (Figure 3-6). They described the elements emitting such energy as radioactive, after radium. **Radioactivity** is the propensity of certain elements to change form and emit rays of energy. The Curies were the first to determine that it was the elements themselves that emitted

unconformity, represents a gap in time during which no sediment layers could form. No rocks exist to yield clues of what happened other than erosion during that time interval. Some time later, the landscape was again submerged in water and the younger rocks were deposited. The youngest rock, the limestone, consists of the shells of tiny marine organisms. Since then, the entire sedimentary sequence emerged again above sea level and deposition ceased. The cause of emergence is the result of a relative drop in sea level; that is, either sea level fell, or the land surface rose, or some combination of the two processes occurred. Modern rivers carved valleys into the upper layers, exposing the underlying unconformity and coal beds.

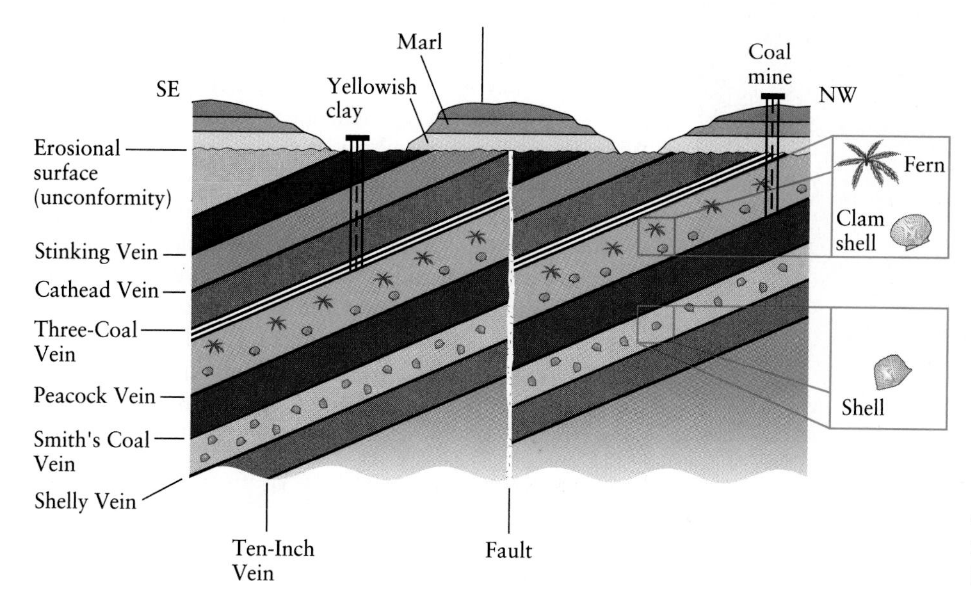

A stratigraphic section for coal-rich Somerset in southwestern England.

the rays, and for their discovery they shared the 1903 Nobel prize for physics with Bequerel. No one yet fully realized the power or the danger of this newly discovered source of energy. Both the Curies began to have recurrent headaches and irritated skin, and Marie Curie later died of leukemia, a cancer that can result from exposure to radioactivity.

Scientists busily set about identifying the properties of radioactive elements. In 1902, two physicists, Ernest Rutherford (1871–1937) and Frederick Soddy (1877–1956), isolated a gas from thorium compounds and studied changes in its activity. They called this gas thorium X; today, we call it radon. Rutherford and Soddy noted that after about 3.6 days, the activity of thorium X had decreased to half its original value, after 7.2 days to one-fourth its initial value, after 10.8 days to one-eighth its initial value, and so forth. At the same time, the activity of the thorium compound from which they had removed thorium X was increasing! They interpreted this to mean that thorium X was being produced by the compound.

Rutherford and Soddy suggested a hypothesis to explain radioactivity:

- An unstable atom decays spontaneously, losing particles and energy from its nucleus, and becomes another element.
- The number of radioactive atoms that decay in a given period of time is proportional to the number of radioactive atoms in the sample: Half will decay over a constant period of time.

FIGURE 3-6 Scientists Marie and Pierre Curie investigating radioactivity in their laboratory.

This pair of statements forms the principles underlying radiometric dating, a quantitative method of dating based on the rate of radioactive decay of unstable atoms. Rutherford and Soddy recognized that some **parent elements**, like those in their thorium compound, decay at specific rates to produce **daughter products**, such as thorium X (radon), which in turn can decay to form other daughter products. Since the work of Rutherford and Soddy, many unstable elements and chains of decay reactions have been identified and now are used as nature's clocks to determine the ages of fossils, minerals, and rocks.

Radioactive Decay of Unstable Isotopes All atoms of a given element have the same number of protons in their nuclei, but most elements occur as more than one variety of atom, each with a slightly different number of neutrons and therefore a slightly different atomic mass. Atoms of an element that differ in the number of neutrons are **isotopes** of that element. Carbon, for example, has 6 protons, but isotopes of carbon can contain as few as 3 to as many as 10 neutrons. As a result, the total number of protons and neutrons in the carbon nucleus, which is equal to the atomic mass, can vary from 9 to 16. In general, isotopes with similar or equal numbers of protons and neutrons tend to be stable, remaining the same over time; those with very different numbers of protons and neutrons tend to be unstable, that is, they decay spontaneously and are radioactive. The most stable and, therefore, the most abundant of the carbon isotopes, with 6 neutrons and 6 protons, is referred to as the ^{12}C (carbon 12) isotope. The next most abundant, with 7 neutrons and 6 protons, is the ^{13}C isotope. Of lesser abundance, the ^{14}C isotope, with 8 neutrons and 6 protons, is the least stable of the three.

The nucleus of an unstable isotope is not held firmly together. Instead, its protons and neutrons break apart spontaneously and continuously—in the process of radioactive decay—until the nucleus reaches a stable configuration. When the nuclei of unstable isotopes decay, they emit subatomic particles and energy; this process explains the source of energy emanating from Earth's interior.

One form of isotope decay occurs when a neutron splits into a proton and a beta particle (an energetic electron, indicated as β^-), and the beta particle is ejected from the nucleus (Figure 3-7). This form of decay is common in isotopes that have excess neutrons, such as the ^{14}C isotope. The chemical equation that illustrates **beta decay** for ^{14}C is

$$^{14}C - \beta^- \rightarrow {}^{14}N$$

The new daughter product, ^{14}N (nitrogen 14), has one fewer neutron than ^{14}C and a new proton. With 7 protons rather than 6, a new element—nitrogen—is formed from the ^{14}C isotope after it decays.

The opposite of β^- decay occurs when a proton transforms into a neutron in one of either two ways: (1) by capturing an electron from the innermost shell of electrons, or (2) by emitting a positron (the antimatter equivalent of an electron, indicated as β^+) from the nucleus. In the nucleus, one proton has been lost and one neutron has been gained, but the atomic mass number has stayed the same. An isotope with a deficiency of neutrons can acquire them via these modes of decay. An example of **electron capture** occurs when an isotope of potassium (^{40}K) transforms to an isotope of argon (^{40}Ar). An example of **positron decay** occurs when an unstable isotope of oxygen (^{15}O) transforms to a stable isotope of nitrogen (^{15}N).

A third type of decay occurs when an alpha (α) particle consisting of two protons and two neutrons is emitted from a nucleus. **Alpha decay** is more common in

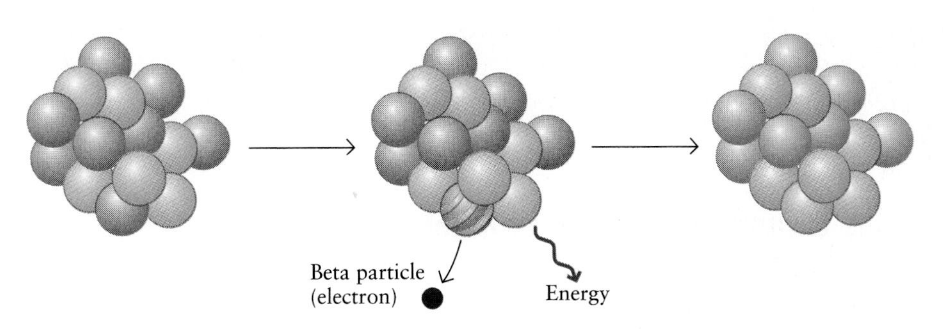

FIGURE 3-7 A carbon isotope, ^{14}C, decays to an isotope of nitrogen, ^{14}N, via emission of a β^- particle (electron).

heavy isotopes, such as uranium (U), for a light isotope is not likely to eject two protons and two neutrons. An example of α decay that is used commonly in radiometric dating is the transformation of ^{238}U to ^{206}Pb (lead):

$$^{238}U \rightarrow {}^{206}Pb + 8\alpha \text{ particles}$$

Decay Rates and Half-Lives Recall that at the beginning of the 20th century, Rutherford and Soddy removed radon (which they called thorium X) from thorium and observed the following: After about 3.6 days, the activity of radon had decreased to half its original value (e.g., from 100 to 50 percent), after 7.2 days to one-fourth its initial value, after 10.8 days to one-eighth its initial value, and so forth. In other words, every 3.6 days the isotope had decayed such that half of the parent isotope was gone. Rutherford and Soddy had discovered a fundamental property of radioactive decay that is the same for all radioactive substances: *The rate of disintegration of an isotope is proportional to the number of parent atoms and occurs at a fixed rate specific to each isotope.* The time it takes for half the atoms of a radioactive isotope to decay is called the **half-life**. Half-lives vary from as short as billionths of a second to as long as billions of years (Table 3-1). The shorter the half-life, the more rapidly the parent isotope decays. Decay of ^{14}C to ^{14}N has a half-life of about 5730 years; decay of ^{238}U to ^{206}Pb has a much longer half-life, of 4.47 billion years.

Rutherford and Soddy's finding gave the half-life of radon as 3.6 days; more recent work indicates that it is closer to 3.8 days. The isotope they studied, radon 222, is now known to pose a possible hazard to people living in houses built above rock types from which it seeps; its decay products can cause lung cancer (Box 3-3).

With each half-life, the number of parent isotopes diminishes by half and the number of accumulating daughter products increases proportionately. Eventually, very little of the parent isotope remains and the amount of

TABLE 3-1 Useful Isotopes for Radiometric Dating

Radioactive parent isotope	Product of decay	Approximate half-life (years)	Useful range for dating (years)*
Rubidium 87	strontium 87	48.6 billion	10 million – 4.6 billion
Thorium 232	lead 208	14.0 billion	10 million – 4.6 billion
Uranium 238	lead 206	4.5 billion	10 million – 4.6 billion
Potassium 40	argon 40	1.3 billion	100,000 – 4.6 billion
Uranium 235	lead 207	0.7 billion	10 million – 4.6 billion
Carbon 14†	nitrogen 14	5730	50,000

*In general, the effective dating range for a radioactive isotope is 10 to 15 times its half-life.

†The carbon isotope is useful for dating organic matter and precipitates, such as bones and shells.

3-3 Case Study

The Environmental Threat of Radon

In the 1500s many silver miners in central Europe were dying young of horrible lung disease. Describing this tragedy in his 1526 textbook on mining, *De re metallica,* Georgius Agricola claimed that "women are found who have married seven husbands, all of whom this terrible consumption has carried off to a premature death." Recent testing has determined that rocks in the mines contain radon, an odorless, radioactive gas linked to lung cancer in humans.

It was the end of the 20th century before the American public recognized in their own houses the threat of radon-induced lung cancer. In 1984, a construction worker at the new Limerick nuclear power plant near Philadelphia set off the radiation detector while checking in at work, before the plant's reactor was even operating. The source of the radiation was a puzzle until an inspection of the worker's small suburban house revealed a level of radon 700 times higher than federal safety standards would allow.

By 1985 the state of Pennsylvania had identified several thousand houses with unusually high levels of radon. All were located above certain types of rocks (granites and shales, in particular) with relatively high amounts of uranium in their minerals. Today maps are available that show where radon gas is most likely to seep from rocks and accumulate in houses. In addition, inexpensive testing kits for radon are available in hardware stores and commercial testing laboratories.

One such testing kit, the alpha-track radon detector, provides some insight into how radon can result in lung cancer. The detector is basically a container with a piece of plastic inside. The container is placed in a house near the suspected radon leak, opened, left undisturbed for several months, then closed and sent to a laboratory for analysis. Radon decays by emitting alpha particles. If radon gas enters the container, the alpha particles it ejects will form tiny pits in the plastic when they hit it. The number of pits in the plastic yields the concentration of radon in the air of the house. Rapidly moving subatomic particles not only from radon but also from its unstable daughter products can cause similar damage to lung tissue and eventually lead to cancer.

The ultimate source of radon gas is the radioactive decay of uranium 238, which has a long chain of daughter products. Uranium 238 typically occurs in small amounts in most rocks, less than a few parts per

A radon map of the United States shows the distribution of rock and soil with uranium levels high enough to produce potentially hazardous concentrations of radon.

million, but is more abundant in some granites and shales (fine-grained sedimentary rock). About midway through the decay chain is radon 222, the form of radon that may occur in houses. Radon 222 is the first gaseous isotope in the chain, and thus the first that can seep easily from surrounding rocks into soil and air. If emitted into open air, radon is dispersed readily, but in an enclosed space, such as a house, radon and its decay products can accumulate to hazardous levels.

Most of the cancer-causing power of radon 222 lies in the emissions from its unstable daughter products. The gas itself has a half-life of only 3.8 days, and if it is inhaled, it is exhaled long before it can emit a significant number of alpha particles. Some of the unstable daughter products of radon 222 have much shorter half-lives, but because they are solids, they can attach to microscopic bits of dust floating in the air. When these unstable daughter products are inhaled, they become stuck in lung tissue and continue to decay from one isotope to another, emitting alpha particles and other types of particles and radiation over many years. Just as particles emitted from radon 222 pit the plastic of a radon detector, particles emitted from its unstable daughter products can destroy the cells of human lungs and cause cancer.

How can you determine if the place where you live has an elevated level of radon gas, and what can you do if it does? You can examine maps prepared by the U.S. Environmental Protection Agency or individual state radon surveys.* However, some houses in an area mapped as low risk may actually have high levels of radon, and vice versa, so it is wise to test your house with an inexpensive detection kit. A high radon level (more than 4 picocuries per liter of air) can be reduced by sealing the floors and walls of the basement, by increasing ventilation from the basement, or by pumping air out of the soil and rock beneath the basement. Each of these approaches generally costs several thousand dollars.

*_Radon: Risk and Remedy_, by David Brenner (see Suggested Readings), contains a list of EPA and state radon program offices, with their addresses and telephone numbers. It also contains information on how to purchase reliable radon detectors.

(a) (b)

(a) In an alpha-track radon detector, sensitive plastic is within the can, which is about 2 to 3 cm in diameter. If radon gas is present, it enters the detector can and its decay products pit the plastic piece inside. (b) In this enlarged photograph, such pits—each from a single alpha particle—can be seen.

daughter product is nearly equal to the original amount of parent isotope. In essence, the process of radioactive decay is like the ticking of a clock. The greater the age of a substance containing radioactive isotopes, the greater the number of ticks that have occurred.

Radiometric Dating

The use of radioactive decay processes to date rocks and organic matter is known as **radiometric dating.** Radiometric dating requires certain information to be useful; in particular, one must know how much of the parent isotope was originally present in a substance at the time of its formation. To date a sample of a mineral or organic matter using a given isotope, the amounts of daughter products and remaining parent isotopes are measured, then the duration of time that radioactive decay has been occurring is calculated from decay curves like those shown in Figure 3-8.

For example, if the remains of a human body found entombed in a peat bog in Norway contain only 1/16 the amount of ^{14}C in the modern atmosphere, it is possible to determine when that person died. (Peat, the precursor of coal, is composed of compressed plant remains buried in a wetland environment.) The original amount of ^{14}C in the body was likely to be similar to that in the atmosphere, because living organisms constantly exchange carbon with the biosphere and atmosphere. Four half-lives have elapsed since the person found in the peat bog died, as the amount of ^{14}C has decayed from its original amount to 1/2, then 1/4, then 1/8, and finally to 1/16 of the original. As the half-life of ^{14}C is 5730 years, the person lived about 22,920 years ago. Although other carbon isotopes are radioactive, they are too rare to use for radiometric dating.

For most unstable isotopes, their use in radiometric dating is diminished after about 10 to 15 half-lives because the number of parent atoms remaining becomes too small to detect. The longer the half-life, the older the sample that can be dated (see Table 3-1). For example, with a half-life of about 5730 years, ^{14}C has a range of about 50,000 years for radiometric dating. This range is most useful for dating the evidence of events that occurred just before and during the most recent time of full glacial conditions on Earth and the evolution of *Homo sapiens*. In contrast, with a half-life of 4.5 billion years, the decay of uranium 238 to lead 206 is most useful for dating rocks that are tens of millions to billions of years old.

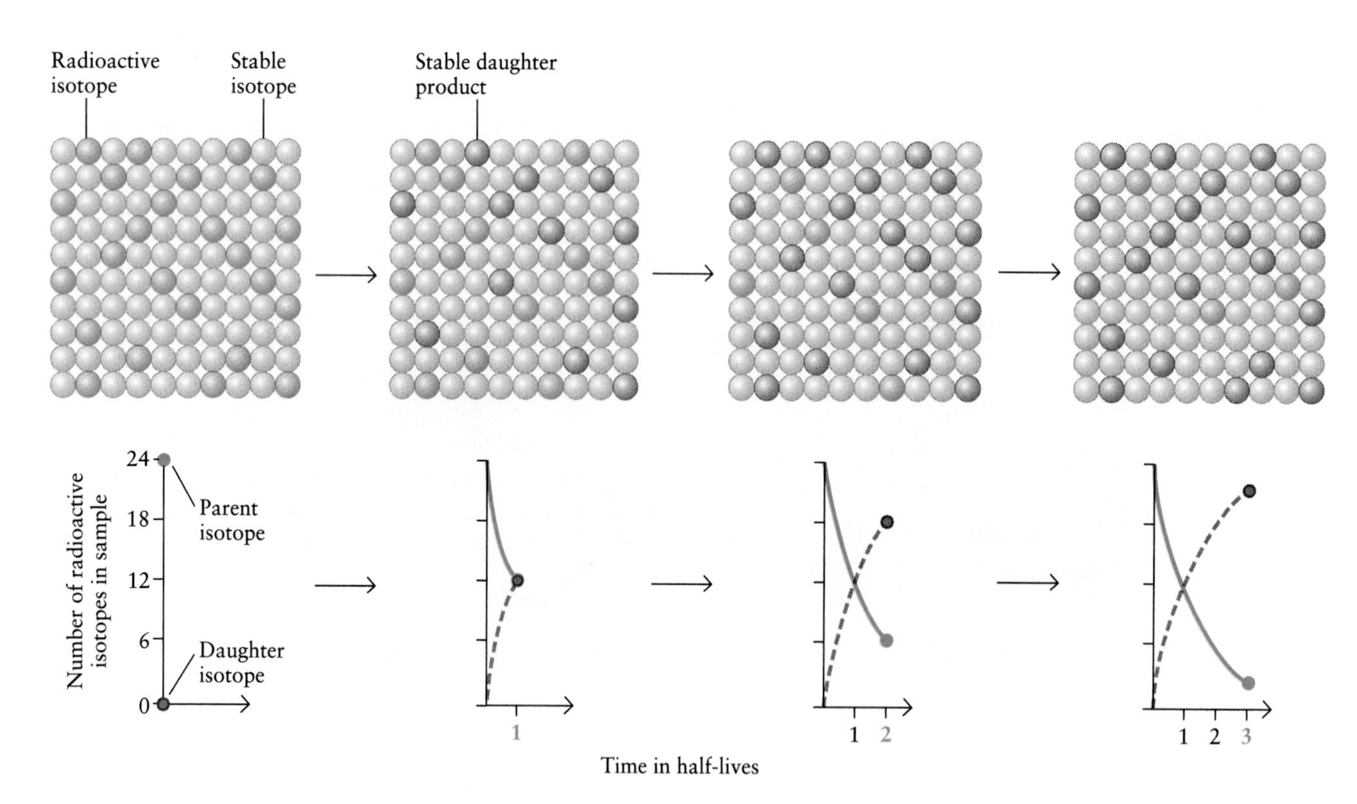

FIGURE 3-8 A radioactive decay curve for a hypothetical group of 100 atoms that originally contained 24 radioactive isotopes shows that, after one half-life, 12 of those atoms have decayed and become stable daughter products. After two half-lives, one-half of the remaining 12, or 6, have decayed, and after three half-lives half of the remaining 6 have decayed. The number of atoms that decays is proportional to the number of parent isotopes present at any given time.

Rutherford was the first to realize the link between his observations and the use of radioactivity to date igneous rocks. In 1905, Rutherford lectured at Yale University, where he used calculations of helium decay to show his audience that a mineral sample was 497 million years old, a substantially older age for Earth than anyone's guess at that time.

Since the work of the Curies, Rutherford, and others, rocks from all over Earth, from the Moon, and from meteorites have been dated. The oldest rocks yet found on Earth are in northwest Canada and are 3.96 billion years old. The oldest minerals, however, are crystals of zircon contained within a younger sandstone in Australia. The crystals are nearly 4.3 billion years old. The oldest fossils are single-celled bacteria found in rocks in Canada and Australia that are 3.5 billion years old (Figure 3-9). Few rocks and fossils of such great ages exist because of the constant recycling of material at Earth's surface. Hundreds of rocks from the Moon have been dated, most at between 3.2 billion and 4 billion years, but some are as old as 4.5 billion years. Many meteorites have been dated, and most are between 4.4 billion and 4.6 billion years old. These ages have helped us to determine the origin of the solar system.

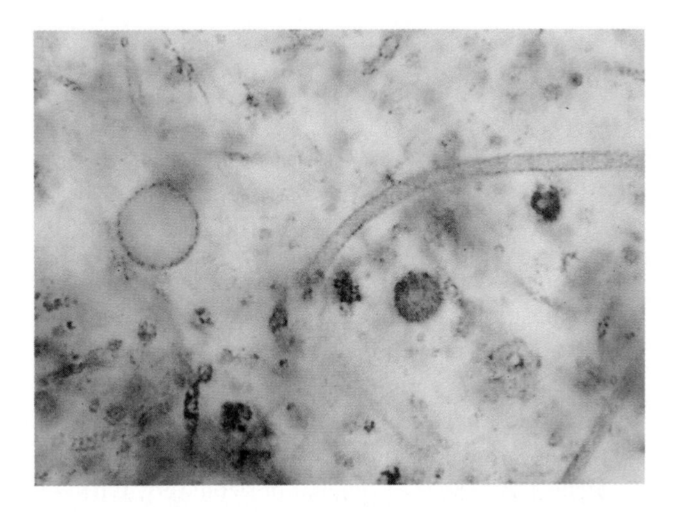

FIGURE 3-9 These fossils of bacteria, probably marine, were found in sedimentary rock in Ontario, Canada. The rocks containing these fossils have been radiometrically dated, and are 3.5 billion years old.

Although it now is known from radiometric dating of meteorites—which formed at the same time as the planets in our solar system—that Earth is 4.6 billion years old, the oldest rocks found on Earth have dates less than 4 billion years old and many igneous rocks are younger than 1000 years old. Radiometric dating yields the time at which the originally molten (liquid) igneous rock cooled below the temperature at which the atoms produced by radioactive decay could no longer escape the increasingly rigid crystalline structure in which they were forming. Because Earth is still hot, and consequently changing all the time because of convection in its interior and the formation of new plates on its surface, the dates of igneous rocks vary from place to place.

As we have seen, the age of organic as well as inorganic material can be found if a suitable isotope is used. The radioactive isotope ^{14}C accumulates naturally in living organisms because it is produced at a fairly steady rate in the atmosphere. As organisms take in gases from the atmosphere, the supply of ^{14}C in their tissues is constantly replenished. Once an organism dies and stops respiring, however, its supply of ^{14}C is halted and only decay to ^{14}N occurs. Hence, in ^{14}C dating, the time of death is obtained from a radiocarbon date. This is how scientists are able to say, for example, that the human remains found in a peat bog are those of someone who died 22,920 years ago.

Unlike igneous rocks, sedimentary rocks do not cool from a molten state and trap radioactive decay products from escape, so radiometric dates cannot tell the time of origin of a sedimentary layer. If a mineral in a sedimentary rock were dated, it might yield the age of the older igneous rock from which it was eroded. Similarly, metamorphosed rocks yield ages that represent the time when the radioactive clock was reset by heating and recrystallization. In fact, radiometric dating can give only part of the story of Earth's history.

Processes and events of past ages are inferred from stratigraphy, not from radiometric dating, but the two can be combined to provide a fuller picture of Earth history. Earlier in this chapter it was shown that layered sedimentary rocks containing fossils enabled geologists to develop a geologic column in which rocks are arranged in chronological order relative to one another. Geologists combine relative stratigraphic ages with radiometric dating by studying localities where molten rock has intruded or buried a sequence of strata. The intrusions and flows—which can be radiometrically dated—are used to bracket the ages of the sedimentary strata above and below them. In this way, scientists have combined stratigraphy and radiometric dating to assess the suitability of Yucca Mountain as a repository for nuclear waste (Box 3-4).

The Immensity of Geologic Time

Stretch out your arms and imagine that the tip of the middle finger of your right hand represents the moment, 4.6 billion years ago, when Earth first began to coalesce from the collision of bits of planetary debris circling the Sun. The tip of the middle finger of your left hand represents the present, and the span of your arms between your outstretched fingers represents Earth's 4.6 billion year

3-4 Case Study

Predicting the Geologic Stability of Yucca Mountain

Congress and the U.S. Department of Energy must be assured that Yucca Mountain in Nevada will be geologically stable for at least the next 10,000 years, undisturbed by volcanic eruptions and earthquakes, if the site is to be approved for the storage of radioactive waste. Applying Steno's principles of original horizontality and superposition in combination with radiometric dates to a geologic cross section of the underlying strata, we can assess the past and probable future stability of this site.

The principle of superposition tells us that the oldest rocks are those at the bottom of the stratigraphic section. At Yucca Mountain, these are ancient limestones that were deposited as fossil-rich mud in shallow seas. Fossils in the limestones are from animals that lived more than 245 million years ago, during the Paleozoic era. The top of the limestone beds, an unconformity, is planed off and mantled with debris that resulted from weathering and erosion when the rock was exposed to the atmosphere, before any of the overlying rocks were laid down. The limestone beds are folded rather than horizontal, indicating that after the muds were deposited horizontally they were squeezed, contorted, and pushed upward by tectonic deformation.

Above the limestones are volcanic rocks laid down during a series of eruptions from Timber Mountain, a large, explosive volcano. Lava and ash spewed from this volcano were so hot when they fell to the ground that they welded together, forming a common volcanic rock known as *welded tuff.* Radiometric dating indicates that the tuff strata—named locally the Crater Flat and Paintbrush tuffs—range in age from 8 million to 13 million years. This means that between 245 million and 13 million years ago, either no strata were deposited above the limestones or, if they were, they have been removed by erosion since then. The older Crater Flat tuffs are slightly folded, and even the younger Paintbrush tuffs are tilted, indicating that some deformation of the crust occurred in the area since 13 million to 8 million years ago. No tuffs have formed since then, and Timber Mountain volcano is now extinct.

All the older rocks at Yucca Mountain are offset by a fault, a break in Earth's crust along which the rocks have slipped past one another. Rocks slipping along a fault can cause earthquakes, or ground shaking. Because the fault cuts all the rocks older than 8 million years, it must be less than 8 million years in age, so earthquakes might have occurred sometime in the past 8 million years.

The fault does not cut across or offset two younger deposits, small cones of volcanic cinders and layers of sand and gravel deposited by recent streams. Volcanic debris in the cinder cones has been dated radiometrically at between 10,000 and 1 million years old. Some of the stream sediments overlie the volcanic deposits, and so are even younger. Fossils of plants and animals in the youngest stream sediments indicate that they were deposited less than 10,000 years ago.

This sequence of events indicates that volcanism is active in the area, but the most recent eruptions occurred more than 10,000 to 1 million years ago and were very small. The older volcanic rocks (8 million to 13 million years old) erupted from a much larger volcano that now is extinct. The fault that cuts the older rocks cannot have slipped in at least 10,000 years, because it does not offset the younger cinder cones or the stream sediments, and more detailed studies of the fault using trenches dug several meters into the ground indicate that the most recent fault movement occurred about 20,000 years ago.

Although no scientist can say with absolute certainty that the site is completely safe from any eruption or earthquake for the next 10,000 years, the probability of an eruption is about 1 in 500 million per year and the probability of a moderate-sized earthquake is similarly small. Furthermore, engineers consider it possible to design a repository for nuclear waste that would withstand any rupture or ground shaking of the size likely to occur in that area in the unlikely event of an earthquake.

Scientists working for the Department of Energy will continue to study Yucca Mountain until sometime after the year 2001. The total cost of all the investigations is likely to exceed $7 billion. If the Yucca Mountain site is deemed unsuitable, the United States will have to look for another place to store its 70,000 tons of nuclear waste.

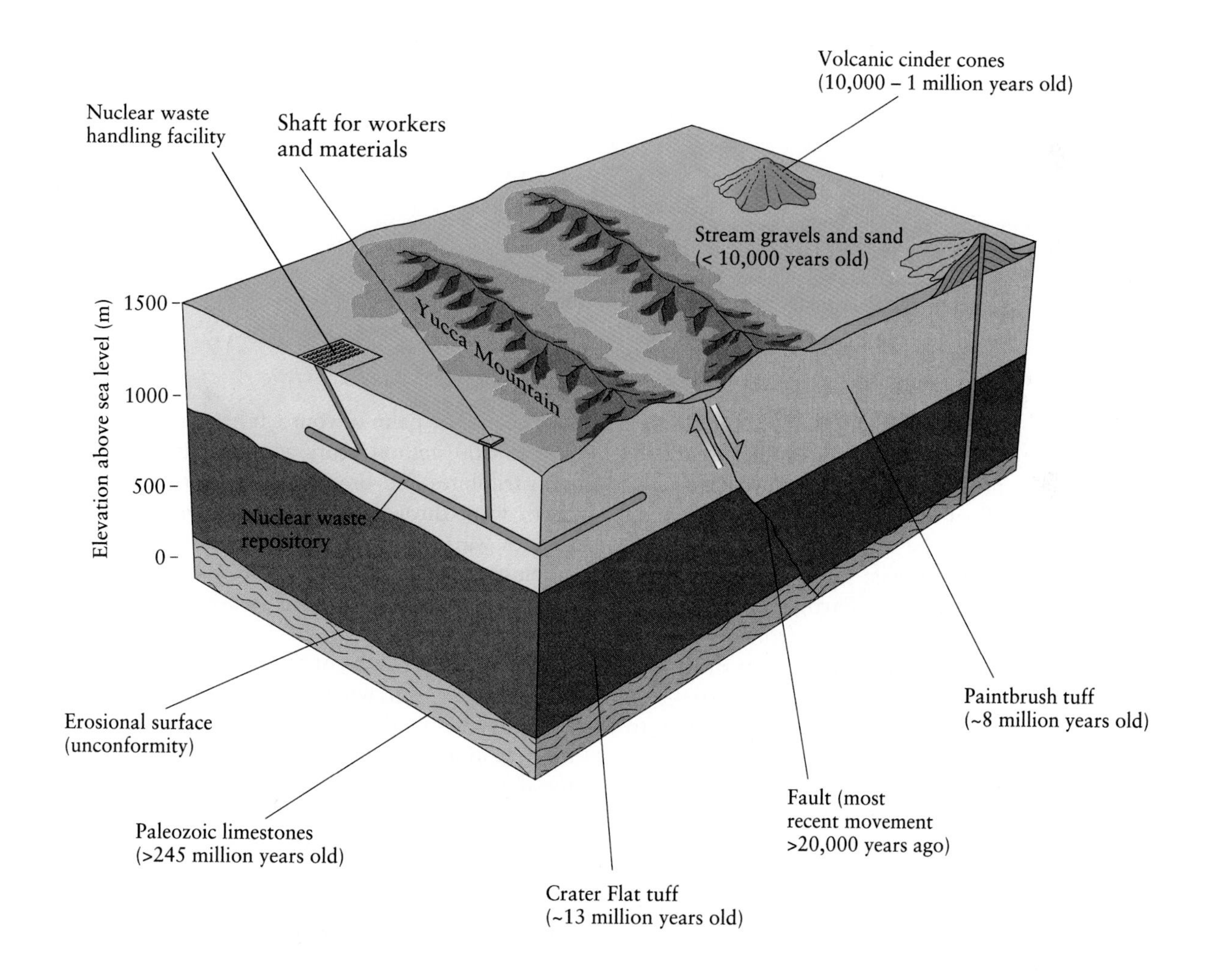

Subsurface geology around the proposed nuclear waste repository at Yucca Mountain, Nevada, suggests that earthquake and volcanic activity are highly unlikely to occur over the next 10,000 years.

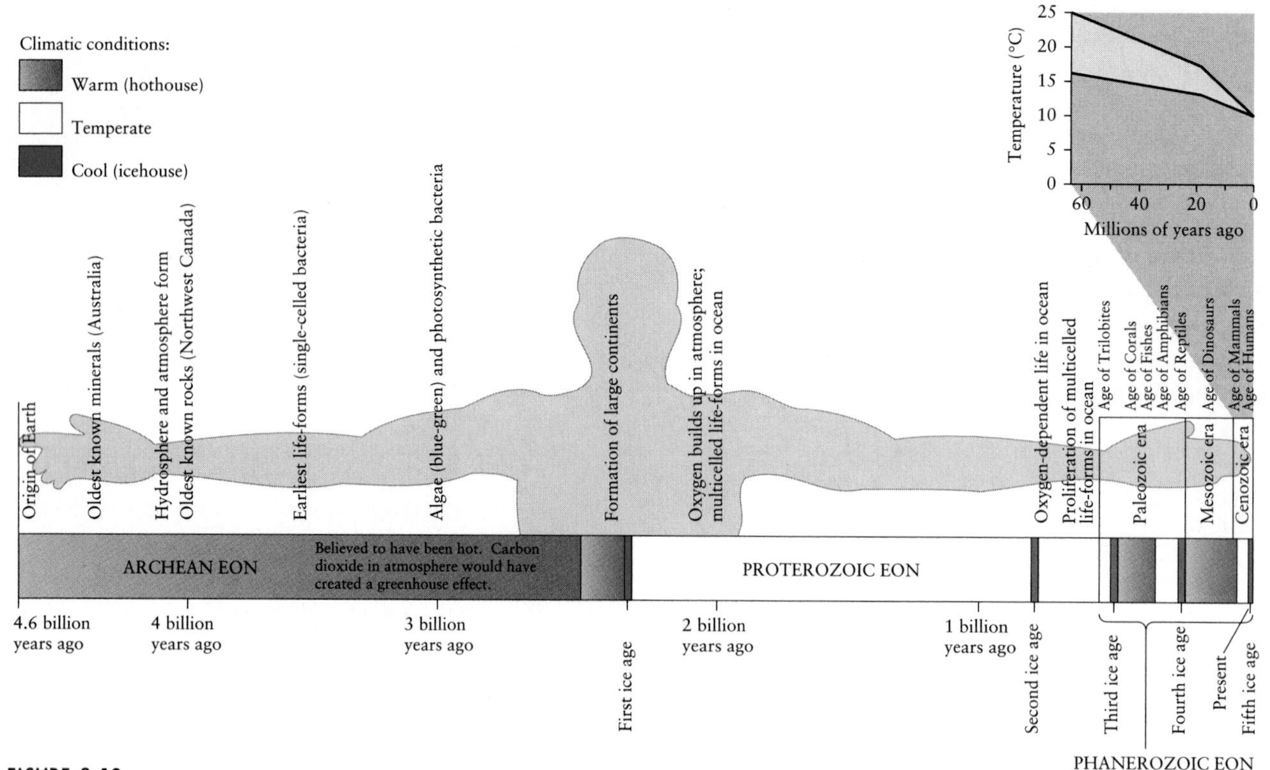

FIGURE 3-10 The immensity of geologic time is shown relative to the human arm span. Generalized global climatic changes ("hothouse" and "icehouse" conditions) are shown, as is the general decline in average global temperature during the Cenozoic era (the past 65 million years). Between icehouse and hothouse conditions, Earth was in transition between the two extremes.

history (Figure 3-10). Now imagine going forward in time (from right arm toward the left) to 3.5 billion years ago when the earliest, most primitive life-form—single-celled bacteria—began to float about in the primordial world oceans. That point would be somewhere near your right elbow, and the entire span from right fingertip to right elbow is a time devoid of life on Earth, a time when the Earth was mostly molten, and gases from its interior were seeping out through volcanoes to form the atmosphere and the hydrosphere.

Continue forward in time from your right elbow toward your left arm, imagining blue, oxygen-rich seas teeming with constantly evolving life-forms. It is thought that Earth was very warm at this time, for substantial amounts of carbon dioxide in the atmosphere would have created a greenhouse effect (see Chapters 12 and 13). Subsequently Earth cooled and entered its first known ice age about 2.3 billion years ago.

Following the arrowhead of time moving across your neck and left shoulder, think of the time 2 billion years ago when the atmosphere became enriched in oxygen, a result of the photosynthesis of the multitudes of plants living in the oceans. Farther along your left arm, past your left elbow, there is a second ice age, and then the time when multicelled life-forms began to proliferate in the ocean. The palm of your left hand represents 600 million to 200 million years ago: Imagine seas filled first with trilobites and brachiopods, then corals and fishes. Earth went through two more ice ages separated by a time of warmth. About 200 million years ago, the first amphibians struggled out of the water and onto an ancient continent still barren of seed-bearing plants such as grasses and flowers.

Move quickly over your left hand, to the beginning of your middle finger, when dinosaurs, birds, and mammals appeared on an Earth green with ferns and coniferous trees. Continue to the small knuckle of this finger, when the dinosaurs disappeared, but birds and mammals remained. Some believe that a large asteroid crashed into the Earth about 65 million years ago, near the present-day Yucatán peninsula of Mexico, and the subsequent envelope of dust that encircled Earth and blocked sunlight might have caused the extinction of nearly 50 to 90 percent of all species on Earth.

Of much interest to humans is their own origin—when did humans arrive on the geologic time scale? Continuing with the analogy of the extent of geologic time and the human arm span, imagine an inch divided into 32 parts, and look at the fingernail of your left middle finger. One shaving with a nail file would remove 1/32 of an inch, and would wipe out the entire 3 million to 4 million year history of hominids on Earth. (Hominids are mammals that walk upright on two legs; humans are included within the *Hominidae* family, as it contains the genus *Homo*. See Appendix 1.) The first upright-walking primates developed during this time period. The famous skeleton of an early hominid in Africa, named Lucy by its discoverers, is estimated from radiometric dating of the deposits in which it was found to be 2.8 million years old. The past few million years also were characterized by rapid global climate changes. In general, a long-term cooling trend was punctuated by warm, interglacial periods, such as the present, alternating with much longer cold, glacial periods.

Geologic time is so immense that the time span of our own species, *Homo sapiens,* is insignificant in comparison. Modern *Homo sapiens* evolved about 50,000 years ago, and only the last 10,000 years are considered the period of modern human history. Called the Holocene epoch, this 10,000 year time span is characterized by a warm interglacial stage that made the Agricultural Revolution possible (see Chapter 1). Human beings domesticated plants and animals, and developed settled agricultural societies. From these arose cities and all the accomplishments of civilization to which we lay claim: irrigation; metallurgy; writing and record keeping; and advances in the sciences, technology, and medicine. The age of discovery and exploration, as well as the scientific revolution, occurred in only the past 500 years. The past 500 years also have been characterized by a rapid escalation in the number of humans populating Earth.

- The absolute age of an Earth material can be calculated by comparing the amount of a particular radioactive isotope in the material with the amount originally present at the time of its formation.
- Radiometric dating of igneous rock yields an estimate of the time when the molten rock cooled to a crystalline state.
- Because sedimentary rocks do not solidify from a molten state, their time of origin is inferred through stratigraphy and radiometric dating of igneous rock that has intruded or buried a sequence of strata.
- Relative and absolute age dating have enabled Earth scientists to reconstruct the long history of changing environments and life forms for planet Earth.

Closing Thoughts

Geologists appreciate the immensity of geologic time relative to the span of the human species' existence on Earth. The geologist typically studies rock, soil, glacial ice, and fossils for clues that will help in reconstructing geological events that occurred over millions, hundreds of millions, or even thousands of millions of years.

Dating Earth and inferring the sequence and rhythm of past events is important for more reasons than determining where to place nuclear waste. Knowledge of rates of different processes helps us trace the cycling of elements among Earth's systems and compare the impact of human activities on the environment with the impact of processes that are independent of human actions and hence less controllable. For example, reconstructions of recent Earth history from ice cores, sediment cores, and tree rings reveal that the amount of carbon dioxide in the atmosphere varied and corresponding changes in global climate occurred long before the Industrial Revolution. Even earlier in Earth's history, global climate fluctuated between warm and cool conditions (see Figure 3-10).

It has taken several centuries for scientists to unravel the details and timing of the history of Earth and the evolution of life on its surface. Only a little more than 100 years ago, few scientists had any notion that a method of dating rocks and fossils such as that provided by radioactivity even existed. In a short time, relative to the tenure of humans on Earth, we have learned much about Earth and human history, including the facts that our home is billions, not thousands, of years old, and that we are just one of many species that have evolved on an ever-changing Earth.

Summary

- An understanding of geologic time and knowledge of Earth history are essential for numerous practical reasons, including safe disposal of radioactive wastes, prediction of volcanic eruptions, and wise use of resources.
- The principles of superposition and original horizontality stated by Steno in the 1600s provide rational means for determining the relative ages of sequences of rocks.

- The principle of faunal (animal) and floral (plant) succession formulated by Smith in the 1700s provides a means of determining relative ages of rocks and of correlating strata from one site to another. In the 1800s, Darwin showed that the succession of species in the fossil record is caused by extinction of species and evolution of new species over time.
- The age of a geologic event or feature expressed by its position in a sequence of events is its relative age. Geologists in the 19th century constructed a "geologic column" establishing the relative ages of rocks and fossils worldwide.
- The age of a geologic event or feature expressed in years is its absolute, or numerical, age. Absolute ages are obtained by radiometric dating techniques.
- The time it takes for half the atoms of a radioactive isotope to decay is called the half-life, which varies from seconds to billions of years for different isotopes.
- Because half-lives of different isotopes remain fairly constant with time, they provide a means of determining the absolute age of a substance from the relative amounts of parent and daughter isotopes present.
- The oldest rocks discovered on Earth have been radiometrically dated at about 3.96 billion years, and the oldest minerals at 4.3 billion years. However, from dating meteorites, which are between 4.4 billion and 4.6 billion years old, scientists have found that the solar system, including Earth, formed nearly 4.6 billion years ago.

Key Terms

relative age dating (p. 66)
absolute age dating (p. 66)
principle of uniformitarianism (p. 66)
strata (p. 69)
stratigraphy (p. 69)
principle of superposition (p. 69)
principle of original horizontality (p. 69)
principle of faunal and floral succession (p. 70)
formation (p. 70)
geologic column (p. 70)
geologic time scale (p. 70)
eon (p. 70)
era (p. 71)
period (p. 71)
epoch (p. 71)
radioactivity (p. 74)
unconformity (p. 75)
parent element (p. 76)
daughter product (p. 76)
isotope (p. 76)
beta decay (p. 76)
electron capture (p. 76)
positron decay (p. 76)
alpha decay (p. 76)
half-life (p. 77)
radiometric dating (p. 80)

Review Questions

1. Why are scientists trying to determine whether or not a volcanic eruption may occur at Yucca Mountain in the next 10,000 years?
2. How do the time and space scales at which fossil fuels form compare with those at which they now are being consumed?
3. How do the time and space scales at which soils develop compare with those at which they are eroding in some places today?
4. How are relative and absolute dating methods used to assess the suitability of Yucca Mountain as a nuclear waste repository?
5. How is the principle of faunal and floral succession related to Darwin's ideas of evolution of species?
6. In what era and period did the anatomically modern human, *Homo sapiens,* appear?
7. In which era did the Age of Fishes and Age of Amphibians occur?
8. How did Rutherford and Soddy discover the half-life of radon?
9. Why is the radioactive decay of radon gas a potential hazard to residents of some buildings? What is the importance of the half-life of radon in understanding this threat?
10. What are three common modes of radioactive decay?

11. Which isotope of carbon is useful for radiometric dating of materials less than about 50,000 years old?

12. Why are radioactive isotopes useful for dating over a range only about 10 to 15 times as long as their half-lives?

Thought Questions

1. Examine Table 3-1 and try to determine which isotopes would be the most useful for dating a rock that you think is about 2.4 billion years old. Explain your reasoning.
2. An isotope has a half-life of about 800,000 years. Over what time range would this isotope be useful for radiometric dating? Would it be useful to help determine the age of Earth? Why or why not?
3. If you learned that an isotope of beryllium (^{11}Be) is susceptible to decay by emission of a β^- particle, would you be able to determine what isotope it becomes? How? You will need to examine a periodic table of elements to help answer this question (see Appendix 2).
4. Why can't Paleozoic coals be dated using radiocarbon methods?
5. Why can't the time of deposition of sedimentary rocks be dated with a relatively common isotope such as ^{40}K, which decays to ^{40}Ar?

Exercises

1. Find out how old the rocks are where you live. Geologic maps have been made for almost every part of the United States. Most libraries have such maps, or you can look in physical and historical geology books for less detailed maps of North America and smaller regions within the United States. If you live in a mountain belt, the rocks are likely to be older than if you live along a large river, such as the Mississippi, or in glaciated areas, such as Illinois, Minnesota, and other northern states. Think about the reasons why.
2. Along the coast of northern California, tidal platforms cut by wave action have been raised several meters out of the water during past earthquakes, and some are now covered with sand dunes. During earthquakes, all exposed, immobile marine life is killed instantly or dies soon after emergence above water. At one location, samples of fossil clam shells found in their original growth position on a platform contained only 1/4 the amount of ^{14}C in the modern atmosphere. How long ago was the earthquake that raised the tidal platform and killed the clams? A burned log left in a Native American fire pit was found in the bottom layer of dune sand covering the platform and shell. It contained 1/2 the amount of ^{14}C in the modern atmosphere. When did Native Americans camp on the dune that migrated atop the newly exposed platform? Do the radiometric ages make sense if you compare them with the sequence of events that you would have reconstructed using stratigraphic principles such as Steno's law of superposition? Why or why not?

Suggested Readings

Alvarez, W., and Asaro, F., 1990. "What Caused the Mass Extinction? An Extraterrestrial Impact." *Scientific American* (October):78–84.

Brenner, David J., 1989. *Radon: Risk and Remedy.* New York: W. H. Freeman and Company.

Coveney, Peter, and Highfield, Roger, 1990. *The Arrow of Time: A Voyage Through Science to Solve Time's Greatest Mystery.* New York: Ballantine Books.

Dalrymple, G. Brent, 1991. *The Age of the Earth.* Stanford, CA: Stanford University Press.

Eldredge, Niles, 1997. *Fossils: The Evolution and Extinction of Species.* Princeton, NJ: Princeton University Press.

Eldredge, Niles, 1994. *The Miner's Canary: Unraveling the Mysteries of Extinction.* Princeton, NJ: Princeton University Press.

Gould, S. J., 1988. *Time's Arrow, Time's Cycle.* Harmondsworth, England: Penguin.

Gould, S. J., 1989. *Wonderful Life: The Burgess Shale and the Nature of History.* New York: W. W. Norton.

Hawking, Stephen, 1992. *Stephen Hawking's A Brief History of Time: A Reader's Companion.* New York: Bantam Books.

Mayr, Helmut, 1996. *A Guide to Fossils.* Princeton, NJ: Princeton University Press.

Stanley, Steven M., 1993. *Exploring Earth and Life Through Time.* New York: W. H. Freeman and Company.

van Andel, Tjeerd H., 1994. *New Views on an Old Planet.* 2nd ed. Cambridge, England: Cambridge University Press.

Whipple, Chris G., 1996. "Can Nuclear Waste Be Stored Safely at Yucca Mountain?" *Scientific American* (June): 17–79.

Whitrow, G. J., 1988. *Time in History: Views of Time From Prehistory to the Present Day.* New York: Oxford University Press.

PART II

Solid Earth Systems

We live at the interface between Earth's outermost solid layer, the lithosphere, and its liquid and gaseous layers, the hydrosphere and the atmosphere. Dynamic interactions among these systems form the pedosphere, Earth's weathered face.

CHAPTER 4

The Arabian Peninsula is surrounded by the Mediterranean Sea to the north, the Red Sea to the south, Egypt to the west, and Saudi Arabia to the east. The Red Sea formed as the Arabian Plate split off from the African Plate and the Indian Ocean rushed in to fill the rift. The peninsula was created as rifting tore open the Gulf of Suez northwest from the Red Sea, and the Gulf of ʼAqaba northeast. The Nile River flows north into a densely vegetated delta and empties into the Mediterranean Sea.

Lithosphere: The Rock and Sediment System

Earth is a dynamic system in which matter and energy are transferred from one reservoir to another along cyclic paths. The satellite image on the opposite page, showing the Arabian peninsula splitting away from northern Africa, reveals parts of the rock cycle. Continental crust is splitting along a rift that deepens and becomes flooded with seawater, forming the Red Sea and its two northern fingers, the Gulf of Suez extending northwest and the Gulf of 'Aqaba extending northeast. Magma rising from the mantle seeps upward along the rift zone and cools to form new oceanic crust. If rifting continues, an ocean basin will grow between the African and Arabian landmasses. Surface processes, too, continuously alter Earth's exposed crustal face and move material through the rock cycle. The Nile River, carrying sediment eroded from northern Africa, drops part of its load to form a marshy delta where the river enters the Mediterranean Sea near Alexandria, Egypt. Winds blowing across the northern Sahara Desert also lift and drop loose sediment, covering the land surrounding the Nile with sand. In this chapter, we examine the materials that form the solid Earth and the processes that produce them, from formation of new crust at rifts to erosion of sediment by rivers and wind and deposition in deltas and oceans. Specifically, we will:

- Investigate the nature and properties of Earth's minerals and rocks.
- Identify some fundamental minerals and rocks.
- Discuss the crustal and surface processes that produce minerals, rocks, and sediments.
- Focus on the role of plate tectonics in altering Earth's surface with time.

Map views like the satellite image of the Arabian peninsula that opens this chapter provide many insights into the processes that shape Earth's crust. Well before satellites, however, map views inspired speculation about what the shapes of landmasses might mean. The discovery and exploration of the Americas in the 16th and 17th centuries required new maps, and Renaissance mapmakers drawing in the eastern margin of the New World noticed its uncanny similarity to the western edge of the Old World, as if the two had once been joined together. Conventional wisdom, however, held that the continents were permanently fixed in place. Finally, in 1912, Austrian meteorologist Alfred Wegener (1880–1930) dared to propose that the continents had indeed been joined in the past, had since separated, and were continuing to drift apart (Figure 4-1).

Geologists as recently as the 1960s scoffed at the idea of continental drift, but a few were persuaded by the lines of evidence Wegener amassed, particularly the near-perfect match of rocks and fossils of similar ages and types on opposite sides of the Atlantic Ocean. No one, however, could explain how thick, rocky continents could plow through solid ocean basins to drift about the world, and consequently his hypothesis had few supporters. Wegener perished in 1930 on an expedition to the Greenland ice sheet, his ideas on continental drift regarded as trifling by much of the geological community.

Decades went by before Wegener was recognized as a genius ahead of his time. In 1925 the German submarine *Meteor* came upon a high, north-south trending mountain ridge in the middle of the Atlantic Ocean, and in 1953 an American scientist, Maurice Ewing, discovered that the center of this ridge is a deep, narrow rift from which lava emerges. Harry Hess, another American scientist, suggested that the lava flowing out at mid-ocean ridges creates new crust, which displaces the existing ocean crust on either side in a process he called **seafloor spreading.** As the opposing blocks of oceanic crust move apart, the continents embedded in them also move. Subsequently, long, deep trenches were discovered elsewhere in the oceans, most of them in the Pacific Ocean, and identified as places where opposing blocks of oceanic and continental crust collide and one dives down underneath the other. This process of **subduction** carries crust back down into the mantle. By the end of the 1960s most geoscientists accepted that Wegener had been right. Continents do indeed "drift": They are the above-water surfaces of rigid, lithospheric plates that move away from one another at spreading centers like the Mid-Atlantic Ridge, which extends the length of the Atlantic Ocean, and collide with one another at subduction zones.

This new view of the way Earth works is the theory of **plate tectonics,** which holds that the lithosphere is broken

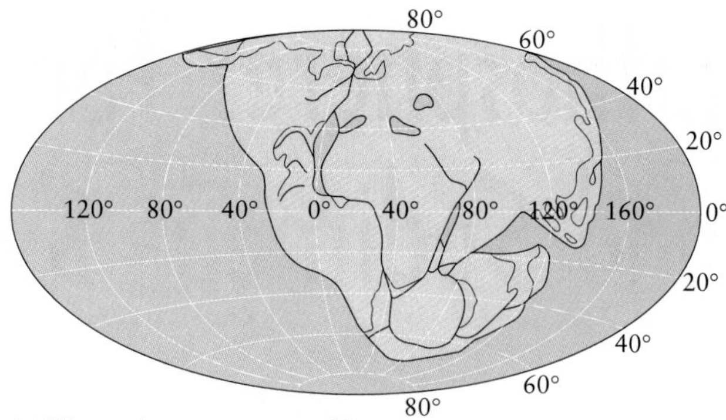

(a) Wegener's reconstructon of Pangaea

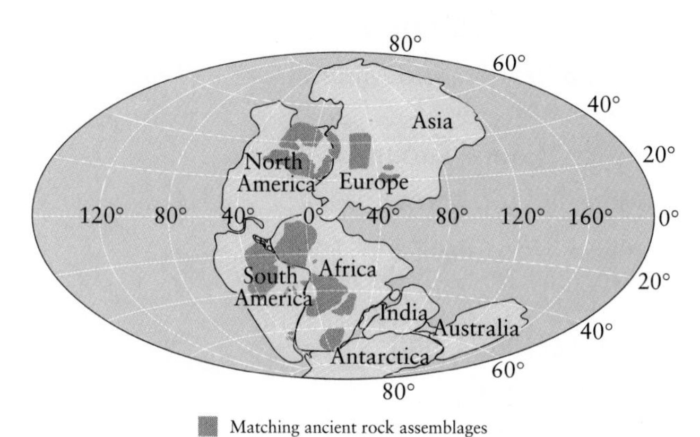

(b) Modern reconstruction of Pangaea

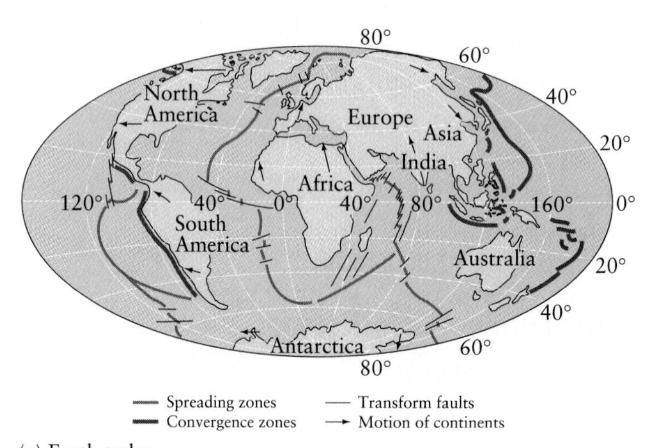

(c) Earth today

FIGURE 4-1 (a) In 1912 Alfred Wegener proposed that Earth's continents were joined about 200 million years ago in a single supercontinent that he called Pangaea. As evidence, he cited matching coastlines and the existence of similar ancient rock assemblages and fossil types on modern continents on opposite sides of the Atlantic. (b) Advances in radiometric dating of rocks and evidence of plate tectonics have enabled scientists to determine that Wegener was correct. (c) About 200 million years ago Pangaea split apart along the Mid-Atlantic Ridge, which extends the length of the Atlantic Ocean.

into rigid plates that move horizontally above the less rigid asthenosphere. Plate tectonics provides a consistent way to explain rock types and their origins, the making of mountains, and changes in global climate. Nevertheless, it was not until 1987 that the actual motion of plates was measured, providing a test of one of the fundamental predictions of the theory of plate tectonics. In that year, scientists at the National Atmospheric and Space Agency used signals sent from satellites to monitor the positions of different parts of Earth's surface. Just as Wegener had argued 75 years earlier, Earth's surface is in motion, with plates moving away from, toward, and alongside one another at rates of up to tens of centimeters per year.

Plate tectonics theory provides a framework for understanding how and where rocks and sediments formed. The heat and pressure associated with different types of plate movements determine the structure and properties of the minerals and rocks that form Earth's crust.

Lithosphere Materials: Elements, Minerals, and Rocks

Majestic Ayers Rock, glowing red and gold in the Australian sunlight, rears up above a vast, nearly flat desert (Figure 4-2). It is no wonder that Native Australians use it as a sacred place of worship. As awe-inspiring as its appearance is its composition: This large outcrop of solid rock is made of pebbles and sand. Ayers Rock is a sandstone, a sedimentary rock formed most commonly from broken bits of the mineral *quartz*. Quartz is made up of silica, a compound of the two most abundant elements in Earth's crust. The silica molecule is created when four atoms of oxygen bond chemically with one atom of silicon (see Chapter 2). If we break any rock down into progressively smaller and smaller parts, we will discover its component minerals and their component atoms (Figure 4-3).

All **rock** is an assemblage of minerals. A **mineral** is a naturally formed, inorganic, crystalline solid with a specific chemical composition or range of compositions. Several thousand minerals and rocks have been identified on Earth. Some minerals, called native element minerals, consist of only one element. Native element minerals include gold, silver, copper, sulfur, and diamond. Most minerals, however, are composed of two or more elements, as is quartz. If Earth's crust contained only oxygen and silicon, it might consist entirely of the mineral quartz. Other elements, though much less abundant, combine with one another and with oxygen and silicon to give rise to different minerals. Because eight elements account for nearly 99 percent of the weight of Earth's crust (Table 4-1), and the range of temperatures and pressures in the crust is relatively small, about a dozen mineral structures and combinations are especially common. These are called the *rock-forming minerals.*

Crystallization and Mineral Structure

Minerals form when atoms bond into orderly, three-dimensional crystal structures in a process known as

FIGURE 4-2 Ayers Rock in Australia is a deep red sedimentary rock made of layered sands and pebbles. Native Australians call the rock Uluru and visit it for sacred rituals.

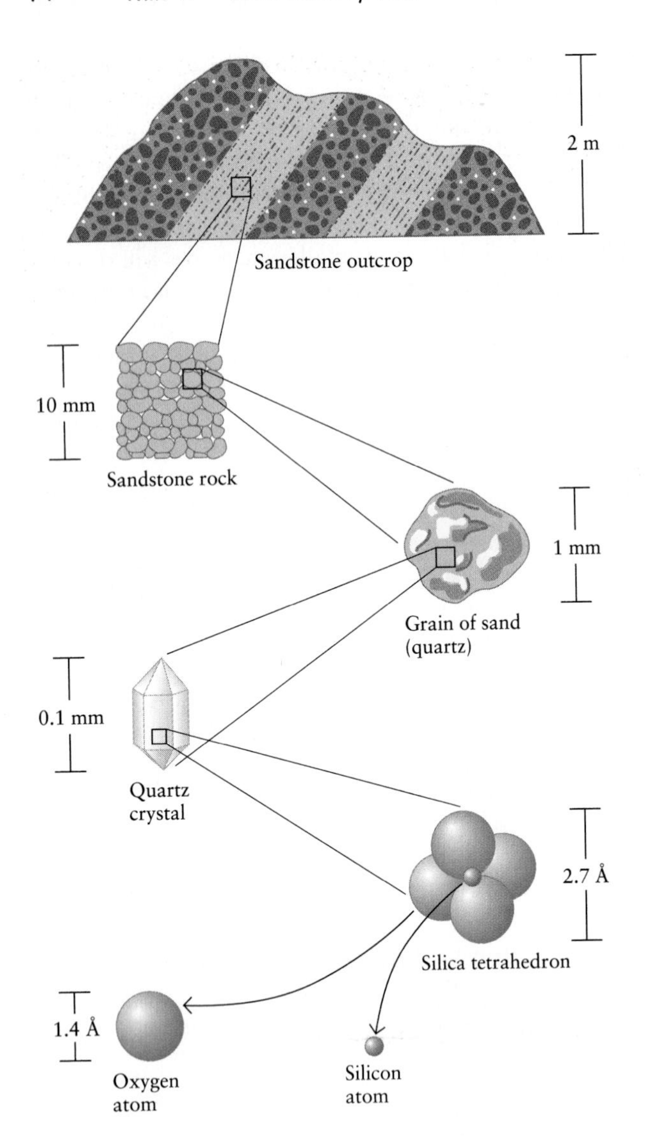

FIGURE 4-3 A piece of sandstone, like that from Ayers Rock, can be broken down into smaller and smaller parts, from sand grains to quartz crystals and then to individual atoms that form the crystals.

TABLE 4-1 The Abundance of Elements in Earth's Crust, by Weight

Element	Abundance (%)
Oxygen (O)	46
Silicon (Si)	28
Aluminum (Al)	8
Iron (Fe)	6
Magnesium (Mg)	4
Calcium (Ca)	2.4
Potassium (K)	2.3
Sodium (Na)	2.1

crystallization. During crystallization, atoms become arranged in regular patterns that are repeated throughout a given mineral. When crystalline substances grow freely in open cavities and spaces, their external shapes develop unhindered and express the internal geometric order of the component atoms. Crystals have regular, planar (that is, flat) faces. Usually, however, minerals growing during crystallization are confined by other crystallizing substances that surround them, so their size and external regularity are inhibited (Figure 4-4). Nevertheless, the orderly framework of atoms within the minerals can be seen from a microscopic view of its internal atomic structure.

The process of crystallization occurs in two ways—through cooling and through precipitation from a solution. Crystallization occurs through cooling when liquids or gases lose enough thermal energy for their atoms to slow down and bond strongly to one another so that they freeze in place, barely able to move, thereby forming solids. Snowflakes, for example, are crystals of ice that solidify from water vapor in the atmosphere when the air temperature falls below the freezing point of water. Similarly, minerals crystallize in magma as it cools on its way up from the hot mantle. The slower the rate of cooling, the more time the atoms will have to form extended frameworks of repeating patterns, and the larger and more perfectly formed the crystals will be, whether they are snowflakes or crystals of quartz.

In contrast, instantaneous cooling freezes atoms in place before they can become arranged into orderly frameworks. The resulting solid is not a mineral because it is not crystalline. Consider obsidian, a volcanic glass (Figure 4-5). Glass formed in volcanoes can be made of silicon and oxygen, as is quartz, but obsidian is not a mineral because its atoms have no internal symmetrical order and do not form a crystalline array. Lacking the planar faces of crystalline solids, obsidian does not break into regularly shaped pieces. Instead, it breaks into flakes along curved, irregular surfaces. It is this characteristic of obsidian that made it valuable for Stone Age toolmaking.

Crystallization through precipitation occurs when atoms or ions left behind in an evaporating solution bond with one another to form solids. A common example is the formation of salt crystals along the rim of a bowl from which salt water has evaporated. Similarly, the mineral used as common table salt, halite (NaCl), forms by crystallization of sodium and chloride ions from evaporating seawater, as along the margins of a shallow bay. As water evaporates, the concentration of oppositely charged ions of sodium and chloride increases and they become closer to one another. Because ions of opposite charge attract each other, they arrange themselves so that each ion is surrounded by ions of opposite charge (Figure 4-6).

(a)

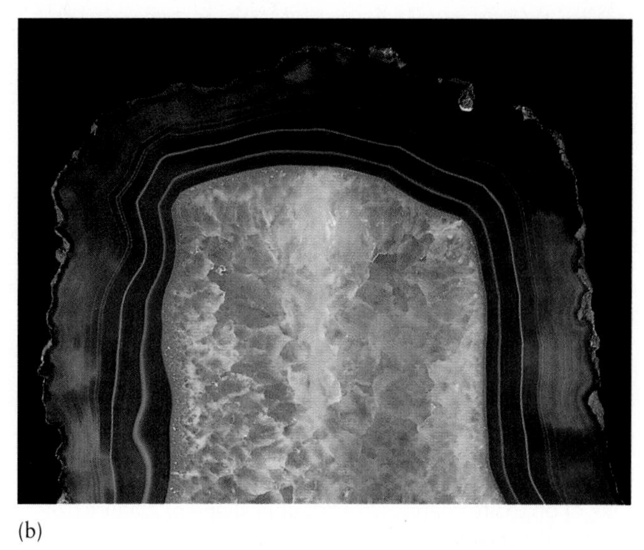
(b)

FIGURE 4-4 (a) During crystal growth, exquisite forms develop if a mineral grows unhindered, as in these quartz crystals formed in open cavities. (b) When growth is hindered, as for these quartz crystals filling a cavity, a solid mass of intergrown particles forms in which crystal faces cannot be seen with the naked eye.

The *structure* of a mineral refers to the way in which its atoms are arranged into crystals. For example, halite has a cubic mineral structure. The arrangement of atoms, in turn, is related to the environment in which the mineral formed. For example, the pressure surrounding rocks increases with depth below Earth's surface; in general, atoms are packed more closely together at higher pressures. Near Earth's surface, where the pressure is low, the element carbon forms the low-density, soft mineral graphite. At depths greater than 150 km, where pressure is much higher, carbon forms the high-density, hard mineral diamond (see Box 2-2). The only cause of the observable differences between the two minerals is their crystalline structure, or the way in which the atoms are arranged.

Major Mineral Groups

Earlier we noted that the abundance of certain elements in Earth's crust (see Table 4-1) and the relatively small range in conditions of pressure and temperature near or at Earth's surface give rise to a few common rock-forming minerals. With oxygen and silicon together making up 74 percent of the composition of Earth's crust, it is not surprising that the most abundant rock-forming minerals in Earth's crust are the **silicates**, a group of minerals composed mostly of oxygen and silicon (see Box 2-2). At least 98 percent of Earth's crust consists of silicate minerals.

FIGURE 4-5 Obsidian is volcanic glass, formed from magma that has cooled very quickly when it reaches Earth's surface. Because it lacks an ordered internal crystalline structure, obsidian does not break evenly along crystal faces. Rather, it breaks along curved surfaces, shaped rather like a shell.

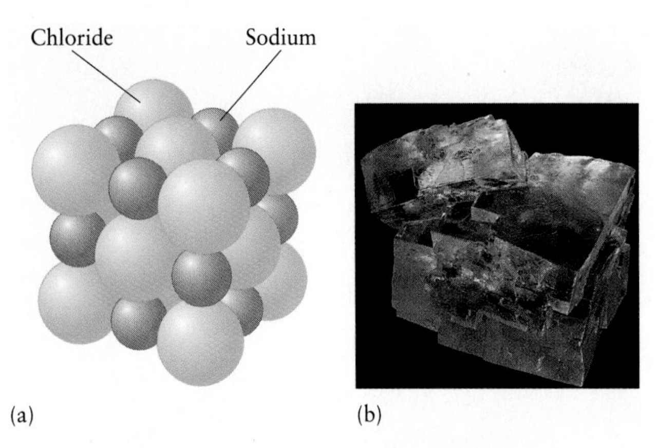

FIGURE 4-6 (a) Sodium and chloride ions form a cubic mineral structure in which each ion is surrounded by six ions of opposite charge. (b) The cubic shape of a halite (table salt) crystal reveals its cubic atomic structure.

The structural unit in silicate minerals is the silica tetrahedron, a molecule in the shape of a pyramid, with four oxygen atoms clustered around a silicon atom. Silica tetrahedra have a large negative charge (−4), which is reduced when tetrahedra link together and share oxygen atoms. For this reason, silica tetrahedra have a strong tendency to link to one another, forming polymers, or chains and clusters of connected molecules, linked by the oxygen atoms at the corners of the tetrahedra (Figure 4-7). In a magma, linkage occurs as the temperature of the magma decreases and the silica tetrahedra move more slowly relative to one another. The arrangement of linked silica tetrahedra results in different silicate structures and resultant minerals, such as olivine, pyroxene, amphibole, mica, feldspar, and quartz. The minerals that form *asbestos* have different silicate mineral structures and are the cause of much recent concern: One type can cause

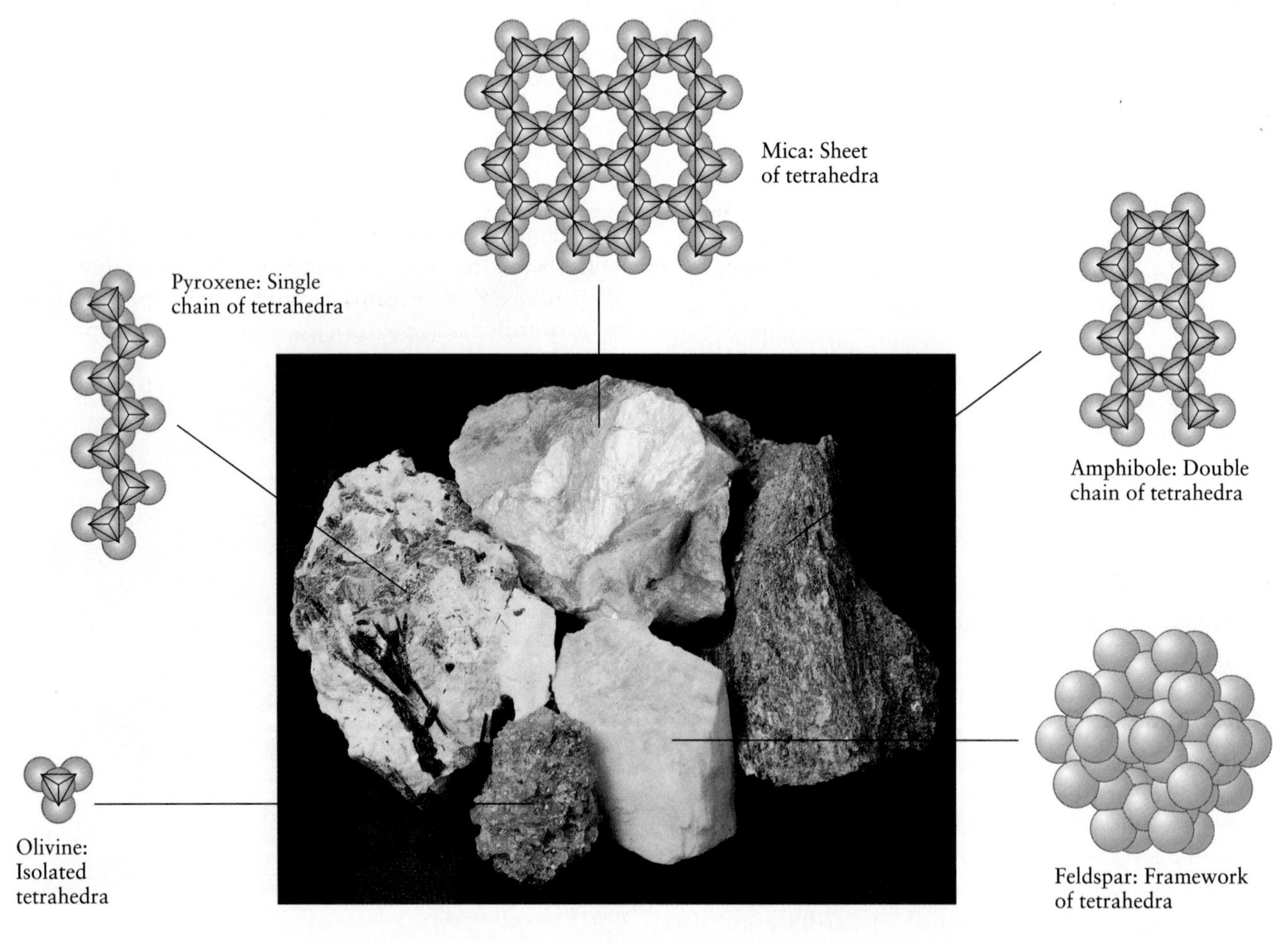

FIGURE 4-7 The silica tetrahedron is the fundamental building block for silicate mineral structures. For example, individual tetrahedra linked by cations can form olivine. Silica tetrahedra link end-to-end to form single chains, as in the common mineral pyroxene, and double chains as in amphibole. With increasing polymerization, tetrahedra link to each other at all their tips, forming three-dimensional frameworks, as in the minerals feldspar and quartz.

TABLE 4-2 Major Categories of Rock-Forming Minerals

Class	Defining anion(s)	Examples*
Silicates	silicate ion (SiO_4^{4-})	quartz (SiO_2), olivine (Mg_2SiO_4)
Oxides	oxygen ion (O^{2-})	hematite (Fe_2O_3), magnetite (Fe_3O_4), ice (H_2O), spinel ($MgAl_2O_4$)
Hydroxides	hydroxyl ion (OH^-)	geothite [$FeO(OH)$], gibbsite [$Al(OH)_3$]
Carbonates	carbonate ion (CO_3^{2-})	calcite ($CaCO_3$)
Sulfides	sulfide ion (S^{2-})	pyrite (FeS_2), galena (PbS)
Sulfates	sulfate ion (SO_4^{2-})	gypsum ($CaSO_4 \cdot 2\ H_2O$), barite ($BaSO_4$)
Halides	ions of chlorine (Cl^-), fluorine (F^-), fluorine (F^-), bromine (Br^-), and iodine (I^-)	halite (NaCl) (table salt), fluorite (CaF_2)
Native elements	none (no charged ions)	gold (Au), copper (Cu)

* The center dot in the chemical formula for gypsum indicates that a calcium sulfate molecule is bonded with a water molecule.

cancer in humans, yet regulatory agencies treat all types as hazardous (Box 4-1).

The remaining 2 percent of Earth's crust consists of nonsilicate minerals: oxides and hydroxides, carbonates, sulfates and sulfides, halides, and native elements (Table 4-2 and Figure 4-8). Most metals that are mined occur in these groups of rock-forming minerals. Each mineral group is defined by the type of anion (a negatively charged ion) in its structure. For example, oxide minerals contain oxygen anions (O^{2-}), whereas sulfides contain sulfide anions (S^{2-}). The anions combine with cations (positively charged ions) to form a variety of minerals within each mineral group. For example, iron (a cation) combines with oxygen to form the oxide minerals hematite and magnetite. If iron combines with the sulfur anion, it forms a sulfide mineral, pyrite (also called fool's gold because of its slight color resemblance to gold).

FIGURE 4-8 Nonsilicate minerals include (clockwise from upper left) halite, spinel, gypsum, hematite, calcite, pyrite, and galena.

4-1 Geologist's Toolbox

Analyzing the Effect of Asbestos on Human Health

Asbestos has caused one of the greatest environmental health controversies of the 20th century, as groups of scientists disagree with one another about the hazard of asbestos. At stake as well as health is billions of dollars of federal and state funding that many experts say is spent unnecessarily.

Six different silicate minerals all have the same commercial name, asbestos. All have a similar, fibrous, appearance, and a similar property of heat resistance. Because they are heat-resistant and durable, asbestos minerals have been used commonly in fire-retardant materials and as insulation in walls, floors, and ceilings of buildings, as well as in coverings for pipes. The structure of one asbestos mineral makes it likely to cause lung cancer, but this mineral is seldom used in commercial products. Nevertheless, the U.S. government classifies all six asbestos minerals as hazardous substances and has passed laws that require their removal from all public buildings.

The most common asbestos minerals are structured in one of two ways. The more common form is as a sheet silicate known as *chrysotile,* which constitutes about 95 percent of the asbestos used in industry. When viewed under high magnification, chrysotile resembles a mass of tiny rolls of wrapping paper. In contrast, *crocidolite* is a double-chain silicate, or amphibole mineral. Under high magnification, its crystals resemble sharp-tipped needles. Crocidolite, which constitutes less than 5 percent of the asbestos used commercially, is thought by most experts to be the only inherently carcinogenic asbestos mineral.

All asbestos minerals form minute, fibrous particles that can fill the air when they are disturbed. Anyone who inhales extremely high levels of them (more than 20 fibers per cubic centimeter of air) for many years is increasingly susceptible to silicosis, a very serious lung disease similar to that caused from exposure to coal dust; both conditions cause the lungs to harden with time. However, only those exposed to moderate to high levels of crocidolite run an increased risk of lung cancer. Such extreme exposure might occur in an asbestos factory.

Workers have been exposed to asbestos minerals in mines and factories, during construction, and, more recently, during attempts to remove all asbestos materials from public buildings. Estimates of the total cost of removing asbestos range from $50 billion to $150 billion. Because a person's chance of getting lung cancer from exposure to crocidolite is less than that of being struck by lightning, many experts think that this money would be better spent for other purposes.

How does the risk associated with exposure to asbestos compare with other threats to human health? First, asbestos is a health hazard only if it is airborne and inhaled. Several studies indicate that skin contact with asbestos is not dangerous and even ingestion of asbestos particles that settle on water or food poses no threat to health. Second, the health risk

Mineral Properties: Hardness and Cleavage

Mineral structure and the strength of atomic bonds in the structure are important determinants of the physical properties of a mineral. In diamond, for example, the carbon atoms are tightly packed and the bonds between the atoms are equally strong in all directions; as a consequence, diamond is extremely difficult to break. In contrast, the carbon atoms in graphite are spaced farther apart, and the bonds between the atoms are weaker in some directions than in others, so this mineral is soft and forms sheets that are readily pulled apart (see Box 2-2). Graphite can be broken so easily along the planes in which its carbon bonds are weakest that it leaves a mark on paper. The "lead" in your pencil is graphite.

In addition to hardness, a number of physical properties can be used to identify minerals, such as color, density, and the way in which a mineral breaks. Hardness and the way a mineral breaks are particularly useful. Each mineral can be ranked based on its hardness relative to other minerals. In the early 1800s Austrian mineralogist Friedrich Mohs (1773–1839) ranked all known minerals by hardness on a scale of 1 to 10, with 10 the hardest. Later workers related the **Mohs hardness scale** to familiar materials such as steel, window glass, and fingernails

from asbestos is small compared with the risk from other common threats. Fewer than 100 people die from asbestos-related lung disease each year, while several thousand die from choking on food or other objects. The calculated risk for asbestos-related lung disease is 1 in 100,000, while it is 1 in 5 for cancer resulting from smoking. Finally, research shows that spending long periods of time in buildings that contain chrysotile asbestos poses little risk. As a result, many geoscientists are urging government officials to reconsider the nature of the asbestos minerals at each site and judge on a case by case basis whether the public is better served by removing the asbestos or leaving it alone.

(a)

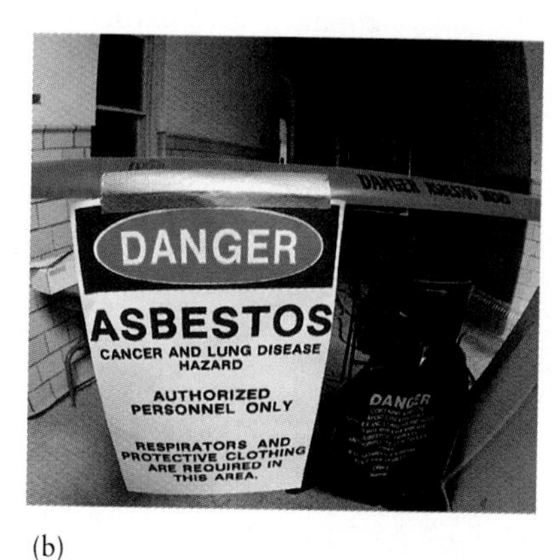

(b)

(a) These microscopic views of two forms of asbestos minerals are magnified 100 times. Both minerals are silicates. Chrysotile (left), a sheet structure mineral, resembles wrapped, tubular bundles, whereas crocidolite (right), an amphibole mineral, has sharp, needlelike forms and is carcinogenic. (b) Asbestos removal is difficult and dangerous. To protect themselves from exposure to the mineral, workers wear protective clothing and breathing apparatuses.

(Table 4-3). Talc, for example, has a hardness of 1 and can be scratched by harder minerals as well as by your fingernail, which has a hardness between 2 and 3. A steel file or knife can scratch feldspar (hardness 6) but not quartz (hardness 7). Of all the minerals at Earth's surface, diamond is the hardest and can scratch all others. This is why jewelers are very careful to keep diamonds segregated from other gems.

Cleavage refers to the manner in which a mineral breaks. Some minerals break along parallel and even planes in one direction only, as mica does (Figure 4-9). Others break along two or more planes that are nearly perpendicular to one another (pyroxene). Still others break along two or more planes that are not at right angles to one another (amphiboles). The planes along which a mineral breaks reflect the internal ordering of its atoms and the nature of their chemical bonds, with breakage occurring along planes with weakest atomic bonds. For example, the weak ionic bonds between adjacent silicate sheets in mica break easily, so mica has perfect cleavage in one direction and poor or no cleavage in any other direction. For this reason, since antiquity—long before manufactured glass became so abundant—thin sheets of mica have been used to make windows for houses and lanterns.

TABLE 4-3 Mohs Mineral Hardness Scale

Hardness*	Mineral	Tests using common objects
1	talc	easily scratched with a fingernail
2	gypsum	scratched with fingernail
3	calcite	scratched with copper penny
4	fluorite	easily scratched with steel knife
5	apatite	scratched with steel knife or window glass
6	orthoclase (feldspar)	scratched with a steel file; scratches window glass
7	quartz	scratches steel file and window glass
8	topaz	harder than most common objects
9	corundum	harder than most common objects
10	diamond	hardest mineral known (forms at depth of more than 100 km, at high pressure)

* Minerals with hardness of 8 and above are rare.

If you know the hardness and cleavage of a mineral, you can make a reasonable estimate of its mineral type and composition (Appendix 4). Consider a mineral that has poor cleavage and can scratch glass and steel but cannot scratch topaz. You might guess that the unknown mineral is quartz, a framework silicate that measures 7 on the Mohs hardness scale and has poor cleavage because of its many, strong covalent bonds that extend in all directions. In contrast, if a mineral is easily scratched with your fingernail and feels slippery when rubbed between your fingers, it is talc (used in talcum powder), a sheet-structure silicate mineral with perfect cleavage in one direction resulting from weak chemical bonds between adjacent sheets of silica tetrahedra. The slippery nature of the mineral is an indication that bonds between sheets are breaking.

Major Rock Groups and the Rock Cycle

Minerals combine to form rocks, and the type of rock formed depends both on the minerals and the rock-forming processes involved. As discussed in Chapter 2, all rocks can be sorted into one of three major rock groups associated with their origin: igneous, sedimentary, and metamorphic. Each type can be identified from clues provided in the rock's mineralogy and texture. *Mineralogy* refers to the relative proportions of each mineral in the rock. *Texture* refers to the different shapes and sizes of the minerals and the ways in which they are assembled.

Igneous rocks are rocks that solidify from hot, molten magma. If magma cools slowly and crystallizes, its atoms form orderly, regular arrays and an interlocking mosaic of crystals (Figure 4-10). Two of the most common igneous rocks are granite, which cools slowly in subsurface pockets of magma and is therefore coarse-textured,

(a)

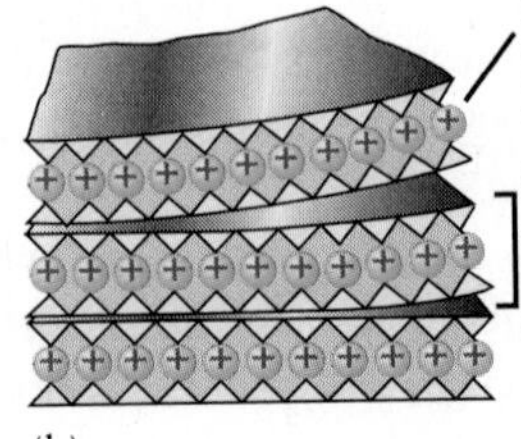

(b)

FIGURE 4-9 (a) Mica breaks into thin sheets. (b) The structure of mica results in pronounced cleavage along a plane between layers of silica tetrahedra.

(a)

(b)

FIGURE 4-10 (a) The crystals in this granite are large enough to be visible even to the naked eye because the magma cooled relatively slowly. (b) When a section of granite is cut very thinly and viewed under a microscope, the nature of each crystal and its faces is easily observed, Most of the brightly colored minerals are mica; the large white crystals, and some of the gray ones, are quartz. The gray crystals with bands, slightly above and to the right of center, are feldspar minerals.

and basalt, which cools more quickly above ground and is therefore fine-textured. Obsidian freezes nearly instantaneously from magma and therefore is glassy in texture with no crystal structure (see Figure 4-5).

Sedimentary rocks form from bits of mineral and organic matter that settle to the bottom of a fluid or move downslope because of gravity to form layers of sediment. If buried, compacted, and cemented together, these loose bits of sediment—which are derived from all three of the major rock groups—can become a solid mass of sedimentary rock, as in the sandstone of Ayers Rock in Australia (see Figure 4-2). Recall from Chapter 3 that sedimentary rocks are the only ones that contain fossils of plants and animals, and thus constitute a record of early life on Earth. Because they contain ancient life-forms, sedimentary rocks are also the only sources of carbon-rich fossil fuels (see Chapter 11). For instance, coal is produced from the burial and compaction of prehistoric swamp vegetation under layers of sediment, and oil develops from the chemical alteration (under pressure) of tiny sea creatures whose bodies at death sank to the ocean floor.

Metamorphic rocks are produced by the transformation of minerals in all types of rock by heat and pressure. However, metamorphic rocks are not heated to temperatures great enough to cause melting. Instead, all changes occur in the solid state. Rocks can be metamorphosed by burial or subduction at great depths, where the internal heat and pressure are sufficient to cause metamorphism, or by intrusion, a process whereby a mass of molten rock heats and recrystallizes closely surrounding rock. In an example of metamorphism resulting from deep burial and pressure, the clays in fine-grained sedimentary rocks alter to mica minerals and grow in a parallel fashion. The new metamorphic rock, slate, is much harder, and the parallel growth of flaky mica minerals makes it possible to split the rock into shingles for roofs and into large sheets for chalkboards.

Each of these rock groups can be viewed as part of the rock cycle. Sedimentary rocks are metamorphosed into metamorphic rocks or are melted to create igneous rocks. All three types of rocks can be uplifted and eroded into sediments. In turn, the sediments can harden into sedimentary rocks (Figure 4-11). Although igneous and metamorphic rocks represent much larger reservoirs in Earth's crust than sedimentary rocks do, they are covered in most places by sedimentary layers. Plate tectonics provides a way to explain the forces in and under Earth's crust that power the cycling of rock among igneous, metamorphic, and sedimentary reservoirs.

- Atoms combine in three-dimensional arrays to form minerals, which in turn combine to form rocks.
- Although a few thousand rocks and minerals have been identified on Earth, only a few dozen of each are common, because Earth's crust is made largely of just a few elements (mostly oxygen, silicon, aluminum, and iron), which are exposed to a limited range of pressures and temperatures.
- Two key properties of minerals—their hardness and cleavage—reflect the internal ordering of their atoms and the strength of the bonds between the atoms. These properties make it possible to identify many minerals. In turn, the types, shapes, and sizes of minerals in a rock make it possible to classify and identify different rock types.

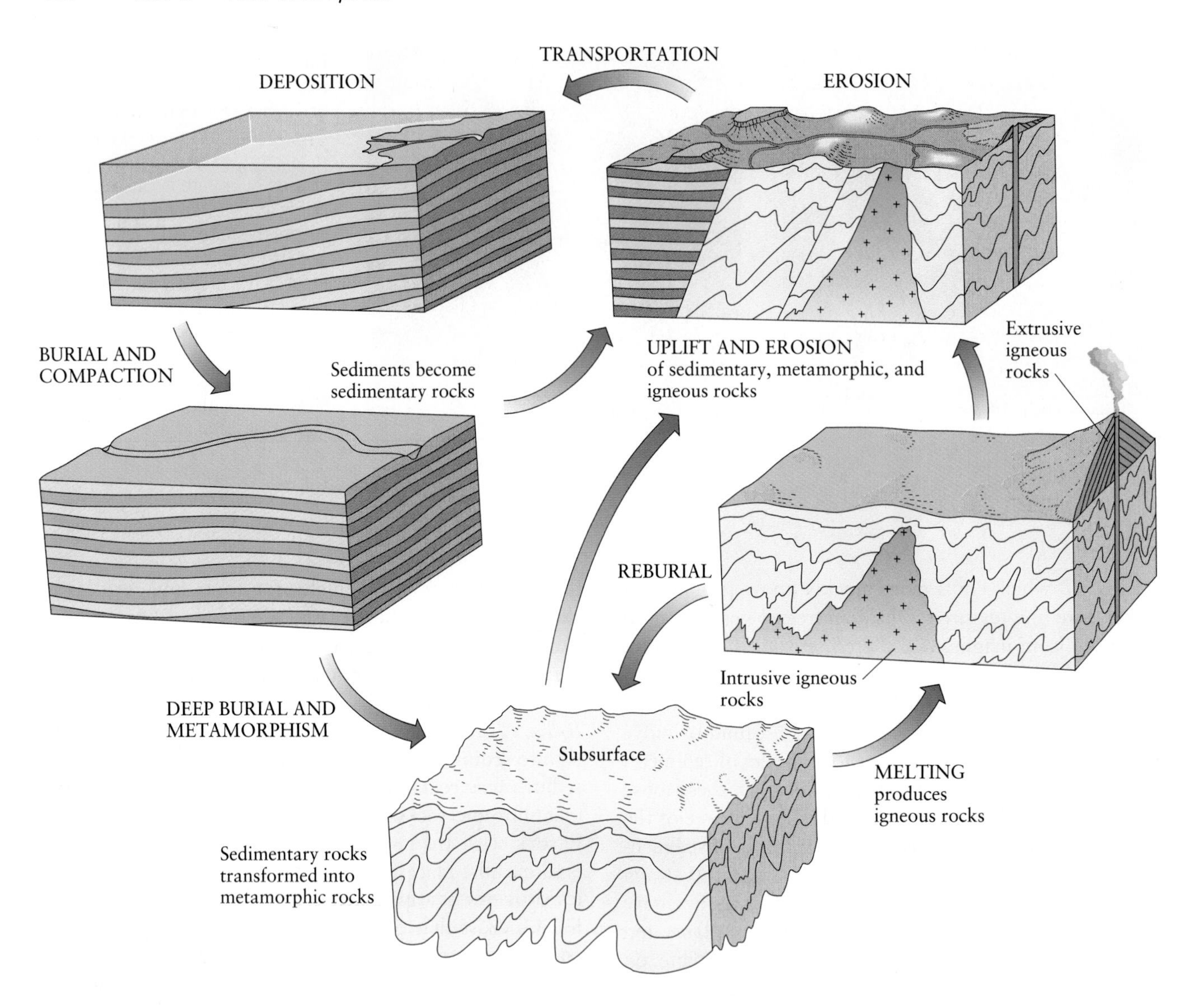

FIGURE 4-11 Cyclical processes link different rock types at Earth's surface. Minerals and rocks are altered by plate tectonics processes that bury rocks deep in Earth and then raise them again during mountain building, as well as by surface processes driven by solar energy, such as weathering and erosion.

Plate Tectonics and the Rock Cycle

As with any dynamic system, continuation of plate tectonics and the rock cycle depends on a source of energy. The two major sources of energy on Earth are Earth's interior heat and solar radiation. Without these energy sources, plates would not move, rocks would not melt into magma, and volcanism, uplift, weathering, and erosion would cease.

Heat from the interior of Earth drives the motion of lithosphere plates. Recall from Chapter 2 that at a depth of 100 km a boundary separates the rigid lithosphere from the underlying, plastic asthenosphere, even though the change is purely physical and no compositional change or melting occurs. From about 100 to 350 km, the temperature is high enough relative to the pressure of overlying rock to make the asthenosphere somewhat ductile, or plastic, just as wax becomes soft and pliable when it is warmed. Because the asthenosphere is a plastic solid, it can move slowly over long periods of time. An analogy with lead and wax explains this motion. A lead sphere set on the surface of a container of wax eventually will sink into the wax and become embedded. Because lead is denser than wax and wax is somewhat plastic at room temperature, the wax flows very slowly around the sphere. Similarly, the asthenosphere can flow very slowly, at rates of several centimeters or so per year, about as fast as the growth of human hair. In places where the lithosphere is densest, it can subduct into the asthenosphere, just as the lead sphere becomes embedded in the wax.

Heat from Earth's interior drives the motion of the asthenosphere, largely by convection. **Convection** is a mechanism of heat transfer in which hot material rises because of its lesser density, while cool material at the surface sinks because of its greater density (Figure 4-12). The convective motion of the asthenosphere sets the lithospheric plates in motion, much as ocean currents set a ship in motion. This constant motion of plates accounts for continental drift, because as plates move they carry along the continents embedded in them.

At some places, magma rises from the asthenosphere and erupts onto Earth's surface, forming new plate material. Only 1 to 3 percent of the asthenosphere is molten, yet it provides most of the magma that rises to the Earth's surface. Elsewhere, denser parts of plates sink into the interior Earth and are destroyed. In this way, lithosphere is formed, destroyed, and recycled, completing the cycle of rock-forming processes. A complete convection cycle in the upper mantle takes about 450 million years. As long as Earth's interior stays hot enough to cause convection in the asthenosphere, plate tectonics and the rock cycle will continue: Continents will drift, join together, and split apart, causing Earth's environments to be ever-changing (Box 4-2).

Types of Plate Boundaries and Plate Interactions

Because of the forces acting on the boundaries between adjacent plates, the rocks there commonly are deformed and broken. Again, wax provides a useful analogy. Imagine placing a block of very hard wax in a vise, then cranking the vise and compressing the wax. As the wax is squeezed, the atoms in it move closer and closer to one another. If the wax is removed from the vise, the pressure is removed and the atoms again move farther apart. The material behaves in an *elastic* fashion, much as a rubber ball does when squeezed and then let go. In a material that behaves elastically, all deformation can be undone and the substance can return to its original shape.

If the wax is squeezed to the point at which the bonds between some of its atoms break, the wax will exhibit *brittle* behavior. Typically, wax, like cold rock, breaks along a plane of failure, a flat surface along which atomic bonds are broken. The plane of failure along which rocks move past one another is a **fault.** The slipping of rocks past one another along a fault plane is the cause of earthquakes. When rocks break and move past one another, energy stored in them is released, sending waves of ground movement outward in all directions from the area of fault-plane rupture (see the discussion of earthquakes in Chapter 5). At greater depths, where rocks are warmer, brittle failure along fault planes is less likely because the rocks are *plastic* (from the heat) and tend to fold rather than crack, just as warm, soft wax would fold if you tried to break it.

Plate borders acquire distinctive characteristics and types of faults depending on activity at the boundaries between plates. Rocks along plate boundaries, like wax, can behave in a brittle manner and break, or in a plastic manner and fold. At **convergent plate boundaries,** where two plates collide with each other, compression occurs and the edges of the plates crumple and break. At **divergent plate boundaries,** where two plates move away from each other, extension occurs and the edges of the plates are thinned and broken. Sometimes two plates merely slide past each other along **transform plate boundaries;** in such cases plate edges typically are not compressed or extended.

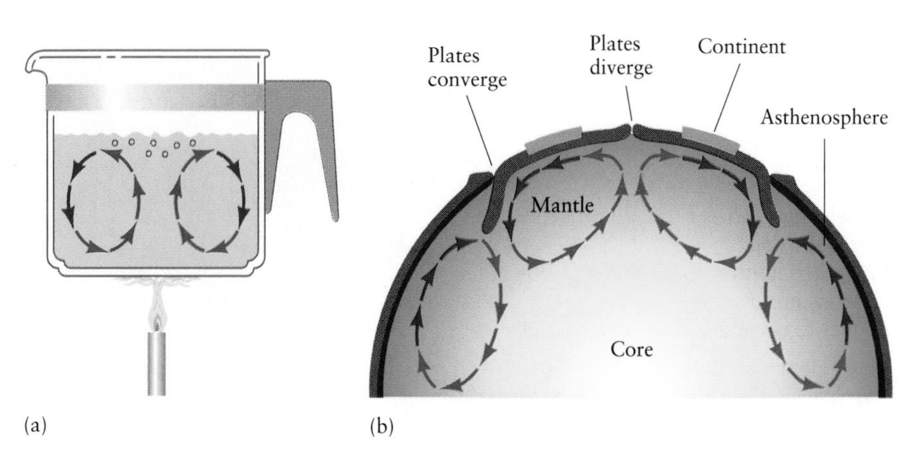

FIGURE 4-12 (a) As water molecules in the beaker absorb energy from the heat source, they expand, become less dense, and then rise. The cooler, denser molecules at the surface sink. (b) Convection currents deep in Earth's interior are believed to be the driving force behind plate movement. As the warm asthenosphere rises under plate boundaries, it flows in opposite directions, dragging the plates along and forcing them to separate. Where the current pushes one plate into another, the cooler and denser rock sinks into the asthenosphere.

4-2 Global and Environmental Change

Plate Tectonics and the Evolution of Species

The lithosphere constantly is changing. It is in motion, and the material that forms it cycles from one Earth system to another with time. Convection arising from Earth's internal heat has caused continuous rearrangement of the lithospheric plates since early in Earth's history. In turn, drifting continents have dramatically influenced the evolution of animal and plant species on Earth. Perhaps the most striking example of this influence can be seen in the fauna of Australia.

About 200 million years ago, from the late Paleozoic era to the early Mesozoic, the continents of Australia, Antarctica, Africa, and North and South America were assembled into the supercontinent Pangaea (see Figure 4-1). About 100 million years ago, marsupials evolved in what is now North America and spread over all of Pangaea. (Marsupials are pouched mammals, such as kangaroos, that give birth to young not yet fully developed. The infants crawl up the mother's abdomen and into a pouch in which they nurse and grow. In contrast, placental mammals lack a pouch and give birth to more fully developed young.)

About 70 million years ago, after South America and Australia had broken away from Pangaea, placental mammals evolved on the remains of Pangaea. Placentals competed with marsupials for food and rapidly drove the marsupials to extinction in what became the continents of North America, Europe, and Asia. Meanwhile, marsupials thrived in South America and Australia, where there were no placentals. When South and North America became joined again approximately 3 million to 4 million years ago, placental mammals migrated south and drove to extinction all the marsupials except opossums. Australia, in contrast, has remained isolated from the other continents and their placental mammals throughout the Cenozoic era, enabling more than 180 different marsupial species to evolve.

In recent history, humans introduced placental mammals to Australia, with adverse affects on native wildlife. From 12 breeding pairs of rabbits introduced in the mid-1800s as a game animal, the rabbit population skyrocketed to more than 20 million individuals in only 6 years. The rabbits lack any natural predators in Australia and are defoliating vast areas of grazing land vital to kangaroos and wallabies.

When a land bridge, the Isthmus of Panama, was uplifted between North and South America, many animals migrated south, including llamas, elephants, bears, deer, dogs, and rabbits. The smaller number of species that migrated north included armadillos, sloths, porcupines, and opossums. For the first time, marsupial mammals in South America came in contact with placental mammals.

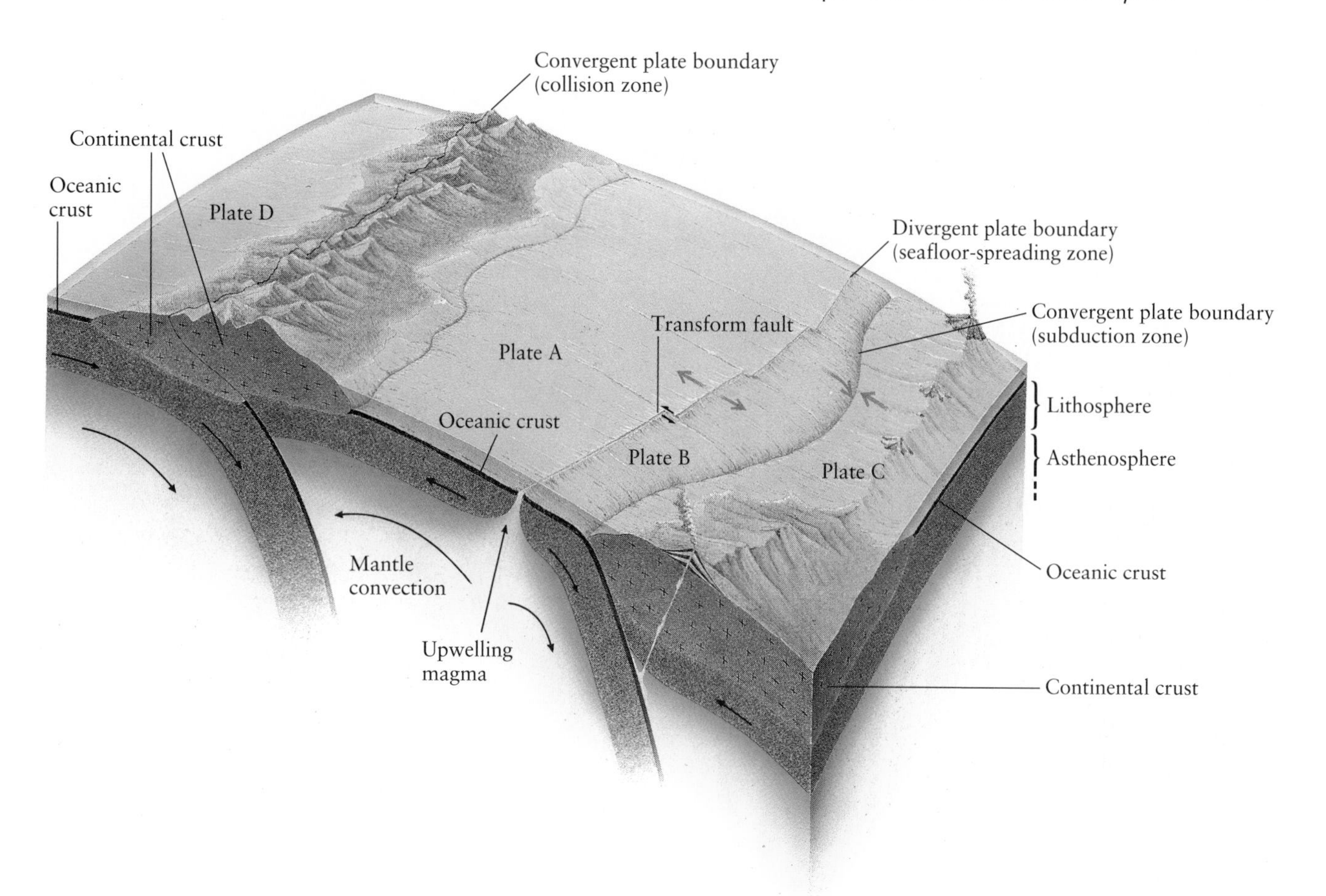

FIGURE 4-13 There are three types of plate boundaries. Plates A and B are moving away from a divergent boundary. They also slip past each other along transform boundaries. The lithosphere grows at the center of seafloor spreading, accompanied by volcanism and earthquakes. Plates B and C are converging along a subduction zone. Plate D is consumed by subduction at the convergent boundary with Plate A. Many geologic processes occur along a convergent boundary, including the creation and upwelling of magma, volcanism, mountain building, creation of deep-sea trenches, and earthquakes. Upwelling of magma, volcanism, and earthquakes also occur at divergent boundaries, but these earthquakes generally are smaller and occur at shallower depths. The type of magma and nature of volcanism at the two boundaries also is very different.

The type of plate boundary affects the nature of the magma, faults, and earthquakes that occur along it and results in different rock-forming environments. Figure 4-13 illustrates the three types of plate boundaries and the types of rock-building or rock-destroying activity associated with each. Figure 4-14 is a world map of plate boundaries and rates of plate motion. Both will be referred to frequently in the discussion of plate boundaries that follows.

Divergent Plate Boundaries, Spreading Ridges, and Normal Faults

Divergent boundaries were first recognized on the seafloor as long chains of volcanic mountains encircling Earth in an arc pattern, similar to the seams on baseballs (see Figure 4-14). Because of their locations and topography, these undersea mountain chains are called *mid-ocean ridges*. The discovery of high mountains on what was thought to be a flat ocean floor occurred just after World War I, when submarines such as the German *Meteor* began roaming the world's ocean basins. Since then, numerous studies—including those that use submersible vehicles to visit the ridges first-hand—have observed long linear depressions, or *rifts*, along the centers of mid-ocean ridges, and have determined that lava erupts from them. These rifts mark the growth of new lithosphere, where new material is added as plates spread away from one another.

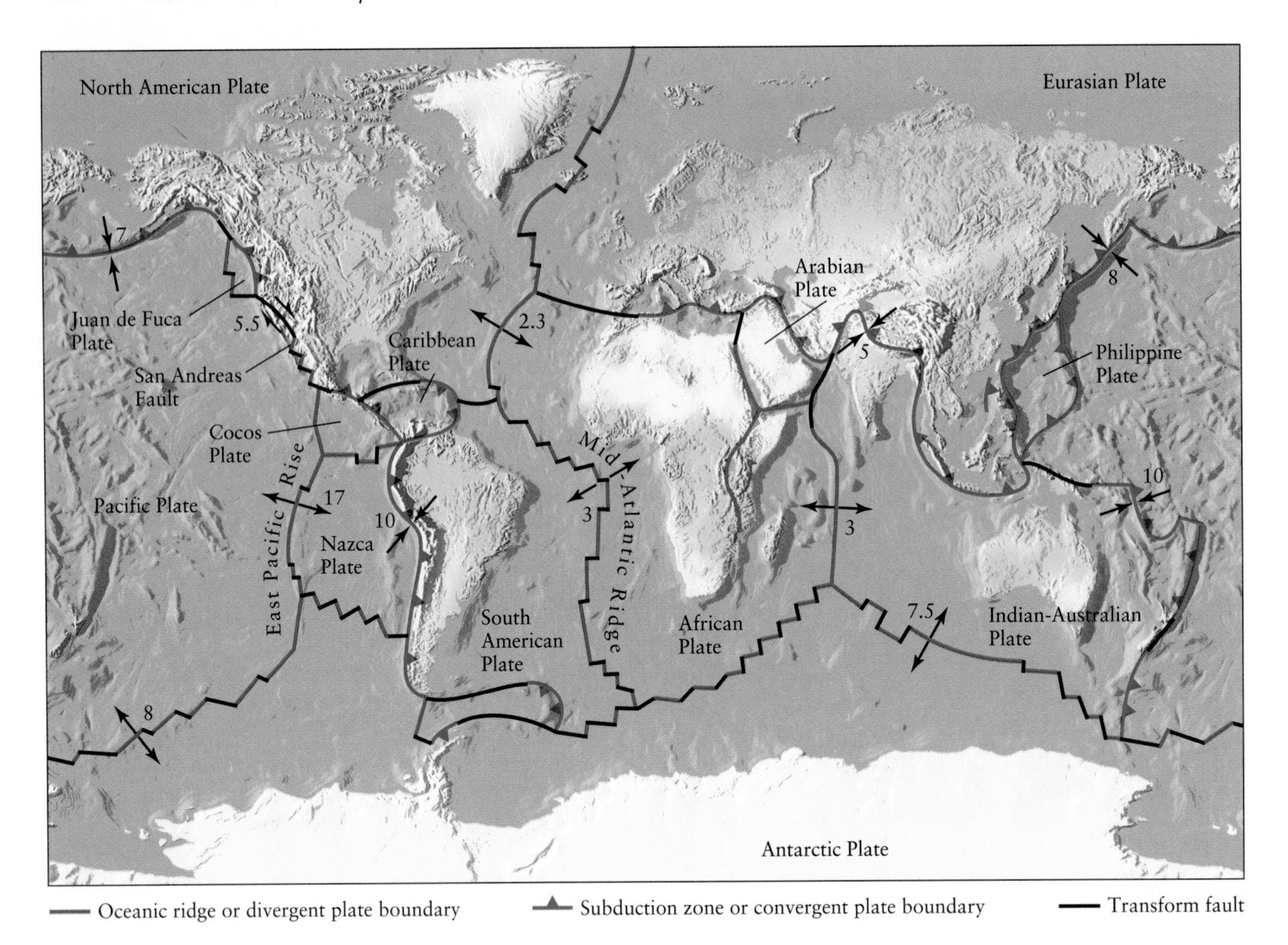

FIGURE 4-14 Earth's tectonic plates move in many directions, at rates up to several centimeters or more per year (see numbers next to arrows). Facing arrows indicate places where plates are converging. Diverging arrows indicate places where plates are separating. The Nazca Plate is converging with the South American Plate at about 10 cm per year, while the Indian-Australian and African plates are diverging at about 3 cm per year. Arrows parallel to a plate boundary, as along the San Andreas Fault in western North America, indicate plates sliding past each other.

At mid-ocean ridges, convection currents move hot, ductile rocks from the mantle slowly upward, and the boundary between asthenosphere and lithosphere rises toward the surface (see Figures 4-12 and 4-13). As warm rocks move upward, the reduced pressure of overlying rocks lowers the temperature at which the material can melt and produce magma. At spreading mid-ocean ridges, some magma seeps out and solidifies, creating new ocean crust. Lithospheric plates on each side of a rift accumulate new material and spread apart from each other atop the slowly moving asthenosphere, forming a divergent plate boundary.

Active divergent boundaries appear above sea level at only a few locations. One is Iceland, as shown in Figure 4-15(a), on the Mid-Atlantic Ridge, a divergent boundary that runs north to south along the length of the Atlantic Ocean and separates the North and South American plates on the west from the Eurasian and African plates on the east (see Figure 4-14). A similar but more ancient rift along which spreading ceased a billion years ago underlies the lower Mississippi valley; it now is buried under hundreds of meters of wet, unconsolidated sediments. A series of very large earthquakes in Missouri in 1811–1812, and many smaller ones in recent years, were associated with this rift. A young divergent boundary is developing in eastern Africa, rifting the continent apart to form a new ocean basin. This new plate boundary is exposed on land, forming the East African Rift and such famous volcanoes as Kilimanjaro. The northern extension of this rift continues beneath the Red Sea, as shown in the satellite view at the beginning of this chapter.

Along all active spreading ridges, crust is thinned and extended and normal faulting is found. In normal faulting, a plane of failure separates rocks that have slipped downward relative to those under the fault plane, as shown in Figure 4-15(b). The term *normal fault* comes from mining terminology. Hot fluids rich in elements such as gold and silver often move upward along fault planes, leaving mineral deposits in the rock cracks. Miners crawling along fault planes in search of minerals described those for which the block of rock above their heads moved downward relative to the block underneath them as normal faults, and the opposite as *reverse faults.* Just as normal faults occur at divergent boundaries, reverse faults occur at convergent boundaries.

Convergent Plate Boundaries, Subduction Zones, and Reverse Faults

New crustal rocks are forming all the time, spreading out from divergent plate boundaries at rates of up to several centimeters or more per year. Yet Earth's size remains about the same. This can only mean that for Earth as a whole, convergent boundaries, where plates collide and lithosphere returns to the lower mantle, must be about as common as divergent ones, and so the rate of creation of new lithosphere is similar to the rate of consumption of older lithosphere at convergent boundaries.

One of two types of plate interaction results when plates collide at convergent boundaries. Subduction occurs when a denser plate collides with and sinks beneath a lower density plate. Density variations are due largely to composition, but temperature also is important. Oceanic crust contains more iron and magnesium than continental crust, so it usually is denser. Consequently, when oceanic and continental crust collide, the oceanic plate tends to be subducted under the continental plate. In addition, crust formed by recent volcanism generally is warmer and less dense than older, cooler, crustal rock. Therefore, when blocks of oceanic crust collide, the lithospheric plate bearing the older crust will be subducted under the plate bearing newer crust. In contrast, little or no subduction occurs when two bodies of continental crust collide. Neither plate sinks because the density of both is about equal. Instead, the edges of the two plates crumple and are pushed upward to form high mountains. These behaviors explain why some continental crust has been dated back almost to the formation of the planet, while the oldest oceanic rocks on present-day ocean floors date back only 180 million to 200 million years.

Ongoing convergence of the Nazca and South American plates is an example of oceanic lithosphere (the Nazca Plate) being subducted under continental lithosphere (see Figure 4-14). The oceanic Nazca Plate consists of denser rocks than the western edge of the South American Plate, which is composed of continental rocks. A long depression, or **ocean trench,** exists along the boundary between the two plates and has filled with

(a)

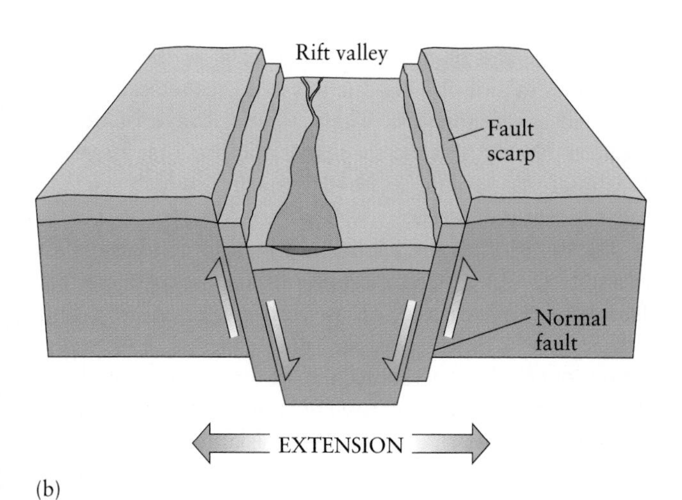

(b)

FIGURE 4-15 (a) Iceland, on the Mid-Atlantic Ridge, is one of the few places on Earth where a divergent boundary can be seen above sea level. (b) Divergent plate boundaries are dominated by normal faults. Rift valleys, such as the one exposed in Iceland, are created when a block of lithosphere drops down between two normal faults.

FIGURE 4-16 The world's highest mountains occur along the convergent margin of two continental plates, where the crust in each plate is too light to subduct. The Himalayan Mountains, along the borders of India, Pakistan, and China, are the roof of an 80-km-thick welt of folded and faulted rocks where the Indian-Australian and Eurasian plates continue to collide with each other.

sediment over time (see Figure 4-13). During subduction, lithospheric material, overlying sediments, and seawater in the sediments are dragged downward into the mantle and heated, sometimes to the point of melting. Magma then rises, forming a chain of volcanoes that parallels the convergent boundary. Because the subducting plate is dipping as it sinks into the asthenosphere, the place where magma rises is inland of the trench and subduction zone.

Because Earth is spherical, the top edge of a sloping subduction boundary is in a bow shape when viewed from above the planet's surface. Likewise, the chain of volcanoes inland of a subduction zone lies in an arc, and is called a **volcanic arc.** The Cascade Range of northwestern North America is a volcanic arc that includes Washington's Mount St. Helens volcano, which erupted most recently in 1980.

The collision of the Pacific and Philippine plates is an example of older, denser oceanic lithosphere (the Pacific Plate) being subducted, along with some sediment and water, under a younger oceanic plate of lesser density (see Figure 4-14). The sediments and water lower the melting point of the subducting crust, forming plumes of magma that erupt into a chain of volcanic islands inland of the subduction zone. Japan, Indonesia, and the Philippine Islands are examples of such volcanic **island arcs.**

The continental crust of the Indian-Australian and Eurasian plates has been colliding over the past 40 million years. As both plates consist of continental rocks that are relatively light, neither is subducted. Instead, the plates remain on the surface and press against each other. This compressive force has deformed large masses of rock and sediment along the edges of the plates and pushed them upward, forming the Himalayan mountain belt and raising the Tibetan Plateau to great heights (Figure 4-16).

Similar thick welts of deformed rock have produced great mountain belts at collision zones throughout geologic history. Between 300 million and 400 million years ago, collision of the North American, South American, Eurasian, and African plates to create Pangaea closed a proto-Atlantic Ocean and formed the Appalachian Mountains. After several hundred million years of weathering and erosion, the ancient mountains still have as much as 1300 m of relief. (Relief is the difference in height between the highest and lowest points in a landscape.) When Pangaea reopened along a new rift about 200 million years ago, the modern Atlantic Ocean was born. The new rift followed the weakened zone of suture—the area where the edges were joined—between the plates, forcing apart the mountain belt. Today, mountainous features equivalent to the Appalachians can be found in northwestern Africa (see Figure 4-1).

Because the rocks tens of kilometers deep in collision zones are under conditions of high temperature and pressure, they often behave as ductile rather than as brittle solids and become intensely folded (Figure 4-17). Oil also tends to form under these conditions. Because light fluids such as oil tend to rise and accumulate under the crests of layers of impermeable rocks in folds, pockets of oil are most common in ancient convergent boundaries between continental plates.

In addition to ductile deformation, brittle deformation (faulting) occurs at convergent boundaries, especially at

FIGURE 4-17 Rocks commonly exhibit evidence of ductile deformation along convergent plate boundaries, as in these folded sedimentary rocks in Alaska. Belts of folded rocks typically mark the locations of ancient convergent plate boundaries. They also commonly are associated with petroleum (crude oil and natural gas), which can be trapped beneath arched layers that are impermeable to the upward flow of petroleum through the rocks.

shallow depths (less than 10 to 20 km) where the rock is cool. The type of faulting generally associated with compressed rocks at convergent plate margins is reverse faulting. As in normal faulting, an inclined plane of failure separates rocks above from those below, but in reverse faulting rocks in the block above the fault plane move upward relative to those below the fault (Figure 4-18). The angle of the failure plane can range from very shallow to very steep. In the case of shallow angles of dip less than about 30°, the reverse fault is called a *thrust fault.* Along a convergent margin, thrust and reverse faults in continental crust typically merge with a master thrust, or megathrust, which forms the actual plate boundary between the overriding and subducting plates. Some of the largest earthquakes known on Earth have occurred along great megathrusts at subduction zones. Recent examples occurred off the coast of Chile in 1960 and southern Alaska in 1964.

(a)

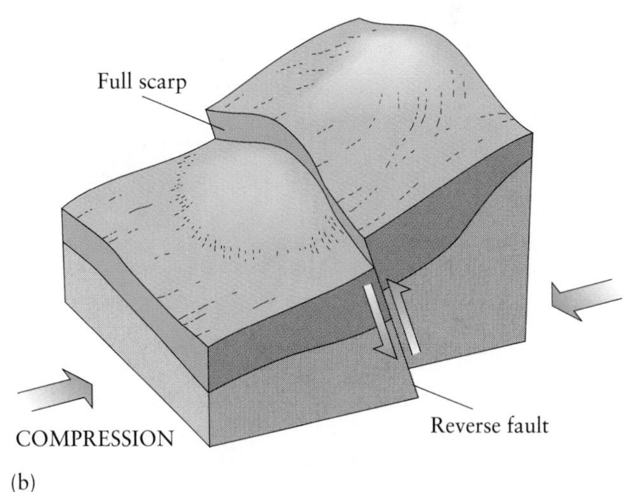

(b)

FIGURE 4-18 (a) In this view of a hillside in northern California, the sharp break crossing the hillside is the trace of a reverse fault. The underlying crust on the upslope side of the fault has moved up relative to the crust on the downslope side. This fault slipped several times in the past few thousand years. A nuclear reactor built over it was closed after the fault was discovered. (b) Convergent plate boundaries, like that between the North American and Juan de Fuca plates along the Pacific Northwest coast, are characterized by reverse faults.

(a)

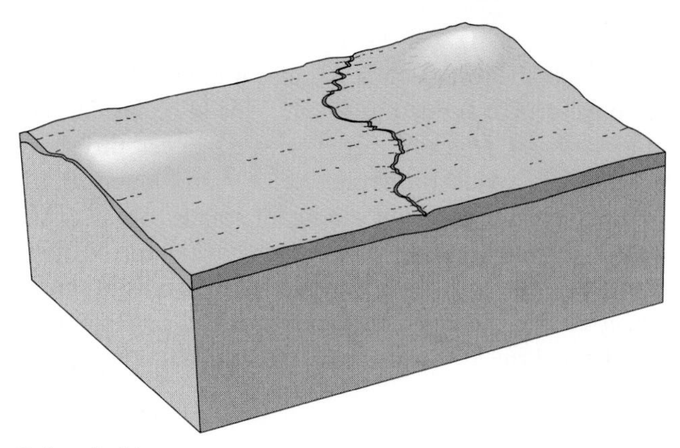

Before faulting

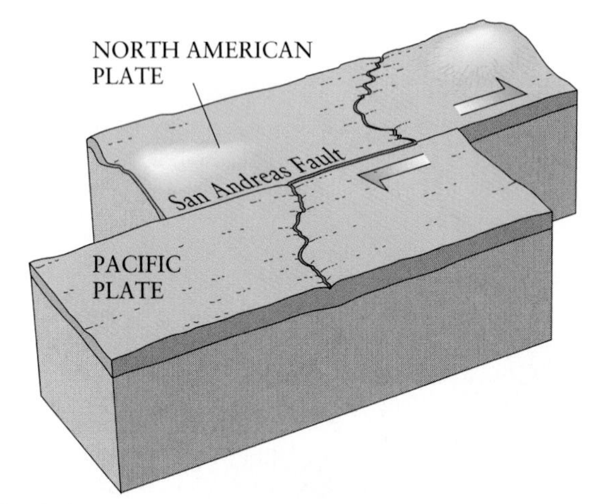

After faulting

(b)

FIGURE 4-19 (a) The transform fault in this view (looking west) is the San Andreas Fault, which extends most of the length of California. The North American Plate is moving southward relative to the Pacific Plate. Note how streams have been offset laterally, with those on the North American Plate (background) moving to the right relative to their counterparts on the Pacific Plate (foreground). (b) Transform plate boundaries, like that in California between the North American and Pacific plates, are dominated by strike-slip faults.

Transform Plate Boundaries and Strike-Slip Faults

A well-known example of a transform boundary is the San Andreas Fault system, where the Pacific Plate is moving northward relative to the North American Plate. The San Andreas Fault system extends more than 1500 km from the Gulf of California to northern California (see Figure 4-14).

Along transform plate boundaries, neither compression nor extension generally occurs, so reverse and normal faults typically are absent. Instead, the type of faulting associated with a transform boundary is called *strike-slip faulting,* and it occurs along a vertical or near-vertical fault plane. The rocks on each side of the vertical fault slip horizontally relative to one another. The orientation of the fault plane with respect to north is called its strike, and slip of the blocks of rock on each side is parallel to the strike, hence the name strike-slip. This type of faulting has interesting consequences. First, because the fault plane is vertical, it appears as a straight line where it intersects the ground surface, and it crosses hills and valleys alike with no change in direction. Second, where features such as streams, fences, and roads cross a strike-slip fault (Figure 4-19), they are offset in a lateral direction, that is, they are displaced on either side of the fault.

- The model of plate tectonics holds that although a chemical boundary exists between the crust and the upper mantle, physically they behave as a unit—the rigid lithosphere.
- New lithosphere forms along rifts at mid-ocean spreading centers on divergent boundaries, and old lithosphere is destroyed at subduction zones along convergent boundaries. The amounts of lithosphere created and destroyed are equal, keeping the lithosphere system in a steady state.
- Plate boundaries are zones of active rock deformation, faulting, and folding. Faulting occurs at shallow depths (generally less than 10 to 30 km), where rocks are cool and brittle, whereas folding generally occurs at greater depths, where rocks are warmer and more ductile.
- The three basic types of faults are normal, reverse, and strike-slip. Although each occurs at all types of plate boundaries, normal faults are dominant at divergent boundaries, where crust is stretched and pulled apart; reverse faults are dominant at convergent boundaries, where crust is compressed and shortened; and strike-slip faults are dominant at transform plate boundaries, where crustal blocks slip horizontally past each other.

Distribution of Rock Types

Once geologists developed the plate tectonics model, the distribution and formation of rocks on Earth could be better understood. For example, igneous rocks with a composition similar to that of the upper asthenosphere are common in oceanic crust adjacent to divergent plate boundaries. In contrast, metamorphic rocks are more common along convergent plate boundaries, where rocks are carried down by subduction or collision and exposed to high pressures and temperatures. With an understanding of how plates interact at the three types of boundaries, we can see how plate tectonics is related to the rock cycle and the occurrence of igneous, metamorphic, and sedimentary rocks.

Igneous Rocks

The composition of an igneous rock is a clue to the plate boundary where it formed, because each such tectonic setting is associated with a characteristic type of magma.

Mafic Magma at Divergent Boundaries The composition of lava emanating from volcanoes at divergent plate boundaries is similar to that of the igneous rock covering nearly two-thirds of Earth's surface as oceanic crust. This lava is rich in iron and magnesium and relatively low in silicon and oxygen (50 to 60 percent SiO_2 by weight). When cooled, this lava forms a dark rock called *basalt* (Figure 4-20). Basalt originates in **mafic magmas,** named for the chemical symbols for magnesium (Mg) and iron (Fe). Rocks that form from lava extruded at Earth's surface, including basalt, are called **extrusive igneous rocks,** or *volcanic rocks.*

Not all mafic magma is extruded at Earth's surface. Some intrudes the crust and cools more slowly than material that seeps from a rift or is erupted from a volcano (see Figure 4-20). Intrusion of mafic magma results in a dark igneous rock called *gabbro,* which has the same composition as basalt but larger mineral grains because of the longer cooling time during crystallization (Figure 4-21). Igneous rocks that form by magma intrusion are called **intrusive igneous rocks,** or *plutonic rocks,* after Pluto, Greek god of the underworld.

A magma that is even more enriched in iron and magnesium, and correspondingly lower in silicon and oxygen (less than 50 percent SiO_2) is described as ultramafic. The rocks that form the asthenosphere are ultramafic in composition and are thought to be the source of mafic (basaltic) magma. When hot, ductile ultramafic asthenospheric rock moves upward by convection at a divergent plate boundary, the pressure surrounding the rock decreases. As a consequence, the temperature at which the rock melts is lowered. Laboratory experiments on the melting of ultramafic rocks at pressures similar to those in the upper asthenosphere indicate that at about 1000°C partial melting occurs. The mixture that forms contains

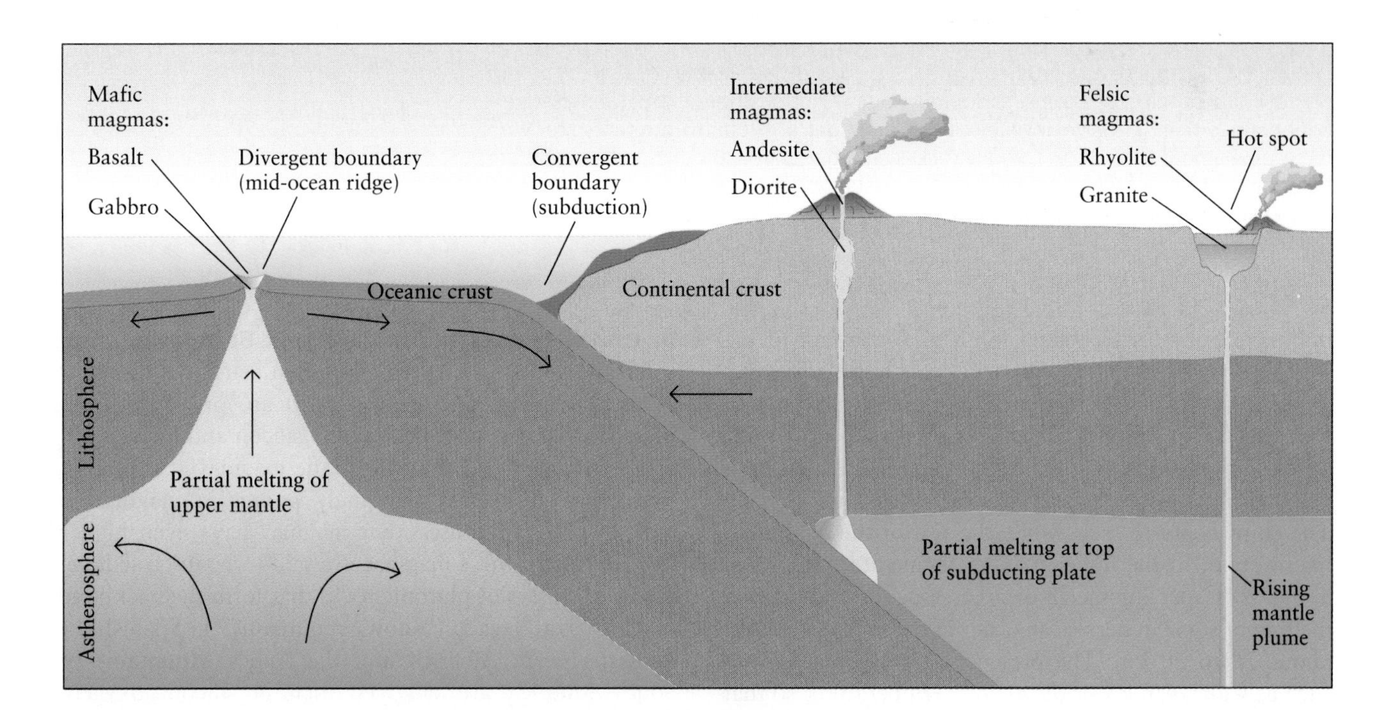

FIGURE 4-20 The distribution and formation of different types of magma are controlled by plate tectonics. Basaltic volcanism occurs at oceanic hot spots and divergent plate boundaries, whereas intermediate volcanism and plutonism occur along subduction zones (convergent plate boundaries). Silicic volcanism occurs at continental hot spots.

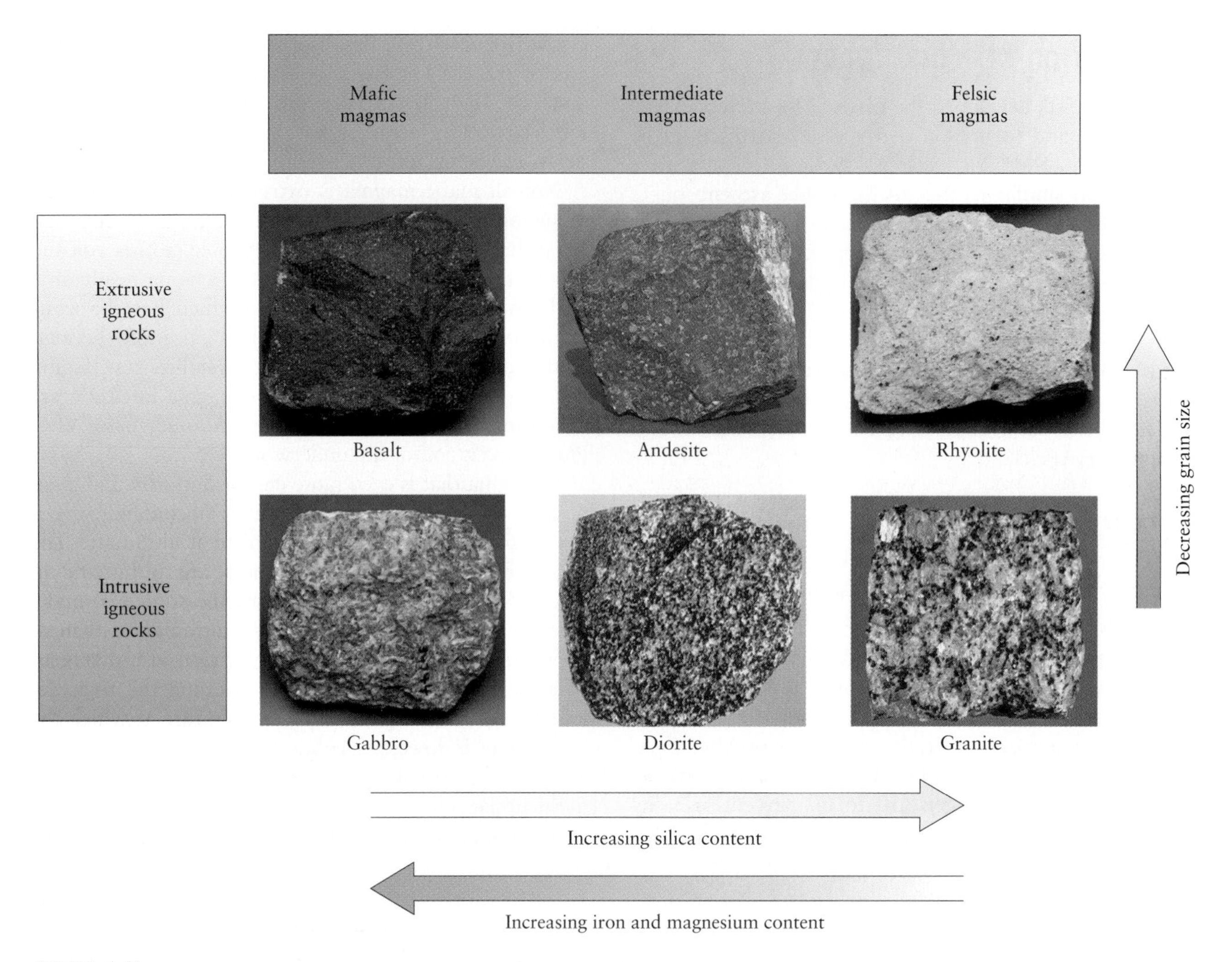

FIGURE 4-21 Extrusive and intrusive igneous rocks are classified according to the size of individual crystals, or apains, and mineralogical composition. Extrusive rocks are finer-grained than intrusive, for they cooled more quickly from the molten state. With increasing amounts of silica, igneous rocks contain less iron and magnesium and are lighter in color as a result. Oceanic crust consists mostly of basalt, whereas continental crust consists mostly of granite, diorite, and andesite.

both solid, unmelted crystals and a small amount of liquid. The first minerals to melt from the ultramafic rocks are those that contain relatively high amounts of silicon and oxygen, so the liquid has the composition of basalt.

This process of partial melting is similar to what happens when a chocolate chip cookie is heated in an oven: The chips begin to melt, while the rest of the cookie remains solid. In the upper mantle, temperature increases about 20°C per kilometer of greater depth, and partial melting of some minerals can occur at depths greater than about 25 to 50 km. The presence of other substances, such as water, can lower the melting temperatures so that even more of the minerals melt. Partial melting of the asthenosphere, which is mostly solid, explains the transfer of rock from the mantle to the oceanic crust reservoir that was described in Chapter 2 (see Figure 2-19).

Intermediate Magma at Convergent Boundaries Plutonic and volcanic rocks in mountain belts at convergent plate boundaries (see Figure 4-20) are produced from magmas that are more enriched in silicon and oxygen (60 to 70 percent SiO_2) than the mafic magmas at divergent boundaries and correspondingly poorer in magnesium and iron. But because other magmas have even higher SiO_2 contents, these magmas are said to be **intermediate.** Plutons (masses of plutonic rock) that form the backbone of the Sierra Nevada ("snowy mountains" in Spanish) of western North America and the Andes Mountains of South America are relatively rich in silicon, oxygen, potassium, sodium, and aluminum. As vast chambers of magma cool to form the plutons, potassium and sodium combine with silica tetrahedra to form feldspar minerals; aluminum and a few other elements join silicate sheet structures to form mica; and the remaining silica tetrahedra link to form relatively abundant quartz crystals. The plutonic rocks in these mountain belts, which cooled slowly over millions of years, have large crystals and are known as *diorite* (see Figure 4-21). Volcanoes and volcanic rocks with similar magma composition occur in association with the plutons, but these rocks have smaller crystals and are known as *andesite,* named for the Andes Mountains.

Much of the material ejected from Mount St. Helens in its 1980 eruption was andesitic in composition. Because andesites and diorites are associated almost exclusively with subduction zones, they are thought to result from partial melting of subducted oceanic crust mixed with weathered, wet sediments derived from adjacent continental crust. Since oceanic crust is basaltic, it is reasonable to assume that the partial melt's composition, if mixed with silica-rich continental sediments, would contain more silicates than a melt of basaltic crust alone.

Felsic Magma at Continental Interiors A third type of igneous magma and associated igneous rock is found on Earth. Magmas that contain high amounts of feldspar minerals and silica are called **felsic magmas,** *fe* for feldspar, *l* for the similar mineral lenad, and *si* for silica. Felsic magmas form *granite,* an intrusive rock, and its extrusive equivalent, *rhyolite* (see Figure 4-21). Granites and rhyolites are common in continental interiors, such as at Yellowstone National Park in Wyoming. Here a plume of hot mantle material rises and heats the underbelly of the North American crust, partially melting some of it (Box 4-3). Because no basaltic ocean crust is involved and mostly wet continental crust is melted, the resultant partial melt is especially rich in silicon and oxygen.

Metamorphic Rocks

Metamorphic rocks result when rocks are exposed to high pressure, or high temperature, or both, but are not melted. Metamorphism will occur, for example, if rocks are deeply buried, compressed, and folded during plate collision and mountain building along a convergent margin. The increasing weight of overlying rocks with depth can increase pressure on buried rocks sufficiently for their minerals to react and recrystallize. If rocks are subducted about 35 to 40 km deep, where temperature is greater than about 650°C, igneous processes of partial melting and magma formation will begin. For this reason, igneous and metamorphic rocks often are found together in mountain belts at convergent boundaries.

Metamorphic rocks also are found adjacent to igneous rocks at both divergent and convergent margins because the hot magmas that produce igneous rocks expose surrounding rocks to high temperatures. Chemists have known for centuries that higher temperatures generally increase reaction rates. To understand how an increase in temperature can cause metamorphism, consider what would happen if you made a clay pot and fired it in a kiln. As temperature increases, chemical reactions begin to occur that make the clay pot harder and firmer. Before firing, the clay softens in water; after firing it can be used to carry water. Similarly, when minerals are dragged down into the asthenosphere, where temperature increases about 20°C per kilometer of depth, the chemical compounds forming the minerals will react with one another and recrystallize, forming new minerals.

The most striking characteristic of many metamorphic rocks is their texture, the characteristic sizes, shapes, and orientation of individual minerals in the rock. Metamorphic rocks often possess directional texture known as **foliation,** in which minerals are arranged in parallel mode. Directional texture results from crystal growth under conditions of immense rock pressure: Crystals grow in preferred directions in which stresses are the least. The stresses stem from continued plate motion and rock deformation.

Metamorphic rocks are classified as foliated and nonfoliated. Foliation is common in metamorphic rocks that originally were fine-grained and contain large amounts of clay minerals, such as the sedimentary rock known as shale. Nonfoliated texture occurs in rocks that consist predominantly of one mineral—such as quartz or calcite—that does not have a tendency to form elongated crystals. The most common nonfoliated metamorphic rocks are *quartzite* and *marble,* which result from the metamorphism of sandstone and limestone respectively.

In general, crystal size and degree of foliation increase with higher pressures and temperatures and continued metamorphism. With increasing degree of metamorphism, foliation is expressed in fine-grained rocks first by closely spaced fractures, resulting in *slaty cleavage;* then by the parallel arrangement of platelike minerals such as mica, resulting in *schistosity;* and finally by alternating bands of dark and light minerals, resulting in *gneissic banding* (Figure 4-22). These textures, respectively, define three types of metamorphic rock: slate, schist, and gneiss.

Sedimentary Rocks

Sedimentary rocks form when sediments are deposited, preserved, and lithified. Sediments are deposited in any environment in which transporting fluids lose energy and are unable to transport their sediment loads (Figure 4-23 and Table 4-4).

Lithification is the process that turns sediments into rocks, compacting the sediments—that is, lessening the space between grains—and cementing them into a consolidated mass. Deposition and lithification occur commonly in *sedimentary basins,* topographically low places where sediments can accumulate to great thicknesses, providing the pressure needed for compaction. Examples of large sedimentary basins include the low areas just inland of volcanic arcs on continents and the deep

4-3 Geologist's Toolbox

Measuring Plate Velocities With Mantle Plumes

A mantle plume, one of the columns of magma from deep in the mantle that jet upward toward the surface, underlies Yellowstone National Park in Wyoming. Also called a *hot spot,* the mantle plume is fixed in position while above it moves the North American continental plate. Heat from this mantle plume is the cause of geothermal activity at the Yellowstone geyser field. More than 120 recently active hot spots have been identified on Earth. Some geologists think that hot spots are the last remaining large pockets of magma in a once completely molten mantle. Plumes may originate as deep as the mantle-core boundary (2891 km), and some are more than 100 km in diameter.

Mantle plumes have been detected at both continental and oceanic plates. One plume under the Pacific Plate forms the volcanic islands of Hawaii. The stationary plume punctures the moving plate, over time creating one volcano after another. Each volcano stays active for about 2 million years and is extinguished when the plate has carried it past its source of magma. The youngest volcano is Loihi, south of Hawaii. Nearly 5 km high, the seamount—the submerged volcanic peak—is within a kilometer of the sea surface. Loihi will continue to grow upward and is estimated to emerge above sea level in a few hundred thousand years.

The resulting linear chain of volcanoes records the Pacific Plate's long-term absolute speed and direction. The bend in the chain of Hawaiian Islands about midway along its length indicates a change in plate direction from north to northwest that occurred about 40 million years ago.

The chains of volcanoes leading to hot spots can be used to calculate the absolute speed and direction at which each plate is moving (as shown by the arrows in Figure 4-14). Volcanic rock in a given island can be radiometrically dated, and this age can be compared to the island's distance from the hot spot that formed it in order to calculate velocity. Some plates move very slowly, less than 1 cm per year. Others move rapidly, more than 10 cm per year. The speed at which a plate moves influences, in part, the degree of volcanic and earthquake activity along its boundaries. In general, the greater a plate's rate of motion, the greater the tectonic activity along its edges.

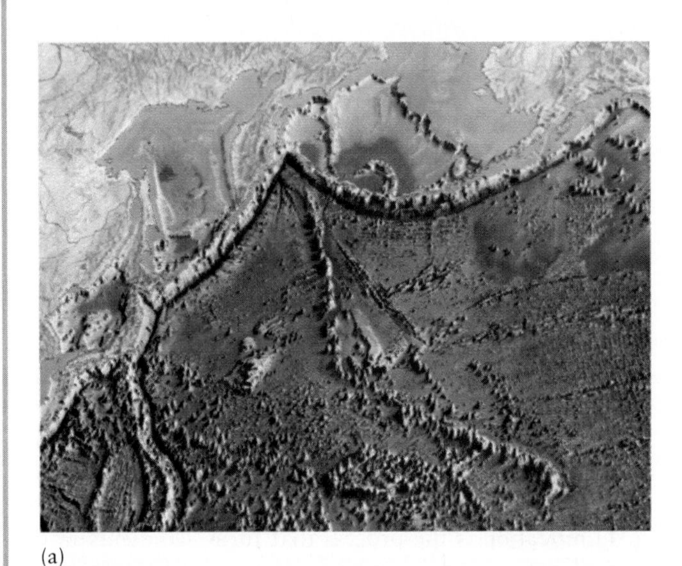
(a)

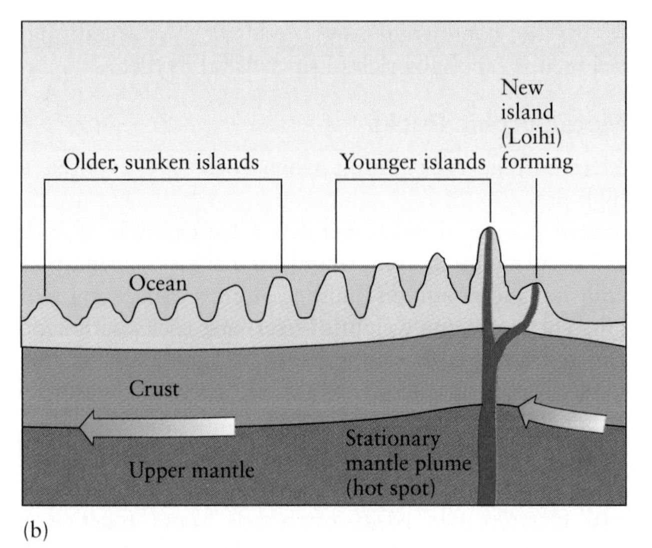

(b)

Of the 120 or more active hot spots on Earth, the largest is the one beneath the Pacific Plate that forms the Hawaiian Islands. (a) Motion of the Pacific Plate is traced by a long chain of islands, with a dogleg to the east where the plate changed directions about 40 million years ago. Thousands of recent earthquakes and a major collapse of the Loihi seamount's peak occurred in the summer of 1996, leaving a crater half a mile wide and 1000 ft deep. (b) As the Pacific Plate moves over a volcanic hot spot, islands rise and sink. The oldest have sunk the most, and are below sea level.

FIGURE 4-22 Foliated metamorphic rocks are named based on the degree of foliation and crystal size, which reflect the intensity of metamorphism. When shale is metamorphosed, for example, foliation occurs first as slaty cleavage in (a) slate, then as schistosity in (b) schist, and then as banding in (c) gneiss.

trenches along subduction zones. Not all **depositional environments,** however, are located on active plate boundaries. For example, in the delta at the mouth of the Nile River, several thousand meters of sediment are accumulating in the Mediterranean Sea (see the chapter-opening photograph). Local depositional environments are various, ranging from river valleys to coastal beaches.

Lithification is distinct from the metamorphic processes of recrystallization and the igneous processes of magma formation by partial melting. The distinction is gradational, however, rather than abrupt. Sediments deposited at Earth's surface can becom105 e lithified at depths of several kilometers and temperatures up to about 200°C, but at higher pressure or temperature metamorphic processes begin (see Figure 4-11).

After deposition, sediments often are buried under yet more sediments. Those at the bottom of the depositional stack are compacted and pore spaces between grains shrink. Furthermore, fluids circulating through the sediments can precipitate iron oxides, calcium carbonate, or silica in the pore spaces. These compounds cement the grains together, producing rock.

Sedimentary rocks are not as closely linked to plate tectonics processes as are igneous and metamorphic rocks. Instead, sedimentary rocks are linked more closely to processes driven by solar energy, including weathering, deposition of sediment, and crystallization from evaporating solutions.

Sedimentary rocks can be subdivided into three categories: clastic, chemical, and biological. *Clastic sedimentary*

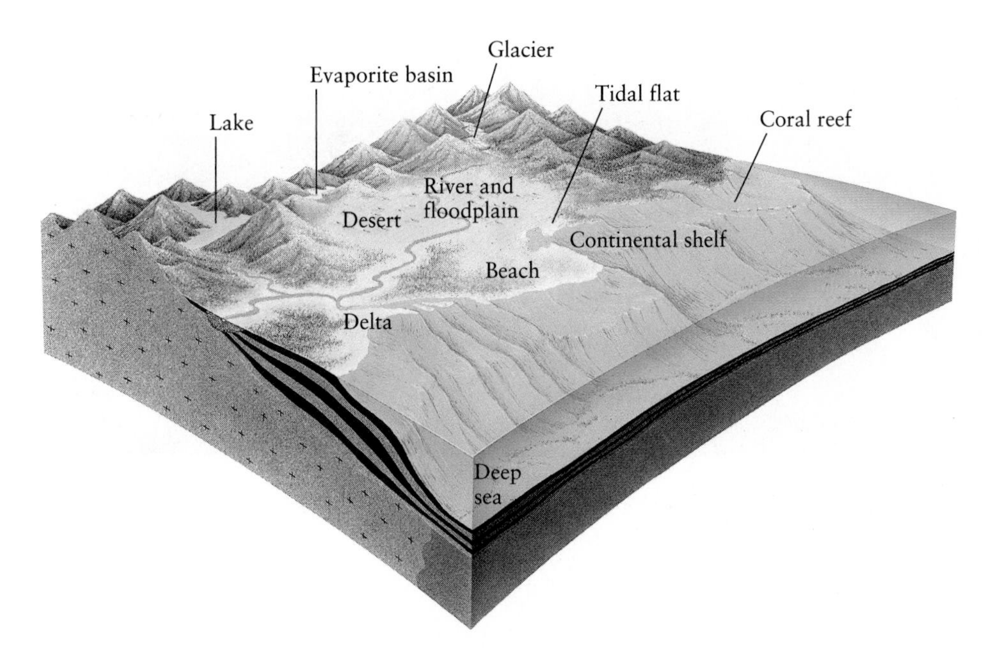

FIGURE 4-23 Sedimentary rocks form in a variety of depositional environments. The nature of a deposit, such as grain size and rounding, provides clues about the environment in which it was deposited. Very rounded and well sorted grains are likely to have come from sand dunes, for example, whereas much coarser and more poorly sorted ones are characteristic of glacial deposits.

TABLE 4-4 Characteristics of Various Sedimentary Environments

Environment	Characteristics of sediments found there
Glacier	ridges of unsorted clay- to boulder-sized particles along glacial margins
Lake	clay- and silt-sized particles deposited in horizontal layers (laminations)
Desert	well-rounded and -sorted sands forming dunes; sand grains appear frosted from frequent impacts with one another
River	sequences of sediments that occur in layers, from cobble to silt size; interbedded with clay and silt deposited in floodplains
Delta	sequence of layers of well-sorted sand, silt, and clay that dip gently seaward and get finer with greater distance offshore
Beach	well-rounded and -sorted sands; no frosting of grains; marine fossils (shells, corals)
Tidal flat	fine silt and clay with mudcracks, animal burrows, and animal tracks
Continental shelf	mixture of muds and sands from continent; shells, skeletons, and fecal pellets of marine organisms
Reef	calcium carbonate skeletons of corals
Deep sea	laminated fine clays and biologic oozes
Evaporite basin	chemical sediments, deposited as salt crystals in horizontal layers in shallow water, often mixed with fine silt and clay washed from surrounding mountains

rocks are formed of fragments of preexisting rocks and minerals that have been cemented together. *Chemical sedimentary rocks* form when dissolved solids derived from other rocks, minerals, and soils crystallize from solution. *Biological sedimentary rocks* are made of the shells and skeletons of various organisms.

Clastic Sedimentary Rocks Clastic sediments (from the Greek *klastos,* "broken") arise largely from the action of physical weathering processes that break rocks into fragments. Ranging in size from tiny clay particles to large boulders, these fragments—or clasts—can be eroded and transported to a new location by wind, water, ice, or gravity (as in a rockslide). If lithified, the sediments become clastic sedimentary rocks. Clastic sediments and sedimentary rocks are classified by the sizes of the clasts forming them (Figure 4-24). For example, a rock formed of clay-sized particles less than 4 μm (micrometers; 1 μm = 0.001 mm = 0.0001 cm) in diameter is called a mudstone or a shale, while a rock formed of silt-sized grains (4–63 μm) is a siltstone and one formed of sand-sized grains (63 μm to 2 mm) is called a sandstone (Table 4-5).

(a)

(b)

(c)

FIGURE 4-24 Classification of clastic sedimentary rocks is based on grain size: (a) shale with plant fossils (clasts are clay-sized, finer than sand and silt), (b) sandstone (sand-sized clasts), (c) conglomerate (rounded particles, including pebble-sized clasts).

TABLE 4-5 Classification Scheme for Clastic Sedimentary Rocks

Particle size (diameter)	Sediment name	Rock name
< 4 μm	clay	mudstone or shale (fine layering)
4–63 μm	silt	siltstone
63 μm–2 mm	sand	sandstone
> 2 mm	gravel	conglomerate

The size of the grains is a clue to the environment in which the sediment was deposited. The larger the grain size or greater the volume of clastic sediment, the more energy is required to transport it. For example, it would be impossible to move a 10-ton boulder with a trickle of water. However, that trickle may have sufficient energy to transport clay-sized particles only 2 μm in diameter. Environments change with time, as when a large flood passes through a river valley or ice sheets wax and wane, and as a consequence the energy available to transport sediments changes, too. When the energy of the transporting medium decreases, it no longer is able to move the largest and heaviest grains and they are deposited.

Chemical Sedimentary Rocks Chemical sediments originate from chemical weathering processes that partially or wholly dissolve and alter existing minerals (see Chapter 6). Have you ever noticed that allowing tap water to dry on dishes can produce white spots? These spots are calcium carbonate that had been dissolved in the tap water and was left behind when the water evaporated. All natural water contains some dissolved ions. In fresh water, the concentration of ions is very small, in the range of just a few parts per million (ppm). In comparison, seawater contains roughly 35,000 ppm salt. Crystalline sedimentary deposits that form from evaporating waters are called **evaporites.** Chemical sediments have a variety of compositions, including calcium carbonate, calcium sulfate, and sodium chloride.

The floor of Death Valley in eastern California is covered with modern evaporite deposits (Figure 4-25), and many ancient evaporites are found throughout the world as buried layers where shallow seas and saline lakes once existed. Examples of older evaporite deposits occur in the coastal areas of Texas and the southeastern United States, Germany, and the Mediterranean region. Most of the world's supply of table salt (sodium chloride) is mined from sediments in evaporite basins.

Sometimes chemical sediments crystallize even before the water in which they were dissolved has completely evaporated. Consider a lake in an enclosed basin like Death Valley. The lake continually receives stream water carrying dissolved ions. Some of the water in the lake is removed by evaporation, but the ions stay behind. If the rates of inflow of stream water and outflow by evaporation balance, the lake maintains a stable size, but the concentration of salts increases year after year. Eventually, the concentration of ions will be so high that the water is saturated with salt and no longer is capable of holding any more in dissolved form. Minerals containing the ions will begin to form by crystallization, as they do in Death Valley, the Great Salt Lake in Utah, the Dead Sea in Israel, and elsewhere. To simulate this process, mix a small quantity of table salt in water. All the salt dissolves in the water. However, if you add more salt, the water eventually becomes saturated with sodium and chloride ions. If you try to add yet more salt, it will not dissolve but instead will accumulate as solid crystals at the bottom of the glass.

Biological Sedimentary Rocks Biological sediments accumulate from the carcasses of organisms. For example, the toppling of trees and the death of small shrubs and grasses in swamps and marshes lead to the formation of peat, and the death of tiny organisms in the oceans

FIGURE 4-25 Death Valley is a deep basin with no streams draining out of it. Instead, streams from surrounding mountains drain water and salts into the basin, where water evaporates and salt crystallizes. Frequent wetting and drying, with associated expansion and contraction of salt minerals, results in polygonal cracks on the desert floor.

produces calcium-rich mud on the ocean floor. Many biological sediments actually are considered chemical in origin because the organisms that secrete carbonate (containing carbon and oxygen) and silica skeletons derive their skeletal materials from the waters in which they live. For example, here is the reversible chemical reaction by which mollusks such as clams extract bicarbonate (HCO_3^-) and calcium (Ca^{2+}) ions from water and combine them into calcium carbonate ($CaCO_3$) for shells:

$$2\,HCO_3^- + Ca^{2+} \rightleftharpoons CaCO_3 + CO_2 + H_2O$$

Notice that in this reaction, carbon dioxide gas (CO_2) and water (H_2O) are produced along with the calcium carbonate. This common biochemical reaction is important to Earth systems. In this way, ions dissolved in seawater, which themselves were weathered and eroded from continental rocks, combine to form a solid and release carbon dioxide, a greenhouse gas. Carbon dioxide is important in the global carbon cycle and directly affects global warming (see Chapter 12). Note also from this reaction that some carbon can become locked into sediments as calcium carbonate, thus removing it from the atmosphere and hydrosphere reservoirs (see Chapter 9). In fact, much of Earth's carbon is stored in these reservoirs.

Many organisms secrete calcium carbonate shells to protect their soft bodies from injury and predation by other organisms. Some of them, such as foraminifera (animals) and coccoliths (plants), are a millimeter or less in diameter and float throughout the ocean while alive. When these organisms die, their bodies sink to the ocean floor and their soft parts decay, leaving their shells behind to form thick oozes of calcium carbonate. If these oozes are lithified, the resulting rock is *limestone*. Limestone formed of foraminiferal oozes also is known as *chalk*, and it is used to make a variety of products, including toothpaste and soft writing tools for slate blackboards (Figure 4-26).

Another animal that secretes large amounts of calcareous hard parts is the soft coral polyp, but it lives in the near shore environment. Coral reefs consist of large colonies of coral polyps and cover extensive areas along continental margins in the tropics, between latitudes of 23° north and south of the equator in today's climate. Many ancient reefs are preserved as limestones in the fossil record and, because of continental drift, can be found today at most latitudes.

Not all organisms living in the ocean secrete calcium carbonate shells. Some create skeletons and shells of silica. As in the case of carbonate-secreting animals, their hard skeletons and shells occur in a variety of sizes. Single-celled microorganisms called radiolaria (animals) and diatoms (plants) create siliceous oozes when their bodies fall to the ocean floor. If lithified, these oozes are called *diatomite* (in the case of diatoms) or *chert* (in the case of radiolaria).

Some terrestrial environments also produce biological sediments. Swamps, wetlands, and bogs frequently produce peat, compressed plant remains containing more than 50 percent carbon, which may eventually turn to coal given sufficient time and depths of burial (see Chapter 11). Although not as rich in carbon as coal, peat is mined and burned to produce heat in many parts of the world.

(a)

(b)

FIGURE 4-26 (a) Layers of chalk, a type of limestone, are exposed in the sea cliffs at Normandy, France. (b) Chalk consists of sediments rich in calcareous microorganisms called foraminifera, like the large rounded shells in this magnified (81×) scanning electron microscope image of sediments from the floor of the Indian Ocean. The coarse-meshed objects are siliceous microorganisms (radiolaria) and the long objects are sponges. The foraminifera and radiolaria float near the ocean surface while alive, and sink to the seafloor after death.

- Magma can erupt at Earth's surface, forming volcanic (or extrusive) rocks, or can cool and solidify within the crust, forming plutonic (or intrusive) rocks.
- Partial melting is the process that explains the origin of basalt from incomplete melting of the asthenosphere. Basaltic magmas are formed dominantly at divergent plate boundaries, whereas intermediate magmas are common at convergent boundaries and felsic magmas are common in continental interiors above hot spots.
- The three types of sedimentary rocks—clastic, chemical, and biological—form in different sedimentary environments, such as floodplains (clastic), desert basins (chemical evaporites), and coral reefs (biological).
- Sedimentary rocks are the only rocks that contain evidence of past life on Earth (as fossils) and fossil fuels such as oil, coal, and peat.

Closing Thoughts

Rocks contain clues to their past: the depths to which they were buried, the temperatures to which they were heated, and the places to which they have traveled. The first step in discovering these clues is identifying the rocks and the minerals within them. An experienced geoscientist can tell much from even small rock exposures. By mapping different rock types and determining their ages, geoscientists have reconstructed much of Earth's long and eventful history. Since the recognition of plate tectonics, a revolutionary concept that forever changed our understanding of Earth, the stories geoscientists discover from rocks are even more fantastic: Rocks from the south pole once were close to those now located in the northern hemisphere, near modern-day Las Vegas; fragments of plates and island arcs once were sutured to the coasts of North America, forming modern-day New Jersey and Alaska. Rocks also contain information about past climates, atmospheres, environments, oceans, and life-forms that existed when the rocks formed. The clues in rocks and the tales they tell also help us to find mineral resources, to understand earthquakes, and to predict volcanic eruptions, as discussed in the next chapter.

Summary

- Minerals are naturally formed, inorganic, crystalline solids with specific chemical compositions or ranges of compositions.
- Crystallization is the process by which atoms combine to form orderly arrays that are repeated in three dimensions.
- The most common minerals in Earth's crust are silicates; they consist of silica tetrahedra that are either isolated or polymerized (linked in the form of clusters, chains, sheets, and three-dimensional frameworks of tetrahedra).
- The properties of minerals reflect the internal arrangement and bonding of atoms. For example, minerals break, displaying cleavage, along planes of weak atomic bonding between layers of atoms.
- Plate tectonics occurs on Earth because its interior is hot enough to cause slow convection of the asthenosphere, and because some parts of the lithosphere are dense enough to sink into the asthenosphere.
- In the cool upper parts of the lithosphere, brittle deformation leads to faults, while in the warm lower parts, plastic deformation leads to folding.
- Magma is described as mafic, intermediate, or felsic, with increasing amounts of silicon and oxygen in each form.
- Mafic magmas result in basalt (extrusive) and gabbro (intrusive), and are most common along divergent plate boundaries. They are the dominant rock types in oceanic crust.
- Intermediate magmas result in andesite (extrusive) and diorite (intrusive), and are most common in continental crust along subduction zones (convergent boundaries).

- Felsic magmas result in rhyolite (extrusive) and granite (intrusive), and are the dominant rock types in continental crust.
- Metamorphic rocks are classified as foliated or nonfoliated. Common examples of foliated metamorphic rocks are slate, schist, and gneiss; common examples of nonfoliated metamorphic rocks are quartzite and marble.
- Sedimentary rocks are classified on the basis of their origin: clastic, chemical, or biological.
- Sediments are deposited in different environments, which can be determined long after the environment is gone from clues such as grain size and shape. These clues help geologists reconstruct the history of Earth's changing environments.

Key Terms

seafloor spreading (p. 92)
subduction (p. 92)
plate tectonics (p. 92)
rock (p. 93)
mineral (p. 93)
crystallization (p. 94)
silicates (p. 95)
Mohs hardness scale (p. 98)
cleavage (p. 99)
igneous rock (p. 100)
sedimentary rock (p. 101)
metamorphic rock (p. 101)
convection (p. 103)
fault (p. 103)
convergent plate boundary (p. 103)
divergent plate boundary (p. 103)
transform plate boundary (p. 103)
ocean trench (p. 107)
volcanic arc (p. 108)
island arc (p. 108)
mafic magma (p. 111)
extrusive igneous rock (p. 111)
intrusive igneous rock (p. 111)
intermediate magma (p. 112)
felsic magma (p. 113)
foliation (p. 113)
lithification (p. 113)
depositional environment (p. 115)
evaporites (p. 117)

Review Questions

1. How did Alfred Wegener determine that all continents once were connected in the single landmass that he called Pangaea?
2. In the early to mid-20th century, what different discoveries were combined into a single model, the plate tectonics model, to explain many diverse phenomena on Earth?
3. Why are silicate minerals so common in Earth's crust?
4. What are some examples of common silicate minerals formed of sheets of silica tetrahedra, and of three-dimensional frameworks of silica tetrahedra?
5. How does crystallization occur in a magma? How does crystallization occur in an evaporating solution?
6. How are mineral hardness and cleavage related to the internal structural arrangement of a mineral's atoms and the strength of the bonds between atoms?
7. What are the three major rock groups, and how are they related to one another by the rock cycle?
8. How do volcanic glasses, like obsidian, form? Why were they so useful for toolmaking during the Stone Age?
9. What is the driving force behind plate tectonics?
10. Along what type of ancient plate boundary did the large Missouri earthquakes occur in 1811–1812?
11. What type of sediments would you expect to find deposited along a large river? In a sand dune? In a lake?
12. Along what type of plate boundary is oil most likely to be found in sedimentary rocks?
13. What is a megathrust? Where on Earth might one exist?
14. What are the names of intrusive and extrusive igneous rocks associated with mafic, intermediate, and felsic magmas?

Thought Questions

1. If you found out that chrysotile (an asbestos mineral) exists in the insulation in your house, should you be concerned about a possible risk to your health?
2. Is obsidian a mineral? Why or why not?
3. Why do most earthquakes occur at relatively shallow depths (generally less than 20 to 30 km beneath the surface)?
4. How might the fact that plates with oceanic crust are more likely to subduct than those with continental crust be related to the composition of oceanic crust?
5. Why don't some metamorphic rocks (e.g., quartzite) display foliation in their texture?
6. How is the formation of calcareous sediments in the oceans related to the amount of carbon dioxide in the atmosphere?

Exercises

1. Use a magnifying glass, hand lens, or microscope to examine closely grains of table salt (halite). These are crystals, and they are broken along cleavage planes. What type of cleavage can you see (e.g., one, two, or three planes broken at right angles to one another, or at another angle)?
2. The island of Hawaii consists of several active volcanoes that are generally less than 1 million years old and still are actively growing. Each volcano north of the main island of Hawaii is progressively older (see Box 4-3), and the volcano located where the chain of islands bends is 40 million years old. Use the distance between these two points—Hawaii and the bend—and the age difference in volcanoes at the two sites to determine the speed at which the Pacific Plate is moving over the hot spot fixed beneath it.
3. One of the largest historic earthquakes in the world occurred along the coast of Chile in 1960. Along what type of plate boundary did it occur (refer to Figure 4-14)? Draw a vertical cross section from west to east that shows the two plates along this boundary, and the nature of their interaction with each other.

Suggested Readings

Bernarde, M. A. (ed.), 1990. *Asbestos: The Hazardous Fiber.* Boca Raton, FL: CRC Press.

Cox, Allan, 1986. *Plate Tectonics: How It Works.* Palo Alto: Blackwell Scientific Publications.

Decker, Robert W., and Decker, Barbara, 1991. *Mountains of Fire—The Nature of Volcanoes.* New York: Cambridge University Press.

Hsu, Kenneth, J., 1992. *Challenger at Sea: A Ship That Revolutionized Earth Science.* Princeton, NJ: Princeton University Press.

Robinson, George W., 1994. *Minerals: An Illustrated Exploration of the Dynamic World of Minerals and Their Properties.* New York: Simon and Schuster.

Scholz, Christopher, 1997. *Fieldwork: A Geologist's Memoir of the Kalahari.* Princeton, NJ: Princeton University Press.

Simkin, Tom, and Siebert, Lee, 1994. *Volcanoes of the World.* 2nd ed. Smithsonian Institution Press.

Skinner, H. Catherine, Ross, Malcolm, and Frondel, Clifford, 1988. *Asbestos and Other Fibrous Minerals: Mineralogy, Crystal Chemistry and Health Effects.* New York: Oxford University Press.

Wallace, Robert E. (ed.), 1990. *The San Andreas Fault System, California.* Reston, VA: U. S. Geological Survey. Professional Paper 1515.

CHAPTER 5

PT Freeport Indonesia, a mining company, is excavating copper, gold, and silver 4000 m above sea level on the island of New Guinea. The yellowish metal-rich rocks stand out in stark contrast to the layered gray limestone. Valley glaciers that recently covered the area have retreated higher into the mountains (upper left), leaving broad, U-shaped valleys carved into the limestone.

Lithosphere: Resources, Hazards, and Change

The highest mountains between India's Himalayas and South America's Andes rise nearly 5000 m above sea level at the equator, forming the island of New Guinea. A Dutch sea captain who sailed past in 1623 reported glimpses of glaciers at the equatorial island, to the scorn of his contemporaries. Yet Captain Carstenz was right. The captain had visited New Guinea during the peak of the Little Ice Age, a few hundred years of global temperatures sufficiently low for ice masses to have advanced. Since then, Earth's atmosphere has warmed and most ice masses have shrunk.

As the ice over New Guinea melted, it revealed the world's largest deposit of gold and third largest of copper. A company called PT Freeport Indonesia uses helicopters, the world's longest aerial tramway, and the world's largest dump trucks to operate the mountain-top mine.

New Guinea's mountains were built as the subduction of the Indian-Australian Plate beneath the Pacific and Philippine plates formed a metal-rich volcanic arc that still rocks the island with earthquakes and volcanoes. This chapter explores the relationship between plate tectonics and mineral deposits, volcanic hazards, and earthquakes. We will:

- Examine the origins of minerals and the issues connected with their extraction.
- Examine the characteristics of volcanoes and ways to predict eruptions.
- Investigate the causes of earthquakes and ways to minimize the associated risks.

Heat from Earth's interior (thermal energy) drives the processes of plate formation, subduction, melting of rocks, igneous intrusion, volcanism, and earthquakes. Earth's heat also helps to concentrate mineral deposits, like the gold and copper in the mountains of New Guinea. The discovery that heat could purify, combine, and temporarily soften metals marked a turning point in human history, from the Stone Age to the Bronze Age.

In *Historia naturalis,* the world's first encyclopedia, the Roman scholar Pliny the Elder (AD 23–79) wrote of the accidental discovery that heat could transform sand to glass. According to Pliny, a Phoenician ship with a cargo of natural soda, the common name for an evaporite that contains large amounts of the element sodium, took harbor along the Mediterranean coast sometime in the second millennium BC. Sodium was, and still is, widely used in soaps and bath salts:

> [The sailors] scattered along the shore to prepare a meal. Since, however, no stones suitable for supporting their cauldrons were forthcoming, they rested them on lumps of soda from their cargo. When these became heated and were completely mingled with the sand on the beach a strange translucent liquid flowed forth in streams; and this, it is said, was the origin of glass. (D. E. Eichholz, ed., 1962, *Pliny's Natural History,* vol. X, Cambridge, MA: Harvard University Press)

As the Phoenicians discovered, glass can be made by melting beach sand, which typically is rich in quartz, a mineral that contains only two elements, silicon and oxygen. During melting, some of the atomic bonds in the quartz minerals are broken apart, enabling the material to flow as a liquid. If it is cooled rapidly, before the atoms have time to rearrange themselves in an orderly crystalline array, the melt solidifies as glass (Figure 5-1). Had the Phoenician sailors not introduced soda into the sand, they would not have produced glass. The melting temperature of quartz is 1610°C, more than twice the temperature of a wood fire, which typically is less than 800°C. By adding sodium—in the form of "lumps of soda"—to the sand, the sailors inadvertently created a mixture with a far lower melting point than quartz sand alone.

The Phoenicians already were familiar with the mysteries of melting minerals and rocks. Their cauldrons, tools, and weapons were made of bronze (an alloy of copper and tin) and iron, metals that can be extracted by smelting certain minerals at temperatures of about 900°C or more. (Smelting is the separation of a metal from other minerals or elements by melting, or by processes that include melting.) The Phoenicians were familiar, too, with the natural melting of rock, because the Mediterranean region has a long history of volcanic activity. Pliny himself died either of asphyxiation from volcanic gases or of heart failure when he ventured into the city of Pompeii in the year 79 to make observations during the eruption of Mount Vesuvius.

Rock and Mineral Resources

Long before the Phoenicians discovered how to make glass, early humans used obsidian, a natural glass pro-

(a)

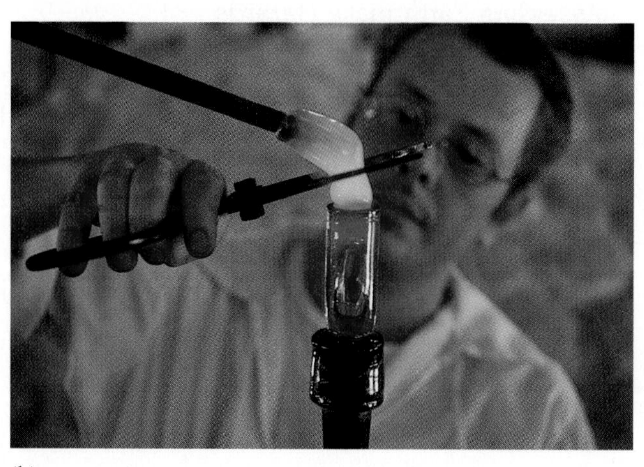

(b)

FIGURE 5-1 (a) Resistant to weathering and to the abrasion associated with transport in water, grains of quartz are common in river and beach sands. (b) Quartz sand or crushed quartz is melted in a furnace, then blown or molded—as shown here—in its molten form to produce glass articles. Glass must be worked quickly because at room temperature it freezes nearly instantaneously into a noncrystalline solid.

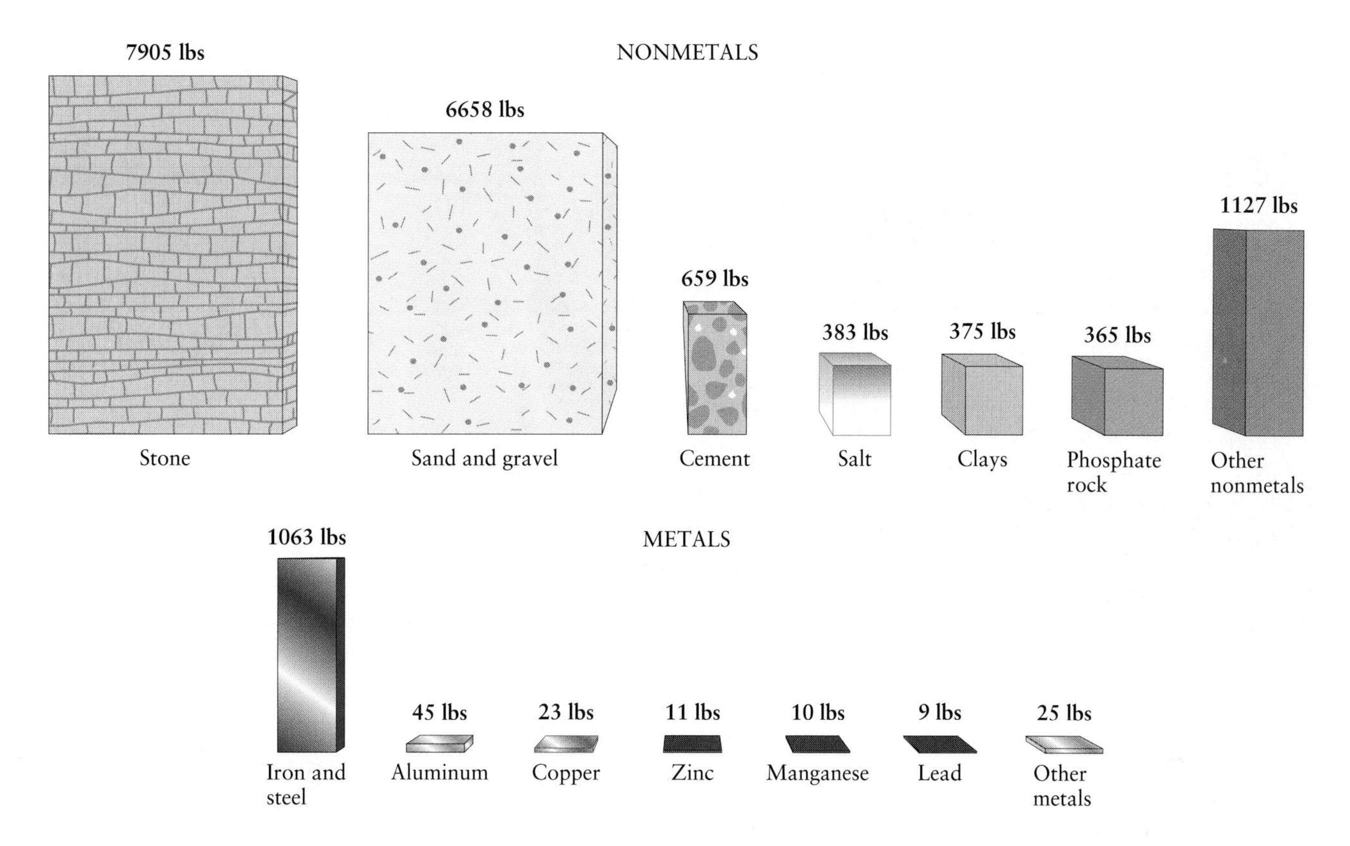

FIGURE 5-2 Use of minerals in the United States per person during the 1980s was more than 18,000 pounds each year. In a single year, an average of nearly 4 tons of stone and more than 3 tons of sand and gravel is used for each U.S. citizen.

duced during some volcanic eruptions, to manufacture blades and arrowheads. The use of different Earth materials by humans through time enabled an early Danish archaeologist, Christian Jurgensen Thomsen, to separate millions of years of human history in Eurasia and Africa into three great technological ages: the Stone Age, which began several million years ago and ended about 5500 years ago; the Bronze Age, which existed from about 5500 to 3600 years ago; and the Iron Age, which began about 3600 years ago. We still are in the Iron Age, because iron remains the dominant metal for industrial purposes. At the same time, so many modern materials are made of glass (from quartz) and ceramics (from clay) that some scientists consider that the present is the New Stone Age.

Today, human beings use large amounts of minerals, especially in industrialized countries. Minerals are needed, among other things, to build roads and farming equipment, to manufacture trucks and trains for hauling food and goods, and to construct homes, appliances, automobiles, and bicycles. With less than 5 percent of the world's population and 7 percent of its land area, the United States consumes about a quarter of the minerals produced. Dividing total annual U.S. nonfuel mineral consumption by the nation's population shows that nearly 10 tons are used per person per year (Figure 5-2). At current rates of use, each newborn American will need a lifetime supply of 619 tons of stone, sand, gravel, and cement, 16 tons of iron, 14 tons of salt, 13 tons of clays, 2 tons of aluminum, almost 1 ton of copper—and much more. Some minerals are mined and processed in the United States, but many are imported. For example, it is likely that some items in your home, school, or workplace are made of copper that was mined from the mountains of New Guinea.

Valuable nonfuel mineral resources include **metals**, such as iron, copper, and aluminum, and nonmetals, such as sand and gravel. Metals form **metallic bonds** in which electrons in outermost shells can flow freely through the array of atoms. For this reason, metals are good conductors of heat and electricity, and are malleable.

Ore Deposits and Ore Minerals

Recall from previous chapters that only three of the industrially important metals—aluminum, iron, and magnesium—occur in proportions greater than 2 percent of Earth's crust. Most others constitute less than 0.1 percent. Copper, for instance, composes only 0.0058 percent of Earth's crust. Fortunately for humans, geologic processes cause minerals to be concentrated in small areas, resulting in rich deposits from which commercially valuable minerals can be extracted profitably. These are called **ore deposits**, and the desired minerals in them are **ore minerals**. All unwanted minerals in the ore deposit are treated as

FIGURE 5-3 In order to mine economically valuable ore deposits, the overburden, or surrounding rock with no economic value, must first be removed. Here at PT Freeport Indonesia's Grasberg mine in Irian Jaya, Indonesia, for example, at least 3 billion tons of overburden must be removed. At modern mine sites such as this one, waste is carefully managed so that disturbed areas can be restored and revegetated when ore removal is complete. Freeport's overburden management program takes advantage of the neutralizing capacity of limestone to minimize the generation of acidic mine waters, saves topsoil for final landscaping, and replants native alpine plants.

waste and left at the mine site as spoil piles, or **tailings** (Figure 5-3).

The abundance of an element in an ore deposit compared with the abundance of the element in the whole crust is known as the element's **economical concentration factor** (Table 5-1). Iron, for example has a crustal abundance of about 5.8 percent, but to be economically recoverable an iron ore must be about 50 percent iron. In other words, the economical concentration factor for iron is about 10 times its crustal abundance. Copper needs to be concentrated in amounts nearly 100 times its crustal abundance in order to be economically recoverable. The global production of copper is almost 10 million tons, which means that almost 100 times that amount, or about a billion tons of rock, probably was mined in order to extract that much pure copper.

Historical images of mining often evoke the picture of a bearded prospector and his mule working their way along a narrow trail in rugged mountains. The coincidence between mountains and ore deposits exists because most ores are associated with the high crustal temperatures found at convergent boundaries where mountain building occurs. Ore deposits along the western coasts of North and South America coincide with convergent plate boundaries and subduction zones, whereas deposits in Europe and Asia coincide with collision zones between the Indian, Australian, and Eurasian plates (Figure 5-4). Ore deposits also are being discovered along divergent boundaries, where seawater is heated at spreading centers on the ocean floor.

Geologists recently discovered evidence that supports their hypothesis about the close association between thermal activity and ore deposition. While drilling deep in the continental crust at the hot spot that stokes the geysers of Yellowstone National Park, they found fresh deposits of sulfide minerals along vents feeding the steaming springs.

Types of Ore Deposits

All metallic and most nonmetallic ore deposits can be classified according to one of four broad types of origin: hydrothermal, igneous, sedimentary, and weathering-related. The first three are discussed here, while the

TABLE 5-1 Economical Concentration Factors of Some Elements

Element	Crustal abundance (% by weight)	Concentration factor*
Aluminum (Al)	8.00	3–4
Iron (Fe)	5.8	5–10
Copper (Cu)	0.0058	80–100
Nickel (Ni)	0.0072	150
Zinc (Zn)	0.0082	300
Uranium (U)	0.00016	1,200
Lead (Pb)	0.00010	2,000
Gold (Au)	0.0000002	4,000
Mercury (Hg)	0.000002	100,000

*The concentration factor is the abundance in the deposit divided by the crustal abundance.

Sources: Data from B. J. Skinner, *Earth Resources*, Prentice-Hall, 1969, and D. A. Brobst and W. P. Pratt, *Mineral Resources of the U.S.*, Prof. Paper 820, 1973.

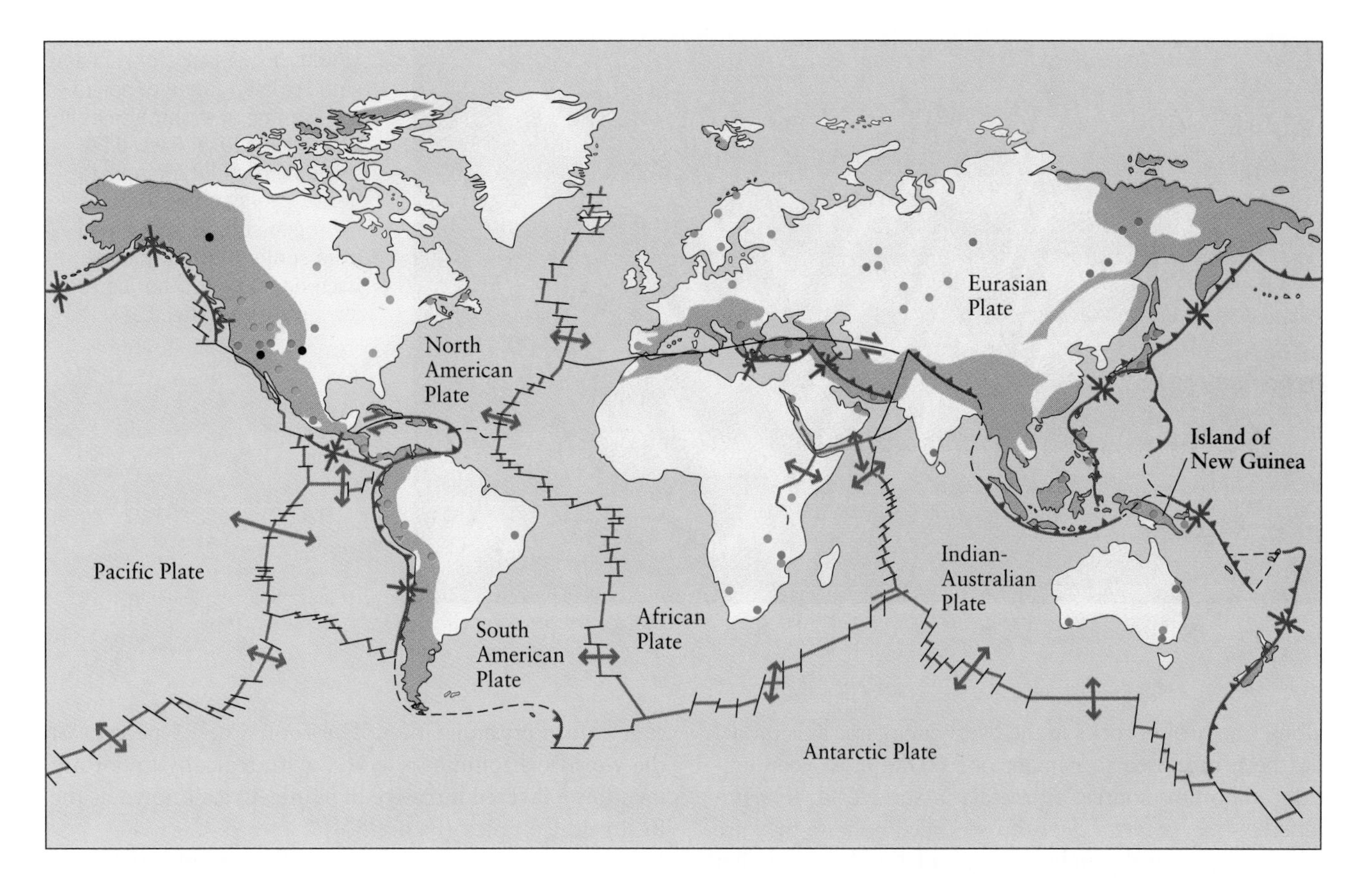

FIGURE 5-4 The distribution of major copper, molybdenum, and gold deposits (blue, black, and red dots) in the world is similar to the locations of Mesozoic- to Cenozoic-age mountain-building activities (pink shading). For example, the island of New Guinea, where the world's largest gold and third largest copper mine is located, occurs along a Cenozoic convergent plate boundary.

fourth, which explains the origin of most aluminum ores and clay, is discussed in the next chapter.

Hydrothermal Ore Deposits In **hydrothermal ore deposits,** mineral ores form by crystallization from metal-rich solutions, or brines, which typically are so hot that they are boiling. The term *hydrothermal* refers to the role of high-temperature seawater and groundwater in the processes of alteration and emplacement of minerals. In the 1960s the richest underwater sulfide mineral deposits known on Earth were discovered during deep-sea drilling in the Red Sea. The Red Sea marks the location of a rift forming along a spreading center between the African and Arabian plates (see Figure 4-14). Sediment cores retrieved from numerous deep basins along the rift indicate that sulfide minerals rich in iron, zinc, and copper are crystallizing from hydrothermal brines at depths of about 1800 to 2100 m below sea level.

In 1979 geologists in a submersible vehicle exploring the East Pacific Rise spreading center had a glimpse of how ore minerals might develop in hydrothermal environments on the ocean floor. They discovered black "smokers"—tall, mineral-encrusted chimneys, up to 10 m high—venting hot metallic fluids at temperatures greater than 300°C (Figure 5-5). Metals accumulate at rifts when seawater moves downward along fractures in the shattered rocks common at mid-ocean ridges, is heated to temperatures of several hundred degrees Celsius when it comes in contact with hot rocks or magmas, leaches metals from surrounding rocks, then rises by convection to the seafloor and escapes, depositing minerals and forming chimneys as it cools.

With the discovery of active mineral deposition in the Red Sea and along the East Pacific Rise, geologists realized that some ore deposits on land must have formed this way in the past and been moved toward convergent boundaries as the seafloor spread. On the island of Cyprus in the Mediterranean Sea, volcanic rocks now recognized as remnants of an ancient seafloor-spreading center were thrust upward during plate collision, pushing rich deposits of copper sulfide and other ore minerals to the surface. Copper takes its name from Cyprus, which holds some of the world's most ancient mines and still earns substantial revenues from its copper, chromite, and iron deposits.

Hydrothermal activity also occurs at subduction zones, where metals can be remelted in the descending oceanic plate, rise as hot fluids (250° to 500°C), and be deposited

FIGURE 5-5 Black smokers and sulfide-rich mineral deposits are forming at a vent along the East Pacific Rise. Bacteria in this environment are able to derive energy from oxidizing hydrogen sulfide rather than from sunlight. In turn, these bacteria support other bottom-dwelling organisms.

along fractures in rocks in the overlying plate. This model has been proposed to explain one of the most economically important sources of metals in the world, *disseminated porphyry metal deposits,* so called because they are widely dispersed throughout mazes of fine fractures that permeate large volumes of porphyritic igneous rocks. Porphyry is a type of igneous rock characterized by a distinctive texture of both large and fine crystals. Porphyritic rocks permeated with fractures containing low-grade (low concentration but minable) copper, molybdenum, silver, gold, lead, zinc, and other metals are common in the mountains of southwestern North America, western South America, the southwest Pacific, Australia, New Guinea, and Indonesia (see Figure 5-4).

An example of a tectonically controlled hydrothermal deposit in continental crust is zinc ore in the central and eastern United States. When North America collided with Europe and Africa about 300 million years ago to form Pangaea, uplift of the Appalachian Mountains forced groundwater to flow west to what is now the Mississippi Valley. The water was highly concentrated in zinc and lead from its interaction with igneous plutons in the mountain-building region. Metals were deposited in cavities within the carbonate rocks through which the groundwater flowed, leaving behind the zinc and lead deposits now being mined in Missouri, Oklahoma, and Kansas.

Igneous Ore Deposits Generally associated with magma chambers along the roots of island and volcanic arcs in zones of plate convergence are **igneous ore deposits**. Igneous ore deposits form in association with igneous processes and rocks. In magma chambers, mafic magmas can segregate into rich layers of nickel, copper, and iron sulfides as well as chromium and platinum minerals. Segregation and layering in a magma chamber sometimes occurs when dense sulfide-rich liquids sink to the bottom before crystallization. The source of 98 percent of the world's chromium reserves, a strategically important metal, is a **layered intrusive** in South Africa known as the Bushveld Complex (Figure 5-6).

Nickel is found in igneous ore deposits of a type that no longer forms today. Nickel-bearing ultramafic volcanic rocks called *komatiites* formed billions of years ago, when lavas from mantle magmas rich in magnesium and nickel erupted on the floors of ancient ocean basins. These lavas no longer form near Earth's surface because the planet now has lost so much heat to outer space that it cannot generate a temperature high enough to melt komatiite rocks at shallow depths. Komatiite formation is an example of a process that was active in early Earth history but which essentially ended, in this case, about 2.8 billion years ago. South Africa, Australia, and Canada are among the few countries where komatiites have been found.

The deepest magmas known to have erupted at Earth's surface are the source of diamond, the hardest mineral. Whereas most minerals crystallize from magmas at shallow depths, diamond is a high-pressure mineral, formed from carbon at depths as great as 150 to 300 km. By some unknown process, ultramafic lavas and diamond crystals are ejected to the surface at extremely high velocities, forming near-vertical pipes, up to 200 m in diameter, known as *kimberlites.* Diamond-bearing kimberlite pipes exist in only a few localities. They were first found in, and named after, Kimberly, South Africa. As pure carbon, the diamond is worth no more than a few dollars per kilogram, but as natural diamond it is one of the most valuable minerals that exists.

Sedimentary Ore Deposits Most sedimentary metallic ore deposits can be traced to mild hydrothermal events, usually at temperatures less than 250°C. Examples in-

FIGURE 5-6 The dark layers in the layered igneous intrusive Bushveld Complex in South Africa are chrome ore.

clude the heating of sedimentary rocks adjacent to igneous intrusions. These low-temperature events sometimes cause fluids rich in metals to migrate through the pores of sedimentary deposits, leaving behind cements and bands of metals that become ore deposits.

Sedimentary ore deposits not associated with thermal events include ore minerals that accumulate in streams and rivers. In India, diamonds have been found in modern stream gravels as well as in one of the rock types drained by the streams, a sedimentary conglomerate 550 million to 600 million years old. The diamonds in the conglomerate presumably were deposited after being eroded either from an even older sedimentary rock or from a kimberlite pipe of unknown location, but the source never has been found. Because diamond is relatively dense (~3.5 g/cm^3) and very resistant to abrasion (hardness of 10), it accumulates in low-velocity pools along modern streambeds. Such accumulations of mechanically segregated ore minerals are called **placers**.

A number of dense minerals, including gold, magnetite, and chromite, occur in placer deposits. Although gold is soft, with a Mohs hardness rating of only 2.5 to 3, it is one of the densest minerals known (~19.3 g/cm^3), and even tiny nuggets and flakes will settle from water flow in streams (Figure 5-7). In comparison, the density of quartz, the most common mineral in stream sands and gravels, is only 2.6 g/cm^3. In addition, gold is very resistant to corrosion, and thus likely to be preserved in a placer deposit. The California gold rush of 1848 started when gold nuggets were found as placer deposits, but miners soon traced them to their sources, hydrothermally deposited mineral veins in the Sierra Nevada that miners called the Mother Lode.

Another type of sedimentary ore deposit that is not associated with thermal events is the crystallization of ore minerals in water at low temperatures (<80°C). These chemical sedimentary deposits are much more common than mechanically segregated placers. Chemical sediments can crystallize from closed lakes or restricted waterways when water evaporates. For example, the floor of Death Valley, California, is covered with salt crystals crystallized from waters draining into the closed lake basin from surrounding mountains (see Figure 4-25). Some of the ions in the water combine to form minerals such as borax, which contains the elements sodium and boron, and is used to make cleansing products and heat-proof glass. Other commonly mined evaporite minerals are

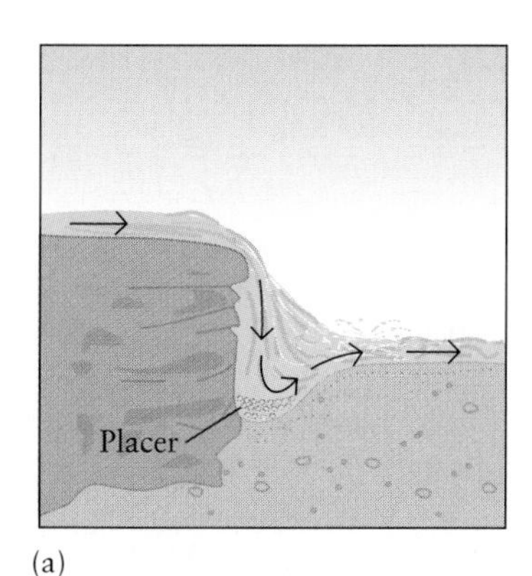

(a)

(b)

FIGURE 5-7 Ore minerals eroded from their source can accumulate as placer deposits at places along streams where velocity is low. (a) In potholes below waterfalls, the water slows and drops high-density particles. Placer activities can be very disruptive to stream ecosystems. (b) During many gold rushes, including those in Alaska and California in the 1800s and South America in the 1900s, miners used pans, sluice boxes, and dredging equipment to search for placer gold in streams. Here in Venezuela, miners are panning for gold.

table salt (NaCl), gypsum ($CaSO_4 \cdot H_2O$), which is used for plaster and drywall, limestone ($CaCO_3$), which is used for cement and building stone, and phosphate (PO_4^{2-}), which is used for fertilizers.

Chemical sedimentary rocks known as *banded iron formations* are the most important source of iron (Figure 5-8). The deposits are so named because the iron occurs as an oxide mineral (hematite) in reddish bands interbedded with light-colored layers of chert, sometimes in formations up to 1000 m thick. The deposits formed when dissolved iron (Fe^{2+}) was washed from continents to shallow ocean basins, where it oxidized to its insoluble form (Fe^{3+}) and crystallized as iron oxides (Fe_2O_3).

All banded iron deposits are more than 2 billion years old, and no equivalent deposits are known to have formed since then. The probable reason is that the composition of Earth's atmosphere changed during its evolution. Two billion years ago the atmosphere contained far less oxygen than it does now, largely because photosynthetic organisms (which release oxygen) had not yet evolved, so iron eroded from continental rocks remained in the reduced state, and thus soluble, until it reached oxygenated ocean water, where it oxidized and crystallized to form banded iron deposits.

Depletion of Mineral Resources

Although black smokers on the seafloor continue to produce sulfide ores and evaporite minerals such as table salt continue to form, we have seen that some ore-forming geologic processes no longer take place because other Earth systems have changed. Hematite no longer collects in banded iron formations because the modern atmosphere is rich in oxygen that reacts quickly with exposed iron. Nickel-bearing lavas no longer occur because Earth's interior has cooled. In general, then, mineral ores are nonrenewable resources. Only a finite amount of them exists, some are no longer forming, and those that still can be produced by geologic processes cannot match rates at which humans extract them.

Known mineral deposits that can be mined profitably with existing technologies are called **reserves** (Figure 5-9). A **resource**, in contrast, is the entire amount of a material known to be available. The boundary between recoverable reserves and subeconomic resources—those with extraction costs greater than their market value—shifts as more efficient mining technologies are developed, existing reserves are depleted, and the price of a particular resource increases. New resources are most likely to be discovered where some mineral resources already are known to occur, so possible ore deposits in such locations are considered hypothetical resources.

Most ore minerals are rare, extraordinarily localized, difficult to find, and used at rapidly increasing rates worldwide. Consider, for example, the production of gold. More gold has been produced in the Witwatersrand basin of South Africa than has been discovered on all the rest of Earth, yet the area of the basin is smaller than that of Los Angeles County. As small areas with rich concentrations of rare ores are mined, is it possible that the world will face a sudden scarcity, or will new resources be found?

Although it is not possible to predict the future with certainty, enough is known about ore minerals to indicate that a major mineral resource crisis is not likely to occur. First, Earth is made of minerals, so a source of minerals will be available as long as the planet exists. Second, the technology for locating and extracting minerals improves constantly, so that deposits once considered subeconomic now are being mined. Finally, when depletion of a mineral seems imminent, ways of using substitute materials, or increasing the use of recycled materials, are developed.

Here are some examples that support these statements. Before 1988 the large gold and copper ore deposit on the island of New Guinea, shown at the beginning of this chapter, was not known to exist, and the technology needed to mine it was not developed until about the same time. Similarly, a large copper ore deposit in Chile was not discovered until 1981. At the time of these two dis-

FIGURE 5-8 Banded iron formations are sedimentary layers of light-colored siliceous chert and reddish iron oxide minerals. They contain the world's largest source of minable iron ore.

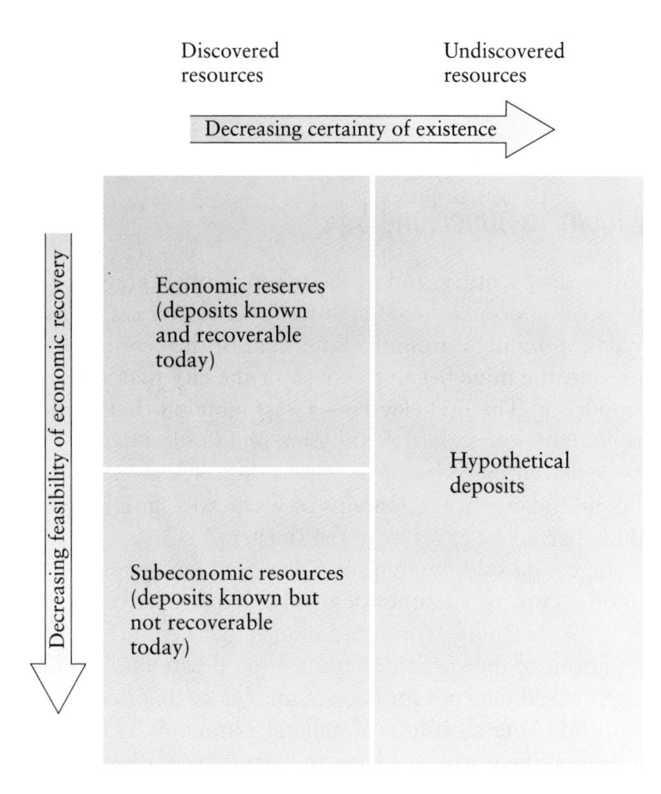

FIGURE 5-9 Anything of use to people is a resource, but some resources are already available to us and can be extracted economically, whereas others have not yet been discovered, or it is too costly to mine them. Resources are classified according to the certainty of their existence and the economic feasibility of their recovery.

coveries, concern over a shortage of copper, which is used for telephone and electrical wiring, already had been alleviated somewhat by the development of fiber-optic cables for communication lines. Fiber-optic cables are made of quartz, and there is plenty of quartz in Earth's crust.

While we are not likely to run out of minerals in spite of the escalating demand, the environmental impact of increased mining is cause for concern.

Environmental Impact of Mining

In the United States, less than 1 percent of the land surface has been altered by mining minerals and coal. More than that area has been affected, however, if consequences such as air pollution from smelting, or water pollution from abandoned tailings and mine tunnels, are included in the environmental assessment (Box 5-1). Spoil piles contain toxic metals and sediments that can enter streams if the piles are not carefully revegetated and stabilized. The environmental impacts associated with mining activities were described in the year 1556 by Agricola in *De re metallica:*

> [T]he fields are devastated by mining operations . . . the woods and groves are cut down. And when the woods and groves are felled, then are exterminated the beasts and birds. . . . Further, when the ores are washed, the water which has been used poisons the brooks and streams, and either destroys the fish or drives them away.

Mining and its environmental impact have changed for the better since the time of Agricola, and there is hope that the tragedies described by him will cease even while mining continues to supply the world with minerals.

Changes in the Mining Industry In many nations, including the United States, the 20th century has been a period of increasing regulation of activities that damage the environment. There are federal and state standards for reclaiming land, controlling emissions to air, soil, surface water, and groundwater, and disposing of hazardous wastes.

The Clean Water Act of 1972 and its amendments (see Chapter 7) require that mining activities discharge close to zero pollutants into surface water. Responsible companies comply. Others have declared bankruptcy, leaving toxic sites in the hands of officials and taxpayers. A recent example is the Summitville mine in the Rocky Mountains of Colorado. In the 1980s a company excavated rock containing small amounts of gold from an open pit, crushed the rock, placed the crushed material on leach pads, sprayed dilute cyanide on the ore to leach gold from it, then collected the gold-laden solution and extracted the gold. This technique is acceptable and widely used. Unfortunately, metal-laden and toxic water leaked from the system at many points, polluting streams that are used for fishing, farming, and drinking water. The federal government is assisting in the cleanup, but the total cost will be hundreds of millions of dollars, most of which will come from taxpayers.

The Clean Air Act of 1970 and its amendments (see Chapter 9) have reduced air pollution from mining activities, but reducing emissions to zero is difficult. Many metals occur in sulfide minerals, and a by-product of separating sulfur from the metals by smelting is sulfur dioxide gas. This gas is vented through smelter stacks to the atmosphere, where it combines with water and oxygen to form sulfuric acid; the contaminated rainwater falls as *acid rain.* To reduce atmospheric emissions, old smelters are upgraded by installation of scrubbers to their stacks. These pollution-control devices contain calcium oxides and hydroxides (from limestone) that combine with and trap sulfur dioxide gas and particles in the stacks before they are released to the atmosphere. New smelters have low emissions compared with older ones. A new smelter built by Utah Copper emits only about 6 pounds of sulfur dioxide for each ton of copper produced. Older smelters emitted more than 100 pounds of the gas per ton of copper.

A lasting problem of mining is *land reclamation*—restoring the land surface, as much as possible, to its

5-1 Case Study

Butte, Montana—From Boom Town to Superfund Site

Butte, Montana, is situated on what may be the richest hill on Earth, a hill underlain by sulfide ore that in places has a concentration of 50 percent copper. Underground veins of copper up to hundreds of meters thick, discovered in the late 1800s, were chipped away and hauled up to the surface along what eventually became almost 16,000 km of tunnels and shafts. As the world's appetite for copper to make electrical wire, pipes, motors, weaponry, and other trappings of the Industrial Age increased, the hill around Butte produced more than 9 billion kg of copper—a third of the U.S. demand and a sixth of the world's—in the first part of the 20th century. The human price was high: In open tunnels where ore once existed, at least 2000 men died in accidents and another 20,000 were seriously injured or disabled. The additional number who died from mining-related lung diseases is unknown.

Anaconda Copper, one of the world's most powerful mining companies, owned the Butte mine during most of the 20th century, and as the mine's tunnels grew deeper and deeper the company shifted to less expensive open-pit mining during the latter half of the century. As a result, the mine began to consume the city that surrounded it. The Berkeley Pit—a vast mine pit that grew to nearly 1 km wide, 1.5 km long, and 0.5 km deep—has replaced one edge of the city of Butte. Surrounding the pit and city are mountains of waste rock and hills made barren by gases from the smelters.

Anaconda sold the mine and degraded landscape to an oil company, Atlantic Richfield Company (ARCO), in 1977. The mine works and pumps that kept groundwater out of the vast pit continued to operate until 1983, when world demand for copper, and hence its price, dropped. After a century of mineral extraction by human endeavor, the mine fell silent, and groundwater began flowing into the maze of abandoned tunnels, flooding them and the Berkeley Pit. The pit will continue filling until its water is level with the local groundwater surface.

As groundwater flows through the maze of tunnels and shafts beneath the Berkeley Pit, it leaches sulfur and poisonous metals from the rock. As a result, the pit's water is extremely acidic and toxic.

Then, sometime just after the year 2021, the water in the pit will be more than 300 m deep. It also will seep outward into the pore spaces of adjacent sediments and rocks that supply streams with some of their flow.

With the Berkeley Pit groundwater comes sulfur dioxide, cadmium, lead, and arsenic leached from sulfide minerals that permeate the hill's bedrock. After combining with oxygen from the atmosphere, the mixture becomes a toxic brew rich in sulfuric acid and poisonous metals. Nothing can be done to stop the groundwater from pouring through the old tunnels or from leaching toxins from the mineralized rock. In the winter of 1995 the lake was so acid—with a pH of 2.3, similar to that of pure lemon juice—that it killed hundreds of Arctic snow geese on their migration to California. Autopsies revealed that tissues lining their esophaguses and digestive organs were burned and blistered.

In 1980 the U.S. government passed the Comprehensive Environmental Resource, Compensation, and Liability Act, commonly known as the Superfund law because of the large fund of money set aside to help finance the cleanup of hazardous waste sites. Butte is the largest of more than 1200 Superfund sites identified by the U.S. Environmental Protection Agency (EPA) under this act.

The EPA is working with state and local officials and ARCO to build equipment that will pump and treat water from the Berkeley Pit as it approaches its full depth, but further remediation of the landscape is not feasible. At this point, scientists and officials are still trying to decide what is best to do. The toxic wasteland that took only a few generations to create will take many more, plus hundreds of millions of dollars, to clean up.

Source: Much of this discussion was drawn from Edwin Dobb's article, "Pennies From Hell," *Harper's*, October 1996.

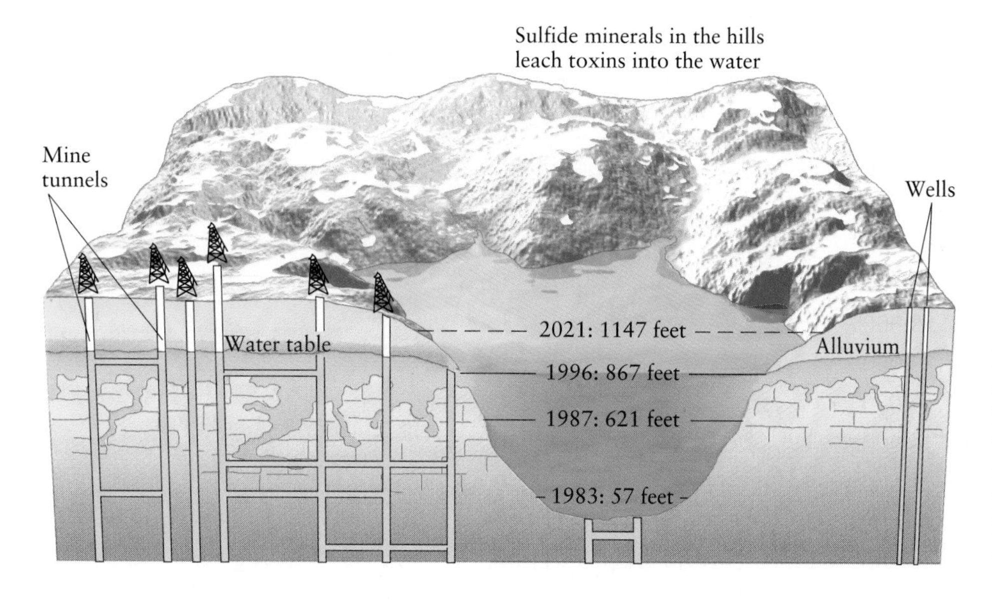

Berkeley Pit, a large open-pit copper mine abandoned in 1983, has been filling with groundwater ever since. The water surface is rising about 3 m per year, and will reach the level of the surrounding groundwater surface about the year 2021.

undamaged state. Removing ore minerals that occur in concentrations of sometimes less than 1 percent of a deposit requires excavation of tremendous volumes of rock, most of which will be waste. In some cases, the mined area is small enough that reclamation is effective. As of the 1990s, about 500,000 acres of land in the United States was affected by mining activities but less than one-third had been reclaimed.

Loading billions of tons of crushed rock back into the voids of large, open-pit mines such as the Freeport copper mine in Indonesia or the Anaconda-ARCO pit in Butte, Montana (see Box 5-1) is not likely to happen. For this reason, there is much hope in new methods, such as *in situ mining,* which extract metals from ore deposits by leaching them "in place," without excavating the ore-bearing rock (Box 5-2).

Using Mercury in Mining Activities Mercury, itself a metal, helps miners extract other metals. Mercury has a strong tendency to gather, or *amalgamate*, gold. Since ancient times, miners have known that blobs of liquid mercury can be used to gather even the tiniest gold flecks. The two metals can be separated afterward merely by squeezing the intermixed mass in a wet leather bag, because liquid mercury escapes through the porous leather, leaving gold.

The drawback of using mercury is that it is highly poisonous to most organisms. In humans, it causes extreme damage to the central nervous system. The residence time of mercury in stream and lake water is short (days to several years), but it has an affinity for streambed sediments and can remain attached to them for many years. Mercury is used so commonly in mining and industrial activities that the flux of mercury from continents to oceans is now 4 times greater and the flux from continents to the atmosphere is now 275 times greater than in pre-industrial times (Table 5-2).

Because mercury is stored and not excreted by living cells, it accumulates at each step in the food chain. Aquatic plants absorb mercury from water and concentrate it in their tissues. Fish that eat the plants further concentrate the mercury. Birds and other carnivorous animals eat the fish, continuing the process. This effect, known as **biological amplification**, means that organisms at the top of the food chain, such as humans, ingest concentrations of mercury that are much higher than those in the water itself.

When environmental regulation is lax or nonexistent, the impact of mining can be catastrophic. Gold rushes in Brazil in the 1980s and Venezuela in the 1990s, like those in 19th century western United States, are marked by dangerous exposure to mercury. South American miners drink from mercury-laden streams and eat mercury-laden fish. On the island of New Guinea, too, tropical streams are clogged with sediment and contaminated with mercury from early years of unregulated mining. There, thousands of local tribal people rioted in protest in 1996; an

TABLE 5-2 Sources of Atmospheric Emissions of Metals

Element	Natural rate (from geologic processes) (metric tons/year)	Anthropogenic rate (from human activities) (metric tons/year)	Anthropogenic/natural ratio
Aluminum (Al)	48,900,000	7,200,000	0.15
Iron (Fe)	27,800,000	10,700,000	0.38
Manganese (Mn)	605,000	316,000	0.52
Cobalt (Co)	7,000	4,400	0.63
Chromium (Cr)	58,000	94,000	1.6
Nickel (Ni)	28,000	98,000	3.5
Copper (Cu)	19,000	263,000	13.8
Zinc (Zn)	36,000	840,000	23.3
Arsenic (As)	2,800	78,000	27.9
Silver (Ag)	60	5,000	83.3
Mercury (Hg)	40	11,000	275
Lead (Pb)	5,900	2,030,000	344

Source: From R. J. Lantzy and F. T. Mackenzie, "Atmospheric Trace Metals: Global Cycles and Assessment of Man's Impact." *Geochim. Cosmochim. Acta* 44: 815–828, 1979.

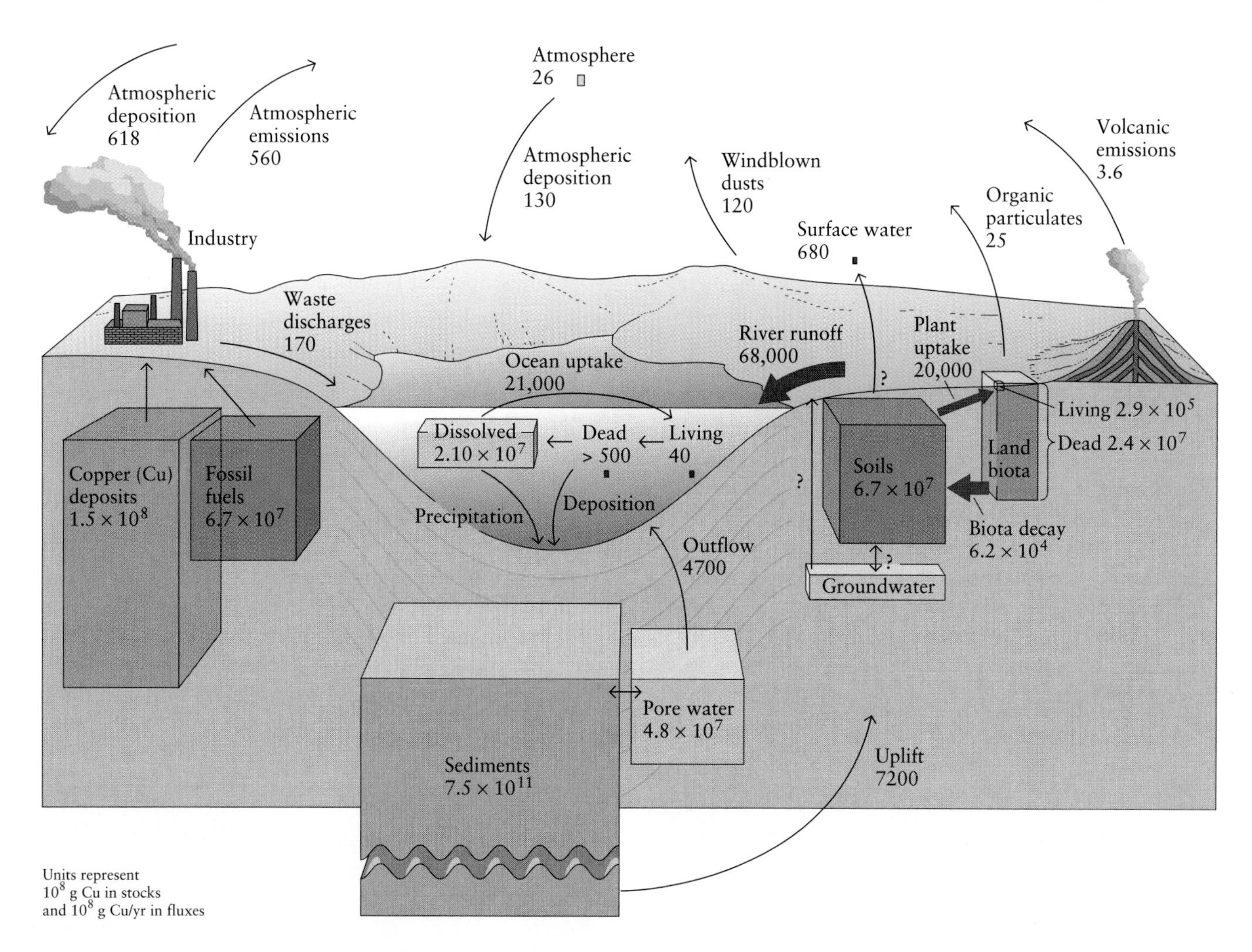

FIGURE 5-10 The biogeochemical cycling of a trace metal—copper—in Earth systems. Stocks (boxes) and fluxes (arrows) are drawn to scale, relative to one another. Unknown fluxes are represented by question marks. Since the amounts of copper range very widely, some of them cannot be represented adequately here. Small stocks of copper are represented by tiny boxes. Fluxes of about 8000×10^8 g/yr and less are represented by lines of the same width because they are very small compared with the largest flux.

Australian human rights group reported that dozens of them disappeared or were murdered. The consequences of despoilment, both environmental and political, can be brutal indeed.

Metals Cycle Through Earth Systems Because of mining, elements that occur in localized concentrations in the lithosphere are *mobilized,* or released, to other Earth systems over wide areas. This is one of mining's greatest effects on the environment. Because metals have a strong tendency to form reversible bonds with many other elements (including carbon), they occur throughout Earth systems, cycling through them in a variety of forms. Elements commonly released to the hydrosphere, atmosphere, and pedosphere during mining include lead, mercury, and chromium. Lead and mercury can damage the central nervous system of humans, while chromium is a potent carcinogen. Increased concentrations of a number of elements, including lead and mercury, have been identified in sediments in lakes and bays and in layers of ice in Antarctica and Greenland. In some cases, global emissions of metals from human activities, such as mining and burning of fossil fuels, have resulted in anthropogenic (human-induced) emission rates that are tens to hundreds of times those of natural background values associated with geologic processes (see Table 5-2).

Copper provides a typical example of how a metal cycles through Earth systems. Like other metals at Earth's surface, copper occurs either concentrated in ore deposits (more than 100 to a few thousand parts per million) or disseminated throughout sediments and rocks in the lithosphere (Figure 5-10). From this reservoir, copper

5-2 Geologist's Toolbox

Mining Methods and Their Environmental Impacts

Human beings have been mining and smelting metals for nearly 6000 years, but with increasing demand for metals and the advent of environmental regulations in the 20th century, the ways that ores are mined and processed are changing rapidly. In conventional mining, ore minerals are excavated, crushed, and then smelted with other compounds in order to produce liquid metal that can be separated from the unwanted residue. Excavation occurs either underground in tunnels and shafts, or above ground in open pits. Because most metals make up less than 2 percent of the ore deposit, the amount of material excavated and left as waste is many times that of the actual metal product. The environmental impacts of conventional mining include:

- A landscape that is difficult and costly to reclaim
- Large amounts of tailings from which metal-rich, acidic waters might drain into groundwater or, if nearby, a stream
- Emission of sulfur dioxide gas into the atmosphere from smelting of sulfide ores

These impacts can be minimized, but at great cost and effort.

A recent mining technique called heap leaching is used to extract metals from oxide and low-grade sulfide ores disseminated over wide areas. In heap leaching, the ore is excavated from open-pit surface mines, crushed, spread on a pad underlain by a protective liner, and

(a)

(b)

In heap leaching (a), ore is excavated, crushed, spread on a protective barrier, and sprayed with solvent. This pile of crushed copper ore is being sprayed with sulfuric acid. (b) The next step in heap leaching is to pump the dissolved copper to a plant where the solvent is removed.

can be mobilized, largely by human activities, weathering, and volcanic activity. Most copper released to the atmosphere is the result of industrial emissions or wind erosion of weathered material. Very little comes from volcanic emissions.

During weathering, oxygen in the atmosphere and hydrosphere bonds with copper to form compounds and ions that are transported readily in solution in water runoff from hill slopes and in streamflow. Some copper is stored in carbon molecules in terrestrial systems,

sprayed with a solvent that leaches the metal from the ore. Cyanide is the solvent used to leach gold. Sulfuric acid is used to remove copper. As sulfuric acid leaches copper from the ore, it forms a blue, copper-rich solution that is sent to a plant where the copper is extracted. As with conventional mining, this technique requires that much material be excavated and left as tailings, but no heating or smelting is required. In some cases, solvents have leaked from storage tanks, or seeped from leach pads, into streams and groundwater, but with care such leaks can be prevented.

A mining method currently being tested requires no excavation at all and might substantially reduce the environmental impact of much mining. In situ (Latin for "in place") mining extracts metal from an ore deposit where it is found, with no excavation. In situ mining uses the same techniques as heap leaching, except that the solvent is injected into the ore deposit through underground wells.

In a pilot project for in situ mining in Arizona, sulfuric acid injected into a deep well travels through the fractures and veins of rock containing copper oxides that occur from 366 to 670 m below the surface. The acid leaches copper from the ore-bearing rock. The copper-bearing solution is drawn up to the surface by other wells, and the copper is extracted from solution by chemical means in a plant. With federal funding to help test this procedure, it might become widely used if fully successful. It leaves no excavation scars or tailings. The only waste is the spent solvent.

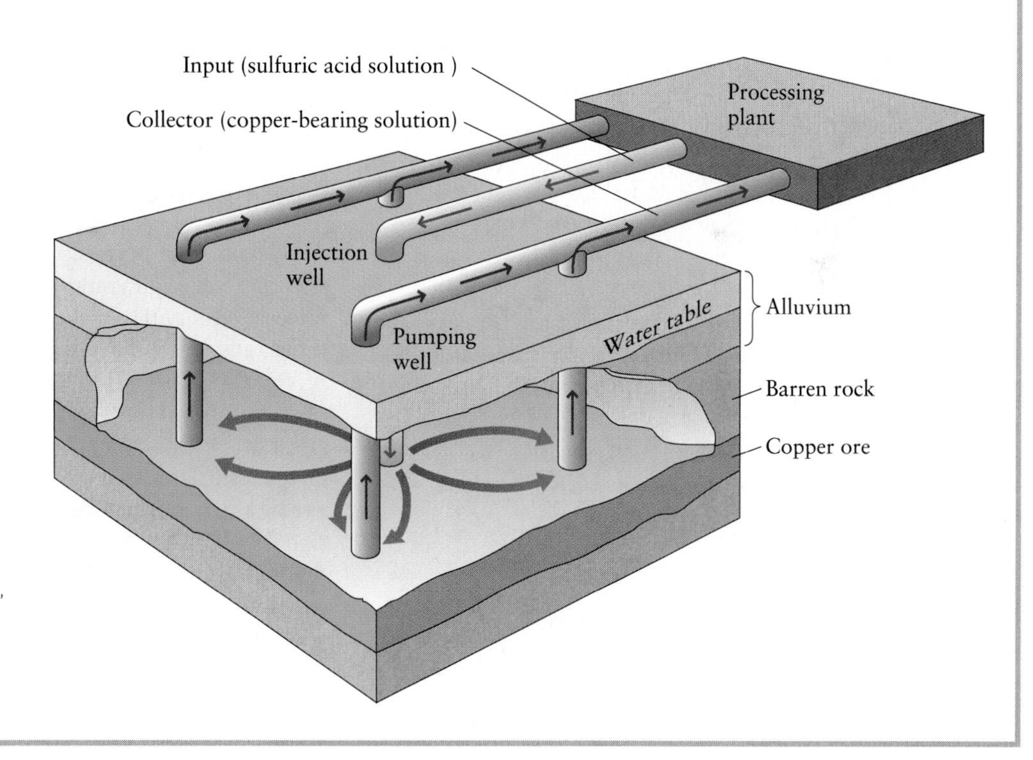

During in situ mining, solvents are injected directly into the ore deposit and the metal-bearing solution is pumped out; damage to land, surface water, and air is minimized. The impact of this procedure on groundwater still is being evaluated.

largely in soils and living organisms. Plants draw copper from soils to build cells, then return it to the soil when they—or the animals that eat them—die and decay. Once released from rock and sediment by weathering processes, most copper is transported to the ocean reservoir in dissolved form or as solid particles in suspension. In the ocean, some copper remains dissolved and cycles through marine organisms, but most is buried with seafloor sediments, thus returning to the lithosphere reservoir. Eventually, because of plate tectonics activities,

FIGURE 5-11 Although the 19th century copper smelter near Copperhill, Tennessee, is gone, its legacy remains in the hillslopes barren of vegetation and in soils contaminated with excess copper.

some of this copper will return to Earth's surface by uplift or volcanism, where it once again becomes mobilized and begins the cycle anew.

▸ Cycling of Copper Among Earth Systems

The residence time of copper in different Earth systems is important to assessing the environmental impact of mining activities. The residence time of copper in Earth system reservoirs can be calculated as:

$$\text{Residence time} = \frac{\text{stock in reservoir}}{\text{flow rate (flux)}}$$

In a steady state system, the flow rate of the copper is the total amount flowing either into or out of a reservoir. For example, the residence time of copper in the atmosphere is small:

$$\text{Residence time of copper in atmosphere} = \frac{26 \times 10^8 \text{ g Cu}}{709 \times 10^8 \text{ g Cu/yr}} = 0.04 \text{ yr} \cong 13 \text{ days}$$

In contrast, copper has a much longer residence time in soils:

$$\text{Residence time of copper in soils} = \frac{6.7 \times 10^{15} \text{ g Cu}}{6.2 \times 10^{12} \text{ g Cu/yr}} \cong 1081 \text{ years}$$

In other words, copper is cycled rapidly through the atmosphere (13 days) and slowly through soils (1081 years). An example of the environmental consequence of this difference in residence times is that even after a smelter reduces its emissions of copper to the atmosphere, it will take many years for excess copper to be removed from soils surrounding the facility (Figure 5-11).

- The most commonly used nonfuel mineral resources are localized in concentrated ore deposits by the following geologic process: mineralization along spreading ridges, mineralization along subduction zones, density layering in magma chambers, formation of chemical sedimentary deposits, and accumulation along stream channels.
- We are unlikely to face a severe crisis in mineral availability, because new ore deposits are still being discovered, new technologies diminish the need for some minerals, and new technologies increase the use of recycled material.
- The total land area affected by extracting minerals from Earth's crust is fairly small, but the impact on surface water and the atmosphere is substantial, for mineral extraction and processing release large amounts of metals and acids into these Earth systems unless substantial efforts to remove the pollutants are made.

Volcanic Eruptions

A volcano is an opening into Earth's hot interior, a beautiful and yet frightening glimpse through the crust to the underlying mantle. In the past few centuries, hundreds of thousands of people have died during eruptions of dozens of volcanoes around the world. In the immediate vicinity of an eruption, death and destruction can result from burial under volcanic debris, high heat, or suffocating blasts of hot, dense air masses laden with rock and ash (Figure 5-12). Farther down the slopes of a volcano, the destruction is often the result of swift-flowing, ash-rich mudflows that overtake villages and their inhabitants, as happened in Armero, Colombia, in 1985, when 25,000 people were killed.

Volcanic eruptions are noisy events, but death sometimes comes silently. Pockets of gas can seep from volcanic craters and surround the unsuspecting, cutting off oxygen and life. In 1986, along the edges of a crater lake in Cameroon, 1700 people were killed this way. Volcanic eruptions in coastal or offshore areas can disrupt the seafloor and displace large masses of ocean water, generating huge waves called tsunamis (see Chapter 10) that inundate coastal areas: 36,000 people were killed by such waves during the 1883 eruption of Krakatoa in Indonesia.

FIGURE 5-12 In 1902 the eruption of Mont Pelée destroyed the port of St. Pierre, Martinique, and all but 4 of the 29,000 inhabitants of the town. A slender spine of the volcano remained afterward, barely visible through the dust in the distance.

More than 1000 volcanoes on Earth are active or potentially active; the greatest activity is in Japan, Indonesia, and the United States (Figure 5-13). The locations of volcanoes are intimately related to plate boundaries. About 80 percent are located along convergent plate margins, in volcanic arcs that form a "ring of fire" surrounding the Pacific Ocean. Most of the other 20 percent are located along mid-ocean spreading ridges. These volcanoes contribute about 20 km^3per year of the world's outpouring of lava, whereas arc volcanoes contribute less than 2 km^3per year. The difference, discussed below, is a result of the type of magma produced in each area. Magma is less likely to reach the surface and erupt from arc volcanoes than from spreading-ridge vents.

Magma Types, Eruptive Styles, and Volcanic Landforms

From extensive studies of volcanoes, geologists have developed a fairly good understanding of how magma rises to the surface from beneath a volcano. A volcano is a conduit for molten rock, gases, water, and heat that rise

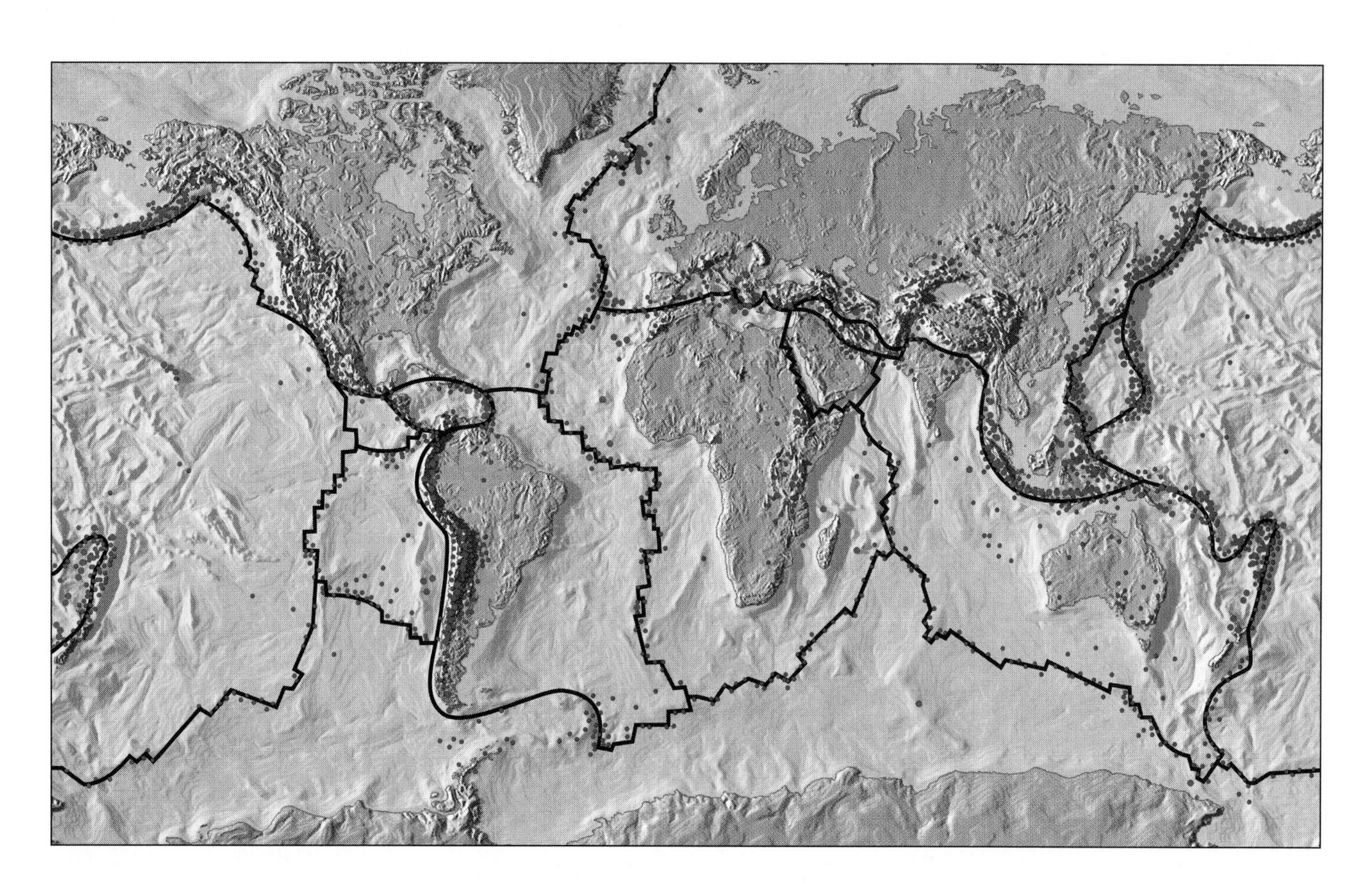

FIGURE 5-13 A global view of Earth and its plate boundaries illustrates the link between plate tectonics, volcanism, and earthquake activity. Most volcanoes (red) are found along convergent and divergent boundaries. The few exceptions are volcanoes at hot spots, such as Hawaii. The Pacific Ocean is rimmed with convergent boundaries and volcanoes. Nearly all earthquakes (purple) occur along plate margins.

(a)

(b)

FIGURE 5-14 (a) When Kilauea volcano erupted in May 1990, low-viscosity, basaltic magma moved rapidly down the volcano's flanks, engulfing buildings and forests. Flames along the edge of the lava tongue are burning organic matter. (b) High-viscosity, andesitic-to-felsic magma erupted from a volcano in Alaska and formed a plug in the volcano's crater.

from the asthenosphere and sometimes from even deeper in the mantle. The general model of a plumbing system for volcanic vents is simple: Magma rises along a thin pipe and fills a reservoir, or magma chamber, in the crust. As the chamber fills, it swells and bulges Earth's surface upward, much like a balloon pumped full of water. After an eruption, the chamber is partly or completely emptied and the crust above the chamber collapses, only to swell again as the magma chamber begins to refill.

Depending on the type of material emanating from a volcanic vent, eruptive styles vary widely and different types of deposits accumulate around the vent. The lower the silica content, the lower the viscosity of the material. The **viscosity** of a fluid is a measure of its resistance to flow—for example, honey has a higher viscosity than water. The cause of differences in viscosity is related to the type and number of molecular bonds in the fluid. Clusters and chains of linked silica tetrahedra increase the viscosity of silica-rich magmas. Basaltic magmas, which contain only about 50 percent SiO_2, have low viscosity, so they are very fluid and easily move upward along magma pipes. As the fluid lava rises, its pressure and temperature drop, and the gases in the lava escape easily into the atmosphere. When basaltic lavas flow out at the surface, they commonly have temperatures between 1000° and 1200°C and can move as rapidly as 100 km per hour, but more typically they flow at only several kilometers per hour [Figure 5-14(a)].

Felsic magmas, with as much as 70 percent SiO_2, have lower melting points and erupt at temperatures between 800° and 1000°C. Felsic magma is very sticky and viscous and flows slowly, often clogging a volcano's plumbing system as it approaches the surface and cools, as shown in Figure 5-14(b). Sometimes felsic magmas cool so much while rising that they don't reach the surface. Eruptions of felsic lava tend to be highly explosive because of the magma's high viscosity and trapped gases. As felsic lava erupts from a volcano, it explodes violently and the shattered bits cool to form debris that ranges in size from ash to building-size "bombs" (Table 5-3). Called pyroclastic debris (from *pyro* for "fire" and *clastos* for "particle"), material ejected from a volcano falls back to Earth's surface and forms layers of volcanic sediments near the volcano. The finest debris, ash, sometimes circles the globe before settling back to Earth.

Intermediate magmas, also known as andesitic because they produce andesite when erupted, contain about 60 percent SiO_2, so are intermediate between basaltic and felsic magmas in their eruptive styles and explosivity. Because andesitic magmas are not as viscous as felsic

TABLE 5-3 Classification of Pyroclastic Debris by Size

Name	Size (mm)
Bomb	>64
Cinder	2–64
Ash	<2

Note: All solid particles ejected from a volcano are called pyroclasts, regardless of magma composition.

magmas, they reach the surface more often, and sometimes flow from volcanoes as lava. However, they move slowly and rarely travel far from the volcano. Much of the material emitted from andesitic volcanoes is pyroclastic debris (see Box 3-1).

For a given magma type, the violence of a volcanic explosion is largely a function of the amount of water vapor contained in the lava. Typically, 70 to 95 percent of volcanic gas is water vapor, with carbon dioxide and sulfur dioxide accounting for most of the remainder. Water vapor inside a volcano is confined at temperatures several times higher than its boiling temperature, resulting in superheated steam that exerts extremely high pressure. Once a magma reaches the surface and the confining pressure of overlying rock is reduced, the water vapor can expand rapidly, resulting in an explosive eruption.

Shield Volcanoes, Stratovolcanoes, and Calderas

The shapes of volcanoes, as well as the force of their eruptions, reflect differences in magma types. Basaltic lavas spread far and rapidly from the vent, building thin sheets of gently dipping flows, one upon the other. With time, a broad dome forms that resembles the convex surface of a shield. The main island of Hawaii is formed by the coalescence of several such **shield volcanoes** (Figure 5-15). In this case, the source of magma is a mantle hot spot. The island—only a few million years old—rises nearly 10 km from the ocean floor, constructed of thousands of lava

(a)

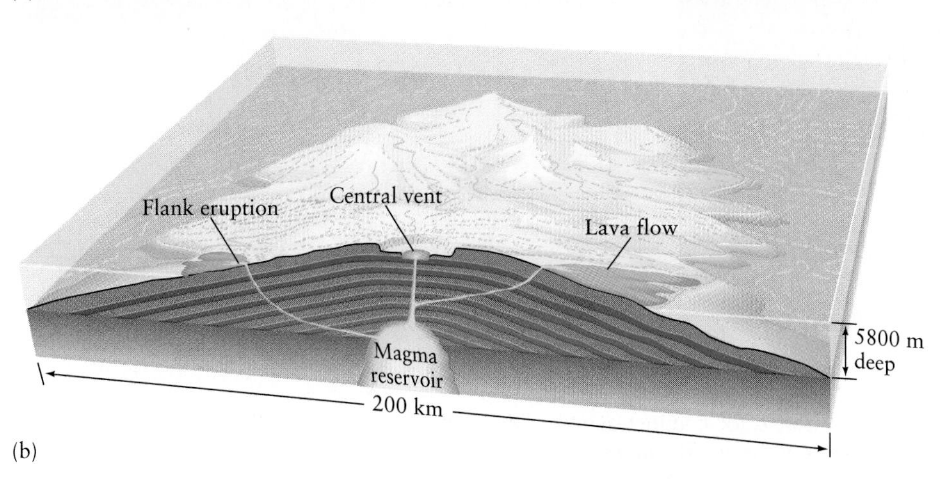

(b)

FIGURE 5-15 Basaltic magma, with its low viscosity, spreads far from where it erupts at the surface, forming thin sheets of volcanic rock. One sheet after another spills down the volcano's flanks, building a thick volcanic dome. (a) The broad, shieldlike surface of Mauna Kea forms the horizon in this view from a nearby cone of recent volcanic cinders. Note the fresh basaltic rock in the foreground. (b) This diagram is representative of the volcanic deposits that have constructed the island of Hawaii.

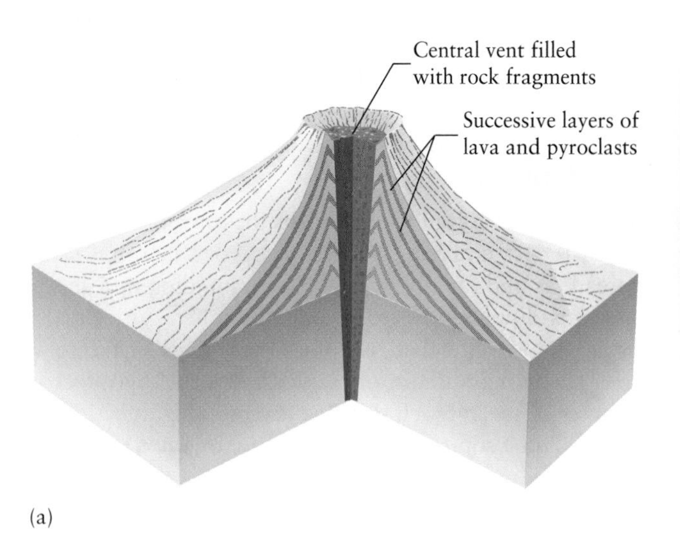

(a)

(b)

FIGURE 5-16 (a) Intermediate (andesitic) magma, with its relatively high viscosity, forms both thick, slow-moving lava flows and pyroclastic debris when it erupts. Together, these deposits build steep, cone-shaped stratovolcanoes, as at Mount Fuji in Japan (b).

flows from the greatest active outpouring of lava on Earth (see Box 4-3).

In contrast, andesitic volcanoes emit both large amounts of pyroclastic material and thick lava flows. With time, the steep flanks of an andesitic volcano are built up by interbedded layers of lava and pyroclasts. The resultant cone-shaped landform is called a **stratovolcano.** Examples of stratovolcanoes include Mount St. Helens, Mount Rainier, and Mount Fuji (Figure 5-16).

A **caldera** is a broad, steep-sided basin from several kilometers to more than 50 km in diameter that forms when so much magma is ejected from the magma chamber that the roof above it collapses. Many oceanic basaltic volcanoes have small craters, fissures, and cinder cones within larger calderas, as at Kilauea (Figure 5-17), but much larger calderas are associated with felsic magmas and continental volcanoes that have not erupted in hundreds of thousands of years. Examples include Yellowstone Caldera in Wyoming and Long Valley Caldera in California (Box 5-3).

Although these caldera-forming continental volcanoes have not erupted recently, they are not inactive. Rather, they are characterized by long periods of quiescence as the magma chamber slowly fills and reinflates. With time,

(a)

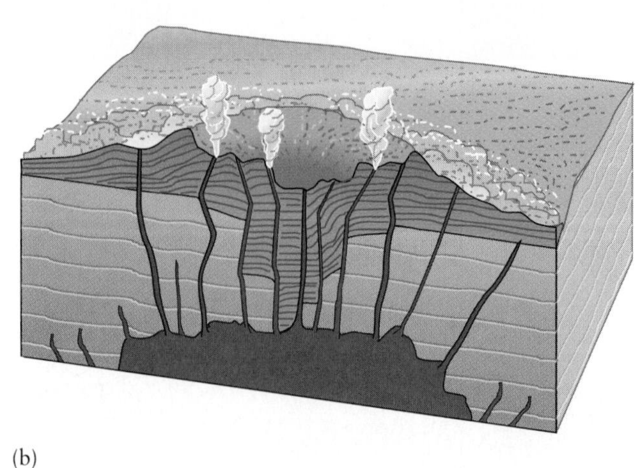

(b)

FIGURE 5-17 (a) The Hawaiian Volcano Observatory is located just outside the rim of a caldera on Kilauea volcano. Scientists from all over the world come here to observe and study active volcanism. (b) Calderas are broad basins that form when so much erupts from a magma chamber that the chamber's roof collapses.

5-3 Case Study

Assessing the Risk of Eruption at Long Valley Caldera

When should the public be alerted about possible volcanic eruptions? In the cases of Mount St. Helens and Mount Pinatubo, the public was alerted and was grateful to be spared the effects of the major eruptions that followed. In the case of Long Valley Caldera in eastern California, however, no eruption followed an alert in the 1980s, tourists were frightened away, and the local economy was damaged. The population was less appreciative of what seemed to them unnecessary meddling.

Long Valley and its resort village of Mammoth Lakes are well known for mountain ski slopes and bike trails, hot springs, and pristine lakes fed with spring snowmelt. The last major eruption, one of the largest in Earth's recent history, occurred 700,000 years ago, forming a caldera 14 km wide and 30 km long. Ash from that eruption is found across much of the United States, and large pieces of debris landed as far away as Missouri. Since then, only small eruptions have occurred, the most recent 500 years ago. Heat flow beneath the surface is high, however, and even provides a source of energy to the local geothermal power plant.

The possibility that the caldera was resurgent, again filling with magma that might initiate collapse, caught the attention of scientists working in the area in the late 1970s. Very frequent earthquake activity began to occur, and the floor of the caldera rose some 40 cm between 1978 and 1982. Obviously, something was stirring inside the volcano. In May 1982, the U.S. Geological Survey (USGS) issued a "potential volcanic hazards" notice, the lowest of three levels of alert at that time. The earthquakes and volcano alert devastated the town's tourist-based economy. In a short time, businesses closed and housing prices fell about 40 percent. In early 1983, earthquake activity and dome uplift continued, and some gases seeped from the ground. Fortunately, no eruption occurred and much of the seismic activity diminished. By the mid- to late 1980s, the region's economy began to rebound.

In 1989, the caldera became active once again, but this time the public was not alerted. Since the previous eruption scare, scientists established a network of instruments in the area to monitor seismicity and ground swelling. Based on the data from monitoring the caldera, scientists concluded that an eruption is not likely to happen before the caldera rises at least another 0.5 to 1 m. They also concluded that the volcano probably will give very strong warnings—such as large earthquakes and substantial tilt of the ground surface—before an imminent eruption. Volcanologists must walk a tightrope between the possibility of failing to warn of a disaster and causing needless alarm.

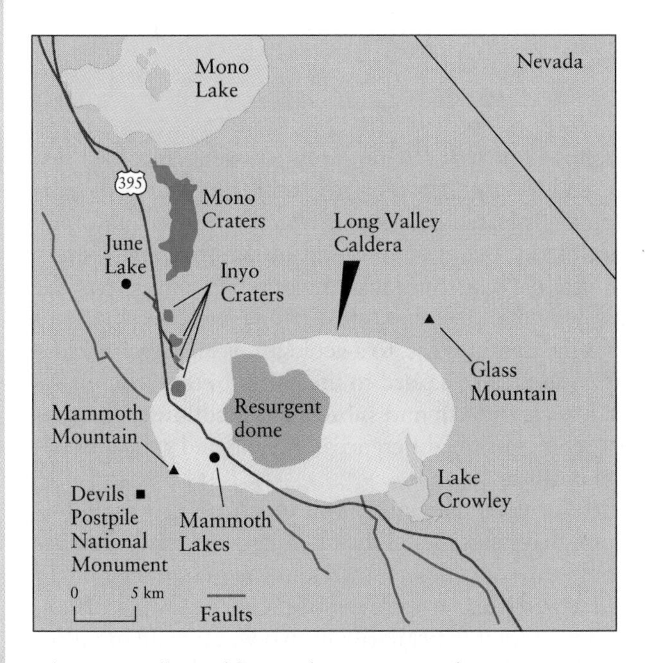

The Long Valley Caldera and its resurgent dome is in eastern California, just east of an ancient volcanic arc that forms the rugged Sierra Nevada. The western edge of the caldera can be seen in the southward-looking view on the right.

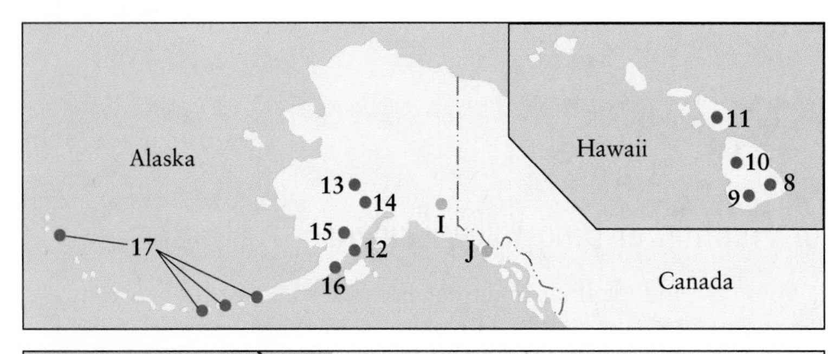

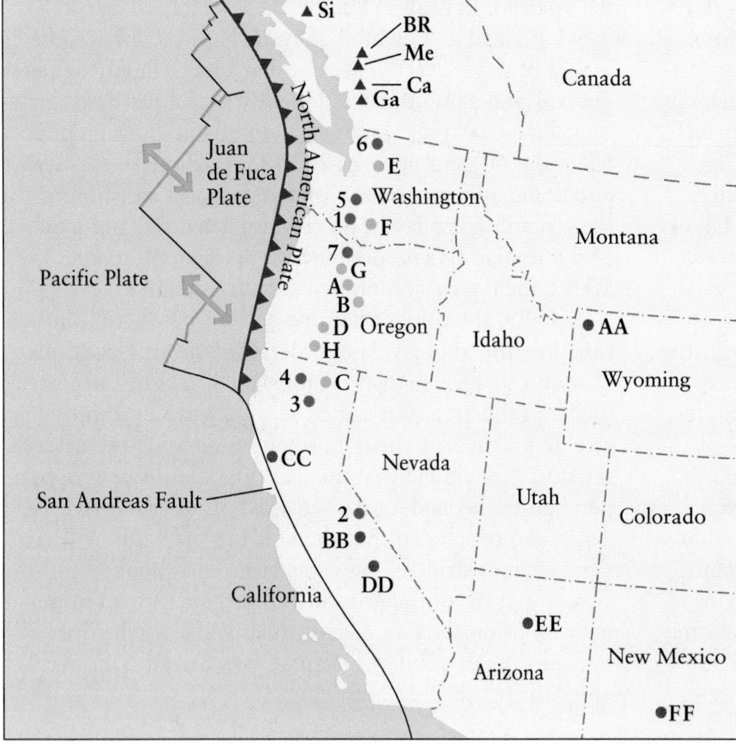

● U.S. volcanoes that have short-term eruption periodicities (100—200 years or less), or have erupted in the past 200—300 years, or both:

Cascades	Hawaii	Alaska
1 Mount St. Helens	8 Kilauea	12 Augustine volcano
2 Mono-Inyo craters	9 Mauna Loa	13 Redoubt volcano
3 Lassen Peak	10 Hualalai	14 Mount Spurr
4 Mount Shasta	11 Haleakala	15 Iliamna volcano
5 Mount Rainier		16 Katmai volcano
6 Mount Baker		17 Aleutian volcanoes
7 Mount Hood		

● U.S. volcanoes that appear to have eruption periodicities of 1000 years or greater and last erupted 1000 years or more ago:

Cascades	Alaska
A Three Sisters	I Mount Wrangell
B Newberry volcano	J Mount Edgecumbe
C Medicine Lake volcano	
D Crater Lake (Mount Mazama)	
E Glacier Peak	
F Mount Adams	
G Mount Jefferson	
H Mount McLoughlin	

● U.S. volcanoes that last erupted more than 10,000 years ago, but beneath which exist large, shallow bodies of magma that are capable of producing exceedingly destructive eruptions:

AA Yellowstone Caldera	DD Coso volcanoes
BB Long Valley Caldera	EE San Francisco Peak
CC Clear Lake volcanoes	FF Socorro

▲ Danger classifications are not available for Canadian volcanoes:

Si Silverthrone	Ga Mount Garibaldi
Ca Mount Cayley	Me Meagher Mountain
BR Bridge River	

FIGURE 5-18 Potentially hazardous volcanoes are found throughout western North America, but U.S. volcanoes have been divided into different categories for danger classification. Some, including Mount St. Helens, Mount Rainier, and Kilauea, erupt every 100 to 200 years, others erupt every 1000 years or more, and some have not erupted in the past 10,000 years.

the chamber might become full enough that the floor of the caldera will bulge upward sufficiently for eruption and caldera collapse to occur again. At Yellowstone, the cycle of caldera collapse and resurgence appears to be between half a million and a million years long. The most recent explosion occurred 600,000 years ago, whereas two previous events occurred 1.2 million and 2.2 million years ago.

Assessing Volcanic Hazards

Most people are surprised to learn that the United States has many volcanic hazards, yet there are 33 active U.S. volcanoes (Figure 5-18). An *active volcano* is one that has erupted during the time of historical record. More than a dozen volcanoes have erupted in the United States in the past few centuries. Consider Mount Rainier, a beautiful, snow-clad volcano that rises several thousand meters above the city of Tacoma near the coast of Washington (Figure 5-19). Although it has been dormant for at least the past 100 years, it is by no means extinct. An *extinct volcano* is one that has not erupted within recorded history, and gives no signs of any future activity. Geologists mapping volcanic deposits in the area have found evidence of volcanic eruptions that occur, on average, every few hundred years. Because this is much longer than the span of human memory, most people consider it safe to live at the foot of what, to a geologist, is an active volcano that poses a clear hazard to life and property from blasts of hot rock and ash and subsequent mudflows. The contrast represents the difference in human and geologic concepts of time.

Geoscientists who study volcanoes can provide information that helps the public to understand how often potential eruptions might occur and how large they might be in a particular area. Scientific understanding of volcanism is acquired primarily in two ways: from monitoring active volcanism, and from studying the remains and deposits of past volcanic eruptions.

Studies of active volcanism are promoted and supported in the United States through federal funding of several

FIGURE 5-19 Mount Rainier—shown here looming behind the city of Tacoma, Washington—is one of more than two dozen volcanoes in the western United States that are potentially hazardous. Mount Rainier is an active volcano along a convergent plate margin. As the oceanic Juan de Fuca Plate subducts beneath the North American Plate, some of the oceanic crust and wet sediment overlying it are melted, producing an intermediate (andesitic) magma. This relatively viscous magma rises through the continental crust of the North American Plate, erupting periodically and building tall steep-coned stratovolcanoes from ash, sticky lava, and other volcanic debris. Mount Rainier erupts every few hundred years or less, and ranks high in terms of cause for concern.

volcano observatories. The oldest, the Hawaiian Volcano Observatory, is located close to the eruptive center of Kilauea on the main island of Hawaii (see Figure 5-17). More recently, observatories have been established in Vancouver, Washington (Cascades Volcano Observatory), and Anchorage, Alaska (Alaska Volcano Observatory).

Because of studies by hundreds of scientists at these sites, as well as elsewhere in the world, we now have a better understanding of the types of phenomena that precede eruptions. These phenomena include heightened earthquake activity and changes in the composition of volcanic gases. The practical applications of this knowledge have included a number of successful eruption predictions in recent years and the development of early warning systems for aircraft. Studying active volcanoes can be dangerous, however, and a number of volcanologists have been killed or injured by sudden eruptions. Dr. Katia Krafft, shown sampling lava in Figure 5-20, was killed along with several other volcanologists including her husband, Maurice, during the eruption of Mount Unzen, Japan, in 1991. Those who do such work accept the risks, and the benefits of their research for humanity have been immense. Successful predictions of major eruptions at Mount St. Helens, Washington, in 1980, and Mount Pinatubo, Philippines, in 1991, saved tens of thousands of lives.

The much safer practice of studying the remains of past volcanic eruptions has led also to greater understanding of volcanic hazards. In particular, it provides a view of processes that occur over geologic time scales. In fact, it was mapping of volcanic deposits around the base of Mount St. Helens in the 1970s that led scientists to conclude that the volcano erupts fairly frequently and regularly, and was likely to erupt again before the end of the 20th century (see Box 3-1).

Typically, those studying past volcanic events examine outcrops of volcanic debris to determine the type of deposit, how far it traveled, how much area it covered, and when it occurred. For example, some deposits are ashes that spread along narrow downwind plumes, whereas others are debris flows that fill valleys and may induce flooding by filling or blocking waterways. Maps of the thickness and extent of each deposit enable researchers to estimate the likely size of the eruption from which they came. Also, organic debris trapped in volcanic deposits, or buried by them, can be radiometrically dated to determine when the events occurred. From such studies, scientists have published hazard-evaluation maps for many volcanoes, and are in the process of completing more (Figure 5-21).

The energy released during major volcanic eruptions is too great ever to be controlled or prevented by engineering methods. However, volcanic deposits associated with active volcanoes are so easily mapped that hazardous areas can be identified readily and avoided. In addition, because they usually are preceded by enough warnings, volcanic eruptions can be predicted to within days and hours before they occur. With appropriate monitoring, it

FIGURE 5-20 Volcanologist Katia Krafft, photographed by her husband and colleague Maurice Krafft, is wearing heat-protective clothing while measuring properties of a lava flow.

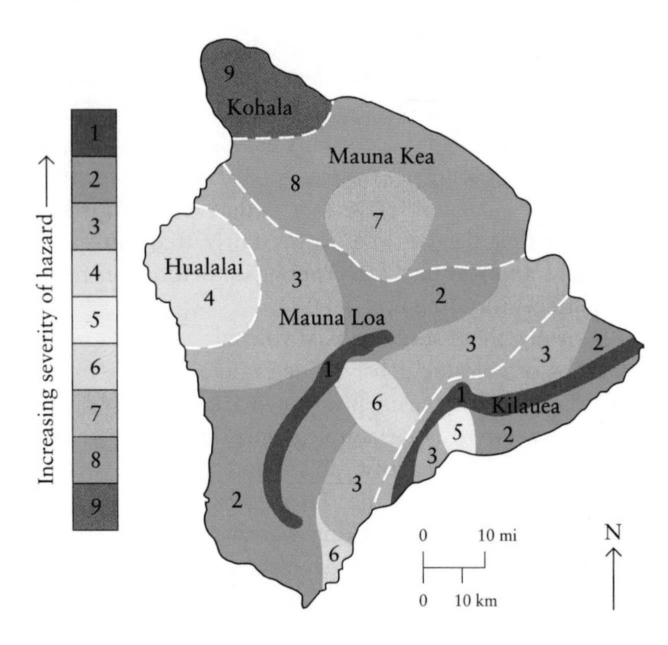

FIGURE 5-21 A map of lava flow hazards for the island of Hawaii indicates the severity of hazard, from low (9) to high (1). The island consists of overlapping shield volcanoes from four sources, including Mauna Loa and Kilauea, which have been the most active in recent times and thus pose the greatest threat.

may be possible to prevent most or all deaths during an eruption, although prediction does not completely alleviate destruction. In the Mount Pinatubo eruption, for example, timely prediction enabled the evacuation of many people, but homes and villages were destroyed under the weight of heavy, wet ash.

The most common types of characteristic pre-eruption activity are related to the movement of magma to shallow depths. Magma movement causes increased earthquake activity at shallow levels, commonly swarms of events, sometimes more than 1000 per day, until the eruption. At the time of an eruption, earthquakes generally occur at greater depths as well, indicating rapid upward flow of magma. Emplacement of magma at shallow depths also results in uplift and extension of crustal rocks, both of which can be measured by surveying the ground surface above a volcano or across a caldera floor.

Minimizing the Risks of Volcanism

The best way to begin to minimize the risks associated with volcanoes is to prepare hazard maps similar to that in Figure 5-21 for all potentially active volcanoes. So far, such maps have been made for about 10 percent of the world's active volcanoes. The United Nations and government agencies have designated the 1990s as the International Decade of Natural Disaster Reduction, and volcanoes are one of the main concerns of their activities. About 20 volcanoes with recent activity and proximity to population centers have been targeted for especially intense studies and monitoring, including Mount Rainier, Mauna Loa, and Long Valley Caldera.

Even the best hazard maps and most careful monitoring are useless, however, if the public is unaware of them and if officials do not use them to restrict settlement or to alert the public to imminent threats. This truth was tragically confirmed in the eruption-induced flow of mud and water that swept over the Colombian town of Armero in 1985. As will be discussed in Chapter 6, officials had a hazard map that illustrated the paths of deadly mudflows that had occurred in the years 1595 and 1845, and the next eruption was likely to take the same path. Geologists had alerted officials to seismic activity at the Nevado del Ruiz volcano above the town. But the public knew little about the map or the alert, and the evacuation order that could have saved 23,000 lives was issued too late, just two hours before 3 m of mud erased the town.

Nevado del Ruiz, like Mount Rainier, is a stratovolcano (see Figure 5-19). Both have steep, icy slopes underlain by loose debris. Deep canyons are eroded into the flanks of both, and sediments in their canyons record dozens of recent debris flows. As at Long Valley, the hazard of Mount Rainier is real, but the timing is uncertain. The hope is that careful mapping and zoning, extensive monitoring, and excellent communication systems will prevent the hazard from becoming a tragedy.

- Basaltic magma generally produces very fluid, fast-moving lava flows that build broad shield volcanoes. Intermediate (andesitic) magma generally produces explosive eruptions with slow-moving lavas and abundant pyroclastic material that build steep-sided cone-shaped stratovolcanoes. Felsic magma generally does not reach Earth's surface, but when it does it is associated with large calderas and very explosive eruptions of pyroclastic debris. Large felsic eruptions have not occurred in hundreds of thousands of years.
- Geologists have predicted many volcanic eruptions because of precursory phenomena that include increased earthquake activity and changes in the composition of gases emanating from volcanic vents.

Earthquakes

Humans long have wondered about the origin of sudden and immense shakings of Earth that mock the phrases "solid earth" and "terra firma." Unlike most volcanic eruptions, earthquakes can happen at any moment with no apparent forewarning. Within seconds, entire buildings and bridges can collapse (Figure 5-22). Some earthquakes are so small they cannot be felt by humans, but

FIGURE 5-22 During the Mexico City earthquake of 1985, about 10,000 deaths occurred throughout the region. Almost 1000 patients and medical staff lost their lives in the collapse of a hospital. Nearby buildings remained standing because they were designed to withstand ground shaking. Most deaths were in the capital city, even though the source of the earthquake along the offshore subduction zone between the Cocos and North American plates was 350 km to the west. Although population density and building construction played a role, some of the greatest damage occurred in Mexico City because it is built upon thick deposits of wet, young lake sediments (mostly silt and clay), which have little strength during shaking.

the largest release 20 orders of magnitude* more energy, an amount equivalent to the annual energy consumption of the United States. What could be happening inside Earth to cause the release of such large amounts of energy in such short times?

The Causes of Earthquakes

We now know that ground shaking during an earthquake is the result of plate motions and slip along faults. Until 1891 most scholars believed that shaking during earthquakes caused the ground to crack and form faults. In that year a large earthquake in Japan formed a **scarp**, or steep slope, up to 6 m high and 70 to 80 km long, offsetting, or separating, many physical features as much as 5.5 m in both horizontal and vertical directions (Figure 5-23). Geologist Bunjiro Koto claimed that offset along the fault had caused the earthquake, not the other way round. His hypothesis was a reversal of the common thinking at the time.

* An order of magnitude is an increase of 10 times; two orders of magnitude is an increase of 100 times, and so on. An earthquake that releases 1000 times more energy than another, for example, would have three orders of magnitude more energy.

FIGURE 5-23 A scarp in Honshu, Japan, formed in 1891 by 5–6 m of offset of Earth's crust along a fault segment 70–80 km long. Geologist Bunjiro Koto examined the scarp and made a revolutionary proposal: that movement on a fault had caused both the scarp and the earthquake, and that the earthquake was an effect rather than a cause.

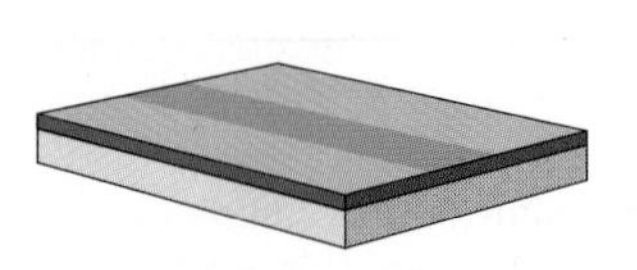

Prior to fault, the road is straight

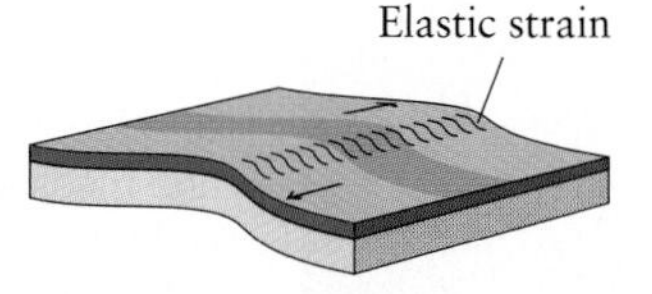

Elastic strain energy accumulates, deforming the crust

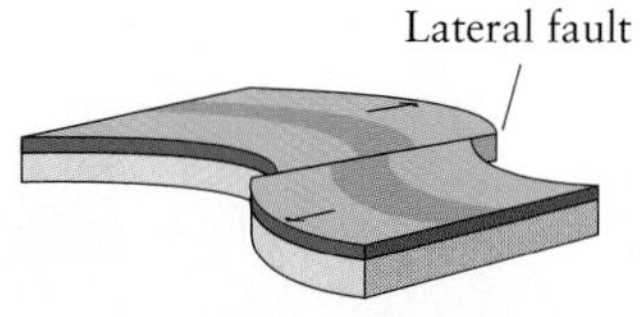

The fault ruptures the crust, releasing seismic energy

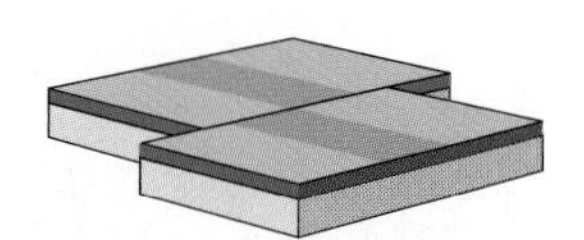

After the earthquake, elastic rebound results in offset along the fault

FIGURE 5-24 The elastic rebound model, the build-up and release of stored elastic seismic energy before and during an earthquake, explains the offset of features such as rocks, fences, and roads after and earthquake. In this example, the road is offset in a right-lateral sense; in other words, one side of the road moved to the right with respect to the other.

In 1906 Koto's idea gained wide acceptance when a large earthquake that destroyed San Francisco was linked directly to slip along a known fault—the San Andreas. Before the earthquake, geologists had noted that the rock types on each side of the fault did not match. Geologists who flocked to the area after the earthquake discovered up to 8 m of horizontal offset of rocks, fences, hillsides, and even walls of buildings along the length of the fault, over a distance of nearly 200 km.

It now is well documented that most earthquakes are the result of plate motions that strain the rocks in opposite directions along a fault plane until they slip. The resultant shaking is due to the release of energy stored in rocks on each side of a fault before they snap and slip past one another. The released energy that causes earthquakes is called seismic energy, from the Greek *seismos*, which means "shock."

An Earthquake Model To explain the fault rupture and release of seismic energy responsible for the San Francisco earthquake, scientists developed the **elastic rebound model**. According to this model, the slow, steady motion of rocks on each side of a fault deforms the crust in an elastic manner, similar to the elastic deformation of a stretched rubber band or a squeezed rubber ball. Markers that cross the fault, such as fences or roads, are distorted by this process. If a rubber band is stretched beyond its elastic limit, it will break and snap back, releasing energy that had been stored. Likewise, with continued stress along a fault, rocks eventually fail in a brittle manner, slipping past one another along the fault plane and releasing stored seismic energy (Figures 5-24 and 5-25). The greater the amount of strain that accumulates in the rocks before they fail, the greater the amount of energy released.

Seismic Sources and Waves Several days after an earthquake struck Boston in 1755, Harvard professor John Winthrop, sitting with feet propped on the brick hearth of his fireplace, was shaken from his repose by an aftershock. One after the other, the bricks under his feet rose up, then dropped back down into place. Winthrop noted with surprise that it looked like "one small *wave of earth* rolling along." His insight was significant: Seismic energy passing through Earth behaves similarly to sound passing though air or a ripple from a dropped stone passing through a still pond of water. In all three situations, energy is transmitted as waves, which in the case of earthquakes are known as **seismic waves.**

During an earthquake, failure begins at a single point on a fault plane named the **hypocenter**, or focus. The rupture rapidly spreads out along the plane until most or all the strain is released (Figure 5-26). Some fault planes rupture at the surface, forming scarps, whereas others do not, resulting in blind faults that can't be seen from the surface. The point directly above the hypocenter at the ground surface is called the earthquake epicenter. At the same time that rupture spreads across the fault plane, seismic waves of several types propagate in all directions, each passing through Earth at a different speed. The result is that an observer at any point on Earth sees batches of different wave types arriving one after the other.

Waves emanating from the hypocenter travel so fast that they can reach the ground surface before the rupture

FIGURE 5-25 This road that was offset in a left-lateral sense by an earthquake that occurred on a strike-slip fault in southern California in 1972. The road on the far side of the fault moved to the left.

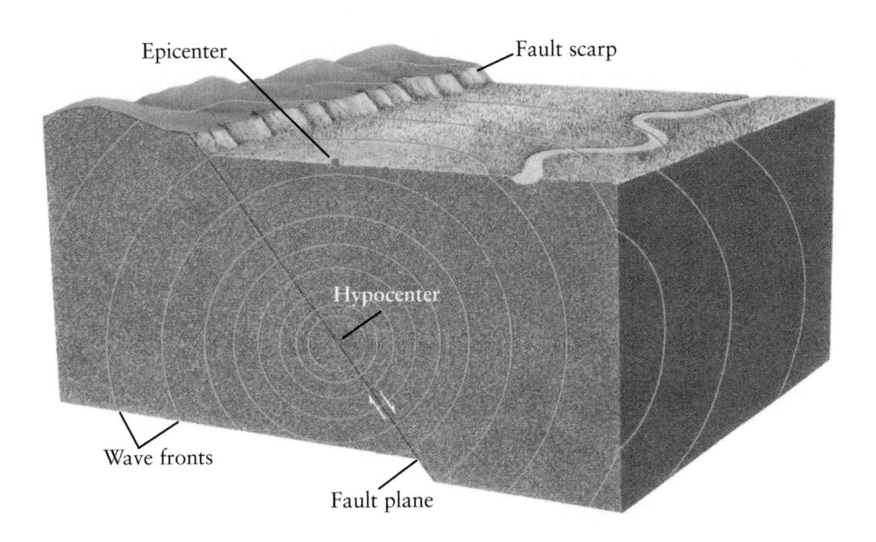

FIGURE 5-26 Seismic waves radiate outward in all directions from the hypocenter, or initial rupture point, along a fault. The waves result in the ground shaking known as an earthquake. As the rupture moves along the fault plane, it reaches the surface and sometimes creates a steep slope, or scarp, like that shown in Figure 5-23.

itself. During a fairly large earthquake in 1983, two elk hunters driving in Idaho suddenly felt dizzy, perhaps because of the arrival of the fastest seismic waves and the rapid shaking of their vehicles. Immediately afterward, they were startled to see a scarp 2 m high emerge 20 m ahead of their vehicle.

The instrument used to detect and measure seismic waves is the **seismometer.** When the ground moves during an earthquake, seismometers fixed to Earth monitor the direction, amplitude (height), frequency (the number of waves per unit of time), and duration of wave motion. A modern seismometer typically is buried up to 3 m underground and connected to a seismograph, an instrument designed to record information regarding ground deformation sent from the seismometer.

Seismic waves can be plotted as a seismogram to show the form of the waves with time (Figure 5-27). The fastest-moving seismic waves are called **P waves,** or **primary waves,** because they are the first to arrive at any station. Like sound waves, they compress and expand the material through which they travel, much like the changing shape of a Slinky toy if one stretches it out across a table and initiates a wave by pushing on one end. Primary waves travel at a rate of about 6 km per second, thus reaching the opposite side of Earth from the earthquake source within about 35 minutes of fault rupture. **Secondary waves,** or **S waves,** are the next to arrive, traveling at a rate of 3.5 km per second. Secondary waves *shear* Earth material back and forth in a sideways motion, perpendicular to the direction in which the wave is moving, much as the shape of a rope stretched between two people is changed when one of them jerks the rope up or down.

When P and S waves reach Earth's surface, they generate lower-frequency **surface waves.** Although their *frequency* is lower, the amplitude of ground motion from surface waves can be quite substantial, resulting in some of the greatest destruction during an earthquake. Surface waves travel about Earth with little impact below ground, as witnessed by miners who have survived earthquakes underground and then climbed to the surface to find widespread devastation. Winthrop probably observed either secondary or surface waves moving through his brick hearth.

Assessing Earthquake Hazards

The energy released during an earthquake is proportional to the size—the length and width—of the fault plane that ruptures and the volume of rock deformed along the fault. The greater the amount of energy released during an earthquake, the greater the amplitude of seismic waves at a given distance from the earthquake source. Likewise, the greater the energy released, the worse the destruction and greater the intensity felt by those close to the source. Both of these measures of earthquake energy—wave amplitude and felt intensity—are used to compare the sizes of earthquakes. The best assessment of the size of an earthquake, however, is based on the area of fault rupture and amount of rock deformed, for this gives a truer estimate of the amount of seismic energy released.

Earthquake Magnitude and the Richter Scale A method that uses wave amplitude to measure earthquake energy was devised in 1934 by the late Charles Richter, then a seismologist at the California Institute of Technology. For years news reporters had asked him to describe the relative sizes of different earthquakes, so Richter decided to invent a scale. The **Richter magnitude scale** uses the amplitude of the largest seismic wave on a seismogram, corrected for distance from the source, to assign a magnitude to the event. The magnitude of an earthquake is the same no matter where it is measured, because the energy released during an individual earthquake is a fixed amount.

The Richter magnitude scale is logarithmic, which means that each unit on the scale is a tenfold—or order-of-magnitude—increase of the preceding one. In other words, the amplitude of the largest seismic wave of a Richter magnitude 5 event is 10 times larger than the amplitude of a magnitude 4 event. Richter used a logarithmic scale because the range in size of earthquakes is so large. Events larger than Richter magnitude 8 are considered to

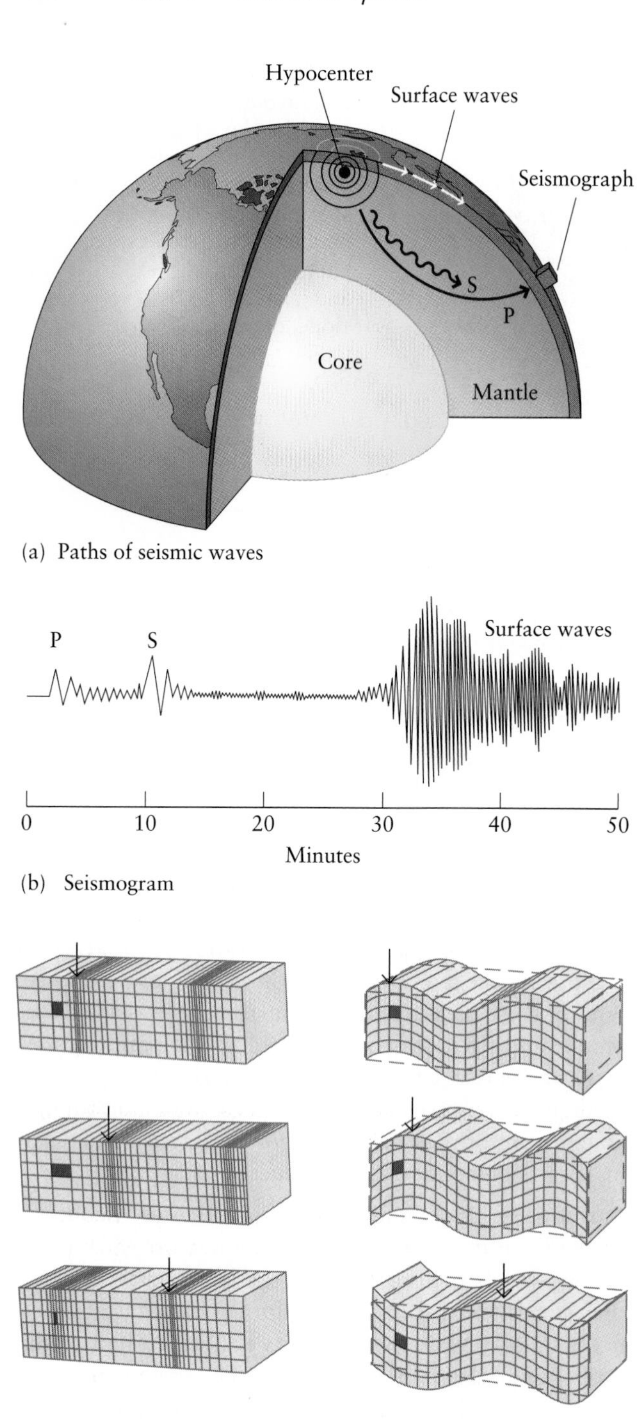

FIGURE 5-27 (a) Because different types of waves radiate outward from the hypocenter, and each travels at a different speed, some waves reach the seismograph station before others. (b) The waveforms on the seismogram are a record of the amplitude and duration of wave motion in the ground. (c) As P waves move through a block of material, pieces within the block alternately expand and compress. The arrow marks the location through the block. (d) As S waves move through a block of material, pieces within the block are shaken side-to-side. The arrow marks the wave crest as it moves through the block.

be "great" earthquakes, of which only one occurs about every decade. More than 100 events greater than magnitude 6 or 7 occur annually. Smaller events are even more frequent, with thousands occurring each year, but those less than Richter magnitude 3 to 4 are barely noticeable.

Earthquake Intensity and the Mercalli Scale Some areas experience far greater damage than others during a given earthquake because of differences in local geologic conditions. For this reason, an older measure of earthquake energy, known as the Mercalli intensity scale, is still useful as a tool for hazard assessment. This scale, developed by Italian seismologist Giuseppe Mercalli in 1902 and since adjusted and renamed the **modified Mercalli intensity scale** (MMI), is based on the effect of an earthquake on human structures and on how people experience it. Each of 12 grades is assigned a Roman numeral, and the scale is not logarithmic: It has been carefully designed so that the increases in intensity from one step to the next are roughly equivalent. While an MMI III event feels like little more than a passing truck, an MMI VIII event can destroy masonry buildings and an MMI XII event results in total destruction.

The intensity of a single earthquake varies with distance from the event and with the nature of geologic materials through which seismic waves pass. Generally, the intensity is smaller with greater distance away from the epicenter. Solid rock dampens seismic waves, so that intensities are lower on bedrock than on unconsolidated sediments. So, unlike the Richter magnitude—which is the same everywhere—the Mercalli intensity of an earthquake varies from one place to another.

Earthquake Moment Magnitude Though news reports still cite "the Richter scale," scientists are discontinuing use of that measure, and rarely use the Mercalli scale. Their reasons are many, but these three are the most important: First, Richter developed his scale in California, and rock types elsewhere are sufficiently different to make the scale difficult to apply. Second, the Richter magnitude scale is a poor measure of the energy actually released by an earthquake. Finally, many people have difficulty understanding a logarithmic scale. As seismologist Thomas H. Heaton recently stated, seismologists "spend as much time explaining . . . what a logarithm is as anything else." Likewise, the Mercalli scale—despite its detailed rankings—is still a subjective measure, and not a good measure of seismic energy at that.

Instead, scientists have developed the **moment magnitude**, which measures the approximate energy of an earthquake source rather than its effects at Earth's surface. Seismic energy can be calculated from the dimensions of the fault plane that broke, as well as the volume of rock deformed along the fault. Seismologists now are encouraging the public and news agencies to accept this measure of the size of an earthquake. Because it takes one or two hours to compute moment magnitude, scientists

FIGURE 5-28 About two-thirds of the deaths during the Loma Prieta earthquake in 1989 occurred in one stretch of Interstate 880 in Oakland, California. Column supports failed when the reinforcing rods within them bent and snapped, and as a result, the upper deck of the freeway dropped onto the lower deck. Nearby, similar structures with support columns wrapped in iron did not fail, but the part of the highway shown here had no such reinforcement.

first release a *preliminary magnitude*, which might be based on ground shaking and hence similar to a Richter magnitude, but soon afterward they announce the moment magnitude.

Local Geologic Conditions Within hours to days, 67 people died in the San Francisco and Santa Cruz areas when bridges, overpasses, and buildings collapsed during the 1989 Loma Prieta earthquake, named for the mountain beneath which the earthquake began (Figure 5-28). Although the epicenter was nearly 75 km southwest of San Francisco, the earthquake wreaked havoc in that city where certain types of sediments existed. In nearly all cases, death was the result of failed engineering structures built on weak, unconsolidated, wet sediments that amplified ground motion as seismic waves passed beneath them (Figure 5-29). Wet sediments behave much the way a stretched drum skin does when tapped, vibrating far more than surrounding rocks and older sediments that are more consolidated, rigid, and resistant. Some sediments have the potential to *liquefy*, or behave more like liquids than solids during intense ground shaking.

One of the greatest contributions of geologists to assessing the hazards associated with earthquakes is geologic mapping. Geologic maps denote the boundaries between different deposits and rock types (see Figure 5-29) and include locations of known and inferred faults. From the experience of Loma Prieta in 1989, as well as other earthquakes, however, we know that maps and geologic information are not enough. In order for death tolls to be lessened and property destruction decreased, such maps must be used during planning and construction phases, not only after structures are completed.

The Potential for Future Earthquakes At present, geoscientists know of no reliable means by which to *predict* an earthquake in terms of its location, exact timing, and size. *Forecasts*, on the other hand, which merely state the approximate location of a future event, are possible because it is known that many faults are active and have a history of recurrent earthquakes. Earthquake forecasts are much like weather forecasts, in that they provide a general idea of what might happen.

One of the most important outcomes of plate tectonics theory is an understanding of the relation between seismicity and plate boundaries (see Figure 5-13). Most faults, and therefore most earthquakes, occur along plate boundaries. Maps of the distribution of earthquakes throughout the world indicate a clear coincidence between plate boundaries and the zones of greatest earthquake activity, with more than 95 percent of all earthquakes occurring at the boundaries between two plates. Most of the other 5 percent occur along ancient plate boundaries, where plates have been sutured together, or rifts are no longer actively spreading. These boundaries are weak spots in Earth's lithosphere.

The depth of an earthquake is related to the type of plate boundary: The deepest earthquakes occur along subduction zones, where plates are carried down into the asthenosphere. The deepest recorded earthquakes have emanated from more than several hundred kilometers below Earth's surface. Earthquakes along transform and divergent plate boundaries are much shallower, generally within a few kilometers of Earth's surface.

It is reasonable to forecast that future events will occur along plate margins, where brittle crust is deforming and elastic energy is accumulating along locked faults, that is, ones that are not slipping (see Figure 5-24). Such a forecast is valuable as a planning tool, ensuring that schools, hospitals, roads, and dams are not built near such features. However, a forecast does not provide a signal for alert or evacuation.

Although scientists cannot yet predict an earthquake, they can estimate the probability that an earthquake of

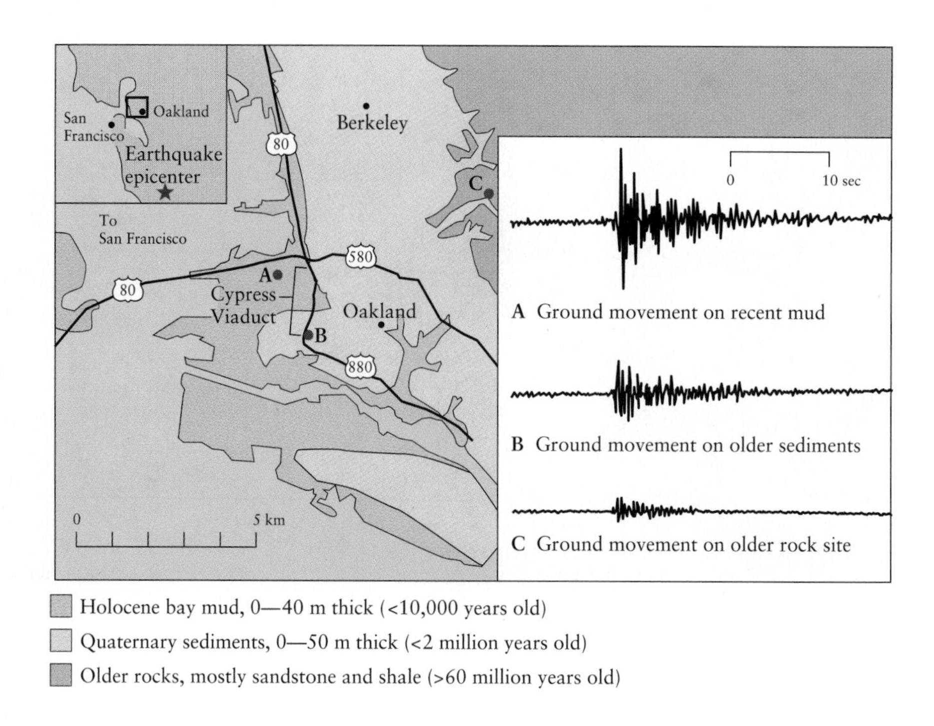

FIGURE 5-29 Much of the failure of the Interstate 880 overpass during the 1989 Loma Prieta earthquake can be attributed to underlying geologic materials that did not receive the necessary special attention at the stage of engineering design. For instance, the highway overpass failed between points A and B, where it is built on young bay muds, but not above older sediments and rocks. During an aftershock, seismograms of ground motion on different geologic materials illustrate the great differences in ground shaking at these different sites. The amplitude of ground shaking is least for the older rocks, and greatest for the young, unconsolidated sediments.

a certain magnitude will occur within a given time period. This analysis is called a *probabilistic assessment*. The assessment is made by using known amounts of slip on mapped faults combined with fault dimensions and long-term rates of strain across the structures. Global Positioning System (GPS) satellites that orbit Earth and monitor its surface provide data on the rates at which plates are moving and crust is deformed along plate boundaries and faults. This method of estimating size, frequency, and probability of occurrence of future earthquakes is based on the elastic rebound theory because it links the release of elastic deformation to rates of accumulation of deformation.

Minimizing the Risks of Earthquakes

Studies during the past few decades have produced a much greater understanding of where and how often earthquakes of a given size are likely to occur, and of the nature of ground shaking that reasonably can be expected in a particular area (Box 5-4). Consequently, it is both possible and prudent for engineers, geoscientists, and politicians to work together to create appropriate building codes for seismically safe structures. Information needed for proper planning includes:

- The location of possible faults, as they are the sources of earthquakes.
- The characteristics of likely earthquakes (magnitude, duration, and type of faulting).
- The type of geologic material between the earthquake source and engineering structures.

In particular, the knowledge of the nature of the geologic material directly beneath a structure is important to good engineering design. Recent building codes in seismically active states such as California require structures to be set back a minimum distance from active faults, and those on sediments that may liquefy must use specific building techniques.

The effectiveness of these approaches is demonstrated by comparison of the Northridge, California, event of 1994 (moment magnitude 6.8) with a similar sized earthquake that occurred in Kobe, Japan, in 1995 (moment magnitude 7.2). Although some aspects of the two earthquakes differ, such as proximity of the epicenter to greatest population densities, in many ways the two events were quite similar. Yet the damage was significantly different. In the Northridge event, 57 people were killed, 9300 were injured, 13,000 buildings were destroyed, and total damage was $13 billion. In Kobe, 5239 people were killed, 26,000 were injured, 50,000 buildings were destroyed, and total economic loss was $200 billion.

Investigators concluded that some of the difference in death toll and destruction was due to the different types of building construction. In Kobe, there were many old buildings that predated modern codes. Masonry buildings constructed of bricks, tiles, and blocks are far less resistant to ground shaking than are wooden frame structures or modern buildings with steel frames and reinforced beams. Because many buildings in Kobe were masonry, destruction was widespread and tragic. In the Northridge area, many buildings are wooden, reinforced, or constructed with steel frames, so they were able to withstand the 7.2-magnitude shock.

5-4 Case Study

Crustal Deformation Associated With Earthquakes in the Cascades Region

Plate tectonics processes explain not only the location and depth of an earthquake but also the nature of crustal deformation before, during, and after an earthquake. The subduction zone between the Juan de Fuca and North American plates is lined with a deep offshore trench and an inland volcanic arc that spans the western coast of North America from British Columbia to northern California (see Figure 5-18). Major cities bordering this margin include Vancouver, British Columbia, and Seattle, Washington.

Because of the motion of the colliding plates, the North American Plate is flexed upward during the period of elastic strain accumulation, when the plate boundary is locked (i.e., not slipping). When the strength of the crust along the plate boundary is exceeded, brittle failure and an earthquake occur. The outermost lip of the overriding plate typically is pushed upward, resulting in coastal uplift that raises tidal pools and beaches up out of the ocean. Inland of the area of uplift, the bulging crust collapses, causing the land to drop and drowning the coastline. Evidence of such submergence exists along convergent boundaries throughout the world, in the form of buried peat and dead trees killed by the exposure of their roots to saltwater or burial by sand and mud.

Although no major earthquake has occurred along this plate boundary in historic time, the native Yurok people of the Pacific Northwest have myths that tell of a god named Yewol (in English "Earthquake") who:

> sank the ground. . . . Every little while there would be an earthquake, then another earthquake, and another earthquake. . . . And then the water would fill those places. . . . (as recorded and translated by A. L. Kroeber, 1978, in *Yurok Myths*, Berkeley: University of California Press).

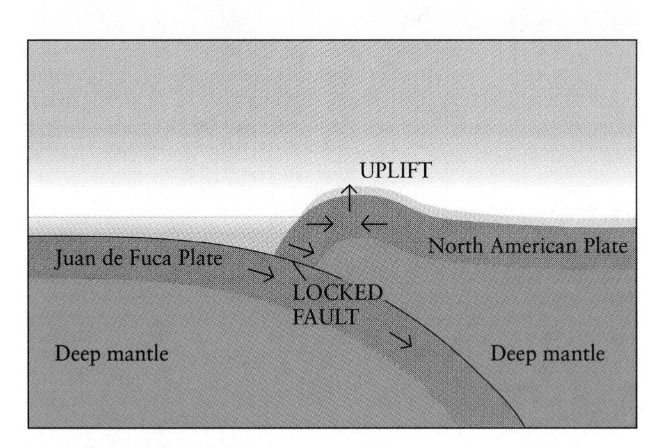

(a) Before earthquake

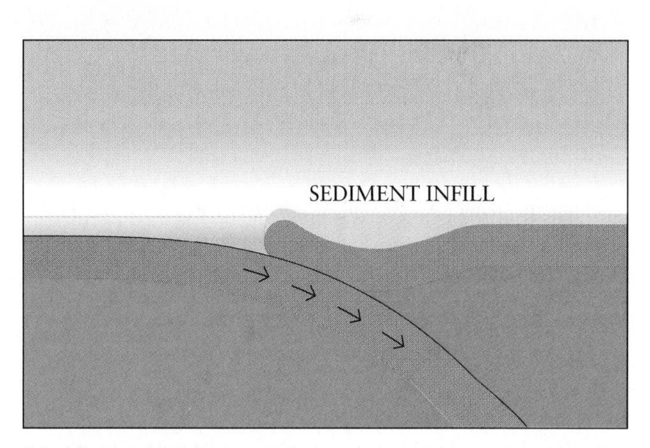

(b) After earthquake

(a) Elastic strain energy accumulates in the overriding North American Plate as it flexes upward before an earthquake. (b) After an earthquake, the outermost lip of the upper plate remains emergent above sea level, while inland the coast subsides to form a trough that fills with water and sediment.

(continued)

5-4 Case Study

(continued)

Geologists also know, from the evidence of dead trees, that an event like that described by the Yurok occurred about 1700, before Europeans settled in the area. The fact that no earthquakes have occurred recently does not mean that faults along the plate boundary are inactive and no hazard exists. Rather, the period of quiescence indicates that elastic strain energy is being stored until the next event.

A ghostly forest of dead red cedar trees in a tidal marsh indicates that the coast of Washington subsided suddenly about 1700. Since the forest was drowned, a meter of mud has buried the bases of the trees.

Restoration Point, near Seattle, is a beach raised out of the water by a major earthquake that occurred sometime between AD 800 and 900.

- Earthquakes occur when rocks slip past one another along a fault plane, releasing seismic energy that had been stored in the rocks. The greater the amount of rock deformed, the greater the amount of energy released and, as a consequence, the greater and longer the ground shaking during the resultant earthquake.
- Most earthquakes occur along active plate boundaries, and the deepest earthquakes emanate from subduction zones, where one plate grinds past another deep beneath Earth's surface.
- When seismic energy is released along a fault plane, it radiates outward through surrounding rock as several types of waves, including primary, secondary, and surface waves. Each travels at a different rate, with primary waves traveling most rapidly.
- Although geologists have an increasingly good understanding of what causes earthquakes and where they are most likely to occur, they have not yet identified any phenomena that consistently precede an earthquake, and so they are not yet able to predict earthquakes before they occur.

Closing Thoughts

Rocks and minerals are economic resources, yet humans derive aesthetic satisfaction from them as well. The Devil's Tower in Wyoming, the Rock of Gibraltar in the Mediterranean Sea, the Giant's Causeway in Ireland, and Ayers Rock in Australia all are large rock exposures that hold a special fascination, evident in their names. Minerals such as gold, silver, and platinum, as well as diamonds, rubies, emeralds, and other gems, have been symbols of wealth, prestige, and status for much of human history and are displayed in museums as objects of interest and beauty. It seems to be part of human nature to treasure Earth materials.

At the same time, however, humans have marred Earth's surface in order to extract the minerals they value. Landscapes scarred and contaminated by mining, such as Butte, Montana, have been described by some officials as "National Sacrifice Areas"—places sacrificed for the sake of minerals and industrialization. Mineral extraction has proceeded despite substantial risks and hazards. Only in the latter part of the 20th century have mining and its environmental impact been regulated, and then only in the most developed countries. As the developing nations become industrialized, world per capita mineral consumption continues to rise while world population continues to grow. Mining has altered the environment for many millennia, but it remains to be seen how governments will respond to the conflict between economic interests and environmental quality in a time of rapidly growing population.

In addition to the human-induced hazards of mining, there are natural hazards associated with the lithosphere. Volcanoes and earthquakes have had much impact on our environment. The 1815 eruption of Mount Tambora, in Indonesia, and the 1991 eruption of Mount Pinatubo, in the Philippines, caused prolonged cooling in the northern hemisphere. Tambora killed tens of thousands of people, but Mount Pinatubo did not, because advances in scientific understanding of the way Earth works enabled geoscientists to warn of the impending Philippine disaster, and thousands of lives were saved.

To date earthquakes cannot be predicted, but greater understanding of the phenomenon has resulted in fewer deaths and less damage in those places where officials have developed and enforced zoning and construction regulations. Moderate-sized earthquakes in California in 1989 (Loma Prieta) and 1994 (Northridge) each caused fewer than 100 deaths. Nevertheless, much more can be done. The Northridge event resulted in few deaths partly because it occurred before dawn, when freeways and business buildings were nearly empty. The actual amount of damage—$13 billion—was greater than in any other natural disaster in U.S. history.

Summary

- In general, metals are disseminated throughout Earth, but some are concentrated locally as the result of geologic processes, especially processes associated with high heat flow and movement of magma, such as intrusion of magma in continental crust above a subduction zone.
- The primary environmental impacts associated with mining and processing minerals are the disturbance of land during excavation, water and soil pollution from metals and acids, and the emission of sulfur dioxide (which forms sulfuric acid) and metals into the atmosphere.
- Human activities have increased the flow rate of metals among Earth systems, to amounts as much as hundreds of times higher than natural rates in the case of metal emissions into the atmosphere.
- Earth has more than 1000 active volcanoes, most of them located along convergent (nearly 80 percent) and divergent (nearly 20 percent) plate boundaries.
- Many of the world's peoples live along the slopes of active volcanoes that erupt only one or a few times every few centuries, such as the residents of Tacoma, who live near the base of Mount Rainier. The recurrence time of eruptions from some volcanoes is longer than human memory, giving residents a false sense of security.
- Geologists study past volcanic eruptions in order to understand volcanism and to prepare maps of different types of volcanic deposits. These maps are used as hazard maps to alert officials and the public to areas of likely future volcanic activity.
- Earthquakes occur if strain built up along a fault is released suddenly as seismic energy when crust on each side of the fault plane slips. The ultimate cause of the build-up of strain is motion between tectonic plates.
- Although in the past the Richter magnitude and Mercalli intensity scales have been used to compare the sizes of different earthquakes, today scientists prefer the more recently developed moment magnitude, as it is a better estimate of the true amount of energy released during an earthquake.
- The damage associated with an earthquake is due largely to ground shaking, which varies substantially depending on the type of geologic material through

which it flows. In general, loose, wet, young sediments are the most susceptible to intense ground vibration, and structures built on them rather than on, for example, granite, are likely to be more heavily damaged.

- Geologists cannot yet predict the exact timing and precise location of earthquakes, but they can provide officials and citizens with maps that locate areas where earthquakes are most likely to occur and where there are geologic materials susceptible to ground shaking and liquefaction. This information can lead to building codes that can substantially reduce the risks associated with future earthquakes.

Key Terms

metal (p. 125)
metallic bond (p. 125)
ore deposit (p. 125)
ore mineral (p. 125)
tailings (p. 126)
economical concentration factor (p. 126)
hydrothermal ore deposit (p. 127)
igneous ore deposit (p. 128)
layered intrusive (p. 128)
placers (p. 129)
reserve (p. 130)
resource (p. 130)
biological amplification (p. 134)
viscosity (p. 140)
shield volcano (p. 141)
stratovolcano (p. 142)
caldera (p. 142)
scarp (p. 147)
elastic rebound model (p. 148)
seismic waves (p. 148)
hypocenter (p. 148)
seismometer (p. 149)
P waves (primary waves) (p. 149)
S waves (secondary waves) (p. 149)
surface waves (p. 149)
Richter magnitude scale (p. 149)
modified Mercalli intensity scale (p. 150)
moment magnitude (p. 150)

Review Questions

1. What is the main mineral ingredient used in the manufacture of glass? What are the two elements in this mineral?
2. What is the process of smelting, and what is its purpose?
3. When did Eurasians and Africans begin to use metals instead of stone to manufacture tools?
4. What three nonfuel materials rank the highest in terms of amount consumed for human activities?
5. What are ore deposits and ore minerals?
6. What is meant by the economical concentration factor of an element? What is the economical concentration factor of copper?
7. Name the four major categories of types of metallic and nonmetallic ore deposits.
8. How are hydrothermal ore deposits formed at spreading centers along divergent plate boundaries?
9. How do diamonds form? What type of ore deposit—sedimentary, igneous, or hydrothermal—are they?
10. What types of minerals are likely to be found in placer deposits? Why?
11. List three reasons why it is unlikely that a major mineral resource crisis will occur in the near future.
12. What are the main environmental impacts of extracting and processing minerals?
13. List different ways in which volcanic processes can damage property and cause death or injury.
14. What parts of the world have the greatest number of active volcanoes? Why?
15. What types of volcanoes are formed by basaltic, intermediate (andesitic), and felsic magmas?
16. Which type of magma (basaltic, andesitic, or felsic) is associated with the most explosive type of volcanic eruption? Why?
17. What two investigative approaches are used by scientists to understand volcanism?
18. What types of phenomena commonly precede volcanic eruptions and can be used to predict their occurrence?
19. How can the risks associated with volcanic eruptions be minimized?
20. How are faults related to the cause of earthquakes?
21. What is a fault scarp, and how is it related to a fault plane?
22. Which type of seismic wave commonly causes the greatest destruction during an earthquake?

23. What is the difference between an earthquake forecast and a prediction?
24. Why do more earthquakes occur along the boundaries between tectonic plates than within plates?
25. What types of information about a given location are needed to minimize the risks associated with earthquakes?

Thought Questions

1. Why are so many ore deposits found along the margins of convergent plate boundaries?
2. How did hydrothermal ore deposits of lead and zinc form along the Mississippi Valley 200 million to 300 million years ago? What did plate tectonics have to do with this process?
3. What type of ore deposit is associated with the copper and gold mines in the mountains of the island of New Guinea? How did it form?
4. Do you think that Australia might be a good place to search for disseminated porphyry metal deposits? Why or why not?
5. Why is the release of mercury from mines potentially more harmful than the release of other metals, such as iron? Why might it be more harmful to a person or fish than to aquatic plants?
6. Why are volcanic eruptions caused by felsic magmas so much more explosive than those caused by basaltic eruptions?
7. Why do basaltic magmas construct broad shield volcanoes, like those that form the island of Hawaii?
8. How do the Richter magnitude and modified Mercalli intensity scales differ from each other and from the newer measure of moment magnitude?
9. Why do some earthquakes occur at shallow depths (several kilometers), whereas others occur much deeper (hundreds of kilometers) beneath Earth's surface? Would you expect earthquakes in Indonesia to be shallow or deep? Why?
10. What causes coastlines to rise out of the water (coastal emergence) or to subside (coastal submergence), killing plant and animal life, during some earthquakes?

Exercises

1. If an earthquake of moment magnitude 7 occurs in San Francisco, how long will it take a P wave to reach Pittsburgh, about 4000 km away? How long will it take an S wave? Is it likely that a resident of Pittsburgh would feel this seismic wave pass through the rock underfoot? Why or why not?
2. Use Figure 5-21, the volcanic hazard map of the island of Hawaii, to determine one of the safer parts of the island on which to live.

Suggested Readings

Bolt, Bruce, 1993. *Earthquakes*. 3rd ed. New York: W. H. Freeman and Company.

Bryant, E. A., 1991. *Natural Hazards*. New York: Cambridge University Press.

California Seismic Safety Commission, 1992. *The Homeowner's Guide to Earthquake Safety*. Sacramento: CA-SSC.

Dobb, Edwin, 1996. "Pennies from Hell." *Harper's* (October): 39–64.

Hyndman, Roy D., 1995. "Giant Earthquakes of the Pacific Northwest." *Scientific American* (October): 68–75.

Nuhfer, Edward B., Proctor, Richard J., and Moser, Paul H., 1993. *The Citizen's Guide to Geologic Hazards*. Arvada, CO: American Institute of Professional Geologists.

Tylecote, R. V., 1976. *A History of Metallurgy*. London: The Metals Society.

U.S. Bureau of Mines, published every 5 years. *Mineral Facts and Problems*. Washington, DC: U.S. Bureau of Mines, Bulletin 675.

U.S. Geological Survey, bimonthly magazine published since 1968. *Earthquakes and Volcanoes*. Denver, CO: USGS.

Wright, Tom L., and Pierson, Tom C., 1992. *Living with Volcanoes: The U.S. Geological Survey's Volcano Hazards Program*. Washington, DC: U.S. Geological Survey Circular 1073.

Young, John E., 1992. *Mining the Earth*. WorldWatch Paper 109. Washington, DC: WorldWatch Institute.

CHAPTER 6

Severe soil erosion is evident on a newly deforested hillslope along Brazil's Transamazon Highway. Soils exposed by deforestation for lumber, farming, and road construction are especially vulnerable to erosion on steep hillslopes.

Soil and Weathering Systems

Scientist Hans Jenny spent his life working with soil and concluded that although it is "not an organism that can multiply, soil on the Earth is a living system." Some narratives of creation view human life as a product of soil. According to the Judeo-Christian tradition, for example, "God Yahweh formed man out of the soil of the Earth and blew into his nostrils the breath of life" (Gen. 2:7). Soil supports life, and the life and death of organisms is vital to soil formation. A single handful of soil contains billions of microbes that decompose plant, animal, and mineral matter and generate nutrients. Fertile soil is one of humanity's most precious resources, a medium for growing food, fiber, and wood. The more that human populations grow and standards of living are raised, the greater the demands we place on our soils. One of the great and paradoxical concerns of modern society is how to conserve soil even as we use it.

Soils occur and soil-forming processes operate in the pedosphere as the result of interactions among the lithosphere, atmosphere, hydrosphere, and biosphere. In this chapter, we will:

- Examine the processes of weathering and soil formation that create the pedosphere and the resources in it that we use.
- Explore the causes of the hazards of soil erosion, desertification, and mass movement, and consider how they can be minimized.
- Investigate examples of changes in the pedosphere over human and geologic time periods.

On Black Sunday in 1935, a massive dust storm darkened the sky over the Great Plains and awakened the United States to the menace of soil erosion. In this semiarid, grassland region, several years of drought in the early 1930s had followed a relatively wet period during which hopeful farmers plowed up the deep-rooted sod to plant crops. The bare, pulverized soil, parched by the subsequent drought and exposed to high winds, began to blow away. On Sunday, April 14, 1935, in the worst dust storm by far, northwest winds whipped millions of tons of loose, dry topsoil into a churning wall of red-brown dust that covered the homes and fields of frightened farmers and destroyed their crops (Figure 6-1). Tens of thousands of impoverished families from Oklahoma, Texas, and other Great Plains states abandoned their farms and migrated west. John Steinbeck immortalized their plight in his novel *The Grapes of Wrath*, and people everywhere lamented the transformation of America's fertile "wheat heaven" into a harsh, barren Dust Bowl.

The finest sediment from the Black Sunday storm was lifted high into the atmosphere and carried eastward by wind currents. In early May, the gritty haze hit the eastern seaboard, and even the politicians in Washington, D.C., found themselves sneezing, coughing, and rubbing the dust out of their eyes. Congress passed the Soil Erosion Act of 1935, and President Franklin D. Roosevelt established the Soil Conservation Service (now the Natural Resources Conservation Service) and put hundreds of thousands of people to work in the Civilian Conservation Corps. The trees these workers planted and the dams they built helped to minimize future soil erosion.

Two of the Soil Conservation Service's early leaders, Hugh H. Bennett and Walter Clay Lowdermilk, were instrumental in impressing on Congress the effects of soil erosion. Giving testimony before a congressional subcommittee, the two men silently laid a towel on the polished tabletop, poured a large glass of water onto it, and watched as the towel absorbed all the water. Then they removed the towel and poured the same amount of water onto the bare surface of the table. The water splashed, splattered, and soon ran off onto the laps of the startled committee members, demonstrating the value of a healthy topsoil that absorbs moisture and protects the ground surface from erosion.

In the modern world, soils are intensively used and altered by humans: Soil loss from erosion and degradation by pollutants, pesticides, and herbicides causes some of the most substantial environmental damage. An understanding of soil-forming processes in the pedosphere is important to assessing the causes and prevention of soil erosion and mass movement, and to determining how pesticides and other chemicals are trapped and degraded in soil.

The Pedosphere: A Geomembrane to Other Earth Systems

The **pedosphere** is the entire layer of disaggregated and decomposed rock debris and organic matter at the surface of the continental Earth. Mass and energy from the lithosphere, hydrosphere, biosphere, and atmosphere are transported across the pedosphere. Because it functions much the same as a biomembrane that surrounds

(a)

(b)

FIGURE 6-1 (a) Walls of churning dust darkened the skies over the Great Plains on Black Sunday, April 14, 1935. (b) Thousands of farms like this one near Guyman, Oklahoma, were rendered infertile; farm families abandoned their land and sought work elsewhere, often as migrant farm workers.

a living cell, the pedosphere can be thought of as the solid Earth's *geomembrane* (Figure 6-2). Dissolved ions, solid particles, and gases move back and forth across this geomembrane, altering the nature of the underlying lithosphere and creating a nutrient-rich environment for plants and animals.

The alteration of rock and mineral matter at this geomembrane by physical, chemical, and biological processes is called *weathering*. In an intricate series of feedback processes that operate at time scales of years to tens of thousands of years, weathered rocks provide essential elements to plants and animals, which in turn act to transform rocks into soil, thus releasing the elements needed for life.

Crust—Earth's weathered face—became susceptible to weathering as soon as it formed at the surface during Earth's early evolution. Rock materials exposed to air, water, ice, wind, acidic solutions, and biological phenomena at and near Earth's surface *disintegrate* physically and *decompose* chemically. These are different weathering processes, but they generally occur in association with each other. Disintegration, or **physical weathering,** refers to the mechanical fragmentation of rocks and minerals. Decomposition, or **chemical weathering,** refers to the chemical alteration of rocks and minerals. Both types of weathering are enhanced by biological processes. Chemical reactions occur on the surfaces of rocks and minerals, so physical weathering facilitates chemical weathering by breaking apart fresh rock to expose greater surface areas (Figure 6-3).

Weathering occurs in place: The weathered rock and mineral fragments remain at their original site. Once rock has been weathered, it is more easily carried away by a moving medium such as wind, water, or ice. The removal and transport of material by these processes is called **erosion.**

Energy is the capacity for doing work, which is the change effected by the application of a force. Weathering, like all processes, requires energy to do work (here, to change rock to weathered debris), and some force must act over a given distance to do that work (see Chapter 1). The force exerted on an area, in this case an area of rock, is known as **stress.** Just as rocks are deformed, or subject to *strain,* by plate tectonics processes, the actions of weathering deform Earth materials. (Stress is a measure of an applied force; strain is a measure of the resultant deformation.) In physical weathering, stresses are created by physical forces such as the thermal expansion of minerals when heated by grassland fires. In chemical weathering, stresses are created by chemical forces such as the incorporation of water molecules in an expandable clay mineral. The greater the stress exerted on rocks and minerals, the greater the work that can be done to disintegrate and decompose them. If the stresses exceed the strengths of the chemical bonds holding the minerals together, then disintegration and decomposition will occur.

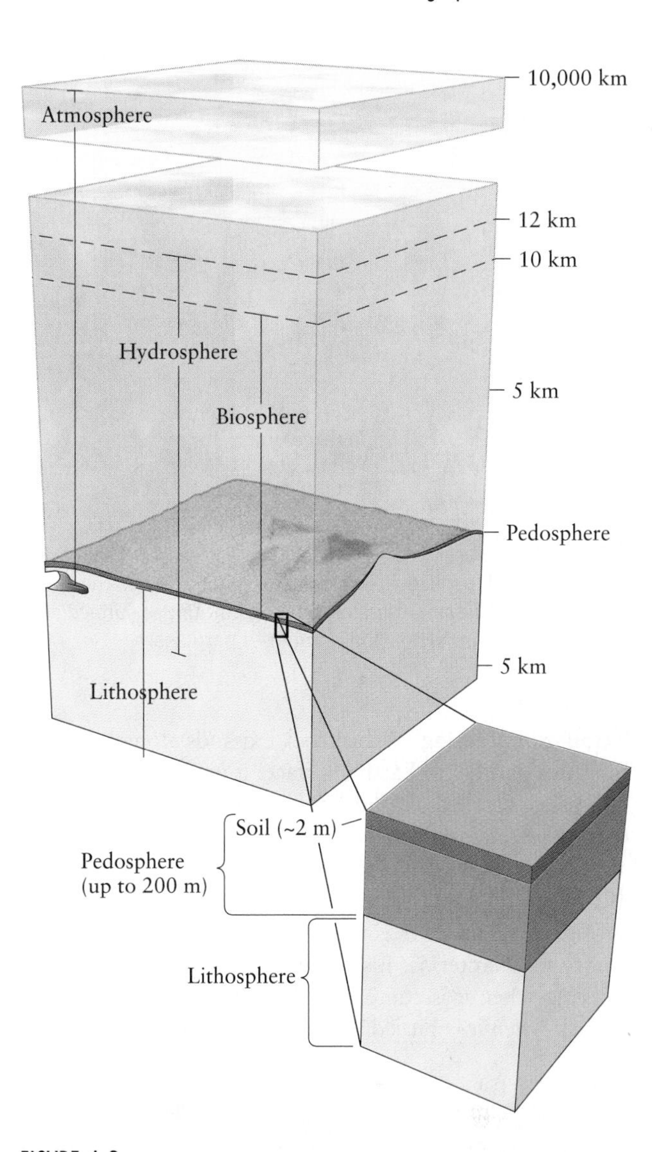

FIGURE 6-2 The pedosphere is the base of continental ecosystems and forms an interface between them and the other Earth systems. Soil, the 1–2-m-thick portion that includes organic matter, forms at the top of the pedosphere, where all Earth's systems interact. Below the soil, weathered rock grades downward to unweathered rock, forming an uneven layer up to 200 m thick.

Physical Weathering

There are many types of physical weathering processes. Several of them are common worldwide and are responsible for most disintegration of rocks and minerals. These processes are exfoliation jointing, thermal expansion, disintegration associated with plant growth and worm activity, and frost weathering.

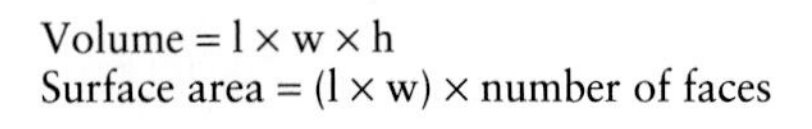

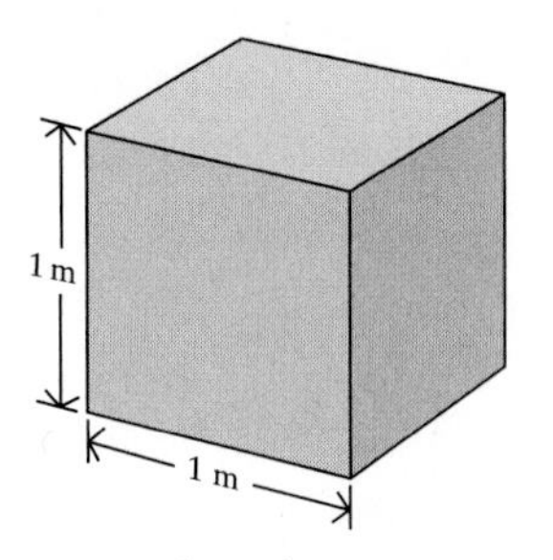

1 cube, 6 faces
Volume = 1 m^3
Surface area = 6 m^2

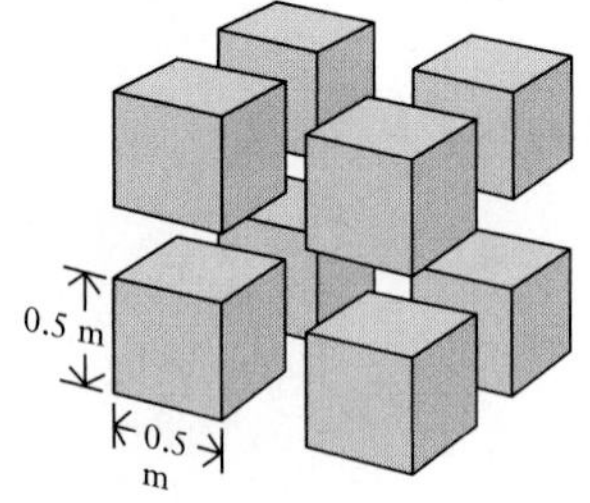

8 cubes, 48 faces
Total volume = 1 m^3
Surface area of 1 cube = 1.5 m^2
Total surface area = 12 m^2

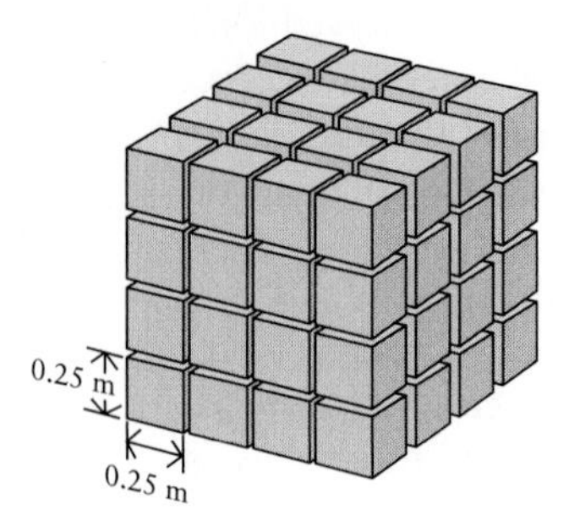

64 cubes, 384 faces
Total volume = 1 m^3
Surface area of 1 cube = 0.375 m^2
Total surface area = 24 m^2

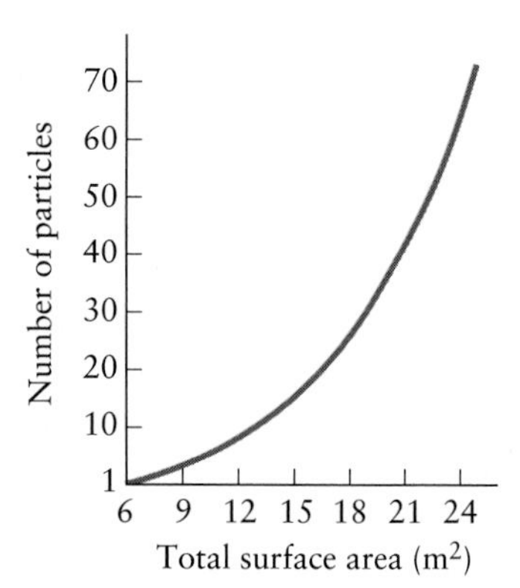

FIGURE 6-3 Physical weathering disaggregates rock into many tiny particles, increasing the amount of rock surface area exposed to chemical weathering. The total volume of the rock remains the same and in place, but the amount of surface area exposed to further weathering increases with the increasing number of particles.

Exfoliation Jointing When rock expands along extensive fractures parallel to Earth's surface, it peels apart in sheets in a process called *exfoliation jointing* (Figure 6-4). Rock that has been confined decompresses as the load of rock above it is removed by erosion. The process is like squeezing an elastic metal spring and then releasing it. Like the spring, rock has some elasticity (recall the elastic rebound theory in Chapter 5). Just as potential energy is stored in a spring when it is compressed, potential energy is stored in rock when it is buried under many thousands of meters of material. As overlying material is removed, the underlying rock expands, generating stresses on adjacent rocks. The amount of stress exerted by rock as it expands along exfoliation sheets is greater than the strength of many rocks. When the stress exceeds the strength of the rock, the rock fractures. The planes along which it fractures, the exfoliation joints, expose many fresh rock surfaces to chemical weathering by water and atmospheric gases.

FIGURE 6-4 Exfoliation weathering in Yosemite National Park, California, produces sheetlike masses of rock with large surface areas that are exposed to other weathering processes.

Thermal Expansion Minerals in rock expand not only when they are decompressed but also when they are heated. In the 1800s, early scientific explorers unfamiliar with hot deserts thought that much of the rubble they saw strewn about the landscapes of Africa, Australia, and North America was caused by intense heating during the day and substantial cooling at night. They thought that extreme daily fluctuations in temperature could cause repeated expansion and contraction of minerals and thereby the disintegration of rock in a process called *thermal expansion.* However, extensive laboratory and field investigations by many scientists have shown no evidence that the temperature ranges typical of hot deserts are sufficient to fracture rock.

Although typical daily temperature fluctuations might not be able to shatter rock in deserts, wood fires have been used in quarries in India to heat rock and loosen slabs up to 15 cm thick from the quarry floor. Some scientiests have proposed that natural fires are likely to reach temperatures high enough to shatter rock as well. Rock is a poor conductor of heat, so the interior of a rock mass receives much less heat than does its exposed exterior surface. As a consequence, outer parts of rock expand in greater amounts than inner parts. This variability in expansion generates substantial differences in stress throughout the mass and causes portions of the rock to break off. Recently, scientists have documented

(a)

(b)

FIGURE 6-5 (a) Rock surfaces that have been spalled by exposure to intense heat, as from (b) a recent range fire in the Sierra Nevada, California. Note the blackened outer surfaces of the spalled flakes from burned organic matter that left a residue.

thin flakes (<5 cm) of rock that have *spalled,* or peeled, off exposed rock surfaces during range fires (Figure 6-5). Fires can occur naturally, as from lightning, and over many thousands of years the process of spalling can be quite effective in causing rock disintegration.

Biological Disintegration The growth and expansion of foreign substances along open spaces in rocks also can cause physical weathering. A common biological example in the upper few meters of the pedosphere is caused by the growth of tree roots. Trees wedge their roots into weak zones in rocks and split them apart as they grow, causing the ever-present bulges and cracks in sidewalks near mature trees.

An even more common, but not as obvious, form of physical weathering caused by biological activity is the continuous disaggregation of rock particles by worms. English naturalist Charles Darwin wrote in the late 1800s:

> Worms have played a more important part in the history of the world than most persons would at first suppose. In almost all humid countries they are extraordinarily numerous, and for their size possess great muscular power. In many parts of England a weight of more than 10 tons [10,516 kg] of dry earth annually passes through their bodies and is brought to the surface on each acre of land. . . . It is a marvellous reflection that the whole of the superficial mould [topsoil] . . . has passed, and will pass again, every few years through the bodies of worms. The plough is one of the most ancient and most valuable of man's inventions; but long before he existed the land was in fact regularly ploughed, and still continues to be thus ploughed by earth-worms.

Since Darwin's astute observations, scientists have documented many instances of physical weathering processes by organisms, including the mound- and tunnel-building of termites, gophers, and prairie dogs.

Frost Weathering Water can split rock. If water in a rock fracture is cooled below its freezing point, it expands as it crystallizes, thereby enlarging fractures and pores in a process called *frost weathering.* The more frequent the cycles of freezing and thawing, the more effective the process of disintegration. Evidence of frost weathering is the accumulation of angular, gravel- to boulder-size debris along slope bases, a common sight in high alpine and arctic areas (Figure 6-6).

FIGURE 6-6 Frost weathering has shaped much of the angular, fragmented rock that has accumulated at the base of this slope in the Sierra Nevada, California. Similar rubble slopes are seen often at high latitudes and in mountainous regions at high altitudes, where freezing conditions are common.

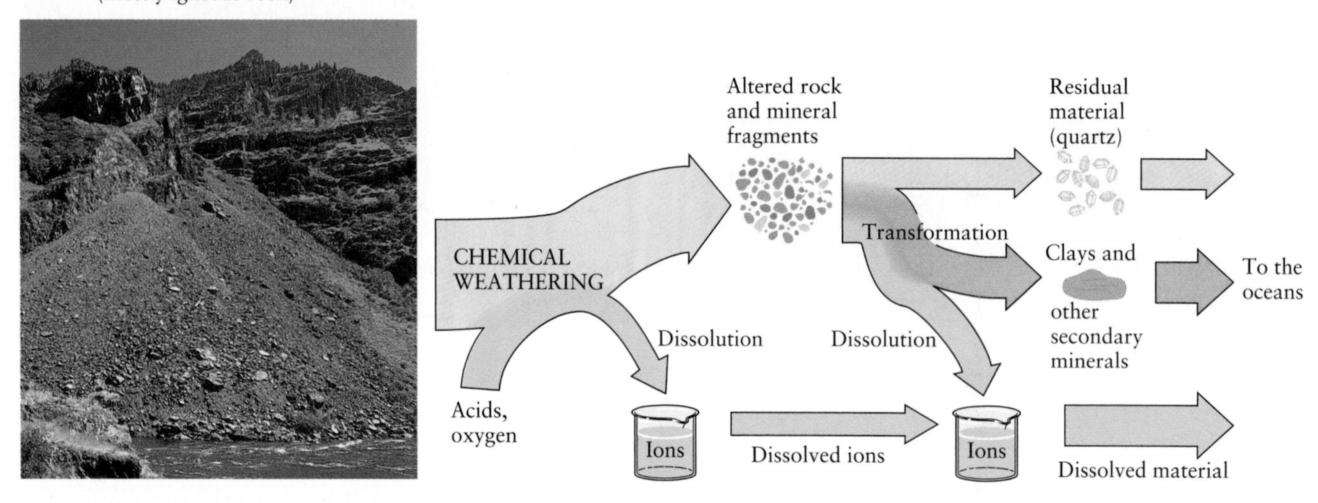

FIGURE 6-7 Chemical weathering decomposes primary minerals in rocks and makes the rock mass vulnerable to fragmentation. Most chemical weathering is caused by the reaction of minerals with acidic solutions and oxygen from the atmosphere, hydrosphere, and biosphere. Products of chemical weathering include fragments of primary minerals and altered (secondary) minerals, like those scattered over these hills in Idaho, and dissolved ions. With continued weathering, residual minerals, clays, and other secondary minerals accumulate in the pedosphere or are washed into the oceans along with dissolved ions.

Chemical Weathering

During chemical weathering, water, ions, and oxygen react with exposed mineral surfaces. The general chemical weathering process is the reaction of unweathered minerals in crustal (mostly igneous) rocks, known as *primary minerals,* with water, acidic solutions, and oxygen from the hydrosphere, biosphere, and atmosphere. The products of these reactions are fragments of primary minerals, newly formed *secondary minerals,* and dissolved ions (Figure 6-7). Together, the residual fragments and secondary minerals released by weathering constitute *sediments.* Mechanical and chemical weathering continues to produce sediments and dissolved ions that eventually make their way to the oceans and are deposited as clastic, chemical, and biological sediments. Most chemical weathering results from exposure to acid, oxygen, or both, as expressed in the following equation:

Primary minerals + acids + oxygen →
sediments (rock fragments + secondary minerals) + dissolved ions

Acid Formation in Soil The effect of acids is a particularly important factor in the decomposition of the upper lithosphere. Acids can partly or even wholly decompose solid mineral structures because they release hydrogen ions in rainwater or groundwater. The small, mobile hydrogen ions replace cations of other elements on mineral surfaces, thus breaking apart the original mineral structure.

The more hydrogen ions circulating in water, the greater the solution's acidity. The acidity of a solution is expressed as its pH or "potential for hydrogen." The pH scale runs from 0 to 14 and is logarithmic, so each unit of pH represents a tenfold change in the concentration of H^+ ions. A pH of 7 indicates a neutral solution; pure water has a pH of 7. A pH value below 7 indicates acidity, and a pH above 7 indicates alkalinity.

The lower the pH value, the more acidic the solution (Figure 6-8). The pH scale is in negative powers of 10, so a pH of 1 indicates that there are 10^{-1} (or 0.1) g of H^+ per liter of solution, a pH of 2 indicates only 10^{-2} (or 0.01) g of H^+ per liter, and so on. Consequently, a solution with a pH of 4 is 100 times more acidic than a solution with a pH of 6, because it has that much greater an amount of free hydrogen ions. While natural rainwater is slightly acidic, with an average pH of about 5 to 6, the pH of rainfall measured in areas downwind of some industrial centers has been as low as 3, an acidity comparable to that of vinegar.

One of the most common acids in the soil environment is carbonic acid, which forms when the water in raindrops combines with carbon dioxide gas in the atmosphere (Table 6-1, equation 1). Because the formation of carbonic acid removes carbon dioxide from the atmosphere, it is a key component of models of global climate

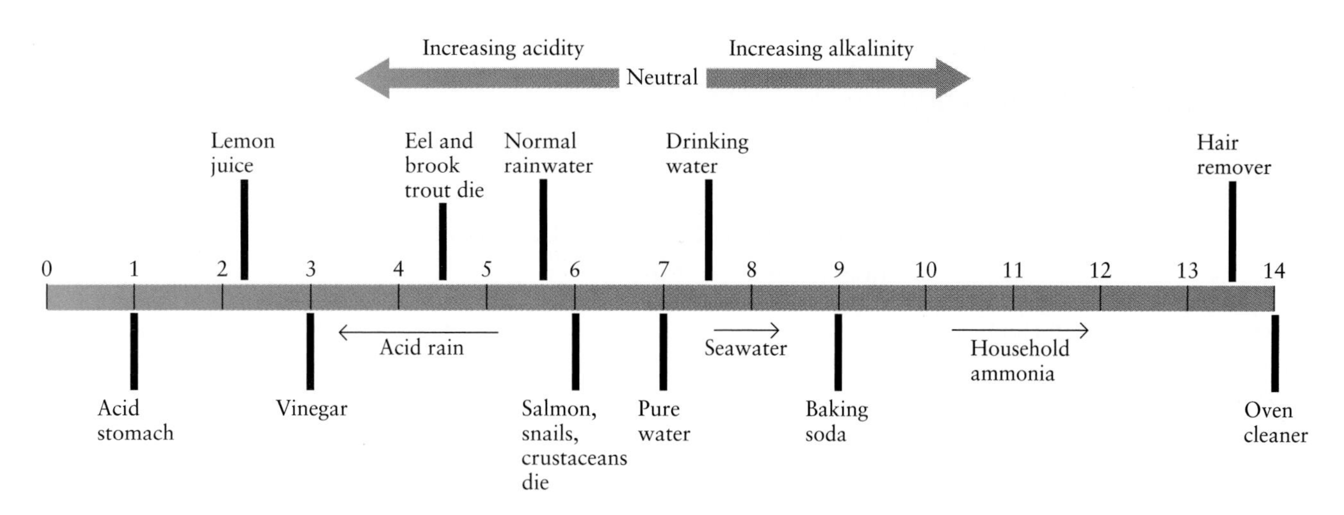

FIGURE 6-8 Common solutions and their acidity or alkalinity on the pH scale and the acidities at which some freshwater life-forms die.

change (Box 6-1). The carbonic acid in unpolluted rainwater gives it a pH between 5 and 6, so it is slightly more acidic than a neutral solution. In the soil zone, however, plants and decaying organic matter release additional carbon dioxide gas that can react with infiltrated rainwater to produce more carbonic acid. As a result, concentrations of carbonic acid are 10 to 200 times higher in soil moisture than in the atmosphere, and the pH of soils is sometimes as low as 4.

Mild as carbonic acid is, it dissolves calcite on contact in a chemical weathering process known as **carbonation** (Table 6-1, equation 2). Calcite ($CaCO_3$) is the mineral that forms the sedimentary rock limestone and its metamorphic equivalent, marble. Carbonation results in

TABLE 6-1 Common Chemical Weathering Reactions at Earth's Surface

Equation 1. Formation of carbonic acid from water and carbon dioxide:

H_2O	+	CO_2	$\leftrightarrow$	H^+	+	$(HCO_3)^-$	$\leftrightarrow$	H_2CO_3
Water		Carbon dioxide gas		Hydrogen ion		Bicarbonate ion		Carbonic acid

Equation 2. Dissolution of calcite in carbonic acid (carbonation):

$CaCO_3$	+	H_2CO_3	$\leftrightarrow$	Ca^{+2}	+	$2\ (HCO_3)^-$
Calcite		Carbonic acid		Calcium ion		Bicarbonate ion

Equation 3. Dissolution of a silicate mineral (sodium feldspar) in carbonic acid (hydrolysis) and formation of a clay mineral (kaolinite):

$2\ NaAlSi_3O_8$	+	$2\ H_2CO_3$	+	$9\ H_2O$	$\rightarrow$	$2\ Na^+$	+	$2\ (HCO_3)^-$	+	$4\ H_4SiO_4$	+	$Al_2Si_2O_5(OH)_4$
Sodium feldspar		Carbonic acid		Water		Sodium ion		Bicarbonate ion		Silicic acid		Kaolinite

Equation 4. Oxidation of pyrite and formation of iron oxide and sulfuric acid (common cause of acid mine drainage):

$4\ FeS_2$	+	$15\ O_2$	+	$8\ H_2O$	$\rightarrow$	$2\ Fe_2O_3$	+	$8\ H_2SO_4$
Pyrite		Oxygen gas		Water		Hematite		Sulfuric acid

6-1 Global and Environmental Change

The Relationship Between Weathering and Long-Term Climate Change

Recent scientific investigations indicate that rates of weathering might contribute to long-term global environmental changes that take place over thousands to millions of years. One possible scenario begins with the rapid uplift of a large landmass, such as the Tibetan Plateau and Himalayan Mountains of Asia. Tectonic uplift of this broad region in the past few tens of millions of years has raised the landmass so high that glaciers have formed at high altitudes and air masses containing moisture evaporated from the Indian Ocean have been blocked. Glaciers carve valleys into bedrock, and moist air masses produce intense storms that contribute to river flow. Both of these processes result in increased rates of landscape erosion, which in turn exposes fresh rock to weathering.

If the amount of fresh rock exposed at Earth's surface increases, the chemistry of the atmosphere could be altered by increased rates of chemical weathering. Water and carbon dioxide in the atmosphere combine to produce carbonic acid, which dissolves silicate and carbonate minerals and releases the bicarbonate ion $(HCO_3)^-$ in solution. Increased chemical weathering could mean an increase in the amount of bicarbonate in solution. Ultimately, bicarbonate ions make their

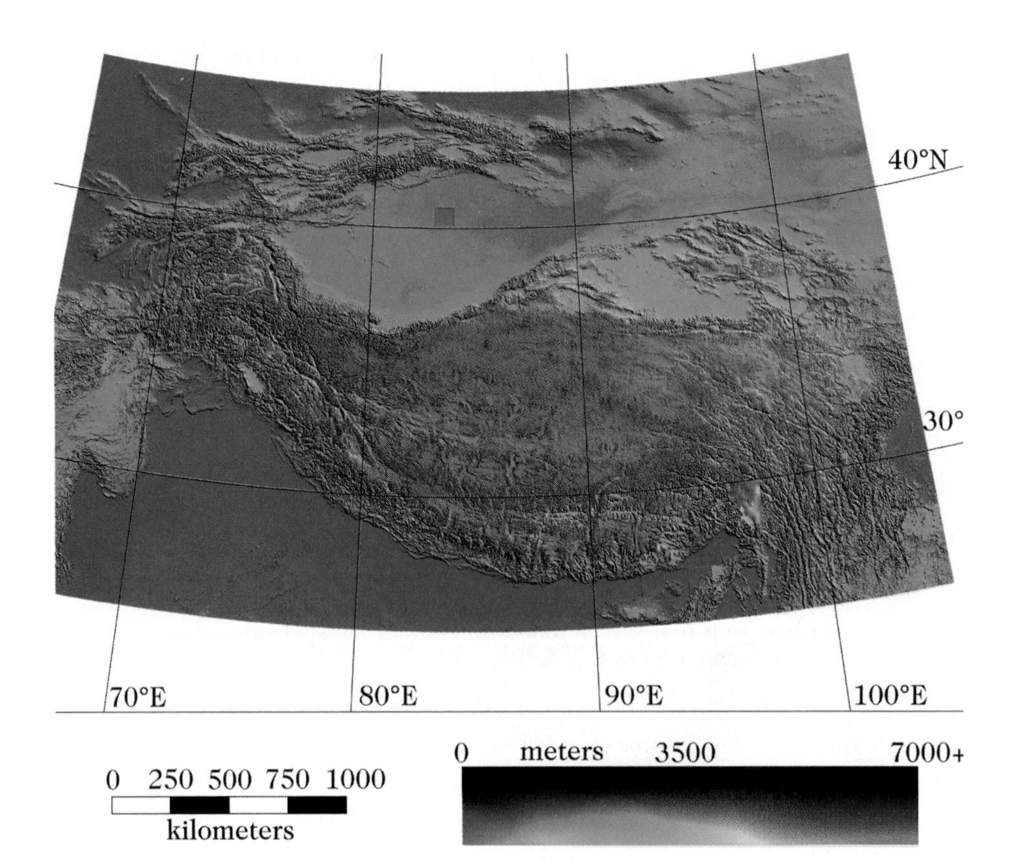

This topographical map of the Tibetan Plateau shows the extensive area of land that has been raised many kilometers above sea level (green, yellow, and shades of red) by recent tectonic uplift.

way downstream to the nearest ocean, where they can be recombined with calcium ions in the calcareous shells of marine organisms or crystallized into limestone. Either way, carbon that had been stored in the atmosphere is moved to the oceans, the amount of heat-absorbing carbon dioxide in the atmosphere is reduced, and global climate undergoes a cooling trend.

Considering all the attention being paid to global warming, you might think that this scenario does not make sense. In fact, the global climate has been cooling for the past few tens of million years and, in particular, for the last few million (see Figure 3-10). The modern warm period is far cooler than the global climates of about 25 million years ago, but it is warmer than the cyclical full-glacial episodes that have marked the past 2 million to 3 million years. In a relative sense, the climate of the past few million years is a time of very cool conditions for Earth, even though slight warming has occurred over the past 10,000 years as Earth has entered one of the warmer interglacial intervals. Even slighter has been the warming in the past century, which scientists estimate to be about 0.5° to 1°C. It is this warming to which scientists refer when warning of "global warming" due to human activities.

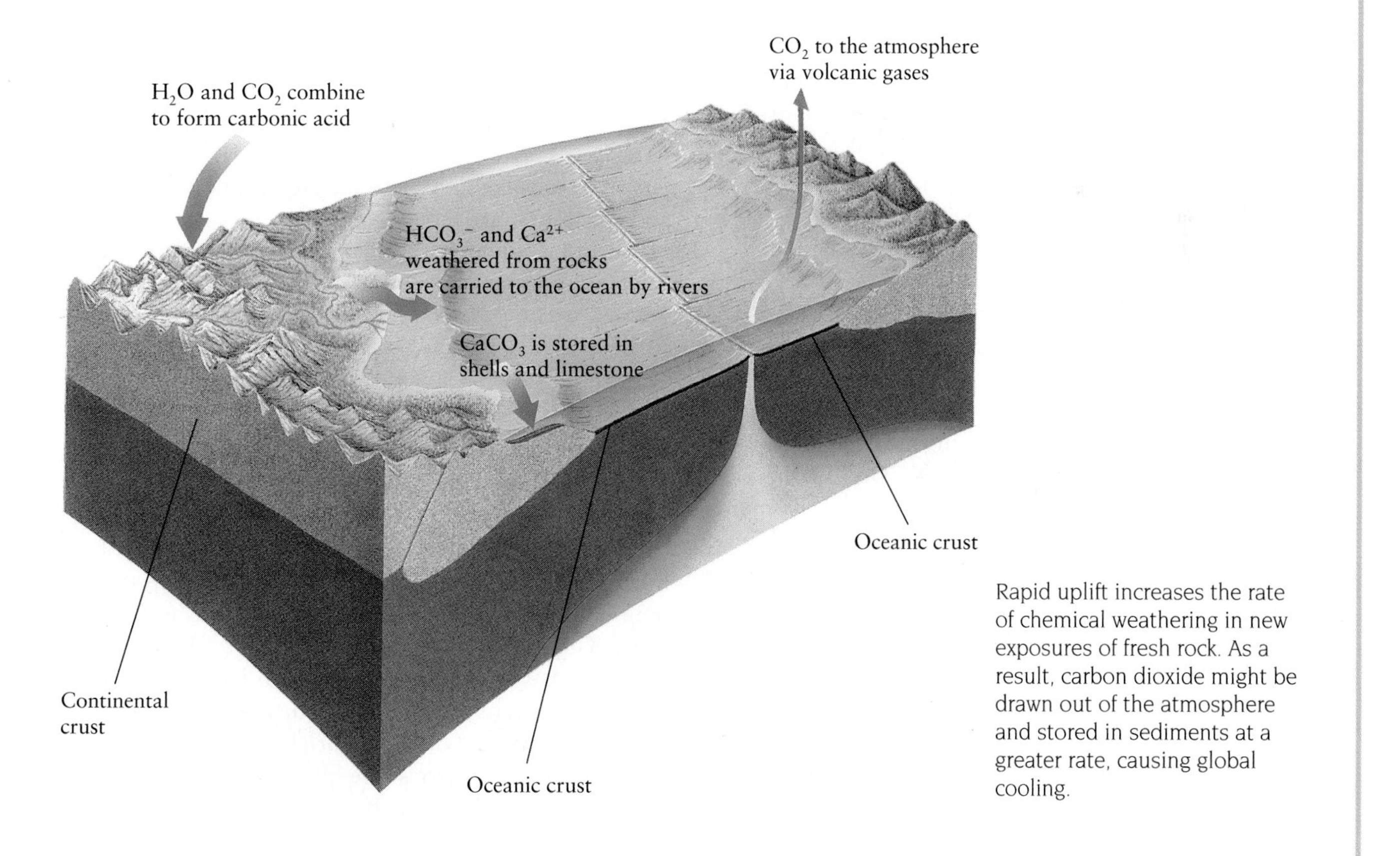

Rapid uplift increases the rate of chemical weathering in new exposures of fresh rock. As a result, carbon dioxide might be drawn out of the atmosphere and stored in sediments at a greater rate, causing global cooling.

dramatically etched landscapes and features such as sinkholes, which form when the rock structure and overlying soils collapse into caverns and solution fractures along which rock has been removed by carbonation (Figure 6-9).

Carbonic acid is not the only source of acid in soil. Organisms produce a number of organic acids. For example, plant roots release citric acid, the acid in lemons and other citrus fruits. The more organic matter in the soil, the more acid the soil will be. Acids less common in most soils are nitric and sulfuric acids, which result from the combination of water with nitrogen dioxide and sulfur dioxide, respectively. Soils developed on rock types that contain sulfide minerals such as pyrite (an iron sulfide; see Chapter 4) sometimes have high concentrations of sulfuric acid, especially where mining activities have crushed the rock and exposed large areas of sulfide mineral surfaces to oxygen.

Carbonic and other acids attack silicate minerals in a reaction called **hydrolysis** (Table 6-1, equation 3). Because most rocks in the lithosphere consist of silicate minerals, hydrolysis is the most common form of chemical weathering. Hydrogen ions from acidic soil solutions replace cations in silicate mineral structures, and the silicate cations in turn are released into solution. Some of the remaining elements and altered minerals form secondary **clay minerals.** Clay minerals are crystalline sheet-structure silicates characterized by small particle size.

Clays are an extremely important component of a soil, and together with organic matter they contribute to properties such as a soil's ability to hold water and nutrients (Box 6-2). When Bennett and Lowdermilk demonstrated to members of Congress the role of a healthy topsoil in absorbing moisture and protecting the ground surface from erosion, they were illustrating the properties of a soil that contains a mixture of tiny weathered rock fragments, weathered secondary clays, and organic matter.

Oxidation The second basic process of chemical weathering common at Earth's surface results from the abundance of oxygen in the atmosphere. During **oxidation,** an element combines with oxygen to form oxide or hydroxide minerals (Table 6-1, equation 4). One of the elements most commonly oxidized is iron, which is abundant in the silicate minerals hornblende, biotite, olivine, and pyroxene. The oxidation of pyrite (FeS_2), the most abundant of sulfide minerals, alters it to an iron oxide called hematite (Fe_2O_3), and the released sulfur dissolves in water to form sulfuric acid. This strong acid is common in the water runoff, or drainage, from mining operations. *Acid drainage* from mines into bodies of water at the surface has caused extreme environmental degradation (Figure 6-10).

Solubility Most elements and minerals are to some extent soluble in soil water, where typical pH values range from 4 to 9. However, solubilities vary enough that some elements are more likely to be removed from the soil and carried to streams by percolating waters, while others accumulate in soil with time. Calcium, magnesium, sodium, and potassium are highly soluble and are the most common cations found in streams, while aluminum, iron, and silica are commonly found in soils. The residual elements left in a weathered soil typically form aluminum and iron oxide minerals, clay minerals, and quartz.

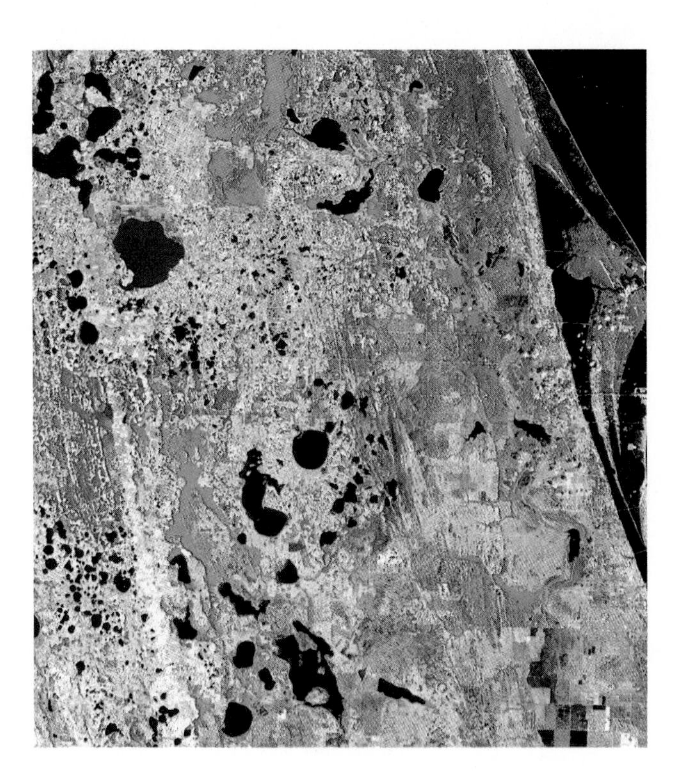

FIGURE 6-9 A satellite view of eastern Florida, where much of the subsurface is cavernous limestone and the surface is marked with sinkholes where cavern roofs have collapsed. Caverns and sinkholes are common in carbonate rocks, such as limestone, which are dissolved easily by carbonic acid. The point of land at right is Cape Canaveral.

Soil Profiles and Soil-Forming Factors

Earth scientists describe **soil** as an internally organized, natural body of weathered mineral and organic constituents arranged in **soil horizons,** zones roughly parallel to Earth's land surface. The vertical arrangement of soil horizons forms a **soil profile,** which often appears as multicolored layers in road cuts and housing foundations. The soil profile varies with climate (Figure 6-11). Soil scientists label soil horizons with letters, starting at the surface. Above all is the O *horizon,* a litter of dead plants and animals over the surface. Below the O horizon, all other horizons are placed in three separate groups known as the A, B, and C horizons. In moist, temperate climates, the *A horizons* form the dark, organic-rich soil, also known as *topsoil.* Below are the *B horizons,* mineral-rich layers that vary in color from browns, reds, and yellows to grays and blues. Clays and other minerals, weathered from upper horizons and

FIGURE 6-10 Acid mine drainage, containing sulfuric and other strong acids, is created by the oxidation of sulfide minerals exposed during mining operations. At this copper mine in Michigan, acidic waters (red) are divided and treated before being released into nearby streams.

transported downward, accumulate in the B horizons. At even greater depths, soils grade into slightly weathered parent material with less distinctive C *horizons,* then to weathered parent material with no horizons, and finally into unweathered parent material.

The total thickness of the pedosphere varies from 1 m to 200 m, with greater depths in areas where rainfall and temperature are high and weathering processes extend deeply into the crust. Horizonated soil, however, which is intimately linked to the plants and animals at the surface, is rarely more than a few meters thick.

Soil profiles in Nebraska and Ethiopia are very different from those in Hawaii, Brazil, or northern Canada, because soil-forming factors vary from one location to another. Studies of soil variations throughout the world point to five major *soil-forming factors:* climate, relief of the ground surface, nature and composition of parent material, amount of time over which the soil has formed, and amount and types of living organisms.

Climate may be the most significant factor in the nature and fertility of a soil. The climate of an area is characterized by its precipitation and temperature. Precipitation

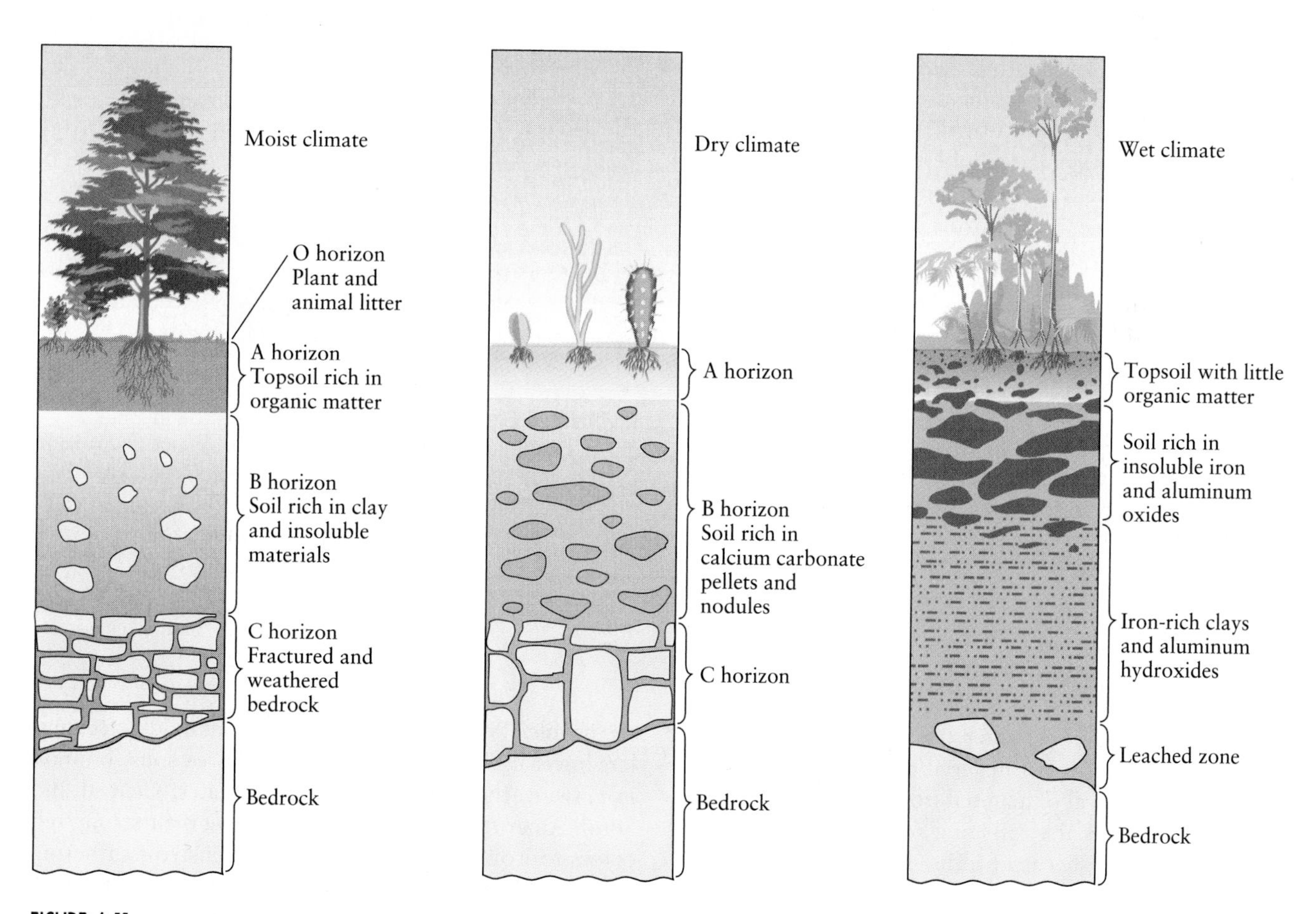

FIGURE 6-11 Soil profiles in moist, dry, and wet climates differ primarily in the amounts of organic matter and the thickness of their A horizons, and in the minerals that accumulate in the B horizons. In very wet climates, a thin topsoil (A horizon) sometimes overlies a subsoil that is so extensively leached that only layers of iron and aluminum oxides and clays remain. Technically, these horizons are all B horizons, but they are harder to define than other soils.

6-2 Geologist's Toolbox

The Unusual Properties of Clay Minerals

Earth scientists spend much time determining the percentage of sand, silt, and clay in a soil. In the field, they do this by wetting small samples of soil with water, then rubbing the soil between thumb and forefinger: Sand is gritty, silt is velvety, and clay is sticky and plastic. Knowing the amount of clay in a soil is crucial to determining its fertility and engineering qualities.

Soil consists of broken bits of primary minerals as small as sand (0.05 to 2 mm in diameter) and silt (0.002 to 0.05 mm in diameter), but its smallest mineral component—clay—makes it quite different from ground-up parent rock. Clay minerals, which are less than 0.002 mm in diameter, are secondary minerals created by the transformation of primary minerals during chemical weathering. In contrast to the larger and relatively inert grains of sand and silt, tiny clay minerals are highly reactive. Clays absorb and expel water, shrink and swell, attract cations, and retain nutrients, compounds that can be used directly by plants, such as carbon dioxide or water. Were it not for the capacity of clay minerals to hold nutrients in soils, land plants would not have been able to colonize the continents during the early Paleozoic era.

More than 20 different types of clay minerals are formed by the chemical weathering of mica, feldspar, olivine, and amphibole silicate minerals. Scanning electron microscopes, with their powerful magnification, give Earth scientists a close look at the structure of clay minerals. Clay minerals are constructed from flat sheets of mostly silicon, oxygen, and aluminum that are bonded to one another by ion-sharing with cations. When arranged in stacks, the sheets produce layered clay mineral structures with extensive surface areas, much like the pages in a book. One gram of clay (about the volume of a pencil eraser) has as much as 800 square meters (about 9000 square feet) of surface area, nearly the area of an Olympic-size swimming pool!

The surface area of clay minerals affects their interactions with ions, other particles, and water. Negative ions—anions—occur along the edges of clay sheets. Oppositely charged particles attract each other, and the positive cations in soil solution are attracted to

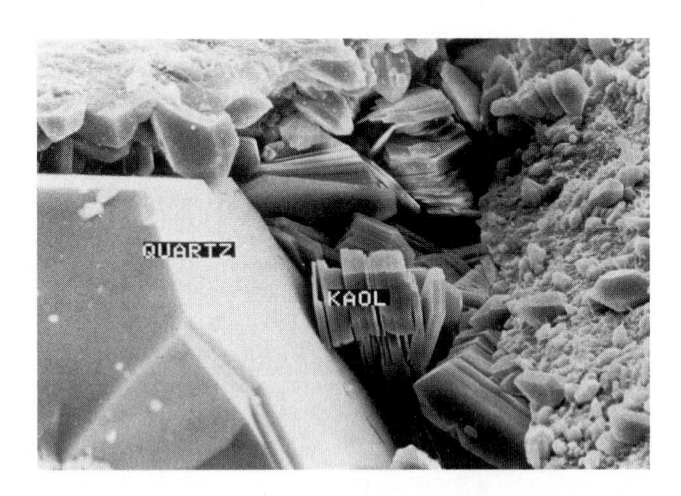

Kaolinite, a clay mineral formed during weathering, has grown in open pore spaces adjacent to the quartz grains seen in this scanning electron micrograph. The booklike structure of the clay mineral is evident.

affects the amount of water available for weathering processes, for transfer of material throughout a soil profile, and for removal of material from a soil. Temperature also affects chemical weathering reactions, which occur more readily at higher temperatures.

In warm, wet climates, the extent of leaching and removal of iron, aluminum, and organic matter from lower A horizons sometimes is so great that a nearly white, quartz-rich *E horizon* forms in the upper part of the profile. Typically, iron and aluminum oxides accumulate lower in the soil profile. The bright red soils common in Hawaii, the southeastern United States, Central and South America, and Southeast Asia owe their strong red colors to iron oxides produced by extensive weathering and oxidation (see the chapter-opening photograph). In contrast, calcium carbonate and some other salts can precipitate in dry-region soils when soil waters carrying dissolved ions are depleted by evaporation and con-

the clay particles, in effect swarming about negatively charged clay mineral sheets much as bees swarm about a hive. From the clay mineral surfaces, cations can move freely to the roots of plants, where they are drawn upward and distributed to stems and leaves. Some of these ions are also plant nutrients. One teaspoonful of soil with abundant clays and organic matter might contain as many as 1.2 quintillion (1.2×10^{21}) sites where nutrients can be held and exchanged with plants. The attraction and retention of nutrients by clay minerals keeps them in the root zone; otherwise they would be washed away with water moving through the soil.

Clay minerals also contribute to the properties of soil *expansion* and *plasticity,* both of which increase susceptibility to irreversible deformation. Soil expansion occurs when water enters clay mineral structures and they increase in volume. Soil plasticity refers to a soil's ability to deform in a pliable, rather than an elastic or brittle, manner. Both these properties are related to water content in the soil. Water molecules are asymmetrical in charge, with positive and negative poles, so their positive poles collect about the negatively charged surfaces of clay minerals in an oriented fashion, much the way iron filings are attracted to a magnet. Clays with weakly bonded sheets can incorporate a lot of water between the sheets, causing substantial expansion of the mineral structure during rainy seasons. Certain *expandable clays* can increase their volume by as much as 2000 percent when wet. Such soils can become too wet and sticky to till, and clay-rich dirt roads can become impassable.

Furthermore, the attraction between the individual clay particles is reduced as the mineral is wetted and swells; thus the particles become more mobile and can slip past one another more easily, resulting in plasticity. Plastic soils lose their rigidity, or their strength, when wet, causing roads and building foundations constructed above them to fail. During dry seasons, expandable clays expel water and shrink substantially. Although soils with expandable clays become more rigid when dry, they also can crack and become too hard to till.

Soils with certain types of clay minerals are very sticky and plastic when wet and can expand, or swell, by absorbing water into their mineral structures. During dry seasons, expandable clay soils shrink and harden; deep cracks result.

sumption by plants. In much of the American Southwest, northern Africa, and the Middle East, desert soils often contain nodules and even thick layers of hard calcite, known as caliche, that have precipitated over thousands of years.

The Interaction of Earth Systems to Form Soil

The pedosphere is an open system with fluxes of matter and energy from the lithosphere, biosphere, hydrosphere, and atmosphere. The interaction of these Earth systems contributes to weathering and the formation of soils through a variety of processes that fall into four categories: *additions, chemical transformations, transfers,* and *removals* (Figure 6-12).

The lithosphere contributes to soil formation by adding mineral matter from rocks and sediments. Soils begin as fresh rock materials, such as layers of volcanic ash and lava, or glacially deposited sands and gravels, or

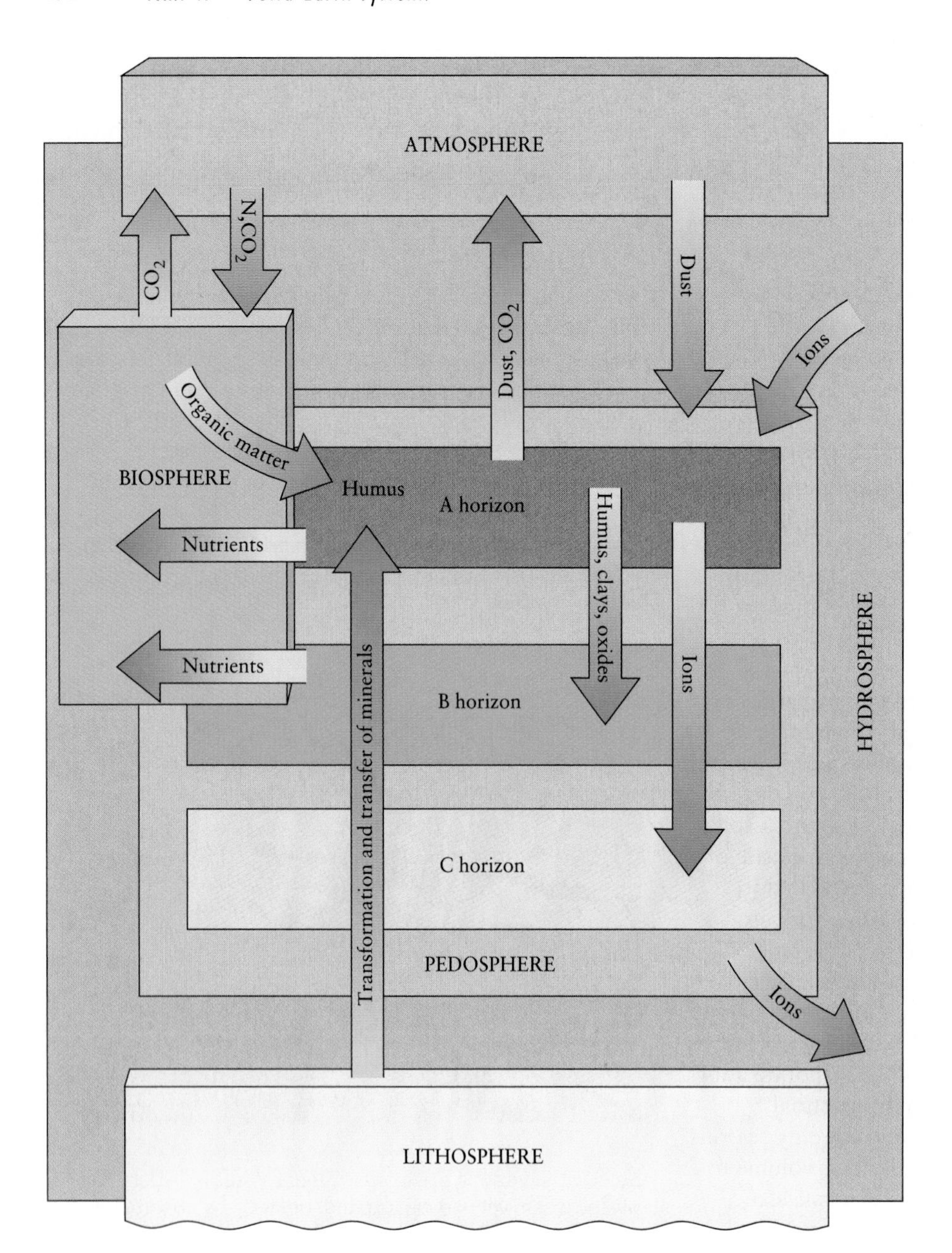

FIGURE 6-12 The interaction of the atmosphere, hydrosphere, biosphere, pedosphere, and lithosphere in soil formation is characterized by the cycling of matter and the development of distinctive soil horizons.

crystalline igneous rocks raised to Earth's surface by mountain-building processes. Stored in minerals are elements that are essential to organisms for growth and sustenance. These elements include phosphorus, calcium, sulfur, magnesium, potassium, and iron. Weathering processes transform primary minerals to secondary minerals, such as clays and oxides, and release ions from minerals to soil water, making them accessible to plants and animals.

The hydrosphere is the Earth system most responsible for the transfer of solid and dissolved substances among soil horizons and the removal of ions from soil profiles. Ions in solution can be removed if soil water continues downward to join with groundwater flow, which transports the ions to streams that drain watersheds. Some soluble substances removed from the upper part of the soil profile are precipitated deep in the soil because of the changing chemical nature of the soil horizons. Insoluble substances (clays and other minerals) added to the soil or produced during transformation can be transferred downward, primarily in water, but the depth of movement generally is limited to several meters by the decreasing size of pore spaces. For these reasons, the lower parts of most soil profiles contain layers of accumulation of clay and oxide minerals (that is, B horizons). Clay minerals contribute to the stickiness and plasticity of B horizons, whereas iron oxides contribute to their typically reddish and yellowish-brown colors.

The biosphere adds organic matter, which enters the soil as litter from dying plants and animals. Decay transforms organic matter and releases carbon dioxide gas, most of which recirculates to the atmosphere, and nutrients, used by soil organisms and plants. Some of the nutrients are synthesized into large, complex, organic molecules with a high molecular weight known as **humus.** Containing carbon, nitrogen, and phosphorus, humic substances are essential to soil-forming processes: They help to bind soil materials together, they help to transfer ions among soil horizons, and they retain ions in the soil. In addition, humic molecules act as soil sanitizers because their large molecular size enables them to trap pesticides, herbicides, and toxic metals (such as lead). Once trapped, pesticides and herbicides are broken down as the soil decomposes.

The atmosphere and hydrosphere interact at the soil surface to add many elements to soil: Sodium, potassium, magnesium, calcium, chloride, and other ions enter the soil in precipitation, particularly near coasts where sea spray is abundant. Pollutants, such as the carbon dioxide and sulfur dioxide derived from power-plant and automobile emissions, also enter the soil in precipitation. Solid particles such as wind-blown dust are added to the soil. One of the most important atmospheric additions is nitrogen, which enters the soil through a series of biological transformations as nitrogen cycles from the atmosphere to the soil and back.

▸ Cycling of Nitrogen Among Atmosphere, Biosphere, and Pedosphere

The **nitrogen cycle** (Figure 6-13), the continuous flow of nitrogen through the atmosphere, biosphere, and pedosphere, is as essential to life as are the water and carbon cycles. Because nitrogen is a key ingredient in proteins and the nucleic acids DNA and RNA, it is a *limiting nutrient* for life—that is, life is limited by the amount of available nitrogen. Although nitrogen is the most abundant element in the atmosphere (79 percent by volume),

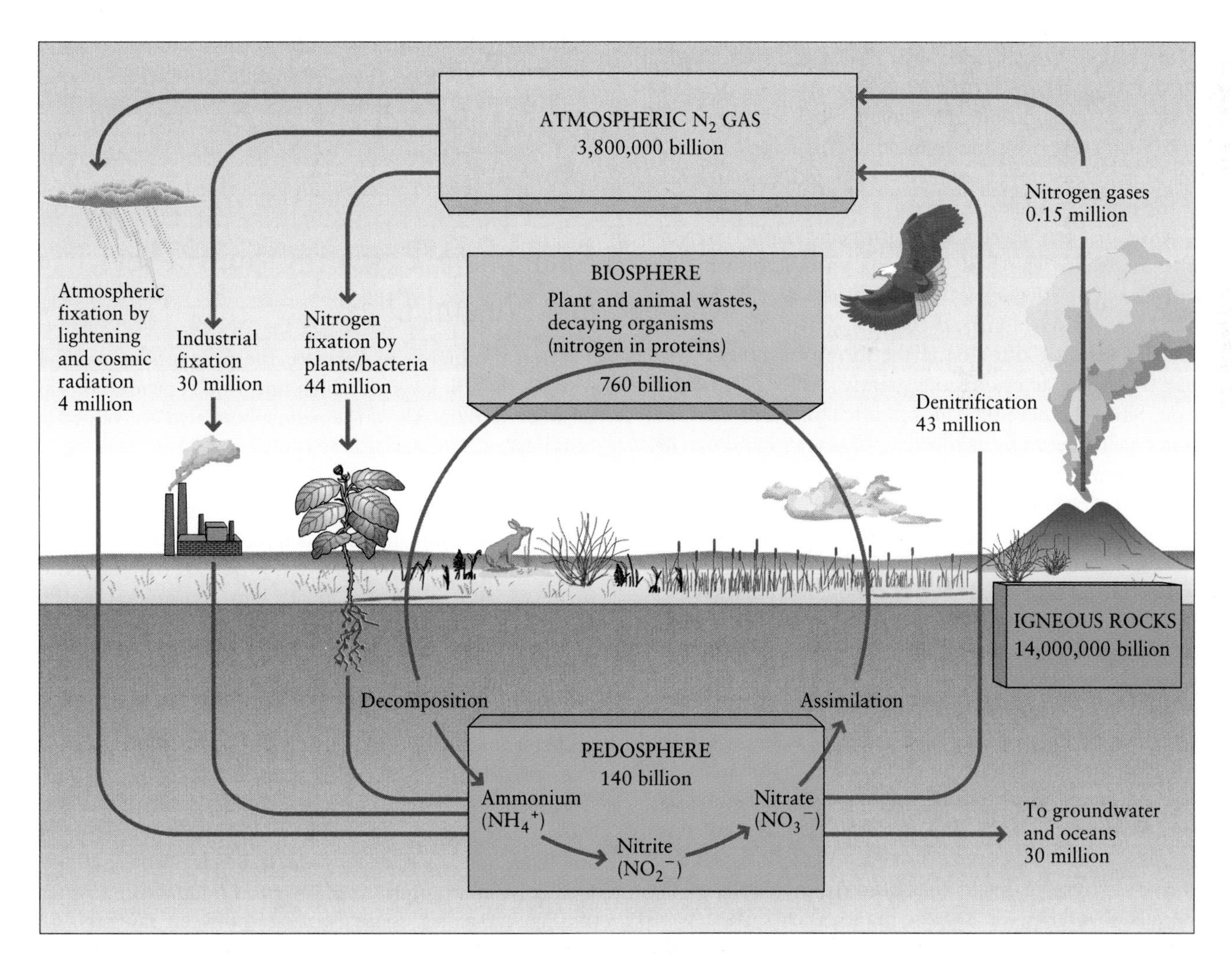

FIGURE 6-13 The nitrogen cycle—with stocks and flows—for continental land areas. Rates of annual flow are indicated by the arrows, and stocks are shown as rectangular boxes. Flow rates are in millions of metric tons per year and stocks are in billions of metric tons.

it is an inert gas, and only a few organisms are able to use it directly. These organisms are valuable for their ability to "fix" nitrogen—to extract the inert gas (N_2) from the atmosphere and combine it with oxygen or hydrogen into a reactive form usable by plants and animals.

A few types of soil bacteria are able to fix atmospheric nitrogen. Many of these bacteria live in a mutually beneficial relationship, or symbiotically, with certain higher plants, commonly in masses of nodules attached to the roots of clovers, bean plants, and other legumes. The bacteria supply the plants with usable nitrogen and feed off the sugars and starches made by the plants. This arrangement enables the bacteria to survive and reproduce and, more important from the plants' point of view, to continue breaking apart N_2 molecules for the plants' use.

As organisms die and litter the soil, their protein molecules and the nitrogen they contain become available to other soil microbes. They convert part of the nitrogen to ammonium (NH_4^+), a usable plant nutrient, and oxidize some of the ammonia to nitrite (NO_2^-) and nitrate (NO_3^-), both of which are also usable plant nutrients. Not all the nitrogen is recycled, however; some returns to the atmosphere by *denitrification,* the natural conversion of organic nitrates into gaseous nitrogen. This entire process, from the fixing of nitrogen gas by bacteria to its release from organisms back to the atmosphere, is the nitrogen cycle.

Humans are altering the nitrogen cycle in two important ways. The first is the increasing cultivation of nitrogen-fixing legumes throughout the world. The second, known as *industrial fixation,* is the use of fossil fuels to produce nitrogen in the form of synthetic fertilizer. Nitrogen fixation now greatly exceeds denitrification. Since the development of agriculture, nitrogen fixation has increased by as much as 10 percent, and much of that increase has occurred in the 20th century.

A direct result of increased nitrogen fixation has been the increased discharge of nitrogen into streams and rivers, loading them with extra nitrogen. Nitrogen is washed from farm fields during runoff and makes its way to lakes, bays, estuaries, and the ocean. In many bodies of water, the excessive supply of nitrogen and other fertilizers in nutrient form intensifies biological activity and can produce algae blooms, the rapid growth of algae in response to an overload of nutrients. When algae and other plants die and settle to the bottoms of lakes and bays, the organisms that decompose them consume oxygen, depleting it from the water. Ultimately, nitrogen-loading results in increased oxygen depletion, which in turn destroys organisms that need oxygen, such as fish. Because fish graze on aquatic plants, their demise worsens the problem of plant overpopulation. This overall process of nutrient-induced oxygen depletion is known as *eutrophication* and has been blamed for the deaths of many fish in aquatic environments.

- The alteration of rock and mineral matter to broken fragments, clays, and ions in the pedosphere is called weathering.
- Physical weathering is the mechanical disintegration of rocks and minerals. Common processes of physical weathering are exfoliation jointing, thermal expansion, biological disintegration, and frost weathering.
- Chemical weathering is the chemical decomposition of rocks and minerals. Carbonation, hydrolysis, and oxidation are dominant processes of chemical weathering.
- Acids in soils contribute to the chemical weathering of minerals into secondary minerals such as clay, a major component of soil that has the ability to hold water and nutrients.
- Soil profiles are vertical arrangements of layered organic and mineral matter in the uppermost few meters of the pedosphere.
- Soil profiles vary throughout the world because of variability in five major soil-forming factors: climate, topography (relief), rock type, types of plant and animal life, and time.

Pedosphere Resources: Soils, Clays, and Mineral Ores

The most valuable resource of the pedosphere is fertile topsoil, the soil layer in which our food is produced. The weathered materials of the pedosphere also are valued as building and industrial materials.

The Fertile Soil

Although all undisturbed soils have topsoil, the relatively organic-rich upper layer of a soil produced by weathering processes and accumulation of plant and animal matter, not all topsoil is especially fertile, that is, suitable for growing plants. The *fertility* of a soil depends on the amount and availability of essential elements such as nitrogen, carbon, potassium, phosphorus, and calcium. Nutrients are cycled from the roots of plants to the stems and leaves and then returned to the soil as litter, maintaining the soil's richness. In a fertile topsoil, the surface soil is dark, contains much humus, and forms stable crumbs that keep the soil loose and facilitate aeration. It also is porous and permeable, qualities that make for good drainage and water-storage capacities. In dry periods between rainfalls, the topsoil slowly releases water to plants.

Soils are classified according to diagnostic characteristics, many of which make them useful for particular purposes. For example, Pennsylvania farmers in the market

TABLE 6-2 Primary Soil Orders According to U.S. Classification

Soil Order (% of world total)	Source of Name	Description
Alfisols (14.7)	"Al" for aluminum; "f" (from Fe) for iron	shallow penetration of humus; accumulation of clay in lower horizons; high in cations†; well-developed horizons
Andisols (~1)*	from Japanese *ando*, "dark soil"	usually in volcanic deposits; dark in color; high content of organic matter and cations (especially phosphorus)
Aridisols (19.2)	from Latin *aridus*,"dry"	all horizons dry for more than 6 months per year; low in organic matter; high in cations
Entisols (12.5)	from "rec*ent*"	mineral matter dominant; no distinct horizons; very young soils (e.g., as on floodplains)
Histosols (0.8)	from Greek *histos*, "tissue"	high content of organic matter (e.g., peat)
Inceptisols (15.8)	from Latin *inceptum,* "beginning"	relatively young soils in humid environments; horizons indicate transformation but little transfer or accumulation; usually moist; typically shallow soils
Mollisols (9.0)	from Latin *mollis,* "soft"	nearly black surface horizons rich in organic matter and bases; common in grasslands and forests with understory plants such as ferns
Oxisols (9.2)	from French *oxyde,* "oxide"	substantially weathered soils typically on ancient landscapes in tropical and subtropical environments; rich in kaolinite (clay), iron oxides, and sometimes humus; thick and deep B horizons
Spodosols (5.4)	from Greek *spodos,* "wood ash"	soils with upper horizon that is light-gray to white and extensively leached (E horizon) and overlying a reddish B horizon with substantial accumulation of iron, aluminum, and clay
Ultisols (8.5)	from Latin *ultimus,* "last"	intensely leached soils, typically in humid, warm climates; B horizons marked by substantial clay accumulation; low content of cations
Vertisols (4.9)	from Latin *verto,* "to turn"	dark, clay-rich soils that shrink and swell, forming wide cracks during dry season (also called "expansive" soils)

* This soil order was added recently for soils formed in volcanic ash. It includes, for example, many of the soils in Hawaii.

† Common cations in soils are Ca^{2+}, Na^{+}, K^{+}, Mg^{2+}, Al^{2+}, and Fe^{2+} and Fe^{3+}.

Source: From information in Hans Jenny, *The Soil Resource,* New York: Springer-Verlag, 1980.

for land would look for Chester silt loam, a well-drained soil that supports especially high crop yields, while real estate developers in Arizona would avoid the Cave silty clay loam, a soil with a thick, hard, calcareous horizon (one containing calcium carbonate) at a shallow depth that is so difficult to excavate that dynamite is occasionally required.

The soil classification scheme developed by the U.S. Department of Agriculture groups all soils in the United States into one of 11 orders (Table 6-2) and subdivides them further into suborders, great groups, subgroups, families, series, and types. Each soil order ends with the suffix "*sol,*" from *solum,* the Latin word for "soil." The soil orders closely correspond to variations in precipitation and temperature, which are important soil-forming factors (Figure 6-14). The division of soil orders into increasingly finer subdivisions is based on variations in other soil-forming factors, such as topography, which affect soil properties such as horizon thickness and clay content.

For about two-thirds of the United States, soil types have been mapped on a mosaic of black and white aerial photographs and published for each county (Figure 6-15). For example, two soil series well suited to cultivation in Lancaster County, Pennsylvania, are the Chester silt loam and the Manor silt loam. Home to the Amish culture, this county has some of the highest crop yields in the world for

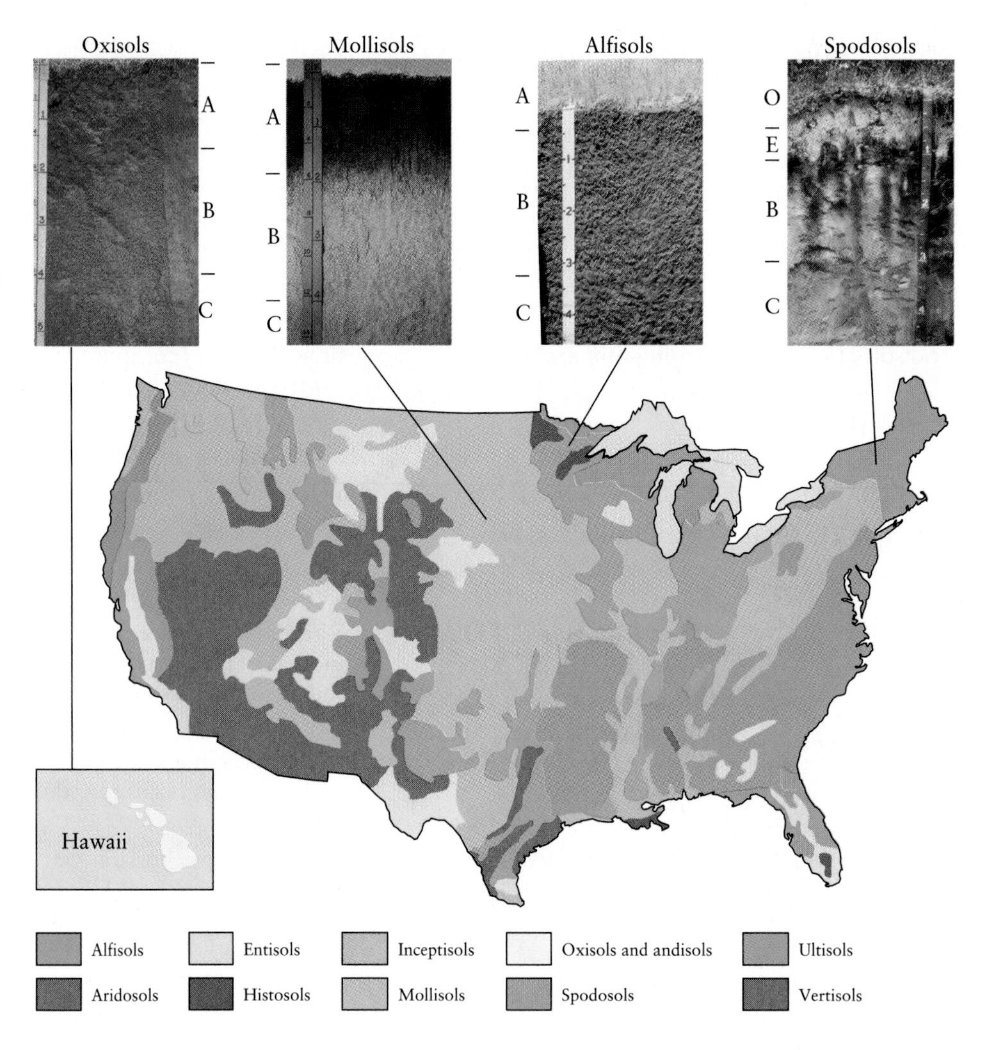

FIGURE 6-14 Soil orders mapped in the United States. Profiles for four soil orders, with horizon designation, are shown above the map. Note the association between geographic location, climatic conditions, and soil order. For example, aridosols are common in the dry Southwest, whereas mollisols are common in the wetter Midwest and ultisols along the humid East Coast.

nonirrigated farmland. Both soil series have developed in weathered metamorphic rocks. The Chester is an ultisol, and the Manor is an inceptisol (see Table 6-2). The thicker Chester soil occurs on broad ridge tops, the Manor soil on valley slopes.

Soil surveys published by the Department of Agriculture are especially useful for environmental planning. Tables and descriptions include information such as depth to bedrock and water table, soil permeability, drainage conditions, and potential for shrinking and swelling—all factors that indicate different uses of land. The suitability of a soil type for cultivation, woodland, range land, wildlife, and recreation is listed as well as soil properties such as color, plasticity, and strength (the ability to support the weight of buildings, bridges, and other engineering structures). Soil maps are valuable for suggesting not only the best use for land, but also the number of people who can be supported by a given area of land (Box 6-3).

Clay, Laterite, and Mineral Ores

Of all the mineral resources used by humans, clay ranks fifth in terms of mass annually produced (see Figure 5-2). In 1988, each person in the United States used, on average, 195 kg of clay. Clay is an essential ingredient in pottery, ceramics, bricks, and tiles. When mixed with water and molded to different shapes, then baked either by the Sun or in a kiln, clay becomes a durable and waterproof material. Many early humans mixed clay with other ingredients, such as plant fibers and animal hairs, to make bricks for constructing houses. Archaeologists have found Sun-dried bricks dating to 4000 BC in the ruins of Sumeria and fired bricks dating to the first century AD in Roman ruins. Ancient brick and tile structures also have been found in Egypt, Asia, and the American Southwest.

One of the most durable brick materials comes from a type of oxisol known as a *laterite,* from *later*, the Latin word for "brick." In wet regions, some oxisols are so extensively leached that the A horizons have little organic matter and the B horizons are composed almost solely of iron and aluminum compounds. Typical minerals in the laterite subsoil are kaolinite (aluminum clay), goethite (iron hydroxide), hematite (iron oxide), gibbsite (aluminum hydroxide), and quartz. The iron oxides and hydro-

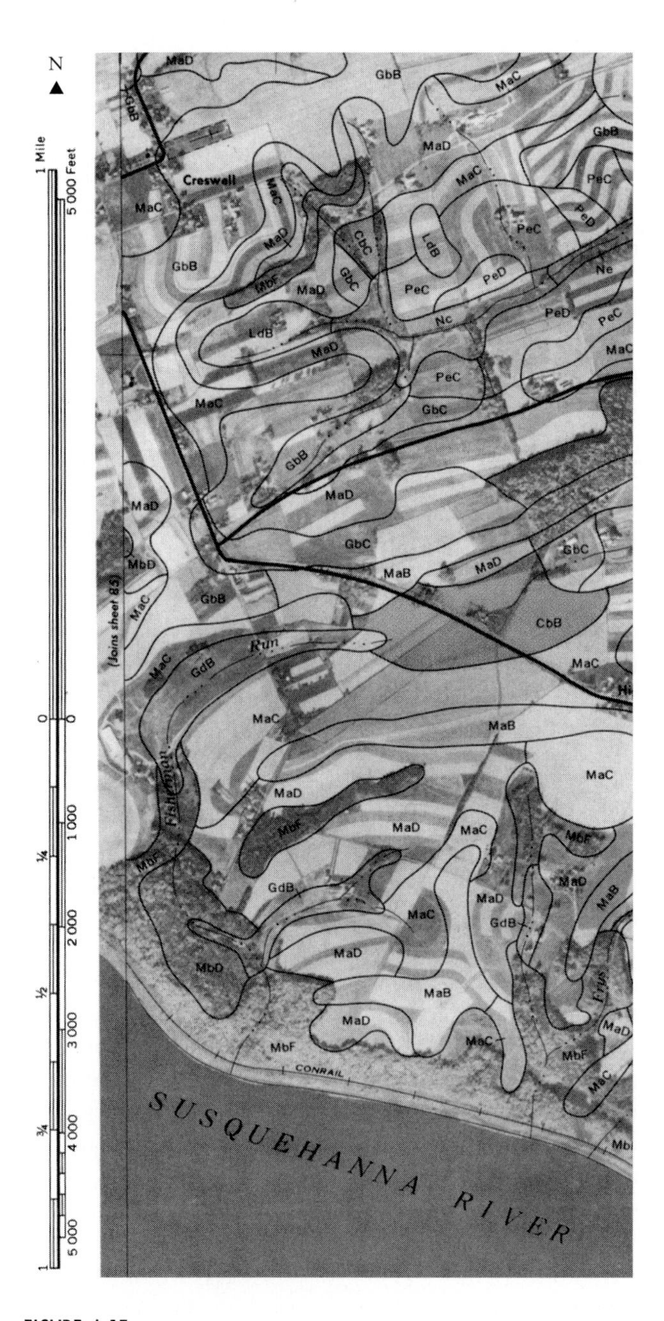

FIGURE 6-15 Detail from soil map for Lancaster County, Pennsylvania, along the Susquehanna River. Most of the area is agricultural or woods. The thin black lines are contacts between soil types; the thicker black lines are roads. Abbreviations represent soil map units: For example, CbB is the Chester silt loam on a slope of 3 to 8 percent. CbC is the same soil type on a slope of 8 to 15 percent.

xides impart a striking brick-red color to the soil. The clay causes the soil to harden to the consistency of brick when exposed to air and sunlight (Figure 6-16).

Fresh exposures of red laterite soils typically reveal a mottled, soft, earthy material that can be cut with a blade. When excavated and dried, however, this clay-rich soil becomes extremely hard and cannot be resoftened with water. For thousands of years, people have used this type of soil to make bricks; ancient laterite temples are found in India, Thailand, and Cambodia. The very properties that make laterites desirable for building materials also make them impossible for cultivation.

The processes of chemical alteration that yield fertile soils and clays also can produce ore minerals in the pedosphere. With time and extensive leaching, weathering processes can remove all but the most insoluble elements in the pedosphere, which accumulate as *residual deposits.* If leaching is particularly extensive, the deposits can become highly enriched in residual minerals. The least soluble minerals in the weathered zone are goethite and two aluminum hydroxide minerals, gibbsite and diaspore. The residual aluminum ores of gibbsite and diaspore are referred to as *bauxite.* Although iron ores occur in both the lithosphere and pedosphere, aluminum ore occurs only in the pedosphere. Large amounts of residual iron and aluminum ore minerals are produced in Australia and Brazil.

- Fertile soil, one of the world's most valuable resources, is characterized by good drainage, relatively large amounts of organic matter (humus), and clay that can hold water and nutrients.
- Weathering processes produce clay minerals that are valuable for many purposes, such as the manufacture of brick and tile.
- Iron and aluminum are relatively insoluble and, as a result, accumulate in deeply weathered soil profiles, forming rich ore deposits in some parts of the world.

FIGURE 6-16 Laterite soil in Brazil quickly hardens and cracks when exposed to air and sunlight after deforestation if the ground is not replanted immediately.

6-3 Case Study

Using Soil Maps to Estimate National Carrying Capacities

Soil is vital to producing food for human consumption and thus is an essential component of any assessment of a nation's carrying capacity—the hypothetical number of people that can be supported by a given amount of land (see Chapter 1). In 1984 the United Nations Food and Agriculture Organization (FAO) used soil maps and climatic information to estimate the carrying capacities of 117 developing nations under three different technological and economic scenarios: low-optimization, which assumes traditional farming practices; medium-optimization, in which some fertilizer is used; and high-optimization, which includes the use of herbicides, pesticides, fertilizer, and improved crop varieties. The results for the different regions are shown in the accompanying table. For example, the value of 3.0 for Africa in 1975, given a low-optimization scenario, means that the region could support three times its population. A value of less than 1 means that the region has insufficient soil resources to feed its populace.

The table shows that the overall average carrying capacity for different regions is cause for both optimism and concern. In the year 2000, most regions will have sufficient resources to feed their populations. However, the potential population-supporting capacity diminishes markedly for every region and every technological scenario between 1975 and 2000, excepting only Southeast Asia. Far more discouraging, however, are the specific values for the individual countries within these regions. Fifty-five of the 117 countries had a value of less than 1 in 1975; the result was famine and thus starvation. By 2000 the number of countries at risk is expected to increase to 64, and much of southwestern Asia could suffer famine.

The FAO study has some shortcomings. On one hand, it does not take into account variables like the availability of nonterrestrial food sources such as fish. On the other hand, it assumes that all the potentially cultivable land is used to produce food rather than cash crops such as tobacco or coffee. Moreover, it does not consider the role of international trading: A nation with plentiful oil resources but little arable land might trade with another for food. In the modern global arena, a variety of resources can be exchanged for food.

Events of the 1980s and 1990s have borne out some of the predictions of the FAO study. A number of nations suffered food shortages, mass starvation, and exodus of refugees. However, the causes have been many and not limited to the extent and fertility of available soil: Catastrophic droughts, civil strife and political unrest that prohibit transport of food, and inequitable distributions of resources all have their effects. Cash crops often are grown on the best soils, while food is grown on marginal lands of low productivity. The result is an even greater acceleration in soil erosion and lowering of fertility of the marginal land. The gaps between having data, interpreting them, and implementing sound land-use policies are dangerously wide.

Potential Population-Supporting Capacities Divided by 1975 and 2000 Populations

	1975 Ratios			2000 Ratios		
	Low	Intermediate	High	Low	Intermediate	High
Africa	3.0	11.6	33.9	1.6	5.8	16.5
Southwest Asia	0.8	1.3	2.0	0.7	0.9	1.2
South America	5.9	23.9	57.2	3.5	13.3	31.5
Central America	1.6	4.2	11.5	1.4	2.6	6.0
Southeast Asia	1.1	3.0	5.1	1.3	2.3	3.3
Average	2.0	6.9	16.6	1.6	4.2	9.3

Source: U.N. Food and Agriculture Organization (FAO), *Land, Food, and People,* Rome: FAO, 1984.

Soil Erosion Hazards and Soil Conservation

Weathering creates soils, clays, and ore minerals from primary minerals in the lithosphere; other processes in the pedosphere move the loosened debris from one place to another. *Erosion* is the general name for all processes that transport loosened Earth material downhill or downwind; it results from the action of water, wind, glacial ice, and gravity. Erosion caused primarily by gravity is called mass wasting. While erosion is a natural process that serves to cycle materials through Earth systems, it also can pose serious environmental hazards to humans. Soil erosion and *desertification,* which is the degradation of land by both erosion and soil deterioration, and soil conservation strategies designed to minimize soil erosion, are considered here; the hazard of mass wasting will be discussed in the next section.

Soil Erosion: A Quiet Crisis

When walls of dust towered above the Great Plains in the 1930s, few doubted that the accelerated erosion of the soil was due to negligent farming practices during a time of drought. In 1996 a drought of equal severity destroyed much of the region's wheat crops, but it did not result in such massive soil erosion. The difference lies in conservation practices learned from experience. Elderly farmers who were children during the "dirty thirties" have lived to see the valuable effects of planting trees as wind breaks and leaving grasses in place to provide some vegetation cover between row crops. Nevertheless, accelerated soil erosion still occurs throughout North America as well as on every other continent on which humans practice agriculture.

In Greece, geoarchaeologists have found sequences of sediments that indicate periods of increased rates of erosion caused by changes in land use. Layers of sediments that were deposited by streams that drained farmed hillslopes are dated to times of deforestation and agricultural activity, both of which accelerate erosion. As erosion intensified, abused sites were abandoned by prehistoric farmers and soils were able to form on the new sedimentary deposits. With time, new waves of expansion and agricultural activity occurred, repeating the sequence and resulting in additional layers of sediment and soil that buried the older deposits (Figure 6-17).

Both soil formation and soil erosion have occurred simultaneously in every soil on Earth since long before humans began using soils for agriculture. In the absence of any disturbance, natural or human-induced, the rates of soil formation and erosion generally balance each other. Typical rates of soil formation are very slow on a human time scale, only 0.02 to 0.11 mm per year. In other words, the formation of a single meter of soil depth requires 10,000 to 50,000 years.

In some parts of the world today, rates of erosion due to human activity are 18 to 100 times greater than the natural rate of soil renewal. In the 1980s the WorldWatch Institute, a research group based in Washington, D.C., projected a 32 percent decline in the amount of topsoil per person between 1984 and 2000.

The environmental impact of soil erosion is felt on the slopes and fields where erosion occurs, as well as

(a)

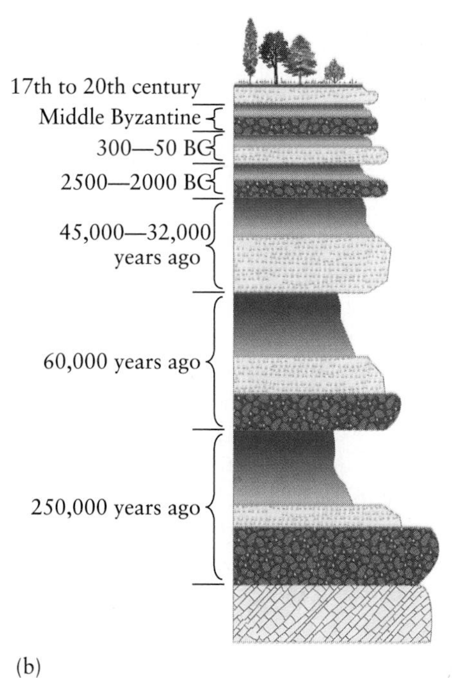

(b)

FIGURE 6-17 (a) Thin, buried soils formed on sedimentary deposits that resulted from accelerated erosion caused by deforestation and poor farming practices in ancient Greece. Each buried soil represents a time of landscape stability, when rates of soil formation exceeded those of soil erosion. (b) Older deposits and soils, dating to past glacial-interglacial cycles between 32,000 and 250,000 years ago, underlie the more recent deposits.

downstream, where increased sediment chokes streams, rivers, and bays. On site, the loss of topsoil reduces soil productivity through the removal of organic matter and clay, both of which retain nutrients and moisture. Degraded soils have lower water-holding capacities and are more susceptible to erosion, as demonstrated by Bennett and Lowdermilk in the 1930s. Downstream, sediment washed into waterways damages fish life, coral reefs, reservoir-storage potential, and navigable waterways.

Worldwide, nations suffer billions of dollars in damage and repairs and unestimated losses in natural resources due to soil erosion. The threat of accelerated soil erosion associated with poor farming practices and rapid deforestation is one of the world's most serious environmental crises. It is, however, a quiet crisis. The development and application of large amounts of synthetic fertilizers, herbicides, pesticides, and improved crop varieties in the 1950s —called the Green Revolution—has increased crop yields in many parts of the world, masking the problem of soil erosion and creating a false sense of security. In fact, this valuable resource is being "mined"—that is, overused—in many places. At some urban construction sites and heavily logged areas, soil has been completely depleted.

Although many factors contribute to soil erosion, the two most important are the degree of slope and the amount of bare soil that is exposed to wind and water. The steeper the slope and the less the amount of cover, the greater the rate of soil erosion and runoff of water. For different amounts of cover, ranging from soils protected by woods to those that are bare, soil losses range from nearly zero to about 70 tons per acre, and the amount of water lost in runoff ranges from less than 1 percent of precipitation to nearly 50 percent. In Nigeria, the rates of annual soil erosion range from 3 metric tons per hectare on shallow slopes of 1 percent to 221 metric tons per hectare on steeper slopes of 15 percent. (Slope percentage refers to the rate of increase in elevation with distance. A 1 percent slope rises 1 m in elevation in 100 m of horizontal distance, whereas a 15 percent slope rises 15 m.)

Soil Erosion by Water Rainsplash and sheet wash erosion are responsible for most soil erosion by water in temperate regions. **Rainsplash** occurs when raindrops hit exposed (that is, unvegetated) soil during intense rainstorms, lifting fine particles high above the soil surface and, if there is a slope, transporting them downhill. If the soil contains much clay, swelling of the soil makes the surface impermeable to additional water, and continued rainfall accumulates as a nearly uniform sheet of water flowing across the ground surface. Close examination reveals that the sheet actually consists of numerous tiny rivulets that form an intricate pattern of joining and splitting "threads." Because of their rapid migration across the soil surface, over time they remove a fairly uniform layer of soil, which is why this process is called **sheet wash erosion** (Figure 6-18).

FIGURE 6-18 In the past half century, the bare, trampled ground beneath this rural village in Ghana has been lowered at least 1 m below the level of the road (the curb is at right) as a result of sheet wash erosion. As a consequence, the roots of trees and bases of buildings have become exposed. Furthermore, concentrated sheet wash has led to the incision of gullies, like that seen beyond the tree.

The amount of water increases toward the bottom of the slope and some tiny rivulets might increase in depth and abrasive power until small, temporary **rills** several centimeters wide and deep are gouged from the landscape. If these incised channels become so large that they cannot be removed by a plow, and become permanently entrenched features, they are called **gullies.** Gullies have dimensions of several to many meters in depth, tens to hundreds of meters in width, and lengths as great as several kilometers (Figure 6-19). Over a small, local region, rilling and gullying may be the most evident cause of soil erosion by water.

Desertification One of the greatest global environmental concerns is the transformation of once productive, or marginally productive, land to deteriorated land and soil unable to support plants and animals. Because the land becomes barren and dry, the process is described as **desertification,** although it occurs in many regions other than true deserts. With increasing population pressure, every bit of available land is used. Much land that once was productive is over-exploited, and other land that once was considered marginally productive is converted for farming and grazing. Many marginal lands have steep slopes, are prone to high rates of erosion, and are more fragile than other lands. The result is widespread degra-

FIGURE 6-19 Rill and gully erosion in the midwestern United States during the years of the Dust Bowl in the 1930s.

dation and deterioration of the soil and land resource base. Desertification occurs as a sequence of events:

- Exposure and excessive stress dries out the soil.
- Native plant species decline, resulting in less production of organic matter.
- As soil fertility is reduced, the soil hardens.
- Because less water can infiltrate the hardened soil during storms, soil drying and hardening are exacerbated in a positive feedback process.
- During rainstorms, water runs off the landscape rather than infiltrating, scouring the soil so that rilling, gullying, and widespread erosion occur.

Depending how far this sequence has progressed, deteriorated land is classified as moderately, severely, or very severely desertified. In cases of moderate desertification, crop yields are reduced by 10 to 50 percent. In severely desertified land, soil is heavily eroded and crop yields can be reduced by more than 50 percent. In cases of very severely desertified land, all vegetation is gone, and crop yields are reduced by more than 90 percent. The United Nations has estimated that 40 percent of Africa's nondesert land, 32 percent of Asia's, and 19 percent of Latin America's is at risk of desertification. More than 50 percent of the nondesert land in Australia is moderately to severely desertified, and 30 percent of the nondesert land in North and Central America and Europe is moderately to severely desertified. In the past few decades, nearly 40 percent of the world's agricultural lands have been moderately desertified or worse (see Chapter 13).

Assessing Soil Erosion

Since the 1940s, many researchers have gathered information on soil erosion in order to develop an equation that can be used to predict rates of erosion under different land-use or soil-conservation conditions. Erosion plots are used to monitor the amount of soil removed from a given area by various factors, such as degree of slope and crop management, which are varied experimentally. From years of work on thousands of plots, scientists have discovered a relationship between erosion and four factors that control its rate: climate (rainfall intensity and duration), soil erodibility, topography (hillslope length and steepness), and land cover (erosion control and crop management). The experimentally determined relationship among these factors is called the **universal soil loss equation:**

$$\underset{\text{Annual soil loss}}{A} = \underset{\text{Climatic factor}}{R} \times \underset{\text{Soil erodibility}}{K} \times \underset{\text{Topography}}{LS} \times \underset{\text{Erosion control} \times \text{crop management}}{CP}$$

Using this equation, it is possible to estimate how much erosion might occur if a given crop type is planted, or if the intensity of rainstorms increases. The soil loss equation has been used in the United States to establish erosion-control and crop-management practices necessary to maintain soil erosion rates below a level of **tolerable soil loss,** which is quantified as a T-value. The tolerable soil loss for a given soil is the erosion limit below which a soil can yield a high level of crop productivity indefinitely—in other words, the T-value is the rate of soil loss at which the crop can be sustained. Depending on the soil type and its rate of renewal, the T-value varies from about 1 to 6 tons per acre per year for different parts of the United States. In the 1980s, erosion rates were as high as 10 tons per acre on cultivated land in many parts of the United States. Small wonder, then, that the U.S. Environmental Protection Agency states that soil erosion is one of the nation's five gravest environmental problems.

Soil-Conservation Practices

To conserve soil and water, the U.S. Department of Agriculture advocates terracing, contour plowing, preserving remnant woodlands as wind barriers that create shelter belts, planting vegetation barriers along waterways, and building structures to trap water and sediment (Figure 6-20).

Terracing, still used throughout the world, is an ancient practice. Evidence of terraces is found at archaeological sites thousands of years old in South America, North America, Africa, and Eurasia. Terraces conserve

(a)

(b)

FIGURE 6-20 (a) Aerial view of a Wisconsin watershed, illustrating different ways to prevent soil erosion, including contour stripping, vegetation barriers, and small reservoirs to trap water and sediment. (b) Handmade terraces have been used for centuries on fields overlying fertile soils developed in volcanic ash erupted from Misti volcano in Peru.

soil and water by reducing the length and steepness of a hillslope by dividing it into relatively short segments that trap or slow the downward movement of sediment and water. By reducing the topography component of the universal soil loss equation, the amount of soil that can be removed by erosion is reduced. Contour plowing is a method of cultivation in which the plows (and other farm machinery) follow contour lines, or lines of equal elevation, rather than moving up and down the farm slopes and crossing contour lines. This prevents the formation of rills and gullies along wheel ruts. Furrows with small ridges created by plowing act as miniterraces, slowing or stopping the downward movement of water and soil.

These conservation strategies have reduced the rates of erosion on much American farmland. However, because of economic pressures and the lack of a strong national policy to protect the soil resource, sound land-use practices are not always strictly adhered to. Since the 1970s, only about 50 percent of farmers have practiced the conservation methods recommended by the U.S. Department of Agriculture, and erosion rates again have been increasing in some areas. Because soil forms slowly, over thousands of years, the result of increased erosion rates will be thinner soil. Eventually, agricultural yields will decrease on those thinned soils, despite increased use of fertilizer.

- Processes that transport loosened earth material downhill or downwind by the action of water, wind, glacial ice, and gravity are called erosion.
- The main processes of soil erosion are rainsplash, sheet wash, rilling, and gullying.
- In recent decades, widespread deterioration and erosion of soil associated with overgrazing and poor farming practices have resulted in desertification in nearly 40 percent of the world's agricultural lands.

Mass Movement Hazards and Their Mitigation

Earth materials can move down a slope primarily as a result of the force of gravity: This type of erosion is called **mass wasting,** whereas the actual process of downslope motion is referred to as **mass movement.** All loose, weathered material on hillslopes is prone to some form of gravitational movement of mass, seen in the slow creep of clay-rich soil on a gentle hillside or the rapid fall of large blocks of rock from a cliff wall. Unlike other erosion processes, mass movement is viewed as a hazard more because of its catastrophic, short-term effects than because it removes soil from the landscape. Like soil erosion, mass movement is a natural process that occurs worldwide, but it has been accelerated by human activities in many places. The hazard of mass movement grows as worldwide population grows and more people settle in areas where risk of mass movement is high.

The downward movement of Earth mass is considered hazardous if it threatens human life, safety, property, or other features of value to humans. The hazard is substantial. In the United States, mass movement causes 25 to 50 deaths and $1 billion to $2 billion in economic losses each year. In China, a single catastrophic event killed more than 200,000 people in 1920. The most recent mass movement disaster took 25,000 lives in 1985, when a small volcanic eruption from South America's northernmost volcano, Nevado del Ruiz in Colombia, generated a large mudflow. The snow and ice cap over the volcano's summit melted during the eruption, providing a source of liquid to mobilize loose volcanic debris from the mountain's slopes. The torrent of mud and water emanating from the summit surged downstream at 30 km per hour, burying at least 22,000 people sleeping in the town of Armero and several thousand others nearby (Figure 6-21.)

FIGURE 6-21 By melting the volcano's ice cap, the small volcanic eruption of Nevado del Ruiz in 1985 resulted in a flow of mud that moved rapidly down the volcano's slope to the town of Armero, Colombia, killing nearly 25,000 people.

The Roles of Gravity and Water in Mass Movement

If mass movement is caused by gravity, which affects the entire Earth, why is it more likely to occur on some hillslopes than on others? And why do some mass movements occur suddenly and unexpectedly, perhaps after several days of rain, with dreadful consequences to unsuspecting residents below? The answer has to do with resistance to the forces of gravity. For a mass of earth to move, the force of gravity must be greater than any resistance of the weathered material. The force of gravity itself does not change substantially at Earth's surface, but the amount of resistance, or *strength*, of the material relative to the force of gravity does vary. It varies from one place to another, and it varies over time at a given location. Variation in space depends largely on the slope of the surface on which the debris rests. The steeper the slope, the more likely it is that mass movement will occur. Variation at a given location over time depends largely on the amount of moisture in the mass.

Consider a flat surface, on which debris remains motionless and at rest. The force of gravity pulls the loose debris downward, holding it in place. On a sloping surface, however, the force of gravity can be separated into two components: (1) a *slope-perpendicular* component that pulls the debris in a direction perpendicular to the slope and helps to hold it in place; and (2) a *slope-parallel* component that acts to move the debris along and down the slope (Figure 6-22). The perpendicular component of gravity provides frictional resistance against mass movement, while the slope-parallel component provides a driving force that favors mass movement.

The steeper the slope, the greater the slope-parallel component of the force of gravity relative to the slope-perpendicular component. At some critical angle of slope from the horizontal, the slope will become unstable. The critical angle above which failure occurs is called the **angle of repose,** and is about 30° to 35° in loose, dry material such as sand and gravel.

Slope steepness cannot be the only criterion for the strength of a material, because some mass remains stable on steep slopes for many thousands of years and then fails suddenly, usually after a prolonged, heavy rain. Water is a second factor in the strength of a material. If water enters the pore spaces between particles in weathered material, the material's resistance to movement changes. A certain amount of water can provide greater resistance because it holds particles together by **surface tension,** a force caused by the attraction of water molecules to one another. Along a water surface, molecules are pulled into the water by the attraction of the molecules underneath, creating a force that can bind water and particles; this effect is seen in the cohesion of moist sand used to build a sand castle. However, if water completely fills the pore spaces surrounding the particles, then the pressure of the water in the pore spaces pushes the grains apart, causing

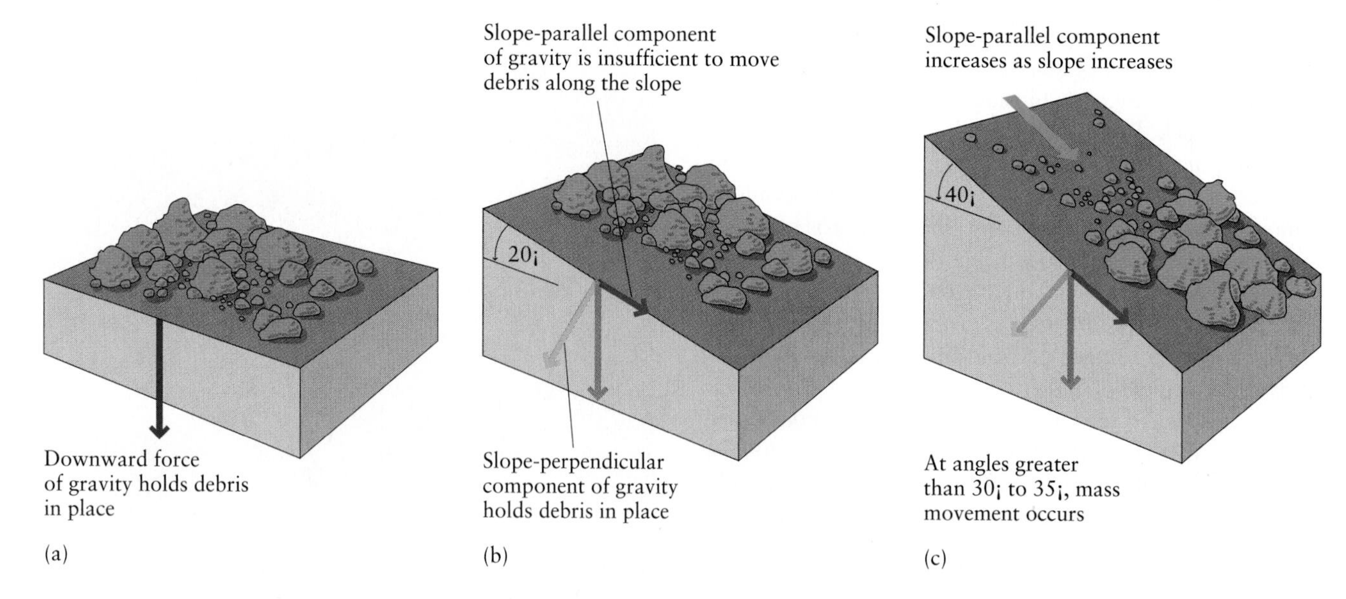

FIGURE 6-22 The effect of increasing slope on the relative resistance, or strength, of loose debris. (a) The force of gravity is perpendicular to Earth's surface and holds debris in place on a flat slope. (b) On a sloping surface, the force of gravity can be separated into slope-perpendicular and slope-parallel components. (c) On a slope steeper than about 30° to 35°, the slope-parallel force is enough to initiate mass movement.

them to move more freely. In this case, water lowers the resistance of the mass to movement, and the mass then behaves more like a liquid than a solid. When too much water is added, the sand castle collapses.

Types of Mass Movement

The phenomenon of mass movement presents a spectrum of many possibilities, from rockslides to debris flows and mudflows. With the exception of pure free fall of rock debris, the three basic mechanisms of mass movement are heave, slide, and flow (Figure 6-23). The **heave** mechanism, caused by alternating expansion and contraction of debris from freezing and thawing or wetting and drying, raises and lowers material in a direction perpendicular to the hillslope. Because of the slope and gravity, some material makes its way downhill as it is dropped back to the soil surface. **Slide** occurs when cohesive blocks of material move, or fail, along a well-defined plane, and thus the term *landslide* is seen to have a very specific meaning. **Flow** occurs when the debris moves like a fluid and, in contrast to sliding, no clear plane of failure exists within or below the moving mass.

All types of mass movement except rockfall can be attributed to one of these three mechanisms, but the type of debris, the amount of water, and the speed at which the debris moves can vary substantially. As a result, many different types of mass movement have been identified, including rockslides, debris slumps, debris slides, and debris flows (see Figure 6-23).

The mass movement most closely associated with heave is *soil creep,* a very slow process that involves varying amounts of water. Evidence of soil creep is the downward curvature of trees and gravestones, and the tilt of power poles. Because of the ubiquity of soil creep, cemeteries usually are located on flat parts of the landscape.

Rockslides, debris slumps, and debris slides can occur with little or no water; they typically move rapidly. Intermediate between slides and a flow of stream water that carries some sediment (which is not considered mass movement) are *debris flows* and *mudflows.* Of the two, mudflows contain more water relative to sediment. Typical amounts of water in debris flows and mudflows are 10 to 30 percent of the total weight of the moving mass. The Armero, Colombia, disaster was caused by a mudflow (see Figure 6-21). Debris flows are tens to hundreds of times more viscous—resistant to flow—than honey, and can transport enormous boulders, houses, and heavy equipment. Debris flows are a common form of mass movement in the mountains that surround Los Angeles, where more than 120 sediment-trapping basins have been constructed at a cost of hundreds of millions of dollars to catch debris on its way downhill, before it reaches buildings and roads.

Causes and Prevention of Mass Movement

All the causes of mass movement can be attributed to one of two factors: Either the driving forces acting on a slope are increased, or the resisting forces are decreased (that is, the strength of mass is reduced) (Table 6-3). Preventing mass movement requires that either the driving forces be reduced, or the resisting forces be increased.

A common cause of both increased driving force and decreased resisting forces along a slope is road-building

TABLE 6-3 Causes of Mass Movement and Preventive Measures

Causes	Preventive measures
Factors related to increased driving forces	
Slope gradient (steeper slopes are more prone to mass movement)	reduce slope (e.g., by constructing benches)
Factors related to decreased resisting forces (reduced strength of mass)	
Lateral support removed by erosion or construction	reinforce base of slope with retaining walls or by grouting
Moisture content	seal surface cracks to prevent infiltration; drain surface water from potential mass movement material with ditches; install subsurface drainage system
Vegetation	replant slopes immediately after removal of vegetation during logging or development; protect slopes with cover or mulch while seedlings become established
Nature of geologic materials (e.g., highly weathered; or rock layers inclined parallel to slope)	construct pilings through mass; avoid building on slopes with rock layers inclined parallel to the slope (they are prone to failure)

(Figure 6-24). During road-building, hillslopes often are cut and steepened, increasing the driving force (the slope-parallel force shown in Figure 6-22) that tends to push material down the slope. In addition, excavation of material that had provided some lateral support against the weight of the material above reduces the resisting forces along a potential failure plane. Similarly, coastal erosion can steepen slopes and remove the lateral support of rock material at the base of a cliff, making mass movement common along eroding coastlines. Preventing such failures requires developers to site roads carefully and to build structures such as benches and retaining walls along road cuts and coastal cliffs.

A common cause of reduced resistance to movement, or diminished strength, is the addition of so much water to the slope material that it begins to behave more like a liquid than a solid. For this reason, many mass movements occur immediately or soon after prolonged periods of rainfall. For example, after hurricanes in the eastern United States, dozens to hundreds of small slides and debris flows occur in hilly areas. Structures that drain water from potential slide masses can prevent or minimize failure. Sometimes the cause of increased moisture in hillslope debris is the concentration of water along newly constructed roads or leakage of water from sewage pipes, canals, or water pipes. In all cases, mass movement can be mitigated by not permitting water to accumulate.

Because the roots of vegetation provide much strength to material on hillslopes, the removal of plants often results in mass movement. Vegetation provides a further stabilizing role in that it absorbs moisture and minimizes the accumulation of water in the soil. In logged areas, the combination of road-building and removal of vegetation leads to conditions very favorable for mass movement. Clear-cutting and associated road-building were blamed for debris slides that killed five people in Oregon in late 1996. Failure is most common within a few years of vegetation removal, after the roots have fully decayed. Mitigating mass movement on devegetated slopes requires replanting as soon as possible, and taking measures, such as mulching, to help establish the plants.

Finally, the nature of the geologic material on a hillslope plays a role in mass movement. For example, if strata are inclined toward a valley bottom such that beds are parallel to the hillslope, mass movement is more likely than if beds are horizontal or dipping into the slope. The reason is that the contacts between each rock layer can become planes of failure themselves, and entire beds of rock can slide downhill along these contacts.

It is the job of geoscientists to map soil, sediment, and rock types, deposits from previous mass movement, and areas of likely instability. Such maps, known as *slope hazard maps,* can be used as planning tools. Potential instability does not necessarily preclude development, for engineering methods to strengthen structures, stabilize slopes, and prevent their failure are known and practiced in many places. In Japan, for example, a country plagued by mass movement, a government effort to reduce landslide losses was initiated in 1958 and has markedly lowered death and property losses.

Debris slide

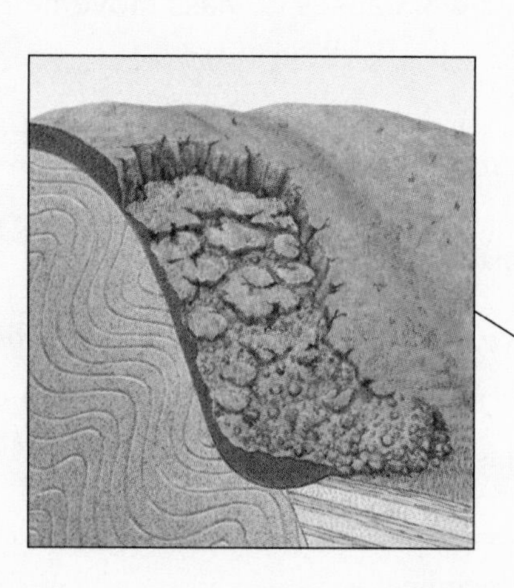

Debris slump

Rockslide

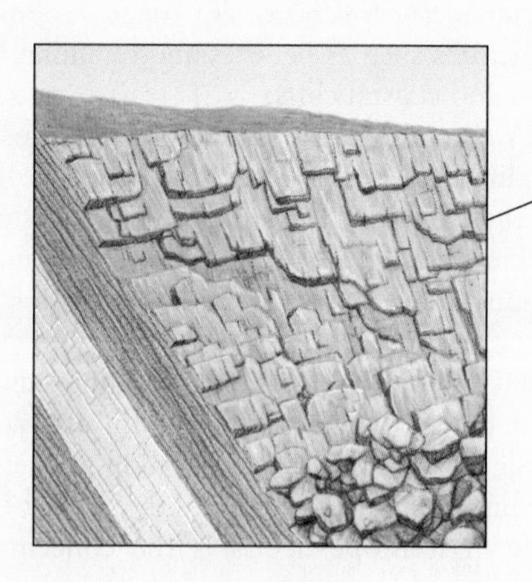

FIGURE 6-23 Different types of mass movement are related to the mechanism of movement (heave, slide, or flow), the nature of the mass, the amount of water, and the speed of movement.

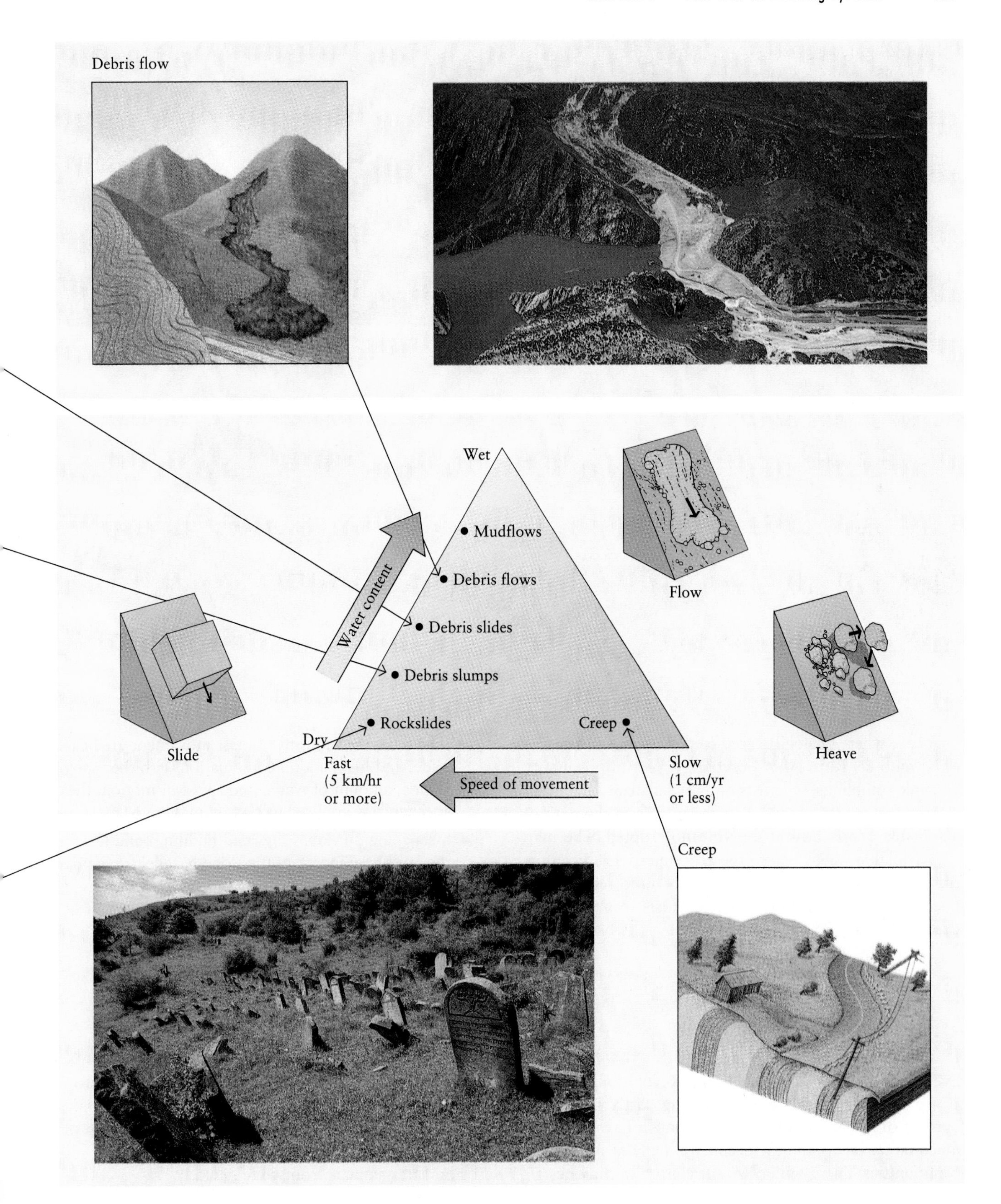
Debris flow
Wet
Water content
• Mudflows
• Debris flows
• Debris slides
• Debris slumps
• Rockslides
Creep •
Dry
Fast
(5 km/hr
or more)
Speed of movement
Slow
(1 cm/yr
or less)
Flow
Slide
Heave
Creep

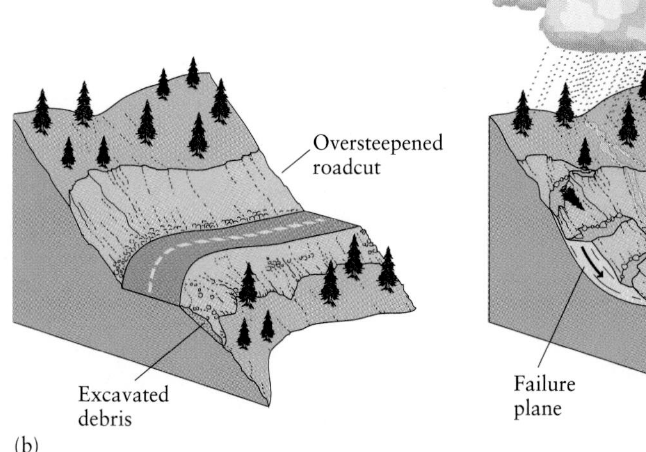

FIGURE 6-24 (a) Along roads such as this one in the Cascade mountains spanning western Canada, Washington, and Oregon, many slumps and slides (white scars) are initiated along fresh road cuts (horizontal benches), where slopes are steepened and rainwater collects and infiltrates the soil. (b) After excavation for road-building, the hillslope is steeper and might fail. Stormwater from rain and snow collects along roads and drains into loose, excavated debris, making that material likely to fail as well. After failure, the angle of the new slope is similar to that of the pre-excavation slope, close to the angle of repose.

A case where geologists predicted mass movement was at Nevado del Ruiz. After observing a year of precursory volcanic rumblings, scientists prepared a hazard map that clearly showed the village of Armero lying in the path of a possible debris flow if the volcano erupted. The map was available weeks before the eruption in 1985; unfortunately, government officials did not urge residents to flee. The decision to ignore the scientists resulted in the deaths of nearly 25,000 people.

- The three mechanisms of mass movement are heave, slide, and flow, and in combination with the type of debris, amount of water, and speed of motion, these mechanisms are used to classify mass movement.
- Steepening hillslopes (by road-building) and leaving the land bare increases the slope-parallel force on the slope and decreases the slope materials' resistance to movement, increasing the possibility of mass movement.

Closing Thoughts

Ever since the start of the Industrial Revolution, around 1760, urbanization has been increasing. With the development of mechanized farming equipment have come much larger farms and the decline of many farming communities. Energy-intensive agricultural techniques since the Green Revolution of the 1950s have raised crop yields nearly four times in some places, so increasing numbers of people can be supported by a farmland base that has grown little. But increasing crop yields can lead to soil erosion sufficient to cause a long-term loss of the resource required to grow future crops.

In urban areas, few people have the opportunity to see much soil. Environmental concerns such as global warming and air pollution easily catch the attention of a largely urbanized society, while the quiet crisis of soil erosion rarely makes front-page news.

Throughout the world, much land is at risk of soil erosion, desertification, gullying, and catastrophic mass movement. Developing strategies to protect Earth's outermost, fertile surface, while the world's human population continues to grow, will be one of the greatest environmental challenges of the 21st century.

Summary

- Physical weathering is the mechanical disintegration of rocks and minerals; decomposition, or chemical weathering, is the dissolution and chemical alteration of rocks and minerals.
- The two main processes of chemical weathering are reactions of mineral surfaces with acids and with oxygen. Acids react with silicate minerals to produce clay minerals and dissolved ions and can completely dissolve carbonate minerals. Oxygen reacts with silicates to form oxide and hydroxide minerals, and when it reacts with sulfide minerals, it also produces sulfuric acid, which leads to acid drainage from mines.
- Clay minerals, a secondary weathering product, are a distinguishing feature of soils; they are chemically and physically different from their parent minerals derived from the lithosphere.
- All Earth systems interact at the surface of Earth to move ions, particles, and gases across the pedosphere, resulting in a vertical arrangement of soil horizons (a soil profile) that varies in composition and appearance with depth in the pedosphere.
- Three primary resources formed by weathering in the pedosphere are fertile soil for agriculture, clay minerals for building materials, and mineral ores (iron and aluminum oxides and hyroxides) used in industry.
- Water can remove substantial amounts of soil if the soil surface is unvegetated or steep. Rates of soil formation are very slow in human terms, and in some parts of the world human activities have increased soil erosion to rates 18 to 100 times greater than natural rates.
- Studies of controlled soil-erosion plots enable scientists to assess erosion and soil-conservation strategies. Fairly simple, and often ancient, practices including terracing and contour plowing are very effective strategies.
- The two primary factors that determine whether or not mass movement will occur on a hillslope are gravity and the amount of water in the weathered rock debris.

Key Terms

pedosphere (p. 160)
physical weathering (p. 161)
chemical weathering (p. 161)
erosion (p. 161)
stress (p. 161)
carbonation (p. 165)
hydrolysis (p. 168)
clay minerals (p. 168)
oxidation (p. 168)
soil (p. 168)
soil horizon (p. 168)
soil profile (p. 168)
humus (p. 173)
nitrogen cycle (p. 173)
rainsplash (p. 180)
sheet wash erosion (p. 180)
rill (p. 180)
gully (p. 180)
desertification (p. 180)
universal soil loss equation (p. 181)
tolerable soil loss (p. 181)
mass wasting (p. 182)
mass movement (p. 182)
angle of repose (p. 183)
surface tension (p. 183)
heave (p. 184)
slide (p. 184)
flow (p. 184)

Review Questions

1. What natural events and human activities caused the Dust Bowl in the Great Plains in the 1930s? How have soil conservation practices changed since that time?
2. How do the growth of plant roots and ice crystals cause physical weathering of rocks and minerals?
3. What types of acids are involved in chemical weathering in the pedosphere? How are they produced?
4. How has the emission of waste products into the atmosphere affected rates of chemical weathering since the Industrial Revolution?
5. How do soils in humid, tropical regions differ from soils in drier and cooler regions?
6. How can soil maps be used to (a) assess the carrying capacity of nations, and (b) make local land-use decisions?
7. How do chemical weathering processes result in the accumulation of rich ores of iron and aluminum?
8. How can buried, ancient soils be used to reconstruct past land-use patterns?
9. Using concepts of force and resistance, explain how deforestation and road-building trigger mass movement.

Thought Questions

1. Living and decaying plants release carbon dioxide in soil, enhancing weathering and producing clay minerals to help retain nutrients and water, leading to greater plant activity. Is this sequence an example of positive or negative feedback? Why?
2. Why can't plants (and hence food) be grown easily in partly weathered bedrock once the upper meter or so of soil is eroded?
3. After heavy rains and pronounced landslide activity in the Pacific Northwest during the winter of 1996–1997, many have blamed logging companies for the deaths and damage resulting from the slides. Some have called for government intervention to regulate, and even stop, logging. The logging companies and some forest officials argue that in a tectonically active area with much hilly terrain, during a year of exceptionally high rainfall, landslides are likely to occur regardless of logging activities. How might scientists go about studying landslides in the area to determine the role of logging in the recent increase in landslide occurrence?
4. Using radiometric and other dating methods, researchers studying ancient soils and sedimentary deposits in Greece have determined that they are of widely variable ages. From this information, they concluded that the changing sequences were not the result of regional climatic changes, such as prolonged droughts, but rather of patterns of land-use change. What might have been their reasoning?

Exercises

1. A scientist samples soil moisture from a field and determines that it contains 10^{-4} g of H^+ ions per liter of solution. What is the pH of this soil? Would it be suitable for agriculture? Why or why not?
2. Using a pile of loose sand, make as steep a cone as possible and measure its angle of repose with a protractor. Why is the angle you measured probably between 30° and 35°? Then add a small amount of water to the sand, make as steep a cone as possible, and measure the new angle. Why is this angle much steeper than the previous one? What force is involved in the change?
3. The average soil erosion rate measured on a hillslope in Kenya is about 5 mm per year, but average rates of soil formation are about 0.02 to 0.11 mm per year. Viewing soil formation as an input and soil erosion as an output, how much will the stock of soil on this hillslope reservoir decrease each year? Assuming that the soil is 2 m thick, in how many years will the soil be completely depleted by erosion?

Suggested Readings

Colman, S. M., and Dethier, D. P. (eds), 1986. *Rates of Chemical Weathering of Rocks and Minerals.* New York: Academic Press.

Evans, A. M., 1987. *An Introduction to Ore Geology.* 2nd ed. Oxford, England: Blackwell Scientific Press.

Hillel, Daniel, 1991. *Out of the Earth: Civilization and the Life of the Soil.* Berkeley: University of California Press.

McLaren, R., and Cameron, K., 1996. *Soil Science: Sustainable Production and Environmental Protection.* New York: Oxford University Press.

Owen, O. S., and Chiras, D. D., 1990. *Natural Resource Conservation.* New York: Macmillan.

Retallick, G. J., 1990. *Soils of the Past: An Introduction to Paleopedology.* Boston: Unwin Hyman Press.

Sidle, Roy C., 1995. *Hillslope Stability and Land Use.* Washington, DC: American Geophysical Union.

Steiner, Frederick R., 1990. *Soil Conservation in the United States: Policy and Planning.* Baltimore: The Johns Hopkins Press.

Troeh, Frederick, and Thompson, Louis M., 1993. *Soils and Soil Fertility.* 5th ed. New York: Oxford University Press.

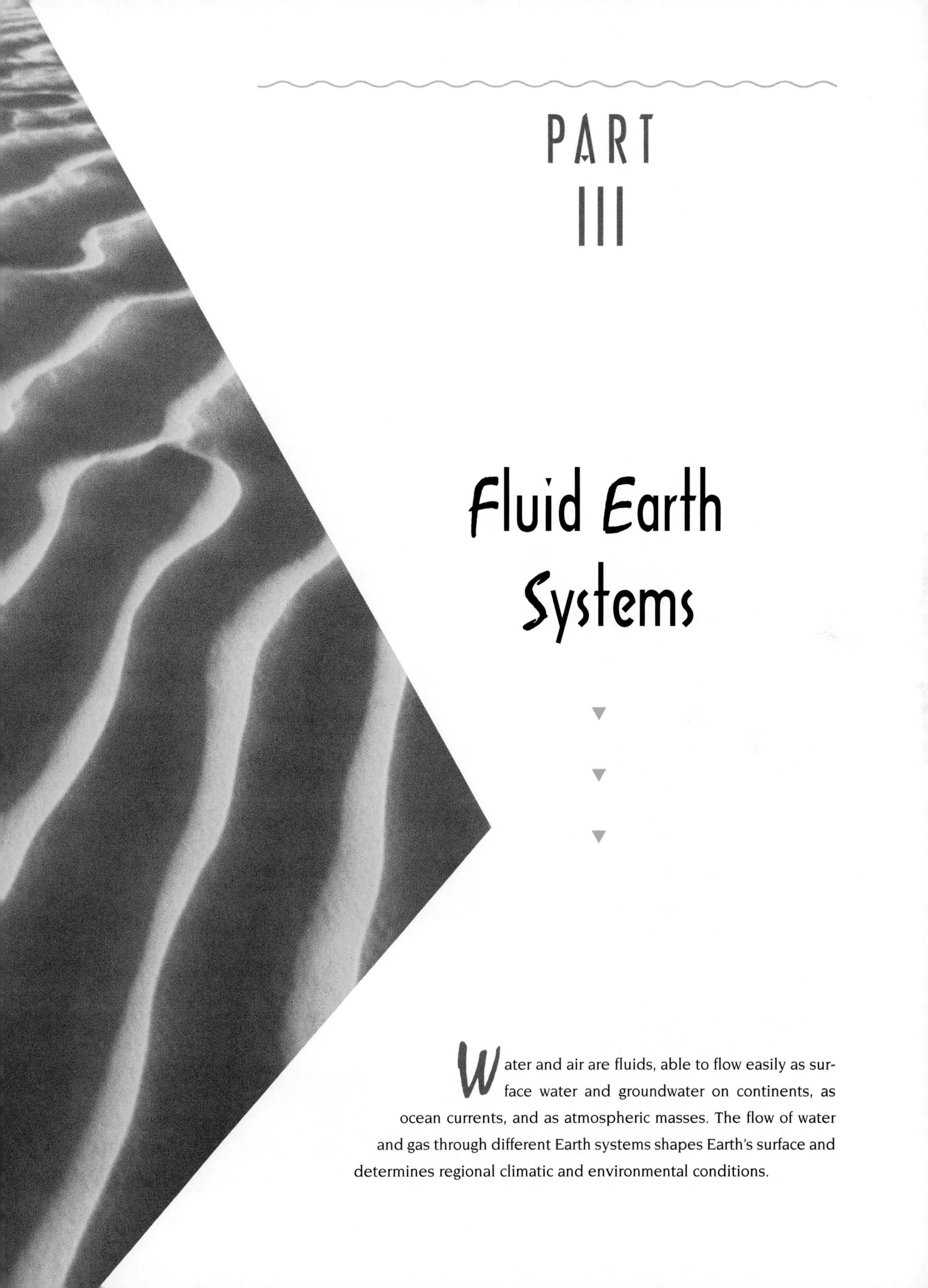

PART III

Fluid Earth Systems

Water and air are fluids, able to flow easily as surface water and groundwater on continents, as ocean currents, and as atmospheric masses. The flow of water and gas through different Earth systems shapes Earth's surface and determines regional climatic and environmental conditions.

CHAPTER 7

To restore sandy beaches in the Grand Canyon along the Colorado River, in 1996 federal officials created an artificial flood by releasing water from four steel drainpipes built into the base of Glen Canyon Dam.

The Surface Water System

Surface Water Distribution

Drainage Basins and Streams

- Runoff Processes
- The Role of Climate in Stream Discharge
- Stream Channel Patterns and Processes

The Hazards of Flooding

- Floodplains and Flood-Recurrence Intervals
- Great Floods
- Minimizing Flood Hazards

Wetlands

- Characteristics and Benefits of Wetlands
- Coastal, Riverine, and Glacial Wetlands
- Wetlands Destruction
- Protecting and Restoring Wetlands

Surface Water Resources and Protection

- Fresh-Water Use
- Surface Water Systems As Waste Disposal Sites
- Regulations to Protect Drinking Water
- Protecting Water Quality in Streams and Rivers

Rivers, streams, lakes, and wetlands store water that precipitates from the atmosphere, supply essential fresh water to the continental biosphere, return some water as vapor to the atmosphere, and drain off the rest to restore the world ocean. Dams now control the flow of most large rivers in the world, storing water, minimizing floods, generating power, and making streams more navigable. However, dams also disrupt the behavior of river water and sediments, and the disruptions affect other Earth systems.

Since 1963, when the Glen Canyon Dam in Arizona was completed, the Colorado River has become a different system. Clear water released from the dam cannot replenish eroding beaches along the river's banks as its sediment-laden floodwaters once did. Native species of plants, fish, and birds lost their habitat and disappeared as beaches and their backwater marshes shrank. The partial success of an experimental flood in 1996, which restored some 50 beaches, has provided guidelines for the timing and manner of flow release from dams along the Colorado.

In this chapter we will:

- See how the surface water system has created an intricate network of streams and storage areas.
- Assess the hazards of flooding, droughts, and pollution of surface water.
- Examine the ways in which environmental changes have affected the surface water system.
- Evaluate current efforts to protect and restore fresh-water systems.

Surface water systems are the drainage channels and enclosed bodies of water that regulate the supply of fresh water on continents. They include rivers and streams, lakes, ponds, and wetlands. Surface water systems store some precipitation and glacial meltwater and carry the runoff to replenish the ocean. Each surface water system supports a complex ecology of interdependent plant and animal life. Altering these systems for human ends, whether by using them as waste dumps or changing their flow patterns, as with the Colorado River, usually has consequences that range from the detrimental to the disastrous.

The inland sea known as Lake Aral, or the Aral Sea, in central Asia, was once the world's fourth largest freshwater lake (Figure 7-1). In the past few decades, water diversions and groundwater pumping for irrigation have reduced its surface area by more than 70 percent, and the environmental and economic impacts are devastating. The salt content of the lake tripled, destroying its fishery and wiping out all native fish. Groundwater levels dropped and soils became salt laden. Plumes of dust and salt 500 km long have blown upward as high as 4 km from the exposed lakebed, then spread downwind and settled on natural ecosystems, crops, and people, wreaking havoc with the health of all. Today, government officials are trying to replenish the lake by allowing less water to be used for irrigation and are considering diverting the course of a river that drains into the Arctic Sea so that it flows into the lake instead.

The Red River, one of the few north-flowing rivers in the United States, originates in South Dakota and flows across the floor of an ancient glacial lake into Canada. Nine thousand years ago, retreating ice sheets and shrinking glacial lakes left behind a marshy, nearly flat landscape. The Red River carved its sinuous channel into the lake sediments, dropping in elevation a mere 3 to 8 cm per kilometer of stream length. In the 19th and 20th centuries, enterprising farmers drained surrounding marshes for farmland. Settlers built levees along the many bends of the Red River to contain it in times of rising waters. In the spring of 1997, snowmelt in the Dakotas swelled the Red River, while farther north in Canada much ground- and surface water still were frozen. Unable to continue its northerly flow, the Red River formed a massive, shallow lake that caused water upstream to back up and overtop its levees. Although the amount of water in the Red River was small in comparison with floodwaters in other rivers, the river's slope is so slight that minor increases in water depth result in widespread flooding. Thousands of people were evacuated, and hundreds lost their homes (Figure 7-2).

The fundamental problem with altering environmental systems to human advantage is that their complexities are not well understood. Scientific models of interdependent natural systems, therefore, do not always recognize all the relevant variables or their relative importance. Engineers and scientists did not foresee, for example, all the negative consequences of damming the Colorado River, reducing streamflow into the Aral Sea, or building levees and houses along the Red River floodplain.

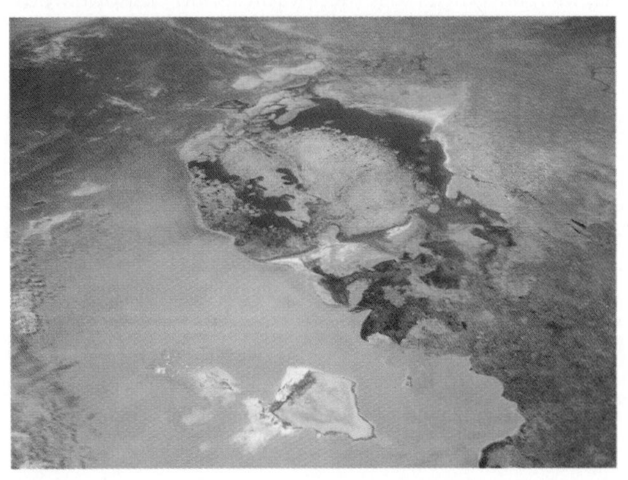

(a)

(b)

FIGURE 7-1 (a) As the Aral Sea shrinks, its sandy, salty bed (seen as white areas in this satellite image) becomes exposed to winds, resulting in devastating salt storms. (b) What once were coastal fishing ports are left high and dry many miles from the present lake shore. All native species of fish have disappeared. A local poet, Mukhammed Salikh, lamented, "You cannot fill the Aral with tears."

FIGURE 7-2 The Red River spilled over its banks and levees in the spring of 1997, as here in Grand Forks, North Dakota, causing widespread flooding and devastation.

Surface Water Distribution

Water, which is everywhere at Earth's surface, would form a layer 3000 m deep if distributed evenly. Only 2 percent of this stock is fresh water; the rest is saline water in oceans and bays. Nearly 87 percent of fresh water is locked in ice caps and glaciers or stored deep underground. Most of the rest is stored as shallow groundwater. A mere 0.01 percent of all water on Earth is available in the form that supplies the needs of more than 90 percent of the world's population—river runoff and lakes.

The total annual volume of fresh water stored on average at one instant in rivers and lakes on Earth is 1250 km^3. However, more water than that is available for human use at any given moment because water flows continuously from one reservoir to another, and from one physical state to another, in the hydrologic cycle. The global average for the residence time of water in the atmosphere is 11 days and in rivers and lakes only 5 days. Such rapid replacement times mean that humans actually have access to more than 40,000 km^3 of surface water each year.

Surface water, the most critical of our freshwater resources, is distributed unequally among and within continental landmasses (Table 7-1). Furthermore, because of rapid population growth and increasing per capita consumption of water, the demand for fresh water exceeds its availability in many parts of the world today (Figure 7-3). Many countries in northern and eastern Africa and in the Middle East already suffer severe to extreme water stress. In the United Kingdom possible climate change is causing water shortages. The years 1995–1997 were the driest in 230 years, and annual rainfall in Essex dropped to 24 inches, as low as that of Middle Eastern countries. The area's climate is now officially listed as semiarid, a change that is difficult to accept for some Britons accustomed to fog and umbrellas. In response, politicians planned to hold a "water summit," and construction began on England's first desalination plant.

In North America the abundance of surface water varies markedly from east to west. Most of the southwestern United States, including Nevada, Arizona, New Mexico, and Colorado, and parts of northwestern Mexico, have suffered water shortages since 1980, and it is projected that nearly all the states west of the Mississippi River will experience shortages by the year 2000. In the eastern United States, only New York and eastern Pennsylvania are expected to face shortages by the year 2000.

The worldwide imbalance between areas of water shortage and water surplus is expected to become progressively greater during the 21st century. Inequities of water resources can lead to political struggles. The drought-plagued African nation of Namibia, for example, recently made plans to withdraw large amounts of water from the Okavango river system. Faced with a growing population, a desire to industrialize, and dams that are only 9 percent full after years of drought, the country is desperate for water. Unfortunately, the river flows into neighboring Botswana, where it is the lifeblood of the world's largest inland wetland, the Okavango delta, which is known as the "jewel of the Kalahari." Without a constant supply of fresh water from the river, the wetland would resemble the neighboring Kalahari Desert itself. The inhabitants of 30 Botswanan villages would face the same deprivations as the people who live along the now-dry shores of the Aral Sea—no water, no fish, no livelihood. Namibia's plan to build a pipeline hundreds of kilometers across the desert to transport 20 million m^3 of

TABLE 7-1 Annual Volume of Fresh Water in Rivers, by Continent

Continent	Volume (km^3)	Percentage of total
Europe	76	3.8
Asia	533	27
Africa	184	9.2
North America	236	12
South America	946	47
Australia	24	1.2
Total	1999	100

Source: *World Water Resources*, World Resources Institute, 1990.

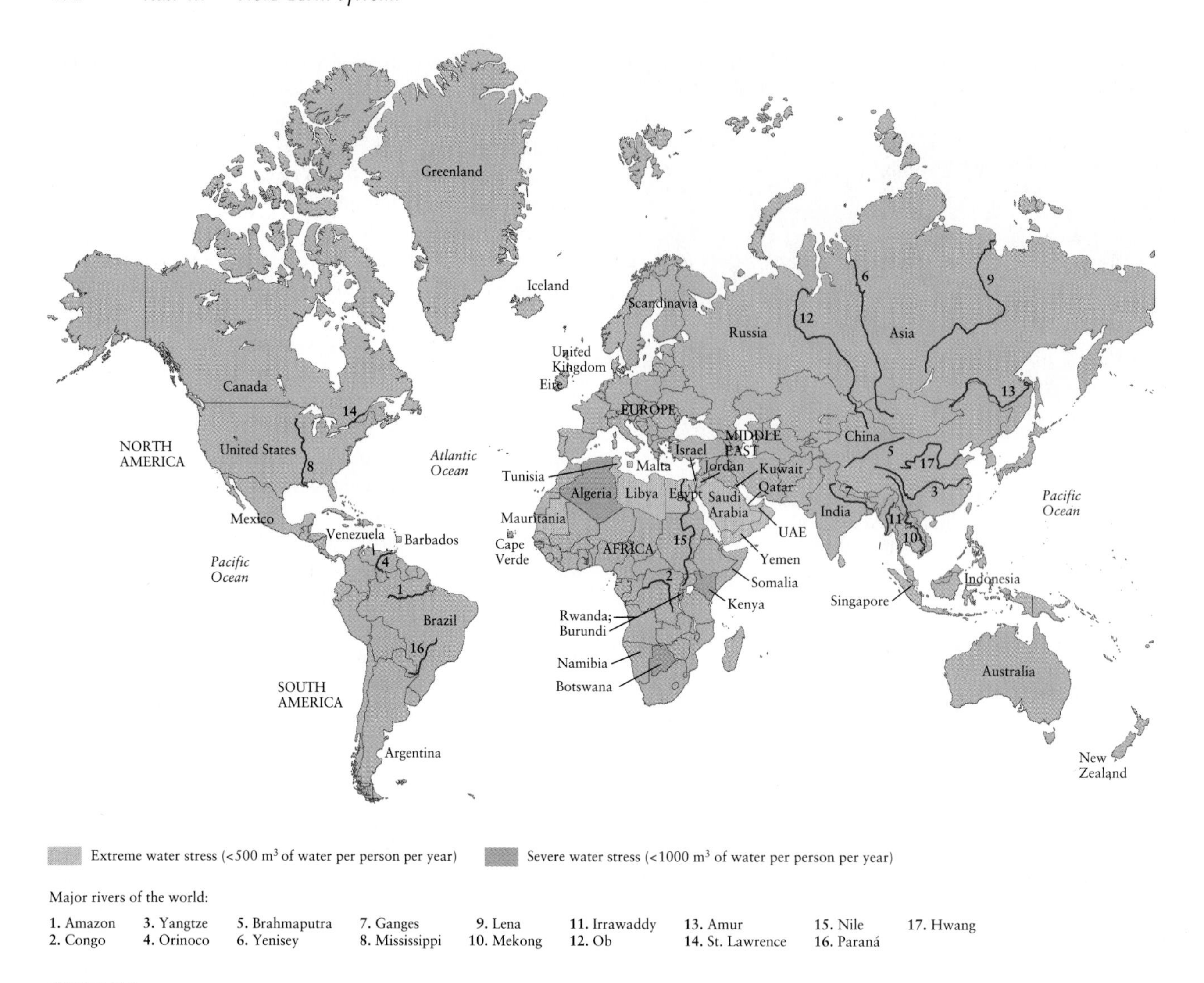

FIGURE 7-3 In countries with scarce surface water resources, extreme water shortages can occur when less than 500 m^3 of water is available per person per year. Severe water shortages can occur when less than 1000 m^3 of water is available per person per year. Although most countries have sufficient water resources, shortages are especially prevalent in Africa and the Middle East. In addition, within a given country water abundance can vary substantially. Shown also are the world's largest rivers (see Table 7-2).

water from the river each year has set off a dispute with Botswana. The case has been referred to the International Court of Justice at The Hague, but Botswana is stockpiling arms in the event that the decision is adverse and only force will protect its interests.

Differences in water supply are related to the distribution of average annual precipitation, **evapotranspiration** (the combination of evaporation from water and land surfaces and transpiration of water vapor from living things), and runoff. If precipitation exceeds evapotranspiration, water is available to fill surface basins and channels and to infiltrate the pedosphere and recharge (replenish) groundwater systems (see Chapter 8). Conversely, if precipitation falls short of evapotranspiration, water deficit will occur.

Values of potential evapotranspiration can be compared with precipitation rates to explain why in the United States most water shortages and *ephemeral streams* (those that do not flow throughout the year) occur west of the Mississippi River. With the exception of mountainous areas that receive snowfall, the average annual precipitation is less than 15 to 20 inches nearly everywhere west of the Mississippi. In many areas precipitation is less than the average annual potential for evapotranspiration. In these semiarid to arid areas, the potential for evapotranspiration is so great that even if more rain were to fall, it might not be enough to exceed evapotranspiration, and no excess water would be available for surface runoff or recharge of soil moisture and groundwater (Figure 7-4).

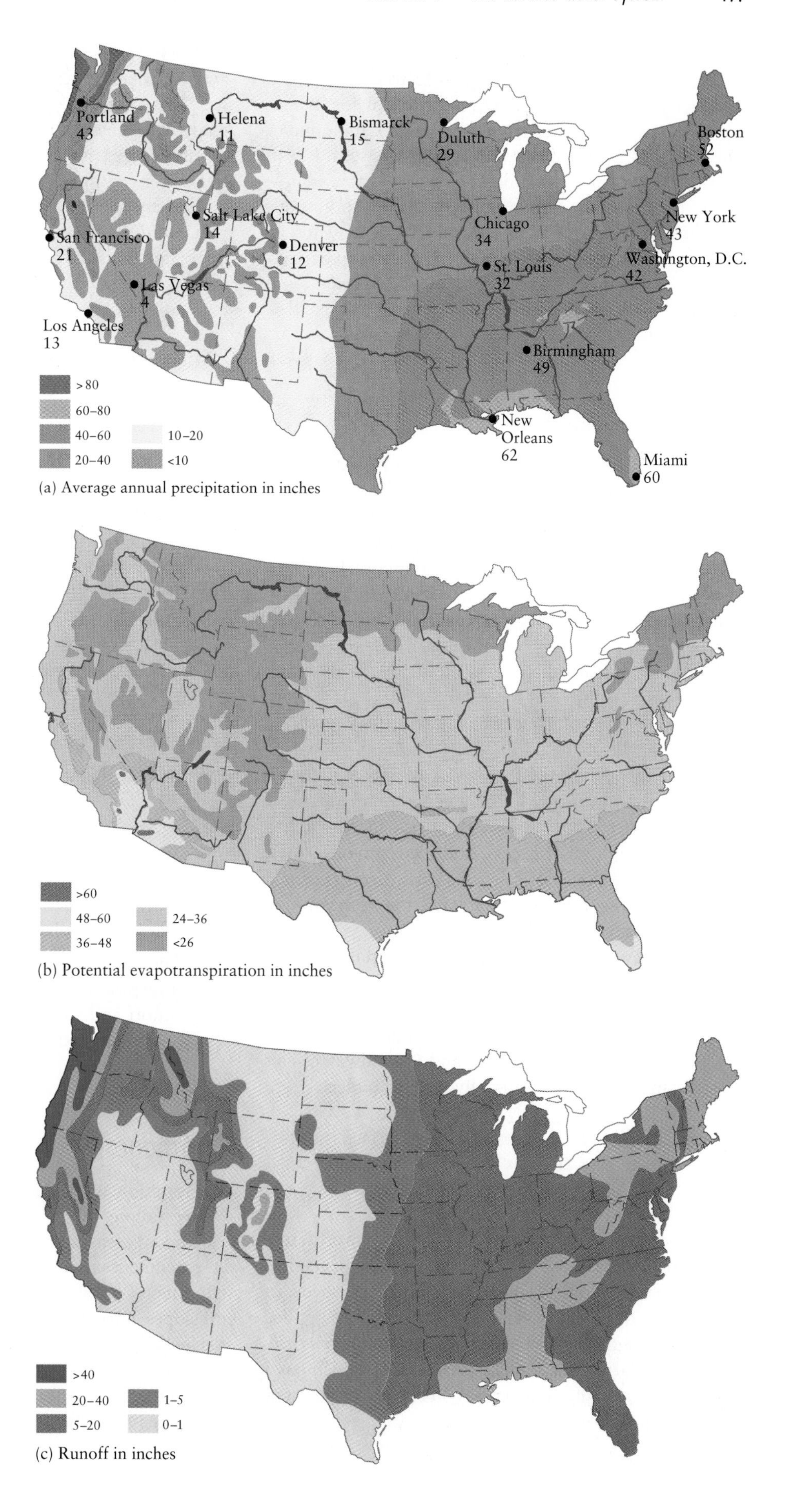

FIGURE 7-4 Rates of average annual (a) precipitation, (b) potential evapotranspiration, and (c) surface runoff for a given area can be compared in order to estimate the availability of surface water. In much of the southwestern United States, where water shortages are common, the potential for evapotranspiration is greater than the amount of precipitation, and thus water is scarce.

East of the Mississippi, the average annual precipitation is nearly equal to or greater than 15 to 20 inches and is greater than the average annual potential for evapotranspiration. Such conditions are typical in semihumid to humid regions. Most of the water used in eastern North America comes from rivers because they are *perennial*—that is, they flow year round. The cause of continuous waterflow between storms is the abundance of water in storage in soil, rocks, and sediments. Long after past rainfalls, this water drains, or percolates, into streams and lakes as **baseflow,** the inflow from groundwater that maintains the average water level between storms.

▸ Cycling of Surface Water Among Atmosphere, Biosphere, and Hydrosphere

Earth's hydrologic system is considered a closed system. With the exception of small amounts of water vapor that escape from occasional volcanic eruptions or geothermal vents, no new water is released at Earth's surface. Likewise, with the exception of small amounts of hydrogen that escape from the outermost atmosphere many kilometers above Earth, no water is lost from Earth's surface. The conservation of water on Earth can be expressed as the **hydrologic equation:**

$$\text{Inflow} = \text{outflow} \pm \text{changes in stock}$$

In other words, if a system is in a steady state, with inflow equal to outflow, no changes in stock occur. If inflow is greater than outflow, the stock increases at a rate equal to the rate of excess, and the reservoir grows. If inflow is less than outflow, the stock decreases at a rate equal to the rate of deficit, and the reservoir becomes depleted.

Water inflows to Earth's surface are precipitation rates and outflows are rates of evapotranspiration and surface runoff. A balance exists between global rates of precipitation, evapotranspiration, surface runoff, and the amount of water that is available to soil and groundwater. This balance can be expressed as the **hydrologic budget:**

$$\text{Precipitation} = \text{evapotranspiration} + \text{surface runoff} \pm \text{changes in soil moisture and groundwater storage}$$

The hydrologic budget equation is a very powerful tool for scientists, environmental planners, and engineers. It is the basis of water-budget studies for lakes, watersheds, irrigation networks, reservoirs, and cities. It can be applied to a small system, such as the soil layer on a farmer's field, or to a much larger one, such as Los Angeles, which has a complex water supply system that comes from as far away as Colorado, Arizona, and Nevada.

If we apply the hydrologic budget equation to the average annual data in Figure 7-4 for the American Southwest, we can quantify the imbalance we have observed already between annual inflows (precipitation) and outflows (evapotranspiration and runoff) in that arid region. Las Vegas, for example, has precipitation of 4 inches per year and potential evapotranspiration of 36 to 48 inches per year. Runoff is less than 1 inch per year in this area. Annual change in storage of soil moisture and groundwater can be calculated by rearranging the hydrologic budget equation:

$$\begin{aligned}\text{Changes in soil moisture and groundwater storage} \\ &= \text{precipitation} - \text{evapotranspiration} - \text{runoff} \\ &= (4\ \text{inches/yr}) - (36\ \text{to}\ 48\ \text{inches/yr}) - (0\ \text{to}\ 1\ \text{inch/yr}) \\ &= -32\ \text{to}\ -45\ \text{inches/yr}\end{aligned}$$

Where potential evapotranspiration is greater than precipitation, soil moisture and groundwater are not recharged and become depleted with time. This is the case in Las Vegas. Little runoff is available except during large rainstorms, streams are dry most of the year, and little or no recharge of groundwater reservoirs occurs. The inhabitants of this region are dependent on deep groundwater and the import of Colorado River water that is impounded by the Hoover Dam for drinking, agriculture, and industry.

- Water shortages occur in some areas because surface water is distributed unequally among and within continents and population growth has exceeded available supplies in more arid regions.
- The balance between precipitation and evapotranspiration determines how much surface water runoff and storage occur at a given place, a relation expressed in the hydrologic budget equation.

Drainage Basins and Streams

Precipitation is the ultimate source of all surface water. Water falling on continental landmasses drains into a catchment area and eventually flows back into the oceans. A catchment area can be imagined as a bowl-shaped continental basin that funnels water and sediment from hillslopes into stream channels. Earth and water scientists—that is, geoscientists and hydrologists—and land-use planners refer to such catchments as **drainage basins,** or *watersheds* (Figure 7-5). The boundaries of a drainage basin are relatively high spots, known as **drainage divides,**

FIGURE 7-5 A drainage basin in Russia showing drainage divides that surround stream channels leading to the Katon River (foreground). Drainage divides separate the waters between drainage basins. During a rainstorm, some waters flow on one side of a divide into a drainage basin, and some flow on the other side into another drainage basin.

that separate the waterflow into adjacent drainage basins. Every continent has major drainage basins separated by a continental divide. In North America, the continental divide follows the crest of the Rocky Mountains and separates basins that drain westward, into the Pacific Ocean, from those that drain eastward, ultimately into the Atlantic Ocean (Figure 7-6).

The entire system of stream channels in a drainage basin forms an intricately linked **drainage network,** similar in function to the human circulatory system or the veins in a leaf: All are branching networks that transport fluids to or from a surface. Examples of drainage networks for the Colorado and Mississippi drainage basins are shown in Figure 7-6. Because many river networks are treelike in form, their patterns are called dendritic, from the Greek word *dendron,* "tree." The major arteries of the river channel network are **trunk streams,** and the much finer channels that feed the trunk streams are **tributaries.**

The primary driving force for circulation in a river network is gravity, and all water and sediment moves in one direction—downslope. A low point that acts as a control on the flow of water and sediment out of a drainage basin is called a **base level.** While the ocean surface is the final base level for all the world's rivers, there are many local and temporary base levels, such as lakes or dams, along most stream networks. Upstream of either a local base level or the level of the ocean surface, the flow of water in a stream cannot drop below that of the base level.

Runoff Processes

If you watch a vegetated hillside during a rainstorm, you may see an occasional trickle of water flowing across the ground. Several hours after the rain stops, the ground may appear nearly dry, yet local streams will be flowing full. During a typical rainstorm, only about 1 to 5 percent of the stormwater falls directly into a river channel or lake. Of the remainder, in a humid-temperate region, up to 60 percent might return directly to the atmosphere by evapotranspiration, leaving perhaps only 35 percent to become **stormflow.** Stormflow is runoff water from a storm that is added to the baseflow, the amount maintained in the stream by groundwater seepage.

If little stormflow travels overland into streams, how does it get to the stream channel? The answer lies in the ground's **infiltration capacity**—its ability to absorb rainwater—and in the types of runoff processes or snowmelt that occur on hillsides.

Soils absorb water until, during prolonged periods of rainfall or snowmelt, their infiltration capacity is exceeded. A vegetated hillslope with a loose, open soil structure has a higher infiltration capacity than does a bare hillslope. The impact of raindrops on exposed soil causes a splashing of clay particles that mats and seals the soil surface, resulting in a progressive decrease in infiltration capacity with time. Rainfall often exceeds infiltration capacity in such conditions, and during a storm the amount of water flowing into channels increases in comparison with the amount seeping into the ground. Surface water accumulates in depressions on the saturated hillsides, then begins to overflow these little basins and run off the surface as *overland stormflow.* Examples of overland stormflow can be observed in semiarid landscapes, on rocky surfaces, and on bare farm soils. Paved surfaces, such as city streets and parking lots, or compacted soil at construction sites, have so little infiltration capacity that even a light drizzle can produce puddles, and a prolonged or heavy rainstorm sends water rushing in streams across the surface (Figure 7-7).

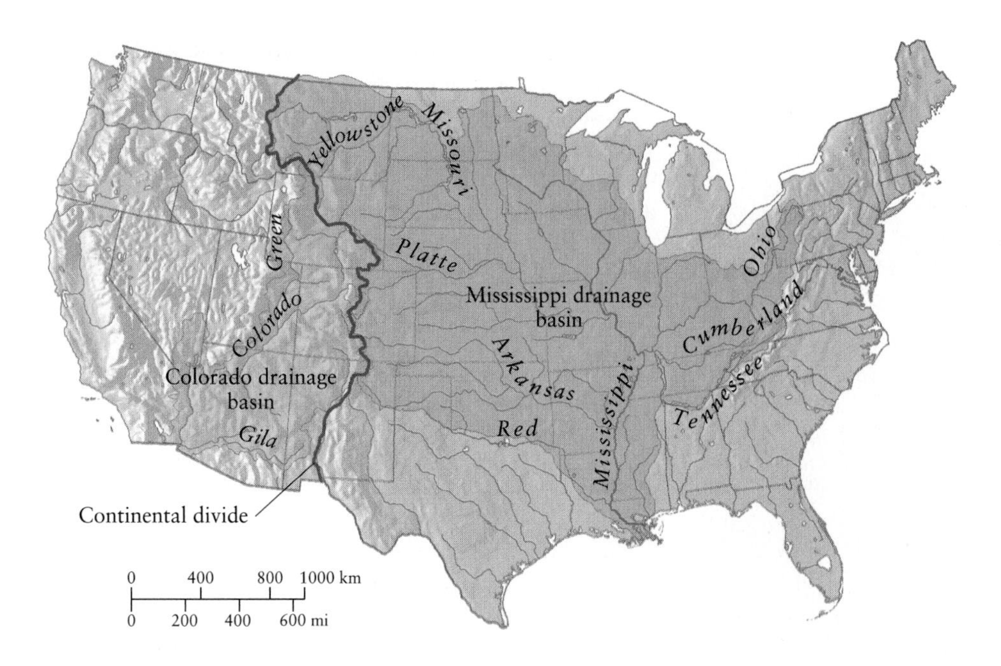

FIGURE 7-6 The North American continental divide in the United States is shown here in red. Drainage networks for the Colorado and Mississippi rivers are blue and their drainage basins are green.

Overland stormflow rarely occurs in humid-temperate or tropical regions on vegetated slopes mantled with organic-rich soils and leafy matter. Infiltration capacities are so high in such areas that most rainfall seeps into the ground. However, some water moves downslope at fairly high velocities just below the ground surface, moving along cracks, animal burrows, and open pore spaces in the soil, sediment, and rock. This *subsurface stormflow* seeps from the ground at springs along the edges of floodplains and stream channels, filling streams soon after a rainstorm.

FIGURE 7-7 A brief but intense rainstorm in Virginia resulted in overland stormflow within minutes at this gravel quarry, where trees had recently been cleared by heavy equipment and burned to expand the mining operation.

Because runoff processes affect other landscape processes occurring in the drainage basin, as well as the response of streams to flooding, it is most important for land-use planners and scientists to know the types of runoff processes that operate in a given area. For example, concentrated overland stormflow on an unvegetated surface, commonly seen on exposed farm soils (see Figure 6-19), is likely to result in rilling and gullying. In contrast, shallow subsurface stormflow on a vegetated slope is likely to be associated with soil creep, giving the hillslope a smoothed, convex appearance. Knowledge of runoff processes is even more important when we consider the negative impact of deforestation and urbanization on flow rates and flooding.

The Role of Climate in Stream Discharge

A region's climate—particularly its precipitation and temperature—determines the variables in the hydrologic budget and consequently the amount of water available for streamflow. A stream's **discharge** is the volume of water flowing past a given point on the stream channel per unit of time. Discharge can be determined by multiplying the average velocity of a stream by the channel's cross-sectional area (Figure 7-8). Discharge depends on three factors:

1. The area drained by the basin upstream of that point.
2. The amount of precipitation falling upon the basin upstream of that point.
3. The loss of water from the basin due to evapotranspiration from soils and vegetation.

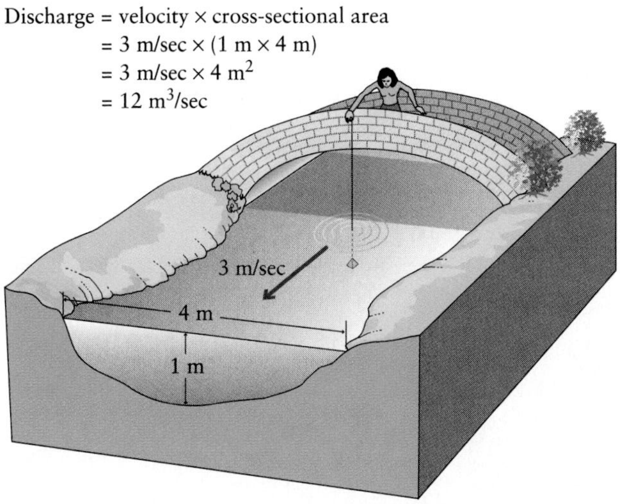

Discharge = velocity × cross-sectional area
= 3 m/sec × (1 m × 4 m)
= 3 m/sec × 4 m²
= 12 m³/sec

3 m/sec
4 m
1 m

FIGURE 7-8 Stream discharge is the velocity of the waterflow multiplied by the cross-sectional area of the stream channel. The depth of the stream can be measured by dangling a weighted measuring line from a bridge, and the bridge's span provides the measure of the width of the stream. To determine the velocity, you can drop a piece of orange peel into the water and clock its travel over a measured distance. In this example the velocity is 3 m/sec. More accurate estimates of discharge are made by surveying the channel geometry and using sensitive current meters to measure the velocity.

Estimating the amount of discharge in a stream is essential to long-term planning of water resources (Box 7-1).

The largest drainage basin in North America is that of the Mississippi River (see Figure 7-6), which drains 3,222,000 km^2 of land east of the continental divide. In terms of total area drained, the Mississippi River's drainage basin is the third largest in the world, after the Amazon and Congo basins (Table 7-2). In terms of the outlet discharge, or volume of water flowing out of a basin at its mouth for a given period of time, the Mississippi drainage basin ranks seventh in the world, with an average annual outlet discharge of 17,300 m^3/sec [611,000 cfs (cubic feet per second); U.S. Geological Survey discharge data is reported in cfs: 1 m^3/sec = 35.3 cfs].

The Colorado River drainage basin is much smaller than the Mississippi basin, but it is of tremendous

TABLE 7-2 The World's Major Rivers in Order of Discharge Volume

River*	Country/continent	Outflow	Average discharge at mouth (thousands of m^3/sec)	Area of drainage basin (thousands of km^2)	Length (km)
Amazon	Brazil/S. America	S. Atlantic Ocean	212.4	5778	6436
Congo	Congo/Africa	S. Atlantic Ocean	39.6	4015	4666
Yangtze	China/Asia	East China Sea	21.8	1943	6299
Brahmaputra	Bangladesh/Asia	Jamuna River	19.8	935	2896
Ganges	India/Asia	Padma River	18.7	1059	2510
Yenisey	Former USSR/Asia	Kara Sea	17.4	2590	5538
Mississippi	USA/N. America	Gulf of Mexico	17.3	3222	5969
Orinoco	Venezuela/ S. America	S. Atlantic Ocean	17.0	881	2735
Lena	Former USSR/Asia	Laptev Sea	15.5	2424	4399
Paraná	Argentina/ S. America	Rio de la Plata	14.9	2305	4878
St. Lawrence	USA, Canada/ N. America	Gulf of St. Lawrence	14.2	1290	4023
Irrawaddy	Burma/Asia	Andaman Sea	13.6	430	1992
Ob	Former USSR/Asia	Gulf of Ob	12.5	2484	5409
Mekong	Thailand/Asia	South China Sea	11.0	803	4344
Amur	Former USSR/Asia	Sea of Okhotsk	11.0	1844	4442
Hwang	China/Asia	Gulf of Chihli	3.3	673	5463
Nile	Egypt/Africa	Mediterranean Sea	2.8	2979	6648

*These rivers are shown on the map in Figure 7-3.

7-1 Case Study

Assessing Water Resources in the American Southwest

Residents of the American Southwest have good reason to worry that drought years may occur: Analysis of tree rings indicates that the original estimate of how much water is annually available in the Colorado River was made during an unusually wet period. Each annual growth ring of a tree reflects the environmental conditions of that year, with wide rings indicating good conditions for growth and narrow rings indicating poor ones. Because plant growth is greatly dependent on the availability of water, wide rings can indicate wet years and narrow rings, drought. Reconstructions of streamflow records from tree rings in the Colorado River region indicate that wet and dry periods have been cyclic over the past four centuries, with average annual flow during wet periods nearly twice that of dry years.

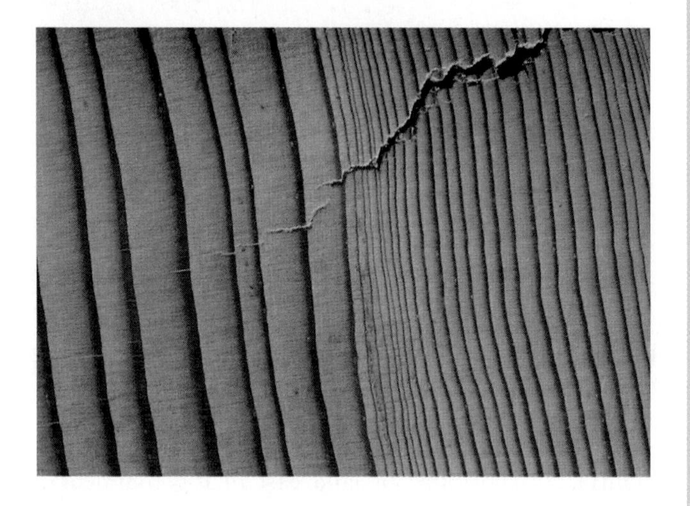

A tree core from the American Southwest showing rings of varying width. Wide rings can indicate relatively wet years, while thin rings can indicate drier ones.

The period of record used to allocate water to users of the Colorado River was the first two decades of the 1900s. This was the wettest period in the entire 400-year record obtained from analysis of tree rings. At that time, mean annual discharge at Lees Ferry in the Grand Canyon was estimated at 22.2×10^9 m^3. The total amount of water entitled to users in the United States and Mexico is about 18.5×10^9 m^3 per year, nearly 4×10^9 m^3 less than the amount estimated to exist in 1922. Since 1922, however, when Mexico and the United States agreed on a distribution of the Colorado River water, the average annual flow has decreased steadily. The long-term average water yield from tree-ring analysis is about 16.5×10^9 m^3, almost 2 billion m^3 less than the amount allocated to modern water users and about equal to the amount consumed in 1978. If the Colorado goes through a period of drought like those revealed in the tree-ring record, the millions of users of Colorado River water will experience a severe water shortage for a number of years.

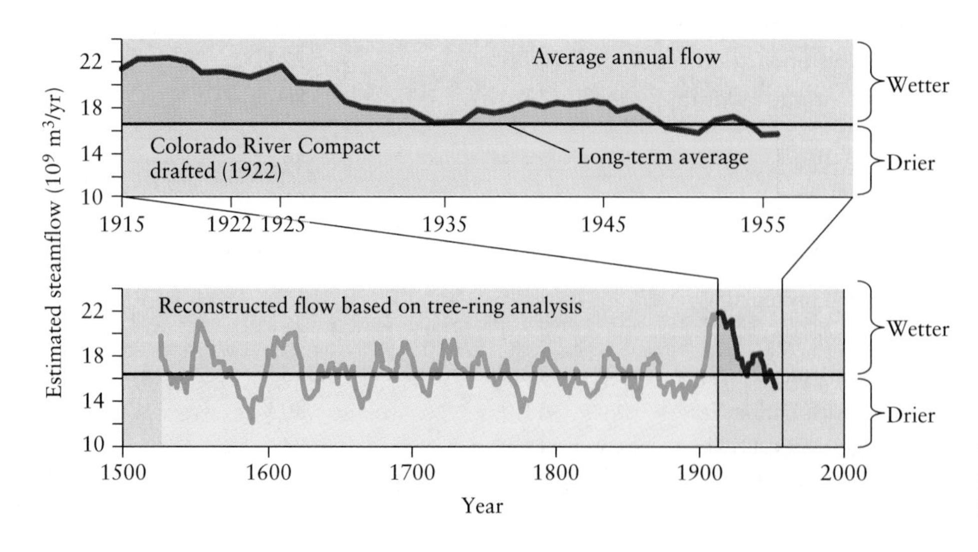

The flow in the Colorado River as determined from tree-ring analysis and historical measurements.

importance to those inhabitants of Mexico, California, Nevada, and Arizona who get water from its river channels (see Figure 7-6). This drainage basin lies west of the continental divide, where mean annual runoff generally is less than 12 inches per year (see Figure 7-4). The trunk stream of the Colorado River starts as a small tributary high in the snowcapped Rocky Mountains. It drains a basin area of 637,144 km^2—about 20 percent the size of the Mississippi basin—before flowing through the desert landscape of the American Southwest and emptying into the Gulf of California off Mexico. Despite its substantial drainage area, the Colorado River has an average discharge at its mouth of less than 425 m^3/sec (15,000 cfs), 2 percent of the flow of the Mississippi River.

The small size of the Colorado's discharge compared with its drainage area demonstrates that climate (as defined by precipitation and evapotranspiration) is more important than drainage area in determining the amount of surface water runoff that flows through a drainage basin and hence is available for surface water use. Even though basin area increases in a downstream direction, the amount of water in a stream in an arid area often decreases because some of it seeps into the channel bed and flows into the underlying porous rock and sediment to recharge groundwater. Such a stream system loses water along the length of its channel. In contrast, in a humid climate where precipitation is greater than evapotranspiration and the excess water infiltrates the soil, high levels of groundwater supply stream channels with baseflow, and the stream network gains water in the downstream direction as the basin area increases.

Stream Channel Patterns and Processes

All flowing water sculpts and changes the landscape by carrying soil off hillslopes and into streams, by gouging canyons and waterfalls, and by transporting sediment from continents to oceans. The higher the energy of a stream, the greater its ability to carry larger sediment particles, a quality known as *stream competence,* and larger volumes of sediment, a quality known as *stream capacity.* The sedimentary particles carried or deposited by flowing water, collectively called **alluvium,** include clay, silt, sand, and gravel (see Table 4-5). Alluvial streams have beds underlain by alluvium, whereas the beds of bedrock streams are relatively free of alluvium.

The typical patterns of flowing streams are braided, meandering, or straight, depending on the energy of the stream system, the nature of the geologic material on which the water is flowing, and the type and volume of sediment being transported. **Braided streams** are so called because they have many branching channels that repeatedly divide and recombine around multiple masses of mounded sand and gravel, called *bars;* thus the stream has the appearance of a braid (Figure 7-9). Generally, braided streams are gravel-bed, alluvial streams that are high in energy, competence, and capacity. They are characterized by a steep stream gradient and have large widths relative to their depths—ideal characteristics for transporting the greatest amount of material under rapidly fluctuating conditions of streamflow. Braided streams are common near glaciers and in disturbed and logged landscapes, where deforestation has resulted in accelerated landsliding (and thus a greater sediment load) and increased stormwater runoff.

FIGURE 7-9 The Skilak River in Alaska, which consists primarily of meltwater from the Skilak Glacier, is a braided stream. The melting ice releases large amounts of poorly sorted sediment.

In contrast, the middle reach of the Mississippi River is a classic example of a **meandering stream** (Figure 7-10). The term is taken from the ancient name of the Meander River—now known as the Buyuk Menderes, in modern Turkey—which was noted for its many bends. This section of the Mississippi River flows upon hundreds of feet of alluvium deposited by ancient braided streams that carried meltwater during repeated glacial stages of the past few million years. The modern river channel—which receives no glacial meltwater during the present interglacial warm stage—is characterized by fine-grained sediment, a gentle gradient, low energy per unit area, and small widths and large depths. Although meandering streams may carry large amounts of sediment, the sediment generally is fine grained and constant in supply. In other words, meandering streams have a high capacity but a low competence compared with that of braided channels.

Just as braided streams are shaped by the bars that divide individual channel braids, meandering streams are shaped by areas of deep water called **pools** and shallow water called **riffles** (Figure 7-11). Pools typically occur on the outside bank of a meander bend, where the energy of swiftly flowing water is concentrated as it moves into the bend. Fast-flowing water scours, or erodes, the outside bank, forming a deep depression. Continued scour shifts

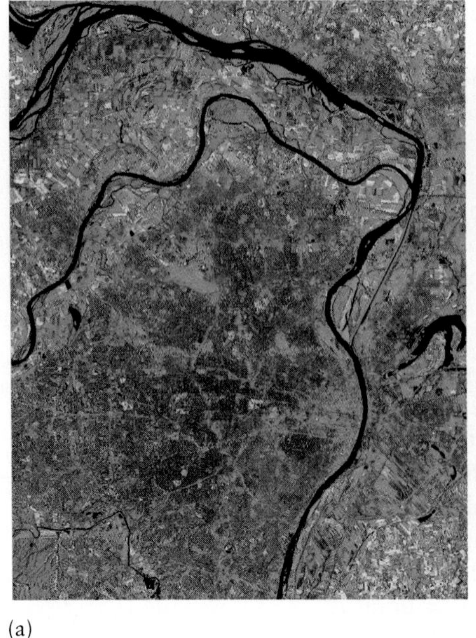

(a)

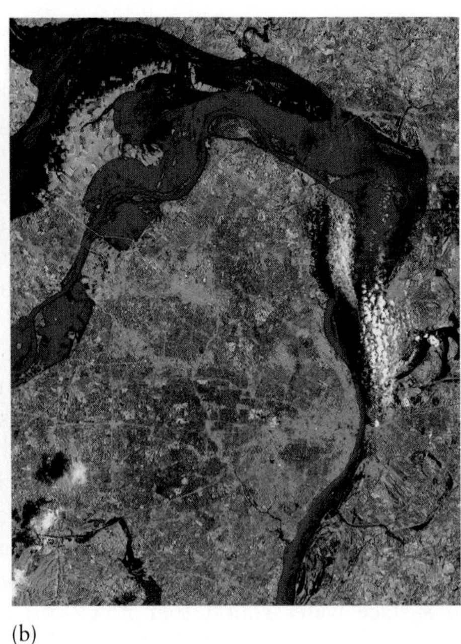

(b)

FIGURE 7-10 (a) This 1988 satellite image of the Mississippi River shows a meandering stream channel. Just upstream of St. Louis (the darkest pink patch at lower right), a large tributary—the Missouri River—joins the Mississippi. (b) Because of the Mississippi's gentle slope, floodwaters in 1993 spread over wide areas at meander bends. Flood-control structures downstream of St. Louis, where the river is less winding, contained the floodwaters.

the stream's position and gradually sweeps it sideways across the valley floor. At the same time, sediment is deposited opposite the pools along the inside banks of meander bends, where the water flows most slowly and has the least amount of energy. Here, silt, sand, and gravel are deposited in crescent-shaped bars, known as *point bars,* that over time migrate in the same direction as the laterally sweeping stream (Figure 7-12).

Between meander bends, sediment scoured from the pools accumulates and forms riffles along straight stretches of the stream channel. Riffles appear as humps in a longitudinal stream profile (a vertical section along the length of a stream), and these humps help to hold water in pools upstream of a riffle during times of low waterflow. For most natural streams in the world, the distance between pools is about five to seven times the channel width. This common characteristic reflects the fundamental nature of the downhill flow of water and sediment and probably is related to the ability of streams to do work with the energy provided them by gravity.

FIGURE 7-11 Geologists (center and right) surveying a steep riffle that separates two deep pools on the Mattole River in California. Like other rivers in northern California, the Mattole has been aggrading, or accumulating sediment, in recent years, most likely because of logging and increased mass wasting. Some scientists blame this process for the local demise of Pacific salmon, which used to come up this and other nearby rivers to spawn. Residents of the Mattole Valley, working with various state and federal groups, are trying to restore the drainage basin by revegetating hillslopes and stabilizing channel banks with resistant materials.

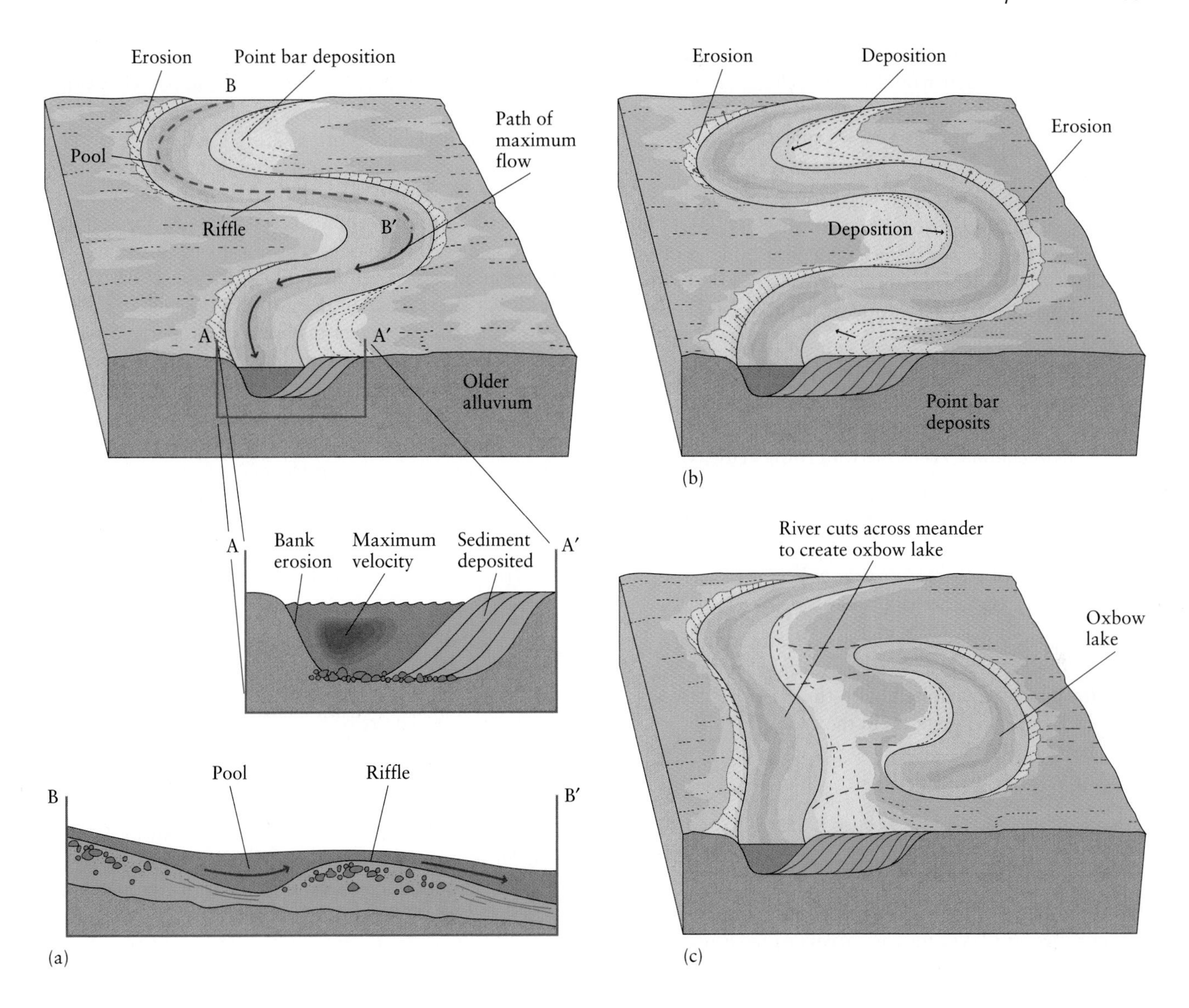

FIGURE 7-12 (a) Meandering streams are characterized by pools and riffles, with pools typically forming along the outside banks of meander bends, where water moving around the bend impinges on the bank and has greatest erosive power. Between pools are steep riffles, which act to pond water upstream in the pools. On the inside banks of meander bends, where stream velocity is low, sediment is deposited as point bars. (b) Meandering streams tend to migrate across their valley floors with time, in the direction of bank erosion, so that meander bends become accentuated and horseshoe-shaped. (c) If a stream cuts across the neck of a meander bend, it abandons its former channel and creates an oxbow lake.

Straight single-thread channels are rare in nature. They can be created in homogeneous sands in artificial streams developed in laboratory sand boxes, but within a very short time, even the artificial channel will alter its shape, creating evenly spaced pools and riffles and a meandering pattern.

The repeating pattern of deep pools always occurs in unaltered meandering streams and rivers and is critical to providing cool, deep water for many river-dwelling plants and animals, including fish such as salmon and trout. Turbulence in the riffles mixes oxygen into the water, a necessity for the health of aquatic organisms. Every good angler knows that the best fishing is along streams with deep, cool pools separated by fast-flowing rapids. Natural stream-channel patterns are essential to minimizing the damage due to flooding as well as to providing suitable habitats for native species.

- All surface water flows within stream networks that drain water from the land; the area of drainage—bounded from adjacent drainage areas by divides, or high spots—is called a drainage basin, or watershed.
- During and soon after a rainstorm, water flows across the land as overland stormflow (especially where the ground surface is impermeable) and beneath the surface as subsurface stormflow, supplying streams with the water that causes them to rise after a storm.
- The volume of water that flows in a given period of time in a stream channel is the stream's discharge, which is calculated as the area of the stream channel times the waterflow velocity.
- Streams are described as braided, meandering, or straight, depending on the nature of their channel patterns. Braided streams typically carry large amounts of coarse sediment, whereas meandering streams carry finer silts and clays.
- Pools and riffles are ubiquitous features along natural stream channels; they are spaced fairly uniformly along channels, with the distance between each pool about six times the channel width.

The Hazards of Flooding

Since the dawn of human existence, people have settled along rivers and lakes for easy access to fresh water and, since the Agricultural Revolution, for fertile soil for crops. The greatest hazard to waterside living, of course, is flooding, and humans have devoted considerable ingenuity and expense to flood control.

The Egyptians have lived along the Nile River in northern Africa since about 4000 BC. Most of the region drained by the Nile is arid or semiarid, so the only part of the drainage basin suitable for large-scale farming is along the floodplain of the main river channel (see the opening photograph of Chapter 4). Like other rivers, the Nile varies in water level throughout the year. Beginning in June the river rises and overflows its banks, flooding for three months. As the river widens and fills its floodplain, its waterflow is retarded and the water drops its load of fine silt. This silt provides continuous enrichment to the soil, which is why the ancient Greek historian Herodotus called the land of Egypt "the gift of the river."

Unfortunately, in the short term rivers are unreliable. Too much water can cause devastating floods, and too little can cause devastating drought. As a consequence, the flow of water is one of the most commonly controlled of natural processes. Rivers have been dammed, diverted, lined with levees, and straightened. When rivers break free from such shackles, the results are disastrous. Recent examples in the United States abound: the upper Mississippi River basin floods of 1993, the Ohio River floods of 1997, and the Red River flood of 1997.

Of all storm-related hazards, floods are the only ones for which the population-adjusted number of deaths in the United States has risen since the 1940s. In addition, flood damage to property has risen throughout this century, topping more than $12 billion for the 1993 Mississippi River floods, one of the greatest disasters in the history of the United States in terms of property damage. In developing countries with high population densities, such as Bangladesh, swelling populations crowd floodplains, and the loss of lives during flooding is far greater than in the United States. Flooding in developing nations is further exacerbated by upstream logging and deforestation, which are associated with rapid economic and population growth.

Floodplains and Flood-Recurrence Intervals

Despite billions of U.S. federal, state, and local dollars spent on flood protection and sophisticated flood-warning systems that use rapid communication and satellite technologies to alert residents to evacuate, flood damages continue to mount. The reason is twofold: imprudent development and management of urban areas along valley floors and the difficulty of identifying floodplains—areas susceptible to flooding.

Floodplains are nearly level, alluvial surfaces that occur alongside stream channels in valley bottoms (Figure 7-13). Floodplains become the beds of channels during high-water stages. In other words, floodplains are a natural control on river systems. The stream channel itself is large enough to hold water for most flows. On average once every 1 to 2 years the flow exceeds the stream channel. Every 2 to 10 years, floodwaters spill over the banks and spread across part of the floodplain. Every 100 to 500 years, large floods can inundate the whole valley bottom.

Floodplains form along both sides of a channel by two processes: lateral migration and overbank deposition. **Lateral migration** occurs when a stream channel cuts away at its concave banks, where stream energy is high, and drops sediment at point bars along its convex banks, where energy is low. This process of simultaneous lateral migration and deposition is responsible for the continuous building and tearing down of a floodplain and constructs most of its height. **Overbank deposition** occurs when fine-grained sediment is deposited in thin, horizontal sheets on the floodplain surface by streams that overflow their banks and spread across the valley bottom. Overbank deposits must be light enough to remain suspended in the uppermost parts of the floodwater. Consequently, they generally are silts and clays and, occasionally, fine sands. Organic matter such as leaves and charcoal are very light and float on floodwater, so they commonly are found at the

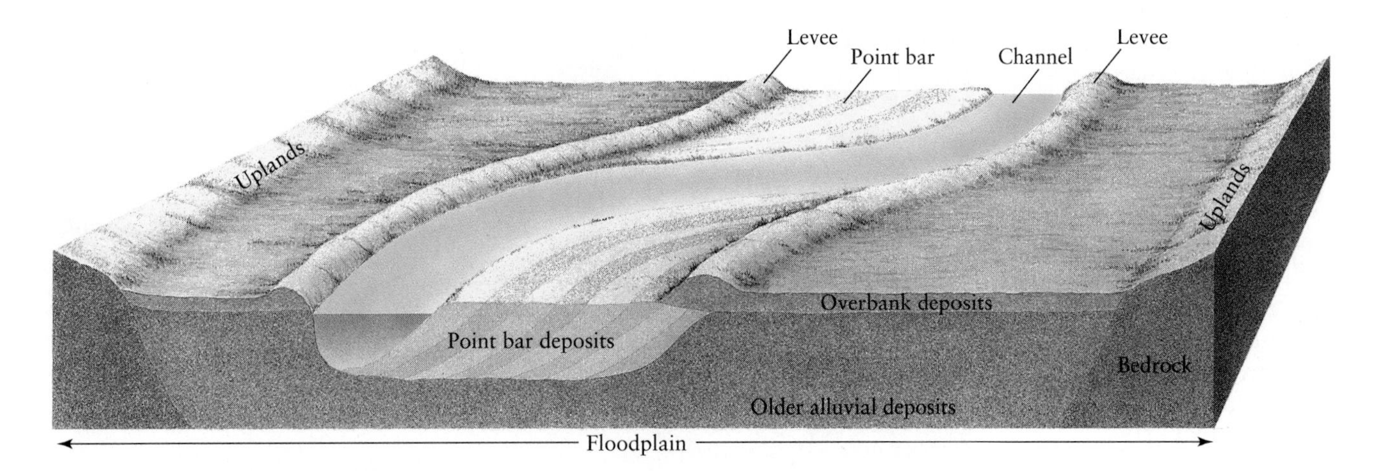

FIGURE 7-13 Floodplains are nearly level surfaces adjacent to stream channels. They are formed by channel migration, which accretes sediment as point bar deposits, and by deposition of sediment during flooding that exceeds the channel's dimensions and flows over its banks. Overbank deposits are thickest near the channels, creating levees that parallel the stream.

top of an overbank deposit, enabling geoscientists to separate different flood deposits from one another in the sedimentary record. Coarsest sediments are deposited immediately adjacent to a stream channel, and commonly form low berms called **levees** that help to contain floodwater within the channel.

Great Floods

Because of the fine-grained, moist nature of overbank deposits and the level nature of the floodplain surface, floodplains are excellent for farming. Nevertheless, early humans avoided settling permanently on them because of the flood risk and in some places constructed high mounds of dirt for temporary refuge. Postindustrial humans have not been so cautious, constructing dams and artificial levees and modifying river channels in an attempt to prevent future flooding, and then settling permanently in flood-prone areas. In consequence, ecosystems such as those along the Colorado River are destroyed, and human dependence on a false sense of security can lead to tragedy when large floods do occur. To minimize the risk of environmental degradation and human disaster, hydrologists and geoscientists attempt to determine where and how often floods happen and advise local governments and engineers on appropriate ways of managing and developing land in flood-prone areas. Two large rivers that have a history of "great" floods—those that cover hundreds to thousands of square kilometers of land—despite human attempts to control them are the Hwang Ho of China and the Mississippi River of the United States.

Hwang Ho The Hwang Ho is called the mother of China for the same reason that Egypt was called the gift of the Nile—the bestowal of annual sediment during flooding. At high-flood stages, however, the Hwang Ho is more lethal than any other river in the world and becomes China's sorrow. Flowing across a vast alluvial plain with a dense population, the river killed 900,000 people in 1887. Less than 45 years later, in 1931, nearly 4 million people drowned during flooding or died soon after from disease and starvation.

It is not the Hwang Ho's water discharge that makes the river so deadly. On average, its discharge is no more than that of the Arkansas River, a tributary of the Mississippi (see Figure 7-6). The Hwang Ho's destructiveness is attributable to two other features: the unconfined nature of the river's natural channel as it spills onto North China's vast, nearly flat Great Plain, and its ever-rising channel bed.

Before heading to sea, the Hwang Ho traverses an alluvial plain larger than the state of Oregon containing thousands of meters of alluvial and coastal sediments deposited over millions of years. The final kilometers of the Hwang Ho cross a **delta** (the term taken from the shape of the Greek letter Δ), a triangular alluvial deposit that occurs where a river empties into the sea or a lake. Most deposition in a delta occurs below the level of the water into which the river flows. The Hwang Ho delta formed over the millennia as sea level repeatedly rose and fell, causing the delta to retreat and advance. Until humans attempted to control the river's location, the Hwang Ho migrated freely in response to changing volumes of water and sea level (Figure 7-14). The loose sediments offer no resistance to channel cutting and widening during times of high flow, and the featureless plain also offers no safe haven to its human inhabitants during flooding.

FIGURE 7-14 (a) The mouth of the Hwang Ho and its delta have shifted many times over the past several thousand years, ranging over a distance of 1000 km on the nearly flat surface named the Great Plain. During great floods along the Hwang Ho, houses and people sometimes are buried under many meters of silt and mud. (b) During a recent small flood, the high concentration of silt and mud is clearly visible in the water pouring through a series of steps and troughs constructed to trap the sediments.

The second reason for the river's lethal nature is related to the amount of sediment it carries and deposits near its mouth, which is the greatest of any river in the world. Downstream of its headwaters on the Tibetan Plateau, the Hwang Ho drains a landscape etched into nearly 150 m of yellow, dustlike sediment known as *loess,* which was deposited by the high winds commonly associated with past glacial ages. The Hwang Ho easily incises the Pleistocene loess, transporting the large loads of yellow, muddy sediment that give the river its name: *Hwang* is the Mandarin word for "yellow." As the river continues beyond the loess plateau and onto the Great Plain, its gradient (drop in elevation over a given distance) decreases to less than 13 cm per km. The stream loses energy and is unable to carry such a heavy load. As a consequence, mud is deposited along the channel bottom during each flood (see Figure 7-14), effectively raising the streambed and hence the level at which water flows.

For thousands of years, people living along the Hwang Ho have raised artificial levees of stones, branches, and rope to check the overtopping waters, but they are caught in a never-ending cycle. As the riverbed continues to rise and flood, they continue to raise the levees. The result is that today, both the river and its levees loom menacingly high above the inhabited alluvial plain (Figure 7-15). Sometimes the river bursts its levees, changing course as much as 1000 km and causing widespread flooding and devastation to those living below the level of the artificially raised river channel.

In the 1950s, in an effort to reduce the amount of silt and water reaching the alluvial and deltaic plains, the Chinese government initiated a massive plan involving the construction of dozens of dams upstream of the Great Plain, reforestation and hillslope terracing throughout the loess plateau, and dredging along the channel. Two other efforts that have been effective are the construction of floodwater-detention basins to hold excess water during high-water stages and settling ponds to trap silt along the deltaic reach of the river, as shown in Figure 7-14 (b). In the long term, the most effective measure is the effort to prevent soil erosion and excess runoff of water in the upper parts of the Hwang Ho drainage basin, because the reservoirs behind the dams rapidly fill with sediment, and levees are likely to continue bursting when overstressed by such a dynamic river.

Mississippi River The Mississippi River drainage basin is considered commonly as two parts, the upper and lower basins. The transition between the two, where the Mississippi trunk stream meets its largest tributary, the Ohio River (see Figure 7-6), is dramatic, marked by the sudden change from a narrow channel cascading over waterfalls to a vast alluvial plain similar to that of the Hwang Ho. Like the Great Plain of China, the lower Mississippi Valley floor formed in the course of millions of years in re-

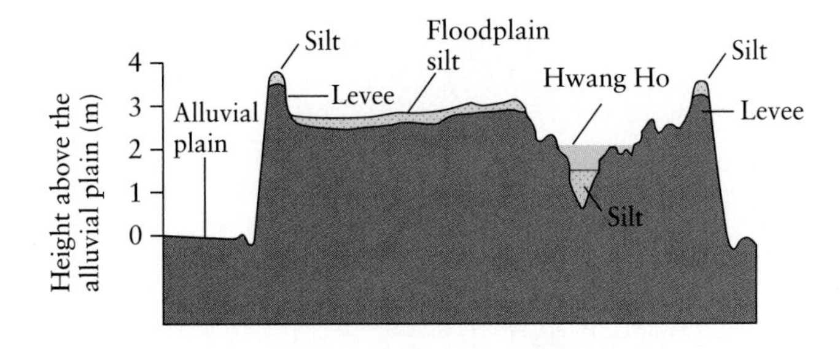

FIGURE 7-15 This cross-sectional diagram of artificially raised levees for the Hwang Ho illustrates the consequences of long-term levee building. With each flood, the river deposits a load of silt that raises its bed and increases the possibility that water will overtop the channel's banks. In response, the inhabitants keep raising the level of the earthen dikes, and the river now stands as much as 3 m above the surrounding alluvial plain in some places. Before the river was restricted by artificial levees, its load was spread across the alluvial plain rather than concentrated in one place.

sponse to changing sea levels and retreating and advancing deltas (Figure 7-16). It now overlies more than 300 m of alluvium, forming a nearly featureless lowland 1000 km long and up to 200 km wide.

Since sea level rose to about its present level some 5000 to 4000 years ago, the river has meandered back and forth across a delta plain at its mouth (Figure 7-17). Rivers shift about on delta plains because of sedimentation. As sediment accumulates at a river mouth, the position of a delta advances seaward and forms a broad, gently sloping surface. Rivers move easily across such nearly level surfaces, typically shifting to locations where the slope is steepest and the distance to the sea is shortest.

When Europeans and Africans arrived in North America, the Mississippi flowed in sweeping, twisting meander bends across its lower valley floor. Today, however, the modern channel of the Mississippi is fettered, constricted, and yoked by thousands of kilometers of artificial levees and hundreds of dams and reservoirs. Unlike the Hwang Ho, the Mississippi River has been substantially straightened and shortened, with nearly 200 km of its original length chopped off at meander bend cutoffs. The call for such shortening came after the levees failed to hold water at more than 120 places during the great Mississippi flood of 1927, which killed several hundred people and left more than half a million homeless. Soon after that flood, Major General Harley B. Ferguson, the newly appointed president of the Mississippi River Commission, declared that levees alone could not control the great river, because the real problem was that the river could not rid itself quickly enough of its floodwaters while navigating such tortuous hairpin turns. "The water wants out," he said. "We will give it out." Less than a decade later, the lower

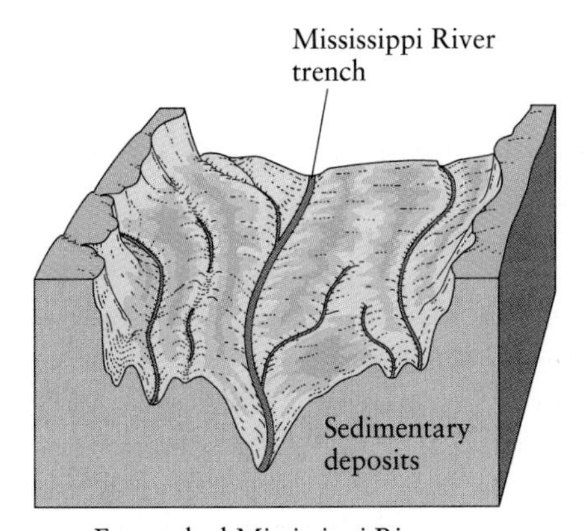

Entrenched Mississippi River
18,000 to 15,000 years ago

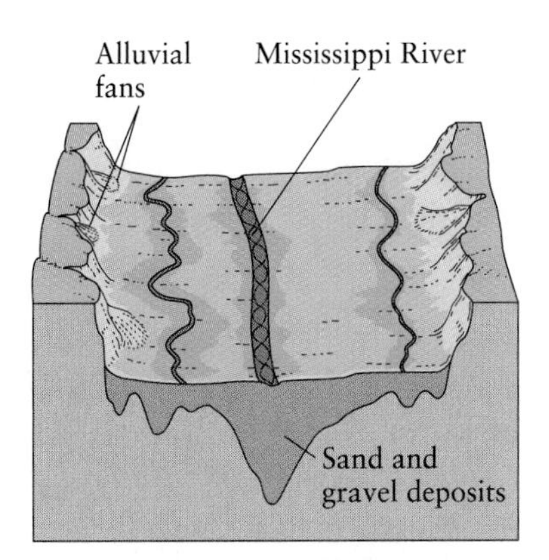

Formation of a braided channel
15,000 to 4000 years ago

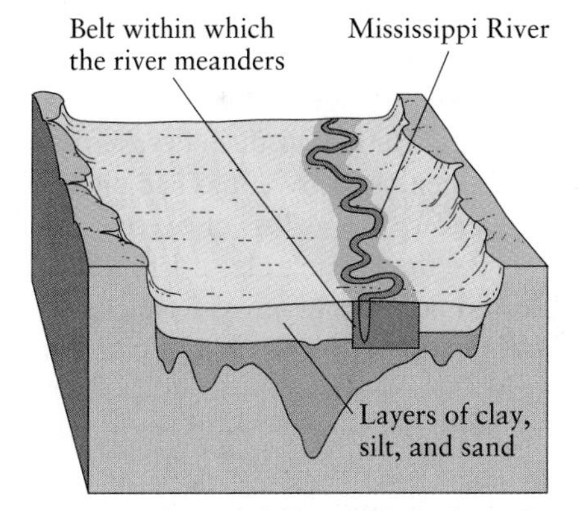

Formation of a meandering channel
4000 years ago to the present

FIGURE 7-16 The changing course of the lower Mississippi River—a cross-sectional view of the changing geometry of the river channel over the past 18,000 years. Approximately 18,000 to 15,000 years ago, advancing glaciers lowered global sea level by about 120 m. In response to the lowered base level at its mouth, the Mississippi River cut deeply into its valley floor. From 15,000 to 4000 years ago, melting ice released large volumes of water and sediment into the Mississippi River system, and worldwide melting of ice sheets raised the sea level. In response to the increased volumes of water and sediment, the Mississippi became a braided stream. The entrenched Mississippi Valley filled with sediment, sea level continued to rise, and the river's slope became gentler in response to the rise in base level. Since about 4000 years ago, the modern Mississippi River carries much less water and sediment than did its braided ancestor and has become a single meandering channel that is better suited to carrying fine sediment on a gently sloping valley floor.

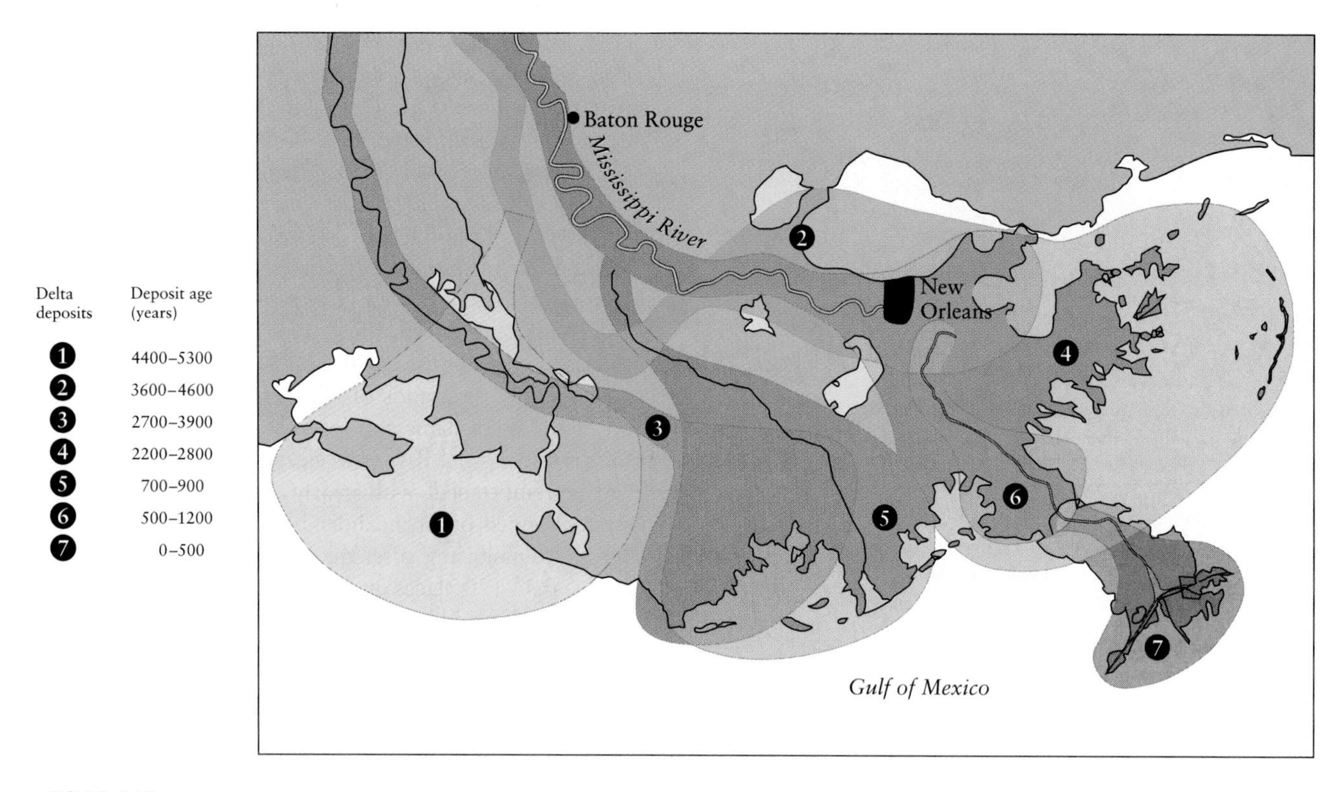

FIGURE 7-17 The Mississippi River deposits sediment as it enters the Gulf of Mexico, forming a delta, but shifts the position of deposition with time. Since about 5300 years ago, the delta lobe has shifted seven times before reaching its present position downstream of New Orleans. In the 19th and 20th centuries, a new delta began to form at about the same location as delta 1, along the Atchafalaya River, but the U.S. Army Corps of Engineers is working to control the river so that it is unable to change direction.

Mississippi River was shorter and straighter, and no flood since has caused as much devastation as that of 1927. But straightening the river has had unanticipated negative consequences such as loss of riverine wetlands.

Like the Hwang Ho, if left alone the Mississippi River would flow freely across its alluvial plain, shifting its path frequently in order to find the easiest routes to the sea. Its levees are designed to keep, forever if possible, the river in its current course, so that riverside cities such as New Orleans will remain just that—riverside. Over the past several thousand years, the Mississippi has followed many different routes to the sea, and as recently as the middle of the 20th century the big river looked as if it would shift to a much steeper and shorter route along the Atchafalaya River (see Figure 7-17). In 1954 Congress authorized the Army Corps of Engineers to do whatever was necessary to keep the river from turning New Orleans and Baton Rouge into backwater swamps. The Corps' main effort was the Old River Control project, which includes a dam that regulates the flow of water from the Mississippi into the Atchafalaya, maintaining for perpetuity the constant value of only 30 percent of the flow of the Mississippi into the Atchafalaya. If this value were to increase—for example, if the Old River Control structures were to fail during a massive flood—the floodwaters could deepen the Atchafalaya stream channel and enable the Mississippi to change course catastrophically.

Efforts to control the Mississippi River were put to an extreme test in 1993, when peak discharges at 42 stream-gauging stations were greater than any previously recorded values. In mid-June through early August, widespread and severe flooding occurred throughout the upper Mississippi River basin, killing more than 50 people, destroying millions of acres of crops, damaging highways and roads, breaking or topping more than 800 levees, severely eroding channel banks and hillslopes, and depositing sediment over large areas (Figure 7-18).

Because of a six-month period of high precipitation, soils throughout the region were saturated when a sequence of intense rainstorms began in late June. With the soils' infiltration capacity already exceeded, most of the rainwater became direct storm runoff, rapidly raising the water levels of many streams that already filled their channels to the tops of their banks. One storm after another pelted the area with rain for nearly two months, and one tributary after another swelled and disgorged into the main Mississippi River trunk stream. With more than 100,000 swollen tributaries covering

FIGURE 7-18 When the Mississippi River flooded in 1993, it broke through or topped more than 800 levees, including this one in Illinois. The levee is the light brown wall just beyond the houses, which were washed in by floodwaters.

about 3.2 million km^2 of land, the threat of severe flooding was great.

In fact, the flooding might have been far worse if not for the storage of large volumes of water in dozens of flood-control reservoirs, most of them within the Missouri River basin (Figure 7-19). A flood-control reservoir is designed to store part of the stormflow during a flood for later release in order to reduce the flood peak downstream. Flood-control reservoirs in the Missouri River basin reduced the discharge at St. Louis by more than 211,000 cfs in July 1993. Computer simulations show that without the reservoirs, many more levees might have been breached, and more cities might have been flooded (Figure 7-20).

Minimizing Flood Hazards

Many scientists scoff at efforts to control the Mississippi River, considering it folly to believe it possible to tame one of the mightiest rivers on Earth for long.* Certainly the amount of death and destruction that occurred in 1993 despite the levees and reservoirs suggests that the only long-term solution to flood hazards is to further dissuade development in flood-prone areas. Some communities, such as Patterson, Kansas, have taken this tack, using federal relief funds to rebuild their towns at higher elevations after the 1993 floods. The U.S. government encourages this type of flood avoidance by the terms of a federal flood insurance program.

Estimating the Probability of Flooding In the early 1960s, after decades of disastrous floods throughout the United States despite more than $1 billion in federal money spent each year for flood losses, the federal government examined alternatives to what engineers call containment approaches, or mechanical flood-control techniques. The result was the National Flood Insurance Act, under which the National Flood Insurance Program (NFIP) was created. The NFIP provides insurance to any

*Dr. Raphael Kazmann has said, "It's going to happen. We can delay it, but ultimately the river will take over. In the long run, man cannot win." Dr. Kazmann, professor of civil engineering at Louisiana State University, was speaking in the 1980s of attempts by the U.S. Army Corps of Engineers to keep the lower Mississippi River from abandoning its present channel in favor of a more direct route to the Gulf of Mexico along the Atchafalaya River.

(a)

(b)

FIGURE 7-19 (a) This flood-control reservoir in Kansas was filled to capacity during the floods of 1993 on the Mississippi River and its tributaries. Excess water was released along the spillway to the right of the reservoir. (b) A bedrock surface scoured and pitted by the enormous volume of swiftly moving water was revealed as the floodwaters receded.

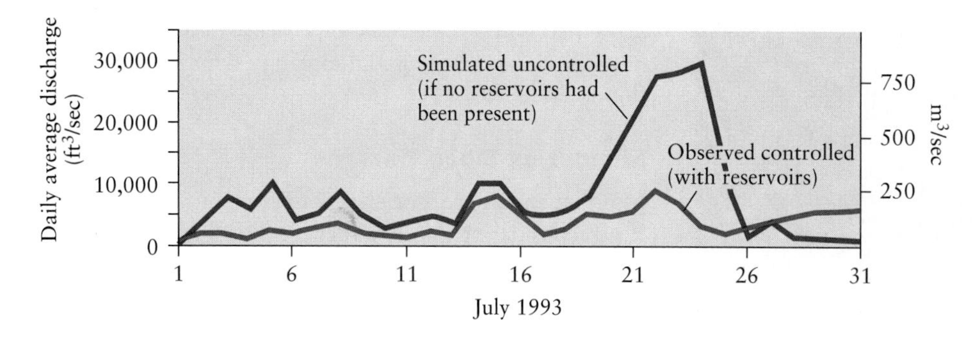

FIGURE 7-20 Engineers with the U.S. Army Corps of Engineers use hydrographs (plots of time versus discharge), as shown here for the Smoky Hill River at Mentor, Kansas, during the flooding of July 1993, to illustrate the effect of detention reservoirs (red) on minimizing the peak discharge during a flood. Without such control structures, peak discharge would be much greater (purple).

community subject to periodic flooding if the community agrees to establish and enforce land-use and flood-control standards for development in flood-prone areas.

Communities that apply for NFIP are sent a team of Federal Emergency Management Agency (FEMA) engineers and geologists. The team defines Special Flood Hazard Areas within a floodplain—areas that would be inundated once water tops the level of the channel banks. The probability of such flooding and the estimate of intervals between floods are based on statistical analysis of the frequency at which the water in a given river reaches a particular stage (Box 7-2). From this study, FEMA provides the community with a Flood Insurance Rate Map, delineating flood elevations and flood-risk zones, that is to be used in local zoning and building code ordinances (Figure 7-21). These maps, available from local city or county planning offices, now exist for more than 10,000 communities.

Controlling Floods Through Channelization Another means of attempting to reduce flooding is **channelization**—the process of replacing natural stream channels with artificial culverts and ditches. The purposes of channelization are to reduce flood hazards and to drain land for agriculture and development. Typically, artificial channels are rectangular or trapezoidal in cross section, are lined wholly or in part with concrete, and are much straighter than the original channel. Roughness of the artificial channel bed is less than that of the original bed, and the gradient is steeper (Figure 7-22). Because the flow of water is directly proportional to the steepness of the channel slope, the steeper gradient results in greater flow velocity.

While speeding the passage of floodwaters reduces flooding in the artificially channelized area, its consequences are not necessarily positive for downstream communities where the stream is not artificially channelized. Flooding can be exacerbated downstream, precisely because upstream floodwaters have not been able to disperse onto a floodplain storage area. Furthermore, the arrival of peak discharge after a storm is hastened by the increased flow velocities.

The first efforts at channelization took place on the Mississippi River in the 1870s, but after passage of the Watershed Protection and Flood Prevention Act in 1954, tens of thousands of kilometers of floodplain channels nationwide were dredged and straightened in order to reduce flood hazards. Simultaneously, riverine wetlands adjacent to streams were drained and filled to increase the amount of land available for cultivation and development. The economic benefits to farmers and developers

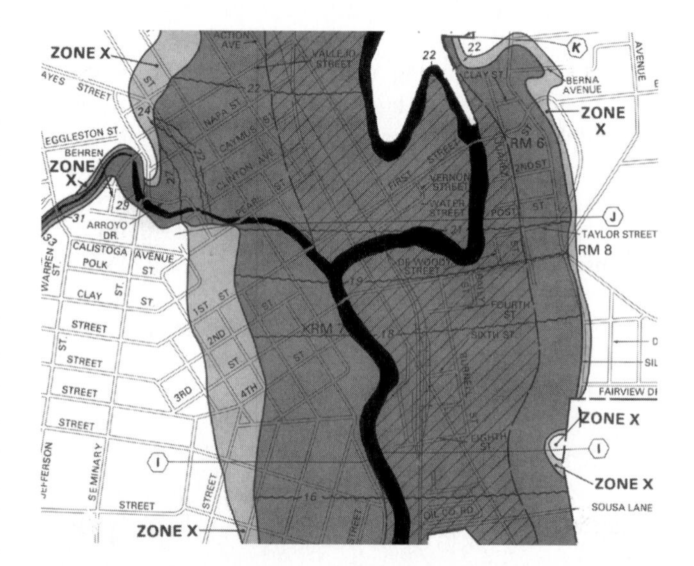

FIGURE 7-21 FEMA has prepared Flood Insurance Rate Maps, like this one for Napa, California, for thousands of communities in the United States. Using historical discharge data and surveyed cross sections of stream channels and floodplains, FEMA scientists and engineers determined the area adjacent to the Napa River (heavy black line) that will be flooded by the 100-year flood (dark gray) and the 500-year flood (Zone X, light gray). Also shown as Zone X, but in white, are areas outside the 500-year floodplain. Within the 100-year flood zone, contour lines with numbers indicate the elevation to which floodwater will rise. At the intersection of Fourth and Burnell streets, for example, 100-year floodwaters will rise to about 18.5 ft. This base elevation is used as a guideline for how high to raise a house if it is built in the floodplain. The first lived-in floor of a house must be at least 1 ft above base elevation.

7-2 Geologist's Toolbox

Flood-Frequency Analysis and the 100-Year Flood

Using geologic and historical records to understand how systems have operated in the past, geologists also try to predict what might happen to a system in the future. An example of such an approach for land-use planning is flood-frequency analysis, in which geologists try to predict how often floods of various sizes might occur. The procedure relies on statistics and probability, combined with a record of past flood discharges. Historical data are obtained from a gauging station, and data from prehistory sometimes can be obtained by radiocarbon dating of ancient flood deposits.

Consider first a simple analogy, the chances of winning a lottery in which six numbers between 1 and 100 will be drawn. The chance of getting one number that matches one of the six is fairly good, but the chance of getting all six numbers is small. Similarly, the chance of a small flood in a given year is large, but the chance of a large flood is small. In other words, small floods are more frequent than large floods and have a greater probability of occurring in a given year.

In flood-frequency analysis, the largest flood for each year of record is ranked by its size (as defined by its discharge). The question addressed by statistical analysis—essentially, probability theory—is how often a flood of that size is likely to occur or be exceeded. Each year of record provides the geologist with a sample, and the more sample points, the more reliable the analysis. Small floods have a high probability of occurring, perhaps a 50 percent probability of being equaled or exceeded in a given year. A 50 percent probability is a 1 in 2 chance, which is the same as the chance of heads appearing on a single toss of a coin. Large floods have a low probability of occurring, perhaps a 1 percent probability of being equaled or exceeded in a given year. A 1 percent probability is the same as a 1 in 100 chance. To estimate uniformly the degree of flooding throughout the country, FEMA adopted the 100-year flood standard. The 100-year flood is not one that occurs only once every 100 years, but rather one that has a 0.01 (1 out of 100, or 1 percent) chance of being equaled or exceeded in any single year. Its probability of occurrence remains the same each year, even if a 100-year flood occurred the previous year. In fact, Houston, Texas, had three 100-year floods in 1979.

The *recurrence interval* is the average period of time that elapses between events of a given size. The greater the probability of occurrence, the more frequent the event, and the smaller the recurrence interval. Thus, the recurrence interval of a flood is the inverse of the probability that a flood of a given size will recur or be exceeded. Use 50 percent probability as an example: The inverse of 50 percent is 2 (100 ÷ 50 = 2). Hence, a 2-year flood has a 50 percent probability of being equaled or exceeded in any given year. In contrast, a 100-year flood has a 1 percent chance of being equaled or exceeded in any given year (100 ÷ 1 = 100). Note that the larger the recurrence interval, the smaller the probability of that event occurring, and the greater the discharge associated with it:

Probability that flood will be equaled or exceeded in a given year	Recurrence interval of flood (years)
0.01 (or 1%)	100
0.02 (or 2%)	50
0.1 (or 10%)	10
0.2 (or 20%)	5
0.5 (or 50%)	2
1 (or 100%)	1

In a complete flood-frequency analysis, either the probability that a flood will occur or be exceeded, or the recurrence interval, or both, are plotted on a graph against the corresponding discharge. A flood-frequency curve can be fitted to the data points in order to show the pattern of the discharge associated with different events. Such curves when projected are essential for estimating the size of large flood events, such as the 100-year flood, because very few gauging stations have records that span more than several decades or preserved ancient flood deposits that can be dated. So the discharge associated with large events must be extrapolated from the limited amount of historical data usually available. The discharge obtained for the 100-year

(continued)

7-2 Geologist's Toolbox

(continued)

flood then is used to determine the stage of the water during that event and to predict how much area will be flooded.

This is the technique that was used in the flood-frequency analysis of the area along the Napa River in California (see Figure 7-21).

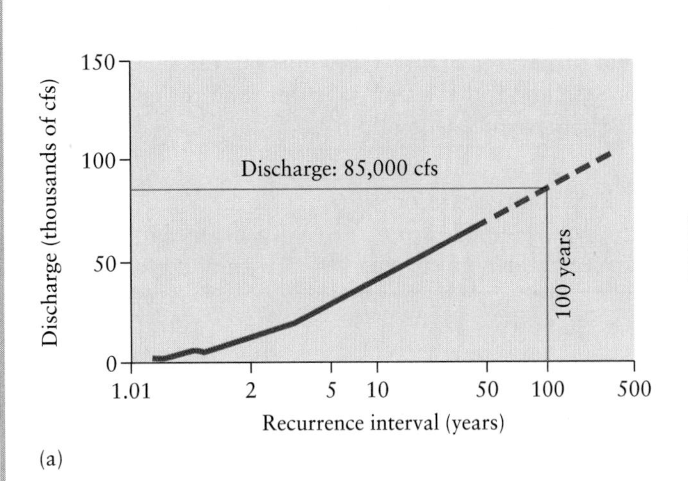

(a)

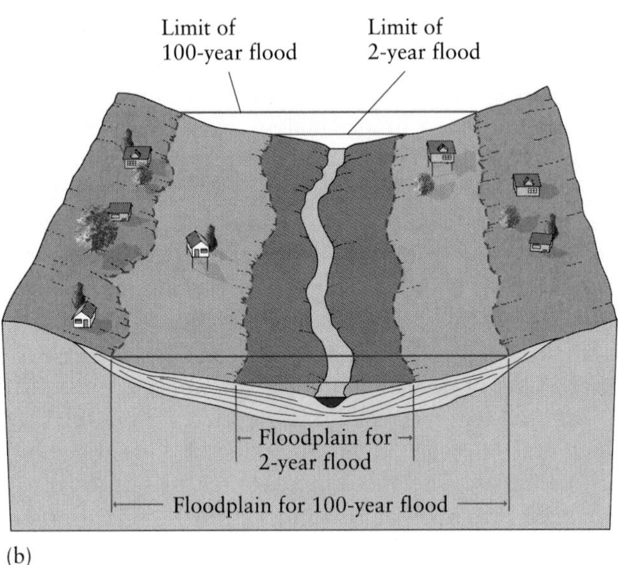

(b)

(a) The flood-frequency curve developed from data (solid line) taken at gauging stations on the Navasota River, Texas. For this river, the 100-year flood is estimated (from the dashed line) to have a discharge of 85,000 cfs, about seven times as large as the 2-year flood (12,000 cfs) and twice as large as the 10-year flood (44,000 cfs). (b) Estimated 100-year-flood levels can be used to identify the height and limits of flooding along the margins of a stream channel. The 2-year flood level coincides roughly with the active floodplain, which is built and maintained by frequent small floods.

are undeniable, but the environmental damages to ecosystems and stream processes are so great that they often outweigh the benefits.

In the 1960s the U.S. government spent $30 million to straighten the Kissimmee River, the main source of fresh water for the Everglades of Florida (Figure 7-23). The U.S. Army Corps of Engineers turned 166 km of twisting, sandy channel beds into fewer than 90 km of straight, concrete-lined channels in an attempt to regulate flooding and make more land available for development. Soon after, however, the wetland ecosystems began to decline and the amount of water seeping into the ground diminished, depleting important sources of groundwater for southern Florida. In 1984, the South Florida Water Management District began to restore the Kissimmee River to its original channel shape and pattern. The projected cost of complete restoration exceeds $350 million. The goal of undoing the straight, concrete channels is to let nature do with flowing water in natural river channels what it does best: recharge groundwater supplies, supply nutrients to

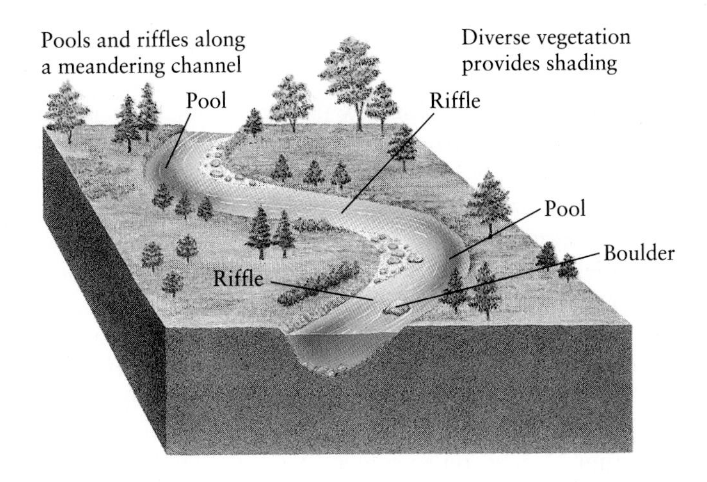

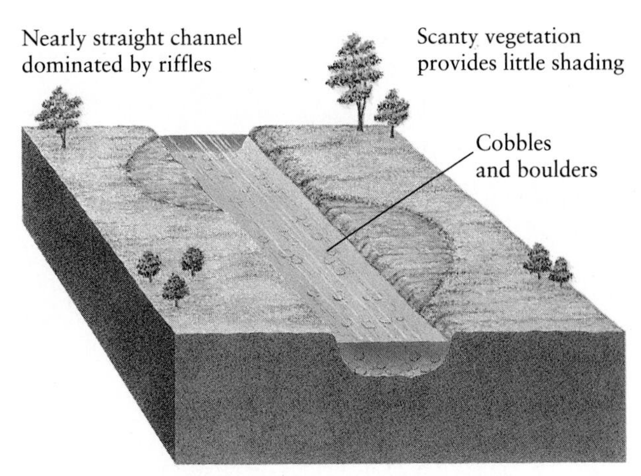

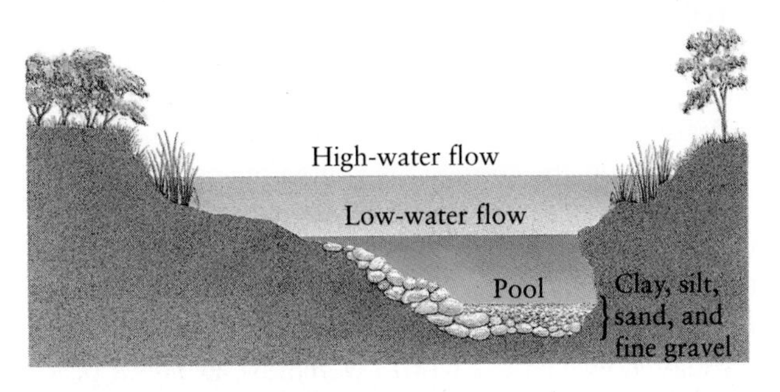

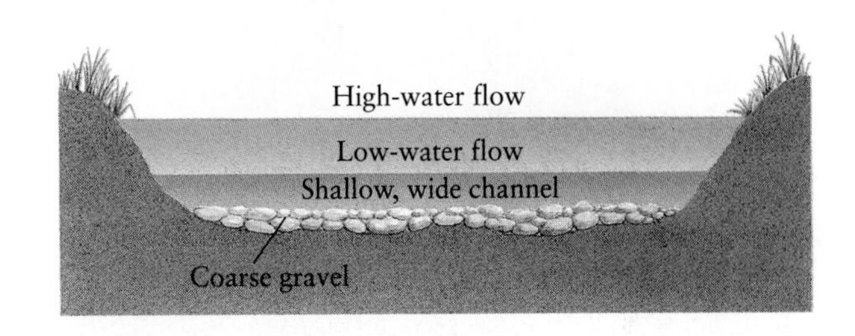

Natural Channel Environment

- Sufficient shading to maintain water temperatures hospitable to aquatic life
- Variety of pool depths and stream gradients; sorted gravels provide habitat for diverse organisms
- Sufficient water depth to support aquatic life during the dry season
- Resting areas for aquatic life behind rocks and beneath undercut banks
- Varied water velocities
- Diversified habitats

Artificial Channel Environment

- Inadequate shading; water temperatures too high to support some aquatic life
- Uniform topography; unsorted gravels provide few habitats; few organism varieties
- Lack of pools results in insufficient depth to support diverse aquatic life during the dry season
- Steep gradient and relatively shallow water
- Fairly uniform water velocity; too fast to support some aquatic life

FIGURE 7-22 Natural stream channels are markedly different from most artifical channels. The importance of meanders, pools, and riffles as habitat for aquatic organisms and for reducing flood hazard is causing scientists and engineers to rethink stream alteration. Increased attention now is given to restoring streams channelized in the past, and to designing new channel alteration projects so that natural characteristics, such as pool and riffle spacing that are five to seven times channel width, are maintained.

floodplain soils, sustain surrounding marshes, and provide habitat for plants and animals.

Today there are federal guidelines for producing channel forms that closely resemble those of natural systems: pools and riffles spaced about five to seven channel widths apart, meandering pathways, and gradients similar to or only slightly steeper than that of the original channel. These design criteria provide resting and spawning places for fish, dissipate flow energy over gravel riffles or sand bars, and minimize degradation of the channel bed.

- Stream channels are able to carry most waterflow most of the time, but about once every two years the water tops its banks and spills onto the floodplain, a nearly level, alluvial surface that bounds the stream along its margins.
- Floodplains are constructed by the deposition of sediment along the sides of channels at point bars, when streams shift position with time, and above the channel level when water overflows its banks in times of flood.

FIGURE 7-23 To regulate flooding and make more land available for development, the Kissimmee River in Florida was channelized and straightened in the 1960s, its length reduced from 166 to 90 km. In the 1980s engineers began to undo the channelization.

- Two of the largest rivers in the world, the Hwang Ho of China and the Mississippi River of the United States, have similar geologic histories. Both are meandering rivers that carry large amounts of fine sediment, and both have deltas formed by the deposition of sediment at the river mouths. Both have frequently (and naturally) shifted their channels until recent times. And, in both cases, for centuries human ingenuity and effort have endeavored to control their flow and position by the construction of dams and artificial levees.
- Flood-frequency analysis, which can yield flood-recurrence intervals, is useful to planning and regulatory agencies that attempt to control floods and provide flood insurance. The concept of the 100-year flood is used as a regulatory standard, and is based on statistical analysis of the historical (and sometimes prehistorical) record of flooding of a stream.

Wetlands

Geoscientists, hydrologists, and ecologists have in the past few decades paid increasing attention to wetlands as their importance to hydrologic processes and ecosystems becomes apparent. To most early Americans, wetlands were reeking pestholes associated with disease.* Doctors in the 18th and 19th centuries observed that repeated plagues of yellow fever and malaria began near wetlands. The true cause of these "swamp fevers," however, was not recognized until a young Baltimore doctor from Johns Hopkins Medical School, Jesse Lazear, willingly let a mosquito in a yellow fever ward draw blood from his hand. Within 12 days, on September 25, 1890, Dr. Lazear died of yellow fever, by his death convincing his colleagues that this dreaded disease was passed from person to person by mosquitoes, which breed in abundance near wetland areas. By the mid-1950s more than half the area of wetlands in the contiguous United States was drained, filled in, farmed, paved over, or developed. Many of the nation's prominent cities, including all of Washington, D.C., and much of New York, Philadelphia, New Orleans, and San Francisco, rest above filled-in wetlands.

Today, however, the word "wetlands" evokes a peaceful image of cattails and mosses, birds and turtles, and perhaps a rustic cabin like that of Thoreau's on Walden Pond (Figure 7-24). This picture is in large part due to the efforts of the environmental movement of the 1960s and 1970s. With this change in attitude have come legislative attempts to protect remaining wetlands from further destruction, yet 121,000 hectares (300,000 acres) still are destroyed each year, with government approval. Effective protection requires the ability to identify wetlands and to understand the factors necessary to their health and well-being.

Characteristics and Benefits of Wetlands

What are wetlands? Scientists, environmentalists, and policymakers have struggled to agree on a single definition of a wetland for legislative and regulatory guidelines. The definition offered here is based simply on moisture content and topography: **Wetlands** are poorly drained, low-relief areas in which the soil is seasonally or perennially saturated or covered with water. Some have shallow ponds, and all support plant life. Unlike lakes, wetlands are not limited exclusively to topographic depressions. They occur anywhere relief is low and drainage is poor, including on drainage divides and hillslopes.

Wetlands are categorized by their sources of water. *Bogs* are sustained by precipitation, while *fens* are sustained by

* In 1766, public opinion forced town officials in colonial Baltimore to pass an "Act to Remove a Nuisance," giving landowners four years to wall off and fill in marshes along the harbor and waterfront, because of their "noxious vapours and putrid effluvia."

FIGURE 7-24 Wetlands provide habitat to nearly half of the species listed as endangered in the United States and are essential to the breeding, nesting, and migration of many birds. This wetland in the mountainous desert east of San Diego provides shelter, food, and water to migratory waterfowl.

the mineral-rich water from subsurface groundwater. Drainage may be impeded if the substrate—the underlying layer of soil or sediment—has a low infiltration capacity, as do the poorly sorted glacial sediments found at the border between the United States and Canada, or if a topographic feature such as a shallow depression traps water and prevents it from flowing away.

Wetlands represent about 6 percent of Earth's total land area. Most wetlands are in temperate regions, where they appear along coastal areas as swamps and marshes. One of the largest salt-marsh wetlands in the world, covering nearly 13,000 km^2, is the Florida Everglades. Because of poorly drained glacial deposits and the broad, low-relief continental feature known as the Canadian shield,* the United States also has an unusually high number of wetlands (and lakes) along its northern border.

A characteristic shared by all wetlands is a low concentration of oxygen in their soil, a consequence of impeded drainage. Oxygen is essential to decay, so decomposition of organic matter in a wetland is less than in other wet areas where drainage is much greater. The anaerobic conditions of a wetland are revealed by a smell reminiscent of rotten eggs that stems from bacterial action in the absence of oxygen. The result of such anaerobic conditions is the accumulation of dead organic matter as **peat,** an unconsolidated deposit of plant remains in a water-saturated environment. Peat has a high moisture content (greater than 75 percent), contains more than 50 percent carbon, and burns freely when dried. Wetland soil usually is dark or mottled. As organic-rich peat accumulates it becomes thicker and the bottommost layers become compressed, forming a substrate with a low infiltration capacity. This peaty bottom perpetuates the life of the wetland by impeding drainage and keeping the wetland moist or saturated.

Wetlands are home to a third of the nation's threatened and endangered species. Along with tropical rain forests, wetlands are the most productive terrestrial ecosystems on Earth (Figure 7-25). In their more commonly known

* So called because of their resemblance to a warrior's shield, low-relief continental shields are vast expanses of ancient, crystalline bedrock that form the cores of continents; the Canadian shield extends across much of eastern Canada and the northern central and eastern United States.

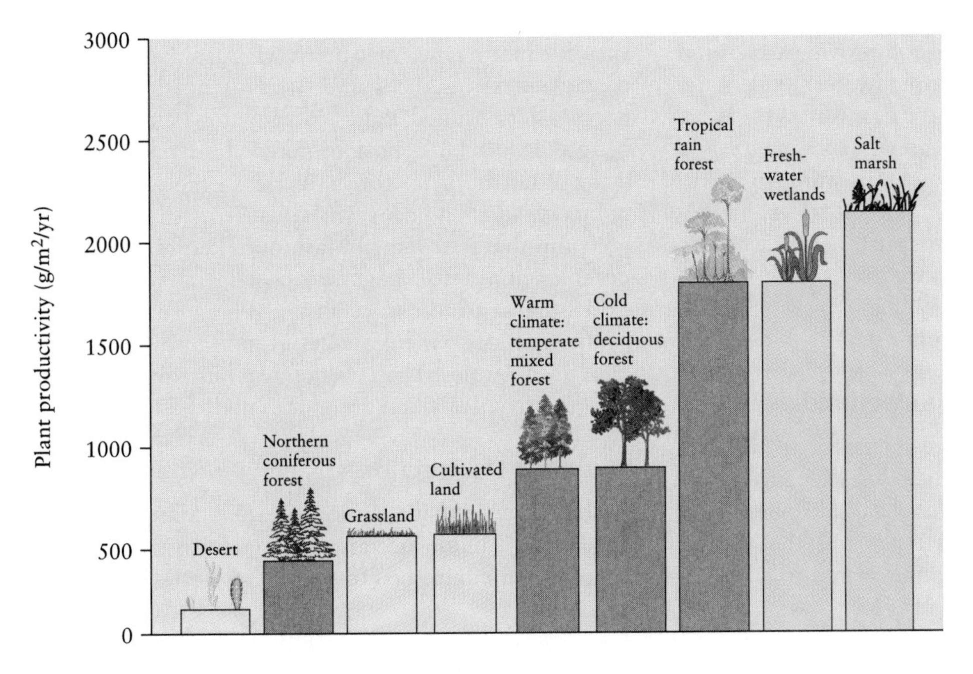

FIGURE 7-25 The net mass of plant matter produced in a given ecosystem over a specific time period—called the net primary productivity—is greater for salt marshes than for any other terrestrial environment. Freshwater wetlands have productivities comparable to tropical rain forests.

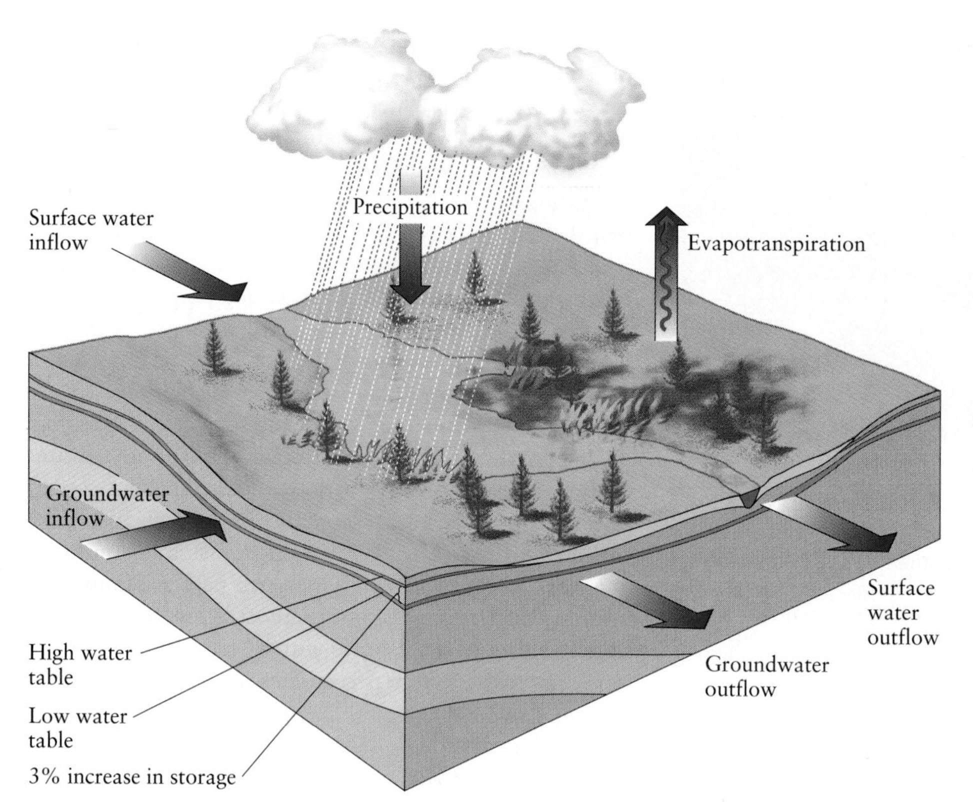

FIGURE 7-26 Wetlands differ from one another largely because of variations in water conditions, which affect wetland soils and vegetation. Components of the water budget for a wetland system include sources of inflow and outflow: Inflow is derived from surface water, groundwater, and precipitation; outflow occurs as surface water, groundwater, and evapotranspiration. This annual water-budget analysis for a fen in Ontario shows a 3 percent gain in water storage (the excess of inflow over outflow) and, consequently, a higher level of the groundwater surface (the water table).

forms as rice paddies, peat lands, mangrove forests, cypress swamps, and salt marshes containing rich beds of shellfish, wetlands have provided humans with food and fiber for millennia. Unless already saturated, as they were during the 1993 Mississippi River basin floods, spongy riverine wetlands with thick grasses act as buffers against seasonal flooding, temporarily storing floodwaters and releasing them slowly. They also act as sources, sinks, cyclers, and transformers of important chemicals and biological substances, such as carbon and nitrogen; they filter pollutants from surface water and provide a plentiful source of purified recharge to groundwater systems. As wetlands commonly are found at the headwaters of rivers, including the Delaware, Susquehanna, Sacramento, and Mississippi, their role in providing clean water to populated areas downstream is critical. Several states, including California, Maryland, and Louisiana, have learned by experience that wetlands can remove toxins from sewage waste more efficiently and cheaply than expensive, high-technology sewage treatment plants.

Coastal, Riverine, and Glacial Wetlands

Wetlands, part wet, part dry, are intermediate between aquatic and terrestrial environments; they exchange water among the atmosphere, hydrosphere, and biosphere. The characteristics of each wetland type, and the way it responds to changed conditions, are related to the wetland's water budget (Figure 7-26). Wetlands located near oceans, bays, or rivers exchange water with these sources. Both tidal-dominated *coastal salt marshes* and freshwater *riverine bottom lands* are fens near open water. Wetlands that are not near open bodies are likely to be small bogs on the order of one to hundreds of hectares. Prairie potholes are a type of bog scooped out of bedrock by glaciers during Quaternary ice ages; today, these rain- and soil-filled depressions that dot the north midwestern United States and Canada provide nesting sites for many types of waterfowl. Also associated with past glaciers are the *kettle bogs* of north-central and northeastern United States. Stagnant blocks of ice embedded in glacial deposits littered the landscape in the wake of retreating ice sheets. As the ice blocks melted, the sediment surrounding them collapsed, leaving wet depressions that look like old-fashioned kettles. Both prairie potholes and kettle bogs are postglacial wetlands that exist because of recent ice ages and because of their temperate latitudes, where annual evapotranspiration is less than precipitation. These wetlands require no water from flooding rivers or fluctuating tides in order to survive.

Wetlands Destruction

At one time, the contiguous United States contained at least 89 million hectares of wetlands. By 1991 that amount had diminished to 42 million hectares. The losses reach 80,000 to 121,000 hectares per year. At that rate, all will be gone in 300 to 400 years.

Wetlands are destroyed in many ways, the two most common of which are draining and filling. Draining is accomplished by digging ditches that connect wetlands to lower elevations or by preventing water from flowing into the wetland. The most common reason for draining a wetland is to convert the land to use for farming or development. Draining is most common for wetlands associated with topographic depressions or for those formed atop impermeable sediments. Along coastal regions, wetlands more commonly are filled with sediment or refuse in order to raise the level of their bottoms. For example, during the 19th century the perimeter of Long Island was extended several kilometers by filling wetlands in order to make more land available for building and development. Box 7-3 discusses further the effects of industry, urbanization, and deforestation on surface water systems.

One of the most acute examples of extreme wetland loss is the Central Valley of California. Before the 1800s, about 1.6 million of the valley's 5.3 million hectares were wetlands. Rivers and streams flowing into the semiarid basin from the snowcapped Sierra Nevada were lined with lush riverine wetlands, and the valley floor was covered with marshes and thick grasses. Several large lakes provided habitat for many migratory birds. In the mid-1800s, farmers began draining the vast wetlands to cultivate the land, and the lakes began to shrink. By 1990, very few of the wetlands remained, and the landscape had been transformed to extensive fields of irrigated crops.

Protecting and Restoring Wetlands

Federal protection came to the wetlands in 1972 when the Clean Water Act was passed. Permits are required now before solid material may be discharged into navigable waters, which are broadly defined to include wetlands. The Act does not prohibit dumping or filling in wetlands; it merely attempts to minimize the adverse effects of filling. Furthermore, it does not prohibit draining wetlands, nor the filling of any wetland less than 0.4 hectare (1 acre) in area.

In 1992 the U.S. Army Corps of Engineers received 90,996 requests for permits to fill in wetlands for development or industrial use; 92 percent were granted. Those are the known losses. Countless other wetlands are destroyed illegally when landowners don't apply for permits. In an effort to keep farmers from draining or otherwise converting wetlands to create agricultural lands, the federal government passed the Food Security Act of 1985 and 1990, also known as "Swampbuster." According to this law, farmers who destroy wetlands are ineligible for most federal farm subsidies. The law was later amended by the Food, Agriculture, Conservation, and Trade Act of 1990 to create the Wetland Reserve Program, which provides farmers with incentives to restore wetlands on their property. The government hopes that more than 400,000 hectares of wetlands on private property will be restored as a result of this program.

Property laws give landowners the right to farm, mine, and develop their property, but the government requires developers to create new wetlands or restore former ones to compensate for those destroyed. This policy is scoffed at by most scientists. According to Ralph Tiner, a wetlands scientist at the U.S. Fish and Wildlife Service, "Trying to recreate a wetland is like taking this vein in your arm and moving it to where there is no vein. It may look like a vein, but it does not function like one."

Creating and restoring wetlands are difficult because the natural systems are complex and incompletely understood, and some wetland types are more difficult to create than others. For example, the Florida Department of Environmental Regulation studied wetland creation projects in 1990 and found that while 45 percent of coastal (that is, tidal) wetland projects were successful, only 12 percent of freshwater wetland projects succeeded. Creating the appropriate balance of inflow and outflow of surface water and groundwater seems to be the critical factor in the success or failure of a project (see Figure 7-26). In most states, more wetlands are lost each year than are successfully restored or created. Restoration is more likely to be successful than new wetland creation, however, because the conditions that can produce and sustain a wetland have existed in the past. In addition, many wetland projects call for mosquito control, typically achieved by spraying insecticides. Although many people are beginning to recognize the values of wetlands, they still prefer not to deal with the problems that made early settlers view wetlands with such disdain.

- Recent environmental activism has challenged the historical view of wetlands. Once regarded as smelly pestholes to be filled in or destroyed, they now are protected and valued because they store floodwater, provide habitat for many organisms, and cycle carbon, nitrogen, and other elements through different Earth systems.
- The amount of carbon-rich organic matter produced and stored in wetlands is prodigious, resulting in some of the most productive ecosystems on Earth and, in some cases, the accumulation of rich stores of peat.
- Despite the change in attitudes toward wetlands, protecting them is difficult because of disagreements over their definition and identification and because the pressure is great to develop land where population is growing.

7-3 Global and Environmental Change

Climate Change, Human Activities, and Surface Water Systems

Surface water systems are affected by changes that occur over both short and long time periods. If you observed the same spot along a river bank over a period of weeks to months, you would see that water levels rise and fall after storms and snowmelt, gradually drop during prolonged dry spells, and rise during prolonged wet periods. Longer-term variations associated with climate change can also be identified. For example, it recently was determined that average water discharge increased in 559 relatively unaltered drainage basins (those with minimal human activity) throughout the United States between the years 1941 and 1988. During the same period, no increase in winter precipitation occurred, but the atmosphere above the United States was cloudier. Increased cloudiness results in less evapotranspiration and, as a consequence, greater amounts of runoff and discharge. The exact cause of the increased cloudiness is unknown, but the phenomenon is probably related to atmospheric changes associated with industrialization since the 19th century.

Another human cause of changes in surface water systems over the past few centuries is the suite of activities typically associated with urbanization. Deforestation, channelization, and levee building affect the local hydrologic cycle by decreasing infiltration and evapotranspiration and increasing stormflow runoff. Conversion of rural land to shopping centers, construction of drainage ways and sewer systems to collect stormwater, and coating of the land surface with impermeable asphalt and concrete alter the local stormflow hydrograph. A *hydrograph* is a plot of discharge in a stream against time. After a storm, discharge increases above baseflow level, then peaks and declines to the preflood level again. As a result of urbanization, the lag time to peak discharge of stormflow decreases, the peak discharge itself increases, and the total discharge for a given storm increases. The changes are largely the result of an increase in the amount of impermeable surface area and overland stormflow after urbanization. Recent laws require developers to detain stormwater in basins and to release it at rates that mimic those of preurbanized conditions. Nevertheless, the speed and scale at which urbanization is occurring throughout the world, and the ubiquity of levees, are such that flood stages once regarded as rare events have become increasingly common along many rivers.

In deforested areas, the pedosphere and surface water systems are greatly altered after trees are removed. The U.S. Forest Service documented the changes in different Earth systems during a well-known experiment carried out during the 1960s at the Hubbard Brook Experimental Forest in New Hampshire. In a series of small valleys along a ridge, all the trees and shrubs in one drainage basin were cut down and kept from regrowing by the application of herbicides. Scientists monitoring the streamflow leaving the valley found that it increased from 25 percent of the rainfall on the drainage basin before deforestation to 65 percent of the rainfall afterward. Scientists also measured the amount of sediment and dissolved nutrients (nitrates and phosphates) in the stream-

Surface Water Resources and Protection

Surface water provides us with most of our fresh water, a resource vital to life. In contrast to other resources, such as various minerals, water has no substitute for most of its functions and cannot be replaced by technological innovation. Because it is recycled continuously through the hydrologic cycle, on a global scale water is considered a renewable resource. Nevertheless, if water use exceeds local replenishment rates, or if water is polluted during use, fresh water can become a limiting resource, placing an ever-lowering ceiling on the number of people that can be sustained, especially in arid and highly populated regions.

Fresh-Water Use

A person needs less than 1 gallon (3.8 liters) of water per day for fluid replenishment; in the United States in 1992, every man, woman, and child used, on average, about 188 gallons of water daily. Obviously this amount was not used only for drinking, but also for cleaning, toilet flushing, food preparation, and watering lawns and gardens.

flow leaving the watershed, and discovered a nearly ten-fold increase. In other words, without plants to retain soil and nutrients on hillslopes and to recycle water to the atmosphere by evapotranspiration, soil, nutrients, and water were lost to the watershed system. This experiment revealed a positive feedback process: The loss of nutrients and soil makes it harder for plants to survive, while the demise of plants increases the loss of nutrients and soil.

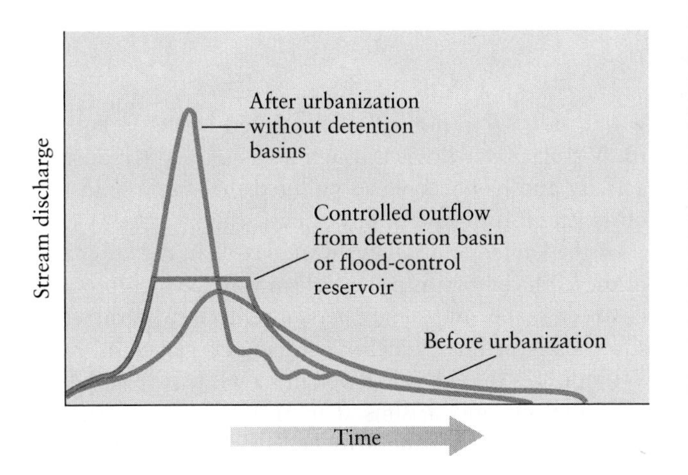

A hydrograph comparing pre- and posturbanization trends. Urbanization results in increased peak discharge and decreased time until the peak is reached during a flood. Engineers attempt to design detention basins so that for a given area, posturbanization hydrographs are similar to those of the preurbanization period.

In the Hubbard Brook Experimental Forest, New Hampshire, ecologists compared rainfall, streamflow, and sediment and nutrients carried in streams for deforested and forested watersheds by stripping an entire watershed (seen at left) of vegetation.

An additional 657 gallons per day per person is used in the United States as irrigation in agriculture, and even more for industrial purposes (cooling, processing, and removing industrial wastes). The total per capita water use in the United States is nearly 2000 gallons per day. The United States is one of the heaviest users of water in the world. In contrast, more than half the world's people use less than 25 gallons of water per day for domestic purposes, because no more is available.

Unfortunately, in much of the developing world water is unsafe to drink because it is polluted during use and returned to its source along with sewage wastes (Figure 7-27). If the waste water is not treated, disease-causing bacteria and viruses from human feces are discharged into streams and lakes. Because the same source of water is used by more people as the water moves downstream, the ratio of waste water to clean water becomes larger and larger. In most developed nations, governments require that all waste water from urban areas be treated in sewage waste treatment plants and that homes in rural areas have septic tanks or other facilities to treat waste before used water returns to the groundwater or local streams. In

FIGURE 7-27 In Jakarta, the capital of Indonesia, raw sewage waste from numerous latrines enters a stream that also serves as a source of drinking water.

many developing nations, however, waste water is not treated, and periodic epidemics of such intestinal diseases as dysentery and cholera result.

In all countries, domestic needs require the smallest portion of water use; agricultural and industrial processes require far greater amounts. Countries in arid regions, such as Iraq, Iran, Sudan, and Egypt, use hundreds to thousands of gallons of water per person per day for agriculture. Furthermore, unlike domestic use, the use of water for agriculture is largely *consumptive:* The water is removed from further use, because much of the water used for irrigation evaporates. This is also true for some water used for industrial purposes because it is stored in ponds or passed through towers for cooling.

Surface Water Systems As Waste Disposal Sites

People realized long ago that rivers are good not only for transportation but also for carrying away sewage and garbage. With the Industrial Revolution, rivers became progressively more important for carrying away different types of waste, including the toxic sludges created as a by-product of new manufacturing technologies. By the middle of the 20th century, many rivers in the United States, cleared of trees and other shade-giving plants, transported reeking mixtures of raw sewage and industrial waste, and many were dead—devoid of aquatic life and unsafe for drinking. As recently as 1968 St. Louis was spewing its raw sewage waste—as much as 300 million gallons per day—directly into the Mississippi River in the hope that the river's great size would make the effects of the dumping inconsequential to those drinking from its waters downstream. Today, the lower Volga River in the former Soviet Union is degraded severely by the large amount of waste dumped into it annually: 300 million tons of solid waste, and 5.5 trillion gallons of liquid waste. In Poland, the Vistula River flows through areas of heavy industrial activity and has become so polluted that more than half of its length is devoid of any aquatic life.

In the United States, laws were passed in the latter half of the 20th century to protect clean water. Much progress has been made, and some rivers once declared dead again show signs of life. Nonetheless, a quick glance at Table 7-3 indicates that all is not well in a typical year in the life of U.S. rivers and streams. The problems that cause fish kills or result in warnings to boil drinking water are no longer the pipes jutting from the backyards of old factories or sewage plants. Many of them are accidents, such as fuel spilled from trucks and river boats (November 1990, Tennessee; January 1991, Pennsylvania; and May 1991, Louisiana), breaks in underground pipes carrying hazardous chemicals (December 1990, Illinois), and unintentional releases of manure or raw sewage (October 1990, New York; October 1990, South Carolina).

Even more alarming than obvious **point-source pollutants**—pollutants discharged directly from pipes or spills—is the steady and growing pollution of the nation's rivers and streams from generalized sources not often recognized as catastrophic. They are the **non–point-source pollutants,** the pollutants in runoff from our backyards, streets, farms, animal feedlots, mines, construction sites, and parking lots (Table 7-4). They include pesticides, herbicides, fertilizers, animal manure, oil, and household chemicals. They are everywhere, and they are not as easily treated as single points of discharge. Non–point-source pollutants are most dangerous during floods, when stormwater surges rapidly across farms and suburban lawns, backyards, and garages filled with household trash.

Two incidents at the Mississippi River illustrate the difficulty of protecting waterways from non–point-source

TABLE 7-3 Sample of U.S. Water-Related Events in an Eight-Month Period

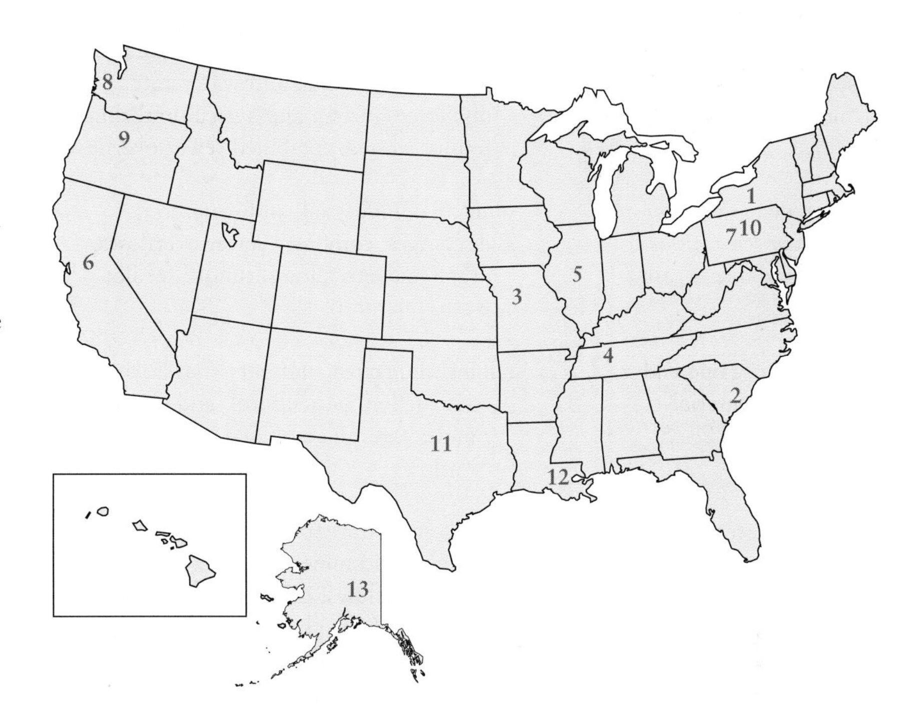

1. October 1, 1990, New York: Pipe at a pig farm ruptured and discharged about 2000 liters (530 gallons) of pig manure into Cryder Creek, which flows into the Genesee River. The Genesee River is the source of the water supply for Wellsville, New York, so residents were asked to conserve water for a few days until the waste dissipated and pumping could be resumed.

2. October 3, 1990, South Carolina: Shell fishing was temporarily banned along the coast because 10 major sewage-treatment plants in the Charleston area were discharging waste into rivers that flowed to the coast. The plants were not yet fully operational because of damage by Hurricane Hugo in late September.

3. November 8, 1990, Missouri: More than 18,000 fish died suddenly in western Missouri after ammonia-laden chicken-manure effluent used to irrigate nearby farmland flowed into a 23-km (14-mile) stretch of streams from a poultry farm as a result of a spraying machine malfunction.

4. November 22, 1990, Tennessee: An overturned truck leaked 32,550 liters (8600 gallons) of gasoline into a creek adjacent to Interstate 40 in Nashville. Homes, businesses, and a kindergarten were evacuated.

5. December 30–31, 1990, Illinois: Nearly 3 million liters (795,000 gallons) of water containing benzene, toluene, and other industrial chemicals spilled into the Kaskaskia River from a broken pipeline leading from a chemical company sludge pit. Water supplies for two communities were contaminated.

6. December 1990, California: Scientists on the California Central Valley Regional Water Quality Control Board reported a link between farm runoff contaminated with pesticides and a rapid decline in the Sacramento River striped bass fishery.

7. January 1, 1991, Pennsylvania: About 37,850 liters (10,000 gallons) of unleaded gasoline spilled into the Monongahela River after a barge hit a bridge and ruptured.

8. January 30–31, 1991, Oregon: More than 15 million liters (3,960,000 gallons) of raw sewage flowed into the Tualatin River because grease and plastic foam combined to clog a 76-cm (30-inch) sewer line and force sewage to overflow into the river.

9. February 8, 1991, Oregon: A tanker truck overturned, spilling nearly 20,000 liters (5300 gallons) of hydrochloric acid into Camas Creek, a tributary to the Columbia River. The spill killed fish over a 19-km (12-mile) distance and people were warned not to fish or eat fish from an 80-km (50-mile) stretch of the river.

10. March 31, 1991, Pennsylvania: A landslide caused a pipeline to split and spill 303,000 liters (80,000 gallons) of diesel fuel, gasoline, and kerosene into a tributary to the Allegheny River. Petroleum products mixed with water formed a 66-km (41-mile) slick that caused seven communities temporarily to discontinue using the river for the public water supply. As many as 1 million people were affected by the spill.

11. April 26, 1991, Texas: A ruptured pipeline spilled 1 million liters (264,000 gallons) of crude oil into the Sabana River, threatening water supplies in Earth and Comanche counties.

12. May 24, 1991, Louisiana: A freighter struck the Jackson Avenue Ferry Landing near the French Quarter of New Orleans and then collided with some barges. An estimated 151,000 liters (40,000 gallons) of fuel oil spilled into the Mississippi River and caused a slick that stretched 100 km (62 miles) downstream.

13. May 29, 1991, Alaska: A tanker train of the Alaska Railroad derailed and spilled 568,000 liters (150,000 gallons) of fuel into the Goldstream Creek valley, one of the richest waterfowl nesting areas on the North American coast.

Source: National Water Summary, 1990–1991.

TABLE 7-4 Pollution Sources and Pathways

Type of waste	Waste-water sources	Pollution indicators
Disease-carrying agents (human and warm-blooded-animal feces)	municipal, industrial, and watercraft discharges; urban and agricultural runoff; feedlot wastes; combined sewer overflows	fecal bacteria and streptococcus, other microbes
Oxygen-demanding wastes (high concentrations of biodegradable organic matter)	municipal, industrial, and watercraft discharges; combined sewer overflows; urban and agricultural runoff; feedlot wastes; natural	biochemical oxygen demand, dissolved oxygen, volatile solids, sulfides
Suspended organic and inorganic material	mining, municipal, and industrial discharges; construction, agricultural, and urban runoff; natural; combined sewer overflows	suspended solids, turbidity, biochemical oxygen demand, sulfides
Inorganic materials, mineral substances (metals, salts, acids, solid matter, other chemicals, oil)	mining, industrial, and municipal discharges; acid mine drainage; combined sewer overflows; urban and agricultural runoff; oil fields; irrigation return flow; natural; transportation spills	pH (acidity, alkalinity), dissolved solids, chlorides, sulfates, sodium, specific metals, oil slicks, toxins
Synthetic organic chemicals (dissolved organic material, e.g. detergents, household aids, pesticides)	industrial, municipal and mining discharges; urban and agricultural runoff; combined sewer overflow; transportation spills	cyanides, phenols, toxins
Nutrients (nitrogen, phosphorus)	municipal and industrial discharges; agricultural runoff; combined sewer overflows; urban runoff; natural	nitrogen, phosphorus
Radioactive materials	industrial discharges; mining	radioactivity
Heat	cooling water, industrial, and municipal discharges	temperature

Source: Council on Environmental Quality, 1981, *Environmental Trends.*

pollutants. In the 1960s fish at the mouth of the Mississippi were so contaminated with agricultural insecticides from upstream farms that one of their predators, the brown pelican, the state bird of Louisiana, disappeared from the state that honored it. It soon was discovered that biological amplification of the insecticide endrin had occurred—in other words, endrin had accumulated in fish to levels so high that it killed animals higher on the food chain that fed on the fish. The insecticide was used by countless farmers on fields far removed from the mouth of the Mississippi River. None foresaw how quickly the toxin would make its way across farm fields with runoff during storms and then downstream along the river, to accumulate in plant tissues and layers of mud along its channel. When the cause of damage was identified, the insecticide was banned.

In the second instance, the U.S. Geological Survey reported in 1991 that it found atrazine, an herbicide that possibly causes cancer, in 141 samples of water taken from the Mississippi River. There is no quick or easy way—other than banning use of the herbicide—to stop the washing of this pollutant into the nation's streams and rivers.

Effects on water quality	Effects on aquatic life	Effects on recreation
Health hazard for human consumption and contact	tainted shellfish	reduced contact recreation
Deoxygenation	fish kills	if severe, recreation eliminated
Reduced light penetration, deposition on bottom, deoxygenation	reduced photosynthesis, changed bottom organism population, reduced fish production and sport-fish population, increased non–sport-fish population	reduced game fishing and aesthetic appreciation
Acidity, salination, toxicity of heavy metals, floating oils	reduced biological productivity and flow, fish kills, tainted fish	reduced recreational use, fishing, and aesthetic appreciation
Biodegradable or persistent synthetic organic material	fish kills, tainted fish, reduced reproduction, skeletal development	reduced fishing, inedible fish for humans
Increased algal growth, oxygen depletion	reduced sport-fish population, increased non–sport-fish population	tainted drinking water, reduced fishing and aesthetic appreciation
Increased radioactivity	altered natural rate of genetic mutation	reduced opportunities
Increased temperature, reduced capacity to absorb oxygen	fish kills, altered species composition	possible increased sport-fishing by extended season for fish that might otherwise migrate

Regulations to Protect Drinking Water

Because of incidents such as the discovery of atrazine and other toxins in river water, the federal government requires all municipal water suppliers to test their water frequently. Safety standards are set below the level at which a contaminant would be expected to cause adverse human health effects after a lifetime of exposure (Table 7-5).

Even if water is deemed safe at the water treatment plant, it might still become unsafe during transit to a domestic tap. For example, houses built before 1930 often had lead plumbing. Lead corroded from the pipes can cause severe brain damage if ingested frequently over a long period. In 1986 the use of lead was banned from new public water systems. Water suppliers now are required to test for lead and take corrective action if more than 15 parts of lead per billion parts of water are found.

Because of water regulations, tap water in the United States could be safer than some bottled water. Federal standards for water quality apply only to commercially bottled water that is sold in bulk and marketed across state lines. Furthermore, the bottlers of commercial water

TABLE 7-5 Selected Federal Drinking Water Standards

Contaminant	Health effects
MICROBIOLOGICAL	
Total coliforms (coliform bacteria, fecal coliform, streptococcal, and other bacteria)	not necessarily disease producing themselves, coliforms can be indicators of organisms that cause assorted digestive system infections (dysentery, hepatitis, typhoid fever, cholera)
INORGANIC CHEMICALS	
Arsenic	dermal and nervous system toxicity effects
Barium	circulatory system effects
Cadmium	kidney effects
Chromium	liver and kidney effects
Lead	central and peripheral nervous system damage; kidney effects; can be highly toxic to infants
Mercury	central nervous system disorders; kidney effects
Nitrate	methemoglobinemia (blue-baby syndrome)
Selenium	digestive system effects
Silver	skin discoloration (argyria)
Fluoride	skeletal damage
ORGANIC CHEMICALS	
Endrin	nervous system and kidney effects
Lindane	nervous system and liver effects
Methoxychlor	nervous system and kidney effects
2,4-D	liver and kidney effects
2,4,5-TP Silvex	liver and kidney effects
Toxaphene	cancer risk
Benzene	cancer risk
Carbon tetrachloride	possible cancer risk
Trichloroethene (TCE)	possible cancer risk
Vinyl chloride	cancer risk

Source: League of Women Voters Education Fund, *Safety on Tap: A Citizen's Drinking Water Handbook,* Washington, DC: League of Women Voters Education Fund, 1987.

are regulated by the Food and Drug Administration rather than the EPA and are allowed to do their own testing. Mineral waters and seltzers are considered "beverages" and do not have to meet EPA standards for drinking water. Bottled water may taste better than tap water, but more than a third of all bottled water in the United States is obtained from the same sources as tap water. At present, bottled-water companies are not required to list the water source on their labels; in fact, they may draw it directly from a tap.

The most common reasons that bottled water tastes slightly better are the dissolved minerals in bottled water obtained from springs and groundwater and the use of ozone as a disinfectant. Municipal water suppliers use chlorine instead, which can leave a noticeable aftertaste. Although bottled water is not likely to be safer or different from tap water (except in taste) and although it costs about 700 to 1500 times more than tap water, the per capita consumption of bottled water is steadily increasing.

Maximum containment level allowed	Sources
1 per 100 ml	human and animal fecal matter
(mg/l)	
0.05	geological, pesticide residues, industrial waste, smelting
1	geological, mining and smelting
0.01	
0.05	
0.05	leaches from lead pipe and lead-based solder pipe joints
0.002	geological, used in manufacture of paint, paper, vinyl chloride and fungicides
10	geological, fertilizer, sewage, feedlots
0.01	geological, mining
0.05	geological, mining
4	geological, additive to drinking water, toothpaste, and foods processed with fluoridated water
0.0002	insecticide used on cotton, small grains, orchards (now banned)
0.004	insecticide used on seed, soil treatments, foliage application, wood protection
0.1	insecticide used on fruit trees, vegetables
0.1	herbicide used to control broadleaf weeds in agriculture; used on forests, range land, pastures, aquatic environments
0.01	herbicide (now banned)
0.005	insecticide used on cotton, corn, grain
0.005	fuel (leaking tanks), manufacture of industrial chemicals, pharmaceuticals, pesticides, paints, and plastics
0.005	cleaning agents and industrial wastes from manufacture of coolants
0.005	waste from disposal of dry cleaning materials and manufacture of pesticides, paints, waxes, varnishes, paint strippers, and metal degreasers
0.002	polyvinylchloride (PVC) pipes and solvents used to join them, industrial wastes from manufacture of plastics and synthetic rubber

Protecting Water Quality in Streams and Rivers

Ecological catastrophes such as the demise of the brown pelican in Louisiana predated the environmental movement of the 1970s and the Clean Water Act, both of which occurred because of growing awareness of the hazards of pollution. The Clean Water Act aimed to make all the nation's surface waters fishable and swimmable by 1983 and to achieve zero discharge of point-source pollutants into those waters by 1985. Neither goal was achieved by the target date; neither goal has yet been achieved. Still, substantial progress has been made on some fronts. In 1987, in recognition that most impaired water sites are the result of runoff rather than direct discharge of toxins, the Clean Water Act was amended to incorporate non–point-source management programs. These "best-management" practices include recommendations for states to develop

(a)

(b)

FIGURE 7-28 (a) By the 1960s, the Nashua River in Massachussetts was a dead, reeking menace, filled with infectious raw sewage, toxic chemicals, and red dye from paper mills. (b) By the 1990s, clean-up efforts paid off, and now the river is inviting and alive with fish that are safe to eat.

plans to prevent soil erosion, reduce runoff from agricultural fields, and limit the spraying of insecticides and pesticides on lawns and golf courses.

In 1990, EPA reported to Congress that of 1.06 million km of rivers and streams tested and surveyed, 80 percent were now fishable, and of 944,000 km surveyed, 75 percent were swimmable (Figure 7-28). However, less than a fourth of the nation's stream miles have been assessed. Non–point-source pollutants keep many of the nation's assessed streams and rivers from meeting EPA criteria for fishable and swimmable waters.

Similar levels of increased awareness and efforts to restore water quality have occurred in other nations. In England, for example, the Thames River flowing through London was used openly for disposal of raw sewage and industrial wastes. Salmon, which are extremely sensitive to pollution, had not been seen in the river since the late 1600s. The filth in the river mounted in the 19th century as the city's population and industry increased. During the second half of the 20th century, however, hundreds of millions of dollars were spent to restore the river's biological quality. By the 1980s the amount of oxygen dissolved in the river's water, an excellent indicator of water quality and ability to support life, had increased markedly. Salmon returned to the river, along with 95 other species of fish.

- Increasing urbanization and greater population density in the past few hundred years have contributed to the degradation of many streams worldwide, because they are used to transport sewage and industrial waste away from urban areas.
- Both point-source and non–point-source pollutants affect streamwater quality: The former is caused by direct discharge of a pollutant from a pipe, leak, or spill, the latter by surface water runoff flowing across farm fields, mines, parking lots, and streets.
- As mandated by the Clean Water Act and its amendments, the U.S. Environmental Protection Agency has established drinking water standards that require that pollutants have concentrations below levels deemed safe for human health.

Closing Thoughts

Each year, the total amount of water used by a growing human population increases. With urbanization and industrialization, the amount of water used per person is increasing in developing nations; it has stabilized at high levels in more developed nations. In February 1996, researchers working to determine Earth's carrying capacity published the results of a study on the availability of fresh water. The researchers concluded that 54 percent of the runoff from continents (river and groundwater flow) that is reasonably accessible to population centers already is being used. They estimate that during the next 30 years, the amount of runoff accessible for human use could be raised by about 10 percent if nearly 10,000 more dams are built. However, population is projected to increase at least 45 percent during that time. As a consequence, the world might reach its carrying capacity with respect to fresh water in several decades.

This grim outlook can be improved only by preventing pollution of available fresh water and using water—especially irrigation water—efficiently. In industrialized nations, laws to protect clean water and facilities to treat waste water have resulted in less-polluted waterways than existed in the first half of the 20th century. It is an open question whether developing nations will be able to protect their water as they build industrial bases.

While water shortages threaten some parts of the world, flooding threatens others. Of all natural disasters, floods cause the greatest loss of life and property, and their toll has not abated despite massive investments of funds and extensive regulation in some parts of the world.

Only recently has it become clear that natural drainage systems—including meandering streams, pools and riffles, and wetlands—evolved in ways that accommodate variable amounts of water. The health of ecological systems depends on their unimpeded operation. For this reason, scientists and land-use planners have turned to natural systems for guidance on how to restore drainage basins, streams, and wetlands to more natural states, while simultaneously diminishing flood hazards.

Summary

- Surface water on continents is a dynamic part of the hydrologic cycle and is distributed through intricate networks of stream channels and wetlands, all of which ultimately drain into the ocean or deep, closed basins in the interiors of continents.
- Human attempts to control surface water include stream channelization, dam control structures, and wetland filling. Consequent changes in the surface water system have resulted in changes in other Earth systems, in particular the biosphere, with dramatic examples provided by the Colorado River, the Kissimmee River (part of a wetlands system), and the Aral Sea.
- Although the amount of water an individual needs to sustain life is small (a few gallons for drinking, cooking, and bathing), the amount of per capita water use is hundreds of times greater than that quantity in industrial nations. In some countries, where annual rainfall is less than about 20 mm per year, water shortages are now occurring or will occur in the near future.
- The quality of much streamwater has diminished since the Industrial Revolution and increased urbanization in the past few hundred years, but environmental laws established in the United States and other nations in the late 20th century have made some progress toward cleaning polluted streams and rivers.

Key Terms

evapotranspiration (p. 196)
baseflow (p. 198)
hydrologic equation (p. 198)
hydrologic budget (p. 198)
drainage basin (p. 198)
drainage divide (p. 198)
drainage network (p. 199)
trunk stream (p. 199)
tributary (p. 199)
base level (p. 199)
stormflow (p. 199)
infiltration capacity (p. 199)
discharge (p. 200)
alluvium (p. 203)
braided stream (p. 203)
meandering stream (p. 203)
pools (p. 203)
riffles (p. 203)

floodplain (p. 206)
lateral migration (p. 206)
overbank deposition (p. 206)
levee (p. 207)
delta (p. 207)
channelization (p. 212)
wetlands (p. 216)
peat (p. 217)
point-source pollutant (p. 222)
non–point-source pollutant (p. 222)

Review Questions

1. Why in 1996 did some scientists urge the U.S. government to create a flood along the Colorado River?
2. What human activities led to the shrinking of the Aral Sea?
3. What is the hydrologic equation? How can it be used to determine the availability of surface water runoff for a given area?
4. What are the main parts of a drainage basin?
5. What are the two major flow paths to streams that rainwater follows after a storm?
6. How is the infiltration capacity of the ground related to the route rainwater follows to streams?
7. How is stream discharge determined? Why is it an important property to measure?
8. Give at least two reasons why the discharge in the Colorado River is so much smaller than that in the Mississippi River.
9. Why are braided streams so common downstream of glaciers and in logged or otherwise disturbed terrain?
10. What are pools and riffles? How do they differ from each other?
11. Compare and contrast the characteristics, benefits, and negative aspects of natural and artificial streams.
12. Why is the U.S. Army Corps of Engineers restoring the Kissimmee River in Florida to its unchannelized state?
13. How are floodplains formed by streams?
14. How have the Chinese tried to control the flow path and flooding of the Hwang Ho?
15. Why is the effort to keep the Mississippi River fixed in a stable position viewed with such skepticism by many geologists?
16. How are floods of different probabilities of occurrence, and recurrence intervals, determined from flood frequency analysis?
17. What characteristics define a wetland?
18. What are several major types of wetlands in the United States? Where are they located?
19. Why has the protection of wetlands been so ineffective in recent years, despite federal laws designed to regulate land use in wetland areas?
20. Describe different ways in which the quality of streamwater is degraded by pollution.
21. Why is bottled water not necessarily as safe to drink as tap water from a public water supplier?

Thought Questions

1. Would you predict that groundwater is or is not being recharged by infiltration of excess surface water in the area of Phoenix, Arizona? Explain your reasoning.
2. Other than building more reservoirs, what can be done to reduce the amount of floodwater that reaches the Mississippi River during a given flood?
3. How has urbanization contributed to increased stream discharge in many streams? (Use a hydrograph in answering this question.)
4. How does the use of artificial levees and straightened stream channels result in greater flooding for areas downstream of such control structures?
5. If sea level rise occurs in the next century, as many scientists predict because of global warming, will this affect the course of the lower Mississippi River? (Think of the concept of base level.)
6. In what ways might global climate change lead to different amounts of discharge in existing streams? Has the amount of stream discharge in different areas varied in the past as a result of global climate

change? If so, give an example, stating when and why the change occurred.

7. Is the 100-year flood a reasonable standard for land-use planners and policymakers to use for zoning and insurance practices, or would you recommend considering another recurrence interval, such as the 200-year flood? Explain your reasoning.

8. What types of changes in land use practices might lead to less non–point-source pollution of streams?

Exercises

1. If a stream has a width of 20 m, a depth of 5 m, and a velocity of 2 m/sec, what is its discharge?
2. If a natural stream has a width of 10 m, what would you predict is the average distance between each pool along its length?
3. A meandering stream flows from an elevation of 1000 m to sea level, and its channel length is 3000 m. The river is straightened for flood control and its final length is only 1500 m. How many times is the slope of the straightened channel increased from its original state?
4. What is the recurrence interval for a flood that has a 0.25 probability of being equaled or exceeded in a given year?

Suggested Readings

Clarke, Robin, 1993. *Water: The International Crisis.* Cambridge, MA: MIT Press.

Collier, Michael P., Webb, Robert H., and Andrews, Edmund D., 1997. "Experimental Flooding in the Grand Canyon." *Scientific American* (January) 82–89.

Collier, Michael P., Webb, Robert H., and Schmidt, J. C., 1996. *Dams and Rivers: A Primer on the Downstream Effects of Dams.* U.S. Geological Survey Circular, vol. 1126.

Dunne, Thomas, and Leopold, Luna, 1978. *Water in Environmental Planning.* New York: W. H. Freeman and Company.

Fradkin, Peter L., 1981. *A River No More: The Colorado River and the West.* New York: Alfred A. Knopf.

Gleick (ed.), Peter H., 1993. *Water in Crisis: A Guide to the World's Fresh Water Resources.* New York: Oxford University Press.

Hillel, Daniel, 1994. *Rivers of Eden: The Struggle for Water and the Quest for Peace in the Middle East.* New York: Oxford University Press.

Lanz, Klaus, 1995. *The Greenpeace Book of Water.* New York: Sterling.

Leopold, Luna, 1994. *A View of the River.* Cambridge, MA: Harvard University Press.

McCully, Patrick, 1996. *Silenced Rivers: The Ecology and Politics of Large Dams.* London: Zed Books, International Rivers Network, and *Ecologist.*

McPhee, John, 1989. *The Control of Nature.* New York: Farrar, Straus and Giroux.

Postel, Sandra L., Daily, Gretchen C., and Ehrlich, Paul R., 1996. "Human Appropriation of Renewable Fresh Water." *Science* 271: 785–788.

Rosgen, Dave, 1996. *Applied River Morphology: A Guide for the Classification, Assessment and Monitoring of Rivers and the Applications for Water Resource Management.* Pogosa Springs, CO: Wildland Hydrology Books.

U.S. Geological Survey, *National Water Summary,* 1986 to present. U.S.G.S. Water-Supply Paper series. Washington, DC: U.S. Printing Office.

CHAPTER 8

A spring-fed oasis in a seemingly dry landscape reveals the presence of underlying groundwater. Rocks and sediments of the lithosphere and pedosphere naturally filter many impurities out of the groundwater and add some minerals that enhance taste, making spring water desirable drinking water.

The Groundwater System

Our lives depend on fresh water. The largest supply of liquid fresh water—32 percent of the total supply on Earth—occurs underground, flowing through spaces between sedimentary grains and cracks in rocks. This groundwater is discharged at Earth's surface into springs, streams, rivers, lakes, and ultimately oceans. Because the growing human population demands ever greater amounts of clean fresh water for agriculture, industry, and domestic use, groundwater is now more heavily used than at any time in the past. Yet human stewardship of groundwater often has been careless, leaving this essential resource exposed to pollution from sewage disposal, leaking underground chemical storage tanks, and chemical spills. As global dependence on groundwater increases, nations worldwide are making efforts to protect and manage groundwater resources and clean up contaminated sites. To illuminate the issues surrounding groundwater, in this chapter we will:

- Find out how groundwater gets underground and flows from one place to another.
- Consider how groundwater is used and the environmental changes that have affected its quality and quantity in recent years.
- Investigate hazards associated with the extraction of groundwater, such as sinkholes, subsidence, and contamination by seawater.
- Examine how groundwater can be cleaned if it becomes contaminated.

Residents of Cape Cod, Massachusetts, rely on groundwater for their water supply. When residents complained about a soapy taste in their water, scientists investigating the problem drilled a well and found the water thick with bubbly detergent lather. After drilling many more wells, they traced the source to a long plume of soapy water emanating from the site of an abandoned military base (Figure 8-1). Otis Air Base had disposed of waste water and sewage in a shallow pit for 50 years, from 1936 until 1986. The wastes seeped downward and outward through sand and gravel, heading south toward the Atlantic Ocean and passing through the groundwater systems of several communities, including Falmouth, some 5 km away.

Until 1964 the personnel at Otis, like everyone else in the United States, used detergents that do not *biodegrade,* or break down naturally in the environment by biological processes. Since 1964 environmental laws in some states mandated that only biodegradable detergents can be used. As a consequence, groundwater in the immediate vicinity of the base no longer contains much detergent, but the groundwater farther downslope will contain detergents for many years unless a cleanup is performed. Some contaminants that entered the groundwater decades ago already have reached public water wells, which have had to be shut down.

Scientists from the U.S. Geological Survey and other institutions have spent years studying the Cape Cod site; they now are discussing how to clean up the contaminated water and prevent it from reaching other public wells. To be effective, this action requires an understanding of where groundwater comes from and how it flows beneath Earth's surface.

FIGURE 8-1 Detergent-contaminated groundwater pumped from beneath Cape Cod nearly 2 km downslope from a sewage treatment field.

Groundwater in the Hydrologic Cycle

Water in the ground, like water at Earth's surface or water vapor in the atmosphere, is in motion because it has energy. Water typically enters the ground as excess surface water, then seeps downward because of gravity and the porous nature of soil, rock, and sediment. Water continues to move underground, flowing eventually to the ocean, the ultimate base level for rivers and groundwater. From the ocean and from the continental surface, water molecules can evaporate into the atmosphere and contribute to rainfall. All the time, water is entering and leaving the groundwater system, forming an essential link in the global hydrologic cycle of water.

The Water Table

Rainfall and snowmelt infiltrate the subsurface and drain downward through any underlying unsaturated material along interconnected pore spaces. At the point where all the voids are saturated, infiltrating water becomes part of groundwater (Figure 8-2). The boundary between the unsaturated and saturated zones forms a surface called the **water table.** Technically, all water below Earth's surface is *underground water,* and only the water below the water table is **groundwater.** Unsaturated rock or sediment above the water table provides an avenue along which surface water can recharge groundwater by gravity drainage.

In areas where the water table rises so high as to intersect Earth's surface, groundwater is exposed in springs. At low places in Earth's topography, such as streams, lakes, and wetlands, the water table is exposed, and groundwater is discharged into these surface water systems (see Figure 8-2). Recall from Chapter 7 that groundwater discharge provides the baseflow to streams between times of rainfall or snowmelt. In addition, groundwater usually is discharged into the oceans along coastal areas. In some coastal areas of the Mediterranean Sea and the Pacific Ocean near Hawaii, large subsurface channels discharge spring water at such high pressures that it rises above the surface of the sea. A poem by the Roman philosopher Lucretius (first century BC) describes such a fountain in the Mediterranean:

> At Aradus there is a spring within the sea.
> The water that this spring pours forth is fresh,
> It parts the salty water all around.
> And elsewhere too the level sea gives help to
> thirsty mariners,
> By pouring out fresh water mid the salty waves.

(*De rerum natura,* Book VI, translated by Alban Dewes Winspear, New York: S. A. Russell, Harbor Press, 1956)

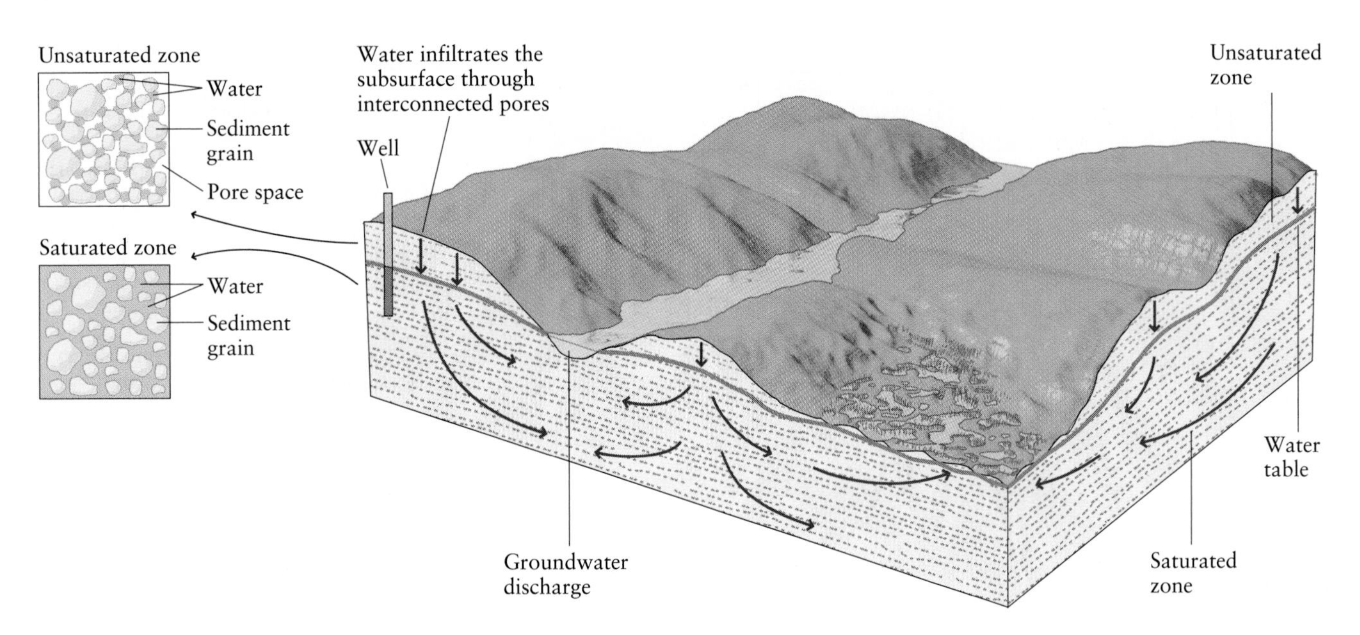

FIGURE 8-2 The water table is the boundary between the zones of unsaturated and saturated pore spaces; its undulations, if any, usually mimic the topography of Earth's surface.

The level of the water table at any location generally changes in response to seasonal and yearly weather patterns. In a humid-temperate region in the northern hemisphere, for example, snow melts at relatively low temperatures during the spring, so the potential for evapotranspiration is low. Recharge of soil moisture, therefore, is likely to be high enough to cause the water table to rise. Conversely, during hot summer months, high temperatures may result in a large potential for evapotranspiration and depletion of soil moisture, causing the water table to fall (Figure 8-3). During prolonged droughts, water tables may fall over large areas, causing wells, springs, and streams to go dry.

Effect of Elevation and Pressure on Groundwater Movement

Elevation and water pressure combine to determine the direction in which groundwater flows. The water table mimics topography and slopes downward toward low points, such as streams, because freely moving groundwater flows from places of higher to lower elevation. Greater elevation provides greater potential energy (see Chapter 1), and water flows from places of high to low potential energy, just as a rock slides from the top to the bottom of a hill.

At the water table, the water pressure in pore spaces is equal to the atmospheric pressure, just as it is at the surface of standing water in a lake or a drinking glass, and so it cannot rise any higher. With increasing depth, the water pressure in pore spaces becomes greater than atmospheric pressure because of the weight of the overlying water column. The greater the height of the column of water, the greater the water pressure. Unlike surface water, groundwater can flow uphill, from a place of low to high elevation, if there is a decrease in water pressure. Deep under a stream channel, where water pressure is high, groundwater flows upward to recharge the stream because water pressure at the stream is relatively lower (see Figure 8-2).

Porosity and Groundwater Storage

It is surprising how much water can be stored underground in the tiny pore spaces between grains of sediment or along cracks in rocks. If all the pore spaces that are filled with water in the subsurface of the United States were connected, they would form one large water-filled cavern, 57 m high, underlying the entire country. The porosity of a rock or sediment reflects the amount of pore space between its grains. **Porosity** is the ratio, usually stated as a percentage, of void space to total volume of rock or sediment in a segment of Earth. The more porous the rock and sediment in an area, the greater the amount of groundwater that can be stored there.

Some pores were originally either spaces between adjacent sedimentary particles, or vesicles—small cavities—that formed in igneous rocks as gases escaped from the cooling melts. In sediments that were compacted and cemented into sedimentary rock, the precipitation of minerals—the cement—into the pore spaces reduced the original porosity of the deposit. This type of porosity is called *primary porosity* because it developed at the same time the rock itself was formed. *Secondary porosity* develops

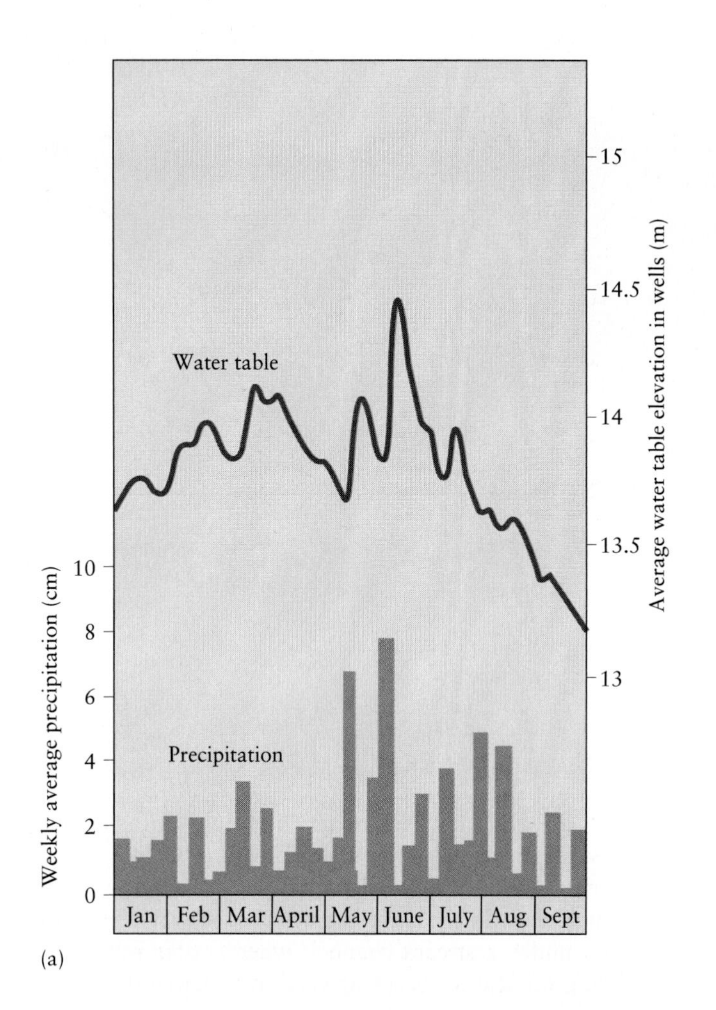

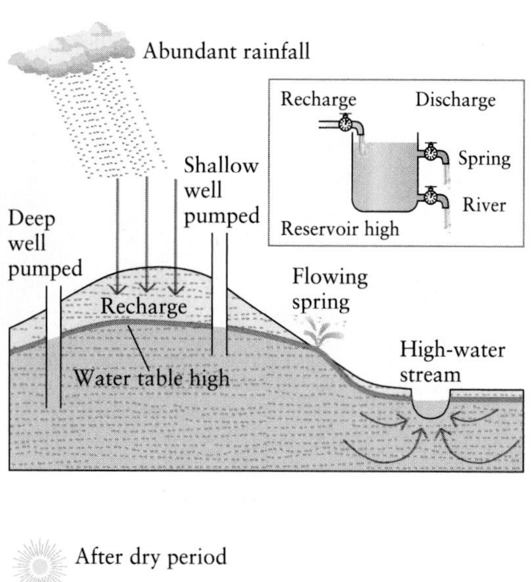

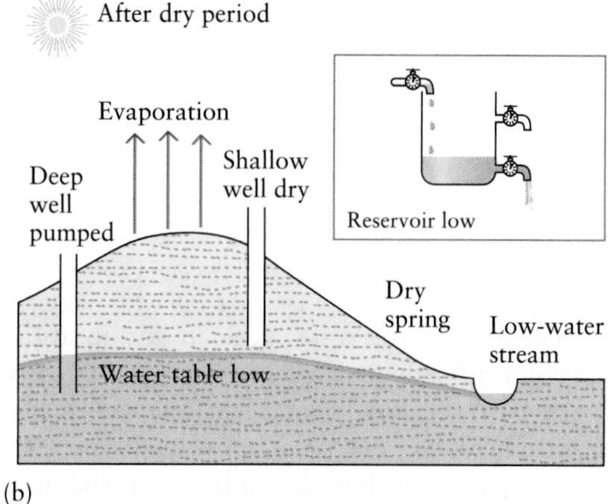

FIGURE 8-3 (a) Monthly trends of precipitation and water table levels in Maryland from January through September. The water table rises shortly after each rainfall. Over the longer term, it rises in the cool, wet spring and drops in the warm, dry summer. (b) The level of the water table varies in response to precipitation and evapotranspiration at Earth's surface. If the groundwater zone is considered as a reservoir, the input is precipitation and the outputs are discharge from evaporation, springs, streams, and wells.

after the rock is formed. Fractures, caused by tectonism, uplift, and erosion, contribute to secondary porosity (Figure 8-4).

In general, primary porosity is highest in sediments and sedimentary rocks and lowest in unweathered crystalline igneous and metamorphic rocks. Primary porosity is higher in volcanic rocks than in plutonic rocks. Some volcanic rocks contain lava tubes, which form when molten material continues to flow beneath cooled, hard crusts during volcanic eruptions. In addition, gases escape and cracks occur in extrusive igneous rocks as they cool, so the tops of individual flows within a stack of lava flows often are vesicular and fragmented, further increasing the primary porosity of volcanic rocks. In metamorphosed rocks, most pore spaces either were squeezed shut by the immense pressures to which the rocks were subjected or filled with crystals that precipitated from hot fluids deep in the lithosphere. Because porosity generally decreases with depth, most groundwater occurs within several kilometers of the surface, although large water-filled fissures have been found at the bottom of the deepest drill hole on Earth, the Kola Hole in Siberia, which was drilled to 12 km.

The size of individual primary pore spaces in a given volume of sediment does not affect the amount of water it can store because the total volume of pore space is the same regardless of the size of the pores. In unconsolidated, uncemented sediment, if the grains are nearly spherical and packed tightly against one another, there is no more than about 26 percent void space between the

(a)

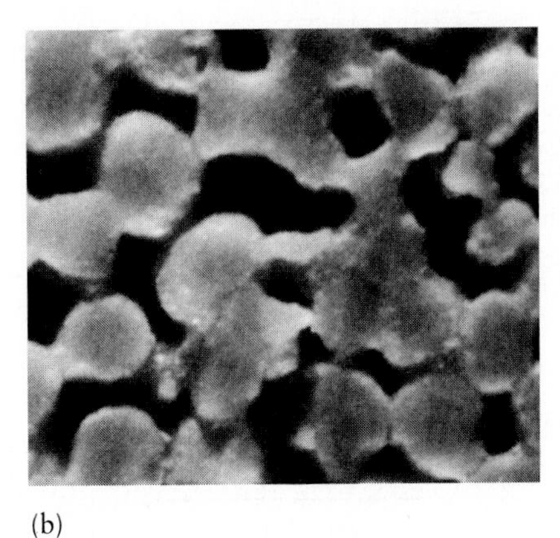
(b)

(c)

FIGURE 8-4 (a) Sand grains without cement, (b) sand grains with cement, and (c) an outcrop of fractured limestone. The porosity of the uncemented sand is about 35 percent; of partly cemented sand, about 25 percent; and of fractured limestone, about 25 percent. Note the spring in the fractured limestone, from which groundwater is emanating.

grains. In fact, all closely packed sediments, regardless of their actual size, that are well rounded and well sorted (similar in size), have a porosity of about 26 percent.

Although grain size does not affect porosity, and thus does not affect the amount of water that can fill the pore spaces, it does affect the amount of water the grains can hold on to by surface tension—the attractive force between the molecules at the surfaces of grains and those at the surface of the enclosed water. Consequently, grain size affects the quantity of water that can flow through the pore spaces. Smaller grains and fractures have more solid surface area per unit volume of sediment and rock than do larger grains and fractures. As a result, the smaller the grains or fractures, the more water that can adhere to the sediment or rock when it drains or is pumped.

The amount of groundwater that can be drained from a porous rock or sediment is called the **specific yield,** which is expressed as the ratio of water drained to total volume of water within the rock or sediment. Specific yield always is less than the porosity (Table 8-1). Because of surface tension, it is not possible to extract all the water from an underground reservoir, just as it is never possible to get all the oil out of the ground by pumping wells. The ratio of the amount of groundwater retained by surface tension to the total volume of water is called **specific retention.** The greater the specific retention, the harder it is to remove every bit of a contaminant from groundwater by flushing and pumping the contaminated area.

Permeability and Groundwater Flow

Note that some geologic materials listed in Table 8-1, such as clay, have high porosity and low specific yield, meaning that little of the water in their pores can move. The ability of rock or sediment to transmit water through its pores is called its **permeability.** Materials with high permeability are those in which pore spaces are relatively large and interconnected. Highly permeable materials

TABLE 8-1 Typical Values of Porosity and Specific Yield for Common Geologic Materials

Material	Porosity (%)	Specific yield (%)
Sedimentary		
Gravel, coarse	24–36	23
Gravel, fine	25–38	25
Sand, coarse	31–46	27
Sand, fine	26–53	23
Silt	34–61	8
Clay	34–60	3
Sedimentary rocks		
Sandstone	5–30	21–27
Siltstone	21–41	15–25
Limestone	0–50	5–45
Shale	0–10	0–3
Crystalline rocks		
Fractured schist	30–38	26
Fractured crystalline rocks	0–10	5
Dense crystalline rocks	0–5	0.1
Basalt	3–35	8
Weathered granite	34–57	40–45
Weathered gabbro	42–45	35–40

TABLE 8-2 Typical Values of Permeability for Common Geologic Materials

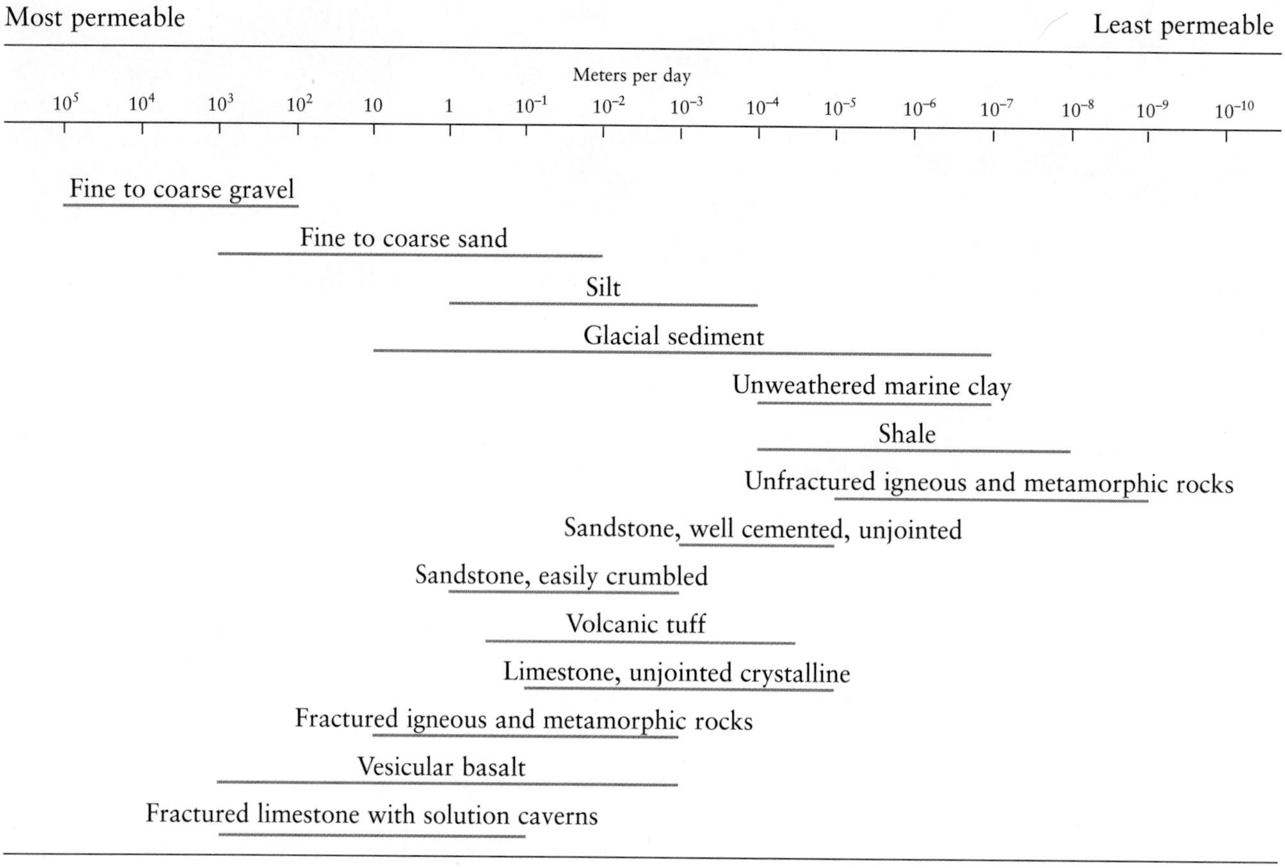

Most permeable — Least permeable

Meters per day

10^5	10^4	10^3	10^2	10	1	10^{-1}	10^{-2}	10^{-3}	10^{-4}	10^{-5}	10^{-6}	10^{-7}	10^{-8}	10^{-9}	10^{-10}

Fine to coarse gravel

Fine to coarse sand

Silt

Glacial sediment

Unweathered marine clay

Shale

Unfractured igneous and metamorphic rocks

Sandstone, well cemented, unjointed

Sandstone, easily crumbled

Volcanic tuff

Limestone, unjointed crystalline

Fractured igneous and metamorphic rocks

Vesicular basalt

Fractured limestone with solution caverns

include well-sorted, coarse sand and gravel, as well as crystalline rocks, such as granite, with extensive fractures. Rocks and sediments with low permeability include crystalline rocks with few or no fractures and very poorly sorted sediments—those highly dissimilar in individual grain size—in which the finer particles block pore spaces (Table 8-2).

A rock or sediment can have high porosity and low permeability: An example is pumice, a glassy volcanic rock formed by the rapid cooling of gas-rich lava. Pumice shows many vesicles from which the gases escaped, but they are largely unconnected to one another. For this reason, pumice is not likely to contain much water in its pores, and even after a soaking rain, the water left in the pores cannot flow through the rock. Clay, a sediment, also can have high porosity, but its pores are narrow and elongate, hindering the flow of water so effectively that the permeability is very low.

Because permeability helps determine the rate at which water—and pollutants—can flow through rocks and sediments, it is a major concern in the siting of facilities for hazardous and radioactive wastes. Perhaps the greatest danger in storing nuclear waste or any hazardous substance underground is that the contaminant could leak into the surrounding host rock or sediment and, ultimately, into groundwater. Once a contaminant reaches groundwater, it can flow with the water toward springs, wells, streams, and lakes. For this reason, hazardous waste-disposal facilities are sited where the water table is deep and the host rock has low permeability. Because igneous rocks typically have low values of porosity and permeability, igneous formations such as the volcanic tuff (rock formed from ash and other airborne debris ejected from a volcano) underlying Yucca Mountain in Nevada have been common sites for consideration of hazardous waste disposal (see Box 3-4).

Aquifers

Rocks and sediments that are porous, permeable and contain water are called **aquifers,** from the Latin words meaning "water" and "carry." Wells sunk into aquifers are used to extract groundwater. Low-permeability rocks and sediments adjacent to aquifers are called *confining layers* because they confine the flow of water through the aquifer. The most efficient aquifers include gravel, sand, sandstone, limestone, and basalt. Less porous and permeable materials such as silt, siltstone, and metamorphic rocks store and transmit relatively little water. Materials that

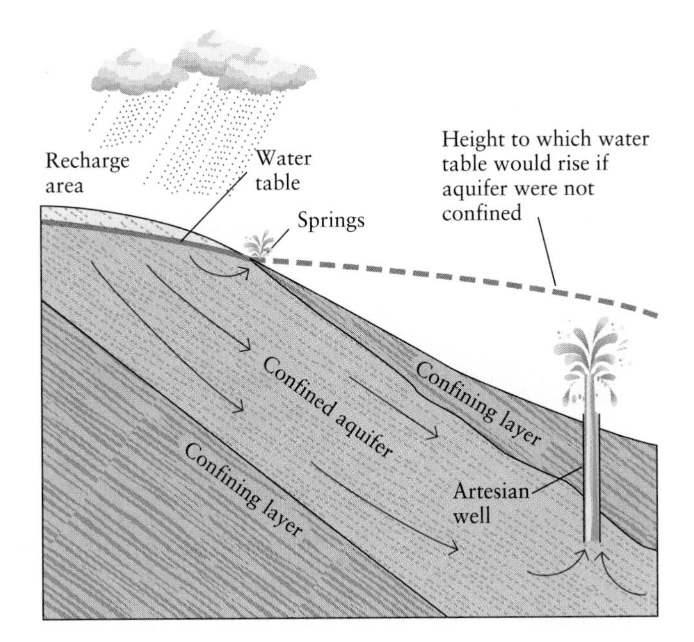

FIGURE 8-5 This aquifer is a sandstone that is inclined downward and confined between two shale layers, so that with increasing distance from the recharge area the water pressure in the aquifer increases. The confined water is under so much pressure that it can flow to the surface if punctured by wells.

make good confining layers enhance the water-bearing quality of the aquifers above and below them; the most common and effective confining layers are clay and shale.

Aquifers are either confined or unconfined. In an **unconfined aquifer,** porous and permeable sediments or rocks are not overlain by a confining layer, and the **hydraulic gradient,** or the slope of the surface along which water flows from an area of higher to lower potential energy, is determined by the differences in elevation from one place to another. Unconfined aquifers can be recharged at any point by precipitation and surface water. In a **confined aquifer,** layers of relatively impermeable rock or sediment occur above and below the aquifer, and water pressure increases with distance from the recharge area at the surface, where the aquifer is exposed between its confining layers (Figure 8-5). As a result, if water can escape from the confining layers, it will rise and perhaps even flow out at the ground surface as an **artesian spring.** These springs are under sufficiently high pressure that water flows out of the ground naturally, without the need for pumping. Similarly, wells that tap water under high pressure in confined aquifers and thus need little or no pumping are called **artesian wells** (Figure 8-6), from the name of the Artois region in France, where many wells of this type were drilled in the 18th century.

In the 1800s and early 1900s, groundwater scientists, or hydrogeologists, explored the United States, identifying and mapping aquifers that were similar in their water-bearing properties. This work resulted in the identification of 12 major groundwater regions for the conterminous United States. These regions were studied in order to develop an inventory of the nation's water-resource potential. Today, the exploration phase of this work is nearly complete, and development, management, and protection of groundwater are the dominant activities of U.S. groundwater scientists (Figure 8-7).

Wells The permeability of an aquifer and the hydraulic gradient of the groundwater flowing through that aquifer affect the flow rate of the aquifer's water. This relationship was identified by Henry Darcy, a French engineer, who expressed it in a formula for groundwater velocity that came to be known as **Darcy's law** (Box 8-1). The rate at which water flows through an aquifer determines how much water can be pumped from a well. When a pump is inserted into a well and begins its pumping action, its suction creates an area of low pressure in the well hole. As water is removed from the well hole, the elevation and pressure of the water in the vicinity of the well become low. Since water flows from areas of high to areas of low elevation and pressure, water flows from the aquifer toward the well. As more water is withdrawn from a well, replenishment comes from farther out in the aquifer. According to the modern version of Darcy's law, the velocity of flow to the well is proportional to the aquifer's permeability and hydraulic gradient. This means that more permeable and higher-pressure aquifers can sustain greater pumping rates and have higher-yielding wells.

FIGURE 8-6 A geologist (left) and a driller are drenched but delighted after drilling into an artesian aquifer in glacial sands and gravels in Massachusetts.

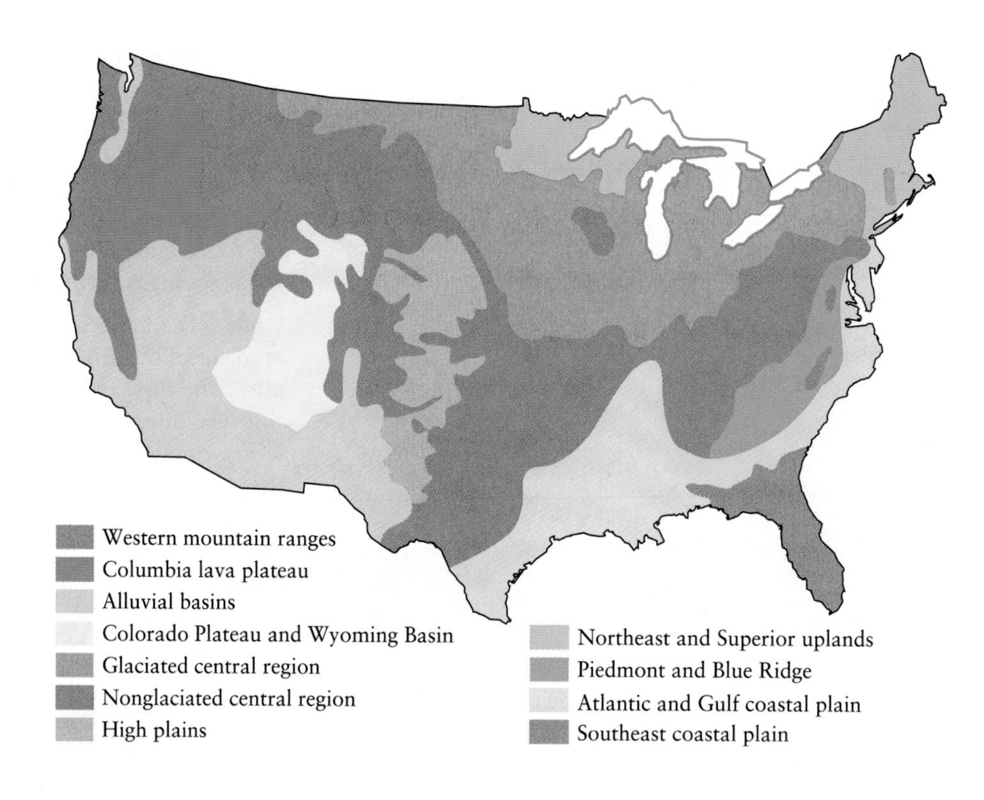

FIGURE 8-7 Eleven of the twelve major groundwater regions of the conterminous United States. Each region is dominated by aquifers of similar character. For example, much water in the semiarid Southwest is stored in alluvial basins, geologically recent sediments derived from surrounding mountains and deposited by streams in the intervening low valleys. In the glaciated central region of the Midwest, much water is stored in geologically recent glacial deposits. Groundwater in the Columbia lava plateau is stored in fractures in thick stacks of basalt flows that spread across the region and formed a vast plateau. The twelfth groundwater region, which is not shown here, is the alluvial valleys found along the floodplains of rivers and streams.

If the rate of pumped discharge is to remain constant, water must flow along steeper and steeper hydraulic gradients as it nears the well, because the area through which it flows becomes smaller and smaller. Eventually, a funnel-shaped area of depressed water level stabilizes around the well (Figure 8-8). If one could look down from above and see this depressed area, it would appear circular, with the well located in the center of the circle, as in a bull's-eye target. From the side, the depressed surface has the shape of a downward-pointing cone, with the tip of the cone at the well intake. This cone-shaped feature is known as the well's **cone of depression,** or, sometimes, *cone of drawdown,* and it exists only while pumping occurs. After the pump is turned off, the cone refills and the original groundwater level is gradually reestablished.

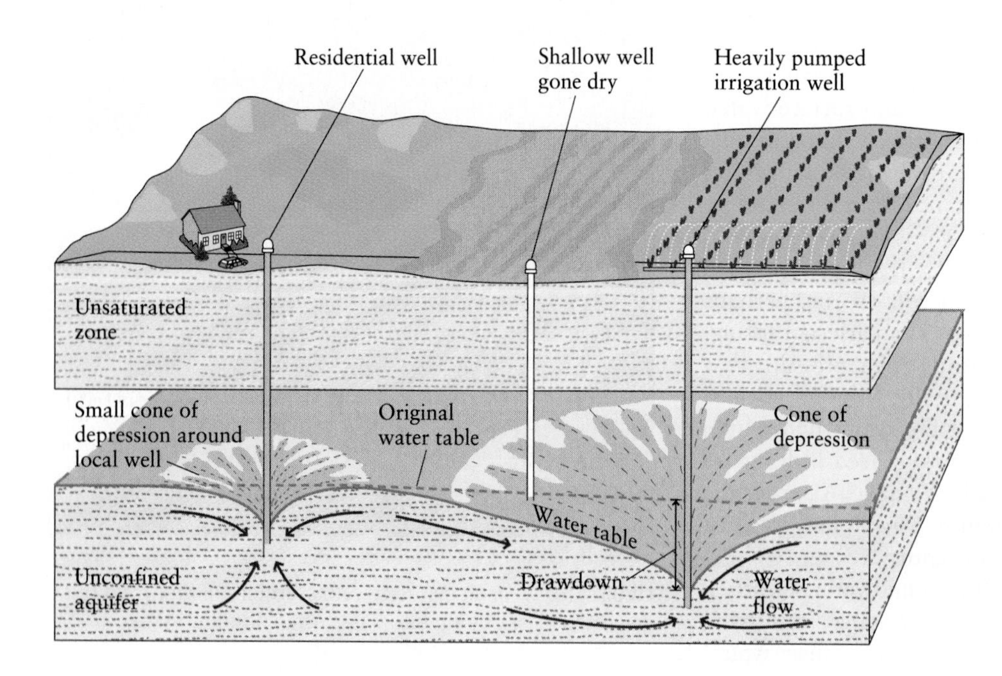

FIGURE 8-8 A three-dimensional view of the cone of depression of groundwater around a pumping well. Nearby shallow wells in the vicinity of a well that pumps large amounts of water can go dry if the water table drops below their intake level.

8-1 Geologist's Toolbox

Darcy's Law and the Flow of Water and Contaminants in Rocks and Sediments

In the mid-1800s, the engineer Henry Darcy was hired by the city of Dijon, France, to help its residents obtain a reliable and clean source of water. Using tubes of sand as water filters, Darcy tried, as he put it, "to determine the law of the flow of water through filters." Darcy's goal was related to a practical objective: to determine the most efficient way to pipe water to the city from a large spring nearly 10 km away and to have it flow by gravity to different supply points. Darcy was successful, and he is credited with giving Dijon a reliable and clean water supply for the first time in its long history.

In the course of his experiments of the flow of water between two points, Darcy discovered that the greater the change in *hydraulic head,* a quantity combining elevation and the water pressure at that elevation, the greater the flow of water from his tubes. The difference in hydraulic head (h) between two points divided by the distance (d) between the points is the *hydraulic gradient* $[(h_1 - h_2)/d]$, and water velocity (v) is proportional to hydraulic gradient:

$$v \propto (h_1 - h_2)/d$$

Darcy also found that water flows more quickly through coarse sands than through fine sands, silts, or clays because the larger pore spaces make the material more permeable, enabling water to pass more readily. Water velocity is thus proportional not only to hydraulic gradient but also to permeability (K):

$$v \propto K$$

Putting the variables of permeability and hydraulic gradient together gives a relationship that came to be called Darcy's law:

$$v = K(h_1 - h_2)/d$$

Darcy's law of water velocity has become the foundation of the science of hydrogeology, the study of groundwater flow, contamination, and well hydraulics.

Darcy calculated the velocity of waterflow from the average rate at which water drained through his tubes of sand. The modern version of Darcy's law goes a step further by taking into account the amount of pore space through which the groundwater is moving. The actual velocity at which water moves from one point to another in porous rock and sediment then is expressed as:

$$v = K(h_1 - h_2)/(d \times n)$$

where n is porosity (the percentage of pore space).

This equation can be used to estimate how long it would take water to travel from one site to another. If its physical characteristics are known, the travel time of a pollutant can also be estimated by Darcy's law. For example, assume that a chemical leak into an aquifer occurs from an underground storage tank and the

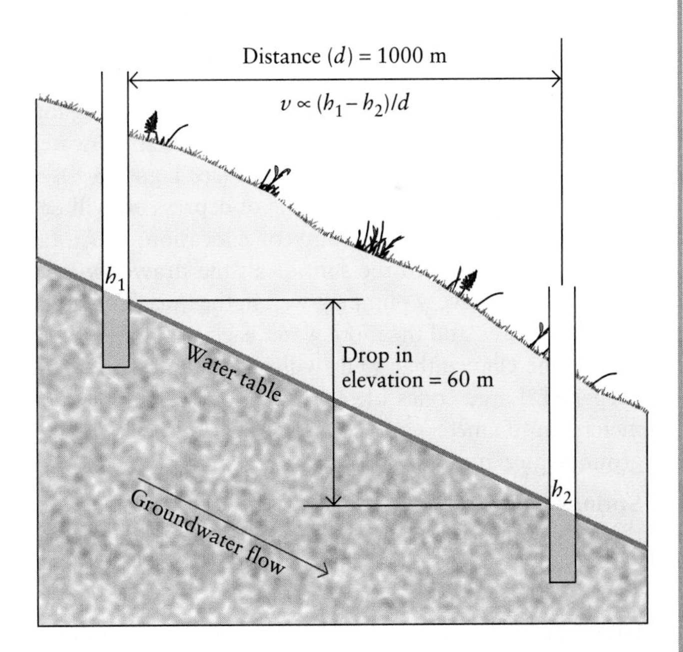

According to Darcy's law, the amount of groundwater flow per unit time is proportional to the drop in elevation divided by the horizontal distance between two points in an aquifer, and proportional to the permeability of the material through which it is flowing. This law can be used to estimate the travel time of a contaminant from its source to points further down gradient. (The diagram is not drawn to scale.)

(continued)

8-1 Geologist's Toolbox

(continued)

chemical moves at the same velocity as groundwater. If a leak occurs in a coarse sand aquifer with a permeability of 60 m per day, a porosity of 30 percent (0.3), and a hydraulic gradient of 1 m per 1000 m, the chemical will travel into the aquifer at the following velocity:

$$
\begin{aligned}
v &= K(h_1 - h_2)/(d \times n) \\
&= (60 \text{ m/day} \times 1 \text{ m})/(1000 \text{ m} \times 0.3) \\
&= 0.2 \text{ m per day, or } 73 \text{ m per year}
\end{aligned}
$$

The direction of travel is from an area of high to one of low potential energy, or high to low hydraulic head. If the leak occurs in a fine-grained clay with a much lower permeability of 0.0001 (or 1×10^{-4}) m per day, a porosity of 20 percent (0.2), and a hydraulic gradient of 1 m per 10 m, the chemical will travel much more slowly:

$$
\begin{aligned}
v &= K(h_1 - h_2)]/(d \times n) \\
&= (10^{-4} \text{ m/day} \times 1 \text{ m})/(10 \text{ m} \times 0.2) \\
&= 5 \times 10^{-5} \text{ m per day, or } 0.02 \text{ m per year}
\end{aligned}
$$

Contaminants can move much more rapidly in materials with high permeability than in those with low permeability; thus, high permeability in an aquifer can pose a considerable threat to drinking water supplies. The faster a contaminant moves through a groundwater system, the greater the amount of the contaminant that will appear in the drinking water supply per unit of time, and the greater the area likely to be affected.

If there is a shallow well within the cone of depression of a deeper and larger well, it will go dry when the water level is drawn down below its bottom (see Figure 8-8). If there is more than one well, a cone of depression will develop around each and, at any one location, the total drawdown is equal to the sum of all the drawdowns at that point from every pumping well in the area. The more wells in an area and the more water each well pumps, the greater the chance that wells will go dry unless they are deepened. Large cones of depression up to tens of kilometers in diameter are common near cities that use groundwater and in heavily irrigated areas.

Springs and Geysers In contrast to wells, which are artificial groundwater retrieval systems that are operated by the water pressure in an aquifer, *springs* are natural mechanisms for relieving water pressure in an aquifer. The larger the groundwater recharge area for a spring and the greater the rate of water recharge to the aquifer, the greater the water pressure in the aquifer and the greater the discharge from the spring.

Although spring water, like most groundwater, generally is filtered by rock and sediment, groundwater from springs and wells near heavily farmed areas may contain large amounts of nitrates from manure and fertilizer. Ingestion by pregnant women of too much nitrate can lead to oxygen depletion in the brains of fetuses and infants, resulting in a serious illness commonly known as blue-baby syndrome. For this reason, the U.S. Environmental Protection Agency's standard for drinking water specifies less than 10 mg of nitrates per liter of water.

Some springs spout hot water and steam above the ground. This phenomenon is caused by superheated steam that builds up pressure in deep subsurface channels. Such springs are known as *geysers,* from the Icelandic word *geysir,* which means to "gush" or "rage." Geysers are common in areas where the heat flow from the crust is high, for instance above magma chambers filled with molten rock. Geysers form when groundwater flows downward to depths of 3000 m or more, where it is heated to extremely high temperatures (Figure 8-9). Because of the great rock and fluid pressures at such depths, the water can be raised to temperatures much higher than its surface boiling point (100°C) before it boils and begins turning to steam. Once steam is formed, it expands and pushes the remaining liquid water upward, where pressures are lower. At reduced pressures, water boils at lower temperatures, so even more steam is produced and more water is forced upward to erupt at the surface. The process continues until the pressure is dissipated and the groundwater cavity begins to refill with water and steam. Such eruptions occur on a periodic schedule at Old Faithful, a well-known geyser at Yellowstone National Park in

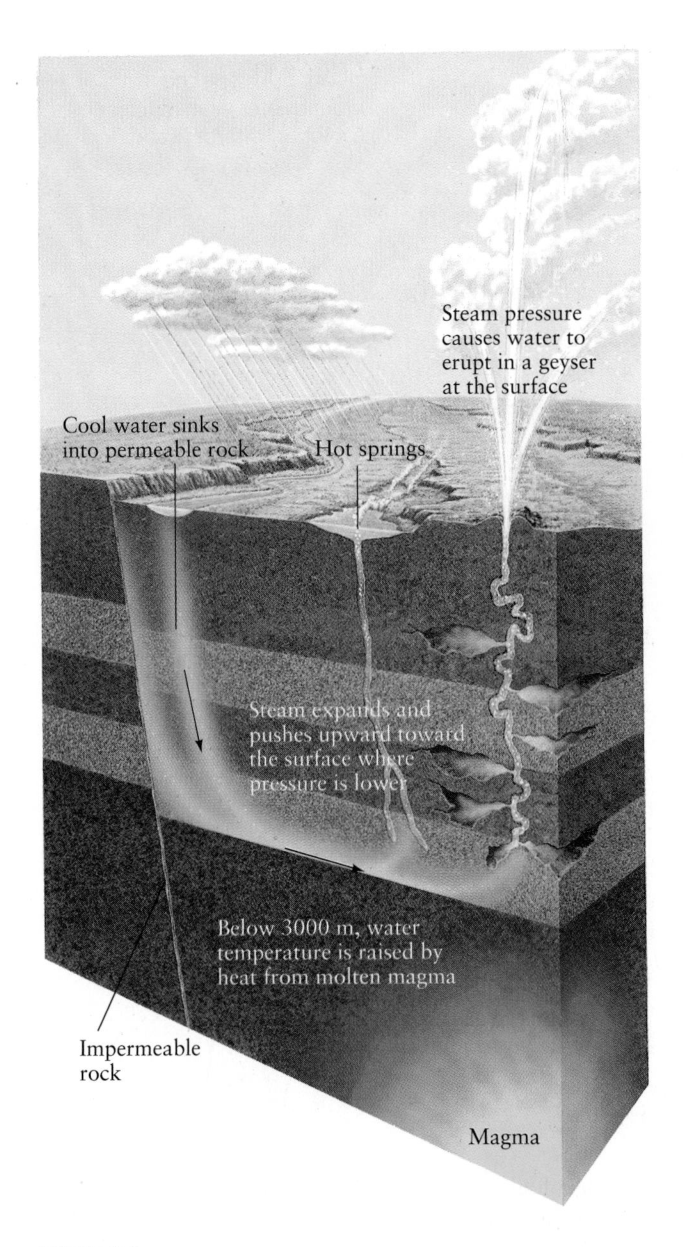

FIGURE 8-9 When groundwater is heated to high temperatures, steam and water erupt periodically with great force, forming geysers. Hot water may seep from the system and pool at the surface as a hot spring.

Wyoming. This geyser is one of many that occur above a large magma chamber supplied with heat from a mantle hot spot.

Groundwater Chemistry

Rain and snow, the sources of groundwater, contain very little dissolved material because the atmosphere acts as a giant still that removes impurities from the system. Human activities such as burning of fossil fuels or wood, and geological events such as the release of volcanic gases, emit compounds into the atmosphere that later are dissolved in falling rainwater, but generally, rainwater contains less than 10 to 20 mg of dissolved substances per liter of solution. By the time precipitation has percolated through soil, sediments, and rocks to replenish the groundwater, it has accumulated many hundreds of times those amounts of dissolved substances. Just as table salt, a mineral, dissolves when you pour water over it, the minerals in rocks and sediments are soluble in water. Each mineral has a different solubility in pure water, and solubility varies with the acidity of the water and the amount of oxygen dissolved in it. Groundwater readily dissolves the most soluble minerals in rocks. With increasing depth, the amounts and types of materials dissolved in groundwater vary, with deeper waters generally having higher concentrations of salts and less oxygen (Figure 8-10).

The composition of rainwater begins to change soon after it infiltrates the soil and becomes underground water. Biological processes remove much oxygen from underground water, for microbes use oxygen to decompose plant and animal matter. With increasing depth, groundwater is depleted of oxygen. Soil-forming processes result in the production of nitrogen (see Chapter 6), which dissolves in water to form nitrogen compounds (NO_2 and NO_3). Nitrogen enhances plant growth and is abundant in manure and artificial fertilizers, both of which farmers and gardeners add to soils in large quantities. The level of nitrates in groundwater has increased rapidly in many farm areas in recent decades.

As water moves through organic-rich soil, much carbon dioxide gas (CO_2) in unsaturated pore spaces is dissolved, and the reaction of CO_2 and H_2O forms carbonic acid (H_2CO_3), a weak acid of about pH 5.7, which in turn dissolves rocks formed of calcium carbonate ($CaCO_3$). Other ions that are common in groundwater from dissolution of minerals are magnesium and potassium; less common are sulfide ions released in the weathering of sulfide minerals.

Together, calcium and magnesium ions contribute to the quality known as **water hardness.** Hard water contains 50 to 100 mg or more of dissolved calcium and magnesium ions per liter of water. These ions can crystallize to form mineral deposits that leave rings in bathtubs and a scaly deposit in water pipes. The ions in hard water also change its flavor and react with soap so that suds cannot readily form. Soft water, with low amounts of these ions, has a much better taste and forms a better lather when soap is added. Many people living in areas underlain by limestone, a calcium carbonate rock, complain about their hard water and often treat it with water softeners, chemicals that react with the calcium and magnesium to reduce the amounts of the free, dissolved cations in the water solution.

The process of dissolution continues as water moves deeper into the subsurface. The total amount of dissolved

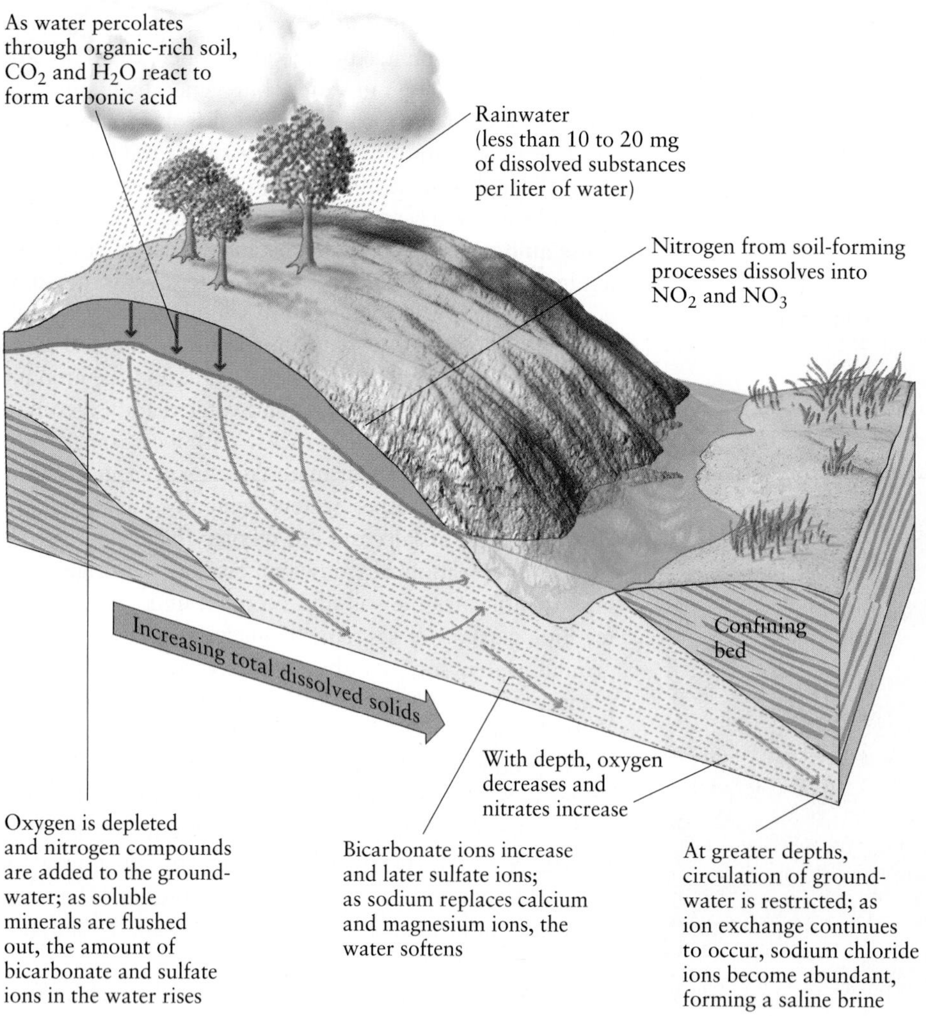

FIGURE 8-10 The chemistry of groundwater in an aquifer and its evolution with time and depth.

solids rises, but there is much exchange of ions between water and mineral surfaces (see Figure 8-10). Clay minerals, common in rocks and sediments, generally are associated with large amounts of sodium during deposition in offshore marine environments, so cation exchange between calcium and magnesium from the water and sodium on exposed surfaces of clay minerals results in water rich in sodium bicarbonate. Because clay particles are small, they have a large total surface area on which are exposed many exchangeable sodium ions.

The age of the groundwater increases with depth: Radiometric dating of carbon in groundwater has shown that its age can be as much as 10,000 to 40,000 years in the deeper parts of some aquifers. The general pattern of groundwater evolution with depth and age is a transformation from bicarbonate-rich water to sulfate (SO_4^{2-})-rich water, then finally to a brine, a water solution rich in sodium chloride. Because the circulation of water at great depths is usually much more restricted than at shallow levels, sodium chloride—which is very soluble—is not flushed from the groundwater. Although some early hydrogeologists thought that the brines pumped from deep wells were ancient marine waters trapped in sediments as they were laid down, it now is clear that groundwater can evolve to become more saline with depth and age. About half of all groundwater is deep and saline and thus not nearly as likely to be used as the fresh water found at shallower depths (see Table 2-1).

- Groundwater recharge occurs at Earth's surface from precipitation that drains downward through the unsaturated zone to the saturated zone. The water table rises as the amount of saturation increases.
- Groundwater is discharged to Earth's surface at springs, wetlands, streams, lakes, and ultimately the ocean.
- Elevation and water pressure are the main determinants of the direction in which groundwater flows. In general, groundwater flows from areas of high to low elevation and high to low pressure.

- The modern version of Darcy's law states that the rate of flow of groundwater is proportional to the change in elevation and pressure between two points and the permeability and porosity of the geologic material through which the water flows.
- Groundwater dissolves minerals during its travels through rocks and sediments. Water that contains relatively large amounts of calcium and magnesium is known as hard water.

FIGURE 8-11 This hand-dug well in the Sinai peninsula is vital to Bedouin and their livestock. The shallow, unconfined aquifer that supplies it is in unconsolidated sediment deposited by an ancient river.

Groundwater As a Resource

Groundwater is not so readily available as surface water, and it has become an increasingly used resource only as new technologies to exploit it have been developed. Today, drilling rigs with diesel engines can bore thousands of meters into Earth's crust. The stock of groundwater, although not so accessible, is much greater than the stock of surface water—at any given time nearly 120 times as much water is stored underground as at the surface. Furthermore, the residence time of water in the groundwater reservoir is much greater than in the surface water reservoir. Much groundwater has been stored in aquifers for very long periods of time with little or no change in that stock until the recent advent of sophisticated pumping technologies.

Gaining Access to Groundwater

In arid regions where year-round surface water is scarce or unavailable, for thousands of years groundwater has been used, reached by shallow wells dug by hand (Figure 8-11). The invention of modern techniques for drilling deep wells has opened many arid and semiarid areas to groundwater exploitation and made water available for farming by irrigation (Figure 8-12). In the United States, groundwater withdrawals have increased almost steadily for decades. Today, about 40 percent of all water used is groundwater, and 14 states get more than half their water supply from the subsurface. About three-fourths of this pumped water is used to irrigate farmlands, largely in the arid and semiarid western states. In rural areas throughout the United States, where public water is not available, groundwater accounts for 96 percent of all domestic

FIGURE 8-12 The sparse vegetation at an oasis (background) in Libya gets its water from a natural seep of groundwater at the surface. Farther from the oasis, where groundwater occurs at a deeper level, dark circles of wheat (foreground) are irrigated by groundwater pumped from wells located in the center of each circle. Sprinklers spread the water around each well.

8-2 Case Study

Groundwater Resource Management in Los Angeles

The metropolitan area of Los Angeles, in southern California, is built on a broad, gently sloping coastal plain made of marine sediments and eroded debris shed from mountains that surround the plain. These deposits have accumulated in the Los Angeles alluvial basin for millions of years, and the upper strata contain large supplies of fresh groundwater. As the climate of the region is semiarid—rainfall is low, only 23 cm per year, and evapotranspiration high, about 76 cm per year—Los Angeles relies heavily on imported surface water and on groundwater from its own basin. Surface water is imported through aqueducts (tunnels or culverts that carry water) from other water basins, including the Mono Lake area 200 km north of Los Angeles and the Colorado River 200 km east of the city.

Groundwater supplies about half the area's water needs. Intensive pumping, however, resulted in falling

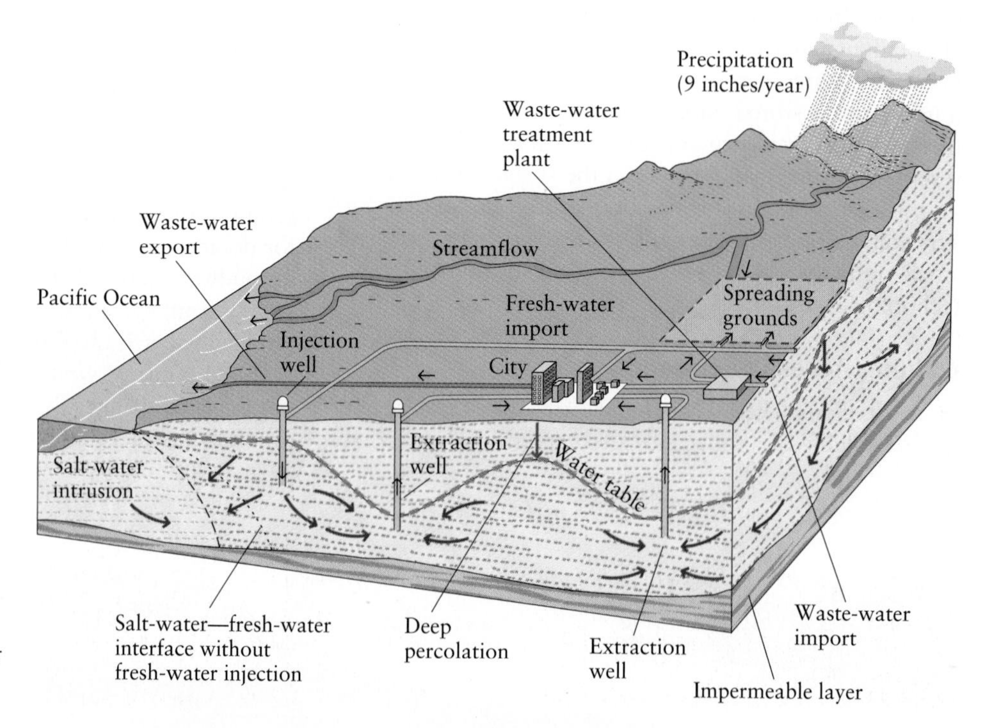

Spreading basins are used to spread water aboveground in order to recharge aquifers under Los Angeles.

water use. Even for towns and cities with public water supplies, groundwater now accounts for nearly half of all water use because many municipalities have installed wells to supplement their surface water resources.

Groundwater Resource Management

Groundwater is considered a *renewable* resource, in contrast to oil and gas, which are *nonrenewable* in the time scale of a human life. A renewable resource is one for which the replenishment rate equals or exceeds the rate of consumption. In some areas, however, groundwater is being withdrawn at rates that exceed rates of groundwater recharge. In semiarid or arid places such as Mexico City and Phoenix, such *groundwater mining* is causing water levels to drop rapidly. Maintaining the groundwater resource so that it will provide a sustainable source of clean water demands an integrated approach to managing water basins by considering their aquifers, areas of recharge and discharge, local climate, and the conditions under which water is used.

water tables that permitted salt water along the coast to flow into the aquifers under the city. In recent decades, the region has experienced a number of severe droughts that have alternated with extremely wet periods, when many floods occurred. In response to drought years, the Los Angeles Metropolitan Water District has developed and maintained a coordinated management plan that makes the most efficient use of what little surface water is available, uses reclaimed waste water, and minimizes negative impacts on groundwater aquifers. The elaborate network of devices to obtain, treat, and transmit water throughout the basin includes the use of spreading basins to recharge aquifers with surface water and injection wells along the coast to maintain a wall of fresh water that will prevent the intrusion of salt water into the groundwater system.

Aqueducts like this one, crossing the desert of Owens Valley, California, transport water to the Los Angeles metropolitan area.

Artificial Recharge If natural recharge rates are low, groundwater management may include artificially adding water to the groundwater system (and returning reclaimed water to it). Recharge of an aquifer can reduce the impact of pumping. For example, in areas of natural recharge of an unconfined aquifer, some precipitation may not move downward into the unsaturated zone and later to the water table if the surface becomes saturated during a storm or a period of snowmelt. In this case the water becomes surface runoff, making its way downslope to streams and leaving the basin. Runoff and streamflow can be collected by diversion, then spread over a large land area to infiltrate in places where the water table is not close to the surface. This practice, known as **water spreading,** is common in many arid and semiarid regions where there is a substantial zone of permeable, unsaturated sediments above the water table (Box 8-2).

Other means of resupplying aquifers are recharge basins and injection wells. **Recharge basins** are less extensive in area than water-spreading grounds, but can hold a greater depth of water; consequently, a large hydraulic gradient is created that helps, in keeping with Darcy's law, increase the rate of recharge. The water added to recharge basins includes stormwater runoff from paved urban areas, reclaimed waste water, and water diverted from streamflow. **Injection wells** are wells through which water is pumped into the ground to maintain the pressure of a reservoir. They are used to recharge water in deep, confined aquifers. Because injection wells have been used for disposal of hazardous wastes, at least one state—Wisconsin—prohibits injection wells for any purpose. Hazardous-waste injection wells are legal in other states.

Siting and Spacing Wells A comprehensive groundwater management plan includes siting and spacing wells so that there is minimal lowering of the water table. Wells should be near areas of natural recharge, so that well discharge and recharge are more likely to balance even under adverse environmental conditions or intense use. If wells are sited near areas of surface water runoff, such as streams, then excess surface water or groundwater already discharged by the aquifer is drawn back into the aquifer; however, pumping near streams, wetlands, springs, and lakes may degrade natural ecosystems, which are sensitive to fluctuations in water amounts, and downstream water users may no longer get sufficient water from the surface water source. Many environmental lawsuits have been brought after supplies of surface water were depleted by the excessive pumping of groundwater.

Clearly, prudent management of groundwater and surface water resources requires knowing the maximum stable yield of water that can be obtained from the basin. The *safe yield* is the maximum possible pumping rate that will neither exceed average annual rates of recharge nor lower the water table to such an extent that pumping becomes too costly or the aquifer quality is degraded.

- Groundwater is considered a renewable resource if rates of recharge are the same as or greater than rates of discharge.
- In areas where rates of discharge exceed natural rates of recharge, water management techniques are used to restore a balance. These include the use of water spreading, recharge basins, and injection wells.

Groundwater Hazards

Like other Earth systems, the groundwater system presents hazards, both natural and induced, as well as providing a valuable resource. One such hazard is the natural occurrence of sinkholes; others include land subsidence and the intrusion of salt water into aquifers because of overpumping.

Solution Caverns and Sinkholes

An aquifer undergoes physical as well as chemical changes as water migrates through rocks and sediments. If the rocks are composed of very soluble minerals, water enlarges the pore spaces to form solution caverns—those formed by the process of solution—and subsurface channels. Limestone ($CaCO_3$), common in many areas, is readily dissolved by the carbonic acid in groundwater. In limestone, much pore space occurs as secondary porosity along fractures, which when enlarged by solution can form mazes of interconnected caverns and tunnels. The terrain in areas of fractured limestone is highly irregular, characterized by tall spires and pinnacles separated by deep channels. Such a topography is called karst, after the Karst region near Trieste, which is dominated by such features. In karst areas water, and its contaminants, can move swiftly because of the high permeability of soluble rocks (see Table 8-2). Furthermore, the direction of groundwater flow is hard to predict in cavernous aquifers.

If the water table fluctuates, the roofs of the caverns can become quite thin, eventually collapsing as the weight of the overlying soil becomes too great for the rock ceiling (see Box 1-1). Such collapsed areas are called sinkholes; they often are aligned in gridlike patterns that reflect the underlying perpendicular fracture zones in the rocks. In some places, the formation of sinkholes coincides with lowering of the water table. In South Africa, groundwater was pumped from an excavation during mining and within two to three years, eight large sinkholes, more than 50 m in diameter and 30 m deep, had formed. Evidently the water pressing against the cavern ceilings had kept the cavern roofs from collapsing.

To determine how fast groundwater moves through cavernous limestones, and to identify the flow paths, in 1996 the U.S. Environmental Protection Agency asked the help of geoscience faculty and students at Franklin and Marshall College in Lancaster, Pennsylvania. Harmless dyes were released into sinkholes and wells in southeastern Pennsylvania, and the arrival of the dyes at various streams and wells was monitored. Within a week groundwater containing the dye appeared several kilometers from one injection site. Such fast travel times pose difficult problems for environment scientists.

Land Subsidence From Groundwater Mining

The pumping of water wells can cause the water table to drop if the rate of groundwater discharge is greater than the rate of recharge. In turn the ground surface may settle and collapse, a phenomenon known as **land subsidence** (Figure 8-13). The effect can be demonstrated by the analogy of a bicycle tire. When air leaks out, the tire deflates, or decompresses, because the pressure of air on the inner surface of the tire is reduced. Similarly, when water is removed from an aquifer, pressure on the inner

FIGURE 8-13 A well casing, once below ground, is now exposed by 7 m of land subsidence in Mexico City that was caused by the excessive extraction of water.

walls of the pores decreases and the porous structure deflates slightly. Subsidence occurs throughout the world, sometimes to great extent: One instance in California has resulted in nearly 9 m of sinking.

Subsidence is especially common above aquifers composed of unconsolidated sediments surrounded by confining layers of clay. As water drains from an aquifer, water from the less permeable, clay-rich layers migrates into the aquifer. Clay layers are especially likely to collapse and contract if they lose water because they are not as rigid as sands and gravel. Although water can be added to sands and gravel after they have been drained (just as air can be pumped into a deflated tire), it is nearly impossible for water to reenter the collapsed pore spaces in clays, so subsidence due to drainage of clay layers above and below drained aquifers is permanent.

Intrusion of Salt Water

Lowered groundwater levels in coastal areas invites inland migration of salty seawater to aquifers, a process known as **saline intrusion,** which makes groundwater unpotable. Saline intrusion destroyed the aquifers beneath the borough of Brooklyn in New York City, when the water table was lowered as much as 15 m in the 1930s.

The causes of saline intrusion are related to the difference in density between fresh water and salt water and to changes in the volume of fresh water when the aquifer is pumped. Seawater is denser than fresh water because it contains more dissolved ions. Fresh water therefore tends to float above seawater as the latter moves inland along coasts (Figure 8-14). At the boundary zone between the overlying fresh water and salty seawater underneath, the two mix. Surface water drains into the sea in a volume usually great enough to push away the denser salt water. When groundwater is pumped, however, the volume of fresh water above the boundary zone is diminished, and the boundary migrates inland and upward. Wells that once pumped groundwater begin to yield useless seawater.

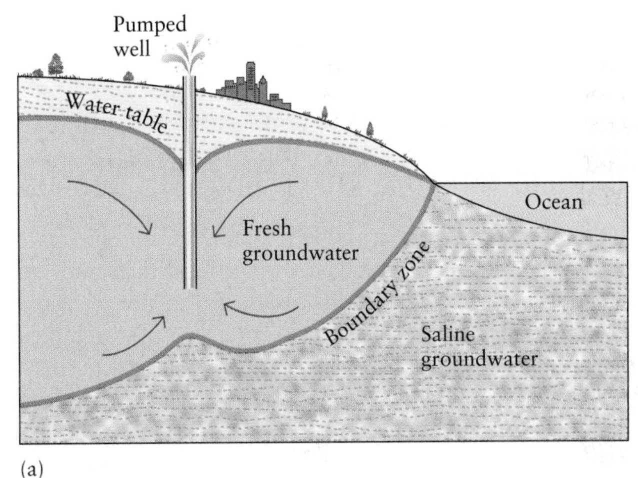

(a)

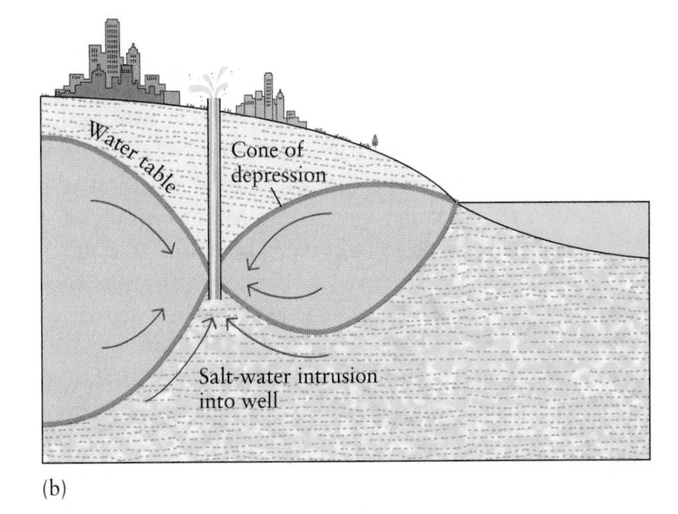

(b)

FIGURE 8-14 (a) The boundary between saline and freshwater zones in a coastal area before groundwater pumping. (b) After prolonged pumping, the water table is lowered and saline intrusion occurs.

Saline intrusion can be prevented or reversed by artificial recharge of the groundwater and by construction of barriers that block the inland flow of seawater. Such efforts were made recently to prevent saline intrusion to the Biscayne aquifer, the sole source of groundwater for Miami, Florida. Today, fresh surface water held in dams is used to recharge groundwater, and the dam walls themselves are barriers to the inland flow of seawater.

- Sinkholes form above soluble rocks, such as limestone, where groundwater has dissolved large amounts of the rock ceiling and caused the roof to collapse.
- Land subsidence occurs where aquifers and confining layers settle because of withdrawal of groundwater.
- Saline intrusion of aquifers, the replacement of fresh groundwater by seawater, occurs when excessive groundwater is removed along coastal areas.

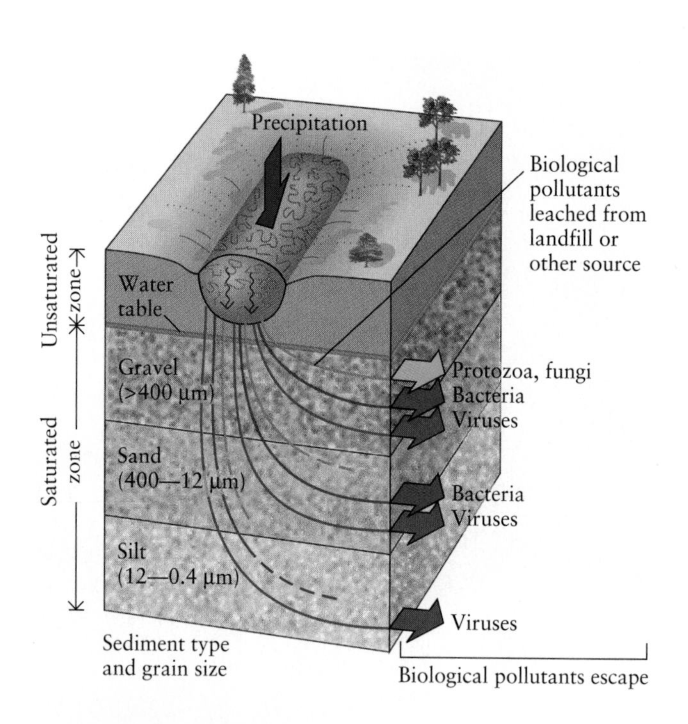

FIGURE 8-15 The smaller the size of the grains, the smaller the pores between the grains and the smaller the microorganisms that can be filtered from water flowing through pore spaces. At 20 to 250 nanometers, viruses are too small to be filtered by silt; bacteria are large enough to be filtered by silt but too small to be filtered by sand.

Groundwater Pollution and Its Cleanup

Pollutants that seep into the ground do not dissipate harmlessly. Rather, they travel with groundwater and can emerge in streams and wells far from the site of initial contamination; the detergent found in contaminated wells in Falmouth, Massachusetts, was released perhaps 5 km away. To understand how to protect and clean groundwater, it is important to know the types of pollutants that reach groundwater and how they flow in water.

Types and Sources of Groundwater Pollution

Groundwater can be contaminated by biological pollutants, such as bacteria, protozoa, or viruses, and by industrial chemicals, such as trichloroethene (TCE). Until the Industrial Revolution, biological contamination was the greatest threat to drinking water. Major outbreaks of bacterial diseases such as bubonic plague in 17th century London and other cities were associated with wells contaminated by human sewage. In general, however, groundwater is less susceptible to biological pollution than surface water because many pathogens are similar in size to grains of silt and sand and thus can be trapped in pore spaces in aquifers or in the unsaturated zone above the groundwater table (Figure 8-15). For this reason, knowledgeable campers and backpackers filter any water collected from streams through fine carbon granules.

Since the Industrial Revolution, the number of chemicals used for manufacturing, farming, and other activities has increased rapidly. More than 70,000 chemicals are used today, many of them distributed throughout the environment, in air, water, and soil, and the effects of many thousands of these chemicals on human health are not known. Each year another 700 or 800 new chemicals are produced. The U.S. Environmental Protection Agency estimates that 55 million tons of hazardous chemical waste are produced in the United States each year. Twenty of the most abundant compounds identified in groundwater at industrial waste disposal sites include TCE, benzene, and vinyl chloride, all of which are known human carcinogens and also can cause liver damage, brain disorders, lesions, and nervous system dysfunction. TCE may be especially prevalent. This toxic chemical is used mainly as a solvent by metal industries and dry cleaners to remove grease. It also is used to extract caffeine from coffee and can be one of the chemicals involved in the manufacture of pesticides, tars, paints, and varnishes. The recommended limit of this contaminant as established by the Clean Water Act is 0.27–27 ppb (parts per billion). Unfortunately, because TCE is so commonly used, drinking water wells in some states, including Pennsylvania, New Jersey, New York, and Massachusetts, contain as much as 14,000 to 27,300 ppb of TCE.

Until the 1970s, no laws in the United States prevented any company or individual from disposing of hazardous wastes in underground storage tanks, surface

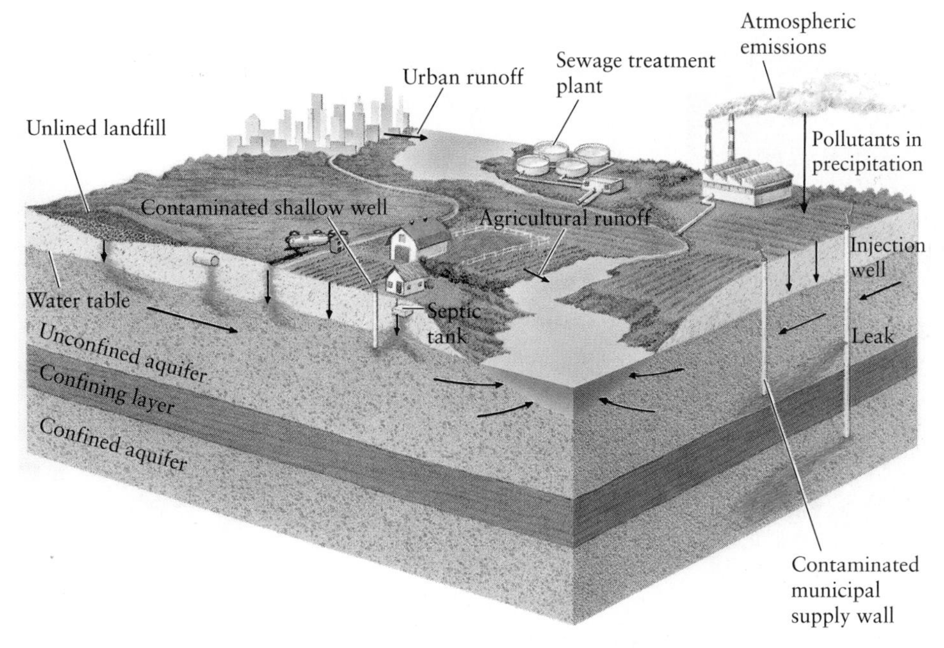

FIGURE 8-16 Corroded and leaking underground storage tanks are one of the main causes of groundwater pollution from industrial wastes, while leaking septic tanks are a main residential cause. Pesticides and fertilizers from farm fields seep underground and can reach the water table, as can metals and other pollutants that move through waste stored in landfills. Depending on the type of waste in a landfill, it may contain metals and chemicals such as solvents and fuel oils that can enter the groundwater. Today, landfills are designed carefully to collect and treat all groundwater before it leaves the site. Although most wells pump water from underground, some are used to inject hazardous waste into the subsurface.

impoundments (also referred to as waste lagoons or ponds) like the one at Otis Air Base, or injection wells. In fact, after passage of the Clean Air and Clean Water acts in the 1960s and 1970s, disposal of wastes to air and surface water was regulated so strictly that even more waste was disposed of underground. Unfortunately, a prevailing misconception was that wastes stored underground could not travel far, if at all. Instead, because the ground is porous, buried contaminants migrate along with the water draining downward in the unsaturated zone or flow down the hydraulic gradient with groundwater in the saturated zone.

A recent assessment of the causes of groundwater contamination in all states identified leaking underground storage tanks and drainage from septic tanks as the top two sources of pollution. Other common sources include landfills, surface impoundments (there are 181,000 in the United States, two-thirds of them unlined), infiltration of pesticides and fertilizers from agricultural fields, and accidental spills from trucks and trains (Figure 8-16).

Many chemicals are stored underground before as well as after becoming waste products, such as the fuel stored at gasoline stations, factories, and other businesses. More than 2 million commercial underground storage tanks are known to exist in the United States. Most are made of metal that can corrode in 10 to 20 years and have supply lines that crack and leak. At least 100,000 are known to be presently leaking (Figure 8-17).

Precipitation or runoff infiltrating downward through landfills and surface impoundments can dissolve metals and chemicals and transport them, along with bacteria, to saturated zones. The sometimes oily, discolored product is known as leachate, which can be seen seeping from the ground near waste sites that are improperly lined and lack leachate collection systems (Figure 8-18).

Migration of Groundwater Pollution

Substances carried by groundwater move at rates that vary with particle or molecular size, solubility, chemical activity, density, and viscosity. If oil is spilled from a truck and seeps downward to the water table, the oil—which does not dissolve in water—floats at the water table above the zone of saturated rock or sediment. It moves along the direction of groundwater flow, giving off petroleum vapors and contaminating wells. In contrast, if salt seeps downward from roads that were salted

FIGURE 8-17 A leaking underground storage tank being pulled out of the ground near Detroit for disposal at a hazardous-waste site. The contaminated groundwater will be cleaned, but the effort will be costly.

(a)

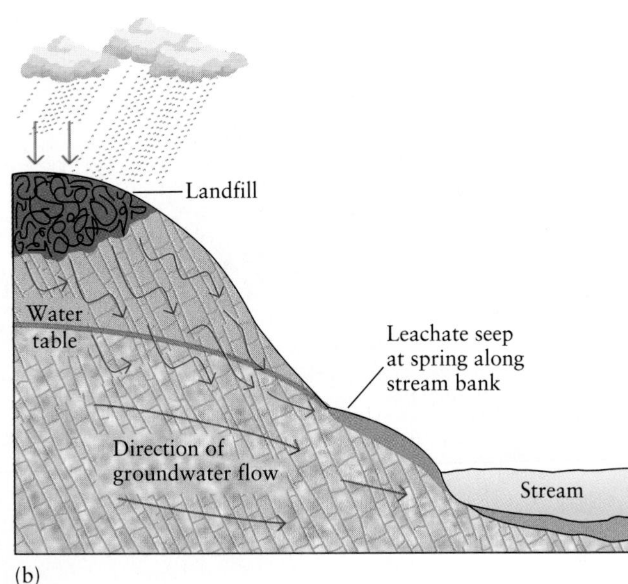

(b)

FIGURE 8-18 (a) Photograph and (b) diagram of leachate from an abandoned city dump, now the Lancaster County Park, Pennsylvania.Infiltrating water flows through the waste and leaches out contaminants. These contribute to the formation of leachate, a liquid that usually is relatively rich in metals and sometimes contains chemicals such as solvents. The leachate flows downward until it reaches the water table. In this instance, the leachate emanates from the hillside at a spring and then flows downhill into a nearby stream.

during the winter, it dissolves and flows into the saturated zone. Both oil and salt are major contaminants of groundwater. In fact, the common use of as much as 2 to 11 metric tons of salt per kilometer of single-lane road to de-ice roads in the course of a single winter has caused extensive environmental degradation of groundwater in northern parts of the United States.

Miscible contaminants flow in the groundwater as dissolved constituents, or solutes, mixing with the groundwater to form a single liquid. Salt is a miscible contaminant. Because most groundwater flows at rates of several meters per year or less and is subject to little turbulence, solutes migrate very slowly from the source of contamination and tend not to become diluted. In fact, recent monitoring and testing of groundwater wells indicate that some contaminants occur in far greater concentrations underground than in localities where contaminated surface water—which moves relatively rapidly—is known to exist.

Miscible contaminants are diluted by streams and rivers, but in groundwater they tend to spread outward in long plumes through the process of *dispersion.* In dispersion, the mechanical combining of contaminated and uncontaminated water, the contaminated water remains at close to full strength as a discrete flow, but is increasingly mixed with uncontaminated water in the downgradient direction. Much of the mixing occurs when contaminants flow around individual grains in paths that cause the contaminated water to diverge from the main flow direction. Dispersion occurs in aquifers because the material through which groundwater flows is not completely homogeneous. Water flows faster through large pores than small pores because frictional resistance with rock surface is less. Also, the water flows more quickly in the centers of the pores than along the pore walls, where frictional resistance to flow is greater. The result is that although the direction of the contaminant flow is that of the regional hydraulic gradient, the shape of the contaminant flow changes. It becomes elongated, and it also may spread out perpendicular to the direction of the groundwater flow as the water follows branching pathways. Dispersion is the primary cause of the characteristic plumelike shape of contaminants in groundwater (Figure 8-19). Contaminant plumes have been identified in many aquifers for all types of miscible pollutants (Box 8-3).

Immiscible contaminants are insoluble, and their fate in groundwater depends partly on their density. Those that are less dense than water float above the water table and move down the hydraulic gradient along the water table's surface. Gasoline is a common example of an immiscible material found floating above groundwater. If any part of the liquid contaminant plume is slightly soluble, it dissolves in the groundwater and moves below the water table. Those immiscible contaminants, such as trichloroethene (TCE), that are denser than water, tend to sink to the bottoms of aquifers. Depending on the geometry of the bottom of the aquifer, the contaminant

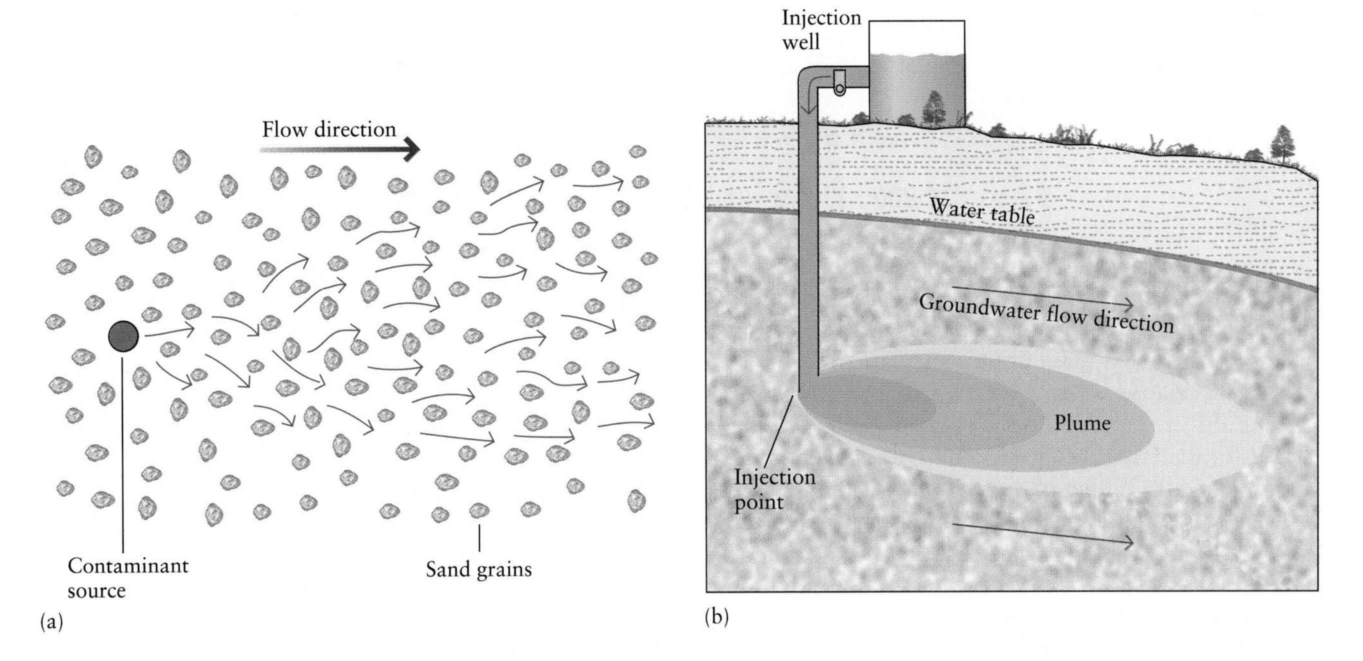

FIGURE 8-19 (a) A primary cause of dispersion is the separation of flow paths around grains. (b) With increasing distance from the source of contamination, such as an injection well at which wastes are injected into the ground, contaminated water is mixed with uncontaminated water in directions parallel to and perpendicular to the flow direction. The result is that the contaminant assumes a plumelike shape.

plume may or may not move in the same direction as the regional groundwater flow. In the case of light immiscible substances, some vapors may be given off that seep into the unfilled pore spaces of the unsaturated zone (Figure 8-20).

U.S. Laws Governing the Quality of Water Resources

Although groundwater is an essential national resource susceptible to contamination from many common sources, there are no federal laws in the United States solely for the protection of groundwater quality. Many aquifers have been seriously polluted, and in some cases the damage is irreparable. In the absence of a specific law, a number of federal laws designed to address other environmental problems have been invoked to protect groundwater. Some of these laws were created to make certain that public water supplies meet established health standards (see the discussion of standards for drinking water in Chapter 7) or to regulate the production, use, and disposal of pesticides, which can affect the quality of underground water.

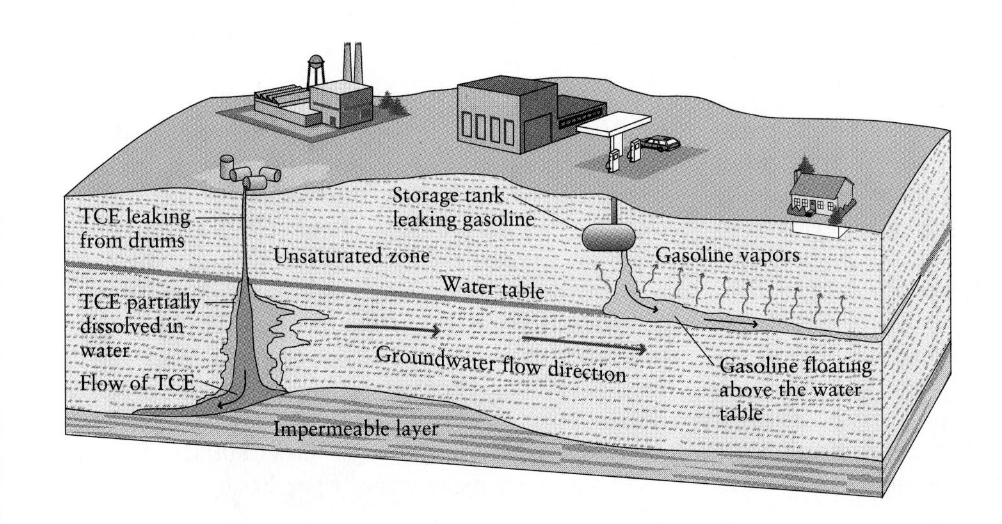

FIGURE 8-20 Light and heavy immiscible contaminants: Gasoline from a leaking underground storage tank floats atop the water table; trichloroethylene (TCE) spilled from drums sinks to the bottom of an aquifer.

8-3 Case Study

Sewage Plume at Cape Cod, Massachusetts

A number of the contaminated sites in the northeastern United States have histories like the one that produced the plume of contaminants emanating from a sewage treatment plant at Otis Air Base at Cape Cod, Massachusetts (see Figure 8-1). Unfortunately, the aquifer at Cape Cod is the only aquifer providing water for the Cape. From 1936 to 1986, Otis treated more than 11.4 billion liters (2.5 billion gallons) of sewage waste and discharged it onto beds of sand to allow it to infiltrate the subsurface. This is a common method of sewage waste treatment and disposal; however, because of the large quantity of contaminants and the permeable, coarse-grained nature of the aquifer, this treatment was ineffective and the degradation of water quality in the aquifer is particularly severe.

Cape Cod is underlain by more than 100 m of sediments deposited by rivers and glaciers. The upper 30 to 40 m are sands and gravel deposited about 10,000 to 15,000 years ago at the front of a melting Pleistocene ice sheet. There is a shallow, unconfined aquifer (the water table is less than 15 m deep) in the upper sediments with a porosity of about 35 percent, permeability of about 116 m per day, and groundwater flow velocity of about 0.3 to 0.5 m per day. Monitoring wells at different sites throughout the area of the plume (see Figure 1-15) led to the discovery that the altitude of the groundwater table drops from 15 m at the sewage infiltration beds to 10 m about 3 km farther south, and the plume of contaminants is moving down the gradient of the water table.

The sediments contain very little clay, which has large surface areas and high ion absorption capacities, and so contaminants move readily and easily through the subsurface at about the same rates as does groundwater. This means that contaminants can migrate downslope from the site at rates of as much as 1 km in 5 to 10 years. For this reason, a well that supplied the town of Falmouth with 25 percent of its water had to be shut down, and many others soon will be affected by the migrating contaminants. One plume of contaminants at this site is more than 3 km long and 0.8 km wide, and about 23 m thick. The plume has contaminated about 7 million m^3 of the aquifer and rendered its water unpotable.

Many contaminants have been identified at this site, including nonbiodegradable detergents used before 1964 and volatile organic compounds such as the carcinogens tetrachloroethene and trichloroethene. Extensive analysis of this site has led some researchers to conclude that cleanup of 11.4 billion liters of sewage waste dispersed throughout an aquifer is not possible. Recent proposals for stopping the spread of the contaminant plume have included the construction of an underground wall of iron compounds designed to absorb many of the chemicals in the plume before groundwater reaches wells farther down the gradient. The cost of the wall would be millions of dollars, and pumping out and treating all the water and contaminants would cost hundreds of millions of dollars. The lessons learned from this site are that despite environmental legislation such as Superfund, the cleanup of aquifers is going to be costly, slow, and sometimes impossible.

One of the most important laws passed for the remediation of polluted groundwater is the Comprehensive Environmental Response, Compensation, and Liability Act of 1980 and its amendments, commonly referred to as Superfund. Many contaminated sites have been targeted for cleanup under the guidelines of this legislation. If the cleanup is not performed by those responsible for the pollution, it is done by the Environmental Protection Agency and billed to the polluters if they are known.

It can be impossible to determine just who released hazardous substances at sites where contamination occurred decades ago and land ownership has changed hands several times. In 1996, when the geology department of Dickinson College in Carlisle, Pennsylvania, investigated possible drilling sites for wells to be used for teaching, they learned that a plume of TCE exists deep beneath the campus and the surrounding area. The plume was traced to its source: an empty lot with no record of past industrial use. No one knows how TCE was spilled or released at the site, and no one can be charged with the costs of cleanup. The law now stipulates that the current owner of the land is liable for the costs of cleaning up any hazardous substances found in the subsurface. This provision has made many wary of purchasing old industrial

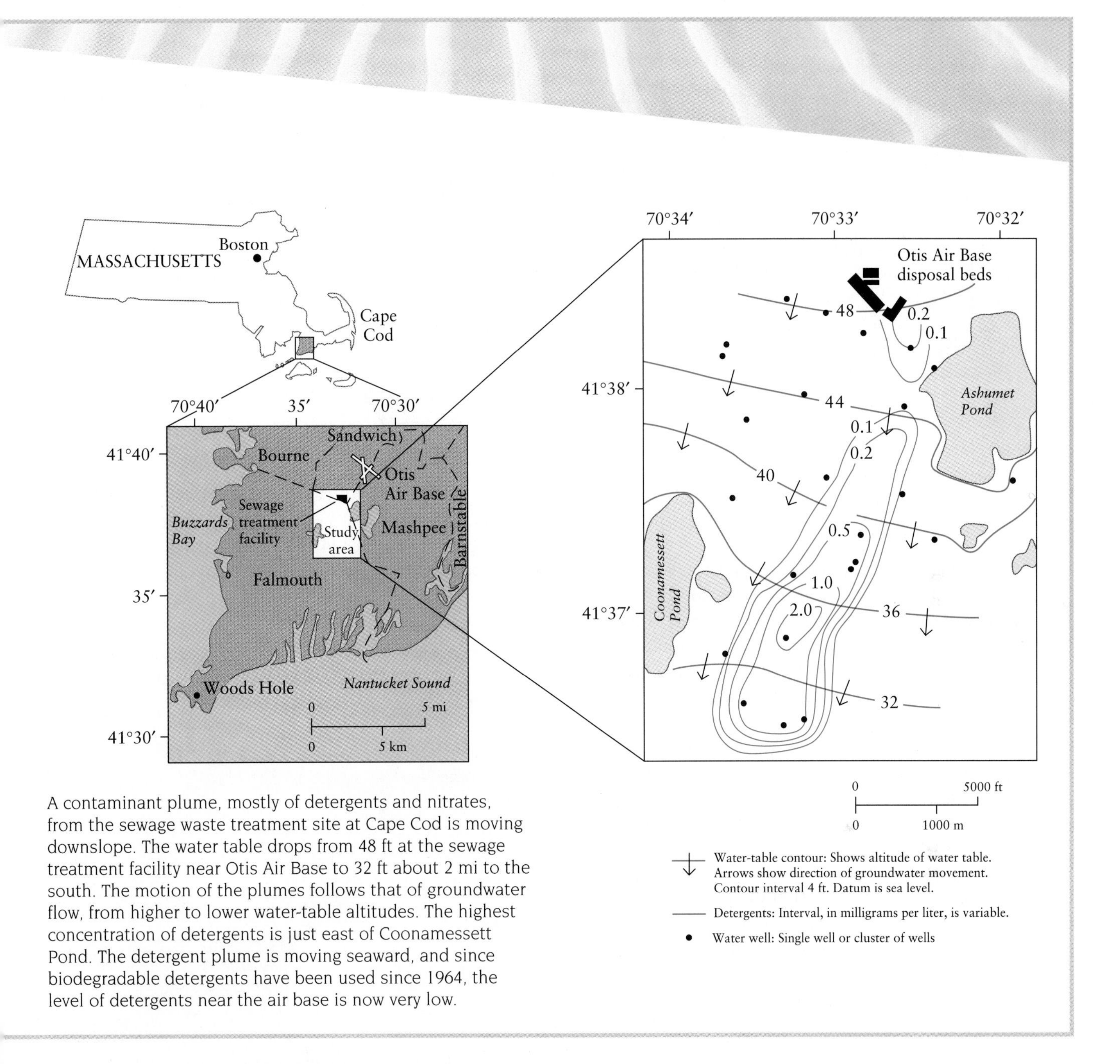

A contaminant plume, mostly of detergents and nitrates, from the sewage waste treatment site at Cape Cod is moving downslope. The water table drops from 48 ft at the sewage treatment facility near Otis Air Base to 32 ft about 2 mi to the south. The motion of the plumes follows that of groundwater flow, from higher to lower water-table altitudes. The highest concentration of detergents is just east of Coonamessett Pond. The detergent plume is moving seaward, and since biodegradable detergents have been used since 1964, the level of detergents near the air base is now very low.

sites—cleanup costs can amount to millions and even hundreds of millions of dollars.

The costs of cleanup are so great that the EPA does not have the funds necessary to remediate more than a few of the sites on its national priority list, let alone the many hundreds of other contaminated sites that dot the country. At a Superfund site in Indiana, for example, the owners of a company that used organic solvents, including TCE, went bankrupt in the 1970s, abandoning 98 tanks and 50,000 drums filled with chemicals. For years, many of these containers leaked chemicals into the ground. Cleanup under the supervision of the EPA began in 1988 and will cost between $25 million and $40 million over the next few decades.

Groundwater and Aquifer Restoration

Remediating contaminated groundwater systems is a booming business that involves scientists, engineers, lawyers, and policymakers. New technologies to treat myriad situations of pollution encountered at hundreds of thousands of sites across the United States are continually being developed and tested. Broadly, cleanup activities can be divided into three major categories: containment,

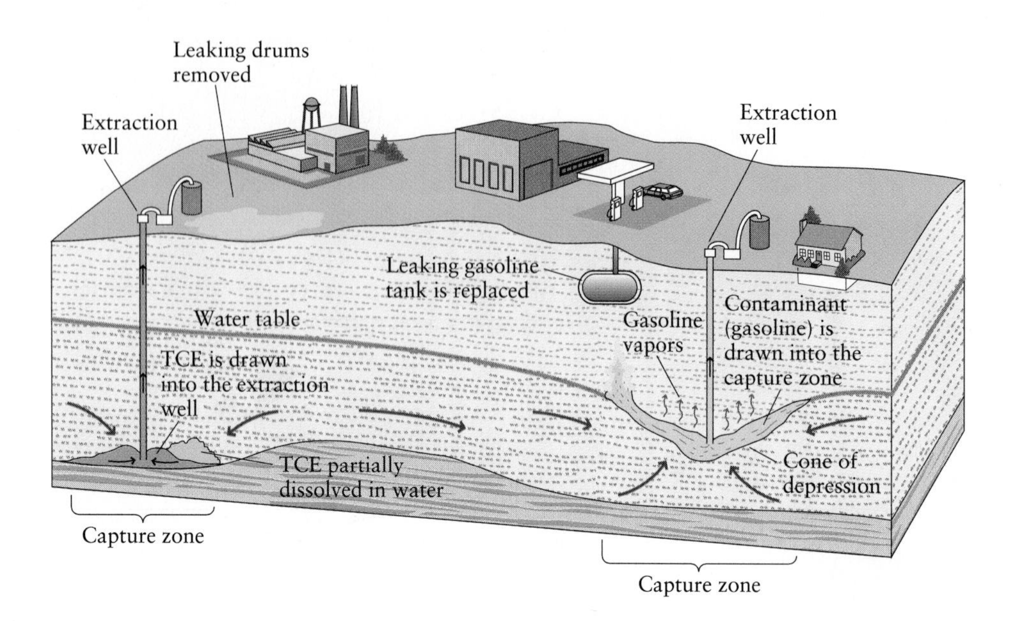

FIGURE 8-21 An extraction well can be used to remove a light contaminant plume resulting from a leaking underground storage tank and a heavy contaminant from drums. Near the extraction wells, pumping causes the groundwater and contaminants to flow toward the well, from which contaminated water can be pumped to the surface and treated.

containment withdrawal, and in situ treatment of contaminants.

Containment, or isolation, is meant to prevent contaminants from leaving the area of the source. The iron wall proposed for Cape Cod is an example of containment. In new solid-waste landfills, wastes and leachate are contained by the use of extensive liners, impermeable caps, and drainage-collection systems to contain and treat waste water at the site.

Containment withdrawal is an enhanced method of containment that uses one or more wells to pump contaminated water from the site (Figure 8-21). In general, the pumped contaminated water is treated onsite or is transported offsite to be treated. Containment withdrawal often is used at industrial sites where underground storage tanks have leaked over a period of years. First the tanks themselves are removed, then the groundwater is withdrawn for treatment. Sometimes remaining soil and sediment must also be excavated and disposed of because surface tension makes it difficult to remove all contaminants from the material.

In situ treatment cleans contaminated water in the subsurface itself without removing the water. Because most hydrocarbons can be decomposed by biological activities in the presence of microbes and oxygen, much effort has been made to develop methods of **bioremediation** (Figure 8-22), in which microbes and oxygen are added to the contaminant site and groundwater is recycled through the subsurface until decontamination is complete.

- Widening plumes of miscible contaminants form in groundwater because the contaminants are dispersed (that is, mechanically combined) with the surrounding water during transport. The amount of mixing increases with greater distance from the source.
- Light immiscible contaminants tend to float above the water table, whereas dense immiscible contaminants sink to the bottoms of aquifers.

FIGURE 8-22 Just beneath the asphalt surface where a gasoline station once stood, microbes injected into the ground through tubes consume gasoline that leaked from an underground storage tank. Bioremediation can occur naturally, but at a very slow rate: It would take 50 to 100 years to clean up a site such as this one near Philadelphia. Intervention shortens the process to less than a year. This technique has been used somewhat successfully to clean up oil field wastes left by the Persian Gulf War. Scientists are doing extensive research on identifying microbes that break down specific contaminants, and they recently identified one type that decomposes TCE. This discovery is significant for those attempting to restore aquifers.

Closing Thoughts

Although the use and contamination of groundwater have resulted in a flurry of scientific activity to understand this important resource and to find ways of restoring groundwater, there is still much to learn about what happens underground. One of the few known facts is that while some water flows in streams through underground caverns, the great majority of groundwater flows through small pores and thin fractures along circuitous pathways. Therefore, the speed at which groundwater flows is much less than that of surface water, except in underground caverns. How contaminants move in groundwater is not well understood, because not all flow at the same rate as their water host and so many different contaminants exist that full understanding of every one seems nearly impossible.

Most scientists have concluded that prevention of groundwater contamination is far more prudent than attempting to clean up after pollution has reached an aquifer. Expensive federal programs such as Superfund deal with wastes that already exist, but for the future it would be wiser to prevent contaminants from entering underground flow paths. The great attention to new technologies to prevent movement of leachate from landfills to groundwater, for example, provides encouraging evidence that progress is being made in avoiding further environmental degradation of aquifers and groundwater.

Summary

- The boundary between the unsaturated and saturated zones of unconfined underground water is a surface called the water table, which mimics Earth's topography and rises and falls in response to changing amounts of recharge and discharge.
- In uncemented, unconsolidated, well-sorted sand, porosity can be as high as 44 percent. In lithified sedimentary rocks and crystalline (metamorphic and igneous) rocks, it generally is much lower (less than 10 percent) unless the rocks are fractured.
- Rocks and sediments that are both porous and permeable make good aquifers, whereas those that are neither porous nor permeable act as confining layers.
- Groundwater contains salts because it flows slowly around minerals, dissolving them and picking up ions in the process. With depth below Earth's surface, groundwater generally becomes increasingly saline.
- Environmental hazards associated with groundwater include sinkholes, land subsidence, groundwater mining, saline intrusion in coastal areas, and groundwater pollution.
- Major sources of groundwater pollution are leaking underground storage tanks and septic tanks, leachate from landfills and unlined waste pits, pesticides and fertilizers from agricultural fields, and accidental spills from trucks and trains.
- The flow of contaminants in groundwater depends on whether they are miscible (soluble) or immiscible (insoluble) in water, and if they are immiscible, on their density relative to groundwater.
- One of the most important laws passed for remediation of polluted groundwater is the Comprehensive Environmental Response, Compensation, and Liability Act and its amendments, commonly referred to as Superfund.
- The three main ways in which contaminated groundwater is cleaned are containment, containment withdrawal, and in situ treatment of contaminants using bioremediation.

Key Terms

water table (p. 234)
groundwater (p. 234)
porosity (p. 235)
specific yield (p. 237)
specific retention (p. 237)
permeability (p. 237)
aquifer (p. 238)
unconfined aquifer (p. 239)
hydraulic gradient (p. 239)
confined aquifer (p. 239)
artesian spring (p. 239)
artesian well (p. 239)
Darcy's law (p. 239)
cone of depression (p. 240)
water hardness (p. 243)
water spreading (p. 247)
recharge basin (p. 247)
injection well (p. 248)
land subsidence (p. 248)
saline intrusion (p. 249)
miscible contaminant (p. 252)
immiscible contaminant (p. 252)
bioremediation (p. 256)

Review Questions

1. List examples of sources of recharge and discharge for groundwater.
2. What are typical values of porosity for sand, sandstone, and fractured limestone?
3. What is Darcy's law? How is it related to the flow of pollutants in groundwater?
4. Would a pollutant migrate more quickly through a well-sorted, coarse, unconsolidated sand or a poorly sorted, fine-grained unconsolidated sand? Why?
5. Why do some springs discharge so much more water than others?
6. How can nitrates from farm areas affect groundwater?
7. How was the groundwater that comes out of geysers at Yellowstone converted to steam below ground?
8. What does "water hardness" mean? Why is some water harder than other water?
9. List common ways that pollutants get into groundwater.
10. What is leachate?
11. In choosing sites for underground disposal of nuclear waste, why is the permeability of the rock in which the waste will be stored a concern?
12. How does dispersion contribute to the formation of contaminant plumes in groundwater?
13. What happens to gasoline that is spilled on the ground or leaks from a tank to the water table?
14. What happens to salt spread on roads that seeps to the water table?
15. How fast is the sewage plume at Cape Cod traveling?
16. What are the contaminants in the Cape Cod plume?
17. Explain how Superfund is dealing with the problem of groundwater pollution.

Thought Questions

1. Which material is likely to have the greater permeability, unconsolidated sand or sandstone? Why?
2. Which of the following is likely to be the most efficient aquifer at yielding water to a well: unconsolidated sandstone, sandstone with cement filling all pore spaces, or crystalline granite with no fractures? Why?
3. Is groundwater in most cases a renewable or a nonrenewable resource? If prolonged drought occurs in a region, why might local groundwater change from a renewable to a nonrenewable resource?
4. What is a cone of depression? If your neighbor installs a well that is deeper than yours and pumps large volumes of water each day, might your well be affected? (Assume that both of you have wells in the same unconfined aquifer.) Why?
5. Suppose you are a hydrogeologist called in to investigate a gasoline smell in someone's house. Gas company personnel already have determined that there is no leak from their equipment to the house. What possibilities would you suspect? How would you investigate them?

Exercises

1. If you live in a humid region, you can learn something about the depth to the water table by examining topographic maps. Points along perennially flowing streams, lakes, and springs are outcrops of the water table surface and their elevations are equal to the hydraulic head of the water table. By contouring the values of hydraulic head (discussed in Box 8-1), you can construct a map of the shape of the water table surface, and from that you can determine how deep it is where you live.
2. You can find out how deep groundwater is in your area, and how much has been yielded from wells drilled there, from state officials. Well-drillers are required by law to file information regarding each well they drill with a state environmental agency. This information, available to the public, records the depth of each well, the depth to groundwater, the nature of the aquifer and its porosity, and the well yield.

Suggested Readings

Fetter, C. W., 1994. *Applied Hydrogeology.* 3rd ed. New York: Prentice-Hall.

Laws, Edward A., 1993. *Aquatic Pollution.* New York: John Wiley.

Heath, C., 1983. *Basic Ground-Water Hydrology.* U.S.G.S. Water-Supply Paper 2220. Washington, DC: U.S. Government Printing Office.

U.S. Geological Survey, 1988. *National Water Summary (1986 to Present): Hydrologic Events and Groundwater Quality*. U.S.G.S. Water-Supply Paper 2325. Washington, DC: U.S. Government Printing Office.

U.S. Water News, a monthly news magazine copublished by U. S. Water News, Inc., and the Freshwater Foundation.

CHAPTER 9

Winter Landscape, painted in 1601 by Pieter Brueghel the Younger, depicts a village in the Netherlands during the Little Ice Age (1450–1890).

The Atmospheric System

Paintings by the 17th century Dutch artist Pieter Brueghel depict a very different Netherlands than exists today. During his lifetime, the world was experiencing a cold snap that began around 1450 and lasted until 1890. In winter, the Dutch canals froze over so solidly that people could skate around town. Elsewhere in Europe the Baltic Sea froze, and glaciers expanded in the Alps. In North America, George Washington crossed the ice-filled Delaware River with his soldiers, mounting a surprise attack that scored a decisive victory in the American Revolutionary War; the Delaware often froze over in Washington's time. This period of global cooling has come to be known as the "Little Ice Age." Hypotheses advanced to explain it include a period of frequent volcanic eruptions that emitted Sun-blocking sulfur dioxide and ash, and a period of decreased solar radiation output. Because glacial accumulation is a destabilizing, positive-feedback process, some geologists suggest that the Little Ice Age might have eventually plunged the planet into full ice-age conditions had it not been reversed by the massive outpouring of carbon dioxide and other greenhouse gases into the atmosphere by the Industrial Revolution.

In this chapter, we will:

- Investigate the impact of the atmosphere's composition and structure on climate.
- Examine the links between atmospheric circulation, regional distribution of moisture, and storms.
- Discuss the role of human-induced atmospheric pollution in climatic and environmental change.

Storyteller Garrison Keillor in his radio show "A Prairie Home Companion" often quips that winter is nature's annual attempt to kill off the inhabitants of Minnesota. Indeed, Minnesotans and their neighbors in the Great Lakes states regularly experience harsh winters with subfreezing temperatures and snow measurable in feet. So in early January 1996, Minnesotans were relieved when nature decided to attend instead to the East Coast. However, even they were surprised by events in eastern states that recalled the Little Ice Age in Europe a century ago. A massive blizzard dumped approximately 50 trillion pounds of snow up and down the Atlantic seaboard, burying many of the nation's largest cities and shattering many historical records (Table 9-1). New York, Philadelphia, Boston, Baltimore, Washington, D.C., and many other cities and towns were immobilized for a week or more as state and local agencies struggled to remove the snowdrifts—some as high as 20 feet (6 m)—that filled every road. Snow removal was particularly difficult in these cities because they are all highly urbanized (Figure 9-1); many houses and businesses lack front yards, and so there was no place to put the snow, which instead had to be loaded into dump trucks and hauled to surrounding fields or thrown into rivers. Many cities used their entire snow-removal budgets for the year in cleaning up the aftermath of this one storm.

The roads were filled with impassable drifts. Travel, including transport of emergency workers, was impossible for several days. Some houses that caught fire burned to the ground because firefighters were unable to reach them. A few babies were born without medical attention and a few heart attack victims died because medical personnel could not get through the snow. Economically, the blizzard was a disaster as well. Stores selling snow-removal equipment experienced a small economic boom, but other retailers lost thousands of dollars as people stayed at home, forgoing shopping, restaurants, and the movies. A dozen or more smaller snowstorms piled up considerable additional amounts of snow. As spring arrived, the massive amounts of water discharging from the melting snow caused flooding in many places. According to the National Climate Data Center, the winter of 1996

TABLE 9-1 Snowfall Summary for the Northeast

		Record Snowfall	
Location	January 1996	Amount	Date
Philadelphia, Pennsylvania	30.7	21.3	February 11–12, 1983
Newark, New Jersey	27.8	22.6	February 3–4, 1961
Washington, D.C.	24.6	22.8	February 10–11, 1983
Providence, Rhode Island	24.0	28.6	February 7–8, 1978
Elkins, West Virginia	23.4	20.7	November 24–25, 1950
Baltimore, Maryland	22.1	22.8	February 11, 1983
Wilkes-Barre/Scranton, Pennsylvania	21.0	21.4	March 13–14, 1993
Charleston, West Virginia	20.5	18.9	March 13–14, 1993
New York, New York	20.2	26.4	December 26–27, 1947
Boston, Massachusetts	18.2	27.1	February 6–7, 1978
Hartford, Connecticut	18.2	21.0	February 11–12, 1983
Bridgeport, Connecticut	15.0	17.0	December 19–20, 1948
Portland, Maine	10.2	27.1	January 17–18, 1979
Pittsburgh, Pennsylvania	9.6	24.6	March 12–13, 1993
Williamsport, Pennsylvania	8.6	24.1	January 12–13, 1964

Note: All values are in inches; U.S. hydrological data are compiled in English units.

Source: Northeast Regional Climate Center at Cornell University.

FIGURE 9-1 Houses in New York City were blanketed by nearly 3 ft (1m) of snow during the East Coast blizzard of 1996.

cost the East Coast 137 lives and $3 billion in lost commerce, unbudgeted expenditures for snow removal, and ensuing floods.

The following winter devastated the western part of the United States. In December 1996 and January 1997, storm after storm hit the West Coast and traveled inland into Idaho and Nevada, dropping rain or snow, depending on the temperatures they encountered. When the ground became saturated and could hold no more water, flooding and mudslides ensued. For weeks, rivers crested above flood stage, forcing tens of thousands of people to evacuate their homes and causing more than 20 deaths and $2 billion to $3 billion worth of property damage. More than 15,000 acres of prime farmland in California's Central Valley were either damaged or destroyed.

Both the eastern snowstorms of 1996 and the western storms of 1997 were caused by atmospheric events that occur infrequently in the span of an average human life but are not at all uncommon in the framework of geologic time. Still, some scientists wonder if the frequency of unusual weather events might be increasing because of human activities. For the last century, people have pumped carbon dioxide, methane, chlorofluorocarbons, and other greenhouse gases into the atmosphere through a variety of activities such as manufacturing, the use of motor vehicles, and cattle ranching (cattle release significant quantities of methane). The impact on the atmosphere is not yet fully understood, although scientists now agree that these gases are causing Earth's average temperature to rise. Computerized climate models predict that the increase in temperature will lead to a much more active hydrologic cycle, with greater storminess in some locations and droughts in others. Whether global warming could also lead to a greater frequency of blizzards is, as yet, an unanswerable question. Nevertheless, the magnitude of the economic and personal devastation caused by severe weather necessitates an understanding of the atmosphere and of the human role in atmospheric change.

The Atmosphere: An Envelope of Gases

An **atmosphere** is an envelope of gases that surrounds a planet. Earth's atmosphere extends from less than a meter below the planet's surface, where gases emanating from Earth's interior penetrate openings in the lithosphere, to more than 10,000 km (6000 mi) beyond Earth's surface, where gases gradually thin and become indistinguishable from the composition of interstellar space (Figure 9-2).

Earth's atmosphere formed early in the planet's history. Light elements such as hydrogen and helium were released during Earth's global meltdown about 4 billion to 4.5 billion years ago. Because of the gravitational attraction of Earth's mass, some of the gases did not escape to outer space but remained near the surface to form an atmosphere (see Chapter 2). In contrast, the Moon and Mercury, with less mass and lower gravitational attraction, were unable to hold onto gases and so have no atmosphere. After Earth solidified, volcanic eruptions continued to add water vapor, nitrogen, carbon dioxide, and other gases to the atmosphere. The evolution of life also strongly influenced

FIGURE 9-2 The atmosphere seen from the space shuttle *Atlantis*. The layer at the base of the atmosphere contains ash and sulfuric acid droplets, which scatter red light. These particles, photographed in August 1992, resulted from the June 1991 eruption of Mount Pinatubo (see Chapter 2). The blue layer above results from the scattering of blue light by atmospheric gases.

atmospheric composition, most notably by adding oxygen. Indeed, the atmosphere owes its present composition primarily to volcanic eruptions and to the evolution of organisms such as plants (Box 9-1).

Present Atmospheric Composition

By volume today's atmosphere is about 78 percent nitrogen, 21 percent oxygen, 0.9 percent argon, and 0.04 percent carbon dioxide, with neon, helium, nitrous oxide, methane, ozone, and other gases in trace amounts (Figure 9-3). The percentage of water vapor varies with atmospheric conditions, from 0.3 percent in cold, dry air to nearly 4 percent in hot, humid air.

The small amounts of carbon dioxide, water vapor, and methane in the atmosphere belie their importance. These greenhouse gases are critical for regulating Earth's temperature and keeping it within a livable range. Without them, Earth would have an average surface temperature of –22°C, too cold for life to have originated and evolved. Water vapor is particularly important because it, along with ocean currents, is responsible for transporting heat from the equator to higher latitudes. Evaporation of water from ocean basins occurs when water absorbs solar energy and changes phase from a liquid to a gas. The energy absorbed, known as *latent heat energy,* is stored in the water vapor and released when the vapor condenses back to water droplets, forming clouds. Energy that water picks up near the equator is released when the vapor reaches higher latitudes and condenses, making the higher latitudes warmer than they otherwise would be. Likewise, removal of heat from the tropics keeps them from becoming unlivably warm. Evaporation and condensation also bring vital precipitation to land and terrestrial organisms.

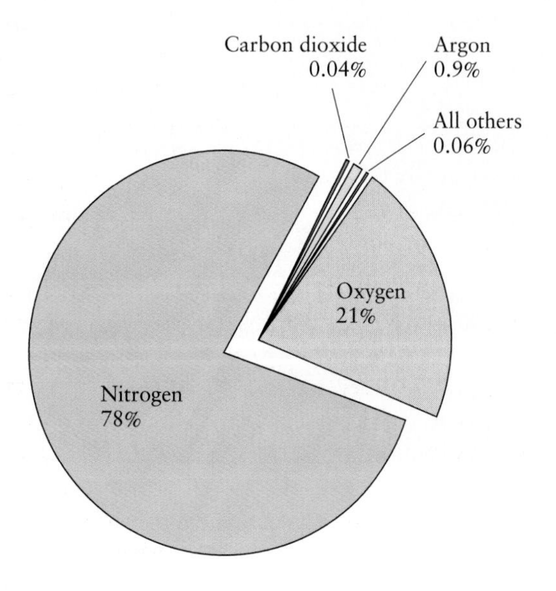

FIGURE 9-3 The composition of the atmosphere by volume. Water vapor is not listed because the amount in the atmosphere is extremely variable.

Ozone also is found only in trace amounts in the atmosphere but is critical to keeping Earth habitable. Ozone at high altitudes absorbs much of the incoming ultraviolet (UV) radiation from the Sun, the same radiation responsible for sunburns, genetic mutations, and skin cancer. Without ozone, incoming UV rays would destroy all life on land, and many geologists believe that land on early Earth remained lifeless until sufficient ozone accumulated in the atmosphere to block the harmful radiation.

The lower parts of the atmosphere also contain some suspended liquid and solid particles, such as acid droplets and dust, which are referred to as **aerosols.** After volcanic eruptions, the concentration of aerosols in the atmosphere increases, but diminishes with time as the particles settle back to the surface. Atmospheric aerosols play an important role in the hydrologic cycle because they provide a surface on which water vapor can condense to water droplets to make clouds. However, some aerosols, introduced into the atmosphere both by natural processes and human activities, are also pollutants responsible for the production of acid rain. (The human impacts on atmospheric chemistry will be discussed later in the chapter.)

Atmospheric Structure

Like Earth's interior, the atmosphere system is organized into layers. These layers have different temperatures (Figure 9-4) because each layer contains a characteristic distribution of gases with varying capabilities to absorb solar energy.

The Troposphere The lower layer of the atmosphere, the **troposphere,** extends from Earth's surface to heights between 10 and 18 km, depending where it is measured. On average, Earth's surface is warm and provides heat to the bottom of the troposphere, but the temperature drops to between –55° and –80°C at the troposphere's upper boundary, referred to as the *tropopause*. Consequently, temperature declines with elevation at a rate of about 7°C per km. You have probably experienced this decrease in temperature with elevation if you have ever driven upward into a mountain range. Some of the temperature decline with elevation also results from the decline of atmospheric pressure (force per unit area) with altitude. As air moves to high altitudes, the atmospheric pressure exerted on it declines, allowing it to expand. To expand, however, the gas molecules in the air must expend energy, resulting in a cooling of the air.

The amount of water vapor that air can hold decreases with decreasing temperature. Therefore, the extremely cold temperatures at the top of the troposphere cause the volume of water vapor there to approach zero and keep water from entering higher levels in the atmosphere.

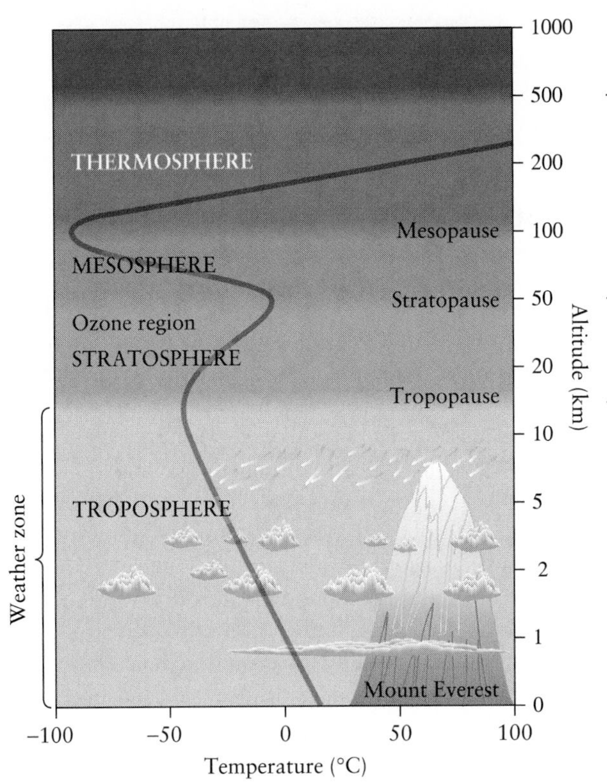

Nearly a vacuum; scarce H, He

Oxygen molecules absorb short-wave solar radiation, resulting in heating

Ozone production zone; UV radiation absorption increases temperature; horizontal mixing spreads pollutant particles

Vertical mixing of air and water vapor results in weather patterns: storms, winds, clouds

Precipitation removes some pollutant particles

CO_2, H_2O vapor, and methane regulate Earth's temperature

FIGURE 9-4 Temperature in the troposphere declines with altitude (which here is measured logarithmically) because the troposphere is heated from below by Earth's surface. Concentration of ozone (O_3) in the stratosphere allows temperatures to rise, but in the mesosphere they fall again. Molecular oxygen (O_2) in the thermosphere absorbs heat energy, so temperature rises with altitude. Clouds depict the location of water in the atmosphere.

The troposphere is marked by strong mixing because of declining density (mass per unit volume) and temperature with elevation. Density at ground level, where the air is compressed by the weight of the overlying atmosphere, is much greater than density at high levels in the atmosphere, which have less mass overlying them. However, when a thin layer of air is warmed near Earth's surface, its density decreases markedly, and so it rises until it reaches a level in the atmosphere with the same low density. If air rises in one place, it must descend in another, leading to vertical and horizontal mixing. The mixing of masses of air and moisture in the troposphere is manifested at Earth's surface as changing weather patterns, storms, clouds, and wind. Because air cools when it rises, its water vapor condenses around dust particles and other aerosols. If these particles fall to the ground in precipitation, the atmosphere is effectively cleaned. This process allows many natural and human-produced compounds to be removed from the troposphere and is one of the key processes involved in the production of acid rain.

The Stratosphere Above the troposphere, at about 10 to 50 km, lies the **stratosphere,** a prominent zone of ozone (O_3) production. Here ultraviolet rays from the Sun break O_2 molecules apart and recombine the O atoms with other O_2 molecules to form O_3. Because ozone is a strong absorber of UV radiation and UV rays enter from the top of the atmosphere, the temperature of the stratosphere increases with altitude. Cooler, denser air cannot rise above warmer, thinner air, so no appreciable vertical mixing occurs in the stratosphere. The stratosphere also contains about 1000 times less water vapor than does the troposphere, and this, combined with the lack of vertical mixing, allows volcanic ash and human-made pollutants to remain in the stratosphere for many years once introduced. Horizontal mixing still takes place, however, allowing these pollutants to be distributed over the globe. In fact, levels of stratospheric ozone are declining over Antarctica because of human-made pollutants produced by industrialized nations far from that continent.

The Upper Atmospheric Layers About 99 percent of the total mass of the atmosphere is found in the troposphere and stratosphere, and all atmospheric circulation occurs below the *mesosphere,* a second layer of mixing that exists at altitudes of about 50 to 85 km above Earth. Above the mesosphere, oxygen molecules (O_2) absorb short-wave solar radiation (ultraviolet light), resulting in heating and increased temperature to about 160 km (96 mi) in a zone known as the *thermosphere.* The mesosphere and thermosphere, while themselves interesting, do not have practical environmental importance. Beyond the thermosphere are layers primarily of helium and hydrogen, the lightest elements in the universe.

9-1 Global and Environmental Change

The Intertwined Histories of Atmospheric Chemistry and Life

Information on the early composition and evolution of Earth's atmosphere is scant, but it is widely acknowledged that initially there was little free oxygen. Early life was adapted to a world without atmospheric oxygen and developed processes to extract oxygen from minerals such as iron sulfate ($FeSO_4$). By 3.5 billion to 3.8 billion years ago, the first photosynthesizing cells, a type of bacteria, had appeared. A by-product of photosynthesis is oxygen, and the oxygen produced by these bacteria began to accumulate because there were few organisms around to use it. By 2 billion (2000 million) years ago, the atmosphere contained about 1 percent as much free oxygen as it does today, and by 700 million years ago oxygen was at 10 percent of its present level. This was enough to enable many invertebrate species to evolve, including soft-bodied sponges and worms and hard-shelled mollusks. At the same time, sufficient oxygen had been produced to allow widespread formation of stratospheric ozone (O_3), a gas that protects Earth's surface from ultraviolet radiation. Such protection permitted life to exist near ocean surfaces and ultimately on land. About 400 million years ago, land plants evolved and diversified. As a result, atmospheric oxygen levels soared, and by 200 million years ago the concentration reached about the value it has today. While the composition of the early atmosphere no doubt had a profound influence on the origins and evolution of life, the biosphere has also had a strong impact on the composition of the atmosphere.

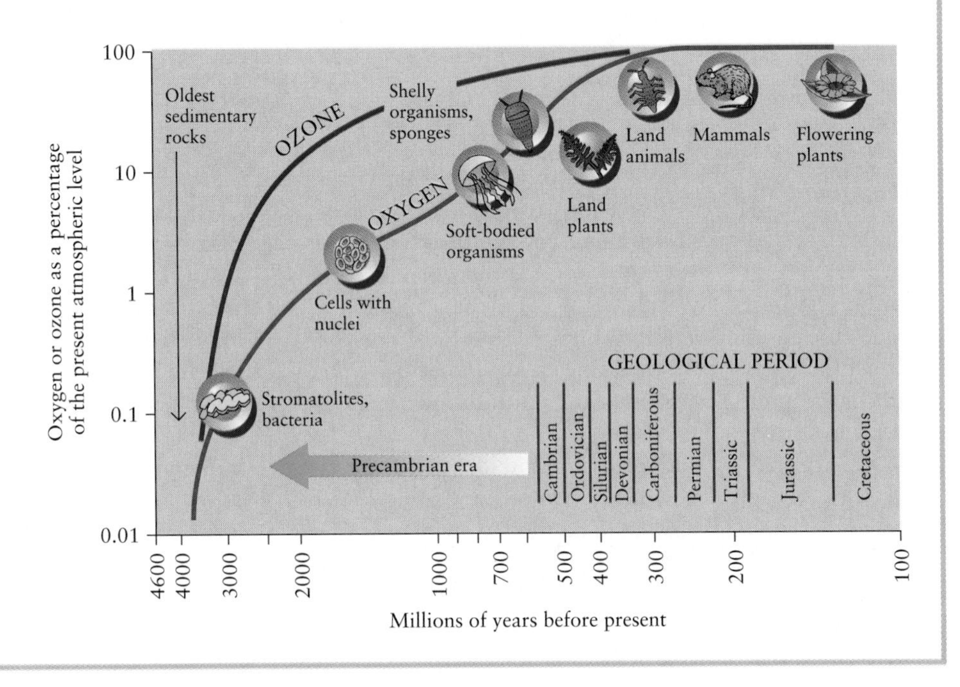

The evolution of atmospheric oxygen and ozone was driven by the appearance of photosynthesizing cyanobacteria. These bacteria, which left a record of their existence in stromatolites, provided oxygen to the atmosphere early in Earth's history. Eventually, the accumulating ozone blocked enough ultraviolet radiation so that plants could live on land, allowing atmospheric oxygen and ozone levels to rise dramatically to near their present values.

The upper atmosphere traps heat and protects us from excess radiation, the lower atmosphere determines our weather.

- Trace gases in the atmosphere, such as carbon dioxide, ozone, and water vapor, are responsible for keeping Earth warm enough to sustain life and for protecting us from UV radiation.
- Vertical mixing in the troposphere produces winds, storms, and clouds.
- Layering of the atmosphere inhibits the vertical transport of material between the troposphere and stratosphere.
- The concentration of water vapor in the atmosphere declines rapidly with elevation and almost all water is confined to the troposphere.
- Pollutants found in the troposphere are easily removed by raindrops, while stratospheric pollutants may linger for longer periods of time.

Climate and Weather

Circulation of the troposphere produces climate and weather. **Climate** refers to the long-term atmospheric and surface conditions that characterize a particular region. **Weather** refers to daily fluctuations in temperature, wind speed, and precipitation. We can think of climate as the long-term average of all daily weather events.

Earth's climate system determines where humans can live for long periods of time. The climate of Antarctica, for example, with its frigid temperatures and high wind speeds, is extremely inhospitable to humans, and farming on ice sheets is out of the question. Although many scientists live on Antarctica for months at a time, they are completely dependent on supplies of food and fuel from lower latitudes.

Weather events may be severe, but they do not in and of themselves restrict human habitation. The East Coast blizzard of 1996, for instance, did not force people to move away from the area inundated by snow because a storm that severe is a singular event from which the population readily recovers. If, however, a change in climate caused blizzards to become constantly recurring events in the northeastern United States, residents would probably migrate to more southerly latitudes, which are not subject to such storms.

The Greenhouse Effect

Every year, Earth's surface receives 1.73×10^{17} W of energy from the Sun. This is about 15,000 times as much energy as the entire human population uses annually. Energy emitted by the Sun consists primarily of ultraviolet radiation, visible light, and infrared radiation—that is, heat (Box 9-2).

As Earth's surface absorbs solar energy, it heats and then emits heat energy as infrared radiation. All this heat would escape into space if Earth had no atmosphere. Instead, some heat energy is trapped by atmospheric gases. Nitrogen and oxygen, the primary constituents of the atmosphere, absorb very little infrared energy. Carbon dioxide (CO_2), methane (CH_4), water vapor, and nitrous oxide (NO_2) are present in only trace amounts but are very efficient at absorbing this energy and redirecting it, some of it to outer space and some to Earth's surface.

This warming process is called the **greenhouse effect** by partial analogy to the way greenhouses maintain heat. In a greenhouse, and at Earth's surface, incoming solar energy is absorbed by the plants, the ground, and other objects. These objects then reemit the energy as infrared radiation, which heats the air. The greenhouse glass traps the warm air, which circulates within the confines of the glass. Although the glass reflects a small portion of the outgoing infrared radiation to the interior of the greenhouse, air retention accounts for most of the greenhouse's heat. In contrast, warm air at Earth's surface rises and loses heat. Much of this heat is absorbed by atmospheric gases, which then reradiate the heat back to the surface (Figure 9-5). In the absence of this so-called greenhouse effect, Earth's average surface temperature would be about 35°C colder than it is today, and the entire planet would be frozen.

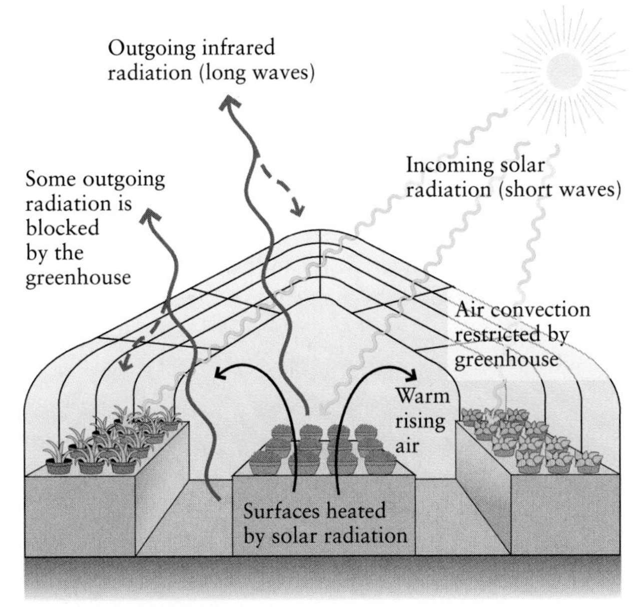

(a) A greenhouse stays warm by allowing in radiation and trapping heated air

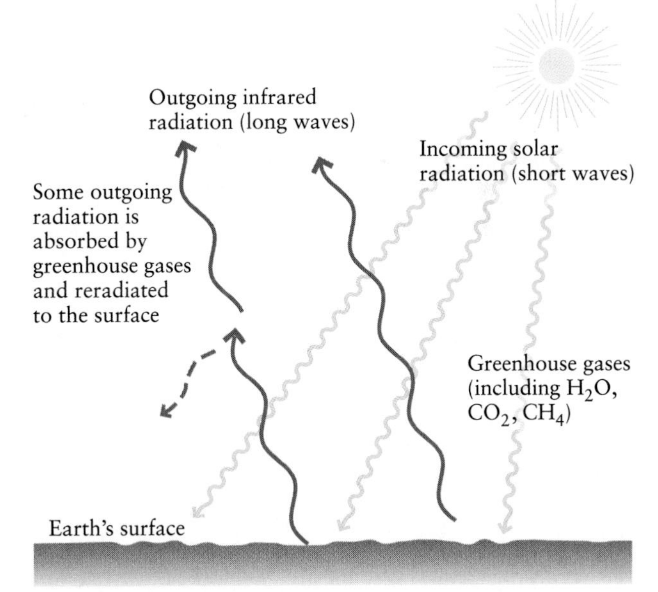

(b) Earth's atmosphere stays warm because greenhouse gases absorb outgoing infrared radiation from Earth's surface and reradiate it

FIGURE 9-5 (a) In a greenhouse, sunlight passes through the glass and is absorbed by plants and the ground within. The ground and plants then emit the energy as infrared radiation that heats the greenhouse air. The glass holds in all the heated air and a bit of the escaping radiation. (b) In the atmosphere, greenhouse gases absorb some of the outgoing radiation given off by Earth and emit part of the energy back toward Earth's surface, thus keeping the planet warm.

9-2 Geologist's Toolbox

Electromagnetic Radiation

All bodies in the universe give off *electromagnetic radiation,* energy transmitted in the form of waves. The wavelength—the distance from the crest of one wave to the crest of the next—of the radiation given off is a function of the temperature of the body; bodies at high temperature radiate at short wavelengths, colder objects radiate at longer wavelengths. The range of wavelengths is the *electromagnetic spectrum,* on which radio waves are the longest and gamma rays are the shortest. The shorter the wavelength, the greater its energy. Much of the Sun's radiation, with wavelengths ranging from 0.4 to 0.7 μm, is in the visible part of the spectrum. Earth, because it is much cooler than the Sun, radiates at a much longer wavelength, in the infrared part of the spectrum. We cannot see infrared radiation, but we can feel it as heat.

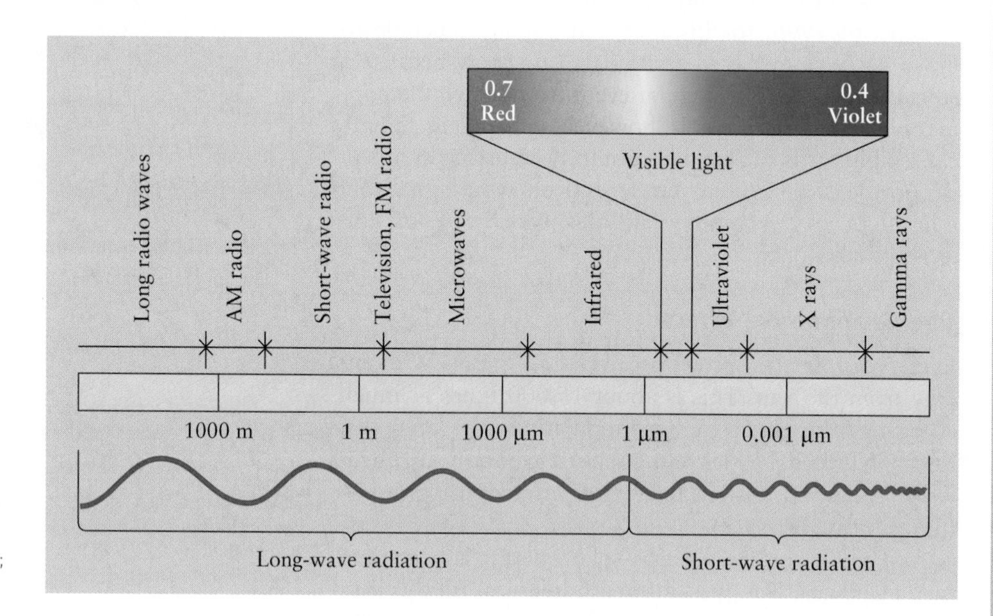

The electromagnetic spectrum. The Sun radiates energy primarily in the visible, infrared, and ultraviolet parts of the spectrum; Earth radiates in the infrared.

Differential Heating of Earth

If Earth were a flat surface oriented perpendicularly to the incoming solar rays, its area would be heated uniformly. Because Earth is a sphere, however, the energy is not evenly distributed (Figure 9-6). Polar locations are farther from the Sun than are equatorial locations; hence polar areas receive less energy than does the equator. However, this difference in distance to the Sun is practically negligible. Of far greater importance is the thickness of the atmosphere through which incoming radiation travels to reach Earth's surface. Atmospheric gases and clouds reflect a lot of the incoming radiation directly back to space. More energy is reflected by the atmosphere at high latitudes, because energy arriving at high latitudes takes a longer route through the atmosphere before reaching the ground than does energy hitting the low latitudes. Finally, the angle of incoming radiation at the poles is lower than at the equator. This angle determines the area over which the incoming radiation is spread. The incoming radiation hitting near the poles is spread over a much greater area than the same amount of radiation hitting near the equator. Because the energy is spread thinner at the poles, they are colder.

Not only does Earth's spherical shape influence how much solar energy it receives during the year, so too does the tilt of Earth's axis of rotation. Summer occurs in a hemisphere when it is tilted toward the Sun, winter when

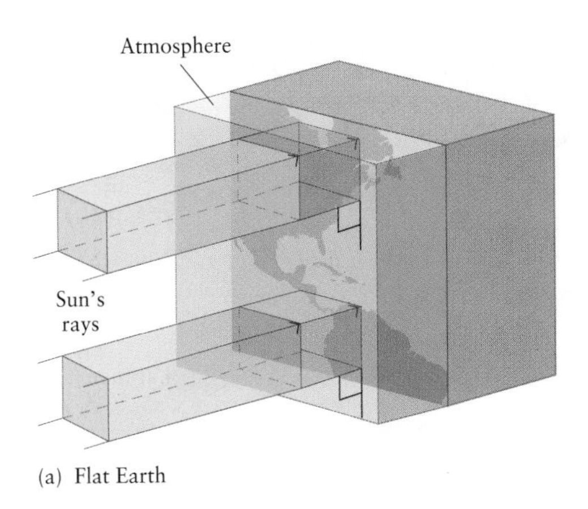

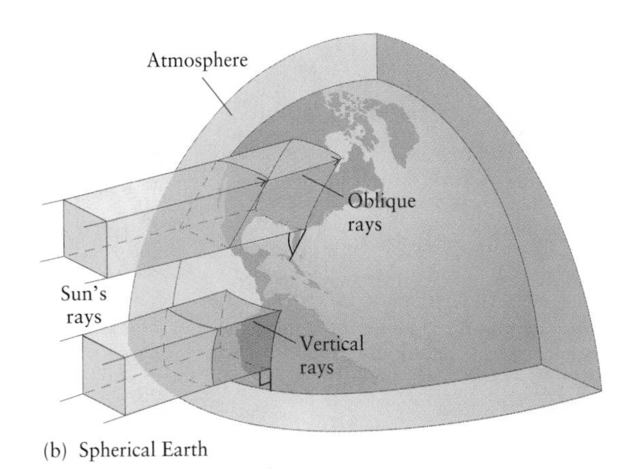

FIGURE 9-6 Radiation striking (a) a flat area is distributed evenly over the surface; radiation hitting (b) a sphere, however, is not. At high latitudes the route taken through the atmosphere by the radiation is longer than at low latitudes, causing more energy to be reflected to outer space. In addition, solar rays hitting higher latitudes make a more oblique angle with the surface and spread out over a much greater area than those at lower latitudes. For these reasons, less energy is absorbed at high latitudes than at low latitudes.

it is pointed away from the Sun (Figure 9-7). This explains why one hemisphere experiences winter while the other experiences summer. The tilted axis also explains why the polar latitudes spend several months each year in total darkness. In contrast, the equatorial regions are illuminated year round, another reason the poles are relatively colder than the equator.

Averaged over Earth's entire surface, incoming solar radiation is balanced by outgoing planetary radiation—heat energy from the radioactive decay of elements in the interior as well as from solar energy absorbed by the surface and reradiated to outer space. Without this balance, Earth would not have a steady temperature but would either heat up or cool down. Despite this overall energy balance, however, individual latitudes of Earth are not in balance. The poles lose more energy than they receive, whereas the equator receives more energy than it loses. Only at about 40° latitude north and south (the latitude of Denver, Philadelphia, Madrid, and Beijing in the northern hemisphere and Wellington, New Zealand, in the

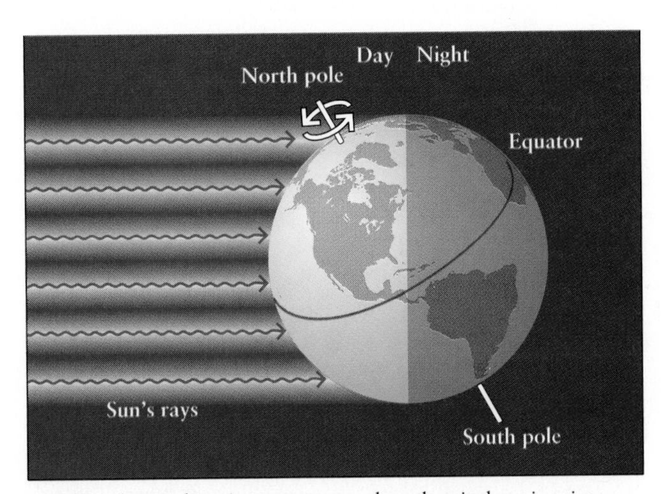

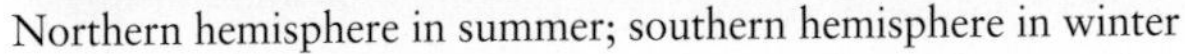
Northern hemisphere in summer; southern hemisphere in winter

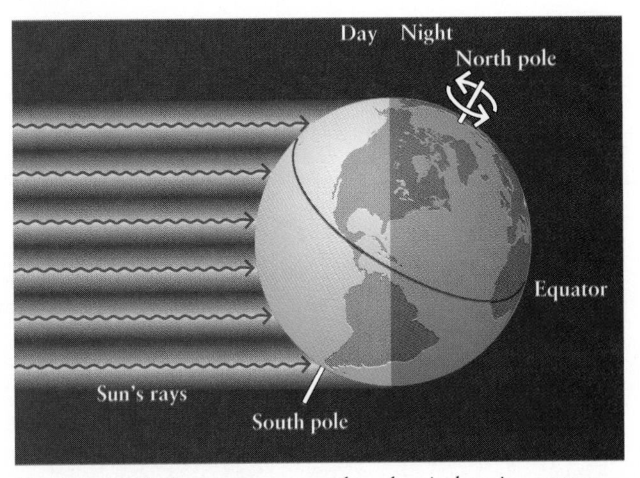

Northern hemisphere in winter; southern hemisphere in summer

FIGURE 9-7 Earth's axis of rotation is tilted. Summer occurs when a hemisphere is tilted toward the Sun. Winter occurs when a hemisphere is tilted away from the Sun. The poles are in complete darkness during winter, while the equator is illuminated year round.

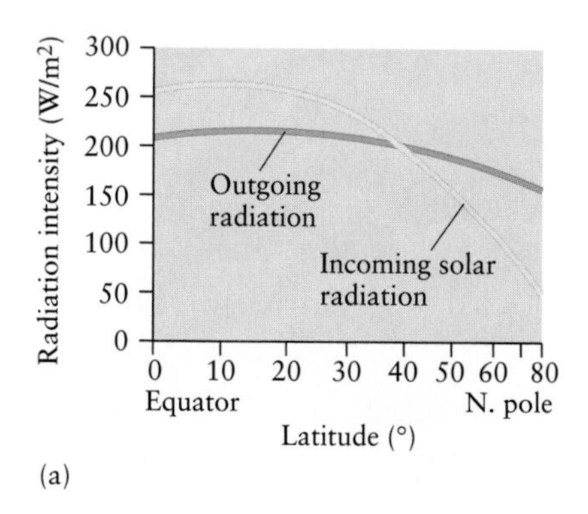

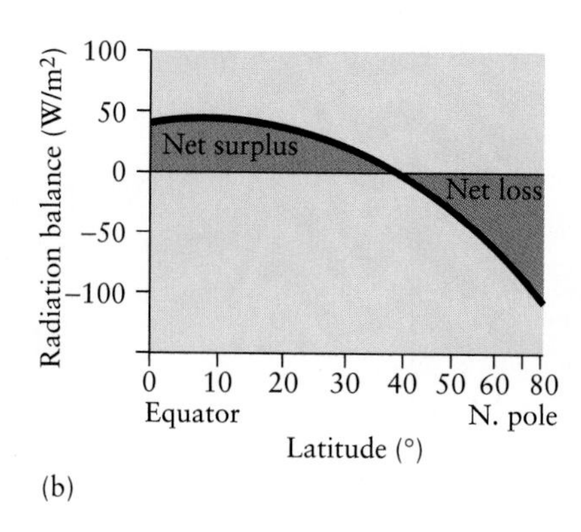

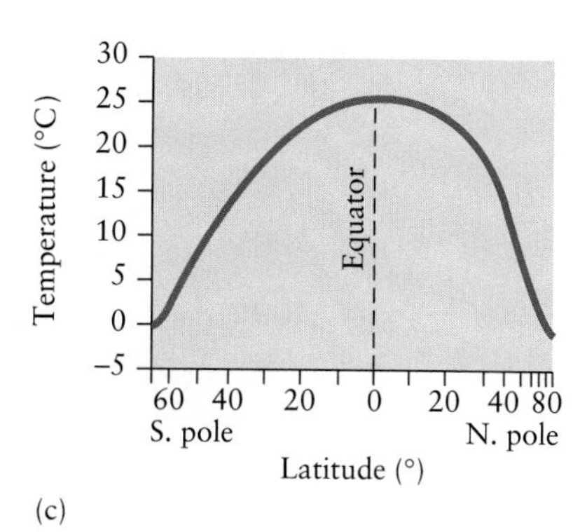

FIGURE 9-8 (a) Incoming and outgoing radiation varies with latitude. (b) The poles lose more energy than they receive, whereas the equatorial regions gain more energy than they lose. Energy gain and loss are equal at about 40° latitude north and south. (c) Temperature varies with latitude. The mean annual temperature of the polar regions is far lower than that of the equatorial region.

southern hemisphere) does the amount of solar energy gained exactly balance the amount of planetary energy given off (Figure 9-8).

If the poles did not receive an inflow of energy from lower latitudes, they would become colder every year. Likewise, because the equator receives more solar energy than it loses to space, it would continuously heat up if there were no way of removing heat energy. These effects do not occur, because the atmosphere and oceans redistribute heat energy through the winds and ocean currents, carrying warm air and water from the equator to the high latitudes and colder air and water from the poles to the equator, and making all parts of Earth more tolerable for life than they would be in the absence of circulation.

Tropospheric Circulation

The mechanism of energy transfer between the equator and the poles is *convection*, the same process you see on your stovetop when you boil a pot of water, and the same

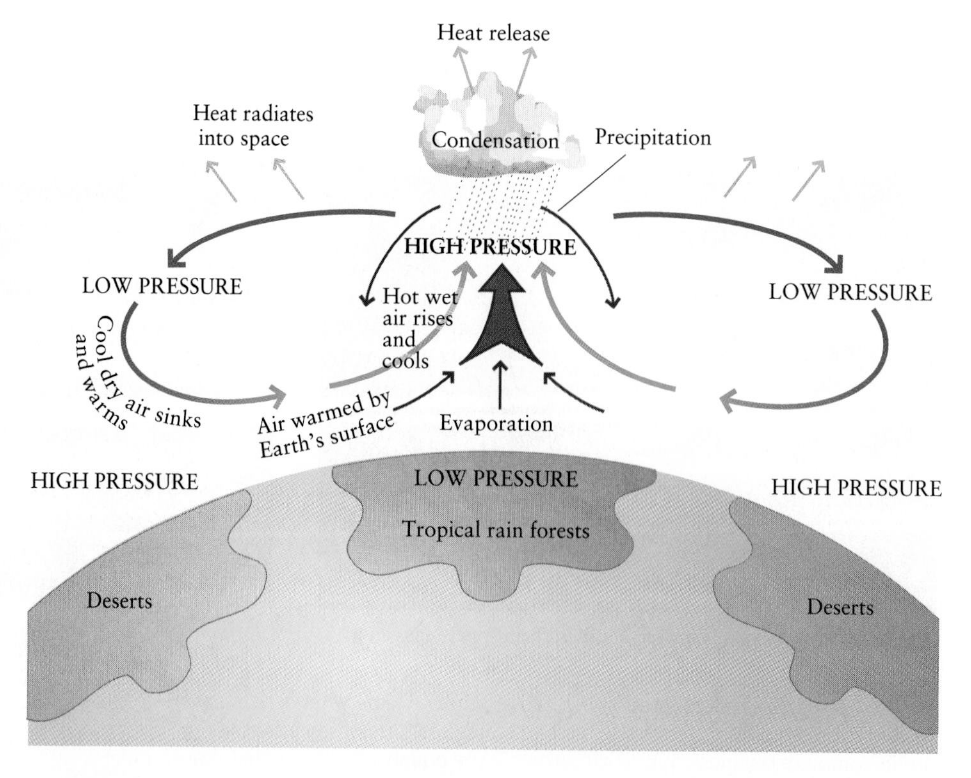

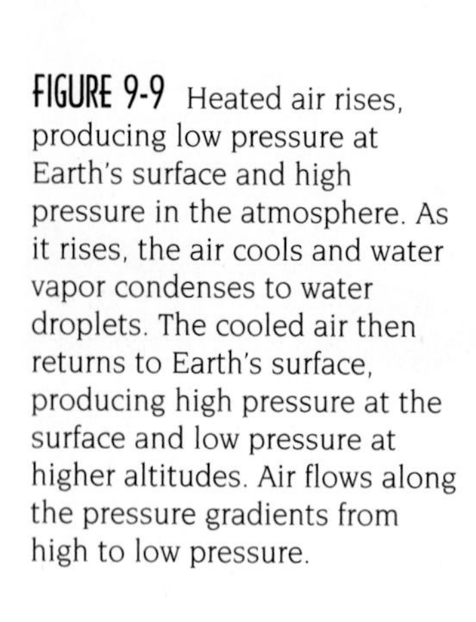

FIGURE 9-9 Heated air rises, producing low pressure at Earth's surface and high pressure in the atmosphere. As it rises, the air cools and water vapor condenses to water droplets. The cooled air then returns to Earth's surface, producing high pressure at the surface and low pressure at higher altitudes. Air flows along the pressure gradients from high to low pressure.

process by which Earth's interior materials circulate and move the lithospheric plates. Convection in the troposphere occurs because the equatorial regions experience a net gain of solar energy every year, heating the air so that it becomes less dense and rises. As the heated air flows poleward, it radiates heat to outer space and becomes colder and therefore denser. It also cools off because of expansion. This cold, dense air sinks at the poles. Because air has mass, it is gravitationally attracted to Earth's surface and therefore exerts a pressure on Earth's surface. When air sinks, the extra mass leads to a higher pressure at Earth's surface than would be experienced if no vertical movement were occurring. Conversely, when air rises, some of the mass is pulled away from Earth's surface, leading to a lower pressure than would otherwise exist. The rising and sinking air masses lead to pressure gradients, or changes in pressure over a distance, in the troposphere (Figure 9-9). Air flows along these pressure gradients, resulting in winds. The steeper the pressure gradient—the greater the pressure difference between two places—the faster the winds will flow.

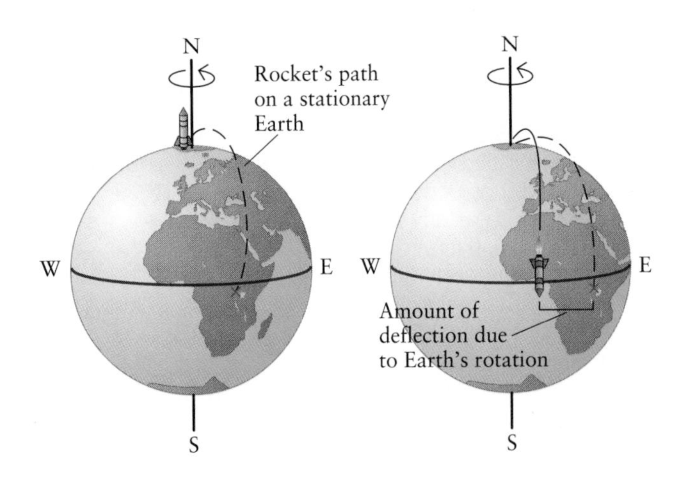

FIGURE 9-10 Coriolis deflection would cause a rocket launched from the north pole directly at Mount Kilimanjaro (marked by the x) to land west of its target because Earth rotates.

The Coriolis Effect and Hadley Cell Circulation Air rises at the equator and then flows toward the poles. Once at the poles, the now-cold air sinks and flows along the ground surface back toward the equator. If Earth did not rotate, the circulation would take the form of two huge convection cells, one for each hemisphere. However, Earth does rotate, making large convection cells unstable. Air and water currents are deflected to the right in the northern hemisphere and to the left in the southern hemisphere. This phenomenon is known as the **Coriolis effect.** To understand its cause, begin by imagining firing a rocket from the north pole toward Mount Kilimanjaro, near the equator (Figure 9-10). The rocket is fired due south, but by the time it arrives at the equator, Mount Kilimanjaro has rotated to the east along with the rest of Earth, and the rocket misses its mark. Relative to its original trajectory from the north pole, the rocket has been deflected to the right, or west.

Earth completes a full rotation in one day, so all points on Earth come back to their starting position in 24 hours. Because of Earth's spherical shape, however, not all points on its surface travel at the same velocity. Velocity varies with latitude, from zero at the poles, where circumference measured along lines of latitude is zero, to maximum velocity at the equator, where the globe is widest. Mount Kilimanjaro, near the equator, moves eastward rapidly, while the pole doesn't move at all, so that by the time the rocket reaches the equator, the mountain has moved far to the east and the rocket, if its course is uncorrected, lands west of it.

Now imagine firing a rocket from the equator toward the north pole. The rocket has a northward velocity from the launch, but it also has an eastward velocity because it is rotating with Earth. As it travels northward, the rocket passes over land moving at slower and slower rotational velocities. Consequently, the rocket "overtakes" the land and again is deflected to the right (this time, east) of its intended trajectory.

Just like the rocket, air currents are influenced by the Coriolis effect. Air flowing poleward from the equator is deflected by the Coriolis effect so that by the time it reaches 30° north or south it is flowing due east. Having lost much of its heat energy by radiation to space and by expansion, the air also has cooled and become denser. The accumulation of cold, dense air at 30° is drawn downward. Some of the sinking air travels back toward the equator, creating a circulation loop known as a **Hadley cell.** The rest of the air flows poleward along the surface of Earth. At 60° latitude north or south, this poleward-flowing air converges with air traveling toward the equator from the poles. The convergence results in uplift of air at 60°. This air splits into two streams when it reaches upper levels in the troposphere. One stream flows toward the equator to form a second Hadley cell between 30° and 60° latitude, and the other flows poleward to create a third cell between 60° and 90° latitude (Figure 9-11).

The tricellular model of tropospheric circulation is a simplification of the actual circulation in the atmosphere but is useful in explaining the distribution on Earth of deserts and wet regions. Hot air can hold a lot of water vapor. As that hot air rises to higher elevations and cools, the water vapor first condenses into clouds and then falls as rain. This process accounts for the enormous quantities of rain falling on the equatorial regions. Having lost much of its water vapor, the drier air then moves poleward. At about 30° it is forced to sink, and as it does, it warms up. The sinking air is now able to hold a lot of water in vapor form rather than in droplet form, and

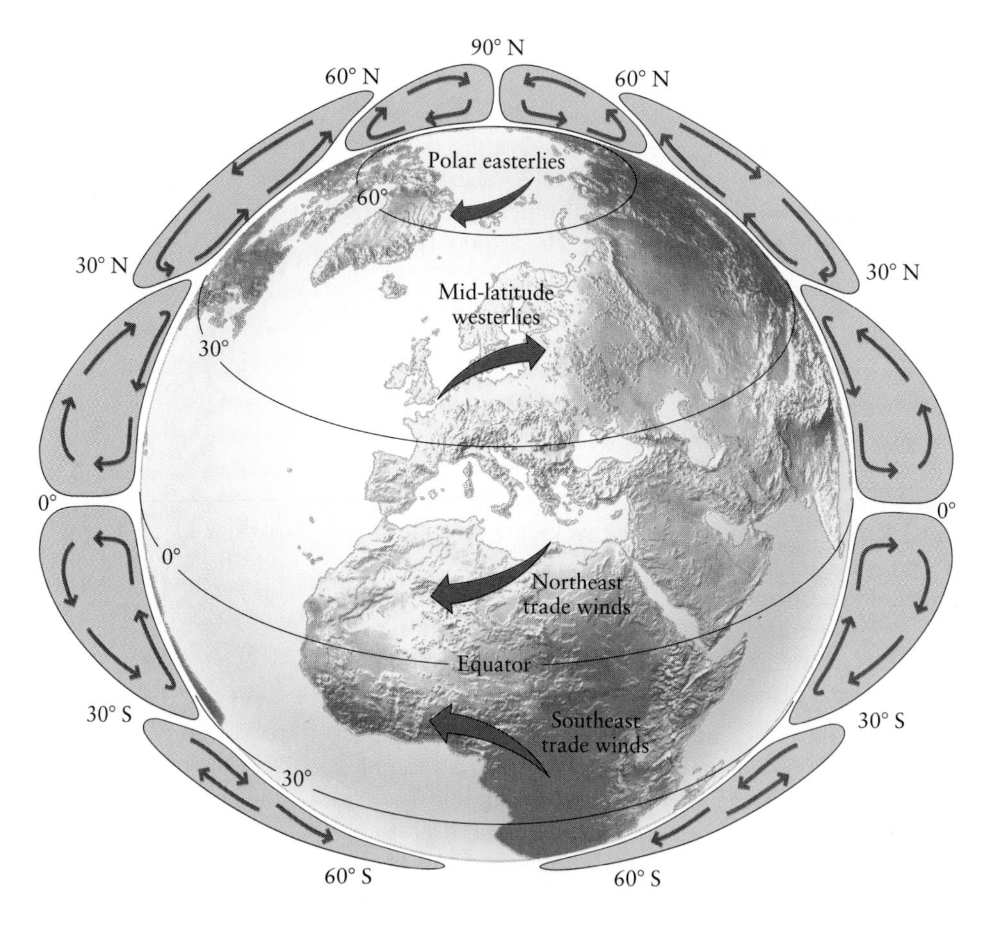

FIGURE 9-11 An idealization of atmospheric circulation on a rotating Earth. Circulation is broken into three large convection cells called Hadley cells. Coriolis deflection of the surface winds results in easterly winds from 0° to 30° and from 60° to 90° latitude, and westerlies from 30° to 60° latitude, north and south of the equator. Rising warm air at the equator and at 60° causes condensation, cloud formation, and precipitation at these latitudes. Sinking cold air at 30° and at the poles causes deserts to form.

water on the surface of Earth quickly evaporates into the warm air. It is no surprise then, that Earth's major deserts are found in latitudinal belts centered about 30° north and south (Figure 9-12).

Although they don't look like conventional deserts, the poles are considered arid lands because they receive very little moisture. The colder air gets, the less moisture it can hold, so air masses moving toward the poles are unlikely to drop much precipitation at high latitudes. Furthermore, air sinking at the poles warms as it sinks, increasing its ability to hold moisture in vapor form.

Influence of Land and Sea on Tropospheric Circulation and Climate The tricellular model assumes a homogeneous, ocean-covered planet. Landmasses complicate that simple picture of circulation for a variety of reasons and so influence local climate and weather patterns. First, air may be forced to flow around or up and over obstructions such as mountain ranges. The air cools as it rises, causing any water vapor contained in it to condense and fall as rain. As the now-dry air flows over the mountain and down the other side, it becomes progressively warmer. Because it can now hold more moisture, evaporation from the ground is greater (Figure 9-13). This is why mountain ranges commonly have a "rainy" side and a "dry" side, and the dry side is often described as being in the "rain shadow" of the mountains.

Second, the very different thermal properties of oceans and landmasses lead to the generation of sea and land breezes. Heat from the Sun is easily carried deep, about 100 m, into ocean water because water is transparent, but on land the Sun's heat is absorbed only by a thin surface layer. Furthermore, heating a unit volume of water 1°C takes far more energy than heating the same volume of rock or soil. As a consequence, the surface temperature of oceans fluctuates much less between day and night than does the surface temperature of land. During the day, the land heats and becomes much hotter than the ocean. The hot land heats overlying air, which becomes less dense and rises, creating a zone of low air pressure on land. When the land is adjacent to an ocean, the low pressure draws in cooler air from over the sea, creating a "sea breeze" (Figure 9-14). At night, the pattern of circulation is opposite: The land cools rapidly, while the ocean maintains a temperature close to that experienced during the day. The ocean is therefore warmer than the land, and air rises over it. Cooler air is then drawn from the land toward the sea in a "land breeze."

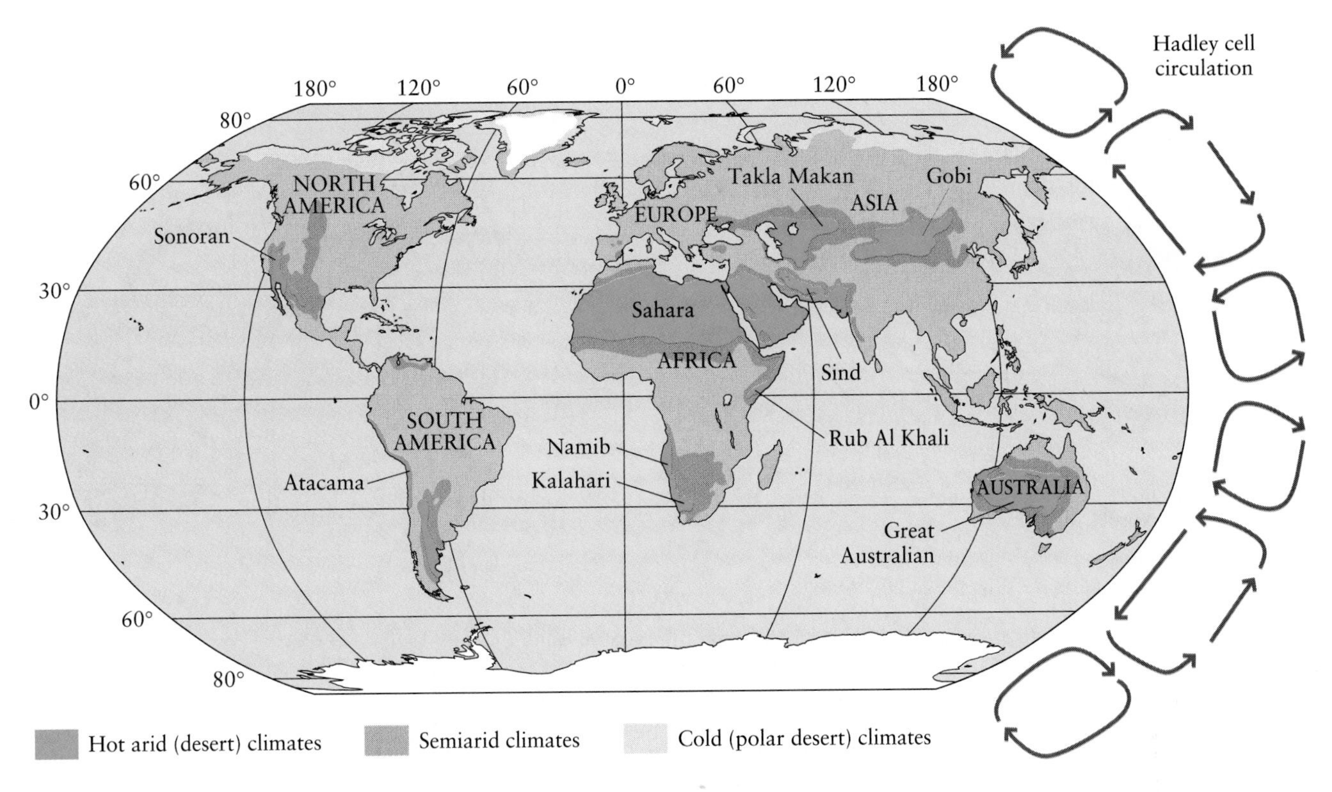

FIGURE 9-12 Most deserts are found in belts centered at about 30° latitude north and south. Deserts are also located at the poles and in the rain shadows of great mountain ranges such as the Andes, Rockies, and Himalayas.

A similar though much larger scale phenomenon produces the **monsoons,** the reversing seasonal winds of India and the southwestern United States. In the summer, the high heating of the Tibetan and Colorado plateaus draws air into the continental interiors from the oceans (Figure 9-15). This air is full of moisture and is quickly forced to higher, colder altitudes by the presence of mountains—the Himalayas or the many mountain ranges of the Colorado Plateau. As the air rises and cools, the water vapor it contains condenses, and therefore these regions tend to receive much rainfall in the summer months. (The summer monsoons in India bring rainfall needed for agriculture, but may also cause floods that drown thousands in low-lying Bangladesh to the east.) The reverse is true in winter, when the land becomes extremely cold. Warm air rises over the oceans while cold air sinks over the plateaus, causing cold, dry winds to blow off the continents toward the oceans.

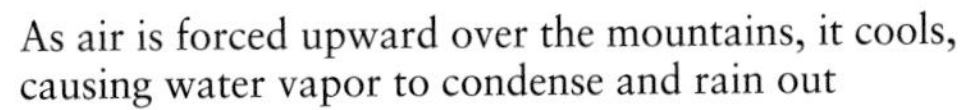

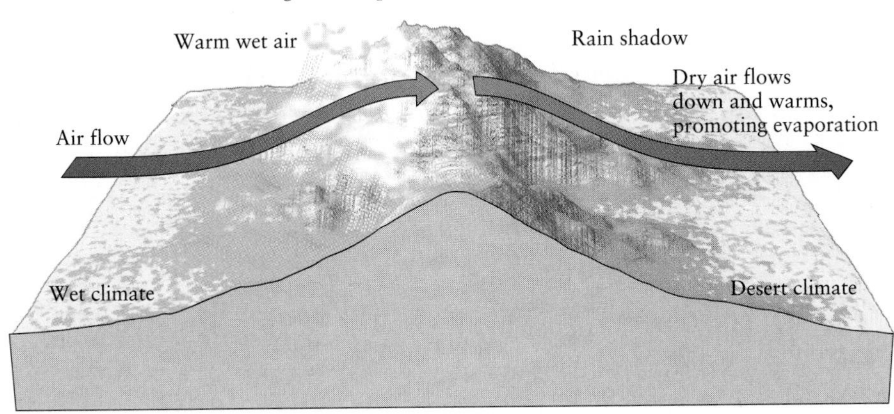

FIGURE 9-13 Moist air on the windward side of a mountain is forced upward, where it cools and condenses. Precipitation strips the air of moisture. On the leeward side of the mountain, the air subsides and warms. The dry, warm air evaporates water from the ground surface. This process creates a "rain shadow" and a very arid climate in the lee of the mountain.

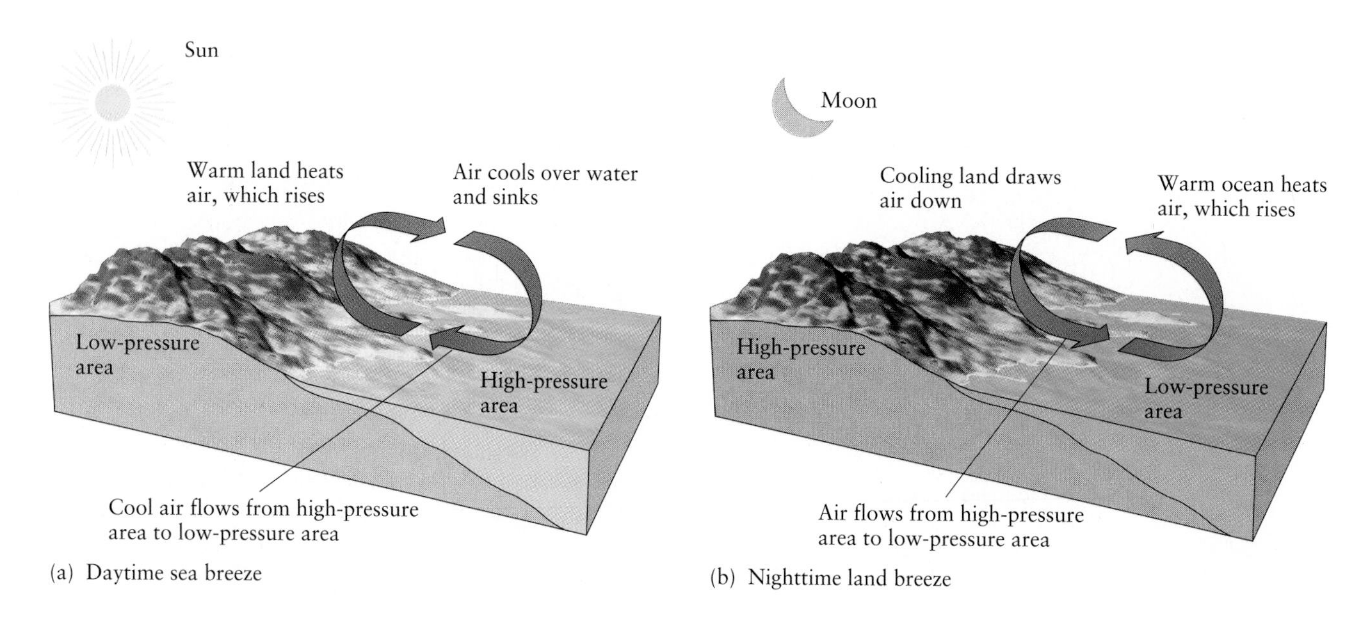

FIGURE 9-14 (a) During the day, the ground heats to a warmer temperature than does the sea. Air over the ground warms, expands, and rises. The low pressure produced by the rising air pulls cooler air in from over the ocean and creates a "sea breeze." (b) At night, the breeze reverses direction, becoming a "land breeze" because the ground quickly becomes cooler than the adjacent sea.

The different thermal properties of sea and land also cause continental interiors to have much more extreme temperature ranges than do coastal regions. Ocean water can absorb huge amounts of energy and slowly release it at night and during the winter; land heats rapidly and cools rapidly. Coastal regions, therefore, experience little seasonal change in temperature. Continental interiors, on the other hand, experience extremes of temperature from one season to the next because they lack the temperature-buffering effect of the ocean. Figure 9-16 shows the monthly average daytime temperature for two locations, Santa Cruz, California, on the Pacific Ocean, and Wichita, Kansas, in the continental interior. Note the much greater variation in temperature in Wichita, which lies about 2500 km from any ocean.

- Earth's climate and weather are produced by solar heating of the atmosphere combined with Earth's rotation and gravity.
- The poles receive less solar energy than does the equator, leading to the establishment of convection in the troposphere.

FIGURE 9-15 Monsoons are winds created by the presence of a high plateau in the interior of a continent. In summer, the plateau heats, causing the air above it to warm and rise. Cooler, moisture-rich air flows into the continental interior from the coasts and upward at the plateau, where it cools and releases abundant rainfall. In winter, the plateau becomes extremely cold, causing the air above it to cool and sink. Cold, dry winds blow outward from the center of the plateau toward the coast.

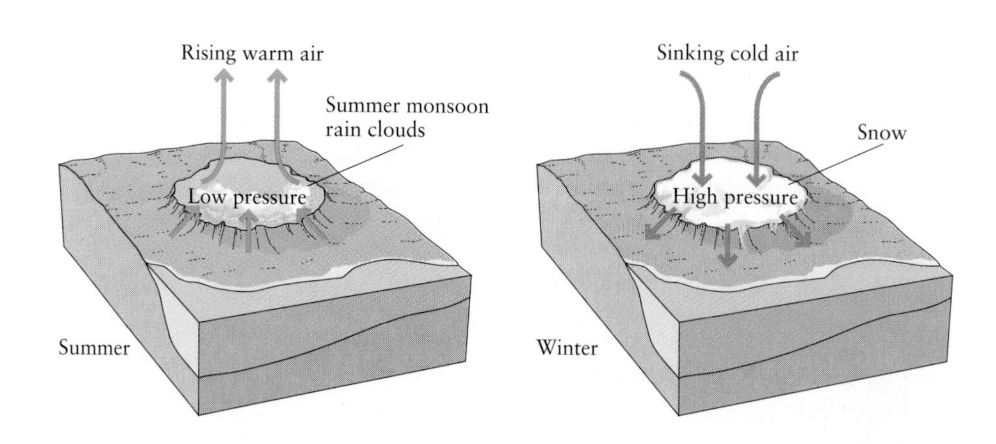

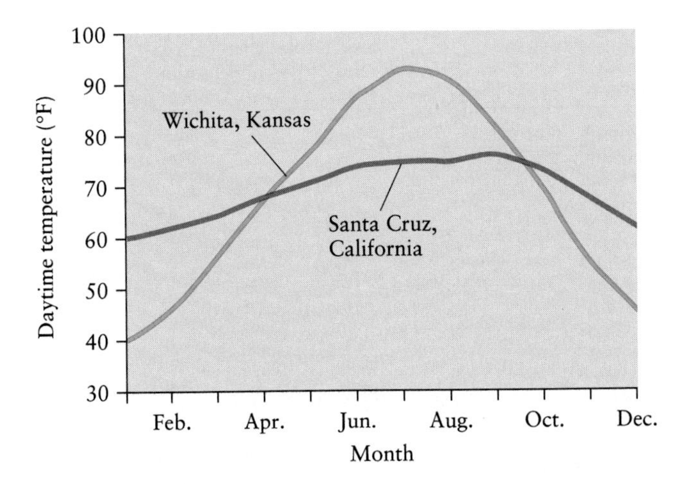

FIGURE 9-16 Average monthly temperatures for Wichita, Kansas, and Santa Cruz, California. The smaller temperature range in Santa Cruz is due to its proximity to the coast.

- The Coriolis effect causes deflection of air currents to the right in the northern hemisphere and to the left in the southern hemisphere. Tropospheric circulation thus deflected breaks into three convection cells, known as Hadley cells, per hemisphere.
- Hadley cell circulation explains the occurrence of deserts at 30° north and south latitudes and in the polar regions.
- Interaction of the atmosphere with the land and ocean influences local and regional climate by producing wet and dry mountainsides, sea and land breezes, and monsoons.

Storms

The temperature and humidity of air depend heavily on its position—over an ocean, over a mountain range, or in the continental interior. Frequently, a large volume of air becomes a distinctive **air mass,** a parcel of air of homogeneous temperature and humidity that moves as an entity. The interaction of air masses along their boundaries is analogous to the interaction of lithospheric plates along faults. The latter cause earthquakes; the former cause storms.

Because they are vital in bringing precipitation to land and because they cost so many lives and so many billions of dollars in damage every year, storms warrant special attention. Storms may be produced anywhere that an upward movement of warm, moist air is initiated. Storms tend to be regional in scale, extending hundreds or even thousands of kilometers, but they may also be rather local in extent if they are produced by uplift of warm, moist air over a mountain range. Likewise, sea breezes may result in the formation of localized storms along coasts.

Development of Air Masses and Frontal Weather Systems

The most common storms are associated with the interaction of air masses along their boundaries. Air masses are produced over oceans or land and tend to come from four different regions in each hemisphere: the Arctic (and Antarctic), polar (55° to 65°), tropical (10° to 30°), and equatorial latitudes. Thus, air masses are classified as maritime polar, continental tropical, and so on (Figure 9-17). Air masses formed over ocean basins are typically more humid than those formed over land. Arctic and polar air masses are much colder and therefore drier than tropical and equatorial air masses. Cold, dry air masses have higher densities than warm, humid air masses because colder temperatures cause gases to contract and because the density of water vapor is less than that of dry air.

The boundary between any two air masses is known as a **front.** Air masses eventually leave their source regions and move in response to the general circulation of the atmosphere. If a cold air mass arrives and replaces a warm air mass, a *cold front* is said to have arrived. Conversely, a warm air mass that replaces cold air is called a *warm front.* Storms tend to occur at fronts because the collision of air masses lifts warm, low-density air over cold, high-density air. As warm air rises, it cools, causing condensation of whatever moisture is present. Condensation releases the water vapor's stored solar energy as heat. This heat energy warms the air again, causing its density to fall

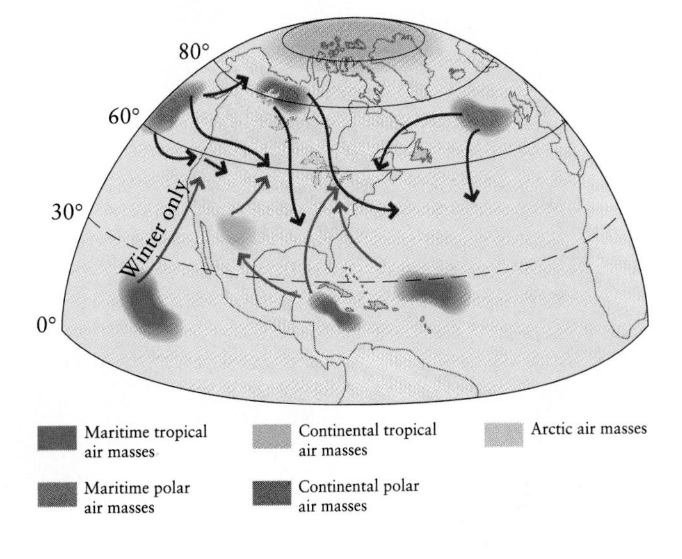

FIGURE 9-17 Source areas and trajectories of air masses that affect North American weather.

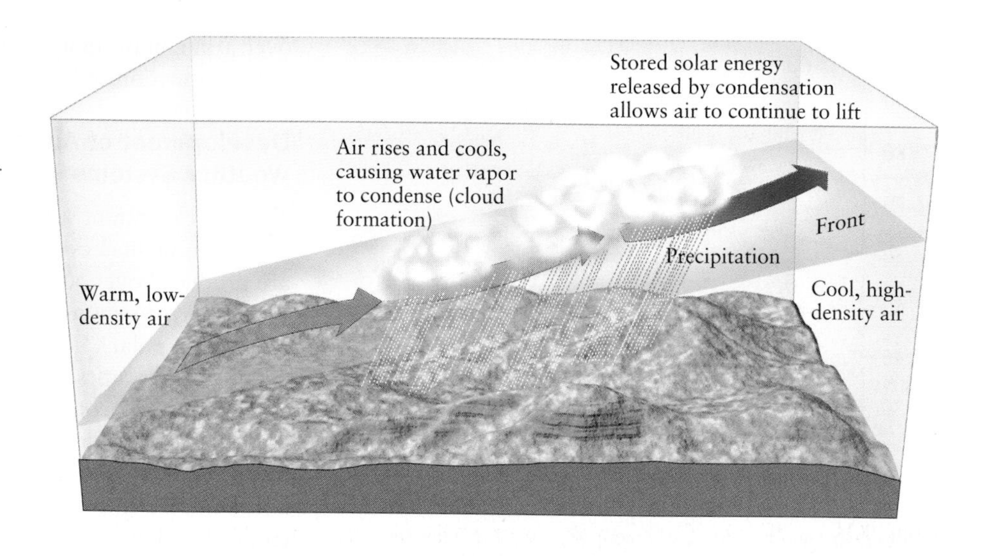

FIGURE 9-18 Frontal lifting of warm air over cold air. As the warm air rises, it cools and the water vapor it contains condenses, leading to precipitation.

even more and the *updraft* of air to continue (Figure 9-18).

Condensation results in the formation of clouds. Initially, the updrafts of air are capable of keeping the cloud droplets suspended. However, as the droplets grow bigger, they eventually exceed the capability of the updraft to support them, and they begin to fall through the cloud. The falling raindrops push air down in front of them and cold air from above rushes down to replace it. The resulting *downdraft* pulls air and precipitation down with it (Figure 9-19). The cold air in the downdraft causes a drop in surrounding air temperature as it moves down toward the ground and cuts off the supply of warm air to the updraft. Eventually, the updraft loses energy and dissipates, resulting in the "death" of the storm.

The East Coast blizzard of 1996 is a perfect example of a frontal storm. A very cold, dry, dense air mass that had originated in the Arctic sat over the Northeast. At the same time, a warmer, moister, less dense air mass traveled up the East Coast from the south along with the prevailing westerlies. The warmer air mass pushing into the Arctic air was forced to slow down because of the greater density of the Arctic air. As the warm air mass slowly slid up and over the Arctic air, it cooled, its water vapor condensed, and precipitation ensued. Because the warmer air was held in place by the denser Arctic air

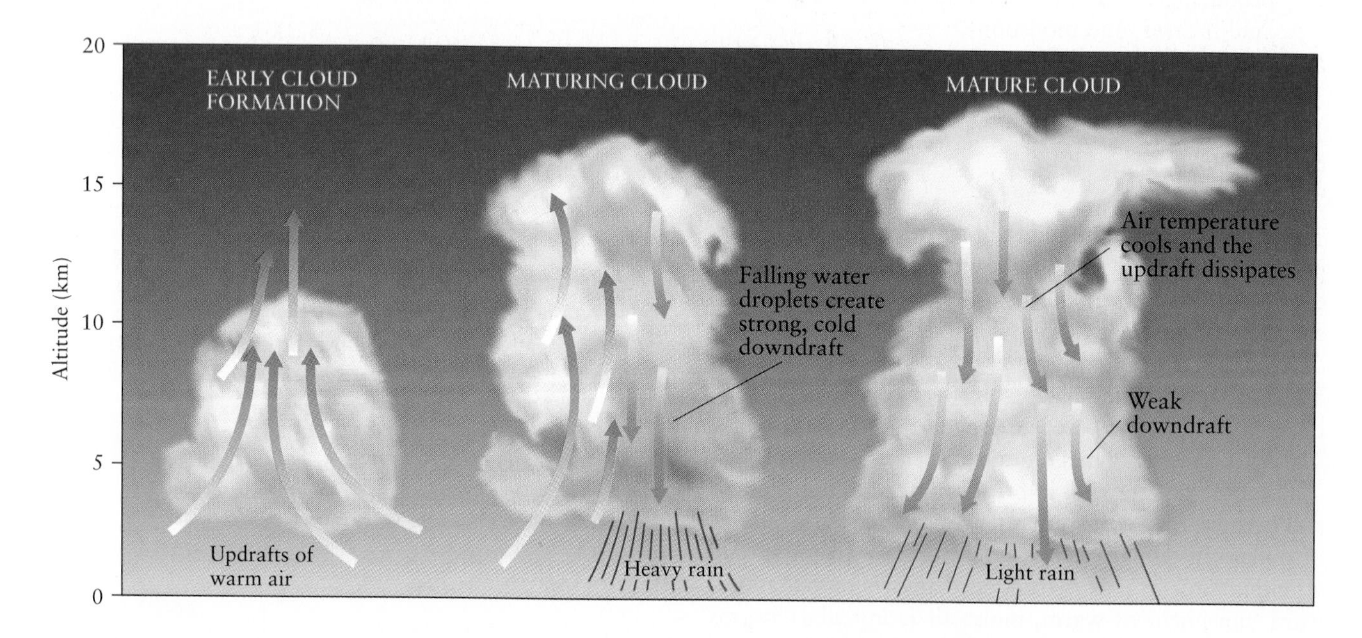

FIGURE 9-19 Updrafts within a storm cloud keep droplets suspended until they are too heavy to be supported. As they fall, they produce a downdraft that pulls cold, high-altitude air down to the ground surface, cutting off the supply of warm air needed to sustain the storm.

mass for a prolonged time, huge amounts of snow accumulated on the ground.

Lightning and Thunder

Sometimes updrafts and downdrafts spawn atmospheric phenomena of extreme violence. Although these phenomena can be very destructive, sometimes, as in the case of lightning, they also play a very beneficial role in biogeochemical cycling.

Convection within clouds results in the development of electrical charges as water droplets, ice and dust particles, and gases rub against one another. For reasons not yet fully understood, the charges that develop in the cloud tend to segregate, the negative charges accumulating near the base of the cloud and positive charges near the top (Figure 9-20). The accumulation of negative charges at the base of a cloud tends to attract positive charges on the ground below. The charges at the surface of Earth "shadow" the charges within the cloud, moving as the cloud moves. Air is a very efficient insulator, so it inhibits the flow of electrons from the cloud to the ground until the electrical potential energy is sufficiently large to overcome the natural resistance of the air to the flow of electricity. At this point, a spark heads from the cloud to the

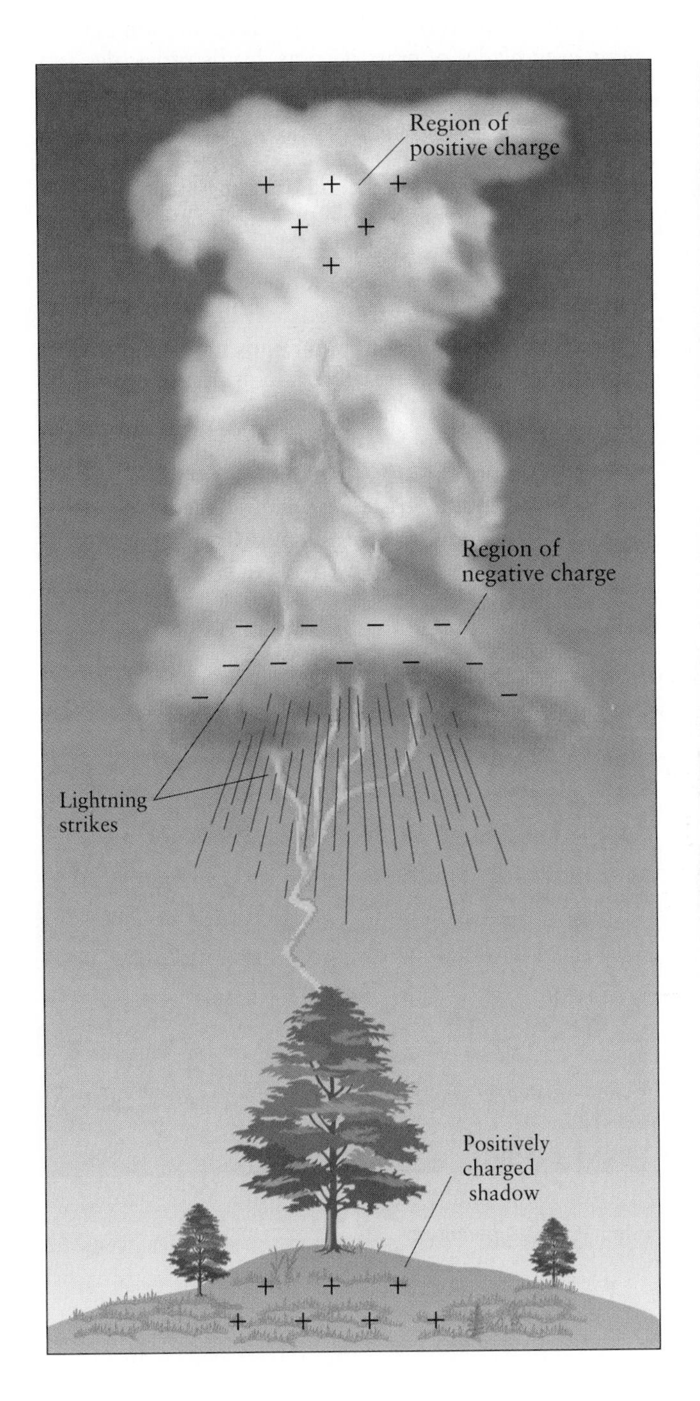

FIGURE 9-20 In a cloud, electrical charges separate and are "shadowed" by charges at the surface of Earth. A spark to the ground leads a massive release of electrical energy, seen as a lightning bolt.

9-3 Case Study

The Andover Tornado

On April 26, 1991, the Kansas sky turned dark greenish purple. A roiling mass of clouds formed and lightning flashed between them. At 6:38 PM, a devastating tornado spawned by these clouds struck the small town of Andover, Kansas, near the state's largest city, Wichita. The twister was on the ground for about 50 minutes and covered 120 km of territory. During its short life, it killed 15 people, injured dozens more, destroyed more than 350 homes (leaving 20 percent of Andover's population homeless), and cost the area millions of dollars in property damage. The Golden Spur mobile home park was particularly hard hit as the tornado picked up mobile homes and slammed them back into the ground as if they were toys. Eleven of the 15 fatalities occurred in the park, and not a single mobile home was left standing. Power lines and trees were snapped in half by winds reaching 420 km per hour. According to weather officials, only about 2 percent of all tornadoes reach wind strengths of this magnitude.

The Andover tornado was only one of 56 tornadoes that struck the Midwest from Nebraska to Texas that day. They were spawned by a storm that formed when warm, moist air moved quickly northward out of Texas toward Oklahoma and Kansas and collided with drier air flowing from the west. The collision of the two air masses caused uplift and condensation of the moist air mass from the south and led to the production of vortices that eventually became tornadoes.

Condensation of water in the updraft is initially what makes the tornado visible. Its visibility increases as the funnel picks up debris, which may be tree limbs, fence posts, and sections of roof. At speeds of 330 km per hour, even blades of grass become deadly projectiles and have been observed protruding from bricks, like darts from a target, after tornadoes subside.

The American Midwest experiences more tornadoes per square kilometer of ground than any other place on Earth and has therefore earned the nickname "Tornado Alley." About 780 tornadoes are reported annually. On April 3, 1974, 148 tornadoes touched the ground over a 12-state area extending from Alabama to Michigan, killing 309 people, injuring thousands more, and causing property damage that ran to hundreds of millions of dollars. This so-called Terrible Tuesday was one of the most devastating tornado disasters in history, but the worst destruction by a single funnel occurred on March 18, 1925, when the "tristate tornado" killed 695 people as it moved from Missouri to Illinois to Indiana. Most of the deaths were caused by flying debris and collapsed walls and roofs.

The Andover tornado was one of the most violent storms ever recorded. It left 20 percent of the small town's population homeless and caused 15 deaths.

ground, blazing a trail for the negative charges to follow. A massive amount of electrical energy, equivalent to a million times the voltage used in an average household, streaks down to Earth. As it travels through the air, the lightning instantaneously heats the surrounding air to temperatures as high as 10,000°C. This intense heating causes rapid expansion of the air and production of a sound wave that our ears register as thunder.

Lightning, which can cause great damage, also serves a very beneficial purpose. Lightning splits nitrogen gas (N_2), which quickly oxidizes to nitrate. This nitrate is then deposited on the ground with the rain and is thereby made available to plants, for which it is an essential nutrient. Every year about 100 million tons of nitrogen are delivered to the biosphere in this manner.

Tornadoes

In the midwestern United States, cold dry air sweeping southward from the Canadian Arctic meets warm moist air traveling northward from the Gulf of Mexico. The interaction of two such dramatically different air masses can produce tornadoes. Frontal lifting causes the warm air to ride up and over the colder Arctic air mass, initiating an updraft. Simultaneously, the air at the front rotates horizontally, creating a vortex parallel to Earth's surface. Water vapor in the updraft condenses as the air rises and cools, releasing heat and maintaining the updraft. Cold downdrafts force the horizontally rotating vortex to bend upward into the updraft, producing a vertically rotating vortex called a funnel cloud. If this cloud thins and stretches out, rotating at higher and higher speeds, it is called a **tornado.** The speed of the wind is what causes tornadoes to be so destructive (Box 9-3).

Tropical Storms and Hurricanes

Rotating over the ocean as tornadoes spin over the land, tropical storms and hurricanes cause death and destruction from a combination of very high winds, torrential rains, and storm surges. **Tropical storms** are the "gentler" of the two, with somewhat slower winds; **hurricanes** have winds of at least 120 km per hour. Both these storm systems tend to be much larger in scale than tornadoes, and therefore more extensive in their impact.

Tropical storms and hurricanes are born near the equator, where solar heating results in uplift of air, creating a low-pressure zone. The low-pressure zone may intensify if water vapor in the uplifting air condenses aloft, because condensation of vapor releases heat energy, which causes more lifting. Air rushes in to fill the low-pressure zone and is very weakly deflected by the Coriolis effect. If the low-pressure zone moves away from the equator and begins to travel to higher latitudes, Coriolis deflection becomes greater and greater. The low thus becomes a spiraling vortex of air that is continually fed by new warm air from below. Average windspeeds typically exceed 100 km per hour, and winds in excess of 300 km per hour have been clocked. These winds rip roofs off of houses and turn tree limbs and other loose debris into deadly missiles. They combine with the low pressure that produces the hurricane to raise the sea level several meters, so low-lying coastal areas are flooded. The torrential rains produced by hurricanes overwhelm the capacity of the ground to soak them up, producing vast amounts of runoff and flooding of river channels. Bangladesh, with much of its land lying within a meter of sea level, has been particularly hard hit by hurricanes throughout its history. Storm surges of several meters have occurred many times in the last 30 years, resulting in flooding and drowning of more than 1 million people. In the United States, damage to life is generally far less severe because of sophisticated weather forecasting, emergency response planning, and evacuation of low-lying areas. Nevertheless, hurricanes in the United States have caused staggering property losses in the billions of dollars (Figure 9-21).

Hurricanes always form over oceans because they require a steady supply of warm, moist air to sustain them. The strength of hurricanes dissipates once they hit land and areas of low temperature. Over land, temperatures rapidly fall below the 26° to 27°C necessary to sustain

FIGURE 9-21 A satellite image of Hurricane Andrew taken August 25, 1992, shortly after it devastated South Florida and one day before it made landfall in Louisiana, shows characteristic swirling hurricane cloud patterns clearly visible over the Gulf of Mexico.

the hurricane, and air over land is also usually not as humid as ocean air. Likewise, as hurricanes move poleward, they move over seas that are much colder. This too tends to shut off their circulation.

- Air masses interact along fronts, which are often the sites of storm activity.
- Water vapor in warm air updrafts condenses as the air travels upward to higher, colder altitudes. Condensation releases stored heat energy that allows the updraft to continue and more condensation to occur. The resulting water droplets fall to the ground as rain or snow.
- Charges develop and separate in clouds when gases, ice crystals, water droplets, and dust particles rub against one another. These charges lead to lightning bolts when the resistance of air to the flow of electricity is overcome by the electrical potential energy developed in the cloud.
- Tornadoes, rotating storms known for their high winds and destructive ability, form at the intersection of air masses with very different physical properties.
- Hurricanes and tropical storms are produced and sustained by a supply of warm, moist air found only over low latitude oceans.

FIGURE 9-22 Oil wells in Kuwait were set ablaze by the retreating Iraqi army during the Persian Gulf War in 1991. The soot from the burning oil was so thick that headlights were needed even for daytime driving. Many children and old people required hospitalization for respiratory problems caused by the smoke and toxic gases released from the burning wells.

Human Influence on Atmospheric Chemistry

The natural processes of tropospheric circulation and the climate and weather produced by it have operated and existed for millions or even billions of years. Increasingly, however, humans are emitting substances into the atmosphere that both interact detrimentally with existing processes and have the potential to alter the processes themselves.

Humans introduce a wide variety of pollutants into the atmosphere through activities such as industry, agriculture, motor vehicle transportation, and war (Figure 9-22). Many of these pollutants are hazardous to the health of animals and plants, while others have the potential to change climate or to destroy Earth's protective layer of ozone in the stratosphere. Amazingly, these hazards are produced by pollutants with concentrations lower than 0.001 percent of the total volume of the atmosphere.

The dispersal of pollutants is driven by atmospheric circulation and the reactions of pollutants with various compounds in the atmosphere. Some pollutants in the troposphere, such as sulfur dioxide and nitrogen oxides, react with water and are quickly removed through precipitation on the ground. Other pollutants, such as chlorofluorocarbons, are chemically more inert and remain in the atmosphere for long periods of time. The residence time of pollutants in the atmosphere determines their environmental impact. Pollutants with short residence times tend to have a local, or at most, a regional impact, whereas those with long residence times may have a global impact. Acid rain and smog are examples of the former; ozone depletion and global warming (discussed in Chapter 12) are examples of the latter.

Acid Rain

Ordinary rainfall is slightly acidic because it reacts with carbon dioxide to make carbonic acid: $H_2O + CO_2 \rightarrow H_2CO_3$. With the advent of the Industrial Revolution, however, the acidity of rain has increased. Combustion of fossil fuels by power plants and motor vehicles and industrial processes such as metal smelting release sulfur and nitrogen oxides into the atmosphere. These gases oxidize to sulfate and nitrate particles, which may be deposited either directly on the ground as dry particles or as rain. Sulfate and nitrates react with water to make sulfuric and nitric acids:

$$\underset{\text{Sulfate}}{2\,SO_4^{-2}} + \underset{\text{Water}}{2\,H_2O} \rightarrow \underset{\text{Sulfuric acid}}{2\,H_2SO_4} + \underset{\text{Oxygen}}{O_2}$$

$$\underset{\text{Nitrates}}{NO_x} + \underset{\text{Water}}{H_2O} \rightarrow \underset{\text{Nitric acid}}{HNO_3}$$

(The subscript x in NO_x indicates that the number of oxygen atoms is variable.)

The pH of rainwater typically falls between 5 and 6, but acid rain usually has a pH between 4 and 5, and values as low as 1.5 have been recorded.

The impact of acid deposition on ecosystems varies. Acid rain falling on land underlain by carbonate rocks is neutralized (see Chapter 6), but rain falling on granitic landscapes is not. When acids react with rocks and soils, they may release toxic metals or strip nutrients from these materials. For instance, aluminum is readily released and transported by acidic waters and both damages roots of plants and interferes with calcium absorption. As a result, plants growing in acidic soils that contain aluminum become sickly or die. Telltale signs of acid-related nutrient deficiency are death of the upper levels, or crown, of the tree, yellow leaves or needles, and excessive dropping of needles (Figure 9-23). Deciduous trees tend to be less vulnerable to acid rain than conifers. Because deciduous trees have a smaller leaf surface area per tree and because they shed their leaves every fall, their exposure to acid rain is limited. Nevertheless they are affected, and both types of trees are also much more susceptible to pests, viruses, and fungi when they are under stress from nutrient deficiency or high concentrations of toxic metals.

FIGURE 9-23 The Black Forest of Germany has been severely damaged by acid rain. Note the yellowing of the needles on these Norway spruce trees.

TABLE 9-2 Effects of Increasing Acidity on Fish

pH	Effects on fish
9.0–6.5	harmless to most fish
6.5–6.0	significant reductions in egg hatchings and growth in brook trout under continued exposure
6.0–5.0	rainbow trout do not occur; small populations of relatively few fish species are found; declines in salmon can be expected; if high aluminum concentrations are present, fish will die; mollusks are rare
5.0–4.5	harmful to salmon eggs and fry, and to common carp
4.5–4.0	harmful to adult salmon, goldfish, common carp; resistance increases with age
4.0–3.5	lethal to salmon; roach, trench, perch, pike survive
3.5–3.0	toxic to most fish; some plants and invertebrates survive

Source: Adapted from D. M. Elsom, *Atmospheric Pollution: A Global Problem,* 2nd ed., New York: Blackwell, 1992.

Toxic metals released by acid rain may be transported by groundwater and surface water into lakes. Aluminum presents a particular problem for fish, causing them to secrete a very thick mucus that clogs gills and results in asphyxiation. In addition, increased acidity of the water inhibits reproductive success; Table 9-2 shows the relationship between pH and toxicity to fish. Of 150 lakes surveyed in southern Ontario, 33 had a pH of less than 4.5 and 32 had a pH between 4.5 and 5.5. Several hundred lakes within 80 km of a smelter in Sudbury, Ontario, have few fish or none at all. In the United States, increased acidity in Adirondack lakes has also led to declining fish populations.

Nor do urban environments escape damage from acid rain. In Sweden increased acidity levels in well water have led to higher concentrations of copper and lead leaching out of plumbing pipes, causing liver and kidney damage in some children. Furthermore, the history of

FIGURE 9-24 Acid rain attacks historic monuments and buildings and literally dissolves them, as shown in this view of York Minster stonework. The central head used to be in the direct path of acidic rainfall, whereas the heads to either side have always been more sheltered.

many countries is slowly dissolving away as acid rain attacks limestone and marble monuments, temples such as the Parthenon in Athens, and grave markers (Figure 9-24).

In an early response to local acid rain, power plants and factories built taller smokestacks. The idea was to emit noxious gases at higher levels in the atmosphere, where they would then be kept aloft by air currents and swept out of the local environment by winds. The problem with this approach is that the acids are transported to new areas where they may cause environmental damage. Sulfur dioxide produced in the Ohio River valley is routinely exported by winds to northeastern states and Canada, and Sweden receives frequent doses of sulfuric acid from eastern European countries. Because of the short residence time of water in the atmosphere, most deposition occurs within 1000 km of the source of the acidity.

In the last 40 years, a brownish acid has been recognized in the Arctic troposphere, previously a pristine environment. This "Arctic haze" is caused by pollutants that traveled great distances from eastern Europe and Asia. In the spring the Arctic haze absorbs so much incoming solar radiation (which otherwise would have been reflected by snow and ice) that it could contribute to global atmospheric warming and climate change. Apparently, the Arctic haze is induced by human activities.

Smog

Industrial activities and automobiles contribute not only to acid rain but also to the production of smog, a common sight in cities. A complex mixture of ozone, nitrogen oxides, and hydrocarbons, **smog** causes respiratory problems, reduced visibility, and damage to vegetation (Figure 9-25). Nitrous oxide gas makes a smoggy atmosphere appear brown because the molecules strongly absorb blue light and scatter yellow and red wavelengths.

All the components of smog have existed in Earth's atmosphere for hundreds of millions of years. Plants produce some hydrocarbons, ozone is manufactured in the stratosphere, and lightning and forest fires generate the heat necessary to produce nitrogen oxides. Industrial activities add significantly to the hydrocarbons and ozone in the troposphere, and the intense heat of automobile cylinders produces a steady supply of nitrogen oxides.

(a)

(b)

FIGURE 9-25 Pasadena, California, (a) on a clear day and (b) on a smoggy day. The San Gabriel Mountains are completely obscured by smog, a thick haze consisting of nitrogen oxides and uncombusted hydrocarbons from motor vehicles, along with ozone.

Solar energy dissociates nitrogen oxides into other nitrogen compounds and free oxygen (O). The free oxygen readily combines with molecular oxygen (O_2) to form ozone (O_3):

$$\underset{\text{Nitrogen dioxide}}{NO_2} + \underset{\text{Solar energy}}{h\nu} \rightarrow \underset{\text{Nitrogen oxide}}{NO} + \underset{\text{Free oxygen}}{O}$$

$$\underset{\text{Free oxygen}}{O} + \underset{\text{Molecular oxygen}}{O_2} \rightarrow \underset{\text{Ozone}}{O_3}$$

(Energy can be described as the wave frequency (ν) times h, a universal constant known as Planck's constant.)

Although in the stratosphere ozone is beneficial, blocking the transmission of harmful ultraviolet rays from the Sun, ozone in the troposphere can cause respiratory irritation and eye damage, may lead to lung cancer, and kills plant tissues. In forests around the Los Angeles basin and in Europe, large stands of trees have been killed by ozone pollution in nearby cities, and California loses an estimated $1 billion annually in ozone-related crop damage.

Smog accumulates in urban areas partly because of the large amounts of atmospheric pollutants generated by motor vehicles, power plants, and industry, and partly because atmospheric temperature can keep the pollutants from escaping. Because temperature generally decreases with altitude above Earth's surface in the troposphere, if a parcel of air at the surface becomes strongly heated during the day, its density declines and the parcel rises until it reaches a level in the atmosphere with the same low density. The upward motion of the parcel of air allows pollutants to disperse. During the night or in winter when surface temperatures are very low, air at the surface is colder and denser than the overlying air and cannot move upward. The result is that temperatures at higher altitudes are higher than at lower altitudes, an inversion of the normal situation. At these times, pollutants produced at the surface remain concentrated there instead of being carried aloft (Figure 9-26). This problem may be exacerbated by geography: If the city is in a basin surrounded by mountain ranges, the dense, cold, pollutant-rich air will not be swept away by winds. Such is the situation of Los Angeles, where the San Gabriel and San Bernardino mountains block winds and trap pollutants.

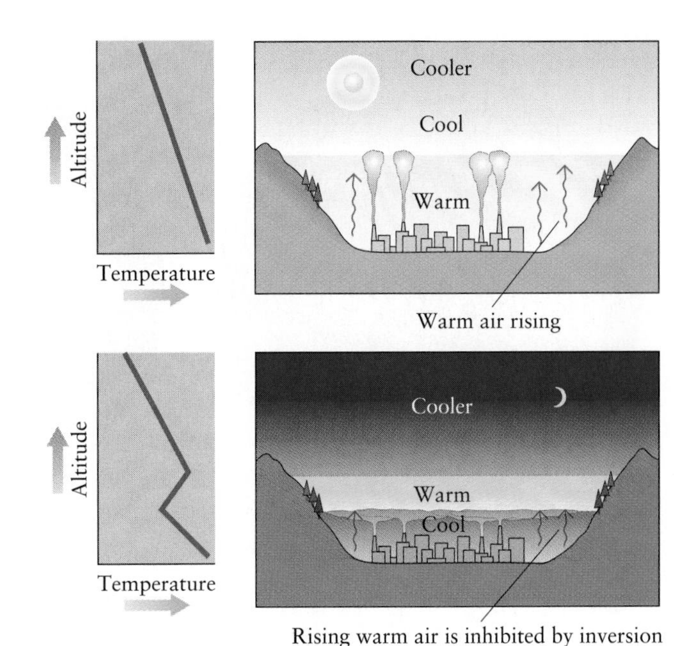

FIGURE 9-26 Atmospheric temperature inversions, which occur at night and in the winter, inhibit vertical mixing of the lower troposphere and lead to high pollutant concentrations.

Ozone Depletion

Ironically, while some urban areas are experiencing smog and excessive amounts of ozone near the ground, stratospheric ozone levels are declining. Ultraviolet (UV) radiation coming from the Sun presents a health hazard to organisms on Earth. While necessary for life (for synthesis of vitamin D in humans, for example), too much ultraviolet radiation can result in sunburned skin, cataracts, damage to the immune system, and genetic mutations that may ultimately lead to cancer. Ozone strongly absorbs ultraviolet radiation and prevents its further penetration to Earth's surface. Ozone is formed primarily in the stratosphere, where ultraviolet rays from the Sun continuously break some oxygen molecules (O_2) into oxygen (O) atoms that combine with other O_2 molecules to produce ozone (O_3). Once created, the ozone itself is dissociated by ultraviolet radiation. Thus, ozone is perpetually forming and reforming, but it is this very dissociation process that absorbs the ultraviolet rays.

In 1974, scientists F. Sherwood Rowland and Mario Molina warned that a class of human-made gases known as **chlorofluorocarbons** (CFCs) produced by industrial processes can destroy ozone in the atmosphere. Only 11 years later, satellite measurements revealed a thinning, or "hole," in the ozone layer over Antarctica (Figure 9-27). In this region, ozone concentrations had dropped to as low as 5 percent of their former levels. Extremely low levels of ozone were then discovered over other parts of the southern hemisphere, such as Australia and New Zealand, and then in the northern hemisphere, over the North Sea. Indeed, ozone levels are declining globally. CFCs, which have been used as refrigerants, as propellants for aerosol sprays, and in the production of styrofoam, are broken down by sunlight to form chlorine monoxide and chlorine. The chlorine reacts with ozone to

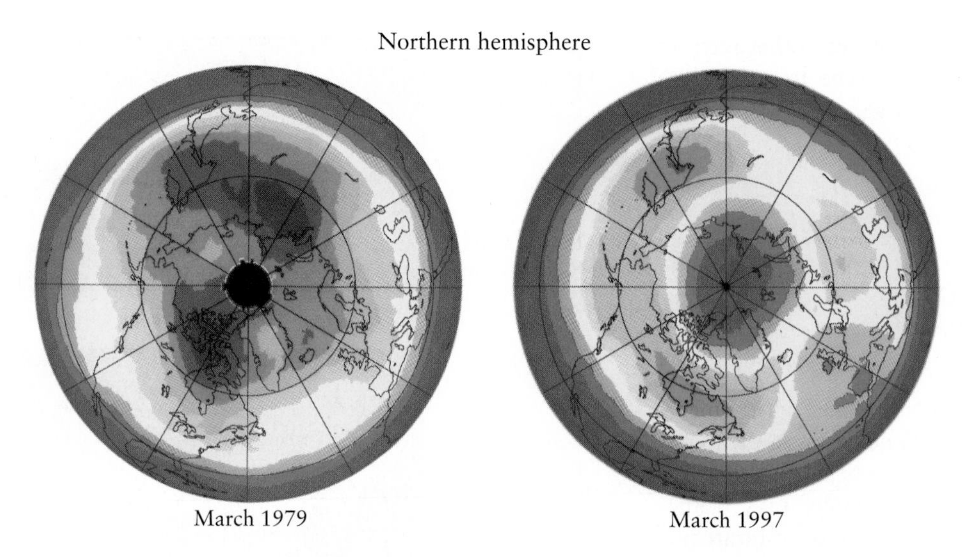

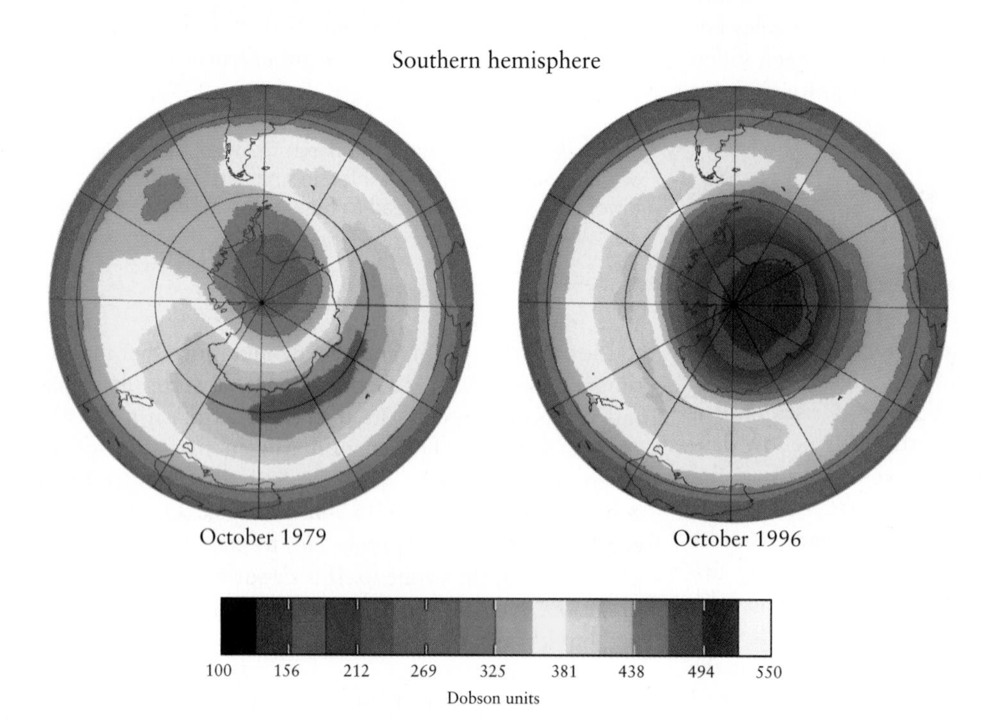

FIGURE 9-27 Ozone concentrations have been declining globally for at least the last two decades, as can be seen in these images produced by the Total Ozone Mapping Spectrometer (TOMS), a satellite in orbit about Earth. Declines have been most dramatic over the poles and high latitudes. Concentrations are measured in Dobson units (1 DU equals an ozone concentration of 1 part per billion), named for George Dobson, an English scientist who designed the first ozone measuring instrument.

produce gaseous oxygen and chlorine monoxide, while the chlorine monoxide reacts with oxygen to form gaseous oxygen and chlorine:

$$\underset{\text{Chlorine dioxide}}{Cl} + \underset{\text{Ozone}}{O_3} \rightarrow \underset{\text{Chlorine monoxide}}{ClO} + \underset{\text{Molecular oxygen}}{O_2}$$

$$\underset{\text{Chlorine monoxide}}{ClO} + \underset{\text{Atomic oxygen}}{O} \rightarrow \underset{\text{Chlorine}}{Cl} + \underset{\text{Molecular oxygen}}{O_2}$$

Because chlorine is perpetually recycled by these reactions, its destructive effect on stratospheric ozone is nearly continuous. The mechanism for removing chlorine from the atmosphere involves a reaction with hydrogen ions to form hydrochloric acid, which is then incorporated in rainfall. Because the concentration of water vapor in the stratosphere is extremely low, a single chlorine atom may reside there long enough to destroy as many as 100,000 ozone molecules. In 1989, 13 industrial nations, including the United States, agreed to stop all production of CFCs by the year 2000. The agreement has been successful, but it will take several centuries before stratospheric ozone returns to its pre-1950s concentration.

- Combustion of fossil fuels, metal smelting, and other industrial processes release to the atmosphere sulfur and nitrogen oxides, which combine with water vapor and droplets to form acid rain.
- The effects of urban smog are exacerbated by the temperature inversion that exists above land at night and during winter.
- Concentration of ozone in the stratosphere is declining because of reaction with chlorofluorocarbons, while concentrations in the troposphere are increasing because of the interaction of solar energy with nitrogen oxides and hydrocarbons given off by automobiles.

Air Pollution and Environmental Management

Clean air became a threatened resource, especially in urban areas, with the Industrial Revolution. Nevertheless, the United States did not enact air quality legislation until 1947.

Legislation

In 1947, California became the first state to enact air pollution legislation. Los Angeles had long been plagued by smog, but its source was unknown. A rubber-making factory was shut down by the city government, but the smog continued. Eventually officials recognized that the primary sources of Los Angeles smog were motor vehicles. The 1947 law established air pollution control districts, but pollution-control devices on vehicles were not required until 1960.

In 1970, the U.S. government passed the Clean Air Act, which established *national ambient air quality standards* (NAAQSs). Standards were set for maximum allowable concentrations of particular pollutants, as measured for a specific period of time at a given distance from the source. Many nations have used a similar approach to controlling air pollution, and ambient air quality standards now are in worldwide use by industrialized nations.

Despite these early efforts, pollution continued to be exported across state and national boundaries, and some pollutants were doing unexpected damage to the atmosphere. For example, the Clean Air Act required many power plants to build tall smokestacks, but the result was that pollutants were emitted higher into the atmosphere and thus enabled to travel even farther from their source. Furthermore, damage to trees by acid rain had become so extensive that entire forests were devastated in some parts of Canada and Europe. In Germany's Black Forest, for instance, one-third of the fir trees have died and another 30 percent are damaged, as are 50 percent of the beech trees and 43 percent of the oaks. Throughout the country, nearly half the forests have been damaged by acid deposition. Likewise, in New York and New England, 20 to 50 percent of red spruce have been killed through a combination of acid precipitation, tropospheric ozone, climatic stresses, and insect infestations.

In response to these new threats, the U.S. Congress amended the Clean Air Act in 1990 to require that annual emissions of sulfur dioxide be reduced by 10 million tons and nitrogen oxides by 2 million tons by the year 2000. In addition, the amendments called for decreased production of CFCs and other gases that contribute to ozone depletion.

The result of the Clean Air Act amendments has been a 96 percent reduction in annual emissions of lead and a 60 percent reduction in emissions of particulate matter. Carbon and nitrogen oxides from new cars have decreased substantially since 1967 (by 76–96 percent), but the number of people and of cars has increased, so the total amount of emissions has merely stabilized or slowed. Nevertheless, air quality probably would have been far worse without the emission controls.

Other industrialized nations also have reduced their atmospheric emissions in the past few decades (Figure 9-28). Unfortunately, little information on air quality is available from developing nations, yet these are the very nations where rapid urban growth is occurring, air quality is severely threatened, and little progress has been made to establish pollution control. The United Nations Environment Programme has established the Global Environment Monitoring System (GEMS) to fill this void, requiring that contributing countries collect data on sulfur dioxide from multiple sites over extended time periods.

Of course, passing laws is not ultimately what reduces air pollution. Cutting down on emissions of various pollutants requires technological innovations.

Cleanup Technologies

The various types and sources of air pollution require different cleanup strategies. A **catalytic converter** in the engine of a motor vehicle turns smog-producing waste products of gasoline combustion into relatively benign gases. The device consists of a steel cylinder filled with porous ceramic that is coated with platinum or palladium. The cylinder resides between the engine block and the exhaust pipe. The ceramic receives the carbon monoxide, unburned hydrocarbons, and nitrous oxides given off by combustion. The platinum or palladium coating acts as a *catalyst* in a chemical reaction that converts carbon monoxide, unburned hydrocarbons, and oxygen into water and carbon dioxide. (A catalyst is a substance that accelerates a chemical reaction but does not otherwise change it.) The catalytic converter also turns nitrogen

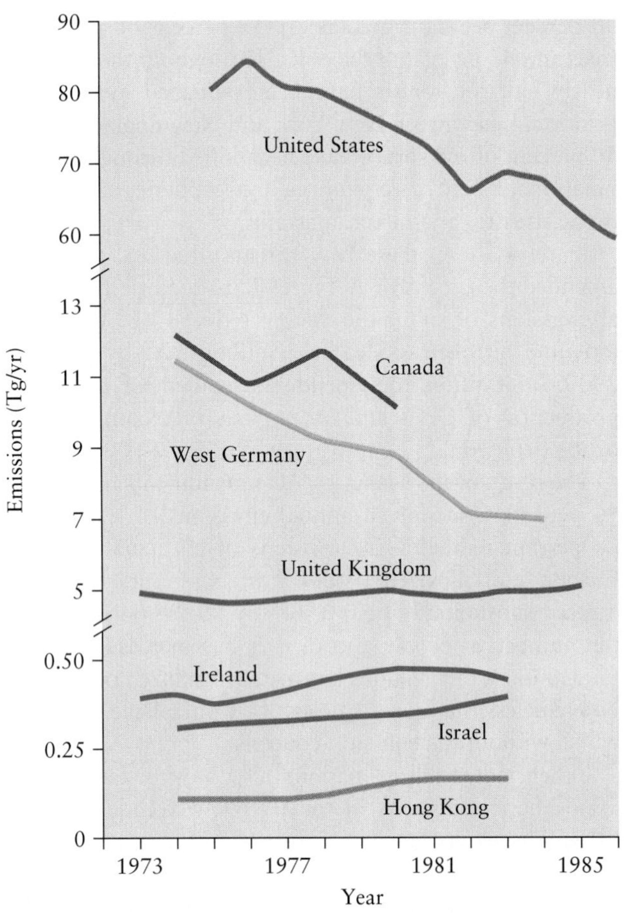

FIGURE 9-28 Emissions of carbon monoxide in different industrialized nations over the last two decades show that some nations have made great strides in reducing their emissions, while others are still experiencing increases. Units are in terragrams (1 Tg = 10^{12} g).

oxides into nitrogen gas, the most abundant gas in the atmosphere. (Another, nontechnological, strategy for dealing with nitrogen oxides produced by cars is the reduction of speed limits. Motor vehicles produce more of these pollutants when their engines are very hot, and speed increases engine heat.)

Power plants that burn fossil fuels give off sulfur dioxide gas. To get rid of sulfur dioxide, calcium oxide or calcium carbonate particles suspended in water, known as scrubbers, are sprayed into the stream of gases coming away from the combustion chamber. The calcium oxide or carbonate combines with the SO_2 to make calcium sulfite or sulfate. Calcium sulfate, commonly known as the mineral gypsum, is used in the manufacture of cement, plaster, and wallboard. Japan, which has little natural gypsum, recycles the gypsum produced by scrubbers. Other technologies turn SO_2 gas into sulfuric acid or into elemental sulfur, both of which can be captured and used for other purposes. Elemental sulfur, for instance, is used in matches.

Another strategy to cut down on the amount of SO_2 gas produced by power plants is the use of coals and oils that are naturally low in sulfur content. This is not always practical, however, so technologies have been devised to remove sulfur from oil and coal. For coal, the technique consists of crushing followed by washing in water to remove the iron sulfide mineral pyrite. Pyrite is much denser than coal and can easily be removed by washing.

Removal of nitrogen oxides from power plant combustion is generally achieved by burning the fuel at a low temperature (700° to 900°C) and by limiting the amount of oxygen allowed for the combustion. This keeps the organic nitrogen in the fossil fuel from oxidizing and can also inhibit the reaction of atmospheric nitrogen gas to nitrogen oxides.

- Enactment of legislation since the late 1940s has led to significant reductions in atmospheric pollution.
- Cleanup strategies depend on the type of pollutant produced and the source of the pollution.

Closing Thoughts

The lives of all of us—those who move south to Florida or Arizona to escape harsh northern winters, Kansas farmers dependent on summer rainfall to grow the crops we eat, firefighters in Los Angeles who know they will be busy every summer when the hot Santa Ana winds blow from the east—are intimately linked to atmospheric phenomena. Yet the atmosphere, like the other Earth systems, is in a constant state of change. It is an open system that perpetually exchanges gases and particles with the lithosphere, hydrosphere, and biosphere.

For many hundreds of thousands of years, the size of the human population remained sufficiently small that human activities had little impact on the overall composition of the atmosphere. In the last two centuries, however, the human population has exploded, and this increase, combined with technological advances, has led to unprecedented changes in atmospheric chemistry. Today we are faced with global warming from greenhouse gas emissions, declining levels of protective ozone, and acidification of forests and watersheds. How we choose to address these problems will determine our future standard of living.

Summary

- Earth's average surface temperature is maintained at a value suitable for life by the presence in the atmosphere of water vapor and trace gases such as carbon dioxide, methane, and nitrous oxide.
- Evaporation of water in the equatorial latitudes and condensation in the higher latitudes transports heat from the equator to the poles.
- Earth's atmosphere is organized into layers identifiable by temperature variations. Temperature reflects the distribution of gases with different capabilities to absorb solar radiation and infrared radiation from Earth's surface.
- Air circulation, and therefore weather, occurs primarily in the troposphere.
- Tropospheric pollutants are easily removed by reaction with water droplets. Stratospheric pollutants have longer residence times.
- Differential heating of Earth's surface produces pressure gradients in the atmosphere along which winds flow. Those winds are deflected by the Coriolis effect, which causes air circulation to take on a tricellular form in each hemisphere and thus explains the distribution of deserts and wetter environments.
- Air masses with homogeneous temperatures, humidities, and densities interact with one another along fronts. Frontal lifting of warm air over cold air often initiates updrafts that lead to storms.
- Human activities have produced pollutants in the atmosphere that are causing environmental and climatic change.
- The impact of a pollutant is determined partly by its residence time in the atmosphere. Acid rain is primarily a local problem because the sulfates and nitrates that produce it react quickly with atmospheric water and fall as rain within a few thousand kilometers of the source. Stratospheric ozone depletion, on the other hand, is occurring globally because CFCs are less reactive and can therefore be mixed throughout the atmosphere before they are removed.
- A combination of legislation and technology has led to vast reductions in pollutant emissions to the atmosphere.

Key Terms

atmosphere (p. 263)
aerosols (p. 264)
troposphere (p. 264)
stratosphere (p. 265)
climate (p. 267)
weather (p. 267)
greenhouse effect (p. 267)
Coriolis effect (p. 271)
Hadley cell (p. 271)
monsoon (p. 273)
air mass (p. 275)
front (p. 275)
tornado (p. 279)
tropical storm (p. 279)
hurricane (p. 279)
smog (p. 282)
chlorofluorocarbons (p. 283)
catalytic converter (p. 285)

Review Questions

1. Why are trace gases in the atmosphere so critical to the existence of life on Earth?
2. Why does vertical mixing occur in the troposphere? Why is it important?
3. What happens to solar radiation as it hits Earth?
4. Explain why the poles receive less solar energy than does the equator.
5. What keeps the poles from getting colder, and the equator from getting hotter, every year?
6. What is the Coriolis effect? What impact does it have on atmospheric circulation?
7. Explain the origin of sea breezes, land breezes, and monsoons. In what ways are these phenomena similar?
8. Why do storms occur along fronts?
9. Why do hurricanes lose their strength when they move over land?

10. Why does acid rain have only a local or regional impact whereas production of chlorofluorocarbons has a global impact?
11. Ozone depletion is occurring in the stratosphere, while too much ozone is forming near Earth's surface in the troposphere. Explain why this is a problem and discuss the causes of ozone depletion and tropospheric pollution.
12. What technologies are used to decrease the amount of acid rain produced in the atmosphere?

Thought Questions

1. Imagine an Earth shaped like a cylinder rather than a sphere. Would the Coriolis effect operate on this planet? Why or why not? How would the atmospheric circulation differ from that which develops on a rotating sphere?
2. Draw a picture of Earth with a rocket at the south pole. If the rocket were fired toward the equator, the rocket would appear to be deflected to the west, or left, relative to an observer in the southern hemisphere; explain this effect.
3. Suppose Earth's axis of rotation were not tilted. Would we experience seasons? Why or why not? Would all parts of Earth receive the same amount of solar energy? Why or why not?
4. Would you expect hurricanes to be more or less frequent if Earth's temperature warms because of increased concentrations of greenhouse gases in the atmosphere?
5. Suppose that the average time a molecule of SO_2 spends in the atmosphere before reacting with water and falling to the ground as acid rain is 4 days. If SO_2 is emitted from a factory smokestack when atmospheric winds are blowing at 15 km per hour, how many miles downwind of the factory will the acid precipitation be felt?

Exercises

1. Construct a barometer to keep track of atmospheric pressure changes. Attach a piece of tubing to a water bottle. Fill the bottle halfway with water and turn it upside down while holding the tubing up so that water doesn't flow out. Water will flow partway up the tube. Tape the tubing to the water bottle and tape a small ruler next to the tubing. Record the level of the water in the tube every day. What do you observe when storms pass through your area? What causes these effects? Hint: How does the air pressure inside the barometer compare with that outside?

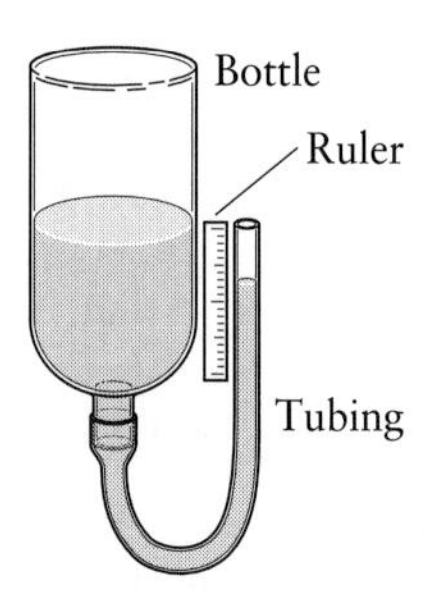

2. To learn more about why vertical mixing occurs in the troposphere but not the stratosphere, fill two clear glasses with lukewarm tap water. Then fill one cake pan with ice and another with hot water. Put one water-filled glass into each cake pan, making sure that the glass is resting on the bottom of the pan. Which glass represents the troposphere, and which the stratosphere? Why? Wait a few minutes for water motions to stop and then drop a little food coloring in each of the glasses. What do you observe? Hint: Think about convection.
3. Determine the pH of rainfall in different local areas by collecting rainwater in a jar or cup and testing it with a piece of pH paper (ask your college chemistry stockroom if they will supply you with a few pieces of pH paper). Try to collect rain from locations near fossil fuel–burning power plants. Does the pH at different locations vary? Do the samples taken near the power plant differ from the others? How?

Suggested Readings

Elsom, D. M., 1992. *Atmospheric Pollution: A Global Problem*. 2nd ed., New York: Blackwell Publishers.

Graedel, T. E., and Crutzen, P. J., 1993. *Atmospheric Change: An Earth System Perspective*. New York: W. H. Freeman and Company.

Graedel, T. E., and Crutzen, P. J., 1995. *Atmosphere, Climate, and Change*. New York: W. H. Freeman and Company.

Lents, J. M., and Kelly, W. J., 1993. "Clearing the Air in Los Angeles." *Scientific American* (October): 32–39.

Parson, E. A., and Greene, O., 1995. "The Complex Chemistry of the International Ozone Agreements." *Environment* 37: 16–41.

Pendick, D., 1995. "Tornado Troopers." *Earth: The Science of Our Planet* (October): 44.

Williams, R., 1996. "The Mystery of Disappearing Heat." *Weatherwise* 49: 28–29.

CHAPTER 10

In 1994, 50 million salmon migrated to Bristol Bay in Alaska, where the Egegik River discharges nutrient-rich sediments, in order to head upstream and spawn. Officials regulate fishing to protect salmon stocks, using undercover agents working as crew, unmarked fishing boats, and aircraft surveillance. In response, fishing companies use increasingly sophisticated technology. Here, the helicopter of a seafood company scopes the water for the glint of salmon scales.

The Ocean and Coastal System

Life on Earth is thought to have begun in the oceans, and for millennia the oceans have helped to sustain human life with abundant harvests of fish. Since the 1970s, however, the stocks of fish have become so depleted by overfishing and pollution that they are losing their potential for renewal. While we draw much of our food from the sea, we also use it as a cesspool for our sewage and other wastes. Nearly 60 percent of humanity lives along coasts. As our numbers grow, so do the amount of ocean pollution and the demand for fish. Countries throughout the world, therefore, are seeking to protect their offshore resources, both fish and the rich stores of petroleum and minerals buried in the seafloor.

The oceans form a dynamic system. Immense currents rise to the surface at some places and sink to great depths at others, carrying water, nutrients, pollutants, and spilled cargos—including shipments of sneakers—from one part of the globe to another. Where deep currents rich in nutrients well up along coasts, as they do off Alaska and Peru, some of the world's most productive fisheries are found. Moreover, ocean currents transport heat and carbon dioxide, forming a coupled atmosphere-ocean system that drives global climatic and environmental change. To understand ocean resources, pollution, and links to global climate and environmental change, we will:

- Examine the nature of ocean basins and their shoreline boundaries.
- Analyze global ocean circulation and its links to other Earth systems.
- Investigate the causes of coastal erosion, sudden flooding, and rise of sea level.
- Evaluate the effects of humans on the seas and efforts to control marine pollution.

The oceans are a great global commons, belonging to no single nation and defying attempts at international governance. In 1609, in a publication entitled *Mare Liberum,* the Dutch lawyer Hugo Grotius articulated the first attempt to manage this commons. Grotius reasoned that, for self-protection, coastal nations were entitled to maintain exclusive access to their offshore waters within distance of a cannon shot from shore. Beyond that distance—about 6 km—all nations were entitled to free access, because the ocean and its resources were too vast to be altered by human activities.

Since then, population growth and the development of technologies for exploiting the resources of the sea make Grotius's reasoning obsolete. At the turn of the 20th century, 2 million to 3 million tons of fish were being harvested from the seas each year, mostly from coastal waters. As the 21st century is about to begin, that amount has increased to more than 80 million tons per year, and many of the fish are caught in deep, mid-ocean waters with the aid of satellite navigation systems, radar tracking, deep-sea sounding devices, and giant nets tens of kilometers long. The 20th century also brought technology that made it possible to drill oil wells and mine ores under water hundreds of meters deep. As demand for marine resources increases, so does international conflict. Who owns the seas? Who has the right to its fish, petroleum, and minerals?

In the 1970s and 1980s, the United Nations convened several conferences on the "Law of the Sea" in order to divide the oceans into areas of national and international jurisdiction. Ultimately the area within 200 nautical miles (370 km) of a country's shore was defined as that country's Exclusive Economic Zone (EEZ). Within their EEZs, nations have exclusive rights to the ocean's natural resources, but the remainder of the sea is open to all. So far, more than 122 nations have laid claim to their EEZs, removing the richest portions of the oceans from the global commons.

In 1994, the Canadian coast guard seized a foreign fishing vessel and arrested its crew 45 km outside Canada's newly defined EEZ but within the Grand Banks fishing grounds on the country's continental shelf. Claiming that the ship's Portuguese operators were pirates who would steal Canada's cod and throw 30,000 people out of work, the Canadian parliament enacted a law authorizing the confiscation of any foreign vessel fishing anywhere on the Grand Banks, even outside the EEZ. By the argument of this law, Canada's jurisdiction extends to its entire continental landmass both above and below the sea, even where its undersea extent exceeds the uniform limit of the EEZ.

While such confrontations can pose an international threat, pollution of the ocean near coastlines can produce a devastating and immediate reality. In 1991, an Asian freighter secretly dumped its sewage in a harbor near Lima, Peru, where beds of shellfish provided villagers with a major source of food. The incident coincided with a festival at which was served ceviche, shellfish marinated in lime juice and served raw. Within days, villagers became violently ill from the Asiatic strain of the bacteria that cause cholera, a life-threatening disease characterized by severe intestinal distress and diarrhea. The bacteria are passed along in human fecal matter. The shellfish in the bay ingested the sewage from the Asian ship and became contaminated with the bacteria. On shore, the disease spread rapidly as water supplies became contaminated because of inadequate sanitation facilities, and travelers carried the bacteria to other countries. By 1993 more than 700,000 people in Central and South America had been diagnosed with Asiatic cholera and more than 6000 had died. As a result of this and similar tragedies elsewhere, many nations now prohibit disposal of raw and treated sewage within set limits of their shores, but complete enforcement is virtually impossible (Figure 10-1).

FIGURE 10-1 A British vessel dumping sewage sludge (treated sewage waste) into the North Sea in 1989. International pressure to ban ocean dumping of both raw and treated sewage is increasing and many nations prohibit the practice.

The oceans are vast but not limitless. According to a report by the WorldWatch Institute, "Rapid population growth, industrial expansion, rising consumption, and persistent poverty have resulted in levels of coastal pollution, habitat destruction and depletion of marine life that now constitute a global threat to the ocean environment." Grotius could not have foreseen or understood today's capacity for overexploiting and polluting the oceans.

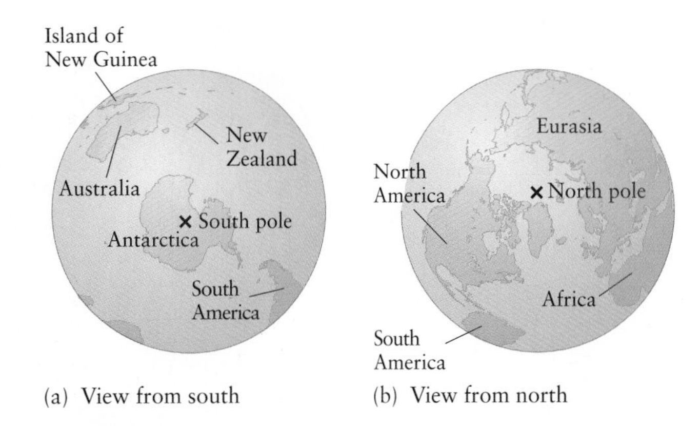

FIGURE 10-2 (a) A view of Earth's southern hemisphere shows mostly water. (b) A view of the northern hemisphere shows approximately equal parts water and land. (× markes the south and north poles.)

The Ocean Basins

If viewed from above the south pole, Earth appears as a vast mass of water surrounding a few relatively small landmasses. Although landmasses in the northern hemisphere are greater, still 71 percent of Earth's surface is covered with ocean water, and all the world's oceans are connected (Figure 10-2). Within this interconnected world ocean are five major ocean basins: the Atlantic, Pacific, Indian, Arctic, and Antarctic oceans. Also connected to the world ocean but largely enclosed by land are smaller bodies called seas. Among them are the Caribbean, Red, North, and Mediterranean seas.

By the end of the 19th century, it was known that continental territory extends beyond the coastline, under the ocean. In the aftermath of World War II, U.S. president Harry S. Truman claimed this territory as it extended from his own nation:

> [T]he Government of the United States regards the natural resources of the subsoil and seabed of the continental shelf beneath the high seas but contiguous to the coasts of the United States as appertaining to the United States, subject to its jurisdiction and control. (Proclamation No. 2667, 1945)

Citing the need for new sources of petroleum and other minerals, and anticipating the energy problems of subsequent decades, the proclamation defined the continental shelf as submerged land that is contiguous to the continent and covered by no more than 100 fathoms (183 m) of water. Canadian officials relied on a similar definition of continental shelf in their 1994 arrest of the Portuguese fishing crew on the Grand Banks.

Features of the Seafloor

At the time of the Truman Proclamation, the submarine weaponry of World War II had given nations ample incentive to understand the topography of the ocean basins. **Bathymetry** is the science of inferring seafloor topography by measuring the depth of the water (from the Greek *bathys*, "deep"). Bathymetric studies, using techniques that range from the ancient line-and-sinker method to modern sonar depth measurement, together with satellite measurements of differences in height of the ocean surface, have provided a notion of what the seafloor looks like and how deep ocean water is at various spots (Box 10-1). A global view of ocean topography from bathymetric studies is provided in Figure 10-3, and continental topography is shown for comparison.

The topography of the ocean floor varies substantially from place to place, but several consistent undersea landforms can be identified. They include the continental shelf, continental slope, continental rise, abyssal plain, mid-ocean ridge, and deep-sea trench (Figure 10-4).

Around every continent is its **continental shelf,** a nearly flat underwater plain, sloping less than 0.5° and covered with up to thousands of meters of layered sand and mud eroded from the continent and transported to the shore by rivers and glaciers. Continental shelves vary in width from a few kilometers along the Pacific coasts of North and South America to more than 1000 km in the Arctic Ocean. The Grand Banks in Canada are part of the broad continental shelf along the northeastern coast of North America, and the variety of hills and valleys on this shelf provides fish with many diverse habitats for breeding and feeding. Off Peru and Chile, in contrast, the continental shelf is narrow and uniformly shallow.

At the outer edges of continental shelves, water depth is as much as 200 m. From this point, the nearly flat shelf gives way to a relatively steep drop called the **continental slope,** which is angled as much as 6° and reaches depths of 3000 m or more. Like continental shelves, continental slopes are covered with thick layers of mud and sand eroded from the continent, and both features are deeply incised by *submarine canyons* that, in some places, dwarf the Grand Canyon in width, depth, and

FIGURE 10-3 A digital shaded-relief map of Earth's surface. Land elevations are taken from topographic maps and ocean floor elevations from depth data acquired by the U.S. Navy and other groups. Color as well as shading is used to make apparent such features as continental shelves, abyssal plains, and oceanic trenches.

length (see Figure 10-4). Through these submarine canyons, sediment is transferred to deep-ocean basins. Sometimes the sediments are spread over broad, gently sloping aprons that form transition zones between the continental slopes and the deep ocean. Such a transition zone, sloping about 1°, is called the **continental rise.**

Together, the continental shelf, slope, and rise constitute the *continental margin.* The Third Law of the Sea Conference, which ended in 1982, allows a nation to extend its territorial claim as far seaward as 350 nautical miles (665 km) if its continental margin extends beyond the 200-nautical-mile-limit of the EEZ, as Canada's does. These limits represent a considerable portion of the ocean floor. If all coastal nations claimed their EEZs, collectively they would control 36 percent of the ocean floor. Furthermore, if all those with especially broad continental margins extended their claims to the 350-nautical-mile-limit, the proportion of the ocean floor controlled by individual nations would rise to 42 percent, about as much area as that which stands above the ocean surface as land.

Continental rises typically grade imperceptibly into a wide, flat **abyssal plain** that forms most of the world's ocean floor and is submerged beneath 3 to 5 km of water. In most places, the abyssal plain consists of a few hundred meters of sediment composed of the shells of microscopic floating organisms that sank and buried the underlying basaltic ocean crust. Although abyssal plains are the flattest features on Earth, they are not completely so. Extinct volcanoes, or *seamounts,* sometimes rise as much as 1 km above the surrounding plain and occasionally break the surface to form islands, as at Bermuda.

The most striking landforms of the ocean floor are the mid-ocean ridges and deep-sea trenches (see Figure 10-3). **Mid-ocean ridges** are submarine volcanic mountain ranges. They form the longest, most nearly continuous mountain belt on Earth, extending over 60,000 km and covering nearly one-third of the ocean floor. Mid-ocean ridges form along divergent plate boundaries, where magma wells up between separating plates and solidifies into new oceanic crust along the ridge crest (see Chapter 4). As new crust solidifies, it also is broken apart, or rifted, by tectonic forces that cause seafloor spreading. *Rift valleys* 1 to 20 km wide form along the crests of mid-ocean ridges. Mid-ocean ridges are the sites of greatest volcanic and earthquake activity on the seafloor.

Substantial earthquake activity also occurs at **deep-sea trenches**—long, narrow depressions at convergent boundaries, where an oceanic plate is subducted beneath a plate of lesser density. There are deep-sea trenches off the western coasts of South and Central America, along the North American Plate from California to Alaska, and adjacent to chains of volcanic islands in the northern and western Pacific Ocean, including the Aleutian, Japanese, Philippine, and Indonesian islands (see Figure 10-3). The

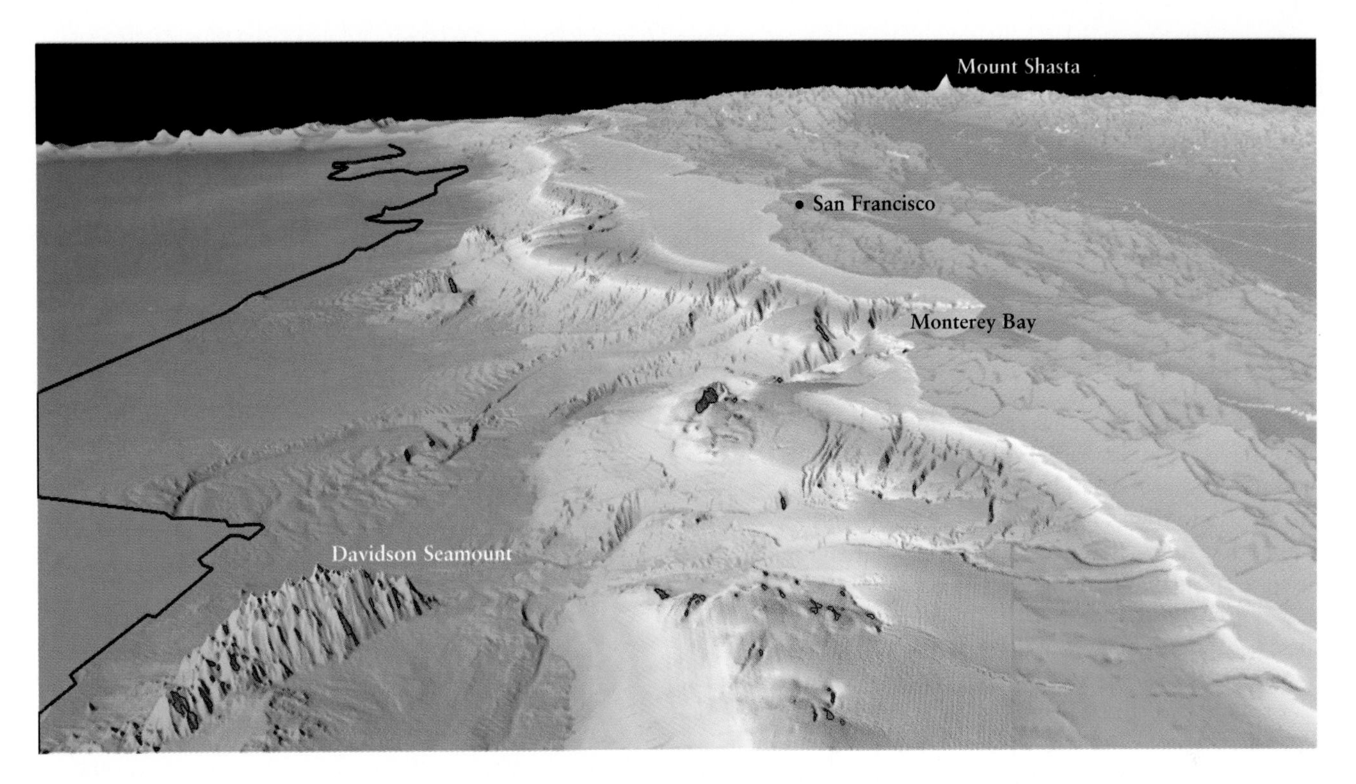

FIGURE 10-4 Multibeam sonar surveys, completed to the right of the black line in this image, provide detailed images of the seafloor. The continental shelf along coastal California, shown here in shades of gray to pink, is especially wide near San Francisco. The continental rise, shown in shades of red to yellow, slopes seaward, in some places more steeply than others, and grades into the continental rise, shown in shades of green to blue. The continental rise merges imperceptibly with the abyssal plain, which is nearly flat except for occasional seamounts and canyons. Rumpled features on the abyssal plain are sediments deposited on extensive undersea fans, places where mass movement has occurred off the continental rise.

bottom of the deepest trench on Earth, the Mariana trench in the western Pacific Ocean, is 11.04 km below sea level.

Ocean Basins, Coastlines, and Plate Tectonics

Even as nations haggle over how best to divide the oceans and their resources, plate tectonics is causing some ocean basins to close and others to widen and is influencing the shape of coastlines. Our maps of the positions of continents and oceans are like snapshots that freeze in place moving continents and oceans. At rates of seafloor spreading and plate convergence up to only 10 m or so per century, however, the change in size of an ocean basin is of little immediate consequence to mapmakers or national leaders.

The Changing Ocean Basins The basin of the Atlantic Ocean has been expanding for some 200 million years. Tectonic forces tear the lithosphere apart along the Mid-Atlantic Ridge, and magma wells up from the mantle to fill the gap with new crust. At the same time, keeping the lithosphere and rock budget in balance, the Pacific Ocean has been shrinking along a vast ring of subduction zones. Although there are spreading centers in the Pacific Ocean, the Pacific Plate is being consumed appreciably faster than new crust is being created. The Mediterranean Sea also is shrinking as the African and Eurasian plates converge and the seafloor between them is subducted.

Juvenile ocean basins are beginning to open in the African Plate at the East African Rift; in the North American Plate at the Gulf of California where Baja California is separating from the Mexican mainland; and between the African and Arabian plates at the Red Sea (see the photograph that opens Chapter 4). The Atlantic Ocean basin must have resembled these rifts when it began to open about 200 million years ago.

Coastlines and Tectonic Settings As ocean basins change shape with time, so do their coastlines. Furthermore, the shape and nature of a coastline are related to its proximity to a plate boundary and to the type of that boundary. Two fundamental sorts of coastlines can be

10-1 Geologist's Toolbox

Exploring the Ocean Floor

How have scientists learned how deep the oceans are? How have they produced detailed maps of seafloor topography? About 85 BC, Posidonius, a Greek philosopher possessed of a great scientific curiosity, sailed to the middle of the Mediterranean Sea to determine its depth. He lowered a rope with a rock attached to one end until it hit bottom; it must have taken a long time, for the bottom was nearly 2 km below the surface. Until the 20th century, that same line-and-sinker method was used to measure ocean depth.

Between 1872 and 1876, during one of the world's first oceanographic expeditions, the scientific crew of HMS *Challenger* took 492 depth soundings of the ocean floor with a weighted hemp line in all ocean basins except the Arctic. Until that trip, it was generally thought that the deep-ocean floor was flat and featureless, but the scientists on the *Challenger* discovered a trench deep enough to contain Mount Everest, as well as mountain ranges that rival continental mountains in their scale.

This vehicle, called the Autonomous Benthic Explorer (ABE) because it travels about the sea bottom, or benthic zone, is set on its preprogrammed course from a support ship. It is shown here during one of its first dives in the Pacific Ocean in 1995. The ABE can go as deep as 5000 m and move into areas, such as hydrothermal vents, that are too dangerous for crewed vehicles. While exploring the Juan de Fuca Ridge spreading center off the Pacific Northwest Coast, the ABE took video images of new lava and measured the temperature of seawater.

Since the *Challenger* expedition, new technologies for exploring the ocean depths have burgeoned, and detailed bathymetric maps are now available (see Figures 10-3 and 10-4). Echo sounders were developed in the 1920s, largely in response to the use of submarines during wartime. Echo sounders mounted on ships send sound waves into the ocean. The waves bounce off the ocean floor and return to the ship. Because the speed of sonar waves is known, the depth of the water can be calculated from the time the wave takes to travel to and from the ship. More recently, ships have towed equipment just above the seafloor that emits bursts of sound in opposite directions. The shape of the seafloor affects the way the energy is scattered, and some of it returns to listening devices also in tow. Called *side-scan sonar,* this technique uses the amplitudes of recorded echoes to produce images that resemble black and white aerial photos. Though useful in the same way that aerial photos help us to study the patterns of the land surface, the sonar images do not reveal the actual depth of the seafloor.

defined based on the tectonic setting (Figure 10-5). A **leading-edge coast** (also known as an *active coast*) forms where one plate, typically continental, overrides a denser, typically oceanic, plate at a convergent boundary. A **trailing-edge coast** (also known as a *passive coast*) is located along continental margins, away from plate boundaries.

The coastlines along the Pacific convergence zones that delineate the "ring of fire" of volcanoes are of the leading-edge type; those set back from spreading centers in the Atlantic are trailing edge. On North America, both leading-edge and trailing-edge coasts are found. The gently sloping eastern coast of the continent is a trailing-edge coast in the middle of the North American Plate. The craggy western margin carries a leading-edge coast that abuts the Cascadia subduction zone.

Each of these tectonic settings determines general coastal characteristics that evolve over a thousand kilometers or more. Leading-edge coasts generally are mountainous, rocky, cliffed shorelines pounded by large waves (Figure 10-6). Mountain building and uplift are common, raising the coastline out of the water and causing streams to incise deep, narrow canyons. Because they typically are bounded by subduction zones and deep-sea trenches, leading-edge coasts usually have very narrow continental shelves and little coastal plain. Sediment eroded from canyons is deposited in deep-sea trenches and offshore

After the designation of Exclusive Economic Zones by the Law of the Sea, coastal nations increased efforts to chart their new territories. In the United States, the National Oceanic and Atmospheric Administration has surveyed more than 200,000 km^2 of seafloor along the Atlantic, Pacific, and Gulf coasts. A new technique, *multibeam sonar*—currently the most sophisticated form of bathymetric seafloor mapping available—was used. Multibeam sonar also bounces sound waves off the ocean floor, but it uses numerous sound sources and listening devices mounted directly on the ship rather than towed behind. Every few seconds, a burst of sound is emitted in a strip perpendicular to the direction the ship is traveling (the red area in the diagram), and listening devices detect sound coming from a narrow strip parallel to the direction of travel (the green area). In this way, the seafloor is mapped in successive swaths as the ship moves back and forth, much as a lawn is mowed. Computers are used to analyze multibeam sonar data and to present detailed maps of the seafloor. These maps are valuable to petroleum companies exploring and drilling for oil and to geologists searching for faults that might cause earthquakes.

A nonbathymetric technology for mapping the ocean floor was made possible by satellites. Signals from satellites to the ocean surface measure subtle changes in the height of the surface. Seafloor topography can be inferred from an analysis of these changes in ocean surface height. Called *satellite altimetry mapping,* this method of seafloor mapping is indirect, but it offers a way to infer seafloor topography even in places inaccessible to ships.

Finally, some ocean-floor exploration is done directly—that is, scientists visit the ocean floor in submersible vehicles or direct uncrewed vehicles from a ship. Crewed submersibles afford first-hand, close-up views of the seafloor, but their coverage is limited because they cannot travel far during a given dive. Crewless vehicles that are remotely controlled or have preprogrammed routes are becoming increasingly common as an effective way to explore either larger or more hazardous areas.

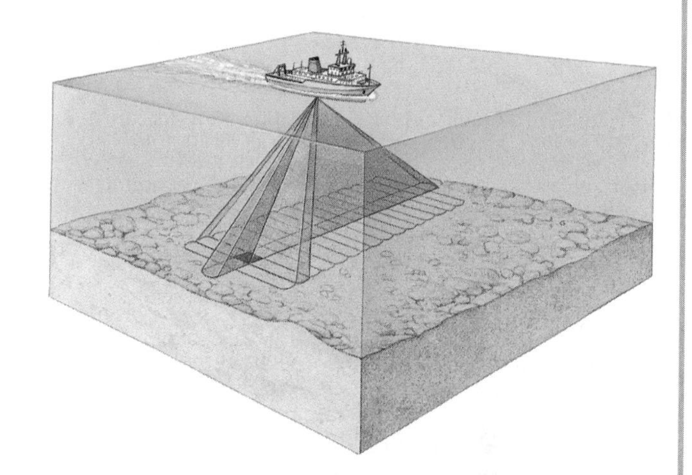

Multibeam sonar instruments on board a ship can map large areas of the seafloor in extremely great detail, as in the map of the western coast of the United States shown in Figure 10-4. The ship's instruments send out sound waves, then record them as they reflect off the seafloor.

basins formed by folding and faulting. Deep water close to shore allows waves to maintain their size even as they approach shore, and hence they hit the coastline with tremendous energy and erosive power. With no shallow nearshore area available to store sediment, and with such high-energy waves to remove any sediment that does accumulate, large deltas cannot form.

In contrast, most of the world's largest deltas occur on trailing-edge coasts, where continental shelves often are broad and extensive and shallow water reduces the size of waves as they approach shore. The extent of the continental shelf is related partly to the age of the ocean basin, older basins generally having wider continental shelves and greater accumulation of sediment. The Gulf of California and the Red Sea—both recently opened ocean basins—are narrow and have little sediment accumulation along their edges. More mature trailing-edge coasts, such as the eastern shores of North and South America and the western shore of Africa, are associated with gently sloping, meandering rivers and broad coastal plains upon which sediment is stored and reworked, typically as sand dunes and coastal islands (Figure 10-7).

Coastlines and Sea Level If sea level rises the coastline retreats, and if sea level falls the coastline advances seaward. Furthermore, a change in sea level alters the nature

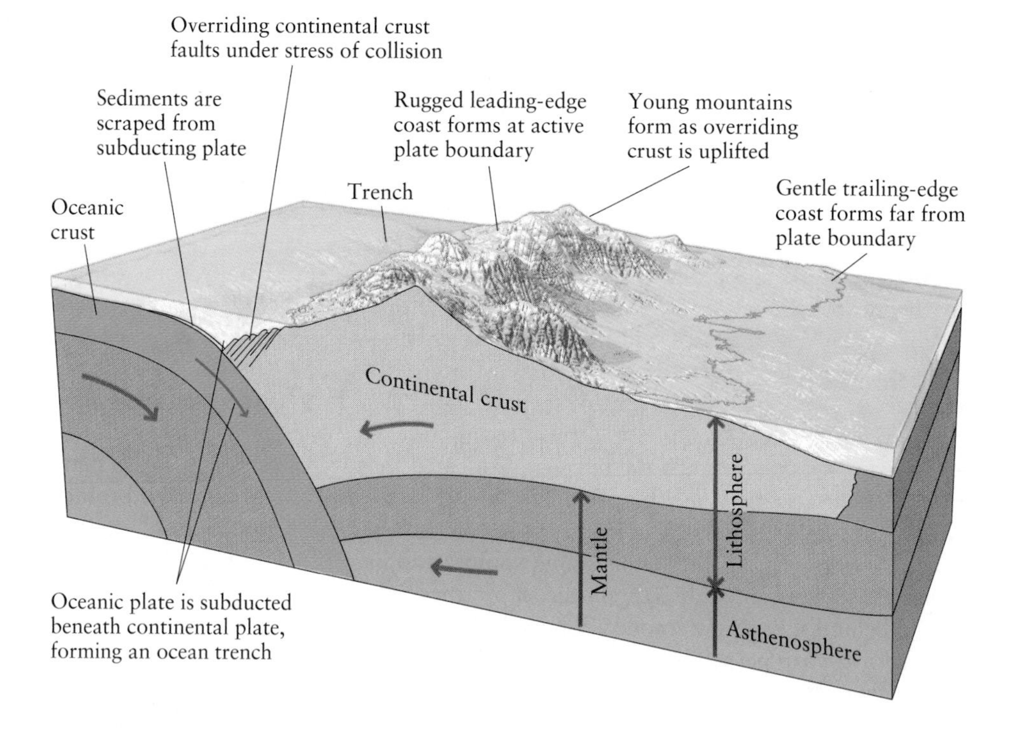

FIGURE 10-5 Tectonically active coasts are associated with active plate boundaries and are classified as leading-edge coasts because they lie at the front, or leading, edge of a moving plate where it comes in contact with other plates. Coasts located away from plate margins, embedded within and riding along with a plate, are called trailing-edge coasts.

of sediment deposition and transport. Plate tectonics can influence sea level through uplift and subsidence. In addition, climate can influence sea level. Global cooling traps water in glacial ice, lowering sea level. Global warming releases water from glacial ice and expands ocean water somewhat, raising sea level.

When sea level falls, substantial parts of continental shelves and sometimes continental slopes are exposed, and rivers cut deep gorges across the surface on their way to the ocean basin. When sea level rises, as at present, the lower reaches of rivers are filled and their mouths drowned, forming estuaries and fjords that are especially prominent on trailing-edge coasts, where little tectonism and uplift of the land occur. *Estuaries* are bodies of ocean water that have encroached inland over the mouth of a river. The constant supply of fresh river water makes the water in estuaries *brackish,* or intermediate between fresh and salt. On the U.S. Atlantic coast, Chesapeake Bay and Delaware Bay are examples of estuaries (Figure 10-8). *Fjords* are drowned coastal valleys carved by glaciers rather than by rivers. Glacial valleys tend to be deeper and more U-shaped than those carved by rivers. Sea level has changed substantially and frequently during geologic history, and the hazards thus presented are discussed below.

FIGURE 10-6 The coast of Peru is a leading-edge coast. Situated on the western margin of the South American Plate, where it overrides the subducting Nazca Plate, it is characterized by steep slopes and a high-energy beach. The steplike benches carved into the coastal landmass are remnants of past shores planed by wave action and since raised out of the water by tectonism.

FIGURE 10-7 The coast of Brazil, along the eastern margin of South America, is a trailing-edge coast, characterized by a broad coastal plain, shallow water, and gentle waves. A flock of scarlet ibises can be seen flying over the mangrove tres.

- From ocean bathymetry, scientists have learned that the ocean floor has substantial topographic relief and consists of six dominant landforms: continental shelf, continental slope, continental rise, abyssal plain, mid-ocean ridge, and deep-sea trench.
- Coastal nations have become increasingly territorial regarding their offshore continental shelf areas because of the wealth of resources—particularly fish, petroleum, and minerals—that are found there.
- According to the Law of the Sea, the area within 200 nautical miles (370 km) of a country's shore is defined as that country's Exclusive Economic Zone (EEZ), within which coastal nations have exclusive rights to the ocean's natural resources, but the resources of the remainder of the sea are open to all nations.
- Two types of coastlines are defined, based on their tectonic setting. A leading-edge coast (or active coast) typically forms where a continental plate overrides an oceanic plate at a convergent boundary; it tends to be characterized by high wave energy and coastal erosion. A trailing-edge coast (or passive coast) forms along continental margins away from plate boundaries and is marked by broad coastal plains and sediment accumulation in deltas and sand dunes.

FIGURE 10-8 This satellite view of the Atlantic coast from New York to southern Virginia was taken during icy, winter conditions. Chesapeake Bay (the larger body of water to the south) and Delaware Bay (to the north) are estuaries that formed about 5000 years ago when rising sea level flooded the lower reaches of many rivers that drain into these coastal indentations. The indentation at Chesapeake Bay is thought to have formed about 25 million years ago, when a meteorite crashed into the continental shelf.

Seawater Properties and Ocean Circulation

Ocean water tastes salty because it contains a relatively large amount of ions from dissolved mineral salts, particularly sodium chloride. Some seawater, such as that in the Dead Sea, is much saltier than ocean water in other, less arid locations, but in general the salinity of ocean water varies only slightly from one place to another and from one year to the next, ranging from 3.38 to 3.68 percent dissolved salts in a given amount of water. This percentage may seem small, but it is enough to make seawater toxic to many living things. The makeup of ancient marine sedimentary rocks shows that 200 million years ago seawater had a composition similar to that of today's oceans, evidence that the amount of the most abundant dissolved substances in the ocean reservoir is in an approximate long-term steady state.

The Salinity of Seawater

The percentage of dissolved salts in a quantity of water is a measure of its **salinity**—its saltiness. How saline is seawater compared with other water? Where do the salts come from?

Seawater contains a total of about 35,000 mg of ions per kg (1,000,000 mg) of water (Table 10-1), so the salinity of seawater is 35,000 mg per kg or 3.5 percent. Of the ions in seawater, the two most abundant are chloride and sodium ions, with average concentrations of about 19,000 mg per kg (1.90 percent) chloride and 10,700 mg per kg (1.07 percent) sodium. Sodium chloride is thus the most abundant salt in the sea, accounting for most of the 3.38 to 3.68 percent of dissolved solids in seawater.

In contrast to seawater, river water is only about 0.01 percent salt (see Table 10-1), and rainwater contains virtually none. If the ocean is fed by rain and rivers, then why is it saline? The answer lies in the flux of ions in the ocean reservoir.

The Ocean Reservoir and Flux of Salt Ions

The salinity of seawater can be explained from a systems perspective (Figure 10-9). The ion input from rivers is relatively small at any one moment, but the ions accumulate in the ocean because they are removed a little more slowly than they are added. Removal of ions may occur through sea spray blown onto the land, through deposition and burial in ocean sediments, or through reaction with volcanic rocks.

Chloride and sodium, the most abundant ions in the oceans, are added slowly but are depleted even more slowly because they have very low chemical and biochemical reactivities in seawater. Only when large evaporite basins form as a result of tectonic and climatic conditions are these ions removed in substantial amounts, and that has happened relatively infrequently in geologic history.

The approximate residence times of the major constituents of seawater can be estimated from the amount of each element in ocean water divided by the rate at which it is added or removed. Those ions with relatively brief

TABLE 10-1 Major Ions in Seawater and River Water

Constituent ions	Concentration in seawater (mg/kg)*	Concentration in river water (mg/kg)*	Mean residence time in ocean (million yrs)
Chloride (Cl^-)	19,350	5.75	120
Sodium (Na^+)	10,760	5.15	75
Sulfate (SO_4^{-2})	2,712	8.25	12
Magnesium (Mg^{2+})	1,294	3.35	14
Calcium (Ca^{2+})	412	13.4	1.1
Potassium (K^+)	399	1.3	11
Bicarbonate (HCO_3^-)	145	52	0.10
Bromide (Br^-)	67	0.02	100
Boron (B^{3+})	4.6	0.01	10.0
Strontium (Sr^{2+})	7.9	0.03	12
Fluoride (F^-)	4.6	0.10	0.5
Total	35,148	89.24	

*Milligrams per kilogram is the same as parts per million by weight and can be expressed as a percentage. For example, the total of 35,148 mg of ions in 1 kg (1,000,000 mg) of seawater can be expressed as 35,148 parts per million or 3.5148 percent.

Source: William H. Schlesinger, *Biogeochemistry: An Analysis of Global Change,* San Diego: Academic Press,1991.

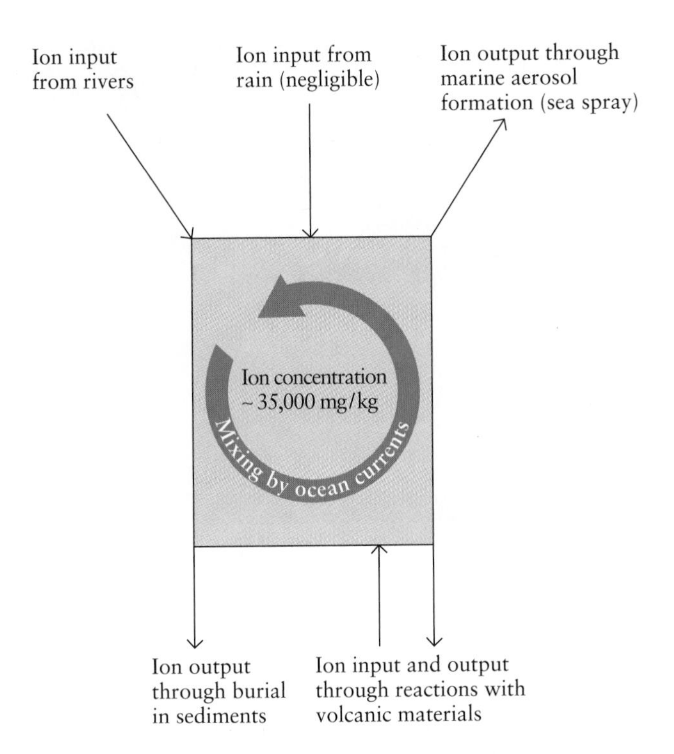

FIGURE 10-9 The ocean can be viewed as a reservoir into and out of which flow dissolved ions and sediment. This model includes only the major constituents of seawater, among them chloride and sodium ions, for which inputs and outputs are approximately balanced over thousands to millions of years. Because the world ocean has no outlet, the only way that ions can be removed is through transfer into the atmosphere by formation of aerosols—suspensions of particles in a gas—or through transfer into the lithosphere by burial in sediments on the seafloor or by reaction with volcanic materials on the seafloor.

residence times, such as calcium and bicarbonate, are depleted rapidly by the biological processes of shell formation. Residence times for the most abundant ions in seawater are extremely long. For example, the residence times of chloride and sodium are respectively 120 million and 75 million years (see Table 10-1). Because these values are much longer than the time periods over which the ions are mixed by ocean currents (1000 to 2000 years or less), their amounts are nearly the same in all parts of the ocean. The effect of the ocean currents is similar to the homogeneity created by a beater that mixes new ingredients being added to a bowl. As long as the beater moves the ingredients around the bowl more rapidly than their rate of addition, whatever is in the bowl will tend to be uniformly distributed.

Nutrients in Seawater

Seawater contains dissolved substances other than salts. **Nutrients** are molecules that provide nourishment to organisms. In the oceans, nutrients nourish life beginning with *phytoplankton*, microscopic plants that float at the surface but are at the bottom of the oceanic food chain. Phytoplankton use nutrients to produce organic compounds and in turn nourish organisms higher on the food chain that eat phytoplankton; these include *zooplankton*, microscopic animals that also float at the ocean surface (Figure 10-10). The main nutrients in seawater include nitrogen (chiefly as nitrate, NO_3^-) and phosphorus (as phosphate, PO_4^{3-}). Carried in runoff from continents, nutrients are abundant along coastlines, especially near the mouths of rivers along continental shelves.

Unlike the salts in seawater, which tend to be mixed uniformly, the amount of nutrients varies markedly in different parts of the ocean. The processes that move nutrients between upper and lower parts of the ocean reservoir occur much more rapidly than global ocean mixing rates. Typically the layer nearest the surface (called the *photic zone* because it gets the most light) is almost completely depleted in nutrients because of the biological activity in this layer. Nutrients are partially recycled within this layer but also are steadily lost to deeper layers through the sinking of dead organic material. Nutrients are recycled from depth to the surface layers where currents well up at coastlines. These areas have high rates of biological activity and are important for fisheries (Figure 10-11).

The Carbon Cycle

In addition to salts and nutrients, seawater contains gases, chiefly nitrogen (N_2), oxygen (O_2), carbon dioxide (CO_2), hydrogen (H_2), and inert, or "noble," gases (argon, neon, and helium). (Noble gases are so called because they "stand alone" and are rarely reactive.) Each of these can be exchanged between the atmosphere and ocean across the air-sea interface. Oxygen and carbon dioxide are highly reactive and are important to photosynthesis. The concentrations of oxygen and carbon dioxide vary greatly in space and time because of differing levels of biologic

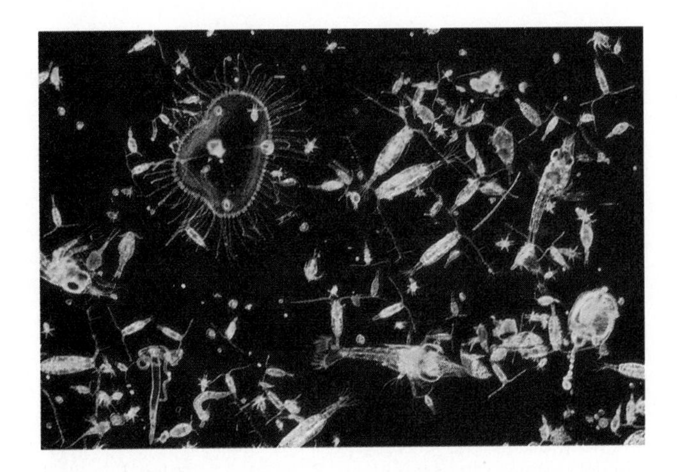

FIGURE 10-10 Passively drifting or weakly swimming, microscopic organisms in the oceans include plant forms called phytoplankton and animal forms called zooplankton.

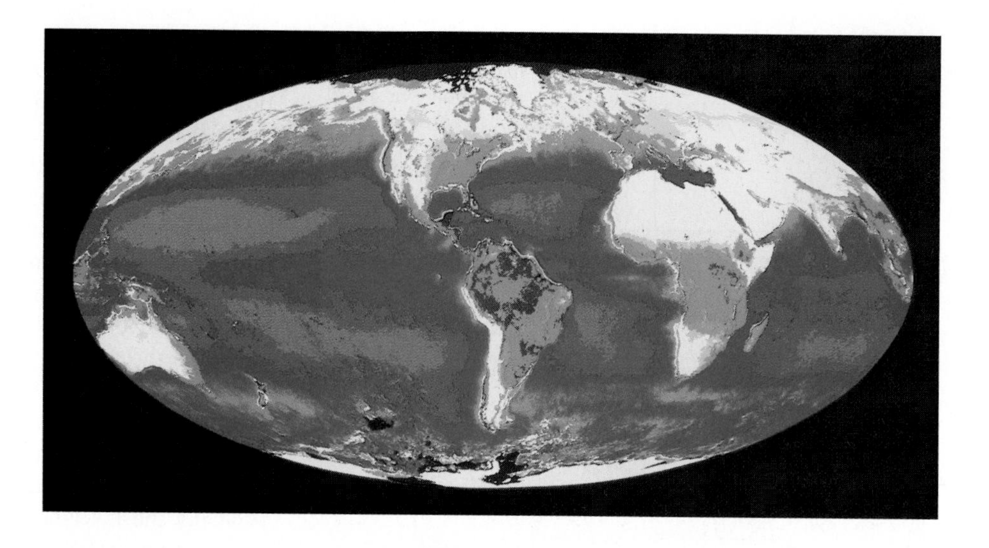

FIGURE 10-11 This false-color satellite image shows the distribution of vegetation on land and phytoplankton in the oceans. Colors represent the density of chlorophyll, from dark green (most dense) to pale yellow (least dense) on land, and from red (most dense) through yellow and blue to pink (least dense) in the oceans. Areas rich in phytoplankton are associated with warm water, broad continental shelves, and upwelling currents.

activity. The amount of carbon dioxide, in particular, has varied as a result of recent human activities. The burning of fossil fuels has caused a steady increase in the rate of addition of excess carbon dioxide to the atmosphere and to the oceans. As a result, the ocean reservoir is not in steady state with respect to this gas.

Carbon dioxide is the main means by which carbon is cycled throughout Earth's system. The cycling of carbon affects the rates of growth of plants and weathering of rocks as well as the temperature of the atmosphere (see Chapter 9). The **carbon cycle** is the continuous circulation of carbon among seven major reservoirs. In order of decreasing stocks, these are the lithosphere (primarily sediments and sedimentary rocks), the deep ocean, fossil fuels, soils, atmosphere, the surface ocean, and plants. A carbon budget for Earth shows the annual quantities of carbon exchanged among these reservoirs (Figure 10-12).

Although the annual flux of carbon through volcanism is small, over geologic time carbon dioxide gas emitted from magma is the chief source of carbon to Earth's atmosphere. Other influxes of CO_2 include emissions from fossil fuel combustion, decomposition of organic matter, and the burning of forests and grasslands.

A conceptual model of the carbon cycle (Figure 10-13) represents the processes by which carbon flows through the Earth system. Carbon is removed from the atmo-

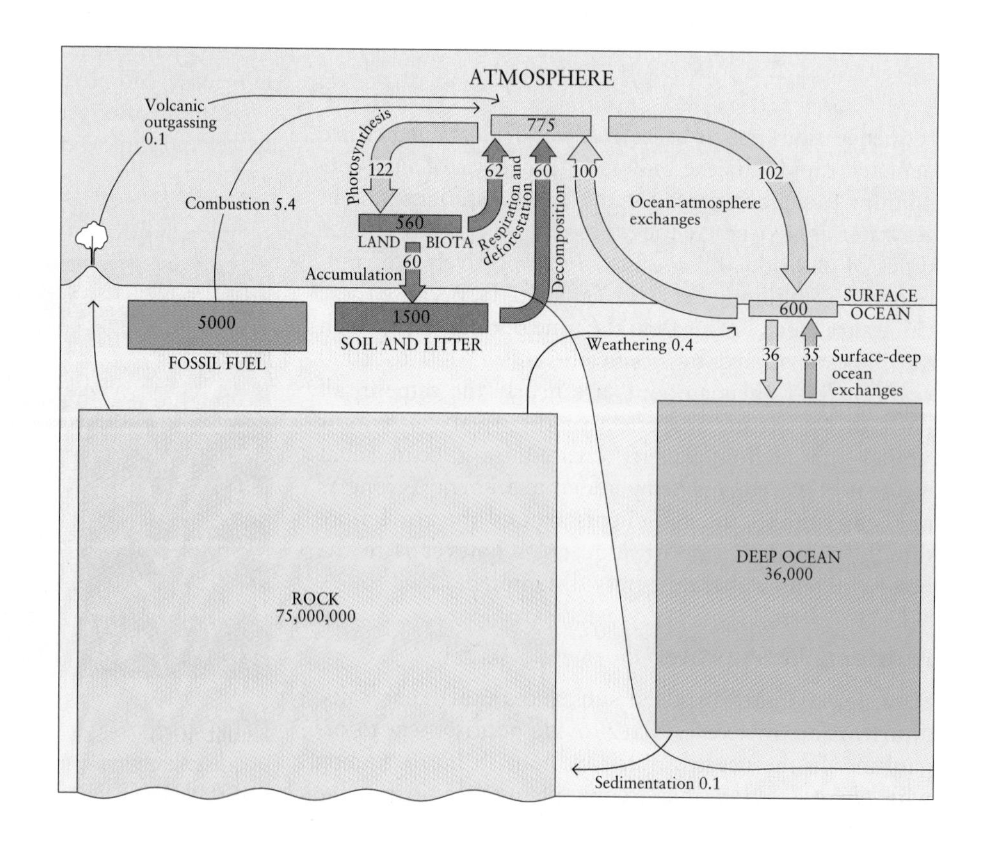

FIGURE 10-12 The carbon budget quantifies the stock of carbon in different Earth reservoirs. Each reservoir is drawn to scale relative to the size of its stock, and arrows representing annual fluxes of carbon from one reservoir to another are shown to scale; the numbers indicate billions of tons of carbon. The largest stock of carbon by far is stored in carbonate rocks in the lithosphere. Of the other stocks of carbon, the largest is the deep ocean. The largest flow of carbon, 122 billion tons a year, occurs between the atmosphere and plants during photosynthesis.

sphere chemically through weathering of rock and biochemically through photosynthesis by plants. Chemical weathering begins when atmospheric carbon dioxide combines with water vapor and falls in rain as carbonic acid (H_2CO_3) (see Chapter 6). This mildly acidic rain chemically weathers rocks, causing some minerals in them to dissolve and dissociate into ions, including calcium (Ca^{2+}) and bicarbonate (HCO_3^-). Ions that run off into rivers and are transported to the ocean are taken up by marine plants and animals. Marine organisms process Ca^{2+} and HCO_3^- ions into hard skeletons and shells of calcium carbonate ($CaCO_3$) (see Figure 4-26). When these marine organisms die, they settle to the ocean floor as carbon-rich, organic sediments.

Tectonic forces may cause some of these carbonate seafloor sediments to be subducted into the mantle. There they melt and react with silicates to form magmas rich in dissolved carbon dioxide. The gas returns to the atmosphere during eruptions at seafloor-spreading centers and volcanoes, completing the cycling of carbon from atmosphere to ocean and back again to the atmosphere.

Carbonate sediments that are not subducted may be buried in the crust, compressed and cemented into sedimentary rocks, then uplifted and exposed during plate collision and mountain-building episodes. Chemical weathering of these rocks releases carbonate ions to solution once again, and the ions are washed into the oceans where they are processed again into biological sediments.

Alternatively, carbon may be removed from the atmosphere through biochemical processes, specifically by plants during photosynthesis. In the oceans, carbon dioxide is used in this way by phytoplankton, which settle to the ocean floor after death. Marine phytoplankton are the dominant source of organic substances in the formation of petroleum (see Chapter 11), so they also contribute to the store of carbon in the fossil fuel reservoir. On land, dead organisms become incorporated into the pedosphere, where they enrich soils, or into the lithosphere as fossil-rich rocks or coal. Decay of organic matter again releases carbon dioxide to the atmosphere.

Two human activities have had great impact on the carbon cycle: combustion of fossil fuels for energy, and burning of forests and grasslands. Both these activities increase the amount of carbon dioxide going into the atmosphere and can lead to global warming, because carbon dioxide gas is a strong absorber of infrared radiation given off by Earth (see Box 12-1).

Scientists are trying to determine exactly where the carbon dioxide in fossil fuels and plants goes after it is released by burning. We know that some of it goes into the atmosphere and accumulates, because the amount of atmospheric carbon dioxide has been rising steadily since the Industrial Revolution. But this amount is only half that known to be released by human activities. Where is the missing carbon?

▸ Cycling of Carbon Among Oceans, Atmosphere, and Biosphere

An Earth system approach can elucidate the fate of the missing carbon. As the carbon budget shows, the annual flux of carbon into and out of the rock reservoir is

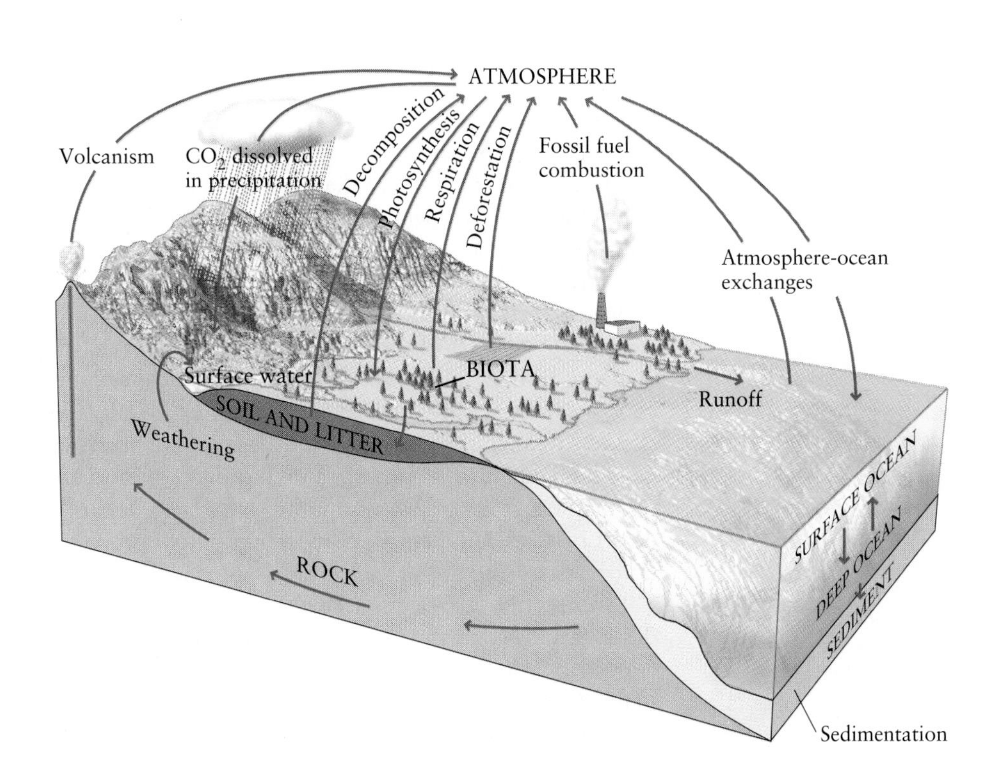

FIGURE 10-13 The carbon cycle is a general model of the sources and flow paths of carbon from one Earth system to another. In this view, carbon contained in surface water that drains weathered carbonate rocks is transferred to the ocean, where it is used by marine organisms to build shells and skeletons. Ultimately, carbon is returned to the mantle by subduction of the seafloor and its sediments, then to Earth's surface by tectonism and uplift.

negligible compared with the fluxes between other reservoirs (see Figure 10-12). Therefore, we can limit our investigation to four carbon reservoirs: the ocean, fossil fuels, biosphere, and atmosphere.

Humans release a total of about 7 billion tons of carbon to the atmosphere annually, with 5.4 billion tons coming from the burning of fossil fuels and 1.6 billion tons from deforestation (Figure 10-14). From studies of bubbles of atmospheric gases trapped in glacial ice cores, scientists have determined that the stock of carbon in the atmosphere is increasing at a rate of about 3.4 billion tons per year, about half the annual amount released into the atmosphere by human activities. Where does the other half go?

Carbon isotopes produced during the testing of nuclear weapons between 1955 and 1967 have provided a clue. These isotopes have made their way from the atmosphere to the oceans, and from them oceanographers have deduced that about 2 billion tons of carbon is extracted from the atmosphere each year by gas exchange with the ocean.

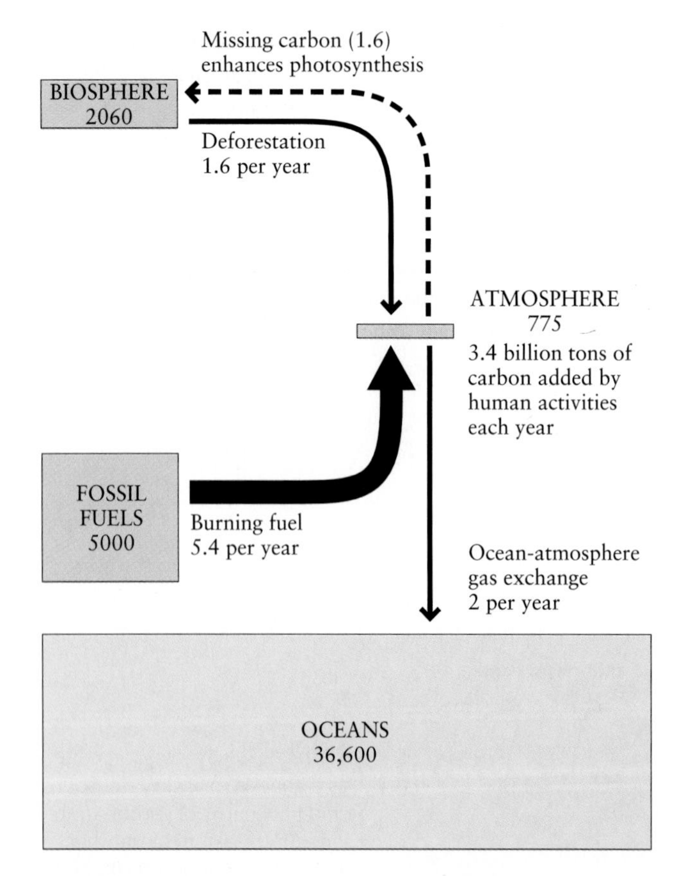

FIGURE 10-14 A carbon budget for the carbon released by human activities. The numbers represent billions of tons. About half the 7 billion tons of carbon released each year, primarily from burning fossil fuels and deforestation, accumulates in the atmosphere. Of the other half, scientists estimate that about 2 billion tons per year goes into the ocean. The remainder, 1.6 billion tons, is the "missing carbon"; some scientists now think that it is used by plants during enhanced photosynthesis.

This means that only 1.6 billion tons is unaccounted for. Where is it going? The only other likely reservoir into which the missing carbon can go is the biosphere, because fairly high exchange rates—or fluxes—are needed to remove the carbon from the atmosphere so quickly. Therefore, scientists suspect that 1.6 billion tons of carbon are cycled back into the biosphere each year through enhanced photosynthesis. This amount is equal to that coming from the biosphere as a result of deforestation. Such an increase in biospheric carbon should be detectable in terrestrial ecosystems, and in fact, scientists studying satellite imagery of Earth's vegetation cover taken over about 10 years reported in 1997 that the amount of photosynthetic activity is increasing.

From many different types of studies, scientists were able to anticipate and tentatively verify the location of the missing stock of anthropogenic carbon. It remains to be seen how Earth systems will regulate the flow of increasing carbon in future years.

Ocean Circulation

The distinct masses of water that flow throughout the oceans and carry matter and energy from one part of Earth's surface to another are the **ocean currents.** They form a pattern of swirling water masses that constitute the ocean's circulation system. Ocean currents depend on the interaction of the ocean with surface phenomena and on the interplay between the water's density and Earth's gravity. The density of water, in turn, depends on its salinity and temperature—the greater the concentration of salts and the lower the temperature, the greater the density of the water. Where there is a density gradient (from different salt concentrations and temperatures), gravity's pull induces movement of water. The ocean has both a horizontal density gradient that runs from the equator to each pole and a vertical density gradient that runs from the surface to the ocean depths. Wind also moves water, but only at the surface.

Mixed by winds and heated through by the Sun, the **surface mixed layer** of the ocean extends to about 0.2 km in depth and is relatively homogenous vertically. Horizontally, however, the surface mixed layer exhibits gradations in temperature, salinity, and density. For instance, near the equator, the rate of rainfall reduces the ocean's salinity, but at about 30° north and south latitude, in the desert belt (see Chapter 9), evaporation outpaces precipitation, increasing the surface layer's salt concentration.

Beneath the surface mixed layer, from approximately 0.2 to 1 km, is an interface where the rate of change of temperature, salinity, or density with depth is at a maximum. Called the *thermocline, halocline,* or *pycnocline,* depending on the physical or chemical property of inter-

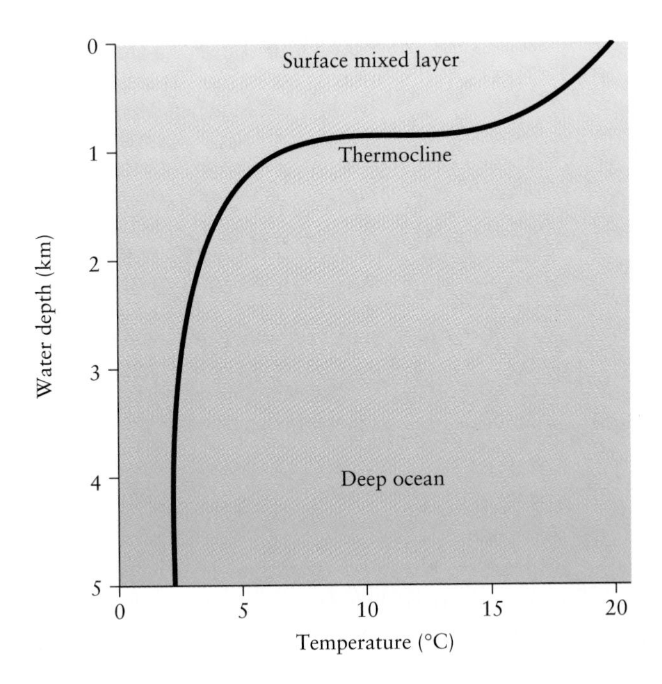

FIGURE 10-15 In mid- to low latitudes, the temperature of seawater decreases with depth, a pronounced drop occurring at about 0.2 to 1 km. Above the temperature drop is the surface mixed layer, which is heated by the Sun and mixed by winds. Below it is the much colder deep ocean, which is not warmed by the Sun or mixed by winds.

est, this interface prevents the surface mixed layer from interacting freely with the **deep ocean,** which lies below 1 km, and so essentially enforces any differences that exist between the two (Figure 10-15). Because the deep ocean is untouched by surface phenomena and cannot readily mix with the surface layer, it is, overall, colder, more saline, and denser. It is also less homogeneous, containing many different water masses, each with distinctive properties. Some of these water masses are extremely cold and saline; others are much warmer and less saline.

Some ocean currents are immense in comparison with masses of continental water. For example, all the rivers flowing into the Atlantic Ocean carry a total discharge of 0.6 million m^3 per sec, but some of the smallest ocean currents, such as that flowing out of the Mediterranean Sea, carry 3 million m^3 per sec, about five times as much water. Larger ocean currents, such as the Gulf Stream, carry many times as much, transporting up to 150 million m^3 per sec. When ocean currents are compared in size with the largest rivers on Earth, it is easy to see why the oceans dominate the global hydrologic cycle.

Circulation of water masses in the surface mixed layer and in the deep ocean is controlled by different processes. The surface ocean moves in response to the winds; the deep ocean is driven by latitudinal and vertical variation in seawater density.

Circulation in the Surface Mixed Layer Winds blowing across the ocean exert a drag on the surface water, pulling and pushing it along. The water in direct contact with the atmosphere moves at the highest speed, and forward velocity decreases with depth in the surface mixed layer until, eventually, no further forward motion takes place.

Because the surface layer of the ocean is driven forward by winds, one might expect that ocean currents would travel in the same direction as the prevailing winds—but they don't. The great Arctic explorer and statesman Fridtjof Nansen of Norway (1861–1930) observed that icebergs floating in the North Atlantic travel at a 20°–40° angle to the prevailing wind direction. Later work by the Swedish oceanographer V. Walfrid Ekman (1874–1954) demonstrated that ocean currents are deflected by the Coriolis force (see Chapter 9) in the same way that atmospheric currents are deflected. Each successively deeper layer of water—moving more slowly than the one above it—is oriented a bit farther away from the prevailing wind direction, producing what is known as the *Ekman spiral.* Averaged over the entire column of water that responds to the winds, the net water movement direction is 90° to the prevailing winds, a phenomenon called the **Ekman transport.** In the northern hemisphere, the net motion is 90° to the right of the winds (Figure 10-16), and in the southern hemisphere, net motion is 90° to the left of the winds.

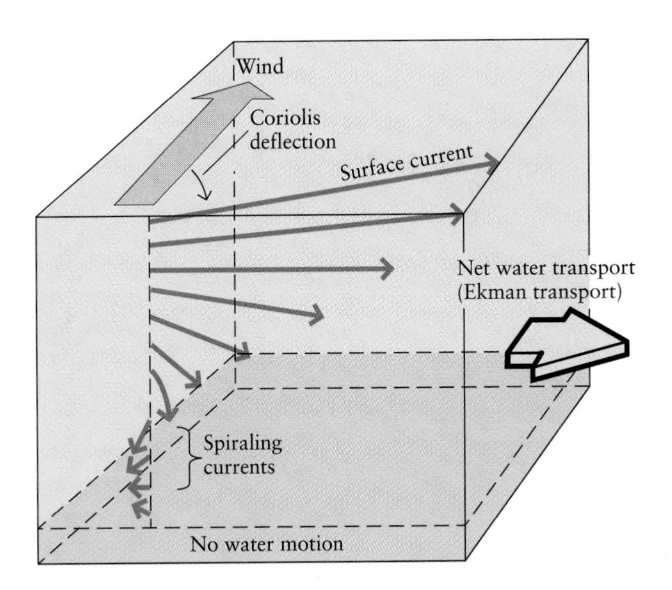

FIGURE 10-16 Ekman transport, shown for the northern hemisphere, results from the interaction of wind friction and Coriolis deflection. Here the Coriolis deflection causes a surface current to flow to the right of the direction in which wind is moving. With depth in the current, each layer of water is deflected a lesser amount to the right of the layer above it, producing a spiraling current called the Ekman spiral. If the flow of all the water depth is averaged, most of it moves 90° to the right of the direction of the generating wind. In the southern hemisphere, Ekman transport is to the left of the direction of the generating wind.

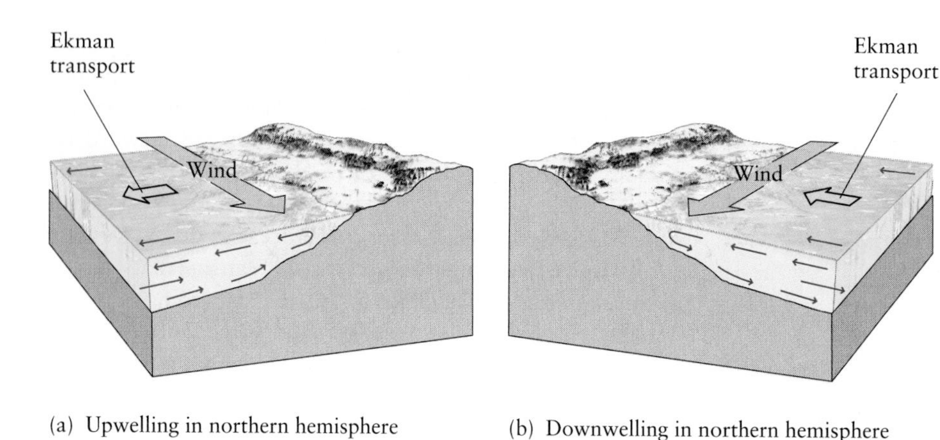

FIGURE 10-17 Upwelling and downwelling ocean currents are caused by winds blowing parallel to a coast and the Ekman transport. These diagrams show the patterns in the northern hemisphere. (a) Where land lies to the east and wind is blowing southward, surface water moves offshore, enabling deep currents to upwell in order to replace the water removed. (b) Where land lies to the west and wind is blowing southward, surface water moves onshore, piling up and then sinking, or downwelling.

The Ekman transport of water plays an important role in the distribution of nutrients and oxygen within the oceans. When winds in the northern hemisphere blow southward along a coastline, the net water transport is to the west because this is the direction 90° to the right of the prevailing wind direction. Thus, if the land lies to the east of the ocean, the net water transport is offshore, and if the land lies to the west of the ocean, the net water transport is onshore (Figure 10-17). In the southern hemisphere, the opposite scenarios occur.

The differences between these two scenarios are important to the locations of the world's fisheries and places of marine biodiversity. When net water transport is offshore, cold water from the deep ocean rushes up to the

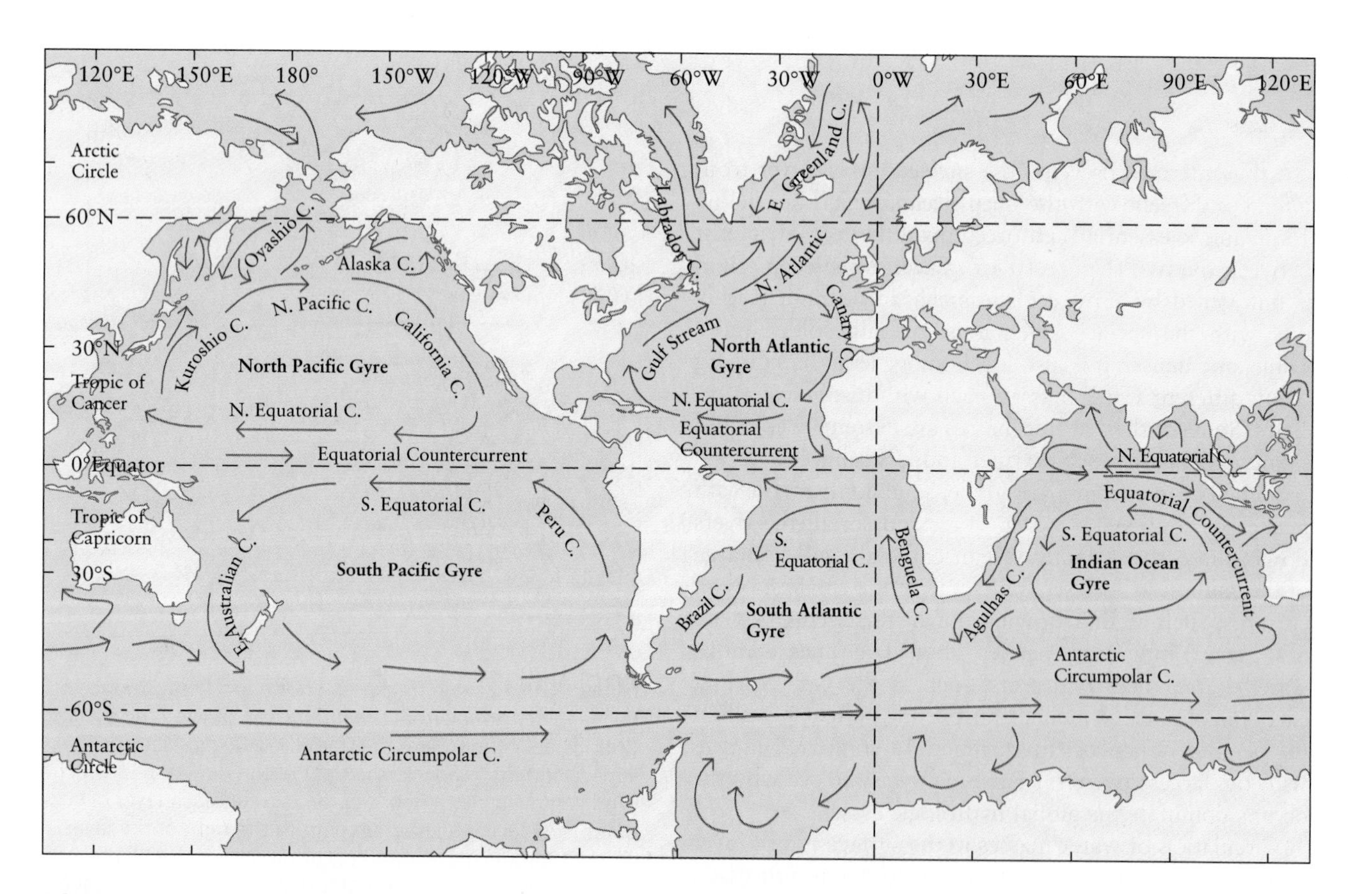

FIGURE 10-18 Surface currents in the world ocean circulate clockwise in the northern hemisphere and counterclockwise in the southern hemisphere. The flow of surface currents is similar to the pattern of atmospheric circulation.

surface to replace the displaced surface water in a process called *upwelling*. The deep water brings with it dissolved nutrients and replenishes the surface mixed layer, which is typically very low in these nutrients because of photosynthesis. In general, zones of upwelling tend to be excellent fisheries, because the high nutrient levels lead to abundant plankton, the primary food for fish. When net water transport is onshore, water is piled up onshore and then sinks, in a process called *downwelling*.

Surface currents result both from frictional drag of wind on water and from slopes in the sea surface generated by winds. Figure 10-18 depicts the system of surface currents. Not surprisingly, this system of currents is in many ways similar to the system of winds described in Chapter 9. Like the winds, oceanic surface currents play a strong part in transporting heat between the equator and high latitudes. In general, in the northern hemisphere water flows northward along the eastern edges and southward along the western edges of continents, forming clockwise currents called *gyres*. The opposite is true in the southern hemisphere, where the gyres run counterclockwise. In the northern hemisphere, the northward flowing currents (the Gulf Stream and the Kuroshio Current) carry heat to the high latitudes, and southward flowing currents (the California and Canary currents) transport cold water to the equator where it can be warmed. Without the warm water of the Gulf Stream, temperatures over northern Europe would be 5° to 10°C lower than they are.

Circulation in the Deep Ocean Deep-ocean currents can be thought of as a conveyor belt transferring water warmed at the equator to high latitudes and cycling water chilled at high latitudes back toward the equator (Figure 10-19). As surface water flows northward from the equator toward Greenland in the northern Atlantic Ocean, it undergoes evaporation and becomes more and more saline. In addition, it loses much of its heat to the overlying atmosphere as Arctic winds blow across its surface. Eventually, this saline water becomes so cold and dense that, near Greenland, it sinks to the bottom of the ocean. From there, the cold, salty water flows southward, crosses the equator, and eventually reaches the southern ocean, extending from 40° to 70° south latitude near Antarctica. From there, it rounds Africa and joins the circumpolar current that surrounds Antarctica.

Surface waters in the southern ocean are bitterly cold and highly saline because of the extensive sea ice formation around Antarctica. When seawater freezes, its dissolved salts are expelled into the surrounding seawater, elevating its salinity. These cold, salty, and therefore dense Antarctic waters sink, displacing, chilling, and combining with the slightly warmer and less dense water that flowed in from the north. This cold, Antarctic bottom water then separates into a series of bottom currents that flow northward into the Atlantic, Pacific, and Indian oceans and well up at the northern ends of these ocean basins, delivering cold, nutrient-rich water to the surface. Surface currents then flow southward in the Pacific and Indian oceans to balance the northward flowing bottom currents. These currents also flow westward, eventually returning to the southern Atlantic, where they start their northward migration toward Greenland again.

The surface part of the "conveyor belt" allows northern Europe to enjoy a fairly mild climate despite its high latitude. The amount of water in this system is immense—20 million m^3 per sec, about the size of 100 Amazon Rivers bundled together—and requires 500 to 2000 years to complete a full cycle.

The conveyor belt image agrees well with much existing data about ocean behavior and has become a powerful tool for modeling climate change. The conveyor belt

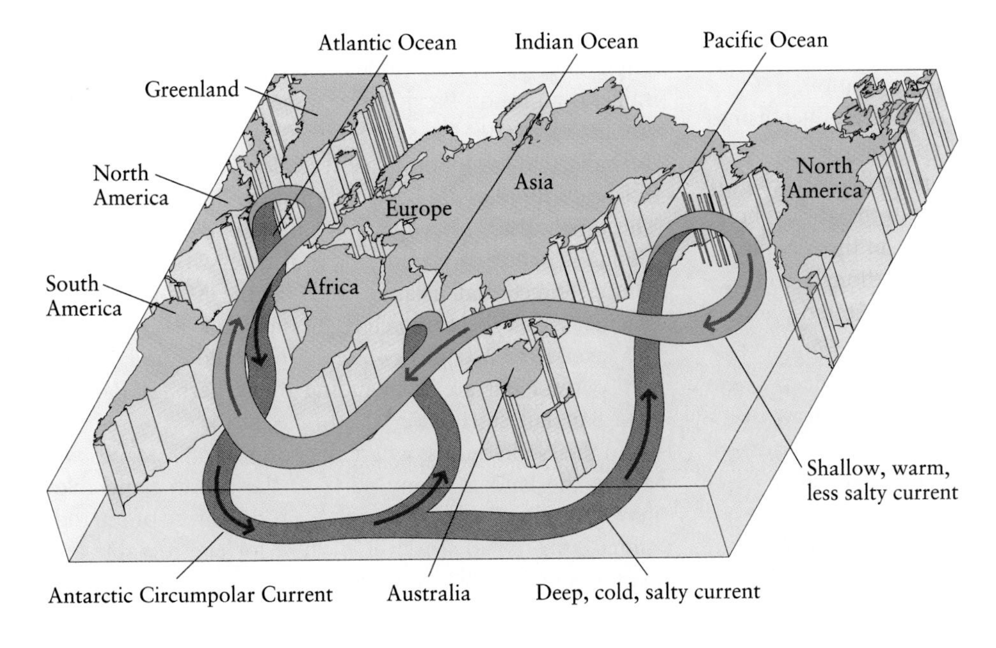

FIGURE 10-19 The model of an oceanic conveyor belt was developed by oceanographer Wallace Broecker of Columbia University's Lamont-Doherty Earth Observatory. The conveyor belt carries warm, salty water into the northern Atlantic Ocean, where it is cooled and becomes dense enough to sink into the deep ocean. From there, it flows southward along the seafloor, draining cold water to the southern oceans at a rate of 10 million to 20 million m^3 per sec.

might be able to start, stop, speed up, or slow down in response to changing amounts of fresh water added at the surface. Some scientists have linked rapid shifts from warm to cool global climate conditions in the past few tens of thousands of years to times when the conveyor belt might not have brought warm water to the northern Atlantic Ocean. Although the ultimate cause of changes in the conveyor belt's pattern are uncertain, there is some concern about how global warming might affect the flow of this great ocean current and, as a consequence, future climate.

- Seawater contains many dissolved ions, making it saline. The salts are derived largely from surface water runoff in rivers and accumulate very slowly over geologically long periods of time. Over a period of thousands to millions of years, the amounts of the major salt ions in seawater—the two most abundant of which are sodium and chloride—are roughly constant.
- The residence times of the most abundant salt ions in seawater are on the order of millions of years, whereas the amount of time needed for ocean currents to mix seawater and its dissolved substances is on the order of thousands of years. As a consequence, the proportions of dissolved salts in seawater are similar throughout the world, on the order of 35 grams of dissolved ions per kilogram of water (35 parts per thousand), or 3.5 percent.
- The most important nutrients in seawater—nitrates and phosphates—are derived from surface water runoff. Their concentrations vary markedly from one part of the ocean to another, largely because they are depleted from the surface layer by biological activity, accumulate in the deep ocean by sedimentation, and are recycled to the surface by upwelling currents from the deep ocean.
- Just as carbon dioxide has accumulated in the atmosphere since the Industrial Revolution, the amount of carbon dioxide in seawater is increasing as a result of the burning of fossil fuels and deforestation. Half the carbon released as a result of human activities is accumulating in the atmosphere, a quarter is stored in the oceans, and the remaining quarter—the "missing carbon"—is thought to be causing enhanced photosynthesis in the biosphere.
- Ocean circulation is divided into two primary types: surface mixed layer currents and deep-ocean currents. In the surface mixed layer, the flow of water results from the movement of air masses—that is, wind—in the atmosphere. In the deep ocean, variations in water density due to differences in temperature and salinity of the water drive the flow of water.

Hazards of the Sea

With population growth and development along coastlines quickening, the hazards of living at the edge of the sea have become increasingly acute. Most are associated with a rising sea level in the face of a developed shore that inhabitants wish to keep fixed in place. A variety of strategies have been devised to protect coasts from the devastation caused by coastal erosion and flooding. Loss of life due to the giant sea waves known as tsunamis that result most often from seismic movement of the seafloor has been reduced substantially in the 20th century because of global monitoring of earthquakes on the seafloor. An occasional hazard of the sea, occurring once every decade or so, is the set of atmospheric and oceanic events known as El Niño (Box 10-2).

Changes in Sea Level

Sea level has been rising about 2 mm a year for the past few decades, but many scientists—and many coastal nations—are more concerned about possible rises of up to several meters in the next century. Sea level varies in response to the amount of water bound in ice stored on land and to the expansion of water when heated. Since calculations show that thermal expansion would raise seawater only several centimeters for a temperature increase of a few degrees Celsius, scientists have concluded that the increased volume of seawater is due primarily to melted ice.

The worrisome thing is that we are spewing greenhouse gases, most notably CO_2, into the atmosphere at unprecedented rates. If the carbon dioxide mounting in the atmosphere causes Earth's surface to heat, then glaciers both large and small might melt and drain into the ocean, raising sea level much higher than it is today. There are today several large ice sheets on Earth: the Greenland, Antarctic, and West Antarctic ice sheets. Numerous smaller glaciers exist in mountainous areas, including the Swiss Alps, Sierra Nevada, Andes, Alaska, and Rocky Mountains. If all ice on Earth were to melt, sea level would rise about 50 to 100 m, submerging the world's coasts and forming shallow inland seas. If sea level rises 5 m, as some scientists think could happen if just the West Antarctic ice sheet melts, then Ho Chi Minh City, Miami, New Orleans, Bangkok, and many more coastal cities would be under water. Some small islands, such as the Maldives in the Pacific, would be completely submerged by a 3-m rise in sea level. Even a 1-m rise in sea level would inundate especially low-lying areas, including many parts of Pakistan and Bangladesh (Figure 10-20).

In the Netherlands, where the ocean surface stands well above the land in many parts of the country and dikes have been built to keep out the sea, the threat of an additional 5-m rise in sea level is cause for considerable concern (Figure 10-21). A break in the dikes in 1953 flooded

FIGURE 10-20 Bangladesh, a lowland nation plagued by repeated flooding, is very susceptible to rise in sea level. If the sea rose just 1 m, the land on which 6 million people live would be submerged.

one-sixth of the country and killed some 2000 people. The Dutch are raising and strengthening their dikes in response to predictions of global warming and continued sea level rise.

Measurements of Sea Level Change What leads scientists to conclude that sea level is rising? Scientists have studied tide gauge records of sea level from many ports around the world, at some locations for more than a century, and the results are complex. For example, it appears that sea level rose some 20 cm in Galveston County, Texas, between 1930 and 1970. In Juneau, Alaska, however, sea level seems to have fallen half a meter in the same time period. Although similar differences are seen around the world, they do not indicate that sea level is rising in some places and falling in others. Rather, they indicate that the land to which tide gauges are attached can move up and down, too.

Several processes can cause land to move up or down. In Galveston County, for example, the ground is subsiding because large amounts of oil and water have been pumped from the ground, and the sediments that contained the fluids have settled and compacted. So, sea level seems to be rising rapidly in Galveston, when in fact the land is sinking. The weight of thick stacks of sediment deposited on continental shelves at the mouths of large rivers also results in slow land subsidence. In contrast, at Juneau, where sea level seems to be falling, the land is rising instead. Juneau is located along a leading-edge coast at an active plate margin where uplift has occurred during earthquakes. In some regions that are not leading-edge coasts, such as Scandinavia and northeastern Canada, the land is rising because it is still springing back from the removal of the weight of massive ice sheets that melted only 10,000 years ago or so.

Because of the many possible causes of apparent sea level change, which occur at time scales ranging from years to millennia, it is difficult to determine the present rate of actual sea level change. The only way to be truly sure of what is happening is to measure the surface of the ocean itself. In the 1990s, the *Poseidon* satellite began doing exactly that, using radar to determine the exact distance between its own position—which is precisely known—and the surface of the ocean below it. These data can be averaged over the whole Earth, helping to sort out the complexities of individual tide gauge stations. Scientists analyzing *Poseidon* data taken over several years confirm that sea level is rising about 2 mm a year.

Few tidal records go back beyond a century or so, and yet geologists are certain that sea level has varied even more in the geologic past than in the last few decades. Marine fossils occur in rocks far inland from the coast, in places where no tectonism is occurring, such as the midwestern United States. Some of these rocks were deposited

FIGURE 10-21 In the Netherlands, hundreds of kilometers of dikes were built to protect low-lying areas from the sea. In some parts of the country, the ocean surface is well above the land.

10-2 Global and Environmental Change

El Niño Climatic Events

Until the 1970s, Peru's anchovy fishery was the largest in the world, but the catch plummeted from some 13 million to less than 2 million tons over a period of 2 years. The primary cause was not overfishing but a major disruption to the ocean current system that sometime appears around Christmas, and is therefore called El Niño, Spanish for "the child." El Niño events, essentially the formation of an unusual warm-water mass off the western coast of South America, occur irregularly and with varying intensity, much the way floods do. Fairly severe events occur about every 7 to 10 years. In 1972 and again in 1982–1983, El Niño events caused extreme droughts in the interior of Peru while coastal areas were hit by storms and floods, and outbreaks of malaria reached epidemic levels. Crops were devastated. The effects of those El Niño events were felt at great distances, as monsoon patterns changed in India, Australia, and southeast Asia, causing flooding and affecting agriculture. In the 1982–1983 event, the largest of the 20th century, economic losses totaled billions of dollars. (As this book was completed, in the summer of 1997, scientists warned that the winter of 1997–1998 might herald the largest El Niño event of the century.)

What triggers El Niño events? Interactions between the oceans and atmosphere exert strong controls on global climate, and the El Niño phenomenon is a manifestation of this connection. During most years, intense trade winds (see Chapter 9) blow from east to west across the Pacific Ocean in response to a pressure gradient between a high-pressure zone in the eastern South Pacific and a low-pressure zone near Indonesia. These winds pile up warm water in the western tropical Pacific, creating an area of high sea level. They also blow water away from the west coasts of North and South America, leading to the upwelling of deep, cold, nutrient-rich currents to replace the surface water blown westward. The upwelling zones support rich biologic communities teeming with plankton and the fish that feed on them. Peruvian fishermen have depended for decades on the vast schools of anchovies that thrive in the upwelling waters.

During El Niño events, the atmospheric and sea conditions change dramatically. The South Pacific high-pressure and the Indonesian low-pressure zones weaken, lessening the gradient and causing the trade winds to relax. No longer confined to the western Pacific by the trade winds, the warm pool of water in the western Pacific migrates eastward and arrives at the west coast of South America around Christmas.

The warm water that flows eastward during El Niño events pumps heat into the atmosphere above the eastern Pacific, resulting in the establishment of a low-pressure zone and associated storms. Heavy rains drench the west coast of tropical South America and flood areas that are usually arid. El Niño events affect not only South America but the rest of the world as well. For reasons that are not yet fully understood, they seem to cause high rainfall rates in western North America and droughts in Africa, India, and Australia.

This explanation of El Niño events is somewhat circular: It remains unknown whether they begin with an atmospheric event or an oceanic event. Does the relaxation of the trade winds allow the Pacific warm pool to migrate eastward, or does the eastward migration of the Pacific warm pool cause a weakening of

during the Cretaceous period, between 144 million and 65 million years ago, when Earth's climate was warmer and more equitable than today. No great ice sheets existed then, and all the water that exists in glaciers today was stored as seawater, adding to the volume of the world ocean. Sea level was as much as 100 m higher than it is today, spilling over the continents and flooding far inland.

The late Cenozoic era has been a time of global cooling over a period of millions of years, and we are in the midst of a major ice age (see Chapter 3). In general, then, sea level has been falling in response to the growth of ice masses. Within this ice age, however, are times of relative warmth and retreating glaciers, called interglacial episodes, and times of relative cold and advancing glaciers, called maximum glacial conditions. As a result of these oscillations in global temperature and ice volumes, sea level has risen and fallen dozens of times in the past few million years (Figure 10-22).

One of the ways that scientists learn about these past changes in sea level is by studying coral reefs submerged by rising sea level. In Barbados, fossils of coral that today is found living only close to the sea surface have been sampled from drill cores more than 100 m below today's sea level. Radiometric dates of these fossil corals have

the trade winds as atmospheric pressures change in response to changing sea surface temperatures? Whatever the trigger, the positive feedbacks between sea-surface temperature and atmospheric pressure amplify the initial disturbance, allowing El Niño to develop. Severe El Niño events appear to be increasing in frequency, with the most recent events in 1997–1998 (predicted), 1994–1995, 1991–1992, 1986–1987, 1982–1983, and 1971–1972.

Many national governments and the United Nations are working to understand the mechanisms that cause El Niño. Global-scale data are needed on ocean circulation and the interactions between the ocean and atmosphere to understand and ultimately predict events. Two new programs are collecting such data. One is the Tropical Ocean Global Atmosphere (TOGA) program, and the other is the World Ocean Circulation Experiment (WOCE).

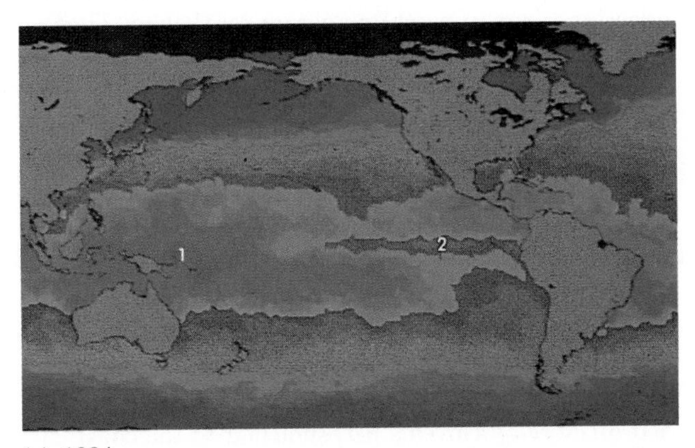

(a) 1984

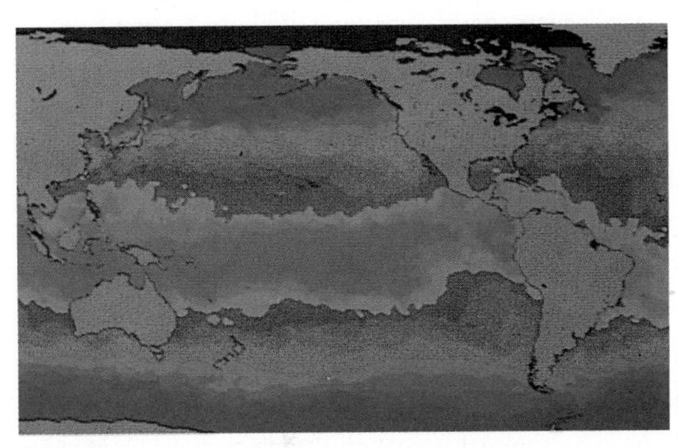
(b) 1983

These satellite images illustrate the difference in sea-surface temperature in 1984, a normal year, and 1983, an El Niño year. (a) During a normal year, the warmest waters occur in the western Pacific (1), and a tongue of upwelling cool water, the Peruvian-Humboldt Current, spreads westward from South America (2). (b) During 1983, an El Niño year, water off the coast of South America was much warmer, and cool water of the Peruvian-Humboldt Current is absent at the surface.

enabled scientists to document the changing position of the sea surface and determine the rate of sea level rise during the last 20,000 years.

About 20,000 years ago, during the most recent glacial maximum conditions, or "last glacial maximum," a portion of the world's ocean water was trapped on Eurasia and North America as huge continental glaciers, resulting in global sea level that was approximately 85 to 120 m below today's level. In places, lowered sea level created land bridges between continents, affecting the migration of many species, including our own (see Figure 1-7). At that time the Atlantic coastline of North America was more than 40 km east of its present position and bays such as the Chesapeake and Delaware were river valleys. Mastodon bones have been dredged up from today's Atlantic continental shelf, where once early hunters probably tracked prey across its gently sloping surface.

The ice sheets and other smaller glaciers began to retreat about 15,000 years ago, and sea level rose rapidly. By 10,000 years ago, sea level was 20 m below present levels, and it continued to rise about 20 to 30 mm per year until between 5000 and 7000 years ago. Since then, sea level rise has been relatively stable, perhaps no more than about 2 mm per year. In the past hundred years,

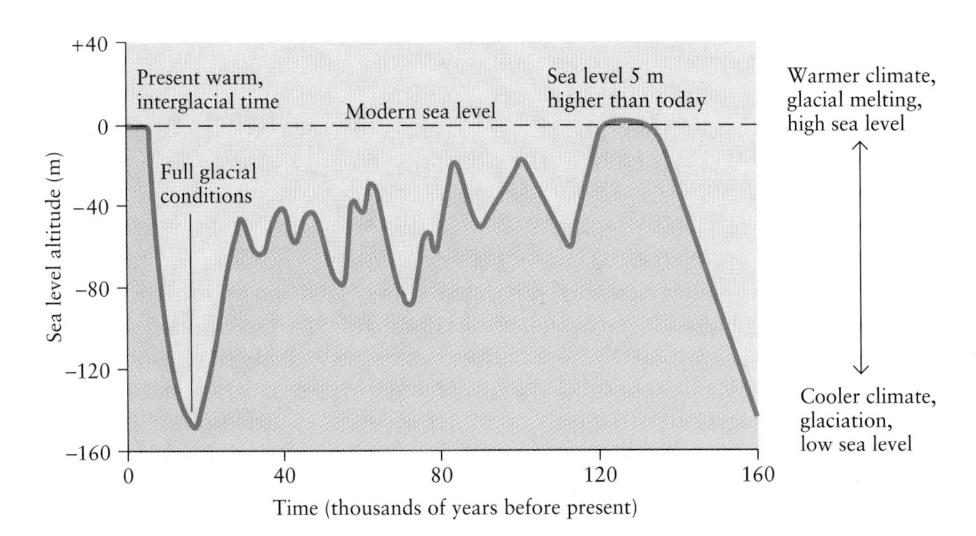

FIGURE 10-22 During the present ice age, which began about 2 million to 3 million years ago, global sea level has fluctuated many times, up to 120 m from its modern level. Scientists have reconstructed this curve of sea level change for the last 160,000 years by radiometric dating of coral in different parts of the world.

however, the rate of sea level rise seems to have increased slightly again, perhaps because of global warming.

How Much Might Sea Level Rise in the Near Future? In 1988, the World Meteorological Organization and the United Nations Development Program established a study group on the possibility of future sea level change. The Intergovernmental Panel on Climate Change assessed the available data and forecast a rise in sea level over the next century of between 20 cm and 1 m. The panel acknowledges, however, a great uncertainty: whether the current ice sheets are shrinking or expanding. This uncertainty can be reduced by measuring changes in elevation of the polar ice caps over a period of several years to decades. The U.S. National Aeronautics and Space Administration (NASA) plans to launch a satellite with a laser rangefinder to map such changes starting in the year 2002. If the West Antarctic ice sheet is shrinking, scientists would be more concerned at the possibility that it might collapse (because it is anchored to bedrock far below sea level), causing a very rapid and worldwide rise in sea level of 5 m.

Coastal Erosion

One of the greatest hazards of living at the edge of the sea is coastal erosion. Coastal erosion is part of the natural geologic process that redistributes sediments along coasts, and occurs primarily through the action of waves and winds. Where coastal erosion poses a threat to human-made structures and recreational beaches, it is regarded as a hazard. But the attraction of **beaches,** deposits mostly of sand that cover the shore where it meets the sea, is undeniable.

In places where sand is abundant and the seafloor slopes gently, barrier islands commonly border the mainland. Separated from the mainland by water in estuaries, bays, and lagoons, **barrier islands** are elongate ridges of sand that parallel the shore. Most formed in the past 5000 to 7000 years as rising sea level submerged elongate ridges and bars of sand on continental shelves. The seaward sides of barrier islands have beaches, whereas the landward sides often are covered with marshes, grasslands, dunes, and even forests. Hatteras Island in North Carolina, Miami Beach in Florida, and Padre Island in Texas are all barrier islands.

One of the most important characteristics of beaches and barrier islands is change: Their sand moves frequently when considered over periods of hundreds to thousands of years. Waves pick up sand grains, and even gravel, and transport them along the shore by *longshore currents* moving parallel to shore. Some sand also is carried offshore by currents perpendicular to the coast, and some is lifted by the wind and transported in dunes. Even as sand is removed, more is added by cliff erosion, rivers, and longshore currents that carry sand into the beaches and barrier islands. These outputs and inputs of sand can be used to assess the stock of sand over time, and to determine whether beaches are growing or shrinking.

In most coastal areas, sand is eroded during winter storms, stored offshore in sand bars, then returned to the beach in the summer. But problems arise when construction begins before the return of the sand, or when sand is removed from underneath houses built on beaches that are stable only on human time scales. Coastal economies depend on summer tourist dollars, and if there is no beach, there are no tourists. In addition, rising sea level for the past 10,000 years has been pushing beaches inland throughout most of the world. If left alone, beaches simply would shift inland with time, maintaining an approximately constant size. Once beaches are developed, however, property owners are reluctant to let them change.

Controlling Coastal Erosion Many schemes have been employed to slow or halt coastal erosion. One of the most common strategies is to build **groins,** low barriers ex-

FIGURE 10-23 Groins protruding from the Miami Beach barrier island were used until the early 1990s in an attempt to protect its beaches from erosion. Although groins trap sand on their upcurrent side, the clear water flowing around the groins is more likely to erode sand on the downcurrent side, so beach protection is minimal. In the mid-1990s, after this photograph was taken, a major sand-replenishment project costing millions of dollars replaced these groins with a beach that extends much farther seaward.

tending perpendicular to the beach to slow the transport and erosion of sand (Figure 10-23). Sand grains move parallel to the beach as well as back and forth across it. Groins block longshore drift, the flow of sand parallel to shore. The longshore current carries its load of sand until it encounters a groin, then the sand is deposited on the upcurrent side of the groin. Water moving around the groin with longshore drift is then relatively clear of sand, with much potential to erode new material, so it erodes the beach on the downcurrent side of the groin. Typically, another groin is built to catch this eroded sand, then another and another farther down the beach. For this reason, groins generally are built in fields, stretching out one groin after another along developed coastal beaches. Costing as much as half a million dollars per groin, the price of beach protection can be substantial.

Another strategy to try and stabilize a beach is to build jetties between barrier islands. **Jetties** are barriers placed to secure inlets and keep them from closing or migrating. When expensive marinas and waterfront developments are built adjacent to inlets, there is a strong incentive to keep the inlet in one place. Even before the U.S. construction boom of the 1970s and 1980s, inlets were stabilized with jetties to preserve lagoonal fisheries behind barrier islands. Seawater exchange through inlets brings in marine water and flushes out the part of the lagoon near the inlet. If the inlet closes or migrates, the salinity and water quality may change dramatically, disrupting commercial oyster, clam, or fishing grounds. Inlet stabilization has been a significant task of the U.S. Army Corps of Engineers. Unfortunately, the same problem that affects groins affects jetties. Longshore drift is impeded by the barrier, and sand builds up on the upstream side and erodes on the downstream side.

Because rivers, which are the ultimate source of sand for beaches, have been dammed upstream on most continents, less and less sand is reaching many coasts. Rising sea level, coupled with ill-planned coastal engineering schemes, has resulted in significant erosion on many beaches in the United States. Beach nourishment, or *replenishment,* is used widely to augment the amount of sand on a beach. Dredges or pumps bring sand from deeper waters offshore or from the upcurrent sides of groins to recreational beaches, where the sand is bulldozed and distributed (Figure 10-24). More than 1000 km of U.S. shoreline were replenished as of 1997, and the practice is increasing.

Replenished beaches are notoriously short-lived and expensive. Ocean City, New Jersey, for example, spent $5.2 million to replenish a small section of beach. Two and a half months and 18 storms later, the beach was gone, and authorities concede that maintaining a beach at Ocean City is too costly a practice to continue for many more years. For densely populated barrier islands such as Miami Beach, Florida, however, there is little choice but to continue to replenish the beach to maintain the flow of tourists that provides the main source of income to the area.

Controlling Cliff Erosion Cliff retreat as a result of the natural processes of coastal erosion is especially common along leading-edge (active) coastlines throughout the world (Figure 10-25). Along the western coast of North America, especially in California, much expensive real estate is in peril of falling into the sea. Many of the cliffs bordering beaches there are composed of relatively soft sediment and rock that is easily eroded. The houses atop such cliffs are susceptible to collapse when coastal erosion undermines their foundations.

FIGURE 10-24 Wrightsville Beach, North Carolina, is prone to excessive erosion, so sand is dredged from the continental shelf offshore, mixed with water, and pumped to the beach to replenish its supply of sand.

Cliff erosion is mainly by direct wave attack, which gradually erodes the rock. Waves also exploit joints and fractures in the rock as planes of weakness, tearing away large blocks at a time. Wind carries salt spray onto the cliffs, and as the water evaporates, salt crystals grow, slowly pushing off small rock fragments. The surface eventually crumbles. Earthquake-induced shaking may bring down unstable portions of the cliff face. Even watering the lawn around a clifftop house may raise the pore pressure in the underlying rock, increasing the chance of failure (see the discussion of mass movement in Chapter 6).

Various engineering techniques have been tried to halt sea cliff erosion, but at best they can only slow the process, and at their worst may actually accelerate it. Covering the cliff face with plastic sheeting is generally futile because during storms, when the sheeting is most needed, it is most likely to tear off. Concrete sheets poured over the cliffs erode as rapidly as the rock behind. At the base of the cliffs, protection in the form of heaps of large boulders, rock-filled wire baskets called gabions, or pilings driven in front of the cliff face, makes **seawalls** intended to break the force of oncoming waves. However, unless the protection covers the height of the cliff and extends for its length, the sea will eventually erode behind it, rendering it useless and increasing the rate of cliff erosion. Creating a buffer zone several hundred meters wide between house and cliff edge, combined with careful management practices, can slow the rate of erosion and limit the need for costly attempts to save properties in peril.

Storm Surges, Flooding, and Delta Erosion

Coastal flooding is the natural hazard deadliest to humans. During storms, strong winds blow water shoreward and the ocean surface rises in response to low atmospheric pressure. The result is an unusually high increase of sea level called a *storm surge.* The greatest known loss of life due to coastal flooding occurred in Bangladesh, on the Ganges River delta, in 1970, when a half-million people died from a massive storm surge that inundated the low-lying delta. In 1991 another storm surge, 6 m high, submerged the outer islands of the delta and flooded the marshy interior; 100,000 people perished.

Most of the damage to delta systems is not so catastrophic. For example, the Mississippi River delta is steadily losing land area, tens of square kilometers per year, because of both natural processes and human activities. The river is the source of sediment to the delta, and natural floods upstream bring new material downstream to build up the delta. Flood-control measures like levees and dams, however, prevent sediment from entering the river, so the delta is not being replenished as much as in the past. The result is that compaction, subsidence, and wave action from the waters of the Gulf of Mexico are relentlessly reducing the delta.

Part of the Mississippi delta that is changing rapidly is the chain of barrier islands called the Isles Dernieres.

FIGURE 10-25 Erosion has eaten away at the cliff beneath these waterfront houses in Maryland, along the western shore of the Chesapeake Bay. Trees and debris are used to protect the shore.

Fragmented into five islands by tidal inlets, the chain is 36 km long. Ongoing erosion by waves, together with rising sea level and reduced sediment supply from the Mississippi delta, has resulted in a net land loss of 78 percent in the 20th century. Hurricanes have had rapid and devastating effects on the islands, opening inlets and submerging parts of the islands. Barrier islands that once were inhabited are now little more than sand shoals, despite the construction of seawalls and other shore protection (Figure 10-26). The ultimate fate of the Isles Dernieres—in English, the "last islands"—is probably reclamation by the Gulf of Mexico.

More than a million people live in the Mississippi delta, which includes about 40 percent of all coastal wetlands in the contiguous United States. One-third of the nation's seafood production originates in the delta marshes. One possible solution to the problem of eroding delta is to destroy the levees south of New Orleans and let the river again overflow into the delta, freely distributing sediment as it did for 10,000 years. This is technically feasible, but probably not politically possible because of the large population living in the affected areas. The U.S. Army Corps of Engineers is proposing several expensive but potentially useful schemes to reintroduce some fresh water and sediment into the delta. The sediment would build about 130 km^2 of land in 150 years, but against a loss of tens of square kilometers per year, the effort will not be sufficient to save the delta from the rising seas.

FIGURE 10-26 The shore of Wine Island, at the eastern end of the chain of barrier islands that forms the Isles Dernieres, has retreated substantially behind a seawall (seen as a dark ring in the water) that was built before Hurricane Andrew in 1992.

Tsunamis

When an earthquake, volcanic eruption, or landslide suddenly displaces a large amount of seawater, it sets in motion a train of waves that radiates from the point of origin across the sea surface. The sea waves thus generated are called **tsunamis.** The phenomenon behaves much like the progressive series of waves that radiates from a pebble tossed into a lake. Tsunamis travel at great speeds, up to 800 km per hr, and the distance between each crest—the *wavelength*—can be as great as 150 km or more. In deep water, the height of the wave might be as small as half a meter, but once the tsunami reaches a shoreline it piles up and breaks into huge waves that inundate coastal areas and cause loss of human life and property damage. By the time a tsunami reaches shore, it can be as high as 40 m, able to carry large ships many kilometers inland.

Tsunamis can occur thousands of kilometers from the original displacing event, catching victims completely unaware of their approach. In 1946, an underwater earthquake in the Aleutian Islands generated tsunamis that spread across the Pacific. Within 5 hours a 15-m wave reached Hawaii, and within 24 hours villages on islands as far away as 8000 km had been destroyed. The number and types of active plate boundaries circling the Pacific Ocean make that region particularly prone to earthquake-generated tsunamis, so in the 1960s seismologists installed a network of seismometers around the Pacific Rim. Now when an earthquake occurs, the path and travel time of tsunamis from that source can be predicted, and warnings of at least a few hours can be given in most events (Figure 10-27).

- Major hazards associated with oceans include rising sea level, coastal erosion and flooding, and tsunamis.
- Sea level has been rising since the time of most recent full glacial conditions, about 20,000 years ago, as a result of the melting of glacial ice, but for the past few thousand years the rate has been very slow, less than a few millimeters per year. Some scientists are concerned that a relatively rapid rise of several meters might occur—inundating many of the world's coastal cities—if global warming causes one of the major ice sheets on Antarctica or Greenland to melt more rapidly.
- Rising sea level has contributed to the accumulation of substantial amounts of sand that form beaches and barrier islands parallel to the coastline. On natural beaches, sand is removed by erosion and replenished by deposition from season to season and year to year, but over a period of hundreds to thousands of years, the rates of removal and replenishment are nearly equal. Recently, however, this long-term flow of sand has been interfered with by efforts to protect property from erosion.
- Many engineered structures, including groins, jetties, and seawalls, are used to stop the erosion of sand from beaches and rock from cliffs, or to maintain inlets to harbors.

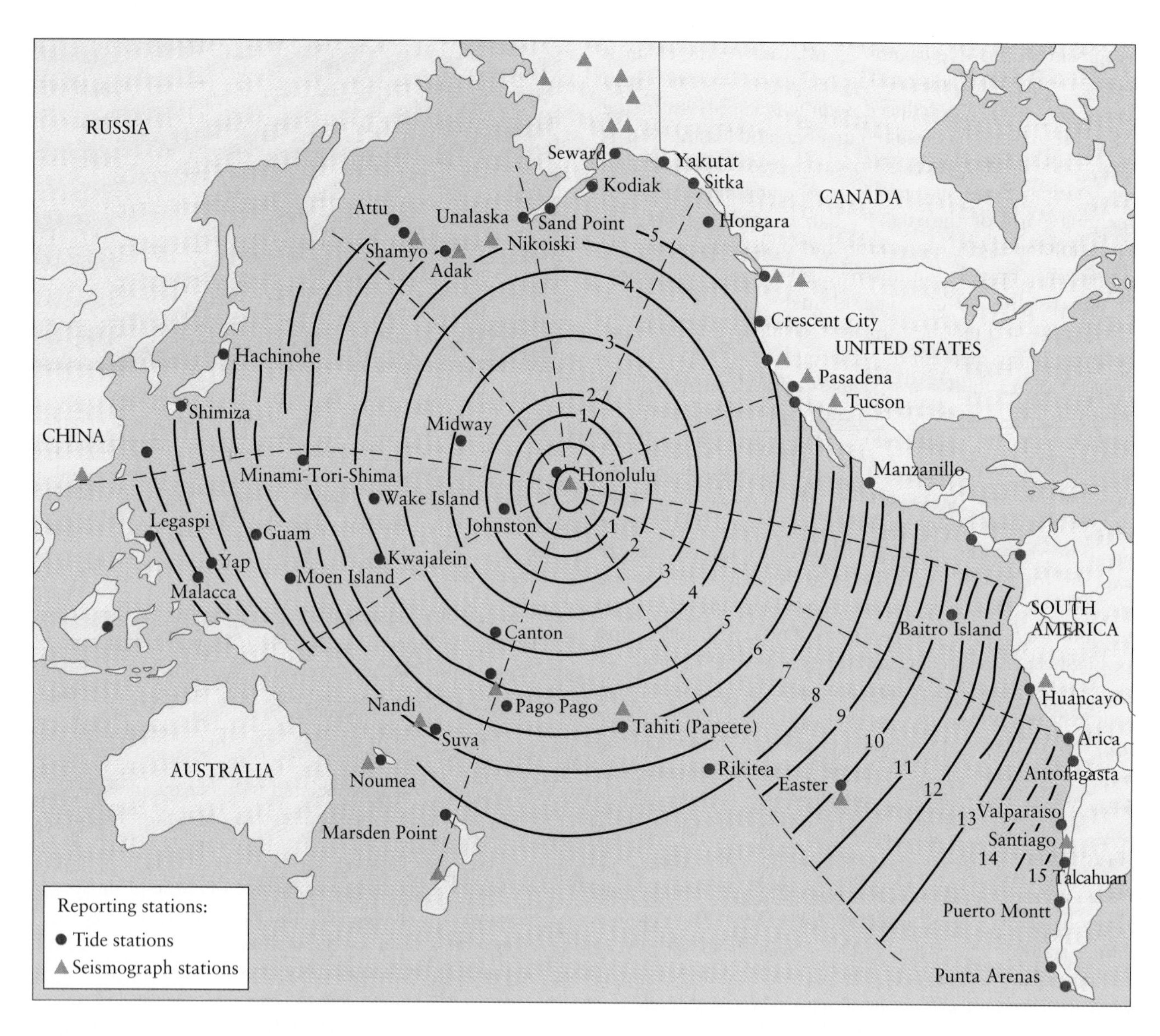

FIGURE 10-27 A warning system established in the Pacific Ocean pinpoints earthquake sources of tsunamis immediately after the event. Concentric lines indicate travel times: An earthquake-generated tsunami originating near Santiago, Chile, would reach Honolulu in about 15 hours, whereas one coming from San Diego would reach the island of Hawaii in little more than 5 hours.

- Delta erosion is common throughout the world as the result of sea-level rise in the past few thousand years and the increased number of dams on streams, the latter of which reduces the supply of sediment to coastlines.
- Tsunamis, sea waves generated by an earthquake, volcanic eruption, or landslide, displace a large amount of seawater and can reach heights of 40 m as they approach shore. A network of seafloor seismometers throughout the Pacific Ocean that pinpoints the sources of tsunamis can provide a warning system that has been effective in reducing loss of life.

Ocean Pollution

Hypodermic needles and other medical wastes washed up on New Jersey beaches in 1988, horrifying the public. The wastes are thought to have come from New York City sewer systems that drained into the sea after heavy rainstorms. Likewise, images of marine birds killed by entanglement in plastic, or of a whale that starved after swallowing a party balloon and ribbon that blocked its digestive tract, have alerted many to the effects of human wastes on wildlife (Figure 10-28). It is estimated that as many as 2 million seabirds and 100,000 marine mammals

FIGURE 10-28 A California sea lion entangled in a plastic fishing net.

die each year because of entanglement in or ingestion of garbage. Still, in their efforts to find places to dispose of unwanted materials, many nations turn to the oceans.

Despite the consequences of ocean dumping, some dumping activities are legal in the United States with a permit and if guidelines for disposal of the waste are followed. The Environmental Protection Agency has designated about 109 approved ocean dumpsites, many of them on the continental shelf and rise of the Atlantic coast. In the late 1980s, New York City legally dumped about 5.3 million tons of sewage sludge—treated sewage waste—per year at one dumpsite. Yet, as the long-term effects of disposal of some types of waste, such as sewage sludge, become apparent, environmentalists and officials are seeking to stop or regulate the dumping. The practice of dumping sewage sludge in the oceans ended in the United States in 1991, and, under pressure from environmentalists and other European governments, it will stop in Britain by 1998. By the Ocean Dumping Regulations of 1989, Canada severely restricts the levels of toxic, carcinogenic, and radioactive substances that can be dumped along coastal waterways.

Effects of Population Growth and Development

According to a National Audubon Society report, the most threatened coastal environments in the world include Los Angeles, Calcutta, Hong Kong, Cairo, Shanghai, and Rio de Janeiro. Coasts are the most threatened part of the ocean system because of their proximity to population centers. More than half of humanity lives within 100 km of the ocean, and this proportion is increasing as the global population becomes more urbanized and migrates to coastal cities. Nine of the 10 largest and most rapidly growing cities in the world are on a coast.

Today, even vacation development away from cities takes its toll on the coastal environment. In the early part of this century small summer villages were established along the coast of New England and the mid-Atlantic region. Colonies of small cottages were set well back from the beach and residents braved mosquitoes and humidity during the summer months. Travel to some sites was by dirt track and boat through snake-ridden wetlands. Now vacationers travel on four-lane highways and stay in air-conditioned condominiums built directly on the beach. Huge shopping malls have replaced the wetlands. Similar development worldwide has placed tremendous pressures on fragile coastal ecosystems.

More than 300 million tons of solid and liquid waste are dumped in the oceans every year, but it is the effect of day-to-day land-based activities, not of isolated oil spills (see Chapter 11), that has the greater impact. From the data presented in Table 10-2, it is clear that dumping and accidental spills hardly cause all of the ocean's ills. In fact, the greatest threat to the oceans is the human wastes running off the land: treated or partly treated sewage, runoff from streets and parking lots, chemical pollutants from industries, and excess lawn fertilizer.

Waste Disposal and Runoff Along Continental Shelves

Continental shelves in many parts of the world, especially adjacent to developed countries, are used for the disposal of a wide variety of wastes, including dredged material, sewage sludge, and industrial waste. Of these wastes, dredged material accounts for 80 to 90 percent of the total volume.

Dredged material may be of various sizes, including clay, silt, sand, gravel, and large blocks of rock. The U.S. Army Corps of Engineers, charged with keeping more than 40,000 km of coastline open to navigation, dredges about 230 million m^3 of sediment from beaches and harbors annually. Habitat is covered when dredged material is dumped, burying and smothering scallop and clam beds and other bottom-dwelling communities. Fine particles can be carried significant distances from dumpsites by currents, and organic debris in dredged material can consume

TABLE 10-2 Pollution in the Oceans

Source	% of total
Runoff and discharge from land	44
Airborne emissions from land	33
Shipping and accidental spills	12
Ocean dumping	10
Offshore mining and oil and gas drilling	1

oxygen from the water as it decomposes. Despite these environmental impacts, if dredged material is dumped suitable distances from shore in places where oxygen-rich water is well circulated, it generally does not cause considerable harm and bottom-dwelling organisms typically recolonize the site within a few months to years.

Sewage sludge, a gummy mixture of organic and inorganic chemicals derived chiefly from human waste, is rich in nitrates, phosphates, and microbes such as viruses and bacteria. Much sewage, treated and untreated, is discharged directly into coastal zones through pipes, and some is carried offshore and dumped by ships (see Figure 10-1). Sewage sludge, containing compounds of nitrogen and phosphorus, adds nutrients to seawater, but there can be too much of a good thing. Nutrients essentially are fertilizers, and when large amounts of them are added to ocean and coastal ecosystems, plants grow more rapidly. Nutrient surges after storms and floods sometimes result in great blooms of algae. But greater rates of plant growth mean that more plants die and decay, and the process of decay requires oxygen. As a result, oxygen is depleted from ocean water and, if depleted enough, invertebrates and fish suffocate and die. Because these animals graze on sea plants, their demise serves as a positive feedback that worsens the situation by enabling even more plants to survive and deplete the water of more oxygen.

This process of nutrient enrichment followed by enhanced plant growth and oxygen depletion is called **eutrophication,** and it is particularly severe in estuaries that are only partly open to the sea and have recirculating water (Box 10-3). Estuarine circulation patterns result in accumulation and retention of suspended sediments and nutrients, giving estuaries their high productivity. But the same circulation patterns trap toxic wastes, metals, and other pollutants in the aquatic system, limiting the ability of estuaries to cleanse themselves.

The Hudson River estuary once supported a thriving fishing and shellfish industry that is now largely defunct because of nutrient loading, overfishing, and pollution. The estuary is the emergency water supply for New York City and a major transportation waterway; it is the sewage disposal system for 16 million people. Because of the Clean Water and Ocean Dumping acts of 1972, raw sewage can no longer be discharged into the estuary, but even treated sewage from such a large population has led to elevated levels of bacteria, organic compounds, and metals, and to eutrophication.

Industrial waste comprises a variety of chemicals manufactured and used in industrial processes, and includes pesticides, petroleum products, and solvents. Although less than in the past, many such wastes are still discharged into the coastal zone through *effluent pipes* that disgorge wastes directly into the ocean (Figure 10-29). The producers of such waste hope—or assume—that rising and falling tides will serve to flush the wastes into the sea, where they will be sufficiently diluted to cause no harm. Instead, the wastes produce a plume, similar to the contaminant plumes formed in groundwater (see Chapter 8). In the case of seawater, effluents discharged from a pipe tend to rise and spread out along pycnoclines with currents. In the downcurrent direction, solid particles settle out and are deposited on the seafloor.

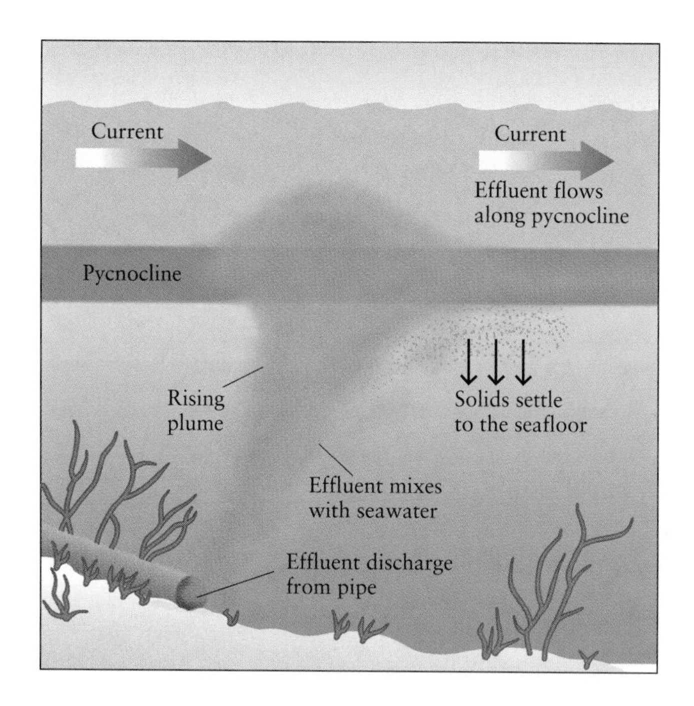

FIGURE 10-29 Effluent discharged from a pipe in a coastal zone forms a plume that rises to the surface and spreads downcurrent in seawater, typically along the pycnocline. Both industrial and sewage wastes have been discharged in this manner. Heavier solid particles settle to the bottom, forming a train of contaminated sediment that spreads away from the discharge pipe.

The U.S. Geological Survey and several other groups are involved currently in a number of environmental lawsuits related to industrial effluents. The lawsuits are being pursued by the Department of Justice on behalf of trustees of the nation's natural resources, including the National Park Service, the Fish and Wildlife Service, the National Oceanic and Atmospheric Administration, and the California Department of Fish and Game. In one case, an industry that was the world's largest producer of dichloro-diphenyl-trichloro-ethane (DDT), a highly toxic pesticide now banned in many nations because of its extreme environmental consequences, discharged large amounts of the substance directly into a sewer system that led to effluent pipes near Los Angeles in the 1950s and 1960s. Another highly toxic substance, polychlorinated biphenyl (PCB), since banned because it is a human carcinogen, was discharged from the same pipes. Both these chemicals resist biodegradation and have a tendency to become attached to the surfaces of silt and clay particles.

As in the general example shown in Figure 10-29, the DDT and PCB effluent was carried away from the pipes by longshore currents, and toxin-laced sediments formed a

10-3 Case Study

Chesapeake Bay and the Problem of Too Many People

One of the most beautiful coastal wetland environments on the eastern shore of North America is Chesapeake Bay, a vast estuary that developed about 5000 years ago as rising sea level drowned the valleys of a number of rivers that drained into the sea (see Figure 10-8). At that time, the estuary's upper reach was near present-day Annapolis. Sea level continued to rise until about 3000 years ago, and since then the estuary has changed little in size. Instead, it has begun filling with the muds and sands of the rivers that flow into its quiet waters. At the mouth of each river, broad flats of silt were colonized by grasses and became marshes.

Some of the Chesapeake's environments are quite unusual. Bald cypress trees—adapted to life in wet, frequently flooded soil, their knobby "knees" protruding above swamp muds—are found at the mouths of Chesapeake rivers, farther north than anywhere else in the United States. Early English settlers described Chesapeake Bay as filled with thick beds of large oysters and teeming masses of crabs, eel, and rockfish. Flocks of waterfowl, including ducks, geese, and osprey, covered its surface as far as the eye could see. Captain John Smith noted that it was a "faire bay encompassed but for the mouth with fruitful and delightsome land." The sight was unlike anything in the settlers' experience in their more densely populated and developed homeland.

Since then, rapid population growth and development along the shores and throughout the drainage area of the bay have placed excessive stress on the bay's habitat, plants, and animals. Waters flowing into the Chesapeake drain some 166,000 km^2 of New York, Pennsylvania, West Virginia, Virginia, Maryland, Delaware, and the District of Columbia, an area encompassing much of the industrial and urban eastern seaboard. Forty-eight rivers with more than 100 sizable tributaries flow into this one body of water, each carrying whatever has washed into it along its way seaward. The bay is an important commercial shipping route and a regional center for the steel, leather, plastics, and chemical industries. Urban growth in the area has almost doubled since 1950, and population has risen to 16 million people.

Between 1960 and 1983 the oyster catch fell by two-thirds, the annual rockfish catch decreased from nearly 3 million to less than 300,000 kg, and wildfowl populations plummeted. Eel-grass beds once covered the bay shallows, forming part of the food web, providing habitat, and protecting the bay's banks from erosion; in the upper bay, the grasses now are almost completely gone, and even carefully planted beds in the middle estuary often don't survive.

The causes of environmental stress in the bay are many and complex, and as a result difficult to remedy. Some have blamed enhanced sedimentation from deforestation, farming, and urban development. Sedimentation rates have increased exponentially. Others have blamed point-source pollution. Although specific sites where pollutants entered the bay have largely been cleaned up since the Clean Water Act of 1972 banned direct waste dumping,

Bald cypress trees, as shown here at Battle Creek Cypress Swamp Sanctuary on Chesapeake Bay, are specially adapted to the conditions of a coastal wetland. Knobby "knees" in the root system protrude above the water and are considered to be a form of anchoring enabling the tall trees to remain stable in wet, muddy soils.

(continued)

10-3 Case Study

(continued)

industries in Maryland still empty more than 2700 tons of metals into the bay each year, and Virginia industries contribute another 400 tons annually. Most experts now agree, however, that non–point-source pollution is the bulk of the problem. Pollutants causing the most damage are eroding soils from farm fields, treated sewage effluent, runoff from streets and animal feed lots, airborne deposits, precipitation, incoming river water, and other diverse sources.

Pesticides and herbicides that run off agricultural fields and chemically treated suburban lawns contain chlorinated hydrocarbons that have a high potential for human toxicity or are probable carcinogens. Many other compounds also enter the bay, and research suggests that almost all of them are either concentrated in the food chain or become entrained in the upper few centimeters of the bottom sediments where they can become deadly ground for burrowing organisms such as worms. Perhaps the most pressing threat to the bay's ecosystem, however, is a lack of oxygen in the brackish water. Sewage and farm runoff contain high concentrations of nitrate and phosphorus. These act as fertilizer for algae and cause algal blooms, the very high growth rates of green algae that lead to eutrophication.

States bordering the bay, in conjunction with the Environmental Protection Agency and strong grassroots environmental groups, have mounted a campaign to "Save the Bay." A voluntary agreement among the neighboring states seeks a 40 percent reduction in total nitrogen and phosphorus levels by the year 2000. Tax incentives are available to help farmers upstream from the bay to implement more rigorous runoff control measures and limit applications of fertilizers, pesticides, and herbicides. Educational campaigns alert the public to the connection between what they do on a daily basis and what happens to the bay downstream. To restore the Chesapeake to its early days of environmental health and resource bounty will demand significantly more compliance and adherence to the agreement than obtains today.

20- to 60-cm thick deposit that now extends at least 12 km along the continental shelf and slope, affecting fish, birds, and other organisms that dwell on the ocean floor and ingest the toxins when feeding. The U.S. Geological Survey is mapping the contaminated sediment by underwater photographs and sediment cores and is using computer modeling of ocean processes to predict what would happen if the system were allowed to recover naturally, with no remedial action. This information is being used in the lawsuits against the industries. As does copper in surface water environments (see Chapter 5), DDT and PCB accumulate in organisms, leading to bioamplification in the food chain. The fatty tissues of animals high on the food chain have been found to contain DDT and PCB in levels hundreds of thousands of times greater than that of the surrounding seawater. When these compounds are found in fish, fishing is prohibited to protect human health.

Wetlands and Their Role in Protecting Coasts

Wetlands bordering estuaries and lining coasts act as filters, absorbing pollutants and nutrients. They are so effective that some communities are turning to natural or constructed wetlands to clean their sewage and trap stormwater runoff and the pollutants it transports before reaching coastal water bodies. The environment of Barnegat Bay, New Jersey, deteriorating as a result of population growth of 700 percent since 1950, exemplifies this trend. In the past few years, a nonprofit land conservation organization in the area has coordinated an effort to purchase 152 km^2 of undeveloped wetlands and their feeder streams along the western shore of the bay. This land, now protected from development, will remain a viable means of minimizing the runoff of additional nutrients and other pollutants.

Yet in many places natural coastal wetlands are dredged for shipping and filled for urbanization. Only about 4 percent of the original 81,000 hectares of wetlands around San Francisco Bay remain. Nearly 60 percent of the wetlands bordering Chesapeake Bay have been destroyed by years of damming, draining, and filling. The percentage destroyed is even higher along smaller, more urbanized estuaries such as England's Thames River. It has been estimated that marshes return $12,000 per hectare per year as commercial fish nurseries; unfortunately, they may return over $250,000 per hectare if sold for development. The pressure for urban and suburban development means that, without strict legislation, coastal wetland loss will continue.

Pollution in the Deep Ocean

The intensity of human impact decreases away from the world's coastlines, and deep-ocean basins are still relatively pristine. However, intermittent pollution occurs in most of the world ocean, especially along major transportation routes. Because of ocean mixing by currents, wastes have been found even on remote islands. Along a 2.4-km stretch of beach on an uninhabited island in the Pacific Ocean, nearly 5000 km from the nearest continent, 950 pieces of trash, including a glue syringe, 6 light bulbs, and a car floormat, were found in 1991.

Plastics are of particular concern because, until recently, they were not biodegradable. Of the approximately 10 million tons of solid waste discharged into the oceans every year, 10 percent are plastics. Plastics in the ocean occur in two forms: plastic pellets which are the raw material from which plastic goods are manufactured, and plastic manufactured objects. Plastics affect the environment by contaminating beaches and killing marine animals through ingestion and entanglement. A recent study suggests that upward of 50,000 northern fur seals are killed through entanglement every year, contributing to a 4 to 8 percent decline in seal numbers per year. Recent bans on the use of plastic drift nets and the disposal of plastic from ships should help in reducing the damage caused by plastics in the ocean.

Protecting the Oceans

Many national and multinational efforts are under way to stop ocean dumping, or to regulate it strictly to protect the ocean and coastal system. In the United States, the Ocean Dumping Ban Act of 1988 prohibited the dumping of sewage sludge and industrial waste in the ocean as of 1991. The London Dumping Convention (1972) banned the dumping of hazardous substances such as mercury, cadmium, oil, and high-level nuclear waste, while dumping of other substances such as arsenic, lead, cyanides, pesticides, and fluorides, requires special permits. The greatest challenge to the world ocean will be the continued growth of the human population, and the greatest hope is that we become better at recycling, reusing, and reducing wastes.

- The greatest source of pollution to the oceans is runoff from continents in the form of treated or partly treated sewage, pollutants from streets and parking lots, chemical pollutants from industries, and excess lawn fertilizer.
- Many human wastes are disposed of on continental shelves, including dredged material (mostly rock and sediment), sewage sludge, and industrial waste. Sewage sludge adds nutrients to seawater, thus contributing to eutrophication, which depletes oxygen from the water.
- Coastal wetlands are an effective pollutant absorber, so some coastal communities are working to preserve remaining wetlands, or to restore those that were destroyed, in order to protect their coastal resources.

Closing Thoughts

Ours is a watery world, yet for most of human history we have been confined to land. At the beginning of the 20th century, the depth of the oceans was known at only the few places where seafarers had dropped and measured the length of a weighted line. In both world wars, submarines traveled along the ocean floors, and in World War II ships used a new technology—sonar—to determine water depth. Until then, the high seas were considered to begin a few kilometers from shore, and were open to any travelers. The oceans were this planet's last frontier.

Soon after World War II, continental shelves became precious for their resources, especially petroleum deposits and fisheries. Nations began laying claim to what had been a global commons, and years of conferences led to the Law of the Sea, an international code that delineates how the seafloor and overlying waters shall be apportioned. Clearly, we live in a time when limits to economic and population growth are being pushed to the extreme. Increasing exploitation of ocean and coastal resources puts pressure on the ocean environment.

But this is an exciting time as well. Until recently, it was possible only to dream of how the seafloor might look. In 1943, the late ocean diver Jacques Cousteau and his compatriot Emile Gagnan, an engineer, invented and tested a device (later to become known as the Aqua-lung) that supplied air from tanks to an underwater diver. The Aqua-lung was the precursor to SCUBA, or Self-Contained Underwater-Breathing Apparatus. Today, it is possible to go much deeper than a scuba diver, if one is willing to enter a small submersible vehicle and sink many kilometers to the ocean floor. Scientists clamor for such opportunities, and those who have the chance to peer out a window and watch giant tube worms swaying on the edge of a hot, hydrothermal vent are considered the lucky ones.

Summary

- The desire to exploit the ocean's resources, combined with an understanding of the connection between those resources and ocean floor topography, has led to changes in territorial claims to the oceans. In past centuries a coastal nation's right to the ocean was based on distance from shore, but modern claims are based partly on depth of water.
- Ocean basins consist of continental margins, abyssal plains that may contain seamounts, mid-ocean ridges along divergent plate boundaries, and deep-sea trenches along subduction zones.
- Most ocean resources—such as oil and fisheries—come from continental margins, where sediment is abundant and water is not very deep.
- The action of plate tectonics is closing some oceans (notably, the Pacific) and expanding others (notably, the Atlantic). Coasts are defined by their tectonic setting. A leading-edge coast forms along a convergent boundary and therefore is mountainous and has a narrow shelf. Erosion is high because deep waters offer no resistance to the progress of high-energy waves. A trailing-edge coast lies far from a plate boundary and so has a broad continental shelf where shallow waters impede wave size, permitting the formation of deltas.
- Ocean water is saline because the rate of addition of ions from rivers exceeds their rate of depletion. Some ions (calcium, bicarbonate) have relatively short residence times in the ocean, while others (sodium, chloride) have very long residence times.
- Nutrients essential to life are dissolved in ocean water and are used by phytoplankton to manufacture food. The phytoplankton are consumed by zooplankton, which in turn sustain other marine life.
- The ocean is a major reservoir of carbon, an element essential to organisms, weathering processes, and global warming. Carbon is cycled among the hydrosphere, lithosphere, biosphere, and atmosphere.
- Wind and gravity act on water, causing it to flow as ocean currents, which transport energy and matter from one part of Earth to another.
- Surface currents are driven by winds and deflected by the Coriolis force. The successively greater displacement of water from wind direction with depth is the Ekman spiral. Ekman transport is the overall water movement 90° to the prevailing winds. Ekman transport is important in distributing the ocean's nutrients because its interaction with land causes either upwelling or downwelling of nutrients along shores.
- Deep-ocean currents occur because gravity pulls water along a density gradient created by vertical differences in temperature and salinity.
- Deep-ocean currents move as if on a conveyor belt, circulating heat and matter around the globe.
- Seacoasts are a pleasant but hazardous place to live, liable to changes in sea level from global warming, tsunamis, and storm surges. Erosion of beaches and cliffs are other hazards stemming from natural processes. Human attempts to ward off these processes are costly and meet with little success.
- Human activities and burgeoning coastal populations threaten coastal environments and deep oceans alike. Changes in landscape, development of beaches, disposal of waste, oil spills, runoff, and destruction of wetlands all affect the land alongside the ocean, the ocean itself, and ultimately our own future.

Key Terms

bathymetry (p. 293)
continental shelf (p. 293)
continental slope (p. 293)
continental rise (p. 294)
abyssal plain (p. 294)
mid-ocean ridge (p. 294)
deep-sea trench (p. 294)
leading-edge coast (p. 296)
trailing-edge coast (p. 296)
salinity (p. 300)
nutrients (p. 301)
carbon cycle (p. 302)
ocean current (p. 304)
surface mixed layer (p. 304)
deep ocean (p. 305)
Ekman transport (p. 305)
beach (p. 312)
barrier island (p. 312)
groin (p. 312)
jetty (p. 313)
seawall (p. 314)
tsunami (p. 315)
eutrophication (p. 318)

Review Questions

1. Briefly summarize the outcome of the United Nations conferences on the Law of the Sea regarding the rights of coastal nations to the ocean's natural resources. Define "Exclusive Economic Zone."
2. What are the major types of resources that nations wish to extract or harvest from the oceans?
3. How have bathymetric methods changed in the past few hundred years? What are the primary techniques used today to determine the topography of the sea floor?
4. Compare the shapes, sizes, locations, and underlying material of the following ocean-floor landforms: continental shelf, continental slope, continental rise, abyssal plain, mid-ocean ridge, and deep-sea trench. Which of these landforms is associated with the highest amounts of volcanic and earthquake activity in the oceans? Why?
5. What are the differences between leading-edge (active) and trailing-edge (passive) coastlines?
6. How do plate tectonics and climate change affect sea level?
7. What are the three most abundant major constituents (dissolved ions) in seawater?
8. What are the two most important nutrients in seawater? What is their source? Why are they so important to the oceans?
9. Why is the amount of nutrients in the surface ocean so much different than in the deep ocean?
10. What is the source of calcium and bicarbonate ions in the oceans? How do these ions become bound in deep-sea sediments?
11. What factors affect the flow of currents in the surface and deep ocean? How do ocean currents in these two zones differ?
12. What is the present global rate of sea level rise, in millimeters per year? How have scientists determined this rate? What might cause it?
13. What processes add sand to barrier islands and beaches? What processes remove sand and transport it along the shore?
14. Describe two ways of attempting to control or halt coastal erosion. How has the damming of rivers contributed to increased rates of coastal erosion on some coasts?
15. Why are efforts to control coastal erosion so costly and so rarely successful over long time periods (tens of years)?
16. Explain the causes of delta erosion in the Mississippi region, the consequences of this erosion, and the nature of attempts to halt it.
17. What is a tsunami? What causes it? How does the tsunami warning system operate?
18. List and briefly describe the major sources of pollution to the oceans.
19. What is eutrophication? How do nutrients contribute to its occurrence?

Thought Questions

1. Why is the extent of the continental margin important in resolving international disputes over rights to marine and seafloor resources? Give specific examples.
2. Which type of coastline—active or passive—is more likely to be associated with coastal erosion and cliff retreat? Why? Which type is more likely to be associated with large deltas? Why?
3. A tide gauge at a site in the state of Washington indicates that over the past few decades, sea level has been rising with respect to the landmass at a rate greater than would be expected from global measures of sea level. What are some possible explanations?
4. How might the Chesapeake Bay area have appeared about 20,000 years ago, when sea level was much lower than today? Was the coastline different then? Would an estuary have existed?
5. Why is the salinity of seawater in the Red Sea so much higher than that in the Atlantic Ocean?
6. Explain how examining glacial ice cores and the testing of nuclear weapons helped scientists determine the fate of the carbon released by human activities.

Exercises

1. Using the carbon budget in Figure 10-12, calculate the residence time of carbon in the atmosphere, in the surface waters of oceans, and in the pedosphere. For which of these reservoirs are the residence times greatest? Least? What would happen to the concentration of carbon dioxide in the atmosphere if the influx of carbon were to increase? What implications might such an increase have for climate?
2. Make a sketch to answer the following questions. In the southern hemisphere, if wind is blowing northward along a coastline with a landmass to the west, is the direction of net water transport toward or away from the shore? Would you predict that upwelling or downwelling would occur along this coast?
3. Use an atlas to determine how much of a coastal city would be submerged if the West Antarctic ice sheet were to melt completely, raising sea level by about 5 m.

Suggested Readings

Carson, Rachel, 1955. *The Edge of the Sea.* Boston: Houghton Mifflin.

Davis, Richard A., 1994. *The Evolving Coast.* New York: W. H. Freeman and Company.

Dolan, Robert, and Lins, Harry, 1994. "Beaches and Barrier Islands." *Scientific American* (July): 68–77.

Flanagan, Ruth, 1993. "Beaches on the Brink." *Earth: The Science of Our Planet* (November): 24-33.

Kemper, S., 1992. "If You Can Fish From Your Condo, You're Too Close." *Smithsonian,* 23: 72–86.

National Research Council, 1990. *Managing Coastal Erosion.* Washington, DC: National Academy Press.

Philander, George, 1989. "El Niño and La Niña." *American Scientist* (September–October): 451–459.

Pilkey, Orrin, and Dixon, Katherine, 1996. *From the Corps to Shores.* Washington, DC: Island Press.

Pinet, Paul, 1996. *Invitation to Oceanography.* Minneapolis/St. Paul, MN: West.

Pivie, R. Gordon (ed.), 1996. *Oceanography: Contemporary Readings in Ocean Sciences,* 3rd ed. New York: Oxford University Press.

United Nations Group of Experts on the Scientific Aspects of Marine Pollution, 1990. *The State of the Marine Environment.* New York: United Nations Publications.

Van Andel, Tjeerd, 1994. *New Views on an Old Planet.* London: Cambridge University Press.

Zaburunov, S. A., 1992. "As the World Breathes: The Carbon Dioxide Cycle." *Earth: The Science of Our Planet* (January): 26–33.

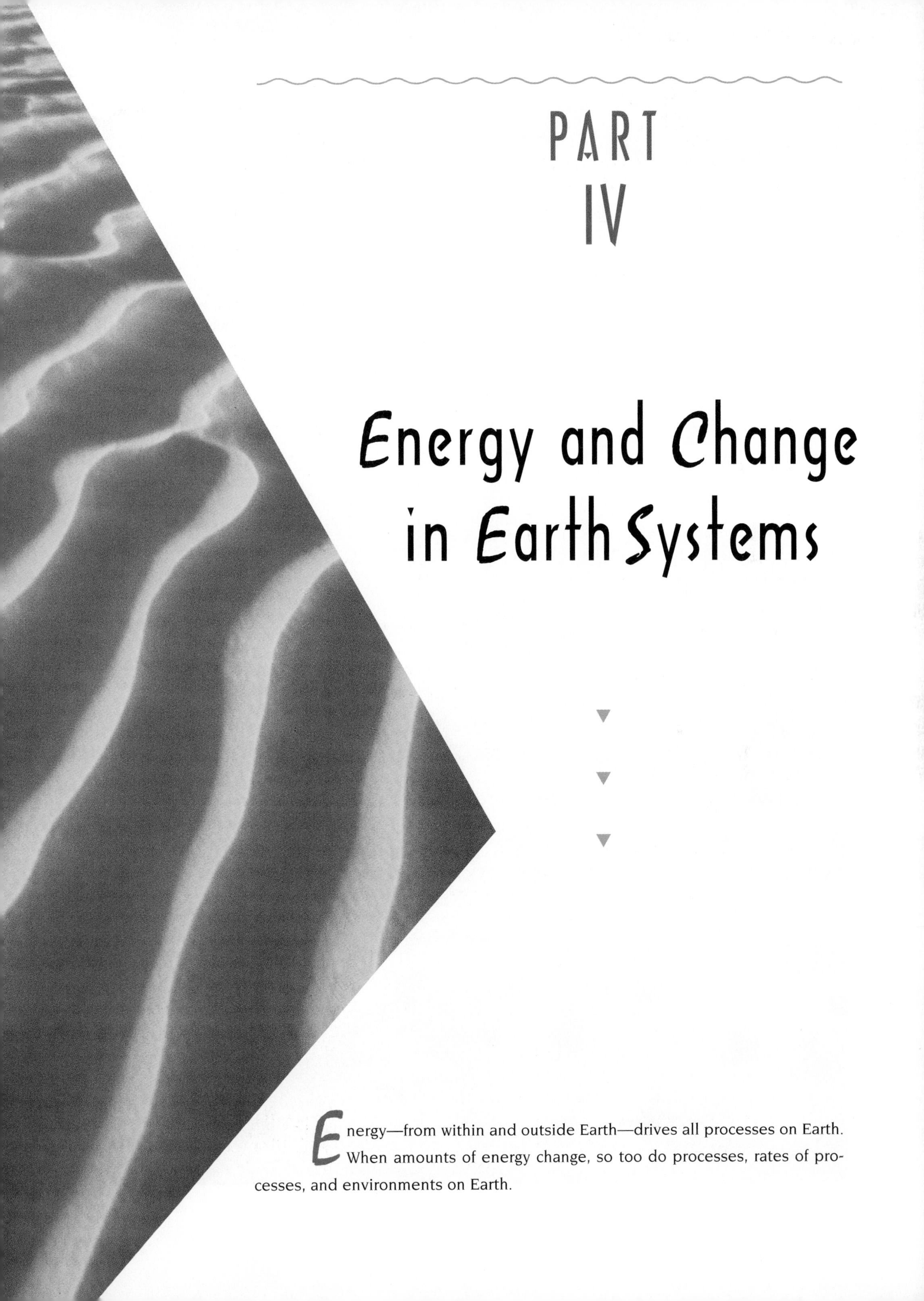

PART IV

Energy and Change in Earth Systems

Energy—from within and outside Earth—drives all processes on Earth. When amounts of energy change, so too do processes, rates of processes, and environments on Earth.

CHAPTER 11

As in this peat bog in Ireland, plants use solar energy and carbon dioxide to make carbon-rich biomass, which can accumulate in the pedosphere and lithosphere as peat, coal, and petroleum. These fossil fuels release their stored chemical energy in the form of heat when burned.

Energy and the Environment

More than 30,000 years ago, our Stone Age ancestors warmed themselves by wood fires, the oldest human use of carbon-rich biomass as an energy resource for heat. Since then, more concentrated sources of carbon have been discovered—ancient biomass buried and transformed into fossil fuels such as peat, coal, oil, and natural gas. Over millions of years, a small fraction of the flow of energy from the Sun has accumulated in aboveground biomass and an even smaller fraction in buried biomass and fossil fuels, giving Earth a very large energy stock.

When burned, biomass and fossil fuels not only release energy but also emit carbon dioxide and other gases which could be contributing to global warming. In our demands for more energy, we could be heating the whole Earth. At the same time, we are using up the fossil fuel supply.

Fossil fuels are nonrenewable resources, but they are not the only energy sources available. For economic as well as environmental stability, the type of energy we use in the 21st century probably will depend more directly on the Sun's radiation and on Earth's internal heat. In this chapter, we will explore the origin, distribution, and environmental impact of the many types of energy resources on Earth. We will:

- Trace flows of energy through Earth systems.
- Investigate responsible ways to recover, use, and conserve nonrenewable fossil fuels.
- Confront the risks involved in producing nuclear energy and disposing of nuclear waste.
- Discover why solar radiation is the only perpetual source of energy safely available worldwide.

The discovery of new sources of energy has resulted in major changes throughout human history, most notably in the 20th century. Until the mid-1800s, more than half the world's energy consumption came from burning wood; much of the rest came from work animals, wind, and water. By 1900, however, coal had become the dominant source of energy. Used in steam engines, coal was pivotal to the early industrialization of Europe and North America. Subsequent development of the internal combustion engine, an engine that could burn liquid fuel, set the stage for major technological changes. The search for underground oil, from which gasoline is produced, intensified, and discoveries outpaced consumption. By World War II, oil was nearly as important as coal to world energy demands, and the importance of oil has increased in the years since then. With these shifts in use of energy resources have come new forms of environmental pollution and the need for environmental regulation (Figure 11-1).

When Adolf Hitler's German forces overran Europe and invaded Russia and Africa in the early 1940s, they used trucks, tanks, and airplanes that depended on petroleum products, but Germany had minuscule domestic supplies of petroleum. Like their oil-poor Japanese allies, who invaded Indonesia to exploit its rich petroleum resources, the Germans sought to gain control of known oil fields in eastern Europe and the Middle East. These efforts failed, and during the Battle of the Bulge in Belgium at the end of the war, German troops abandoned, on the battlefield, brand-new tanks that simply had run out of gas.

Population, mass transportation, and industry all have grown rapidly since then, and by the 1970s oil provided the world with more than 40 percent of its energy. Then OPEC (Oil Producing and Exporting Countries, a cartel formed in 1960) created a series of energy crises in oil-importing nations by repeatedly raising its oil prices. It suddenly became clear that much of the world's oil, perhaps one-half to two-thirds, was located in what generally had been considered a resource-deficient, barren desert—the Middle East. Furthermore, the world's rate of oil consumption was approaching that of new discoveries.

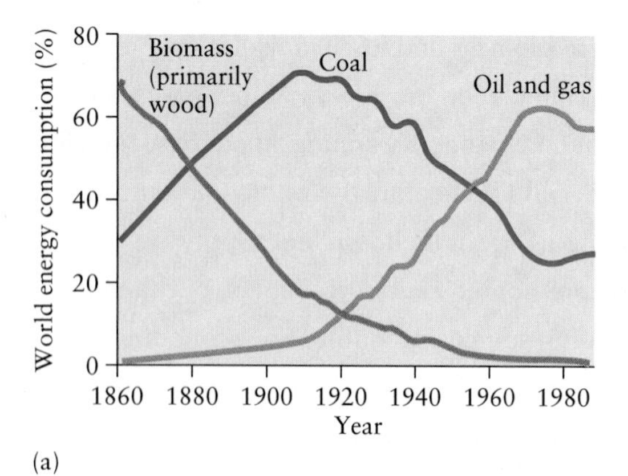

(a)

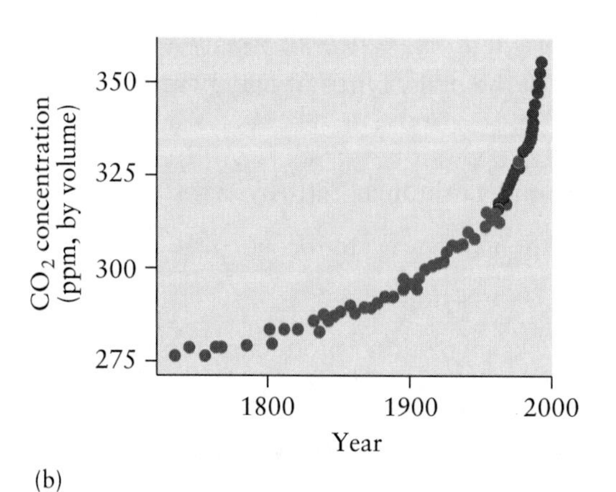

(b)

(c)

FIGURE 11-1 (a) Since 1860, world energy consumption has increased markedly and shifted from burning biomass (mostly wood) to burning fossil fuels. As a result, amounts of atmospheric carbon, nitrogen, and sulfur have increased measurably. (b) Concentrations of carbon dioxide at sites far from industrial centers have risen sharply since the Industrial Revolution, as measured in cores drilled in Antarctic ice (red data points) and in the air over Mauna Loa, Hawaii (blue data points). (c) A modern electrical power installation on the Susquehanna River in Pennsylvania is designed to control damage to the environment. The plant with the tall, brown stack burns both coal and oil, monitors emissions continuously, and traps dust; the adjacent plant uses water power. To protect the endangered shad on the river, a fish lift was constructed in 1997 to transport the fish upstream to their spawning grounds.

As industrialized nations strove to become less dependent on foreign oil, they increased their reliance on natural gas, sought new sources of domestic oil (particularly in the Gulf of Alaska, the Alaskan North Slope, and the North Sea), made some attempts to develop alternative sources such as solar energy, and increased efforts to conserve energy. Consequently, oil prices dropped, and by the 1980s OPEC was contributing a smaller share of the world's energy consumption. Nevertheless, when Iraq invaded Kuwait in 1990 and took over its oil fields, many nations, including the United States, sent troops to defend the Persian Gulf and their interest in its fossil fuels. As summed up in a report by energy analysts Christopher Flavin and Nicholas Lenssen of the WorldWatch Institute, "Not only is the world addicted to cheap oil, but the largest liquor store is in a very dangerous neighborhood."

Some experts think that reliance on oil has reached its peak and that consumption of natural gas will rise until sometime between 2030 and 2050. Although coal could provide us with energy for centuries, its use has much greater environmental impact than other fossil fuels. Many scientists predict that nonfossil energy resources—especially solar energy—will become increasingly important in the 21st century.

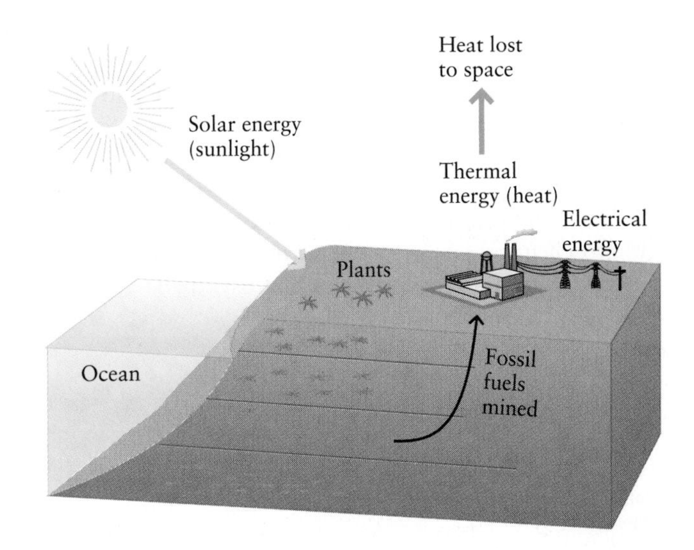

FIGURE 11-2 Matter flows cyclically through Earth systems and little enters or leaves the planet, but the energy flux is one-way. Sunlight reaches Earth and is used by plants to produce organic matter, which stores the solar energy as chemical potential energy. If buried, the organic matter can be transformed into fossil fuels, which when burned release their stored energy as heat. This thermal energy can be used to generate electricity to provide heat or lighting, but ultimately the energy given off as heat is lost to space.

Earth's Energy System

The flow of energy through the lithosphere, pedosphere, hydrosphere, atmosphere, and biosphere drives the movement and transformation of matter that make Earth a dynamic planet capable of sustaining life. Just as matter flows from one reservoir to another, so does energy, as in the transfer of electricity or heat. External energy from the Sun and internal energy from Earth's interior are transferred through Earth's material systems and may be stored there temporarily before being lost to space.

Solar energy, for example, is stored as chemical potential energy in plants, which can be buried and altered into fossil fuels. When plants or fossil fuels are burned, chemical energy is transformed to thermal energy, which can be used to generate electricity and run automobiles (Figure 11-2). In another example of energy transfers, thermal energy from the interior of Earth is stored in elastically strained rocks before an earthquake. When the rocks snap, not only does an earthquake occur, but also potential energy in the rocks is converted to kinetic energy (specifically, seismic energy) and finally thermal energy as a result of friction. Ultimately, all the heat released during any type of energy transfer is dispersed into our surroundings and then to space, lost from the Earth system.

Over a short time scale, Earth approximates a closed system: Energy is exchanged freely across its boundary, but matter is not. The exchange keeps the whole Earth system in an approximate steady state with respect to heat, because the energy inflow from the Sun is about equal to the energy outflow to space.

Sources of Energy on Earth

The greatest source of energy at Earth's surface is electromagnetic radiation from the Sun. This external energy source provides an inflow of $173{,}000 \times 10^{12}$ W, or 173,000 tW (terawatts; see Appendix 3), per year and makes up 99.99 percent of the energy flow on Earth. Solar energy is the ultimate origin of nearly all other energy sources on Earth, including the energy stored in wood, coal, oil, natural gas, falling water, and wind.

Earth's internal heat, which supplies only 32 tW a year to the surface, is responsible for driving the processes of volcanism and plate tectonics. Earth's heat energy also can turn groundwater to steam, producing geysers and geothermal fields.

Gravitational attractions among Earth, Moon, Sun, and other bodies in the solar system supply a mere 3 tW a year of potential and kinetic energy. Nevertheless, this small source of energy causes both ocean water and Earth's crust to change shape with time in a regular fashion, resulting in ocean and rock tides.

Energy Transfers and Photosynthesis

The flow of energy through the pedosphere and biosphere is of special human concern because these systems are the sources of food and fuel energy. The biosphere depends on energy stored during **photosynthesis**, the process by which chlorophyll in green plants converts solar energy into chemical energy in the form of organic (carbon-based) molecules. The simplest photosynthetic reaction produces a carbohydrate called glucose:

$$6\,CO_2 + 6\,H_2O \xrightarrow{\text{sunlight used}} C_6H_{12}O_6 + 6\,O_2$$

$6\,CO_2$	$6\,H_2O$	sunlight used →	$C_6H_{12}O_6$	$6\,O_2$
(Carbon dioxide removed from atmosphere, biosphere, and pedosphere)	(Water removed from hydrosphere)	(Solar energy stored as chemical energy)	(Glucose available to biosphere)	(Oxygen released to atmosphere)

The energy stored in glucose and other organic molecules can be released if the carbon bonds are broken in a reaction with oxygen. *Respiration* is the process by which animals oxidize organic molecules to convert their chemical energy to heat, releasing carbon dioxide and water as by-products. This process is vital to animals, for they are not able to use solar energy directly to synthesize their own carbohydrates. Decomposition and combustion also are processes that release heat energy by oxidizing organic molecules. *Decomposition* is the relatively slow oxidation of dead organic matter by microbes that take in oxygen from the atmosphere. This process is essential to soil and peat formation (see Chapter 6). *Combustion* is the much more rapid oxidation that occurs when organic matter ignites and burns, generating both heat and light:

$$6\,O_2 + C_6H_{12}O_6 \xrightarrow{\text{chemical energy released}} 6\,H_2O + 6\,CO_2$$

$6\,O_2$	$C_6H_{12}O_6$	chemical energy released →	$6\,H_2O$	$6\,CO_2$
(Oxygen removed from atmosphere, hydrosphere, biosphere, and pedosphere)	(Glucose removed from biosphere)	(Heat released to organism by respiration or to environment by decomposition or combustion)	(Water returned to hydrosphere)	(Carbon dioxide returned to atmosphere, pedosphere, and biosphere)

The energy released by this reaction is substantial. Oxidizing 1 kg of carbon—whether by respiration, decomposition, or combustion—releases 33 million joules of stored energy, enough to drive an efficient subcompact car 15 km or more.

Because of the reciprocal reactions of photosynthesis and oxidation of organic matter, the energy inflow into the biosphere is nearly equivalent to the outflow. Of the 173,000 tW of solar energy that reach Earth's surface annually, only 100 tW are stored in the biosphere through photosynthesis of organic molecules, contributing to the formation of soils and peat on land and organic oozes on the ocean floor. Most of the carbon in the biosphere, about 99.9 percent, is returned to the atmosphere as carbon dioxide after organisms die and decay as a result of microbial activity requiring oxygen. A tiny remainder, 0.1 percent, is buried in sediments and protected from oxidation, becoming part of the lithosphere. There, organic matter may be transformed by pressure and heat into fossil fuels, including petroleum, coal, oil shale, and tar sand.

Energy Resources

Extracting and burning fossil fuels releases stored energy that originated as sunlight and was used by organisms for photosynthesis long ago. Although the rate at which solar energy becomes stored in fossil fuels is small, it has accumulated in the lithosphere over periods of millions to hundreds of millions of years, and so the total stock of fossil fuels is very large.

The modern world currently relies on nonrenewable fossil fuels for more than 75 percent of its primary energy needs (for heating and electricity). Renewable biomass and hydroelectricity supply most of the rest. Nuclear power accounts for only 4 percent of energy use. Although the uranium oxides from which nuclear power is derived contain the largest stock of nonrenewable energy on Earth, the risks of radiation exposure inhibit the development of this energy resource. Geothermal energy from Earth's interior heat flow is essentially a perpetual resource at a few locations but is insignificant in the world energy tally. Direct solar energy and wind energy are perpetually available and environmentally benign but are little used (Table 11-1). The origin, economic potential, and environmental effects of each of these energy resources, as well as others of local importance, are described in the following sections.

- Three sources of energy are responsible for all processes on Earth: electromagnetic radiation from the Sun, heat from the interior of Earth, and gravitational potential energy associated with the Earth-Sun-Moon system. Solar radiation accounts for more than 99 percent of Earth's energy budget, and heat from Earth's interior accounts for most of the remaining 1 percent.
- External energy from the Sun and Moon and internal energy from Earth's interior flow through Earth systems and sometimes are stored temporarily before being lost to space as waste heat.
- Energy flows into the biosphere through photosynthesis and is temporarily stored in biomass and fossil fuels in the form of organic carbon compounds. Respiration, decomposition, and combustion all release stored energy by oxidizing organic molecules.

TABLE 11-1 Nonrenewable and Renewable World Energy Resources

Nonrenewable energy resources	Total amount of energy (J)	Annual primary energy use (%)*
Uranium oxides (for nuclear power)	1.6×10^{23}	4
Coal	1.4×10^{23}	25
Syncrude (tar sand and shale oil)	1.4×10^{22}	insignificant
Petroleum (crude oil)	1.0×10^{22}	36
Petroleum (natural gas)	8.0×10^{21}	17
Renewable and perpetual energy resources	**Annual influx of energy (J/yr)**	**Annual primary energy use (%)***
Direct solar energy	4.0×10^{24}	insignificant
Biomass†	$\sim 2 \times 10^{21}$	>13
Hydroelectric energy	9.2×10^{19}	5
Geothermal energy	8.0×10^{19}	insignificant
Wind energy	no global estimate	insignificant

* "Primary energy" refers to energy used for heating and electricity.

† Calculation of biomass stock for energy resources is based on the total amount of new plant growth each year. This is far more than the stock actually available as an energy resource because substantial amounts of plant matter are needed for food and construction products.

Sources: From C. D. Masters, D. H. Root, and E. D. Attanasi,"Resource Constraints in Petroleum Production Potential," *Science* 253 (July 12, 1991): 146–152; and Chauncy Starr, Milton F. Searl, and Sy Alpert, "Energy Sources: A Realistic Outlook," *Science* 256 (May 15, 1992): 981–987.

Petroleum

The 20th century has been called the Petroleum Age, a time of remarkable technological advances made possible largely by the energy stored in petroleum. **Petroleum** is a liquid or gaseous form of organic matter composed of **hydrocarbons**—compounds consisting of carbon atoms and hydrogen atoms. Commonly, "petroleum" means any hydrocarbon mixture that can be extracted through a drill pipe. The main forms of petroleum are **crude oil,** the liquid part of petroleum (sometimes called "black gold") and **natural gas.** Natural gas contains less carbon and more hydrogen than does oil, making it less dense (Table 11-2). Both forms of petroleum are easier to transport and cause less pollution than does coal.

Origin of Petroleum

Scientists who examined the oil produced from early wells were puzzled by its origin. Its composition indicated an organic origin, but the hydrocarbons in petroleum are chemically somewhat different from carbohydrates, proteins, and other organic molecules.

The key difference is the abundance of oxygen in other organic compounds and its nearly complete absence in petroleum (see Table 11-2). Nevertheless, many studies have shown that hydrocarbons are produced by the transformation of different kinds of organic matter; during transformation, oxygen is lost from the organic molecules, commonly in the form of water vapor.

The most common source of petroleum by far is phytoplankton, the single-celled marine plants that live along continental shelves (see Chapter 10). When phytoplankton and other marine organisms sink to the ocean floor after death, most are eaten or decomposed, but in still water depleted of oxygen, scavengers seldom venture and decay is unlikely. Such environments exist in deep basins along continental shelves, where fine-grained sediments (silt and clay-rich mud) and organic matter accumulate and are

TABLE 11-2 Average Elemental Composition of Fossil Fuels (% by weight)

	Natural gas	Crude oil	Coal
Carbon (C)	76	84.5	81
Hydrogen (H)	24	13	5
Sulfur (S)	trace –0.2	1.5	3
Nitrogen (N)	0	0.5	1
Oxygen (O)	0	0.5	10

Source: From John M. Hunt, *Petroleum Geochemistry and Geology,* New York: W. H. Freeman and Company, 1995.

FIGURE 11-3 A well blowout in 1988 on an offshore drilling rig in the North Sea caused a fire that killed 160 workers and locally polluted the ocean with spilled petroleum. Blowouts occur because petroleum exists under conditions of very high fluid pressure, which is suddenly released during drilling.

buried over prolonged periods of time. The petroleum-rich Monterey Formation on the California coast (see the opening photograph of Chapter 3) is the uplifted remnant of one such sedimentary basin. The thick sediments at river deltas also favor hydrocarbon formation. The Mississippi delta and its offshore extension into the Gulf of Mexico have yielded abundant supplies of petroleum.

How are substances such as aquatic plants transformed to the oil and gas we use in our houses, automobiles, and power plants? As mud and organic matter are deposited in sedimentary basins, the deeper and older layers are subjected to increasing pressure from overlying sediment and increasing temperature from Earth's internal heat. A number of changes occur in the sediments as a result. First, the pressure expels water from the pore spaces, just as squeezing forces water from a sponge, and compacts the sedimentary particles into an increasingly firm mass. Below about 1000 m, the sediments are compacted into mudstones and shales. Some clay minerals contain water, but below about 3000 m, water is squeezed from clay particles and into adjacent pore spaces, generating very high pressure in the sedimentary rocks. If a well is drilled to these depths, the localized release of pressure can push the water and any oil and gas floating above it more than 1000 m above the ground surface. Such blowouts are extremely dangerous to workers on oil rigs, as the hydrocarbons are highly combustible. (Figure 11-3).

While water is squeezed out of pore spaces and migrates upward through the cracks in the rock, organic compounds remain behind, stuck to clay grains. They, too, are heated and squeezed during burial. The chemical bonds between their atoms begin to break at depths greater than about 1000 m, depending on the local rate of temperature increase with depth below Earth's surface. As the bonds break, lighter organic molecules are formed that become the components of hydrocarbons. At temperatures between about 40° and 100°C, oil is generated, and at temperatures greater than 100°C, natural gas forms. For average geothermal gradients (between 1.8° and 5.5°C per 100 m), petroleum is most likely to form at depths between 1000 and 8000 m. However, because oil and gas are less dense than the *source rock,* the parent rock where the hydrocarbons were generated, the buoyant fluids will rise and seep out at Earth's surface unless trapped by a relatively impermeable layer, called *cap rock.* Typical cap rocks are fine-grained or crystalline, and include shale, mudstone, and salt.

Petroleum Traps

Any geologic structure that confines the flow of petroleum, causing it to pool beneath cap rock, is a **petroleum trap.** Petroleum migrates upward from the source rocks where it formed, generally dark marine shales and mudstones. These same rocks can act as traps to hydrocarbons that migrate from other source rocks.

Three types of petroleum traps are shown in Figure 11-4. The most common, the **anticlinal trap,** forms where sedimentary layers of variable permeability are folded into a dome or arch called an *anticline.* Petroleum moving up through an anticline pools in the structure's more porous and permeable rock layers. These *reservoir rocks* generally are sandstones and fractured limestones. A **fault trap** occurs where the movement of rock along a fault juxtaposes permeable reservoir rocks with impermeable

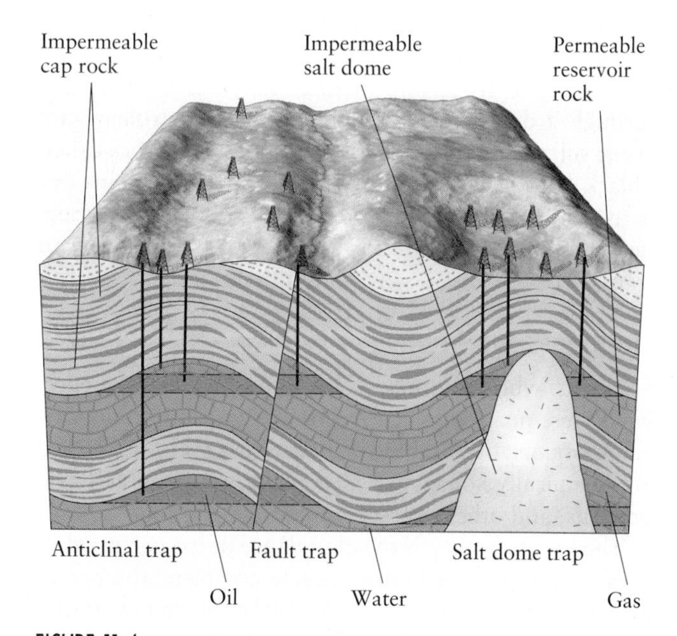

FIGURE 11-4 Three common types of petroleum traps are those found associated with anticlines, faults, and salt domes.

cap rocks. A **salt dome trap** forms when halite (NaCl), which accumulates in sedimentary deposits as low-density evaporites, flows upward through surrounding rock and forms large, impermeable, dome-shaped masses. Halite flows upward for the same reason that water and hydrocarbons do—it is less dense than surrounding rocks. It is also somewhat plastic because layers of sodium chloride molecules glide past one another under pressure. When subjected to tectonic forces and the weight of overlying sediments, the salt flows, pushes against the denser and more rigid surrounding rock, and causes it to fold into broad, dome-shaped structures with impermeable salt cores. Petroleum pools between the outside walls of the salt core and the inner walls of the surrounding rock.

Both oil and natural gas are less dense than water (1 g per cm^3), with average densities of about 0.7 g per cm^3 for oil and 0.1 g per cm^3 for natural gas. As a result, the three fluids are stratified in petroleum reservoirs, the oil located above the water, and the gas located above the oil (see Figure 11-4). The occurrence of natural gas is often the first sign to drillers that they have reached a petroleum reservoir. Unfortunately, because gas is riskier and more difficult to transport than liquid petroleum, the gas is sometimes burned off and wasted before the oil is pumped from a well. In recent years, however, the use of natural gas has increased, and some analysts predict that it will play a more important role in the world's energy future.

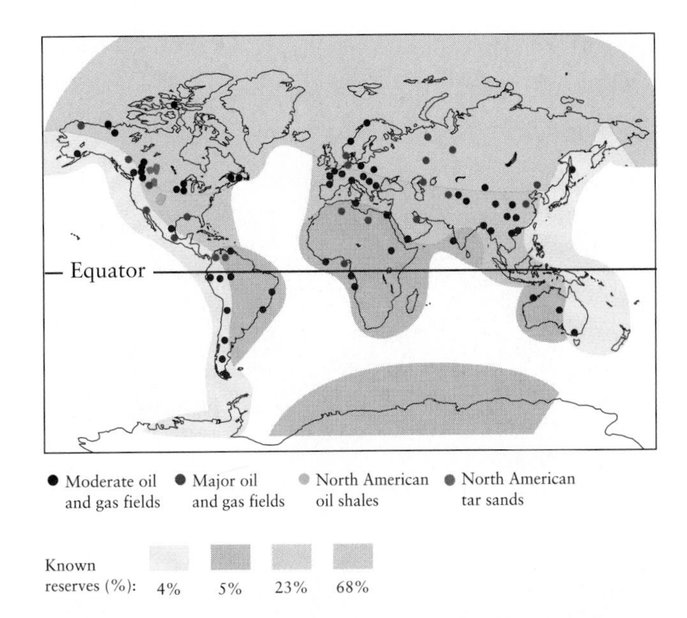

FIGURE 11-5 About two-thirds of the world's petroleum-producing areas, including the oil-rich Middle East, are located at low latitudes (warm-water regions) in areas that are now, or were recently, active convergent plate margins. Percentages are relative to the total amount of oil and gas so far discovered in the world.

Finding, Extracting, and Refining Petroleum

At the turn of the 20th century, wells were drilled where petroleum was found seeping out at the surface. However, oil seeps are shallow and their yield is typically small. For deeper deposits, prospectors would try drilling at the top of the highest hill, because petroleum tends to collect beneath arched rocks. High hills, however, do not necessarily have arched rocks located beneath them. A few geologists began to suspect that the location of oil hundreds of meters below the ground could be inferred from mapping the types and orientations of rocks at Earth's surface, a method called *structural mapping*. They combined their maps with information about subsurface rocks from existing drill holes. Structural mapping provided a more reliable means of finding arched rocks than the occurrence of hills. By 1916, oil companies had hired teams of geologists to map areas with oil potential, commonly associated with folded or faulted rocks. Geologists have remained in charge of oil exploration decisions ever since, predicting and finding oil both onshore and offshore throughout the world, in Saudi Arabia, Alaska, Venezuela, the North Sea, Indonesia, and the region of the Gulf of Mexico and the Caribbean.

As we have seen, petroleum originates with the accumulation of organic-rich sediments and their subsequent burial and deformation, which transform the organic matter to hydrocarbons as a result of increased heat and pressure. Such conditions are most common along plate boundaries. In all, about 68 percent of the world's petroleum reserves are located where convergent plate boundaries either exist now or have existed in the recent geologic past (Figure 11-5). These areas include the productive eastern Mexican coast and Caribbean and Indonesian regions as well as the Middle East. The western coasts of North and South America have been sites of subduction for hundreds of millions of years, accounting for another 5 percent of the world's petroleum. Most of the remainder is found at modern or ancient divergent boundaries, where deformation also occurs.

Finding new petroleum reservoirs is increasingly difficult because the most obvious traps have already been exploited. Drilling a single well can cost millions of dollars, and geologists use a variety of sophisticated geophysical and computer tools to assess Earth's subsurface features and their likelihood of bearing oil (Box 11-1).

Oil can be found and produced on land and at sea, but ocean exploration and production are more expensive, requiring the use of large drilling platforms (Figure 11-6). The use of highly sophisticated technology has led to extraction of oil from undersea areas where water depths exceed 2000 m. However, extracting and transporting petroleum in the ocean can damage the marine environment greatly if an accident occurs.

Once extracted, petroleum is transported to refineries by ocean tanker or pipeline. Oil refineries process the crude oil into a number of compounds, boiling it and condensing

FIGURE 11-6 Drilling into the ocean floor in an area of icebergs and sea ice is challenging: This drilling rig is in Lower Cook Inlet, Alaska. Nevertheless, offshore drilling is increasingly common as the more accessible oil fields on land are depleted.

its vapors in a series of stages so that progressively lighter and smaller hydrocarbon molecules can be collected and separated. From heavier to lighter carbon compounds, petroleum products include lubricating oils, diesel fuel, kerosene, gasoline for automobiles, and natural gas.

Will We Run Out of Petroleum?

The geologist M. King Hubbert once said that he found it hard to decide which is more remarkable, "that it took 600 million years for the Earth to make its oil, or that it took 300 years to use it up." How much of this precious resource exists, and how much of it already has been used? Scientists at the U.S. Geological Survey estimate that 950 billion barrels of oil can be recovered from known fields (Table 11-3). This amount is called the oil *reserves,* because its existence is known and current technologies make its recovery economically viable (see Figure 5-9). Another 665 billion barrels or so exist in those same reservoirs, but it is held tightly to the sedimentary rocks by surface tension and other forces and as yet cannot be pumped out economically.

Still more oil is likely to exist in places where the geologic conditions are right, such as much of the Andean mountains along the convergent plate boundary of western South America, but in some of these areas little exploration or drilling has been done. Scientists speculate that undiscovered, recoverable oil might amount to another 500 billion barrels, bringing the total of how much oil still exists to 1450 billion barrels. In comparison, total global production of oil as of 1990 has been 650 billion barrels.

If the estimated amount of recoverable oil is only a little more than twice as much as has been used in the past century, how much longer will supplies last? The world consumes 20 billion barrels of oil each year. At that rate, it would take 70 years to use up the estimated 1450 billion barrels of oil remaining. However, the rate at which the world uses oil has increased steadily since the first wells were drilled, because the global population is increasing and because developing countries are striving to raise their standard of living to match that of the industrialized nations. If the rate of consumption continues to increase as much as it has in recent years, the amount of recoverable oil left might be used up in as few as 30 years.

Balancing the alarm over oil depletion are "enhanced recovery" technologies, such as horizontal drilling, which have made it possible to extract more oil from old reservoirs. In addition, the use of the lighter component of petroleum, natural gas, is increasing. Gas reserves are less well known than those for oil, but present estimates indicate upward of 100 years of reserves at current rates of increase of consumption each year.

Natural gas is the cleanest fossil fuel, emitting 30 percent less carbon dioxide per unit of energy than oil and 43 percent less than coal. Natural gas also contains little or no sulfur and produces smaller quantities of carbon monoxide and nitrogen oxides than does oil or coal. In the 1950s, when a thick sulfur-rich smog engulfed London and more than 4000 residents died from respiratory ailments, the city deliberately began converting from coal to natural gas to improve its air quality.

TABLE 11-3 Estimated Oil Reserves and Resources (billions of barrels)

	United States	World
Reserves (1991)	25	~950
Undiscovered recoverable	~40	~500
Anticipated reserve growth	~20	unknown
Total oil left for future	~85	~1450
Past production (through 1990)	55	650

Source: From C. D. Masters, D. H. Root, and E. D. Attanasi, "Resource Constraints in Petroleum Production Potential," *Science* 253 (July 12, 1991): 146–152.

11-1 Geologist's Toolbox

Computer Modeling of Petroleum Resources

Geologists use many sophisticated techniques to find potential petroleum reservoirs. One such exploratory method is the use of seismic reflection experiments. In this technique, energy is produced at Earth's surface, usually by explosives on land or a sound-generating device offshore. The sound waves produced from the energy source radiate outward through the underlying material, reflecting off different layers and returning to the surface at different times. The arrival times of the reflected seismic waves are recorded by special receivers called geophones. Combining the data from an array of geophones, geologists can construct a three-dimensional picture of the material beneath Earth's surface, helping them to locate potential petroleum traps and reservoirs. The amount of data collected is immense—as much as 1 billion data points per square kilometer of land surveyed—and even with the largest computer systems it can take up to a year to analyze the data. More digital data exist from the seismic reflection work done in the petroleum industry than for any other scientific activity.

Seismic reflection data and computer technology have been used to find petroleum in many types of sedimentary basins. Much oil has been found in sands deposited in ancient deltas similar to that of the modern Mississippi. In a delta, water and sediment from a continent are transported by stream channels in which the coarsest debris, typically sand, is deposited along the channel beds and the finest debris, mud, is carried offshore as plumes of sediment that slowly settle to the ocean floor. As stream channels cross a growing delta, they separate and branch into many distributary channels. Deposits of sand in distributary channels provide good reservoirs for petroleum rising from organic-rich muds below. When a delta sinks under its own weight, or if sea level rises, its shoreline shifts and channel sands are buried by mud, which later becomes cap rock if petroleum migrates into the sand. Evidence of such shifts in depositional environments is common in the geologic record and mirrors the repeated rise and fall of sea level through geologic time.

Three-dimensional models generated by seismic reflection experiments can reveal the location of ancient deltaic distributary channels made of sand. Each of these might be a potential petroleum reservoir to be discovered during drilling. With this information, exploration geologists recommend where and how deep to drill, then wait anxiously as drillers approach that

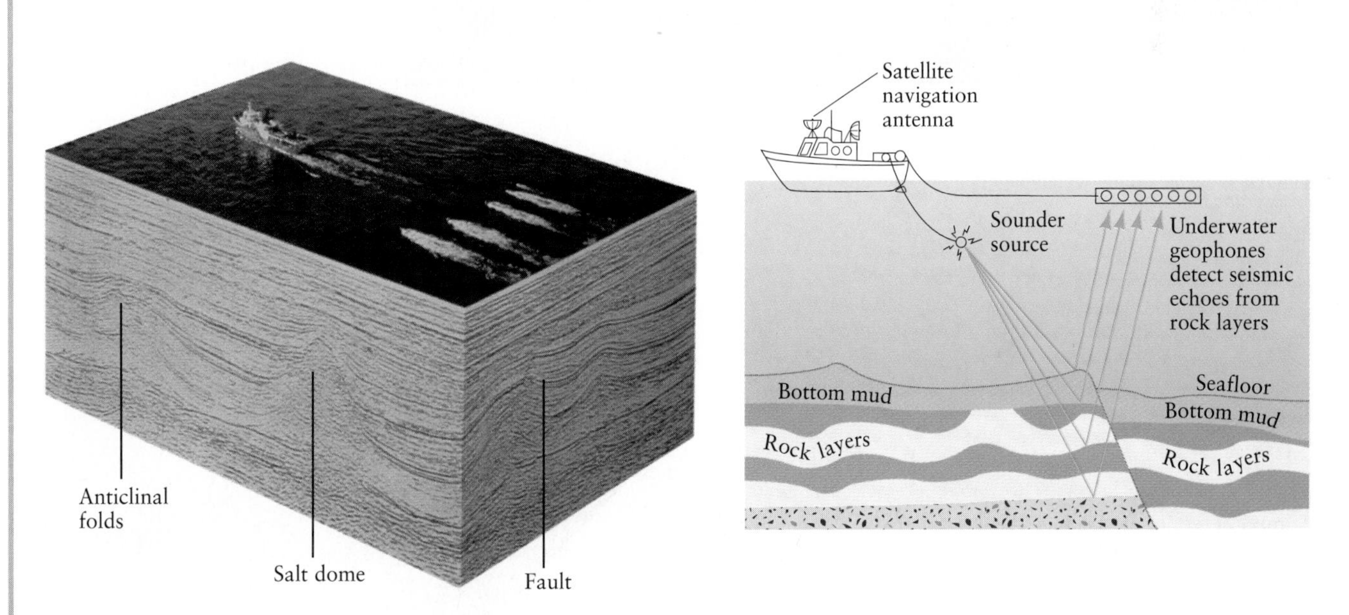

A ship towing a sound source and underwater geophones used to produce three-dimensional seismic reflection diagrams depicting the structure and layering of Earth beneath its surface. Anticlinal folds, a possible salt dome, and a fault are potential petroleum traps.

(continued)

11-1 Geologist's Toolbox

(continued)

level to see if their reasoning leads to another source of oil or if they have a dry hole. In the sediments of the Gulf of Mexico, one out of six holes bears oil and the rest are dry. Even with greater understanding of sedimentary basins and advanced computer technology, finding oil still requires a bit of luck.

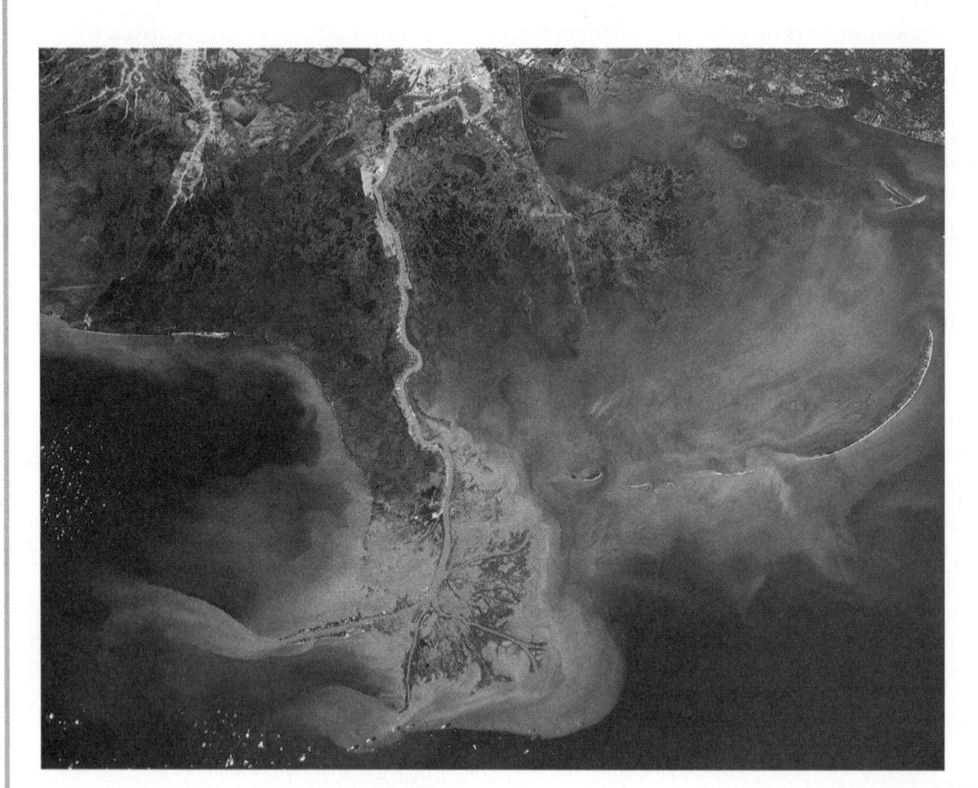

A high-altitude photograph of the lower Mississippi River and its delta. Plumes of muddy water extend outward across the delta forming along distributary channels (lower left). Sediment from the North American continent has accumulated in the Mississippi delta for nearly 150 million years. Offshore drilling has exploited the hydrocarbons in organic-rich sediments buried beneath the delta and its offshore extension into the Gulf of Mexico.

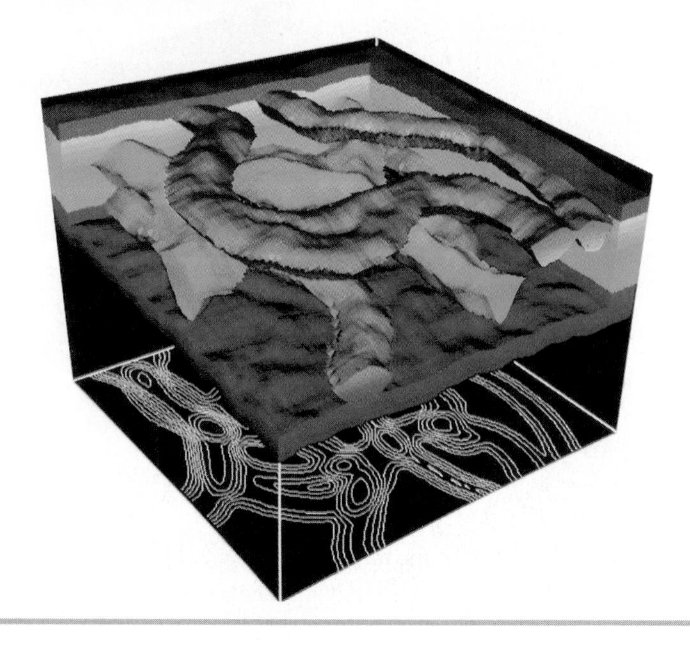

Computer models of seismic reflection data enable scientists to generate colorful three-dimensional models like this one of the Mississippi delta. The shapes of the channels are very similar to those of distributary channels. Numerous channel sands typically are stacked one upon the other, separated by layers of mud. The total thickness of the sands is computed and shown as a contour map at the base of the image.

Natural gas provides hope for a cleaner environment even as developing nations become more industrialized and increase their demand for energy. New designs for power plants that use natural gas are more efficient and less polluting than existing coal and oil plants and have spurred construction of gas plants worldwide. Since the 1980s, gas production has increased 30 percent worldwide, and natural gas dominates the market for new power plants in the United States and Europe.

Economists claim that we will never run out of oil because before that happens we will have switched to cheaper, cleaner, more reliable substitutes. New technologies are likely to make the use of substitute power sources increasingly feasible. According to Christopher Flavin and Nicholas Lenssen of the WorldWatch Institute,

> Just as the world shifted early in this century from solid fuels to liquid ones, so might a shift from liquids to gases be under way today—thereby increasing the efficiency and cleanliness of the overall energy system. . . . [Furthermore, natural gas is the] logical bridge to what . . . [might] become our ultimate energy carrier—gaseous hydrogen produced from solar energy . . . because these two fuels are so similar in their chemical composition . . . and in the infrastructure they require.

Environmental Impact of Using Petroleum

Aside from the toxic gases that fossil fuel combustion releases, the environmental concerns about petroleum use center on the influx of crude oil into the oceans and the proximity of drilling sites to sensitive ecosystems. The very areas that contain rich stores of petroleum are some of the most productive in terms of modern biological activity.

Ocean Pollution Most of the oil pollution in oceans is derived from ships that flush their oil tanks at sea and from industrial and urban runoff from continents. Large accidental oil spills from tankers transporting oil across oceans contribute only 5 percent of the estimated 2.3 million tons of petroleum hydrocarbons entering the oceans every year, but they can cause severe damage to local marine or coastal ecosystems (Figure 11-7).

Most oil spills occur when tankers run aground and their storage compartments are pierced. The first response must be to prevent as much oil as possible from escaping. Large, absorbent buffers are placed around the ship, and oil may be siphoned from its tanks. Oil already on the water must be isolated and kept from washing up on the coastline. Dispersants are sprayed on the oil, or the oil may be burned off. Over time, gases escape from the remaining oil, leaving a dense, viscous residue that sinks to the bottom of the ocean and may smother organisms on the ocean floor. The residue persists for many years, hampering recovery of the ecosystem. If the oil does reach shore, the cleanup is more difficult because waves distribute the oil across tidal zones.

Evidence is accumulating that the cleanup of spilled oil is better left to nature. Oil degrades through the action of sunlight and microorganisms. Past cleanup efforts often have resulted in increased damage to the local ecosystem. After the *Exxon Valdez* oil spill in Prince William Sound, Alaska, in 1989, pressurized hot water was used to clean some of the oiled beaches (see Figure 11-7). Years later, while the beaches left alone had rebounded naturally, those beaches were still barren. Furthermore, biologists

FIGURE 11-7 Coastal oil spills have devastating effects on wildlife and habitat. A generally unsuccessful attempt was made to rescue seabirds, other marine wildlife, and habitat after the wreck of the *Exxon Valdez* in Prince William Sound, Alaska, in 1989. Cleanup efforts included the use of hoses and pressurized hot water to blast oil off the beaches.

who monitored oil-soaked wildfowl that had been cleaned and released found that few survived. The birds had ingested so much oil that cleaning their feathers was not enough to save their lives.

Habitat Disruption Highly publicized disasters such as the *Exxon Valdez* spill have made the public wary of efforts to increase oil drilling in coastal regions. The demand for energy resources is in direct conflict with the desire to preserve habitat and protect wilderness resources. An area of much contention is the Arctic National Wildlife Refuge along Alaska's North Slope. Teeming with caribou, wolves, polar bears, snow geese, and other wildlife, the area currently is protected, but some oil companies would like permission from the U.S. government to drill along the coastal plain. The continental shelf has all the conditions favorable to formation and preservation of petroleum, and the companies argue that tapping this field could reduce American reliance on imported oil. Environmentalists counter that the wilderness is irreplaceable. Furthermore, the oil produced might meet domestic demand for several months to years, but the same amount could be saved by increasing the miles-per-gallon fuel requirements for new automobiles or making power plants more energy efficient.

- Petroleum, found in pore spaces in sediments and sedimentary rocks, occurs in two forms: crude oil and natural gas. The latter is less dense and typically occurs above crude oil in a petroleum reservoir.
- Petroleum forms largely from phytoplankton, single-celled floating marine plants that sink to the ocean floor upon death, accumulate in sediments, and become hydrocarbons at the high pressures and temperatures associated with deep burial and tectonic deformation.
- Major petroleum deposits are associated with recent or ancient convergent plate boundaries where conditions favorable to hydrocarbon formation (deep burial), migration, and trapping (folding and faulting) have occurred. All these conditions have existed in the area that now forms the rich oil fields of the Middle East.
- Crude oil reserves are estimated to last for 30 to 70 years, and natural gas reserves for about 100 years.
- The environmental impacts of producing, transporting, and burning petroleum for energy are many and include the emission of carbon dioxide —a greenhouse gas—into the atmosphere, ocean pollution, and habitat disruption.

Other Fluid Hydrocarbons

The Athabasca Tar Sand is a 5000-km^2 area in Alberta, Canada, that contains enough hydrocarbons in its thick layer of sand to yield 600 billion barrels of oil—almost as much oil as humanity has consumed so far. But there is one drawback. The hydrocarbons permeating the sand are too viscous to be pumped from a well.

Another 2000 billion barrels of oil could be produced from a single deposit in the southwestern United States, the Green River Oil Shales spanning Colorado, Wyoming, and Utah. The hydrocarbons in this deposit are even more viscous than those in the tar sand; in fact, the oily material is nearly solid. Although the environmental and economic costs are great, both tar sands and oil shales can be mined and processed to yield oil. Both tar sands and oil shales are hydrocarbons, but they differ from petroleum in origin and the locations at which they occur.

Tar Sands

Neither tar nor necessarily a sand, **tar sand** is a sedimentary deposit saturated with an asphaltlike organic substance made up almost entirely of dense hydrocarbons. Called *bitumen,* this substance forms where migrating organic compounds come in contact with the atmosphere, pedosphere, and groundwater, but the details of formation are not clear. Recent studies indicate that bacteria in the pedosphere may degrade liquid petroleum as it migrates upward, releasing gaseous hydrocarbons and leaving behind the viscous substances found in tar sands. Alternatively, tar sands may contain hydrocarbons in early stages of formation.

Currently, tar sand is mined at the surface, much the same way as coal. The entire excavated mass is heated with pressurized steam until the bitumen softens and rises. The oily substance is collected and treated to remove sulfur and other impurities, then hydrogen is added to upgrade it to a synthetic crude oil that can be refined. Mining and processing tar sand requires nearly half as much energy as the end product can yield. Nevertheless, the Athabasca Tar Sand produces nearly 16 percent of Canada's oil.

There are significant environmental drawbacks to obtaining oil from tar sand. The amount of rock and sediment that must be mined can lead to extensive land disturbance. Large volumes of water are necessary for processing the hydrocarbons. The contaminated wastewater and sediment accumulate in toxic disposal ponds. Nonetheless, because there are large deposits in Canada, Venezuela, and elsewhere, and because the technology exists for the successful extraction and production of oil from tar sands, these deposits are likely to be exploited more fully as known petroleum reserves are consumed.

Oil Shales

The estimated 2000 billion barrels of oil thought to be recoverable from the Green River formation in Colorado, Wyoming, and Utah (Figure 11-8), is about as much as the sum of all petroleum-derived oil used to date and estimated to remain. **Oil shale** is a fine-grained clastic sedimentary rock, generally similar to shale, but instead of oil it contains a solid, waxy substance called *kerogen*. Kerogen forms from buried plant matter at temperatures and pressures too low to produce liquid hydrocarbons. In other words, oil shales are exposed source rocks that have not yet simmered enough for their organic remains to be converted to oil and migrate away.

The Green River Oil Shales formed during the Eocene epoch (57 million to 34 million years ago), when many large lakes existed in the area. This one formation contains half the world's known supply of oil shale and is capable of producing 10,000 barrels of *syncrude* (synthetic crude oil) each day. Other deposits exist in Brazil, Australia, South Africa, and China, and the world's reserves of oil shale are thought to be 200 times larger than those of conventional oil.

Kerogen can be extracted from oil shale by both in situ and surface mining methods. Below ground, heated air is injected into the rock to vaporize the kerogen, which then is collected as it rises to the surface and condensed to form shale oil. Above ground, oil shale is mined similarly to tar sand, and the kerogen is extracted by heating the entire excavated mass. For production of shale oil to be economically viable, it must yield more energy than the amount required to mine, extract, and process the oil shale. With current technologies, most oil shales are not worth mining while other sources of energy are cheaper. Nevertheless, some interest persists because of the immense volume of oil that potentially is available from this one source.

Producing oil from oil shale is fraught with the same problems as producing oil from tar sand: widespread land disturbance associated with surface mining, the disposal of waste rock, and the need for large quantities of water. In one estimate, if all known U.S. oil shales were mined, the amount of crushed waste rock would be sufficient to cover Earth with a layer 3 m thick. The need for water to process the Green River shales is an especially significant problem because the region is semiarid.

- The tarlike bitumen in tar sands could be either bacterially degraded petroleum or hydrocarbons in an early stage of formation. The waxy kerogen in oil shales formed where temperatures and pressures were not high enough to produce liquid petroleum.
- In contrast to the hydrocarbons in petroleum, which are produced largely from marine phytoplankton, the solid organic substances in oil shale are deposited in continental lakes.

FIGURE 11-8 Oil shale contains large amounts of kerogen, a waxy substance that is similar to oil when extracted and processed.

- Tremendous amounts of syncrude are available from tar sands and oil shales, but exploiting them is costly and potentially very damaging to the environment, so at present they contribute little to the world's energy supply.

Coal

Of the fossil fuels, the largest reserves by far are contained in coal deposits (92 percent). Like petroleum, coal is composed primarily of carbon and hydrogen compounds (see Table 11-2), and most coal is contained within sedimentary rocks. Unlike petroleum, however, coal is a solid rather than a liquid or gas and usually contains elements besides hydrogen and carbon. These include sulfur and nitrogen, which combine with oxygen and water in the atmosphere and hydrosphere to produce the sulfuric and nitric acids that contribute to acid rain and acid mine drainage. Because of coal's sulfur and nitrogen content, the drainage waters from coal mines and gaseous emissions from coal-burning power plants contain more pollutants than the drainage and emissions associated with petroleum extraction and combustion. On the positive side, however, coal is far more widely distributed and abundant than the relatively localized and rare petroleum reservoirs. Most coal formed from land plants that lived in swamps, marshes, and coastal plains. Most petroleum originated from marine phytoplankton that lived along and near continental shelves. Because petroleum migrates after it is formed and collects in permeable rocks adjacent to traps, the occurrence of petroleum tends to be more restricted than the occurrence of coal, which is found in the broadly distributed areas where it formed.

Origin of Coal

Layers of coal buried in Earth's crust are the tombs of ancient wetlands. Early miners noted a variety of plant fossils in coal beds, mostly the remains of primitive ferns and spore-bearing trees that grew in coastal swamps, estuaries, and swampy river floodplains, much like the modern Great Dismal Swamp in coastal Virginia and North Carolina and the Okeefenokee Swamp in coastal Florida.

Wetland environments are characterized by poor drainage, wet soils, and stagnant, oxygen-depleted water. The lack of oxygen preserves organic plant matter. The accumulation of substantial thicknesses of organic matter in a wetland can lead to the development of **peat,** a soft, unconsolidated deposit of compressed plant remains. Peat has a high moisture content, contains less than 50 percent carbon, and burns freely when dried. Any mineral matter contained in the peat will not burn, but instead forms small gritty particles called *ash*.

Depending on the pressure and heat exerted on peat, it may be transformed into any of three grades of coal (Figure 11-9). The softest is **lignite,** a brown sedimentary rock that is about 50 to 60 percent carbon and gives off much more heat than does peat when burned. If deep burial further compresses lignite, thus removing more water from it, it becomes a harder, brown-black sedimentary rock known as **bituminous coal,** which is about 60 to 90 percent carbon. If bituminous coal is heated enough to recrystallize, it becomes a very shiny, dense, hard, black, metamorphic rock known as **anthracite coal.** Anthracite coal is more than 90 percent carbon and has the highest heat value of all coals. If anthracite were to metamorphose even further, it would ultimately become pure carbon, in the form of the mineral graphite. As with peat, mineral matter in any of these three types of coal is much less combustible than is carbon and becomes ash when burned.

The thickness of a coal bed is about one-tenth the original thickness of the peat from which it formed. A 3-m coal bed, for example, represents an original peat thickness of 30 m. Modern peat accumulates at a rate of about 1 m in 1000 years. At this rate, a 3-m coal bed represents about 30,000 years of peat accumulation.

Coals commonly occur as numerous layers, which coal miners call seams, ranging from one to several meters thick and separated by beds of sandstone, siltstone, shale, and limestone. The layers of clastic and chemical sedimentary rocks separating the coal seams formed when the ancient coal-forming swamps were submerged periodically. One cause of submergence in a coastal plain environment is a relative rise in sea level, which occurs either because landmass is sinking or ocean surface is rising (see Box 3-2). Another cause is lateral migration of river channels.

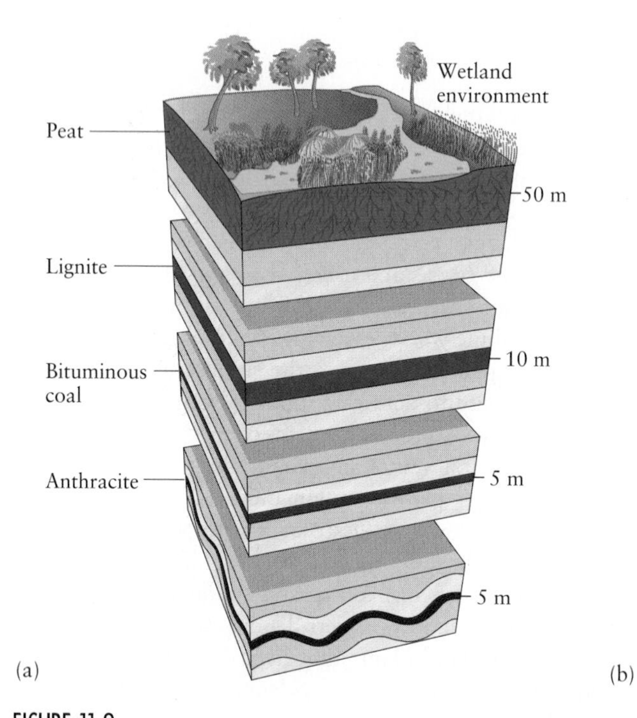

FIGURE 11-9 (a) As organic matter is buried and compressed, the amount of water, oxygen, and hydrogen decreases, producing an increasingly carbon-rich material that grades from peat to lignite, then bituminous coal, and finally anthracite. (b) Samples of peat and coals, arranged in order of increasing degree of transformation and carbon content: peat (top left), lignite (top right), bituminous coal (bottom left), anthracite coal (bottom right). The higher the grade of coal, the greater the density, hardness, and shine.

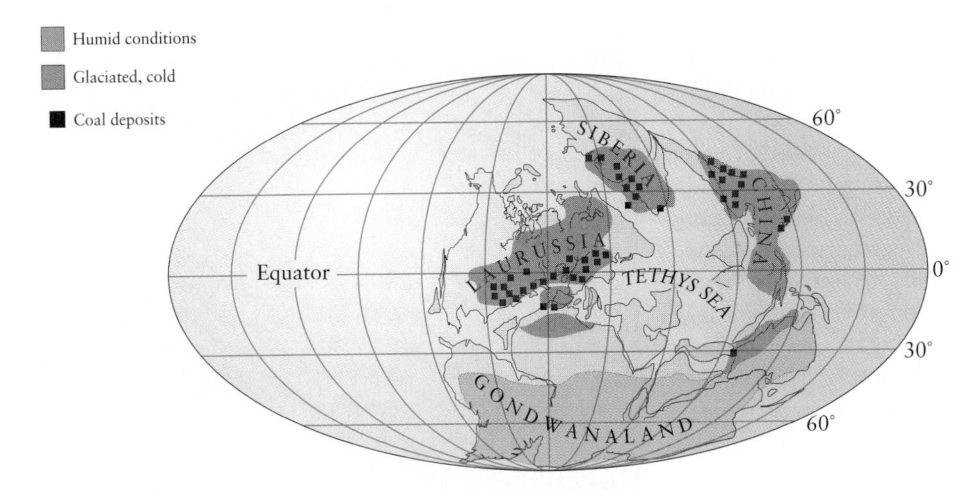

FIGURE 11-10 When continents are arranged in their Carboniferous positions of about 360 million to 290 million years ago, the modern coal beds are located along or near the equator, surrounding a large body of water known as the Tethys Sea. A polar ice cap over Gondwanaland repeatedly grew and shrank throughout Carboniferous time, providing the fluctuating sea level that enabled numerous swamps and coal seams to develop one atop the other around the coastal margins of the Tethys Sea.

Global Distribution of Coal Deposits

Consider the following facts about the global distribution of coal:

- The largest known coal deposits occur in sedimentary and metamorphic rocks of about the same age—360 million to 290 million years old—in the United States, Europe, the former Soviet Union, and China. For this reason, that time is named the Carboniferous period.
- Fossils in these coal deposits indicate they formed in tropical and subtropical coastal swamps through which large rivers flowed seaward.
- All these deposits contain numerous coal seams, sometimes as many as 100, typically separated by marine shales and limestones. The sequences of coal, shale, and limestone are cyclic, repeated over and over in the thick deposits.

What do these clues tell us about the origin and current distribution of much of the world's coal? How does a geologist interpret these facts?

If the areas that now contain Carboniferous coals once were tropical swamps along the ancient coastlines of North America, Europe, and Asia, these landmasses must have been much farther south, closer to the equator than they are now. Continental drift is the explanation: These continents in the Carboniferous period were much closer to one another and straddled the equator (Figure 11-10). As for the number of sequences of coal, shale, and limestone, these indicate that sea level rose and fell many times during the Carboniferous period, inundating continental lowlands and enabling wetlands to cover vast areas. With each rise in sea level, coastal swamps would have been drowned and buried by marine mud that later became shale and limestone. With each fall in sea level, the swamps and river floodplains would have extended out from the continents, forming new swamps in which peat accumulated and later became coal. Although some of the Carboniferous coal deposits now are located in mountain belts, such as the Appalachians, they mark the locations of ancient lowlands periodically inundated by shallow seas.

Why would the Carboniferous sea level have risen and fallen so many times? The most likely cause is an ice age in which ice masses grew and melted over and over again. Geologic evidence indicates that when the northern continents were coming together along the equator, the southern continents (South America, Australia, India, Africa, and Antarctica) already were assembled as the continent Gondwanaland, which was positioned over the south pole. Glacial deposits and rocks scoured by glacial ice provide a picture of an ice sheet much like the modern Antarctic ice sheet. Just as ice sheets and smaller glaciers repeatedly have advanced and retreated for the past few million years, causing sea level to rise and fall more than 100 m, the Gondwanaland ice sheet varied in size and affected the Carboniferous sea level.

Few coal deposits are older than Carboniferous because land plants evolved just 400 million years ago, but coals only somewhat younger exist. Some of these are located along the low interiors of continents that were inundated by high sea levels since Carboniferous time. Global sea level was especially high during the Cretaceous period 146 million to 65 million years ago, when global temperature reached some of the highest values in Earth's history. In North America, a large Cretaceous sea spread throughout the lowland west of what are now the Rocky Mountains. Just as in Carboniferous time, fluctuating sea level favored the development of many layers of coal that now are mined throughout the region.

Will We Run Out of Coal?

Because much coal is exposed at Earth's surface, geologists can more easily estimate the recoverable reserves of coal than of petroleum. By measuring the thickness of different coal seams and the area they cover, geologists estimate that world reserves are sufficient to last for at least 220 years at current rates of use, but only 65 years if the use rate increases 2 percent each year (a little more than the current rate of population growth). Undiscovered coal resources might—once found—supply the world with coal for 900 years at current rates of consumption, or 149 years if the rate increases 2 percent per year. As with petroleum, advances in mining technologies may raise the estimate of how long we can rely on coal. For instance, deep *underground mining* in the eastern United States once provided the nation with most of its coal, but now nearly 70 percent of it is produced by surface mining in western states. In *surface mining*, large machines strip away layers of sandstone, shale, and limestone to uncover coal seams at Earth's surface. Where coal seams are relatively thick, close to horizontal, and less than about 100 m below ground, surface mining is much cheaper and safer than underground mining. It also is more efficient at removing most of the coal; in underground mining, nearly half the available coal must be left as roof support to keep mine openings from collapsing.

FIGURE 11-11 These long, linear pits are open surface mines near Tremont, Pennsylvania, in the Appalachian Mountains. Miners excavated seams of coal sandwiched between layers of tilted sedimentary rocks that form the backbones of the prominent ridges.

Environmental Impact of Mining and Burning Coal

Because there are much greater reserves of coal than oil, and because new mining methods can recover greater amounts than in the past, the use of coal could grow in the future. At present, it is the second most important global primary energy source, providing 25 percent of all energy consumed. Certain countries, such as China, are particularly dependent on coal for their energy needs (China burns coal for 75 percent of its commercial energy) and probably will increase their use of this resource in the future. Coal, however, is arguably the most environmentally damaging energy source.

All stages of the coal energy cycle, from mining to processing and combustion, affect the environment. Coal fires are a problem particularly with underground mines because so much coal is left in place. Hundreds of underground fires are burning throughout the United States alone, and many cannot be extinguished. The mazes of abandoned tunnels and shafts below ground provide ample air flow and oxygen to keep fires burning for years (Box 11-2). Underground mining also results in surface subsidence above collapsed tunnels, thereby damaging buildings, roads, and railways.

Surface mining is safer than underground mining for mine workers, but it leads to slope instability, unsightly and dangerous open pits, and soil and sediment erosion (Figure 11-11). Reclamation of operating coal mines now is required in many nations, including the United States, but old, unreclaimed mines still present a significant hazard to surrounding communities. Furthermore, it is costly and laborious to reclaim surface-mined land in arid and semiarid regions where vegetation and water are scarce.

Sulfur is a persistent environmental concern associated with coal use. Coals formed in coastal swamps, typical of the eastern coal deposits of the United States, have a higher sulfur content than do coals formed in freshwater swamps typical of western deposits. The reason is that the amount of sulfate ions is much higher in salty coastal waters than in fresh water on continents. The sulfate ions combine with the other elements, such as iron, to form sulfide minerals, such as pyrite, in the coals. When sulfide minerals in coal are exposed to the atmosphere, as in tailings, they react with oxygen and water to form iron oxides and hydroxides and sulfuric acid (see Chapter 6). In addition to causing acid mine drainage, sulfur released by human activities is the main contributor to acid deposition in the form of acid rain.

Acid mine drainage increases the acidity of local streams, thus increasing their metal content and affecting the biota. Aquatic plants and animals are sensitive to the acidity of their habitat, and organisms begin to die as the pH of water falls below their tolerance levels (see Figure 6-10). Furthermore, iron oxides and hydroxides precipitate out of the water as a yellow-orange mass called yellowboy that coats the stream bottom and smothers aquatic life. The problem is prevalent in the eastern United

11-2 Case Study

Underground Coal Mine Fires

In the Appalachian Mountains, beneath the now abandoned coal-mining town of Centralia, Pennsylvania, a fire in a single mine pit almost 40 years ago still burns in a honeycombed subterranean world that covers tens of square kilometers, and it could go on burning for centuries. Residents using garbage as landfill for an abandoned surface mine pit decided to burn the waste before their 1962 Memorial Day celebration. The fire ignited an exposed coal seam. More than $10 million has been spent to fight the fire, but it has continued to spread underground. Burning coal in its path, the fire moves uphill, sideways, and downhill and leaps from place to place by hydrogen explosions. The extreme underground temperatures boil off groundwater, creating steam that pinpoints the locations of fires.

State and federal scientists drilled dozens of holes into the ground to monitor temperature and air quality around Centralia. Temperatures have exceeded 500°C, and the fumes contain a variety of deadly substances, including carbon monoxide and carbon dioxide. Both gases are odorless and nearly impossible to detect without instruments, prompting Congress in 1984 to appropriate $42 million for voluntary acquisition and relocation of affected businesses and residences. Later, all remaining structures were condemned.

Where more than 1000 people lived and worked is an evacuated landscape that looks as if it had been bombed. Its surface is marked with deep holes above collapsed tunnels, roads riddled with long cracks, and gases rising from the holes and cracks. While roads are repaired and houses relocated, the fire continues to burn, now too deep and extensive to be controlled. And the fire continues to spread, moving toward the next small town.

(a)

(b)

(a) Roads in Centralia, Pennsylvania, have collapsed above a burning mine fire, marked by steam escaping from cracks in the roads. (b) Nearby, vapors that include carbon monoxide, carbon dioxide, and sulfur dioxide seep from deep collapse pits. In places, the air and ground are so hot that dead trees catch fire frequently, and their charred remains are abundant. Where metal-rich fumes seep out of the ground, exquisite crystals, much like those found at volcanic vents, form thick crusts about the pit rims.

(a)

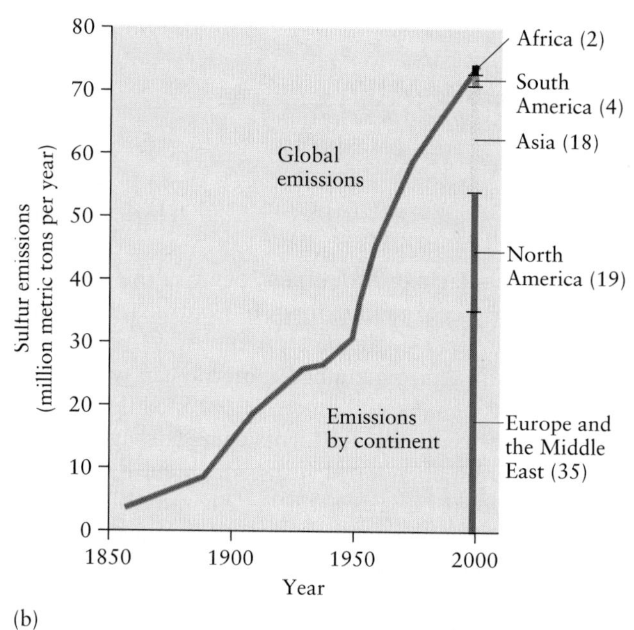

(b)

FIGURE 11-12 (a) Uncontrolled coal combustion from household stoves and small boilers pollutes the air in urban areas throughout northern China. Coal combustion is the single largest source of sulfur emitted to the atmosphere and biosphere by human activities. (b) Europe, the Middle East, North America, and Asia emit the greatest amounts of sulfur to the atmosphere. As developing countries industrialize, their per capita demand for energy increases, and so does their emission of sulfur if they use high-sulfur fossil fuels. Combining this effect with increasing population, some estimates of global sulfur emissions for the year 2020 are as high as 210 million tons per year, three times as much as 1990 rates.

States and northern Europe, where precipitation is high and coals and coal-bearing strata contain more sulfides. Although acid mine drainage has been vigorously dealt with in many areas of the United States, over 9000 km of waterways and tens of thousands of square kilometers of land are still affected.

Coal extraction and combustion are the single largest source of sulfur released to the atmosphere and hydrosphere by human activities and emit more carbon dioxide, a greenhouse gas, per unit of energy than any other fossil fuel. Approximately 18 million tons of sulfur are emitted into the atmosphere every year in the United States from coal-burning power plants. Even more sulfur is emitted from Europe and the Middle East (Figure 11-12).

Sulfur released into the atmosphere returns as acid deposition, but can be reduced by limiting the amount of sulfur emitted when burning high-sulfur coals or by using low-sulfur coal. Advanced combustion technologies in which coal, limestone ($CaCO_3$), and other calcium oxides are mixed together can cut SO_2 emissions significantly and reduce NO_x emissions by reducing combustion temperatures. Once gases leave the combustion chamber, SO_2 can be extracted by a variety of methods, of which one of the most effective, but also most expensive, is the use of scrubbers. The flue gases produced during combustion are injected into a scrubber unit in which a limestone slurry combines with the sulfur to form a calcium sulfite ($CaSO_3$) slurry. Sulfur dioxide is removed and disposed of as a solid waste product contained in the slurry rather than emitted to the atmosphere as a gas.

Scrubbers can reduce sulfur emissions from coal-burning power plants by up to 90 percent, but reduction comes at a significant cost. Scrubbers consume up to 2 percent of the electricity generated by a power plant. Carbon dioxide emissions increase by 4 percent because of the increased consumption of coal and because CO_2 is generated by the reactions in the scrubber. Scrubbers also generate huge quantities of solid waste, which must be disposed of in landfills that may then pose a threat to local groundwater.

Finally, combustion of coal produces other wastes, including ash from mineral matter and small amounts of a variety of metals. Ash is collected in power plants and disposed of as solid waste. Some forms of ash are reused, for example as a component of cement. Some metals that in large amounts can be toxic to organisms are emitted to the atmosphere during combustion. These include selenium (Se), mercury (Hg), and arsenic (As). Those coals with relatively large amounts of these substances are less desirable for burning.

- Coal forms from land plants that are deeply buried. The greater the depth and higher the temperature of burial, the higher the grade of coal and carbon content, ranging from low in peat and lignite to high in bituminous and anthracite coal.
- Major coal deposits are associated with recent or ancient convergent plate boundaries where conditions were favorable to the formation of peat in extensive coastal wetlands, the burial of layers of peat under sediments associated with changing sea level, and the deformation of layers of peat and coal during plate collision. All these conditions existed during and after the Carboniferous period some 360 million to 290 million years ago, when the continents that composed Pangaea were drifting together and sea level fluctuated in response to polar glaciation.
- There are large reserves of coal on many continents, particularly North America and Asia, which are estimated to be sufficient to last for up to 220 years or so, and even more coal might exist as undiscovered resources.
- The environmental impacts of mining, transporting, and burning coal for energy are many; two of the greatest are the emission of carbon dioxide and sulfur dioxide into the atmosphere, the latter contributing to acid rain. In addition, sulfur dioxide that reaches the hydrosphere from mine tailings and abandoned mine tunnels produces acid mine drainage.

Geothermal Energy

The energy available from Earth's interior is called **geothermal energy.** In general, because the average heat flow from Earth's interior is small (0.06 W per m^2 compared with about 1200 W for solar energy), geothermal energy does not represent a significant energy resource. However, heat stored in deeply buried rocks is an energy stock estimated at four times the energy value of the world's fossil fuels. The difficulty lies in extracting the energy.

Geothermal energy has been made use of only where heat flow is high and very hot or molten rock occurs within several kilometers of the surface, conditions primarily found along plate margins. The concentration of volcanoes around the Pacific Ocean known as the "ring of fire" marks convergent boundaries, which yield substantial geothermal energy for Japan and New Zealand. Spreading centers along the Mid-Atlantic Ridge supply most of Iceland's electricity, and some Icelandic cities are heated entirely by geothermal energy; geothermal energy has the potential to be harnessed along the East African Rift. In the United States, areas with high geothermal gradients are concentrated in the western states, where volcanic and tectonic activity occurs. The Geysers, a geothermal field in northern California, is the largest supply of natural steam in the world and provides San Francisco with half its electricity. Hot spots are ideal locations for geothermal energy resources, and the Hawaiian Islands have much geothermal potential. At present, the United States uses nearly half the geothermal electricity generated worldwide.

Geothermal energy can reach Earth's surface through natural convection processes. Groundwater that percolates downward through porous and permeable rocks until it touches hot rock is heated, expands, and circulates back toward the surface as hot water or steam. Just as wells are drilled to extract groundwater, geothermal wells are drilled into rock to extract hot water and steam from its pore spaces (Figure 11-13). Hot water can be used directly for heating and steam can be used to turn a turbine and generate electricity.

The environmental impacts of using geothermal energy are minor relative to those of other energy resources, but the contribution to the world's energy future is likely to remain small. Current costs of energy production and transportation are too high for the heat and steam from geothermal sites to be carried more than about 200 km. Some geothermal fields are showing signs of depletion after several decades of use, apparently because the rate at which heat is extracted from the reservoir exceeds the rate at which heat flows through rock and can be replenished from Earth's interior.

FIGURE 11-13 Geothermal energy tapped from steam or water heated underground by Earth's internal heat is a valuable source of energy at a number of places on Earth, primarily along active plate boundaries. This geothermal power plant is located in Kenya, along the East African Rift.

- Geothermal energy is a renewable resource along active plate margins where heat flow is high and especially where steam is abundant.
- The environmental costs of geothermal energy production are minimal, but so also are recoverable geothermal reserves, and the costs of transporting heat and steam from geothermal sites are too high for this resource to make a substantial contribution to global energy needs.

Nuclear Energy

The greatest known stock of nonrenewable energy on Earth is concentrated in uranium oxides (see Table 11-1), rare minerals containing an even rarer unstable isotope, uranium 235. Uranium 235 is one of the radioactive isotopes in Earth's interior, contributing to the planet's internal heat through spontaneous nuclear decay. Volcanic processes bring uranium minerals to Earth's crust, where they remain in igneous rocks or become incorporated into sandstones. Whether through spontaneous decay or decay induced in a nuclear reactor, uranium 235 has the potential to release 2,500,000 times the energy released by the combustion of an equal amount of carbon.

The aftermath of the first practical use of nuclear power was shocking. In August 1945 atomic bombs devastated the Japanese cities of Hiroshima and Nagasaki. Those who survived the initial blast and the fires that followed were left with severe radiation burns, blindness, debilitating sickness, and cancers, and many died within weeks. Radioactive materials released by the explosions affected people hundreds of kilometers away. Since then, nuclear power has been harnessed for electricity as well as for weapons, but the dangers of radiation leakage have kept uranium from replacing fossil fuels as the world's major source of electricity.

Fission and Fusion

Two types of nuclear reactions can produce energy: *fission,* a type of nuclear decay, and *fusion,* the combining of atomic nuclei. Both reactions release energy, but so far only fission has proved practical for generating substantial amounts of electricity.

Nuclear fission is the splitting of a large nucleus into smaller nuclei by bombarding it with neutrons. Nuclear fission energy has its origin in the formation of Earth and the solar system. During the explosions of massive stars called supernovae, high temperatures and pressures fused smaller nuclei into larger nuclei of the atomic mass of iron or greater, including uranium. Large atomic nuclei were preserved in Earth, but some are unstable and spontaneously decay to smaller nuclei, providing natural clocks by which to date minerals (see Chapter 3). The nuclei of other heavy isotopes are stable under natural conditions, but can be induced to break apart and release energy when bombarded by neutrons.

In a fission reaction, some mass is converted to energy, and the amount lost is proportional to the amount of energy released. Neutrons ejected from the uranium nucleus can cause further fissions in other uranium 235 atoms in a *chain reaction* (Figure 11-14). In addition, some fission products themselves are naturally radioactive and will continue to decay from one radioactive isotope to another.

Nuclear fusion is the welding together of atomic nuclei, a reaction that occurs naturally only in stars, where temperatures exceed 100 million °C and pressures are extremely large. In our Sun and other stars, hydrogen atoms fuse to form helium atoms and release large amounts of energy in the process (see Chapter 1). Hydrogen is abundant on Earth in water, which can be dissociated easily with an electric current, and fusion is what produces the destructive power of the hydrogen bomb. However, achieving the temperatures and pressure necessary to initiate fusion requires more energy than the reaction tends to generate. Although scientists at Princeton University's fusion research laboratories have exceeded the energy break-even point, a safe and economical commercial fusion power plant is unlikely to be a reality before the year 2050.

Problems With Nuclear Energy

Fission of uranium 235 in the cores of some 400 nuclear reactors now provides 17 percent of the world's electricity in 25 countries. In the United States, 109 nuclear power plants provide the nation with 20 percent of its electric energy; in France, nuclear power furnishes almost 80 percent. Japan and some European countries intend to increase their reliance on nuclear energy because they have only small amounts of fossil fuels. There is no shortage of uranium in the world, and new reserves have been discovered on nearly every continent. Yet no new nuclear power plants have been built and operated in the United States since the 1970s, and most nations have slowed the rate of installation of new plants. Nuclear power is not likely to solve the world's energy needs in the future. Safety and disposal concerns have dashed high hopes for nuclear energy as a cheap, clean alternative to fossil fuels. Concerns about nuclear power focus on three issues: extraction of nuclear fuel at mine sites, nuclear power plant safety, and nuclear waste disposal and weapons proliferation.

Radioactive Mine Wastes Like most mining, uranium mining produces residual material, which is disposed of as tailings. The volume of material is not large compared to that of other mining operations, but it is far more dangerous because the tailings are radioactive. Although most of the uranium is extracted from the ore deposit, its

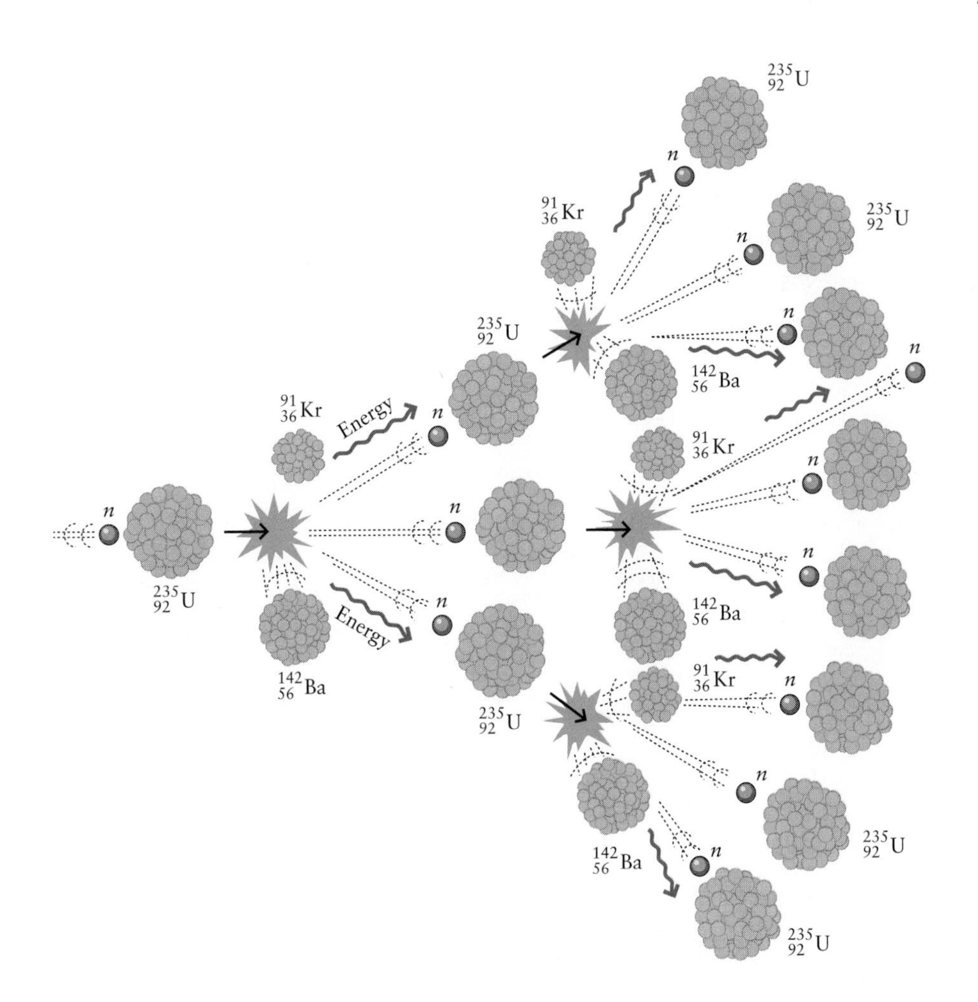

FIGURE 11-14 When a ^{235}U nucleus is bombarded by a neutron, it splits into two smaller atoms (krypton and barium), and several neutrons are ejected from the nucleus. These neutrons then bombard other ^{235}U atoms, triggering a chain reaction that will continue until most atoms are split or the neutrons are absorbed by a nonfissionable substance. At nuclear power plants, the chain reaction is kept under control with boron solution, a substance that absorbs neutrons. (The lower number shown for each nucleus is the number of protons.)

radioactive daughter products are not, and the radioactivity levels still are many times the background levels that we are exposed to in everyday life. In the United States, more than 20 million tons of abandoned radioactive tailings are exposed to the atmosphere, hydrosphere, and biosphere. Of the 24 uranium tailings sites in the United States, approximately 13 were remediated as of 1994. In a typical remediation method, the tailings are moved to a multibarrier repository designed to reduce emissions of radioactivity, prevent infiltration of water, and prevent erosion for up to 1000 years. Unfortunately, at least 10,000 years may be needed for the tailings to dissipate enough radioactivity to be at levels considered safe.

Nuclear Power Plant Safety In a typical commercial nuclear power plant, the induced decay of uranium 235 releases heat, which boils water into steam that turns a turbine to generate electricity (Figure 11-15). A nuclear power plant is similar to other electric power plants run by steam except that the heat comes from nuclear reactions. Keeping the heat under control is critical to keeping the plant intact and preventing radiation leaks.

Two severe accidents at commercial nuclear power plants have helped turn public opinion against nuclear power. One occurred in the United States in 1979, at an installation on Three Mile Island near Harrisburg, Pennsylvania. Loss of cooling fluid in the reactor core resulted in its partial meltdown. A **meltdown** occurs when the temperature in the reactor becomes too high, melting the uranium-laden fuel rods in the core and destroying the reactor container. Once the protective container is breached, radioactive material is released to the environment. As engineers at Three Mile Island struggled to control the temperature in the core and prevent a complete meltdown, the governor called out the national guard and ordered the evacuation of surrounding areas. Fortunately, the engineers were able to gain control of the reactor temperature in time to avoid catastrophic failure. Some radioactive gases were released at the time of the accident, but subsequent studies indicate that no statistically significant increases in mortality occurred in the local population. Nevertheless, residents remain frightened and suspicious, and litigation continues against the utility that owns the reactor.

Far more chilling—and destructive of human life—was the Chernobyl nuclear accident in the Ukraine in 1986. Telltale signs of radiation in the atmosphere alerted

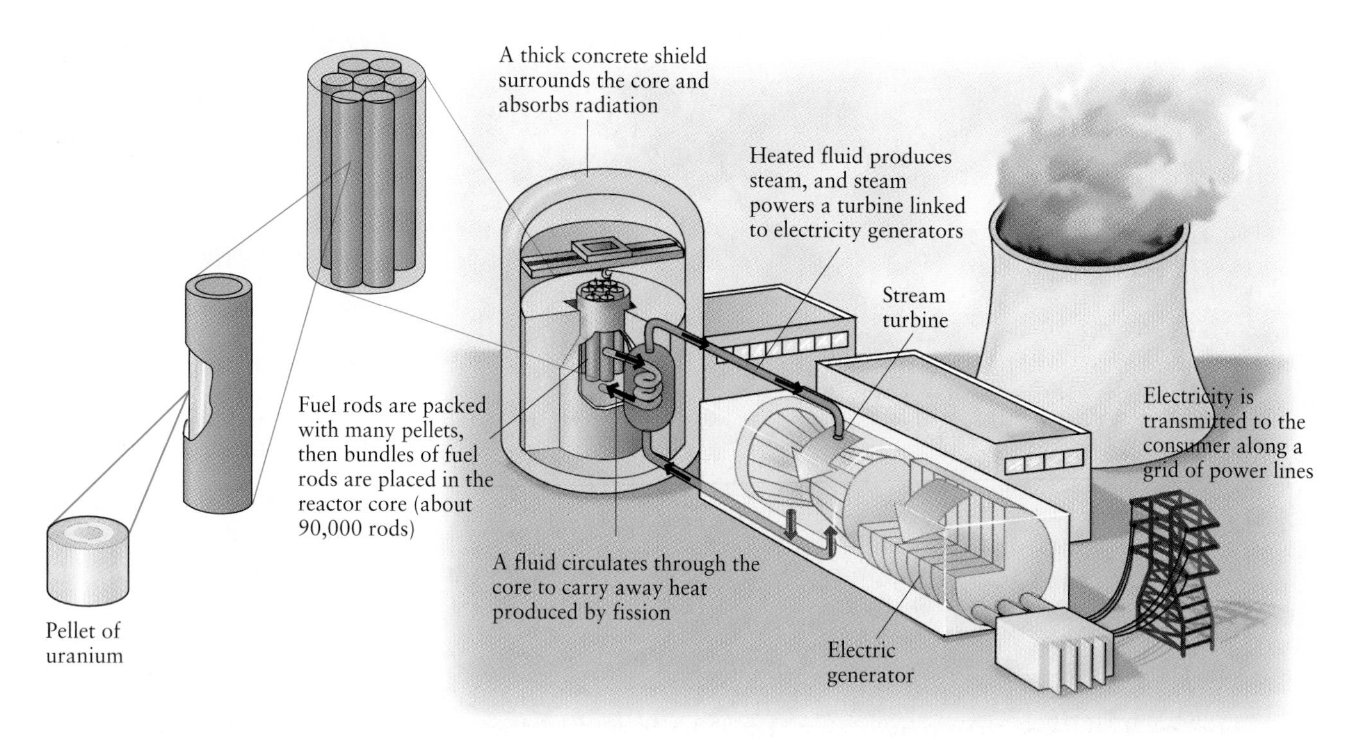

FIGURE 11-15 A nuclear power plant has a reactor core that produces energy from a bundle of metal fuel rods. Each fuel rod contains many small pellets of uranium oxide enriched in uranium 235. Some rods contain boron, which controls the reaction rate by absorbing neutrons. Nuclear fission produces heat, which is used to turn a turbine and generate electricity. If insufficient coolant is present to remove heat, a meltdown will destroy the reactor container and release radiation to the environment.

workers at a Swedish nuclear power plant that an accident had occurred upwind of their own facility. Not until Sweden confronted the Soviet government was news of the meltdown released to the world.

The Chernobyl accident was far worse than the one at Three Mile Island. At Chernobyl, the core reached temperatures as high as 3000°C, completely destroying the reactor container and releasing as much as 80 percent of the radioactive fuel in a giant explosion. The exact cause of the accident is unclear. The reactor was on standby status at the time, yet somehow it surged to 50 percent of its capacity in a short period, resulting in an uncontrollable nuclear chain reaction. The accident caused 31 immediate deaths from massive radiation exposure. Within a week, radioactive gases and particles drifted downwind and spread over Scandinavia and western Europe, contaminating the landscape with radioactive fallout (Figure 11-16). Thousands of reindeer ate contaminated plants and had to be destroyed, and to this day soils in the immediately surrounding area are contaminated. Most of the several billion people in the northern hemisphere were exposed to small amounts of radiation from Chernobyl over the next week, and those living within sever-

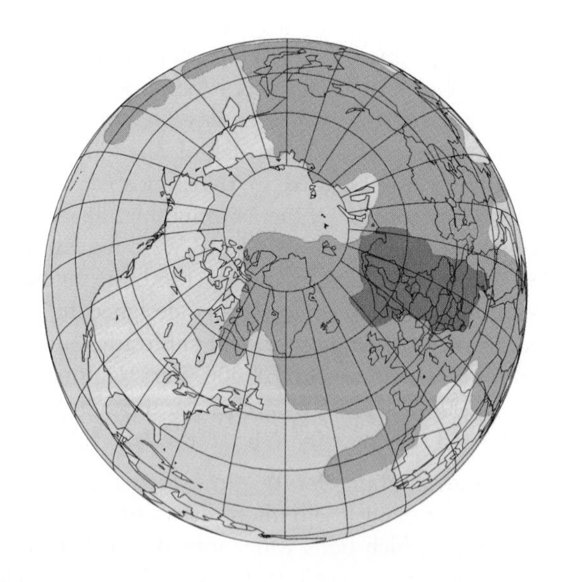

FIGURE 11-16 Dark red shading shows where radioactive particles and gases spread downwind from Chernobyl immediately after it exploded and released the contents of its reactor core on April 26, 1986. Within a week, radioactive material had spread over more than half the northern hemisphere (light red).

al tens of kilometers of the reactor were exposed to extremely high doses. Long-term exposure to radiation in the areas of greatest fallout could cause as many as 30,000 deaths over the next 40 to 50 years.

Disposal of Nuclear Waste One of the most challenging problems associated with nuclear energy is the disposal of its radioactive waste products. Nuclear waste classed as *low-level* includes the machinery, equipment, cleanup residues, and protective clothing used at nuclear power plants, hospitals, and research laboratories. Nuclear waste classed as *high-level* includes the highly radioactive liquid and solid residue from spent nuclear fuel rods and weapons production. After several years, fuel rods in a reactor core must be replaced, because although still radioactive, they contain insufficient uranium 235 for fission reactions. The rods, with their load of radioactive decay products, must be removed and replaced. Some daughter products decay into stable atoms within a few years, but others remain radioactive for many thousands of years.

The U.S. government requires states to provide for the timely disposal of low-level radioactive waste generated within their borders. All are looking for suitable sites—often outside their state lines—but only one site, Barnwell in North Carolina, still accepts waste from other states. The first instance of international nuclear waste trade occurred in 1997, when North Korea signed an agreement with Taiwan to store some of its low-level radioactive waste. Desperate after several years of crop failure and famine, North Korea will accept waste in return for cash to buy food.

After 20 years of scientific investigation, a site near Carlsbad, New Mexico, was approved as the first long-term underground repository for low-level military radioactive waste. The Carlsbad site is scheduled to begin accepting military waste—gloves, aprons, and lab equipment—by 1998. The site is underlain by thick beds of salt, some of which have flowed upward and formed salt domes. Salt is a good repository rock because it is ductile and does not break when deformed, so it has few or no fractures along which wastes might leak. As salt flows upward, it will entomb the waste, sealing it off from the surroundings. In addition, salt is highly soluble in groundwater, so its very existence as a solid indicates that water—which can corrode radiation-proof canisters—is not likely to exist in the repository.

At present, approximately 70,000 tons of high-level radioactive waste from spent fuel is in temporary storage under water at more than 60 nuclear power plant locations across the United States (Figure 11-17), and another 3000 tons are generated each year. An additional 8000 tons of high-level waste from military facilities awaits disposal. Spent fuel rods and waste water from the 300 or so nuclear power plants in other nations also require permanent disposal. Disposal must isolate the waste for approximately 10,000 years, by which time the amount of radioactivity will be comparable to that of a similar size uranium ore deposit. Isolating something from the biosphere, atmosphere, and hydrosphere for that long is difficult on a dynamic planet. Furthermore, disposal sites must be in remote locations that can be protected from sabotage.

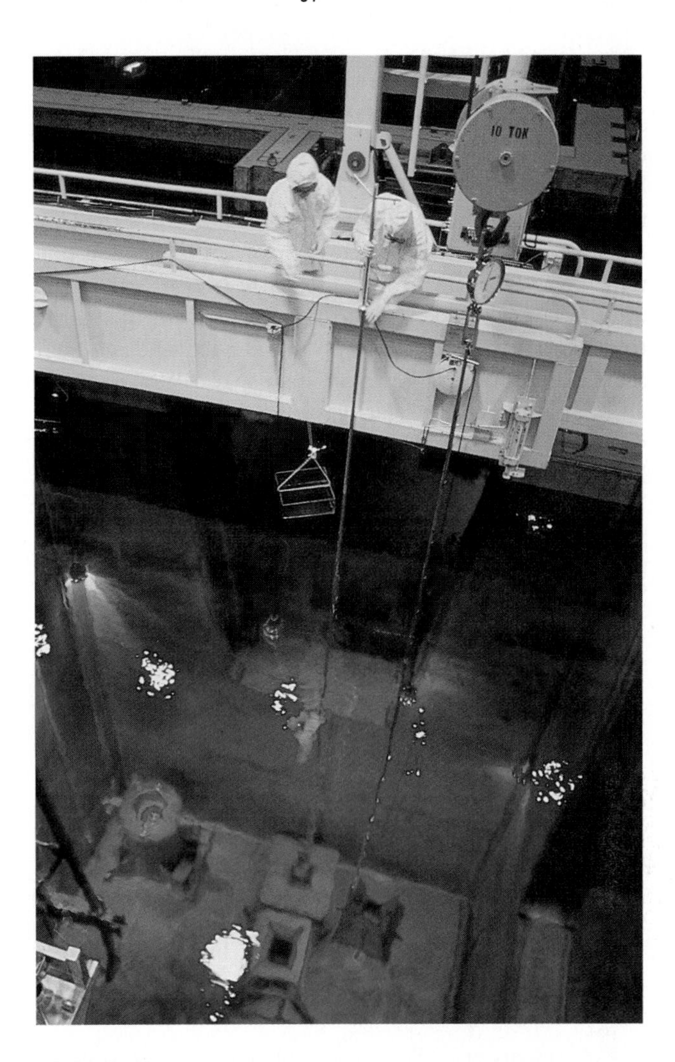

FIGURE 11-17 Spent fuel rods stored temporarily under water at a nuclear power plant. Many nuclear power plants around the world rely on temporary storage while nations decide how and where to store their high-level radioactive waste permanently. Although the fuel rods no longer have enough uranium 235 to be useful in a reactor core (unless processed), they still are hot enough to glow under water and will remain radioactive for thousands of years.

Permanent subterranean storage is the disposal scheme of choice for several nations, including Germany, Canada, and the United States. Only Germany, however, has a permanent storage facility, in salt beds, in operation. In Canada, scientists are investigating a site at Lac du Bonnet in Manitoba, an area underlain by ancient, crystalline bedrock. Although the U.S. government has approved a disposal site in Nevada, at Yucca Mountain (see Box 3-4), scheduled to open in 1998, the current governor of Nevada actively opposes the project, and the facility might not be completed until 2011 or later. In the meantime, more than 30 utilities have filed a joint lawsuit against the U.S. Department of Energy for breach of contract, asserting that they had been promised a waste site by 1998. The utilities contributed funds to the Department of Energy to construct the site by the promised date, and many of them are out of temporary storage space. While storing all high-level U.S. nuclear waste at a single site concentrates the risk, the Yucca Mountain area is highly unlikely to experience any geological disturbance over the next 10,000 years, and having dozens of temporary aboveground storage sites scattered about the nation is viewed by some as an even more dangerous prospect.

- Nuclear energy plants use neutrons to split the nuclei of uranium 235 atoms. In this process, called fission, energy and more neutrons are released, and these neutrons split other atoms to generate more energy in a chain reaction.
- Uranium 235 is obtained from uranium minerals, typically oxides, and is used to produce fuel pellets that are inserted into the core of a nuclear reactor. After several years, the fuel rods are spent and must be replaced. Because they are still highly radioactive, long-term storage is a scientific challenge and a political problem.
- Accidents at nuclear power plants such as those at Three Mile Island and Chernobyl occur if the reactor is not cooled sufficiently to prevent a meltdown in the reactor's core and the radiation-proof casing around the core is damaged as a consequence.

Solar Energy

Solar radiation is, from a human perspective, esentially undepletable. It also is the most abundant energy source on Earth. The planet receives as much solar energy in 20 days as is stored in all its fossil fuels. Yet the solar revolution in power delivery that energy analysts have been predicting for years has not yet arrived.

A major limitation of using solar energy for heat and electricity is its low concentration. Even though a lot of solar energy reaches Earth's surface, its concentration at any one spot is small. Furthermore, the amount of incoming direct solar radiation is concentrated in the world's middle latitudes, the southwestern United States, Australia, and the Sahara and Kalahari deserts in Africa (Figure 11-18). The United States, for example, receives solar energy equivalent to 500 times the country's energy consumption, but the hot, dry southwestern states receive more than twice as much sunlight as the chilly, damp northeastern states. This uneven distribution means that areas where energy is used for heat much of the year get less solar energy than areas where heat is not needed. Furthermore, receipt of solar energy in cool, moist areas is frequently hampered by heavy cloud cover. A cloud-covered area intercepts little solar energy.

Solar radiation is an energy flow and must be intercepted in order to be useful. Under ideal atmospheric conditions, a square that is 1 m on a side laid flat on the ground at noon at mid- to low latitudes would receive a maximum of 1123 W of direct sunlight and 70 W of diffuse—that is, scattered and reflected—sunlight. If these conditions prevailed for an average 12 hours per day, the total amount of energy intercepted by this square yearly would be 5220 kilowatt-hours (kWh). However, because sunshine hours typically number less than 90 percent of the ideal and particles in the atmosphere absorb and scatter solar radiation, the actual total amount of energy received is likely to be less than 2550 kWh per year. This energy receipt is equal to the amount of energy needed to power three 100-W light bulbs for a year, or approximately 4 percent of the average per capita electricity consumption in the United States in 1988. In other words, it would take 25 panels each 1 m^2 to collect enough energy at mid-latitudes to satisfy daily electricity consumption for a single inhabitant of the United States.

The primary technological challenge of using solar energy is to devise solar collectors, or panels, that can intercept enough of the Sun's rays to meet energy demands. The second is to store enough solar energy to provide a sufficient and uninterrupted supply when little solar energy is intercepted.

The Potential of Solar Energy

Despite current technology's limited ability to capture and store solar energy, solar radiation for heat and electricity is being harnessed on a small scale, usually for individual homes. The obvious advantages of solar energy are its overall amount and its accessibility. In contrast to fossil fuels and nuclear fission, solar recovery and power-generating systems produce essentially no pollutants and disturb little land. Furthermore, once a solar energy system is installed and has operated long enough to pay back the initial cost of construction, the supply of energy is free and the only costs are those of maintenance.

Development of larger-scale solar energy plants would particularly benefit more than half the world's population—the 2.5 billion low-income people who live in villages and rural areas in developing nations. Typically, developing nations do not have the resources to build and operate petroleum, coal, or nuclear power plants and run power lines to consumers, nor do most of the potential

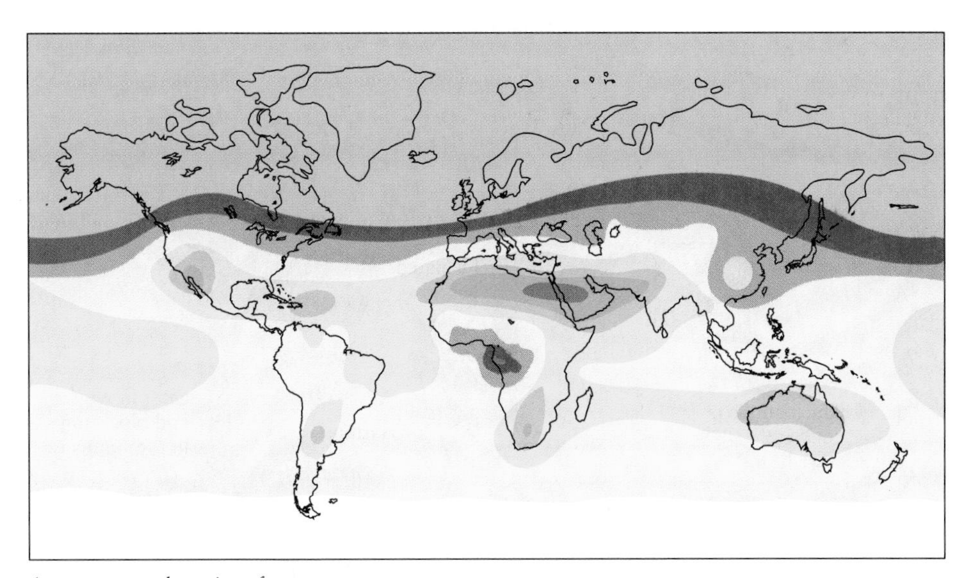

FIGURE 11-18 The global distribution of average annual solar radiation that reaches a horizontal surface on the ground, from low values of 1.16 tWh per km^2 per year to high values of more than 2.55 tWh per km^2 per year. Areas not colored receive less than 1.86 tWh per km^2 per year, but specific values for these areas are not available.

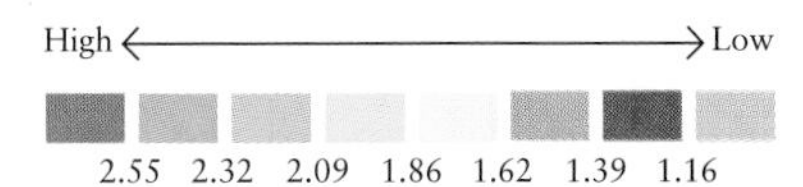

consumers have the resources to pay for the service. In areas where widespread energy infrastructures do not yet exist, solar energy is a practical means for providing heat, electricity, and other services (Box 11-3). India reports that at least 6000 villages already use solar energy systems, and the country has received financing from the World Bank to install solar systems in an additional 38,000 villages. Other nations striving to increase the use of solar energy systems in villages are the Dominican Republic, Sri Lanka, Zimbabwe, and South Africa.

If current trends continue, solar energy could supply 17 percent of the world's electricity by the year 2010, and perhaps 30 percent by the year 2050. World leaders in solar research and development are Germany, Japan, Italy, and the state of California. Since 1990, the U.S. government has increased federal funds for research and development in solar energy technology and now spends more money in this area than any other nation. Clearly, the technology is promising, and the global market ($1 billion in 1993) is growing.

Using Solar Energy for Heating Solar energy can be used to provide heat and hot water through passive and active solar heating systems. A *passive solar heating system* is, essentially, a greenhouse. It intercepts solar energy and allows it to pass through to indoor surfaces that convert the energy to low-temperature heat, but it does not allow the heat energy to pass back out very readily. Any ordinary window can function as a passive solar heating system, but a deliberately designed passive system generally is structured with energy-efficient windows strategically placed and insulation to keep heated air from escaping. Buildings are oriented to maximize the amount of solar radiation that can be collected and are built with thick concrete or adobe walls and floors and other massive materials that can store a lot of heat and release it slowly (Figure 11-19). Salt has a high heat capacity and sometimes is impregnated in timbers used for construction to enhance their heat-storage capacity. Windows, shades, and sometimes shutters are opened and closed throughout the day to regulate the flow of heat into and throughout a solar-heated house. In most modern designs, components of a passive solar system are wired to a computer that monitors and controls the actions of each part of the system. Insulated tanks of water mounted with solar collectors on rooftops can supply most or all of the typical hot water needs of a house.

More than 250,000 North American houses are fully heated by passive solar systems (Figure 11-20) and another million have some passive heating features. Although adding the passive solar features increases the cost of construction by 5 to 10 percent, the energy savings can recover the initial costs in 3 to 7 years.

In *active solar heating*, solar collectors typically are mounted on rooftops and connected to pipes and pumps that circulate heated air or water throughout a building. Excess heat can be stored in tanks of hot water or in other materials, such as rocks. Active solar heating has not been as widely used as the passive mode, largely because of the extra costs and maintenance. As new technologies are developed and tested, however, active solar heating is likely to become more common because of its ability to store excess heat.

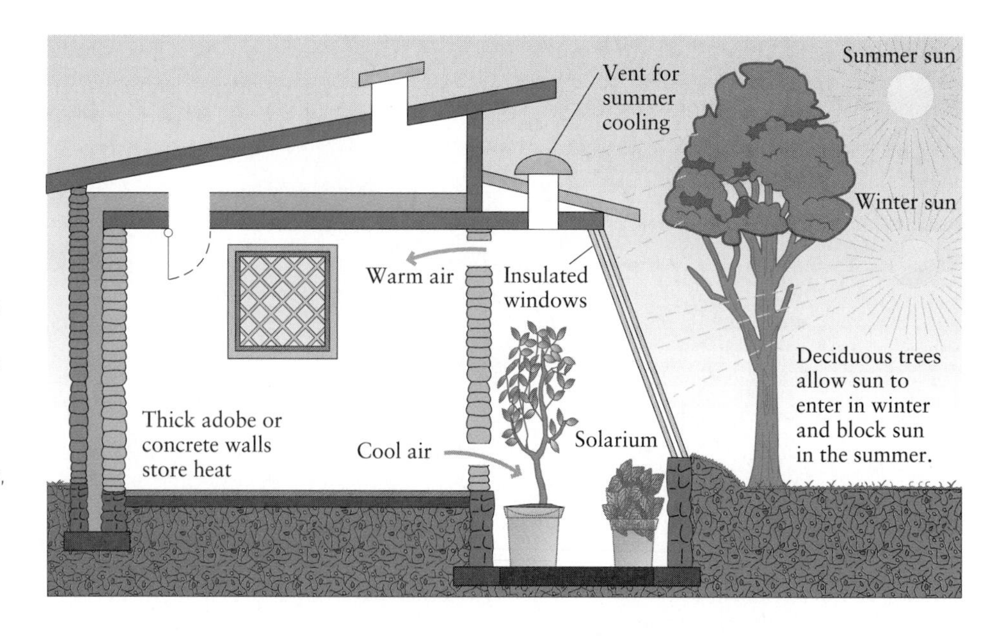

FIGURE 11-19 A common design for a passive solar house uses a sunspace, for example, a greenhouse or solarium, with insulated windows facing the direction of greatest incoming solar radiation to collect solar energy that heats the inside air. Vents enable the warm air to circulate, return to the sunspace when cooled, and be rewarmed. The angle of the Sun is higher in summer and lower in winter, so trees that have summer foliage can be used to block solar radiation during warmer months, and additional vents can be opened for greater cooling.

Using Solar Energy for Electricity One of the primary uses for energy in industrialized nations is the generation of electricity to run appliances and provide light. Solar energy can be converted directly to electricity in devices called **photovoltaic** (PV) **cells.** Photovoltaic cells are made of a *semiconductor* material, so called because at high temperatures it conducts electricity almost as well as metals do but at lower temperatures its conductivity practically disappears. One of the most popular semiconductors is silicon, because it is abundant, cheap, and easy to manipulate. Molten silicon is poured into sheets thinner than paper, so they heat rapidly in sunlight. The cooled and solidified sheets are cut to a variety of shapes and sizes and coated in glass or plastic for protection. Sunlight shining on a photovoltaic cell excites electrons in the semiconductor material, which then flow and create an electric current. Each cell produces just a tiny amount of energy, so many of them are wired together to form a *solar panel.* A number of panels are mounted either on a rooftop or on a structure that can follow the Sun throughout the day. Current technology relies on banks of batteries to store electricity and supply it to a building as needed. Photovoltaic cell technology is still too expensive to be economically feasible in places where electricity already is

FIGURE 11-20 This structure in Snowmass, Colorado, houses the Rocky Mountain Institute, an energy and conservation organization founded by Hunter and Amory Lovins. The southern part of the facility is partly built into the ground for additional insulation and summer cooling. Nearly all heat, hot water, and lighting are provided by the Sun, and the energy savings are expected to pay for the cost of constructing the facility within 40 years.

11-3 Global and Environmental Change

Shifting Energy Pathways

For millennia, people relied on biomass (primarily wood) for most energy needs. Rapid shifts from one dominant energy source to another began just recently, in the 1800s, and since then we have shifted from wood to coal and then to oil, the transitions occurring at about 50-year intervals. We might be in the midst of a shift to natural gas. Some energy analysts predict yet another shift—to solar energy and hydrogen gas.

Future energy sources depend on the world's energy needs and the ways in which energy industries respond to those needs. Two global trends are occurring that result in different energy demands. Rich industrialized nations contain only one-fifth of the world's population but consume more than four-fifths of the world's energy supply. Per capita energy consumption increased rapidly throughout the 20th century, and huge increases in fossil fuel consumption enabled some of their people to reach unprecedented levels of affluence. At the same time, their population growth rates dropped. In contrast, population has expanded rapidly in countries where the economic standard of living is far below that of the industrialized world. While struggling to feed and support their burgeoning numbers, poor nations also are striving to modernize and raise their standard of living. Most new demand for energy comes from these countries. As a result of these two global trends, it is possible that the demand for additional energy will increase 50 percent by the year 2010.

Throughout the 20th century, as industrialized countries increased their energy consumption and their wealth, so too did they increase their emissions of carbon dioxide and other pollutants. If developing nations were to follow a similar path to affluence and modernization, carbon dioxide emissions could double in 20 to 40 years and much energy would be wasted. In modern power plants that run on nuclear energy or fossil fuels, 40 to 86 percent of the energy is wasted and never used by any consumer. In fact, the industrialized world wastes nearly as much energy as the developing world uses.

The future is likely to be different for each country and region. Decentralized, renewable energy sources such as solar power probably will play an important role in the rural areas and villages of developing nations. Nations that already rely on large, complex, centralized distribution systems might continue to search for a renewable, fluid fuel source that can replace fossil fuels. Some analysts predict that this fuel will be hydrogen gas (H_2). Hydrogen is abundant at Earth's surface, occurring primarily in the hydrosphere in water. An electrical current passed through water can split water molecules into gaseous hydrogen and

In the town of Villaseca, Chile, settled by squatters in 1975, a woman uses solar energy to cook food. Earlier inhabitants removed all the trees and shrubs from surrounding hillsides for fuel, so solar kitchens are essential to the villagers.

(continued)

11-3 Global and Environmental Change

(continued)

oxygen. When hydrogen is burned for energy, it reacts with oxygen in the atmosphere to produce water vapor. Hydrogen is thus a renewable fuel source.

Hydrogen gas can be used for energy just as can natural gas. In fact, hydrogen gas *is* natural gas, without the carbon. Prototypes for hydrogen-powered cars, submarines, and buses already exist. The major limitation at present, however, is that energy is needed to generate electricity to split water in the first place. Using fossil fuels or nuclear power to produce hydrogen gas would counteract the reasons for converting to hydrogen-powered systems. The only reasonable long-term solution is to use solar energy to dissociate hydrogen from water. Currently, several nations, including Japan, Germany, and the United States, are researching and developing new solar technologies that can produce hydrogen gas. In the next 50 years, fuel stations may replace their pumps with hydrogen gas canisters.

available from other energy sources (Figure 11-21). However, the cost is dropping and might be competitive by the year 2000.

Alternatively, electricity can be generated in a **solar-thermal power plant.** One design, called a solar power tower, uses reflective solar collectors that track the Sun and concentrate heat and light on a central receiver mounted atop a tower (see Figure 1-10). Fluid in the receiver is heated to the boiling point, producing steam that is used to generate electricity by turning a turbine in much the same way that steam-driven turbines are used in other types of power plants. Solar power plants require about 1 km^2 of land for every 20–60 million watts of electricity generated.

Wind Energy From the Sun

Uneven solar heating of Earth's surface generates winds that can be used as an energy resource (see Chapter 9). For centuries, people have harnessed wind power to do mechanical work, using windmills to lift water from wells, grind grain in mills, and propel sailboats and sailing ships. Like direct solar radiation, wind is an energy flow and needs to be intercepted in order to be converted into a usable form of energy. At present, most wind energy is converted into electricity by wind turbines, a modern adaptation of ancient windmill technology (Figure 11-22). Wind-generated electricity is becoming economically competitive with conventional sources of electricity. More than 17,000 wind turbines in California are completely integrated into the state's electricity grid and generate 1 percent of California's electricity. Worldwide, more than 20,000 wind turbines have an installed capacity of over 2200 million watts. Current technology, however, can recover wind energy economically only from areas of consistently high wind speed (at least 20 km per hour). Only about 13 percent of the contiguous United States is sufficiently flat and free of barriers for winds to reach such speeds. Such areas include the Great Plains and coastal regions.

In principle, the available wind energy could supply most of the world's electricity demand. The amount of energy dissipated as wind flowing over the United States alone each year is about 40 times the country's annual

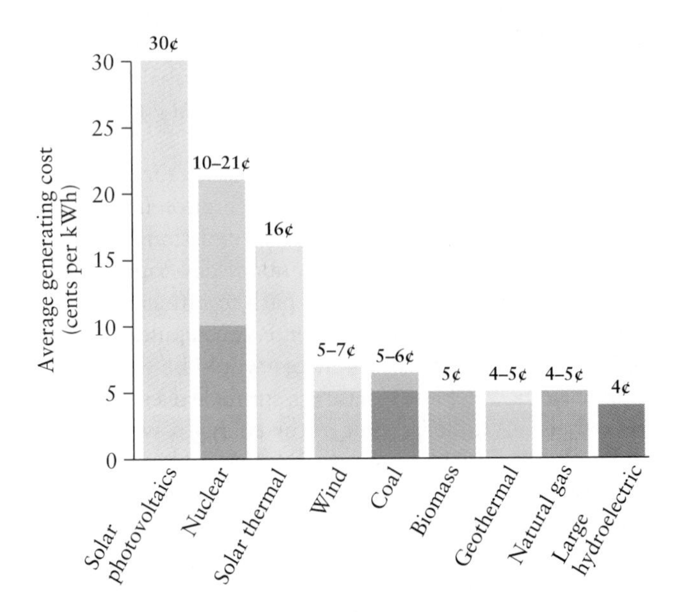

FIGURE 11-21 Different technologies generate electricity at costs that range from as low as 4¢ per kWh for hydroelectric dams and natural gas to as high as 30¢ per kWh for a solar photovoltaic facility. As new technologies develop, the cost of electricity from photovoltaics is expected to go down. The second most expensive source of electricity is nuclear power.

FIGURE 11-22 Wind farms use turbines to generate electricity. In Holland, a pastoral country long associated with windmills, modern wind generators are a common sight.

consumption of energy. Wind farms pose small threat to the environment except where turbines are located along migratory bird routes or near the habitat of hawks, eagles, and other birds of prey. Birds are killed when they fly into the wind blades. High bird mortality associated with wind farms is unacceptable to much of the public and is likely to slow the development of wind power dramatically unless the problem is solved. Aesthetic objections are another hurdle to using wind power. A proposed wind farm in California was defeated partly because of objections from developers and land owners who claim that wind farms deface the landscape.

Water Energy From the Sun

Energy from flowing water, like energy in wind, comes ultimately from the Sun, as solar energy drives the hydrologic cycle. Flowing water has been used for thousands of years to turn water wheels and more recently to turn turbines that generate electricity in hydroelectric power plants. Nevertheless, there are environmental drawbacks to hydropower. Because of rainfall variability and the need to control the rate of energy production, dams are built to store water in reservoirs and release it over waterfalls, generating electricity as needed (see the beginning of Chapter 7).

Hydropower generates about 20 percent of the world's electricity, or 5 percent of the world's total energy production. The estimated world resource base is 9.2×10^{19} J per year (see Table 11-1), of which only about 14 percent has been developed. Asia and Africa have the largest resource bases, with 30 percent and 19 percent of the world total, respectively. North America and Europe have developed the largest proportions of their resource bases (48 percent and 27 percent, respectively).

Creating new large hydroelectric dams is expensive and causes profound damage to river systems. Dams impede the flow of water and change flow rates, trap sediment upstream, cause erosion downstream, drown river valleys, provide breeding grounds for mosquitoes, and block fish migration patterns. In addition to energy production, the benefits of dams to humans include flood control, water storage, and recreational facilities, but these benefits do not always outweigh the environmental costs, and new dam construction in the United States and most of the developed world is likely to be minimal in the future. Dam construction probably will continue in the developing nations, despite the environmental consequences, because of the ever-pressing need for energy. Asia and South America are likely to double their hydropower capacity over the next 50 years.

To build the Three Gorges project, the largest and most complex hydroelectric dam in the world (Figure 11-23), China sacrificed hundreds of kilometers of riverside ecosystems. The benefits, however, include staving off a severe energy shortage and slowing the expansion of more coal-burning power plants, which already supply three-quarters of China's commercial energy. Were China to increase consumption of coal, global efforts to reduce CO_2 emissions would be compromised.

In the future, solar energy may also be extracted from the oceans by tapping their heat differential and by harnessing ocean waves. Solar radiation heats surface waters, but the slow turnover rate of the oceans (see Chapter 10) typically keeps water at depth 4°C cooler than at the surface. Cold water at depth can be used to condense a fluid such as ammonia, which changes readily in density with relatively small changes in temperature. The chilled, condensed fluid is then pumped to the surface, where warm surface water heats the fluid, which expands and drives a turbine. This technology, called *ocean thermal energy conversion,* is in the experimental stage. Several pilot plants have been constructed around the world, but the total generating capacity is negligible.

Biomass Energy

Plant matter and animal waste, the **biomass,** can be burned directly for its energy, or it can be converted to gaseous or liquid biofuels to be used in the same ways as fossil fuels. An example is the use of fermented corn to make an alcohol-gasoline fuel mixture called gasohol. Because alcohol can be mixed with gasoline in a 1:9 ratio in automobile fuel tanks, it can extend the lifetime of nonrenewable petroleum reserves. At present, however, gasohol costs about twice as much to produce as gasoline. Worldwide, the amount of energy provided from biomass has declined over the past few centuries, but wood, animal dung, and other biological materials still are used for heating and cooking by more than 1 billion of the world's population. Globally, biomass supplies 14 percent of total

FIGURE 11-23 The flow of the Yangtze River, shown here near Sichuan, soon will be blocked when the Three Gorges Dam is completed early in the 21st century. The Chinese government has been severely criticized for building the Three Gorges Dam because it will inundate 22,800 hectares of cultivated land, displace 1.1 million people, and disrupt the natural flow of the Yangtze River. However, it will reduce the country's need to burn coal.

energy demand. In countries such as Canada and the United States that rely heavily on fossil fuels, biomass accounts for only 4 to 5 percent of the energy use, but in developing nations, it accounts on average for 36 percent of energy use.

In the United States, biomass energy supplied 4 percent of energy demand in 1990, or 3.12×10^{18} J. Of this amount, 2.87×10^{18} J was generated from wood and wood waste, and the remainder from agricultural and industrial waste and municipal solid waste. At present most wood waste in the United States already is used to generate electricity, and there is little potential to increase this source. On the other hand, the use of municipal waste as an energy source could be expanded dramatically. About 144 million tons of municipal waste are generated in the United States every year, representing a potential energy resource of 1.7×10^{18} J. Of this amount, 10 percent is incinerated, 10 percent is recycled, and the remainder is placed in a landfill. If all nonrecyclable municipal waste were incinerated to generate electricity, this resource could supply 8 percent of U.S. electricity consumption. The major drawback is that burning organic waste generates CO_2 and other potentially harmful emissions. Without stringent emission controls, a large increase in the use of biomass energy would have increasingly severe environmental penalties that would at some point overwhelm all the environmentally positive aspects of this energy source.

Biomass is a renewable energy stock as long as its supply is replenished at the same rate that it is consumed. This requires at least one tree to be planted for every tree cut down, one hectare of crops to be planted for every hectare harvested, and no net loss of soil organic matter. In practice, maintaining equal rates of replenishment and depletion is difficult, especially in those parts of the world where fuel wood is the primary energy resource and its supplies are scarce (Figure 11-24).

- Solar energy is the greatest source of energy available, but its widespread use is currently limited by its low concentration in any given area, especially at high latitudes, and by its unreliable availability because of cloud cover.
- Solar energy can be used to heat the air and water in buildings, and it can be converted directly to electrical energy with photovoltaic, or solar, cells.
- With current technologies, solar electric power plants have the highest costs for producing electricity but are likely to become economically viable in the next few decades.
- Solar energy is the ultimate source of wind power and water power, which are economically viable ways to generate electricity at wind farms and hydroelectric dams. The environmental impacts of power production at such facilities are minimal compared to those of fossil fuels and nuclear power, although hydroelectric dams significantly affect local stream systems and their habitat.
- Solar energy is the ultimate source of chemical energy stored in biomass, which provides nearly half the world's people with energy in the form of plant matter (primarily wood) and animal waste. The problems in using it include sustaining it as a renewable resource and limiting CO_2 emissions.

FIGURE 11-24 Needing wood for fuel, villagers in Mali strip wood from a living tree. In this way trees and shrubs are destroyed faster than they can regrow, and what could be a renewable resource is depleted, rendering the use of biomass unsustainable.

Energy Efficiency and Conservation

The greatest hope for the world's energy future is in improving the efficiency of energy use. Because of physical laws governing energy conversions, some energy waste is unavoidable. However, about half of all energy lost as waste heat in industrial nations could be conserved. The populations of industrialized nations waste three to five times as much energy as they use. Much of the waste is avoidable, but the portion that is not results from two fundamental scientific laws of energy.

The first law of energy, or the law of conservation of energy, states that in all physical and chemical changes, energy is neither created nor destroyed, but it may be converted from one form to another. By the second law of energy, when energy is converted from one form to another, some of the energy always is degraded to lower-quality, more dispersed energy that has less ability to do work.

A simplified example of the proportion of energy wasted in order for a coal-burning power plant to provide electricity illustrates the nature of these two energy laws (Figure 11-25). Chemical energy stored in coal is converted to

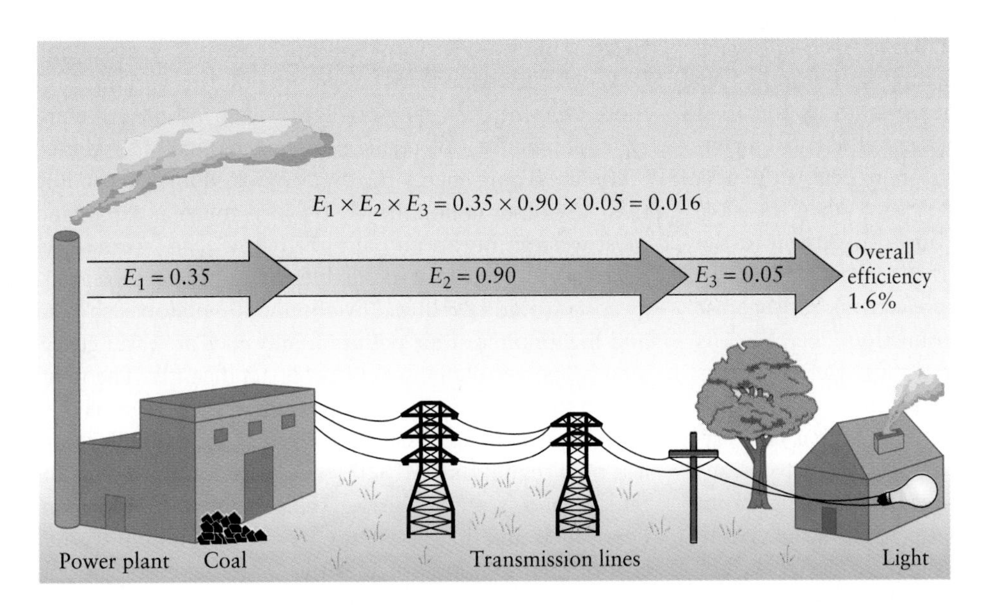

FIGURE 11-25 Fuel that burns at a high temperature—as much as 600°C—is used to generate electricity to burn incandescent light bulbs at temperatures of less than 100°C. In the process, much energy is lost as waste heat. To find the total efficiency for the conversion process shown, the efficiencies (E) of each step are multiplied.

FIGURE 11-26 All power plants that use nuclear energy or fossil fuels produce waste heat that is emitted to the surrounding environment, either to the atmosphere from cooling towers (as at the Three Mile Island nuclear power plant) or to nearby surface water bodies (as at the Brunner Island coal-burning power plant). Waste heat (red) discharged to streams and rivers causes thermal plumes that can be identified from satellite data of thermal radiation emitted from Earth. Pinkish regions, particularly in the upper right, are urban areas.

heat (thermal energy) when it is burned, but only 35 percent of the heat energy is captured and used to make the steam that turns a turbine in a generator to produce electricity. The rest is *waste heat,* heat that is given off to the surroundings and dispersed by the random motion of air and water molecules. The conversion from chemical to electrical energy, therefore, is 35 percent efficient. The **energy efficiency** of an energy conversion process is the percentage of energy not lost as waste heat. When the electricity is transmitted along power lines, 10 percent of the original energy captured is converted to heat because of frictional resistance as the moving electrons rub against other materials in the lines. Transmission, therefore, is 90 percent efficient. When the transmitted electricity is converted into light in an incandescent bulb, 95 percent of the electrical energy that made it to the bulb is converted to waste heat. The conversion from electricity to incandescent light, therefore, is only 5 percent efficient. From the first conversion to the last, 98.4 percent of the energy that went into this system was not used to do work, but instead was converted to waste heat. Only 1.6 percent of the energy input did work. Therefore, 0.016 is the measure of the efficiency of this whole energy conversion system. A system's energy efficiency is equal to the percentage of total energy input that does work, and can be calculated by multiplying the efficiencies of each step in the energy conversion process.

All power plants waste energy in producing electricity. The wasted energy is heat. Most power plants are only 30 to 40 percent efficient, thus converting 60 to 70 percent of their initial energy input to waste heat that is dissipated into the environment. Roughly half the waste heat from fossil fuel power plants is released directly into the atmosphere through tall stacks. The remainder is released to cooling waters, which then transport the waste heat to the environment, either into the atmosphere through cooling towers or into a lake or river. The U.S. Environmental Protection Agency regards cooling towers as the best available technology for the disposal of heat. Heat dissipated in surface water bodies is responsible for fish kills and may have other negative impacts on the biota; its release from factories and plants is strictly controlled by the EPA. Combustion-based power plants are not the sole culprits in heat pollution of surface waters. Nuclear power plants also generate waste heat that is carried away by water (Figure 11-26).

Of the 84 percent of commercial energy that is wasted in the United States, 41 percent is the unavoidable result of the second law of energy. The other 43 percent of wasted energy is avoidable. It dissipates from poorly insulated homes and buildings that lose energy, from "petropig" vehicles that swill gasoline because they get few miles per gallon, and from careless habits such as leaving windows open when air conditioning or heating is in use. Photographs of houses taken with infrared-sensitive film can reveal "hot spots," places where heat is lost to the surroundings. In many cases such total losses are astoundingly high, equivalent to losses that would occur from large holes in the walls or from windows left open year-round.

Why don't Americans invest in better insulation, drive more efficient cars, turn off lights when rooms are empty, and practice other energy-saving habits? The United States uses more energy per unit of gross domestic product than any other nation, nearly twice as much as Japan and most western European nations. If Americans were to reduce the 43 percent of avoidable energy waste, their energy costs would be reduced by about $300 billion each year, and the amount of air pollution and nuclear waste generated by power plants would drop significantly. The main reason that Americans seem unaware of their waste is they have cheap sources of energy. Gasoline, for example, costs less now than it did in 1920 when price is adjusted for inflation, and is much cheaper than in Europe and Japan. Fossil fuels and nuclear energy would cost more if the full cost of their use, in particular the "external costs" of pollution, were accounted for in market prices.

Closing Thoughts

The world's energy future is difficult to foretell. Danger signs are clear, but there are also signs of hope. Finite fossil fuels are being depleted and nuclear technologies not yet 50 years old leave us with a legacy of radioactive wastes that will last thousands of years. Yet new technologies are being developed for using hydrogen gas and solar energy resources, which would produce little pollution and make abundant energy available nearly everywhere on Earth.

Industrialized nations are increasing environmental regulations, requiring emission-control devices on power plants, and encouraging energy conservation, and are pressing less developed nations to do the same. But nations just beginning to build an industrial base resist spending resources on pollution control, claiming that it is an unfair burden. The industrialized nations, after all, didn't start worrying about pollution until after they became rich. Furthermore, conservation cannot be a priority in drought-ravaged areas where people are forced to destroy the last of their trees to get any energy at all.

One aspect of the future is possible to predict with some certainty: Although at current rates of consumption most of the world's fossil fuels could be used up in the next few hundred years, consumption rates are likely to change. Conserving energy, using it efficiently, and increasing the use of solar energy and other environmentally safer substitutes for fossil fuels would allow these nonrenewable energy resources to last until they are no longer needed.

Summary

- Energy enables matter to move and do work. Thus, all flows of matter between reservoirs in the Earth system imply flows of energy as well.
- Earth is a closed system in which matter moves in cycles from one reservoir to another, but energy enters and leaves Earth; all energy on Earth is lost ultimately to space as heat.
- Respiration, decomposition, and combustion all release energy and carbon dioxide and consume oxygen, essentially reversing photosynthesis, which uses energy and carbon dioxide and releases oxygen.
- Fossil fuels form when carbohydrates are buried in sedimentary layers, where they are subjected to enough pressure and temperature to transform them into hydrocarbons.
- Petroleum is a fluid source of hydrocarbons formed from marine phytoplankton and other organisms buried in sedimentary ocean basins primarily along continental shelves. Almost three-fourths of the world's petroleum reserves are found at current or geologically recent subduction zones, mostly in warm climates near the equator where plankton are abundant. Natural gas, the gaseous form of petroleum, contains almost no potential pollutants and burns more cleanly than crude oil, the liquid form.
- Coals and the peat from which they derive are solid hydrocarbons that develop in wetlands and are buried. The greater the conditions of temperature and pressure, the more carbon in the coal.
- Anthracite, the metamorphic form of coal, contains more than 90 percent carbon and therefore generates more heat and burns more cleanly than do peat and the sedimentary forms of coal.
- Environmental problems associated with fossil fuels include landscape defacement and water pollution during extraction, atmospheric pollution as a result of combustion, and release of waste heat into the atmosphere and surface water bodies. Oil released into oceans threatens ocean and coastal habitat. Economic problems include the depletion of these finite resources and the amount of energy dissipated as waste heat when these fuels are converted to heat and electricity.
- The only practical means of generating nuclear energy with current technology is the nuclear fission of uranium 235. Although rare, this isotope represents the greatest stock of nonrenewable energy on Earth. Nuclear energy is relatively little used, however, because of the threat of radiation leaks from reactor meltdowns and the extreme difficulty of storing nuclear waste safely.
- Geothermal energy is environmentally harmless and constantly replenished from Earth's interior heat, but it is a practical energy resource only in the vicinity of tectonically active areas with high heat flow.
- Direct solar energy is inexhaustible as long as the Sun shines, and it has virtually no negative impact on the environment. Passive solar collection is feasible in areas where sunlight is relatively abundant and is currently being used on a small scale for home heating and hot water. Large-scale collection for generating electricity is possible with photovoltaic cell technology

or with solar-thermal (steam) technology, but so far it is economical primarily in less developed countries without sufficient energy infrastructure to operate large power plants and to carry electricity to villages and rural areas.

- Solar energy powers the flow of air and water. High winds and fast-flowing waters can be harnessed to drive electricity-producing turbines. The environmental impacts lie in destruction rather than pollution. Wind farms cause high bird mortality when birds are caught in collector blades. Dams necessary to control surface water flow affect stream patterns, blocking fish migration paths and obliterating spawning grounds.
- Developing direct solar radiation as a major energy resource, conserving energy, and improving the efficiency of energy conversion systems can counterbalance the depletion of fossil fuels. As long as energy-recovery technologies continue to be developed and the Sun continues to shine, the world is not likely to run out of energy.

Key Terms

photosynthesis (p. 330)
petroleum (p. 331)
hydrocarbons (p. 331)
crude oil (p. 331)
natural gas (p. 331)
petroleum trap (p. 332)
anticlinal trap (p. 332)
fault trap (p. 332)
salt dome trap (p. 333)
tar sand (p. 338)
oil shale (p. 339)
peat (p. 340)
lignite (p. 340)
bituminous coal (p. 340)
anthracite coal (p. 340)
geothermal energy (p. 345)
nuclear fission (p. 346)
nuclear fusion (p. 346)
meltdown (p. 347)
photovoltaic cell (p. 352)
solar-thermal power plant (p. 354)
biomass (p. 355)
energy efficiency (p. 358)

Review Questions

1. When did the dominant energy source for human endeavors change from wood to coal to petroleum? Why?
2. What are the three main sources of energy on Earth? Of these, which is the largest? Which is responsible for earthquakes? Which is responsible for the energy released when fossil fuels are burned?
3. What compounds are used by plants during photosynthesis? What compounds are released?
4. What are the similarities and differences between respiration, decomposition, and combustion?
5. What is the difference between carbohydrates and hydrocarbons?
6. In what two main forms does petroleum occur? Why?
7. What type of organism is thought to be the major source of petroleum? What particular geologic conditions are conducive to its transformation to petroleum?
8. Name and describe three types of petroleum traps.
9. List and describe the unique set of conditions and geologic events that led to the development of more than half the world's petroleum within the Middle East.
10. List three environmental drawbacks associated with extracting, transporting, or using petroleum and coal.
11. Why is natural gas considered to be the cleanest fossil fuel?
12. Why is it unlikely that the hydrocarbons extracted from oil shales and tar sands will supplant petroleum as a major source of energy in the near future?
13. From what does coal form? What geologic conditions are most conducive to its formation?
14. Why did the Carboniferous period result in thick deposits of coal on several continents?
15. What is acid mine drainage? How can it be controlled?
16. How are emissions of sulfur dioxide gas from burning coal reduced at power plants?
17. Why is geothermal energy practical only within limited areas?
18. What is the nature and source of the fuel used in the core of a nuclear reactor?
19. List three environmental drawbacks associated with using nuclear energy.

20. What are two major technological challenges to utilizing the full potential of solar energy?
21. List three major environmental advantages associated with using solar energy rather than fossil fuels.
22. List the advantages and disadvantages of using water energy and wind energy to generate electricity.
23. What are the first and second laws of energy? What is energy efficiency?
24. What percentage of the energy used to generate electricity at a coal-burning power plant is lost as waste heat? Why is the amount so large?

Thought Questions

1. Make a list of all the energy transfers that occur from the time that solar energy enters Earth's atmosphere, to when it is stored in water vapor that falls on a continent as rainwater, to when it becomes part of a river along which hydroelectric plants use the falling water to generate electricity, and finally to when it is used to light a home.
2. How are the processes of photosynthesis and respiration by plants examples of transfers of solar energy from one Earth reservoir to another?
3. If the annual amount of solar energy stored in plants that become buried to form fossil fuels is very tiny, why do such large accumulations of fossil fuels exist on Earth?
4. What types of changes in our lifestyles might occur if we were to switch from oil to natural gas as a primary energy source?
5. Predictions of the configuration of tectonic plates and continents for the next few hundred million years indicate that a new landmass will form as the Pacific Ocean closes and the Americas collide with Asia. Explain whether this configuration is or is not conducive to the formation of thick, extensive seams of coal.
6. If you were a political advisor to a developing nation in an arid region with few surface water resources and a largely rural population, what arguments would you present to encourage development of solar energy resources?

Exercises

1. A chunk of anthracite coal about the size of two fists weighs approximately 0.5 kg. How far could you drive an efficient subcompact car if you could burn the coal's carbon in the car's engine?
2. If a solar power plant loses 15 percent of its energy in waste heat while boiling water, 30 percent in running a turbine, and 10 percent in transmitting the electricity through power lines, what is the energy efficiency of the system?

Suggested Readings

Bass, Richard, 1995. *Oil Notes.* Dallas: Southern Methodist University Press.

Hunt, John M., 1996. *Petroleum Geochemistry and Geology.* 2nd Edition. New York: W. H. Freeman and Company.

Johansson, T. B., Kelley, H., Reddy, A. K. N., and Williams, R. H., eds., 1992. *Renewable Energy.* Washington, DC: National Academy Press.

National Research Council, 1992. *Nuclear Power: Technical and Institutional Options for the Future.* Washington, DC: National Academy Press.

Ogden, J. M., and Williams, R. H., 1989. *Solar Hydrogen: Moving Beyond Fossil Fuels.* Washington, DC: World Resources Institute.

Scientific American, September 1990. *Special Issue: Energy for Planet Earth.* New York: W. H. Freeman and Company.

Tester, J. W., Wood, D. O., and Ferrari, N. A., 1991. *Energy and the Environment in the 21st Century.* Cambridge, MA: MIT Press.

CHAPTER 12

The Cowlitz River in the state of Washington quickly became choked by uprooted trees and mudflows of volcanic ash and soil in the wake of the disastrous 1980 eruption of Mount St. Helens.

Understanding Environmental Change

Mount St. Helens, in the state of Washington, erupted on May 18, 1980, with a huge, sideways-directed blast, sending an enormous avalanche of rock and volcanic ash sweeping down over the landscape, destroying vegetation and damming streams. Coldwater Creek, for example, turned into Coldwater Lake, surrounded by a blanket of ash that was meters thick and studded with charred trees. Over the next several years, ash carried downslope by runoff would wash into the lake.

To collect the sediment as it washed in, a team of geologists installed traps in the lake. Every 15 days, the traps automatically released plastic beads on top of the sediments to mark the time of deposition. Comparison of the collected sediments with meteorological records revealed that each identifiable sedimentary layer was deposited by runoff from a specific storm. Furthermore, each layer's thickness corresponded directly to the intensity of the storm, the thicker layers indicating heavier storms.

Of course, during most of Earth's history there were no instruments to measure environmental events, but observations at Coldwater Lake confirm that sedimentary layers are good "proxies"—reliable indicators of environmental conditions. To reconstruct Earth's environmental history and predict future change, scientists study ancient sedimentary layers and other proxies for environmental events. In this chapter, we will:

- Examine the origin of Earth's climate.
- Analyze some of the causes of climatic and environmental change.
- Discover how proxies such as sediments, tree rings, pollen grains, and glacial ice are used to unravel Earth's history of environmental change.

Between 1986 and 1994, California experienced the worst drought in recorded history. Eight years of below-average rainfall might have devastated the state's enormous agricultural and industrial economy were it not for the vast reservoirs constructed in the Sierra Nevada in the late 1800s to supply fresh water to the semiarid lands west of the mountain range. Coastal California, where San Diego, Los Angeles, and San Francisco are located, is home to 80 percent of the state's population but receives only 15 percent of its precipitation.

To cope with the drought, coastal cities imposed strict water rationing. Residents were prohibited from watering their lawns and washing their cars, restaurants served water only upon request, hotels and developers of new homes were required to install low-flow shower heads and toilets, and water-wasters were fined by water companies. With no end to the drought in sight, the cities made plans to construct desalination plants, such as those used in Saudi Arabia and other arid Middle Eastern countries, to extract drinking water from the ocean. Then, as abruptly as it began, the drought ended, and many of these plans were left on hold. State and local governments have realized, however, that they are at the mercy of the climate and are planning to build new reservoirs and aqueducts. So far, only Santa Barbara has built a desalination plant. Although presently not in use, the plant is ready for the next drought.

The drought was not an isolated incident. Tree stumps discovered at the bottom of modern marshes, lakes, and streams in the Sierra Nevada (Figure 12-1) indicate that twice in the geologically recent past these areas were dry long enough for trees to grow before they subsequently refilled with water. Radiocarbon dating of these stumps, coupled with counting of tree rings, indicates that one drought lasted longer than 200 years, ending in AD 1112, and the other, 140 years, ending in AD 1350. Rising water levels drowned these forests as the droughts ended. Such a drought today could destroy California's economy.

A centuries-long drought may be responsible for the downfall of the Mayan civilization in Mexico in the period AD 750–900. Sediments deposited in Lake Chichincanab on the Yucatán peninsula revealed oxygen isotopes and organisms that indicate an episode of extreme drying from AD 800 to 1000. This extended drought could have caused famine and persuaded the Maya to migrate to more hospitable environments.

As these examples illustrate, human societies benefit greatly from studying the climatic and environmental history of their local regions, as well as of the whole Earth. In particular, we need to understand how past changes in climate affected such environmental factors as water resources and vegetation distribution. If climate changes are found to be cyclical, their recurrence may be predictable, a circumstance that would be of great help in land-use planning and management. Even if these events are random and therefore unpredictable, understanding them can help us develop contingency plans. In the same way that people in earthquake-prone areas prepare for the next big earthquake, someday people in drought-prone areas may prepare for the next severe drought.

Climate on Terrestrial Planets

Why do environmental conditions suitable for life exist on Earth but not on Mercury, Venus, or Mars, the other terrestrial (that is, Earthlike) planets in our solar system? For many years conventional wisdom held that Earth's temperature and climate were hospitable to life because the planet happened to be at the "right" distance from the Sun. By this reasoning, the frigid –60°C average surface temperature of Mars occurs because that planet formed too far away from the Sun, while the blistering 460°C average surface temperature of Venus is attributable to its closeness to the Sun.

This hypothesis, however, is flawed. For the first 2.5 billion years of Earth's history, the Sun gave off so little radiation that the planet would have frozen if there had not been some other temperature-regulating mechanism.

FIGURE 12-1 A drowned forest in the West Walker River in July 1991 is evidence of a drought in California within the past 1000 years that lasted more than a century, long enough for trees to grow in the dry riverbed.

That mechanism was an extreme greenhouse effect caused by the predominance of carbon dioxide in Earth's atmosphere at the time (Box 12-1). Likewise, satellite imagery of Mars shows that it was not always a frozen wasteland. The most likely explanation for the deep canyons covering much of its surface is that liquid water—and, therefore, temperatures greater than 0°C—must have existed at one time (Figure 12-2). These river channels subsequently were bombarded by meteorites that left visible impact craters. Scientists have counted the craters superimposed on the river channels and, using typical rates of bombardment, have estimated that Mars had liquid surface water for about 800 million years, from the time the solar system formed about 4.6 billion years ago until about 3.8 billion years ago. Clearly, then, distance from the Sun is not the sole factor in determining a planet's temperature. A study of Earth processes suggests that carbon dioxide plays a major role in planetary climate.

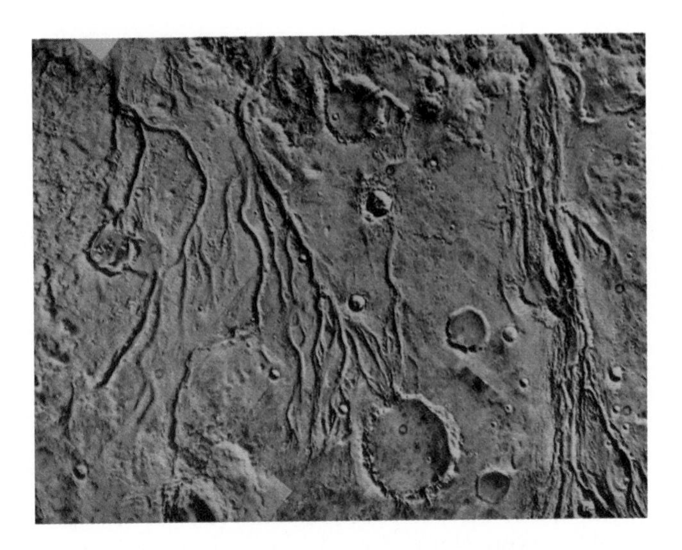

FIGURE 12-2 A satellite image of the surface of Mars. Deeply incised river canyons and meander loops indicate that water once flowed on this now-frozen planet.

Mars: A Frozen Planet

As discussed in Chapter 2, the heat energy responsible for the development of plate tectonics on Earth resulted from the accretion of the planet and from radioactive decay of elements in minerals. Mars was formed by the same accretionary process, but the planet has about one-tenth the mass of Earth and about twice Earth's ratio of surface area to volume. With its heat thus dissipated over a comparatively larger surface, ancient Mars would have cooled much more quickly than Earth and probably never fully developed plate tectonics. At the same time, weathering processes comparable to those on Earth would have stripped carbon dioxide from the Martian atmosphere and converted it to bicarbonate ions. Calcium and bicarbonate ions in solution would have crystallized into calcium carbonate in Martian oceans and drifted down to the ocean floors. The carbon was thereby locked into ocean sediments, but the lack of subduction kept it from being returned to the atmosphere.

Initially, some recycling of carbon had to occur on Mars—otherwise, the carbon dioxide in Mars's atmosphere would have been completely depleted by weathering in about 10 million years, not the 800 million years indicated by the river channel data. Satellite photographs suggest that voluminous volcanism occurred on Mars, and some scientists think that volcanic materials buried carbonate sediments to depths at which pressure and temperature metamorphosed the sediments, releasing CO_2. This recycling has been shown by computer models to be effective for up to a billion years. By then, the planet would have lost too much internal heat for volcanism to take place, carbon dioxide would no longer have been recycled into the atmosphere through volcanism, and the planet's surface would have cooled along with its interior. In addition, Mars may have lost some of its atmospheric CO_2 to outer space because of its relatively small mass and weak gravitational attraction.

Scientists estimate that the initial concentration of atmospheric carbon dioxide on Mars was equivalent to about 10 bars of pressure and subsequently dropped to the present value of 0.006 bars. In comparison, Earth's atmospheric CO_2 concentration is equivalent to 0.0003 bars. Although the Martian atmosphere still contains more CO_2 than does Earth's, the amount is not high enough to keep Mars as warm as Earth through a greenhouse effect because Mars is much farther from the Sun and receives less solar radiation than does Earth.

Venus: A Runaway Greenhouse Effect

At present, 99.9 percent of the carbon on Earth is tied up in rocks, whereas on Venus, the same proportion is found in the atmosphere. In fact, Venus's atmosphere is 97 percent carbon dioxide and is about 100 times denser than Earth's. The reason for the superabundance of carbon dioxide on Venus—and the resulting runaway greenhouse effect—is the planet's lack of water. Venus may initially have had liquid oceans, but today it is nearly dehydrated, with only small amounts of water existing in the Venusian atmosphere, either as vapor or as a component of sulfuric acid.

How did Venus lose its water? Again, a process active on Earth may indicate the answer. Strong ultraviolet radiation coming in from the Sun causes water vapor in the upper levels of Earth's atmosphere to break down into hydrogen and oxygen gases, a process that, because it is caused by light, is called *photodissociation*. The oxygen remains in the atmosphere, but the hydrogen drifts to outer space because its small mass allows it to escape Earth's gravitational pull. However, because very little water ever reaches the upper atmosphere of Earth, our planet's loss of water by photodissociation is minimal.

12-1 Geologist's Toolbox

The Faint Young Sun Paradox

According to well-accepted models of stellar evolution—the way in which stars are formed, develop, and die—the amount of light given off by the Sun has increased through time. It is probable that the output of radiation, or luminosity, of the young Sun at the time of Earth's formation was 30 percent lower than the present value. If early Earth had an atmosphere of today's composition, a 30 percent reduction in the Sun's luminosity would have produced an average surface temperature on Earth of about −22°C, far too cold for liquid water to exist, and as the Sun's luminosity increased, Earth's temperature would not have climbed above freezing until roughly 1.6 billion years ago. Yet geologic evidence suggests that the early Earth was quite warm. Sedimentary rocks reveal that Earth had oceans by 3.8 billion years ago, and the oldest fossils discovered to date are 3.5 billion years old. Furthermore, lack of evidence for glaciation suggests that Earth may well have been warmer than at present for much of its early history.

What enabled Earth to maintain such warm surface temperatures during a time of low solar output? Geologists refer to this puzzle as the *faint young Sun paradox*. The solution lies in the evolution of Earth's atmosphere. If the early atmosphere had a higher concentration of carbon dioxide than is present today, the greenhouse effect would have been correspondingly intense, and the heat retained in the atmosphere would have kept Earth's oceans liquid. Later, as luminosity increased, the increased energy would have promoted evaporation from the ocean basins. The additional water entering the atmosphere would have combined with atmospheric carbon dioxide, fallen to Earth as carbonic acid rain, and caused an increase in rock weathering. The resulting decrease in atmospheric carbon dioxide would have limited the greenhouse effect. In other words, as solar luminosity increased, heat-retaining carbon dioxide decreased and kept Earth at a fairly stable temperature. A temperature drop sufficient to inhibit evaporation would lead to less weathering and a consequent buildup of carbon dioxide in the atmosphere. The resulting increase of the greenhouse effect would cause warming. Thus, despite changes in the input of energy over geologic time, Earth's surface temperature has remained remarkably constant at values between 0° and 100°C, the range necessary for life as we know it.

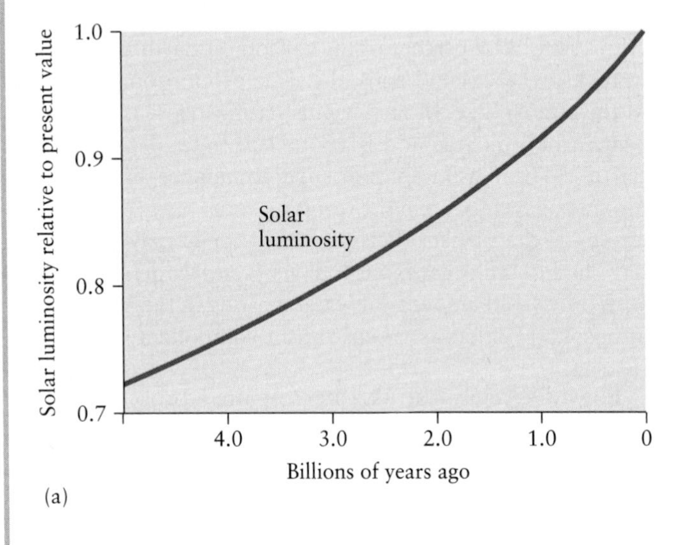

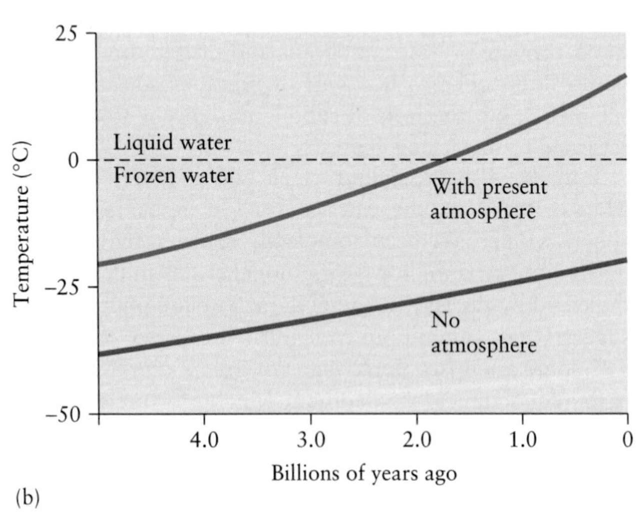

The hypothetical increase in solar luminosity (a) over time and (b) the corresponding increase in Earth's surface temperature, assuming an atmosphere of today's composition. In that case, Earth would not have become warm enough to support liquid water until 1.6 billion years ago. If there had been no atmosphere at all, Earth would still be frozen today.

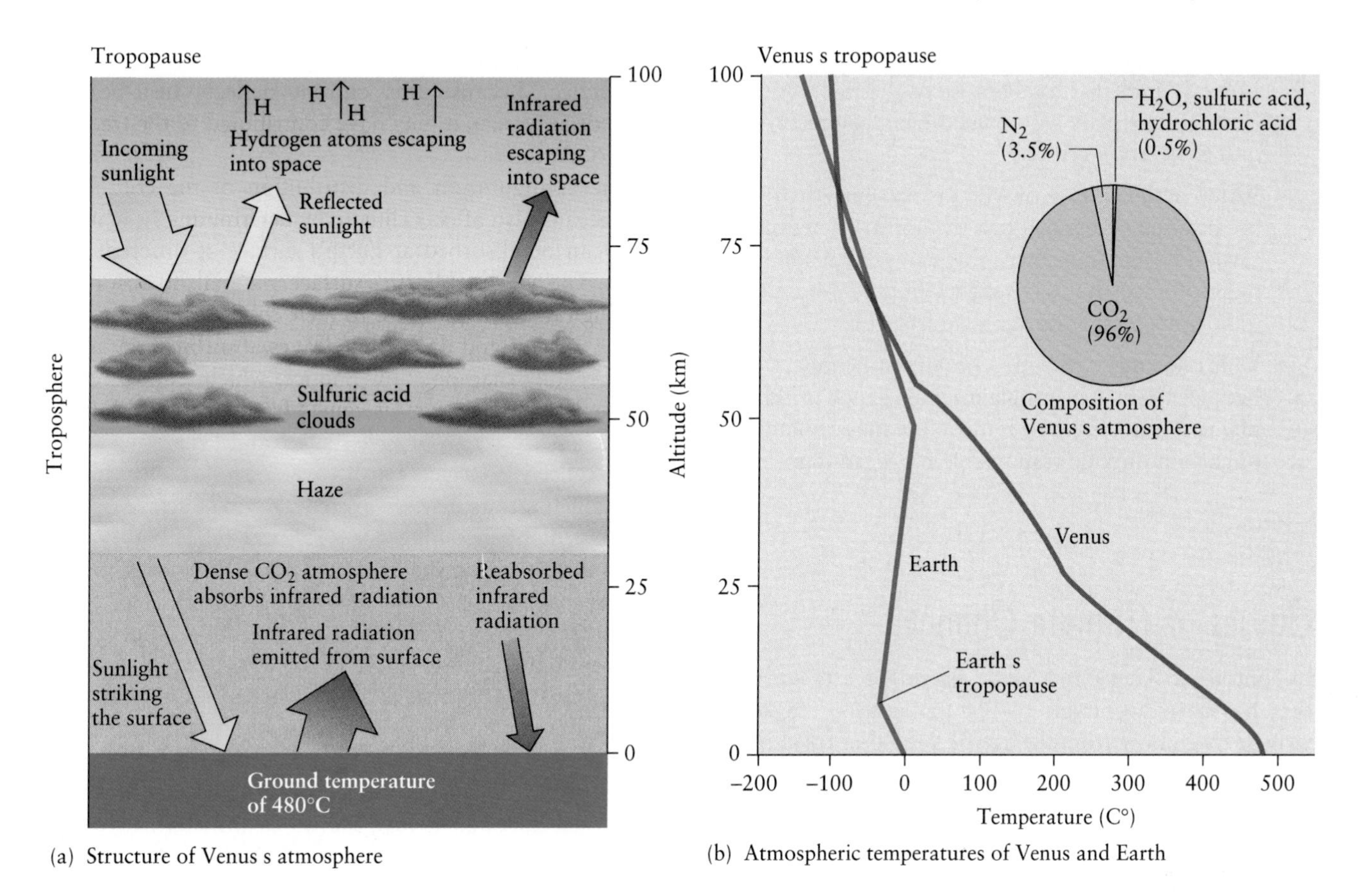

(a) Structure of Venus s atmosphere

(b) Atmospheric temperatures of Venus and Earth

FIGURE 12-3 The greenhouse effect on Venus. (a) Thick sulfuric acid clouds block much of the incoming solar radiation, but dense carbon dioxide clouds underneath them trap most of the outgoing infrared radiation emitted by Venus's surface, keeping the planet hot. Water is lost through the photodissociation of vapor in the upper atmosphere and the escape of hydrogen to outer space. (b) As a consequence of Venus's exaggerated greenhouse effect compared with the Earth's, the average surface temperature of Venus is about 30 times that of Earth.

Earth's water vapor tends to lie low because of the thermal structure of Earth's atmosphere. Recall from Chapter 9 that temperature falls with elevation above Earth's surface and declines to a minimum value at the tropopause. The rate at which temperature declines with elevation depends partly on how much water vapor the atmosphere contains. As air parcels move upward through the troposphere, they cool rapidly and any water vapor they hold condenses to liquid droplets. Condensation releases the water vapor's stored solar energy so that the decline in temperature with elevation is more gradual than if the atmosphere were dry. A dry atmosphere cools at a rate of ~10°C per km of elevation. A saturated atmosphere cools at a rate of 5.5°C per km. When the droplets gather sufficient mass so that they no longer are supported by atmospheric turbulence, they fall as rain. Consequently, little water vapor can be passed beyond the tropopause to levels at which photodissociation occurs.

Its nearness to the Sun gave Venus a much higher surface temperature than Earth's, and therefore surface water on Venus would have evaporated early in the planet's history. The early atmosphere of Venus may have held 20 percent or more water by volume than did the atmosphere of Earth. Heat and humidity probably caused Venus's atmospheric temperature to drop very slowly with altitude compared with the temperature decline in Earth's atmosphere. Therefore, the tropopause on Venus would have been found at about 100 km elevation (Earth's tropopause is at ~10 km). At so high an elevation, water vapor would have photodissociated easily to oxygen and hydrogen, and the hydrogen would have escaped the planet's gravitational pull. Without liquid water, weathering reactions involving carbon dioxide would have been impossible. Thus, CO_2 accumulated in the Venusian atmosphere and generated an intense greenhouse effect (Figure 12-3).

- Earth's surface temperature has remained within a range suitable for life because of the interaction of three cycles: plate tectonics, the hydrologic cycle, and the carbon cycle.
- Water at the surface of Mars is presently frozen because the planet has lost the ability to recycle carbon to its atmosphere through volcanism and therefore does not have sufficient greenhouse gases to keep its temperature above freezing.
- Venus has a runaway greenhouse atmosphere because there is insufficient water vapor in the planet's atmosphere to remove the massive amounts of carbon dioxide responsible for warming.

Causes of Climate Change

In contrast to Venus and Mars, Earth has a temperature that has been hospitable to life for at least 3.5 billion years, as we know from the age of the oldest fossil organism yet found on Earth. But Earth's temperature has not remained static since the planet's formation. Earth's climate has changed continuously in response to complex interactions among interior processes, the pedosphere, biosphere, hydrosphere, and atmosphere, Earth's distance from the Sun, and variations in the intensity of solar radiation. Change in any one of these variables can lead to change in climate. Furthermore, climate changes occur on a variety of time scales. Volcanic eruptions may cool Earth for a few years, whereas plate tectonics movement can lead to climatic changes that are played out over millions of years.

Influence of Plate Tectonics on Climate

Plate tectonics influences climate by changing the distribution of continents between equator and poles, by uplift of land to higher altitudes, and by seafloor spreading and volcanic eruptions. Recall from Box 6-1 that the collision of India and Asia, which began 40 million years ago and is continuing today, uplifted land to elevations that permitted glaciation, which in turn promoted physical and chemical weathering that may have removed CO_2 from the atmosphere and caused global cooling. In addition, the uplifted lands act as barriers to the flow of moisture-laden air from over the Indian and Pacific oceans, which over time have caused large parts of interior China to become desert.

Plate tectonics has also influenced the circulation of the oceans over geologic time by affecting the configuration of the continents and the ocean basins. As an example, recall from Chapter 2 that the Antarctic ice sheet developed 30 million years ago when Antarctica and South America began to split apart, establishing a flow of water around Antarctica. Because this current deflects heat-bearing equatorial waters, it may have contributed to the freezing over of Antarctica.

The configuration and distribution of the continents and oceans also affects climate by determining how much light can be absorbed at Earth's surface at different latitudes. The ability of Earth's surface materials to absorb incoming radiation is related to their color. Dark soils, deep blue seawater, and dark green leaves absorb energy much more readily than do snow and ice, which tend to reflect most energy. Reflectivity is valued from 0 to 1, with 0 denoting a perfectly absorptive surface and 1 denoting a perfectly reflective surface. Earth's average reflectivity value is 0.3, which means that, on average, Earth's surface reflects 30 percent of incoming energy. Local reflectivity values differ depending on the degree of cloud cover and the type of material making up the terrain.

Latitude affects reflectivity because of the angle of incidence of the Sun's rays. Have you ever been in a boat or on a beach at sundown and seen glare from the Sun coming off the water, although around noon you noticed no such effect? When solar radiation strikes the water from a high angle, the water absorbs most of it and gets warm, but when the angle is low, the water absorbs little of the incoming radiation, which instead glances off the water into your eyes. Similarly, at low latitudes where sunlight strikes from a high angle, close to 90°, ocean water absorbs the light almost entirely, reflecting only 3 to 5 percent of the energy, for a reflectivity value of about 0.03 to 0.05. The same ocean water at high latitudes would reflect about 25 percent of the incoming solar radiation, for an average reflectivity of 0.25. Meanwhile, the average reflectivity of land is about 0.2, comparable to the reflectivity of water at higher latitudes. Thus, land areas concentrated at both poles with a large ocean circling the globe at low latitudes would lead to higher average temperatures for Earth than would the reverse situation (Figure 12-4). In both cases, the intensity of radiation received from the Sun is greatest in the low latitudes (where solar radiation strikes the planet's surface at nearly right angles), and is much lower at high latitudes (where incoming radiation is inclined relative to the planet's surface). If the ocean encircles the equator, the water absorbs nearly all the incoming high-intensity solar radiation, while the higher latitude landmasses reflect a substantial amount of the low-intensity radiation. In the reverse situation, however, with land encircling the equator and oceans at the poles, much more of the high-intensity radiation hitting the equator would be reflected back to space, leading to lower average temperatures on Earth. In addition, the polar oceans reflect as much of the low-intensity radiation as would polar land. Consequently, more of the Sun's energy is retained if the ocean encircles the equator between two polar continents than if a continent encircles

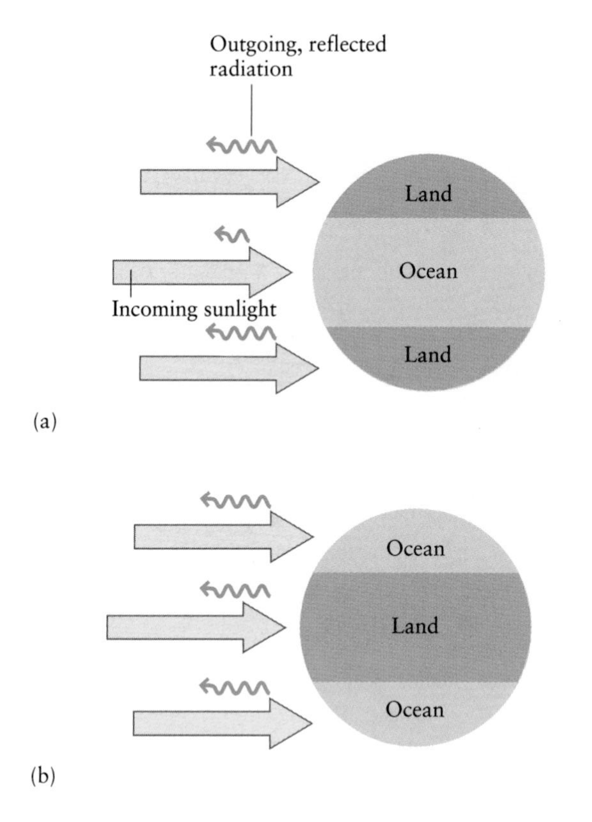

FIGURE 12-4 The distribution of land and ocean affects global temperature and climate. (a) On average, lands at the poles with an ocean circulating between them would warm the planet because, compared with land, the ocean absorbs more light at low latitudes, where the greatest amount of solar radiation is received by the planet. (b) Oceans at the poles with land between them would, therefore, have the opposite effect. Arrows striking the planet represent incoming solar radiation; arrows leaving the planet represent reflected radiation. Shorter outgoing arrows denote that more energy has been absorbed by the planet's surface than longer outgoing arrows, which represent more reflection.

the equator between two polar oceans. This means that Earth's average temperature would be higher if the ocean encircled the equator, although the landmasses themselves, being polar, might be cold.

Seafloor Spreading The Cretaceous period was a time of more rapid seafloor spreading rates than are observed today. This increase in rates is thought to have led to an enormous volume of carbon dioxide in the atmosphere as upwelling magmas along oceanic ridges released dissolved CO_2. Scientists believe that Earth's atmosphere during the Cretaceous period contained 7 to 10 times the CO_2 that it contains today, leading to an enhanced greenhouse effect and warm temperatures globally. No evidence of glaciation exists in any rocks from this time, suggesting that all the ice sheets and glaciers must have been melted. Furthermore, the rapid spreading rates caused a decrease in the depth of the ocean basins because young, hot crust is more buoyant than old, cold crust (see Chapter 4). Ocean water therefore flooded continents and covered much of the midwestern United States, Africa, and Europe with shallow, warm seas (Figure 12-5).

Volcanic Eruptions While increased volcanic activity ultimately acts to increase the concentration of greenhouse gases in the atmosphere and warming, in the short term volcanic eruptions may cause a cooling of Earth by injecting ash, sulfuric acid droplets, and sulfate particles high into the troposphere and stratosphere where they reflect incoming solar radiation (see the discussion of Mount Pinatubo in Chapter 2).

Influence of the Oceans on Climate

Some scientists now conceive of oceanic circulation as comparable to a large conveyor belt (see Figure 10-19). In this model, warm water from low latitudes flows northward in the Atlantic as a surface current. Through evaporation, the salinity of this surface current increases, and through northward migration, the temperature of the current decreases. At high latitudes, this cold, salty water sinks, forming a mass of water called North Atlantic Deep Water, which flows as a bottom current back toward the equator. This circulation is thought to be responsible for maintaining relatively mild winter temperatures in western Europe through sea-air heat exchange.

Occasionally in the past, this oceanic conveyor belt appears to have slowed or stopped entirely, shutting off the transport of heat to high latitudes. Between 12,500 and 11,000 years before the present, for example, a time known as the Younger Dryas cold period, western Europe experienced frigid temperatures and expansion of glaciers.

Influence of Earth's Orbital Parameters on Climate

The amount of solar energy received by Earth changes over geologic time because of variations in Earth's orbital parameters: the shape of Earth's orbit about the Sun and the tilt and wobble of Earth's spin axis.

Earth's Elliptical Orbit The shape of Earth's orbit influences the amount of solar energy the planet receives at different times of the year. A perfectly circular orbit would show no change in radiation receipt with time of year. Earth's orbit is slightly elliptical, however, and Earth is currently closest to the Sun in January and farthest away in July. Therefore, Earth receives more energy on average in January than in July. The **eccentricity**—the deviation from perfect circularity—of Earth's orbit has not remained constant throughout geologic history but undergoes a 100,000-year cycle between ~0 (nearly

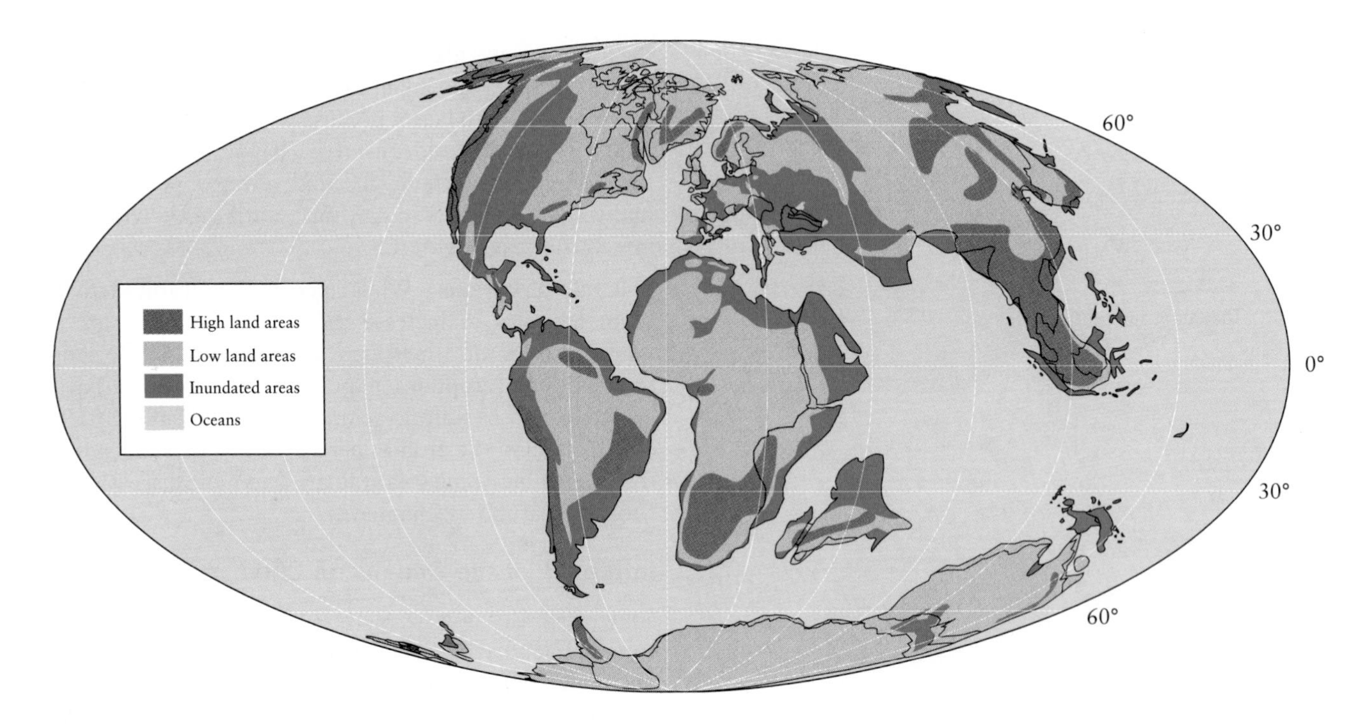

FIGURE 12-5 The distribution of land and sea during the late Cretaceous period, about 95 million years ago. Extensive shallow seas inundated much of the continental landmass. Sea level was very high during the Cretaceous partly because of accelerated rates of seafloor spreading, which created young, buoyant oceanic crust that displaced ocean water onto the continents. In addition, an enhanced greenhouse effect, brought about by increased CO_2 outgassing to the atmosphere, led to melting of ice sheets, contributing to sea level rise.

circular) and 0.05 (slightly elliptical), causing subtle changes in the amount of radiation received over time (Figure 12-6).

Tilt of the Spin Axis The orientation of Earth's spin axis also influences the distribution of solar radiation. As discussed in Chapter 9, Earth's spin axis is tilted relative to the incoming rays from the Sun, and this tilt is what causes Earth to experience seasons. The angle of tilt has changed over geologic time and continues to change today. In a cycle lasting about 40,000 years, the tilt varies between ~21.5° and ~24.5°. The greater the angle, the greater the temperature range between summer and winter at any location on Earth's surface (see Figure 12-6).

Wobble of the Spin Axis Like a top that is winding down, Earth wobbles on its spin axis, and so the direction of tilt of the spin axis changes. The spin axis wobbles through one full rotation about every 23,000 years, during which period the timing of the seasons changes. Recall that winter occurs when a hemisphere is tilted away from the Sun and summer when the same hemisphere is tilted toward the Sun. Currently, therefore, winter occurs from December to March in the northern hemisphere and from June to September in the southern hemisphere. However, 11,500 years from now, when Earth's spin axis is midway through its wobble cycle, winter will occur from June to September in the northern hemisphere and from December to March in the southern hemisphere. In 23,000 years, the spin axis will have completed a full rotation, and winter and summer will fall in the same calendar months as they do today. This phenomenon is known as the **precession of the equinoxes** (see Figure 12-6).

The importance of the wobble in determining the amount of solar energy received by any location on Earth depends on the degree of ellipticity of Earth's orbit. Winter in the northern hemisphere currently occurs when Earth is closest to the Sun in its orbit, and summer occurs when Earth is farthest from the Sun. The result is that winters and summers in the northern hemisphere are relatively mild today because the seasons are offset somewhat by Earth's proximity to the Sun. Eleven thousand years in the future, northern hemisphere winter will occur when Earth is farthest from the Sun, and summer when Earth is closest to the Sun. In this instance, because Earth's proximity to the Sun and hemispheric tilt will amplify each other's effects, summers will be hotter than they are today and winters colder. The southern hemisphere will experience just the opposite effects.

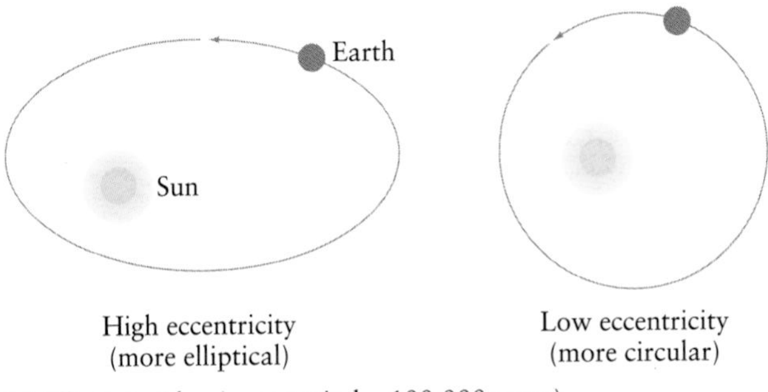

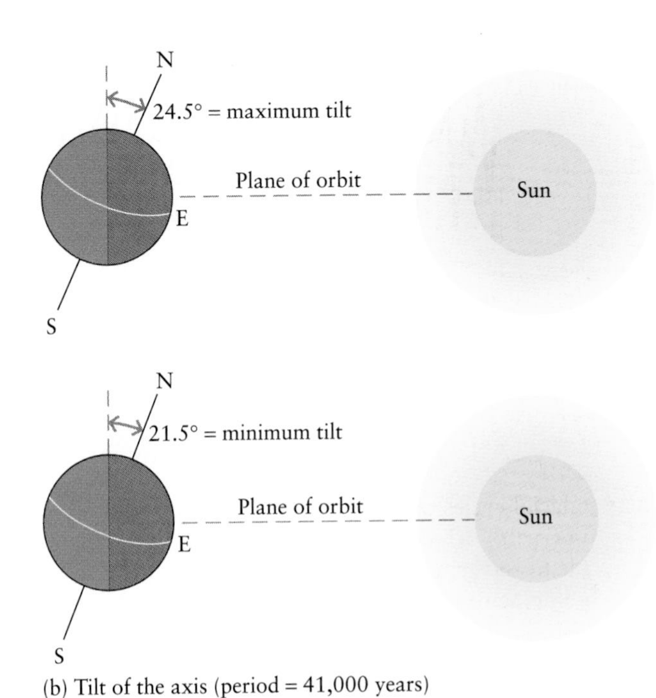

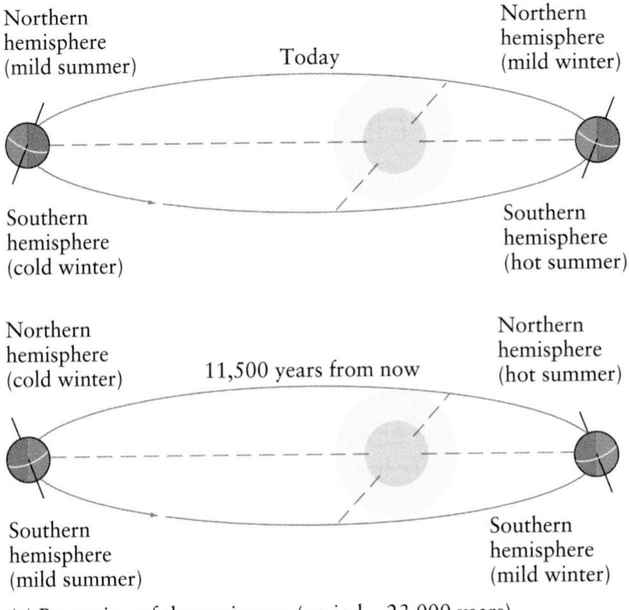

FIGURE 12-6 Changes in Earth's orbital parameters cause changes in the amount of solar radiation the planet receives. (a) Orbital eccentricity determines Earth's distance from the Sun at any given time of the year. (b) As the tilt angle increases, the temperature range between summer and winter increases at any given location. (c) Precession, or wobble, of the spin axis together with orbital eccentricity determines whether the seasons will be extreme or mild.

Milankovitch Cycles The variations in Earth's orbital parameters have come to be known among paleoclimatologists—scientists who study past climate and the history of climate change—as **Milankovitch cycles,** after the Serbian mathematician Milutin Milankovitch (1879–1958), who built on the theory of Scottish scientist James Croll (1821–1890) that orbital variations could lead to varying receipts of solar radiation over time that would affect climate. Combining the effects of the eccentricity, tilt, and precession cycles, Milankovitch calculated the *insolation*—the amount of incoming solar energy—hitting all latitudes of the surface of Earth over the last 600,000 years (Figure 12-7). He surmised that episodes of minimum insolation would have led to cold climates, and perhaps glaciation, and episodes of maximum insolation would be characterized by very warm climates.

Milankovitch considered the amount of energy reaching 65° north latitude in June to be of primary importance in determining whether glaciation will occur in the northern hemisphere, because this is the latitude at which ice sheets begin to grow. He theorized that glaciation would occur during times of relatively mild winters and cool summers. Warmer air in winter can transport more moisture as snow to the ice sheets, and cooler air in summer can minimize meltoff. By identifying periods of particularly low values of summer solar radiation over the last 600,000 years, Milankovitch suggested dates for periods of glaciations evident in the Alps; his correlations are shown as the shaded areas in Figure 12-7(b). Milankovitch thereby became the first person to suggest absolute dates for these glaciations. These dates have since been confirmed through radiocarbon and other dating techniques that had not yet been developed when Milankovitch did his work.

Climatic Feedbacks

While changes in any of the Earth systems can cause changes in climate, positive feedbacks within the climate system can amplify and negative feedbacks can reduce climatic variability.

The tendency of ice to perpetuate itself is a positive feedback mechanism. As ice caps grow, they reflect increasing amounts of energy (see Figure 2-17). As a consequence, less incoming solar radiation is absorbed by

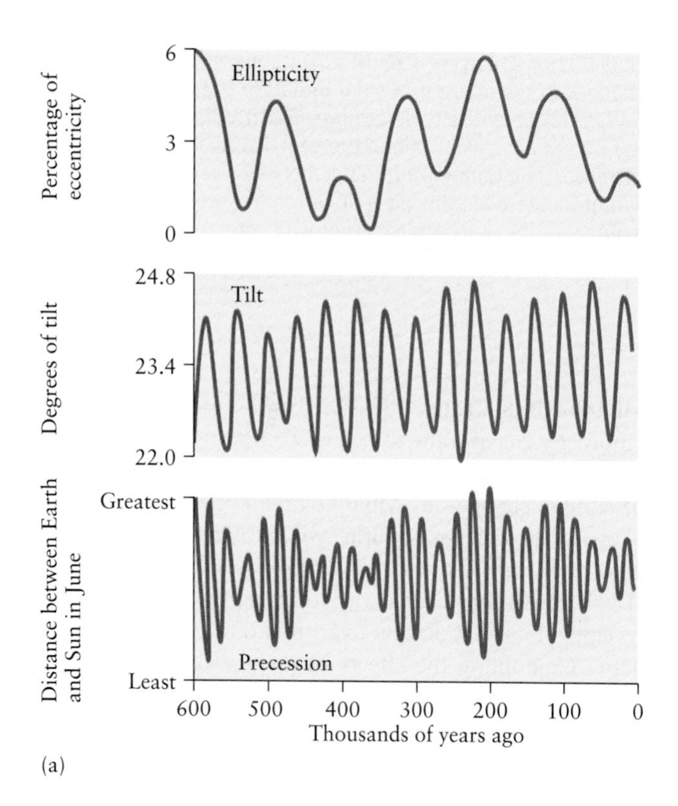

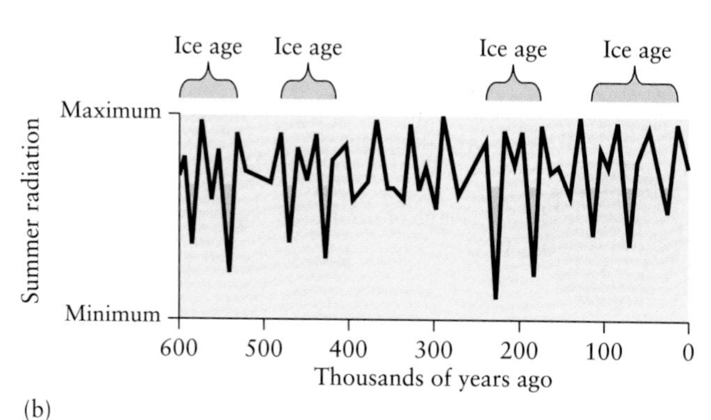

FIGURE 12-7 (a) The eccentricity, tilt, and precession cycles for the last 600,000 years. Note that the eccentricity cycle has a 100,000-year period, the tilt cycle a 40,000-year period, and the precession cycle a 23,000-year period. (b) Solar radiation in June at 65°N latitude over the last 600,000 years, calculated by Milutin Milankovitch from the eccentricity, tilt, and precession cycles shown in (a). The shaded areas represent minima in incoming solar radiation—times when Milankovitch believed the Earth was glaciated.

Earth's surface, resulting in declining temperatures and more ice growth, a continuously self-amplifying cycle. That Earth has never completely frozen over suggests that there is some sort of brake on this positive feedback process.

A negative feedback mechanism affecting Earth's temperature arises from phytoplankton photosynthesis. Phytoplankton are microscopic, aquatic plants that produce dimethyl sulfide gas during photosynthesis and release it to the atmosphere. Like all sulfide gases, in the atmosphere dimethyl sulfide attracts water droplets and forms clouds. The clouds block incoming solar radiation, simultaneously cooling the atmosphere and decreasing the ability of the phytoplankton to photosynthesize. As a result, less dimethyl sulfide is produced, and the clouds begin to dissipate, allowing Earth to warm. Phytoplankton photosynthesis, then, may help to maintain Earth's temperature at a fairly uniform value and reduce climatic variability.

How do we know that these processes were going on in the past? How can we learn what climate and environment were like at different times in Earth history? The answers lie in clues contained in the geologic record.

- Processes that cause climate change operate on distinctive time scales: Volcanic eruptions can cause climate to change for a few years whereas rearrangements of continents and ocean basins by plate tectonics cause climate to change over millions of years.
- Glacial cycles on Earth probably occur when cycles in Earth's orbital parameters, such as tilt and wobble of the spin axis and eccentricity of the orbit, combine to produce insolation minima.
- Positive feedbacks such as ice reflectivity amplify climatic variability, while negative feedbacks such as the radiation-blocking effect of dimethyl sulfide produced during phytoplankton photosynthesis tend to stabilize climate.

Indicators of Environmental Change

Scientists have identified many climatic and environmental indicators in the geologic record that provide information about temperature, precipitation, and vegetation distribution at different times in the past. These indicators fall generally into one of two categories, geologic or biologic. The sediment layers deposited in Coldwater Lake, described at the beginning of this chapter, are geologic indicators of rainfall intensity. Tree rings, used to infer the length of droughts in California, are an example of biologic indicators.

Geologic Records of Climate and Environment

Earth's environmental history is written in deserts and coastlines, glacial landforms, ice, and dry lakebeds.

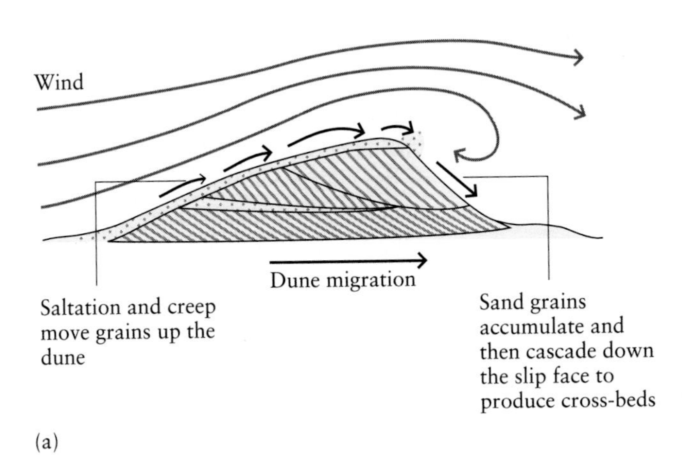

FIGURE 12-8 The formation of cross-beds in dune sands. (a) Grains are blown up the "stoss" (upwind) side of the dune and are redeposited on the "lee" (downwind) side of the dune, leading to distinctive layering that indicates the direction of the prevailing wind. (b) The shape of this dune in Death Valley, California, reveals that the prevailing wind direction is from left to right as this photo is oriented. The upwind side of the dune (left) has a low slope, up which grains are blown. At the crest of the dune, they avalanche down the steep downwind side.

Arid Environments Dunes are a common feature of today's deserts. Sculpted by winds that blow sand grains up one side of the dune and down the other (Figure 12-8), they move forward over time in the direction of the prevailing winds. If you were to take a shovel and dig into the dune, you would see the results of the forward motion as a series of inclined layers of sand known as **cross-beds** that point in the direction of the prevailing wind.

If you were to find 350-million-year-old rocks made of cross-bedded sands that had become cemented together, you might assume that these sands indicate a past desert environment. However, dunes, like many other sedimentary deposits, can form in a variety of environments. All that is required to form dunes is a supply of sand and active winds. Therefore, while often found in deserts, dunes are also common along coasts. To establish whether the dunes formed in a desert or a coastal environment requires additional information and a good understanding of modern depositional environments. If the dunes contain shells of marine organisms such as clams and oysters, and are deposited in a sedimentary environment of beach sands, muds containing marine fossils, and limestones produced by the accumulation of shells of dead marine organisms, the dunes indicate coastal rather than arid conditions (Figure 12-9). If, however, the dunes occur with muds and salts deposited in ephemeral lakes that fill during desert downpours and dry up within a few days to weeks, then the dunes probably indicate arid conditions.

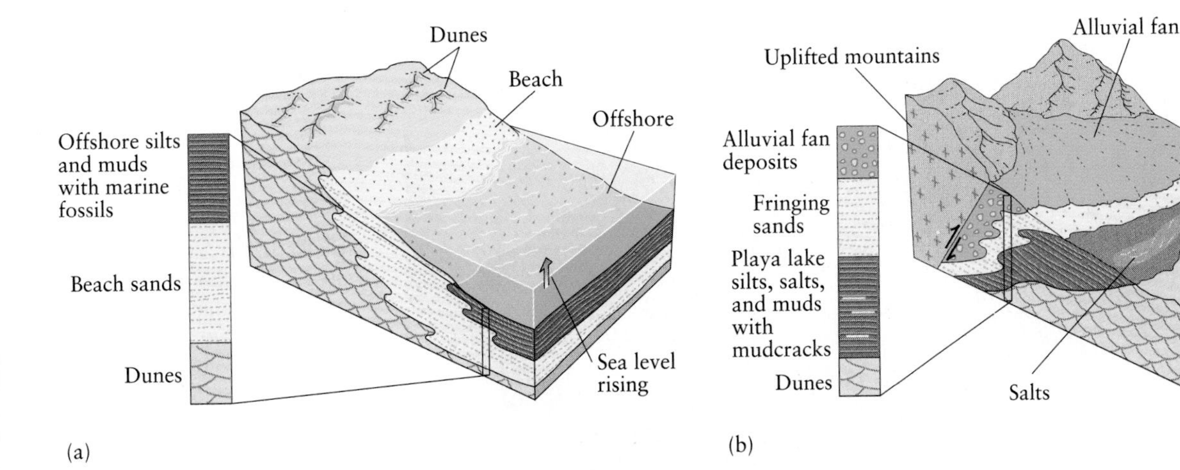

FIGURE 12-9 Dune formation in (a) a beach environment and (b) a desert environment.

Further evidence could be found in polygonal **mudcracks** that form when the lake dries up and subsequently become filled in with sediment. Only after we identify the depositional environment in which the dunes formed can we make climatic interpretations.

Desert or coast, cross-beds that develop in dunes contain an accurate record of past prevailing wind direction. In addition, the shapes of the dunes themselves often give directional information. Dunes were used to determine wind directions during the last global glaciation 20,000 years ago in the area that became the United States. The study revealed that although today winds blow across much of the United States from the southwest, during the last glacial maximum winds blew from the northwest, off the ice sheet that covered North America at the time.

Glacial Landforms Evidence of past glaciation, which indicates colder temperatures than prevail today, is much less ambiguous than that presented by dunes. Valley glaciers (those confined by valley walls) and ice sheets (which may cover whole continents) produce a variety of distinctive landforms that are not reproduced in any other environment. **Moraines** are ridges of boulders and sediment left behind as a glacier melts. They form in a variety of ways at the front and sides of a glacier and underneath the ice. Most commonly, moraines form at the front end of glaciers when debris-laden ice flowing within the glacier reaches the end of the glacier and melts, dropping its sedimentary load (Figure 12-10). Front-end moraines also can form when a glacier pushes rocks and soil ahead of it in the manner of a snowplow. If the climate becomes warmer, the glacial ice may melt, causing the glacier to retreat and leave the moraine behind. Glaciers also drag rocks embedded in the ice over bedrock. Sharp points on the moving rocks scratch into the underlying bedrock a series of grooves known as **striations,** which indicate the past flow direction of the glacial ice (Figure 12-11). The continual grinding of rock against rock is a very effective erosional process. As a result, valley glaciers quickly deepen their valleys, forming steep-sided U-shaped valleys. The thin spine of rock left between two glacial valleys, known as an *arête,* and the spires (or *horns*) found near the headwalls (or *cirques*) of several glaciers are also clear evidence that a landscape was once glaciated (see Figure 12-10).

Ice sheets leave behind somewhat different landforms than do valley glaciers because they are unconfined. **Glacial erratics** are boulders that have traveled hundreds and sometimes thousands of kilometers from their source regions, frequently over hills. Glacial ice is the only medium on Earth dense and viscous enough to carry them. Ice sheets also leave behind vast deposits of **till,** a mixture of particles ranging in size from clay to boulder. Chunks of ice left behind in the till sheet as the ice retreated melt to form *kettle lakes* (Figure 12-12). Minnesota, the "land of 10,000 lakes," owes its geography largely to this process. Debris-laden meltwater streams flowing at high pressures underneath glacial ice deposit their loads of sediment in long, winding ridges called *eskers.*

Glacial Ice A wealth of information regarding past climate can be found in glacial ice. Bubbles trapped in the ice record past atmospheric composition, including concentrations of greenhouse gases such as carbon dioxide and methane. Glacial ice forms in distinguishable annual

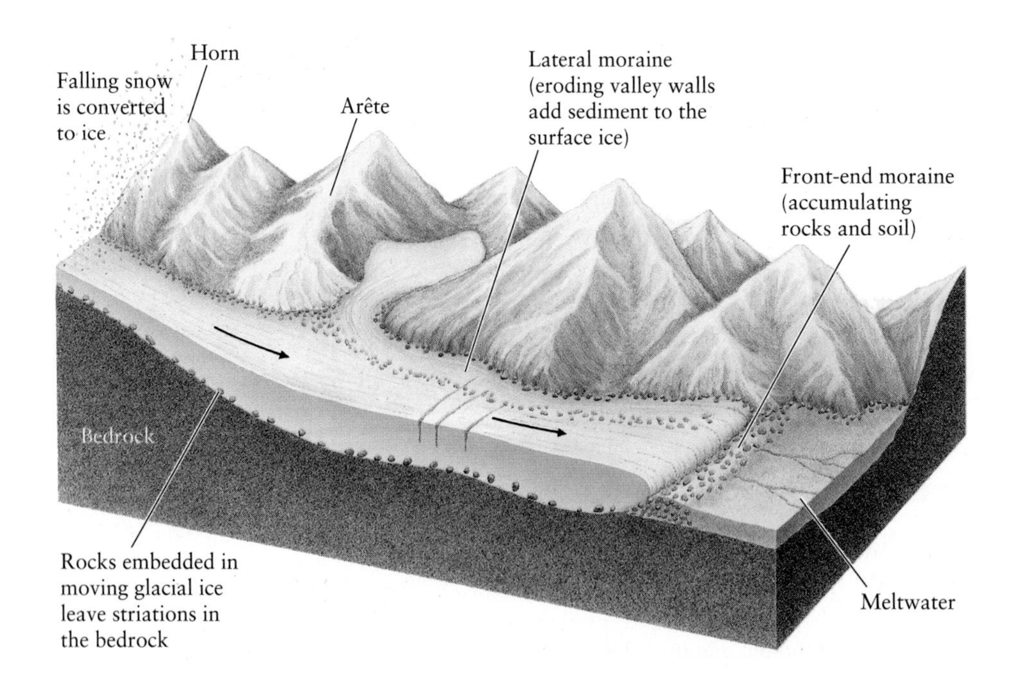

FIGURE 12-10 A typical glaciated landscape. The advance and retreat of a glacier within its valley organizes debris into ridges called moraines, which consist of a variety of grain sizes from small clay particles to large boulders, and carve the valley into characteristic forms.

FIGURE 12-11 Glacial striations in Glacier Bay National Park, Alaska, were created when an angular rock embedded in the glacial ice was dragged over the underlying bedrock. Such striations can be used to infer the direction of past ice flow.

layers, so developing a chronology for climatic events is relatively easy. These layers vary in thickness from year to year in response to temperature variations in the atmosphere. Cold air cannot hold nearly as much moisture as can warm air, so the amount of precipitation falling on an ice sheet in a given year is directly related to atmospheric temperature. During cold years, very little deposition occurs, whereas warmer years are marked by increased precipitation as moisture-laden air masses travel poleward.

Glacial ice also contains a record of past atmospheric temperature in its oxygen isotopic composition. There are three isotopes of oxygen: ^{16}O (8 protons and 8 neutrons), ^{17}O (8 protons and 9 neutrons), and ^{18}O (8 protons and 10 neutrons). Of these, ^{16}O, the isotope with lowest mass, is by far the most abundant, constituting ~98 percent of all oxygen. The heaviest isotope, ^{18}O, is second most abundant at nearly 2 percent, and ^{17}O is quite rare. Water molecules may contain any of these oxygen isotopes.

When water evaporates from ocean basins, molecules containing ^{16}O evaporate preferentially because less energy is required to evaporate molecules of lower mass. Atmospheric water vapor, therefore, is enriched in ^{16}O and depleted in ^{18}O compared with the ocean water from which it was evaporated. If atmospheric water vapor condenses to form clouds, any molecules containing ^{18}O condense first and rain out of the air more readily than molecules containing ^{16}O. The colder the atmosphere, the more likely that the heavy ^{18}O water molecules have precipitated out, leaving the resulting cloud mass made of lighter isotopes. Consequently, storm systems may have lost much of their ^{18}O by the time they reach high latitudes, and the colder the atmosphere, the less ^{18}O in the precipitation. Warmer temperatures would allow more ^{18}O to travel to high latitudes.

Paleoclimatologists use an instrument called a *mass spectrometer* to measure the ratio of ^{16}O to ^{18}O atoms in glacial ice. The lower the proportion of ^{18}O in a layer of glacial ice, the colder the atmosphere when that ice was deposited as snow. Scientists have discovered a relationship between the proportion of ^{18}O and temperature by which they can calculate the atmospheric temperature prevalent when an ice layer was deposited. These calculations reveal that the air over the Greenland ice sheet was approximately 10°C colder 20,000 years ago than it is today (Box 12-2).

Glacial ice contains a record of past winds as well as records of atmospheric temperature. Atmospheric dust settles on glacial ice, and the grain size of the dust particles provides information about wind strength. Faster winds are capable of transporting larger grains, whereas

FIGURE 12-12 A kettle lake in Minnesota, so called because of its round shape, formed when a chunk of glacial ice melted on a till sheet.

12-2 Case Study

Interpreting Greenland Ice Cores

Two teams of scientists, one from the United States and one from Europe, have drilled into the Greenland ice sheet to obtain climatic records for the last 250 thousand years. By studying layer thickness, dust composition and concentration, gas bubble composition, and oxygen isotopes in the ice, these scientists have identified very abrupt, large swings of temperature of ~10°C at the end of the last ice age and at many other times. Some of these temperature changes occurred over time spans as short as 10 years; in other words, temperatures changed by >1°C per year in some instances. By comparison, for the past 10,000 years Earth's climate has been very stable.

What caused those frequent and abrupt changes? They certainly could not have been produced by variations in Earth's orbital parameters because those changes occur much too slowly. The ending of the events about 10,000 years ago suggests that they were in some way related to the presence of ice sheets in North America and Europe, which had retreated almost completely by then. Sediments from the floor of the North Atlantic tend to confirm this observation because they show layers of debris occasionally interrupting the ordinary pattern of sedimentation. These layers suggest that the North American ice sheet sometimes launched armadas of icebergs into the North Atlantic, which carried loads of debris far into the ocean as the icebergs slowly melted.

If the iceberg armadas were sufficiently large, they may have significantly freshened the surface waters of the North Atlantic (see Chapter 10), thereby shutting off the formation of North Atlantic Deep Water. As a consequence, the global ocean conveyor system would have ground to a halt, discontinuing the transport of warm water to high latitudes. In this scenario, northern Europe would have experienced a rapid drop (perhaps over a mere decade) in atmospheric temperature of roughly 10°C, so that the climate of the British Isles would have become similar to the modern climate of Svalbard, an island just north of Norway, within the Arctic Circle.

Many of the climate shifts recorded in the Greenland ice cores are synchronous with the periods of ice-rafted debris deposition in the North Atlantic. However, because some others do not correlate with changes in marine sedimentation, the climatic shifts remain somewhat of a mystery. In addition, if the shifts were caused by releases of icebergs to the North Atlantic, scientists must next learn the mechanism that produced the iceberg flotillas and answer such questions as why the North American ice sheet only occasionally released large quantities of icebergs to the ocean.

(a)

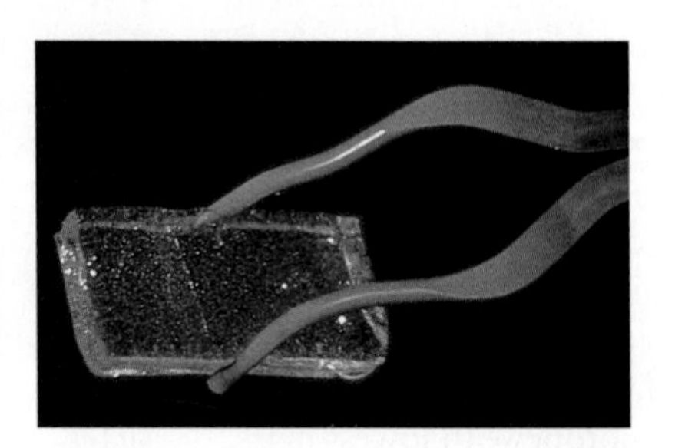
(b)

(a) Scientists removing an ice core from the drill string.
(b) An ice core from Greenland. Note the annual layering and the bubbles and dust in the ice.

gentler winds may transport only the finest particles. Similarly, glacial ice contains a record of past volcanic eruptions (Figure 12-13). Acid aerosols and ash particles may be deposited thousands of kilometers away from the eruptive source if volcanic products are injected high into the atmosphere and stratosphere.

Lake Shorelines In the western United States are basins that are presently arid but which show evidence of once having contained large lakes. This evidence consists of two types of shoreline development, constructional and destructional (Figure 12-14). **Destructional shorelines** are produced when waves strike the same location over and over again, cutting away at the land and forming a wave-cut bench. These benches are commonly covered with a beach deposit of rounded gravels. **Constructional shorelines** are usually identified by an algally-produced $CaCO_3$ deposit known as *tufa,* which is made when springs of fresh water empty into the margins of a saline lake. In either case, both these shorelines are evidence of a previously wetter climate.

Calcium carbonate precipitating in lakes contains a record of the oxygen isotopic composition of lake water that also can be used to infer lake level. Evaporation preferentially removes $H_2{}^{16}O$ from the lake because it is lighter than $H_2{}^{18}O$. Therefore, the water in lakes that are subject to high rates of evaporation tends to become heavier with time. When a lake receives runoff from its surrounding drainage basin or precipitation on its surface, that runoff and precipitation are usually very light isotopically. If there is more inflow to the lake than outflow through evaporation, the lake will grow bigger and contain lighter water. If, however, there is more outflow from the lake than inflow, the lake will shrink and contain heavier water. Calcium carbonate crystals in the lakebed record these isotopic variations and, therefore, also record fluctuations in the size of the lake.

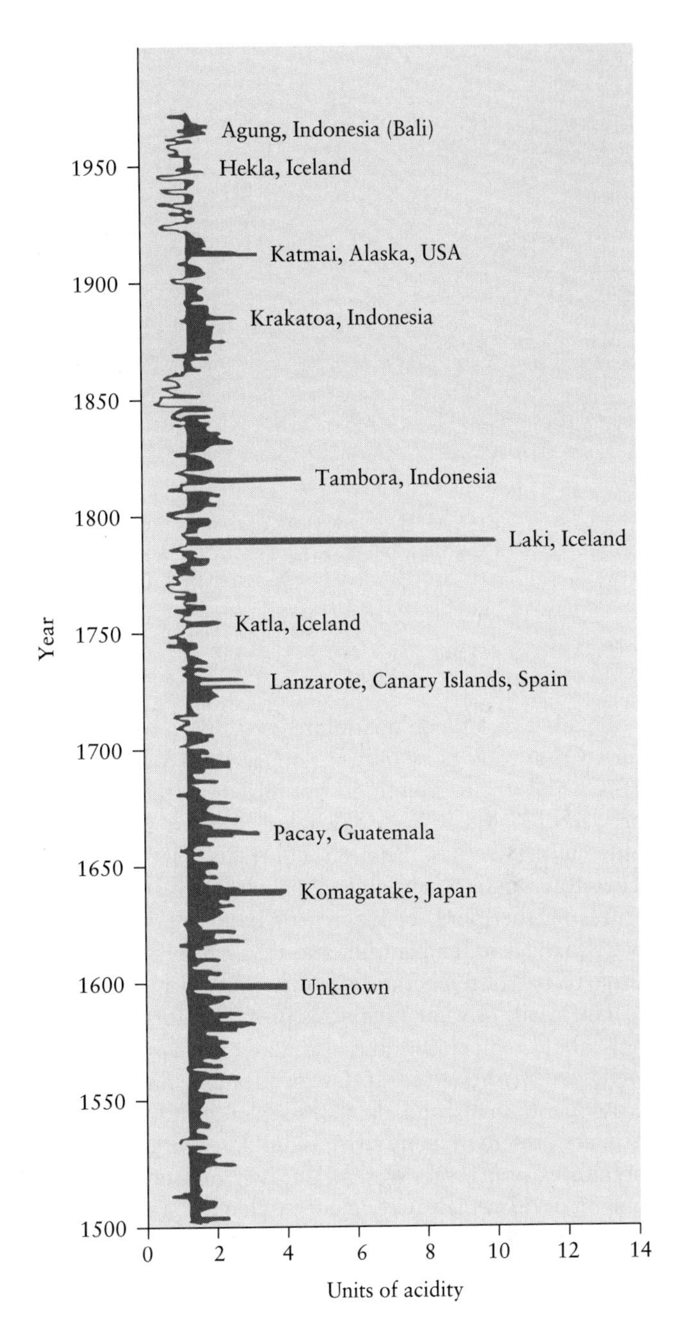

FIGURE 12-13 The acidity of glacial ice. High values correspond to periods of volcanic eruptions.

Biological Records of Climate and Environment

Organisms grow larger and reproduce more when environmental conditions are favorable to them. An analysis of fossils and living organisms thus contributes to an understanding of environments past.

Fossils The presence of specific fossils indicates when organisms flourished, suffered, or became extinct. Crocodile bones dating to the Eocene epoch (57 million to 34 million years ago) have been found buried in sediments in Utah, Colorado, Wyoming, and the Dakotas. Crocodiles are reptiles—cold-blooded animals, requiring temperatures warmer than freezing in order to survive. Currently they are found in parts of Africa and the Americas that are warm year round. In the United States, for example, they are now found only in states along the Gulf Coast, but the states in which *fossil* crocodiles have been found presently experience harsh winters with very cold temperatures. Therefore, paleoclimatologists believe that temperatures in these northern states must have been substantially warmer in the Eocene, perhaps more akin to temperatures observed today in Florida and Louisiana. Furthermore, crocodiles live in swampy habitats quite unlike modern conditions in Utah, Colorado, Wyoming, and the Dakotas. Not only were these states warmer in the Eocene, they must also have been wetter.

(a)

(b)

FIGURE 12-14 Destructional and constructional shorelines at Pyramid Lake, northwestern Nevada. (a) Destructional wave-cut benches in a hillside. (b) Constructional tufa towers formed at the margins of the lake.

Fossils can provide quantitative estimates of temperature. Coral reefs grow only at temperatures warmer than 18°C, and their growth is optimized at temperatures between 23° and 25°C; today, these conditions are found only in the tropics, between 30°N and 30°S latitude. Therefore, fossil corals indicate past warm water temperature. Furthermore, coral growth bands vary in width with changes in temperature and thus can provide a sensitive record of temperature change in tropical waters.

The fossils of some organisms provide information not only about temperature but also about the salinity of the environment they lived in. Ostracodes and diatoms (microscopic animals and plants), as well as molluscs and fish, are present in many streams and lakes. Some of these organisms prefer very fresh water, whereas others thrive only in more saline environments. Often the species found in a lake can be used to infer changes in lake level. Lake water becomes salty when there is more evaporation than inflow, just as it becomes isotopically heavy. Thus, changes in salinity, like changes in isotopic composition, reflect variations in lake size.

Packrat Middens Packrats are small rodents that construct nests, called **middens,** out of twigs, leaves, and bark from their immediate surroundings. Because they are made of biological materials, they contain carbon, and therefore can be dated by radiocarbon. Furthermore, identification of the twigs and other plant debris can be used to determine what climate existed in the region at the time the midden was constructed because different plants require different climatic conditions for their existence. Cacti, for example, are found only in deserts.

Pollen Pollen grains from different plant species have very distinctive shapes and can be used to infer what plants were present at different times in the past. *Palynologists,* scientists who study pollen, typically extract a core of sediments from the floor of a lake into which pollen has been deposited. They then use radiocarbon methods to determine the age of the sediment at different depth horizons in the core and identify the pollen grains to assess what the climate was at each depth or age horizon. Using pollen from many different locations in the northeastern United States, scientists with the Climate/Long-Range Investigation, Mapping and Prediction project (CLIMAP) have shown how forests responded to the retreat of the North American ice sheet from 18,000 years ago (the time of the last glacial maximum) to the present (Figure 12-15). Cold-loving species were initially found in southern states, but migrated northward as the ice sheet retreated. Today, species such as spruce and fir are found in abundance only in Canada and the northern United States.

Tree Rings Trees growing in a climate that varies seasonally go through a growth phase and a dormancy phase each year. During the growth phase, new wood is added to the outer layer of the tree (Figure 12-16). This **early wood** is composed of large, spongy cells and is very low in density. As fall approaches, the tree starts to become dormant, but before it does, it lays down a layer of **late wood,** which consists of very densely packed, smaller cells.

The widths of the early and late wood rings are directly tied to climatic conditions. If growing conditions are optimal, the tree will show a fat growth ring, but if growing conditions are poor, a narrow ring will indicate that very little growth occurred in that year. Many factors influence growth, including cloudiness, availability of nutrients in the soil, temperature, precipitation, and the amount of food stored within the tree. Of these, the two most important factors are usually temperature and precipitation.

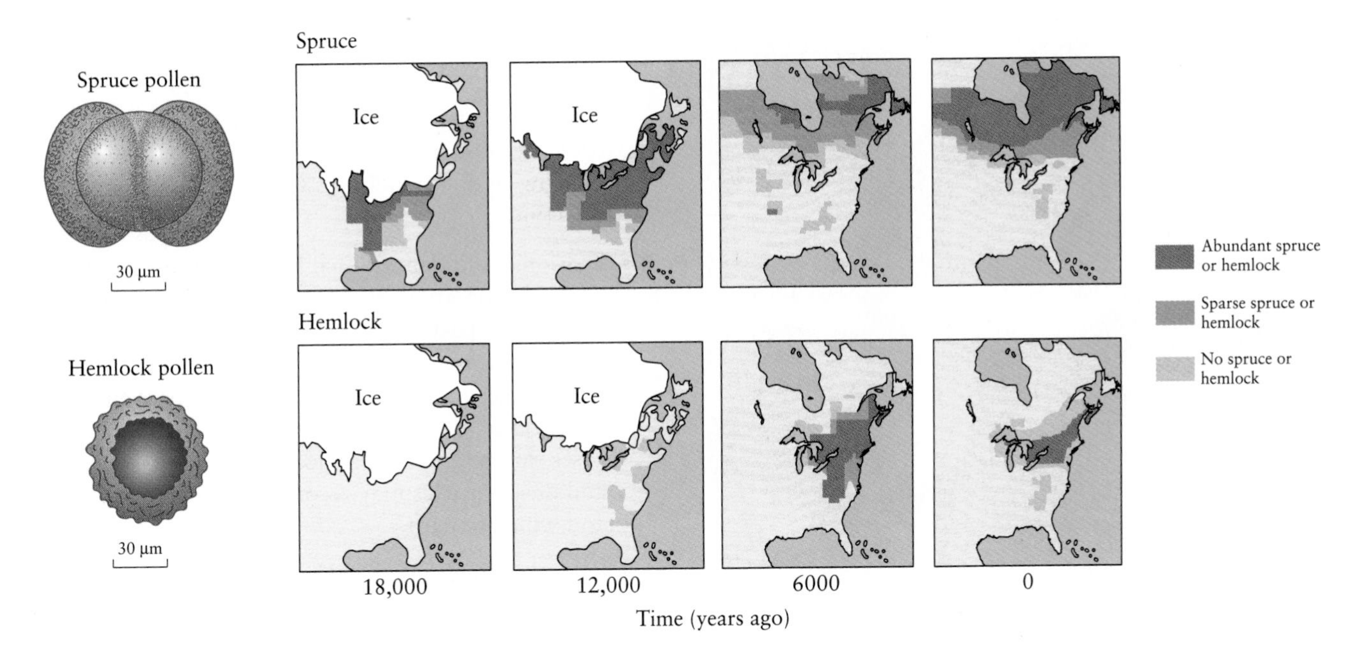

FIGURE 12-15 Spruce and hemlock pollen have unique shapes that allow them to be readily identified. Maps of pollen distribution determined by CLIMAP show the progress of these species as climate warmed and the large ice sheet that blanketed much of North America during the last glacial maximum retreated. Spruce migrated northward, seeking a colder climate, and hemlock developed in areas previously too cold for its survival.

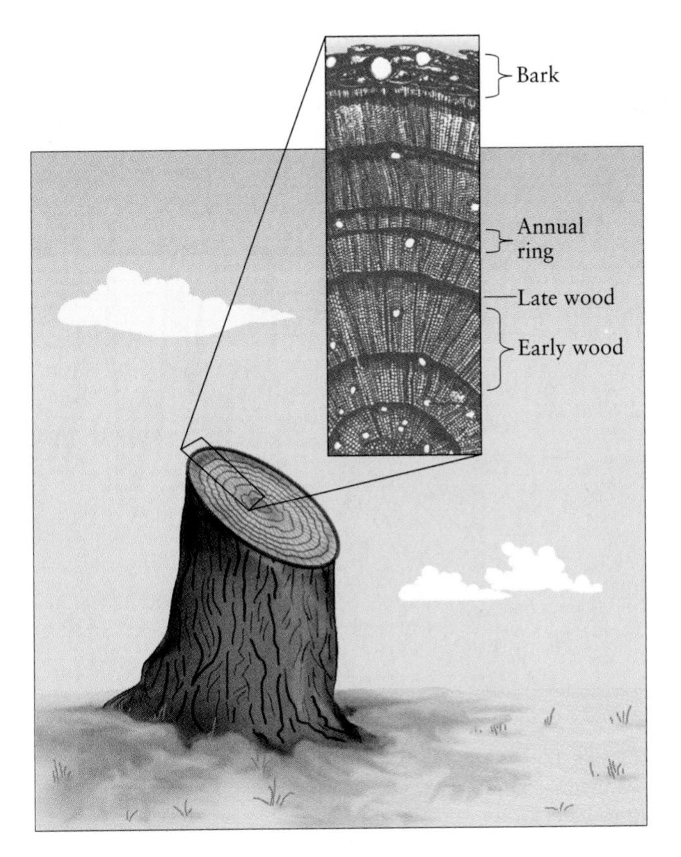

FIGURE 12-16 A cross section of a conifer tree trunk showing early- and late-wood rings. Each early wood–late wood couplet represents one year of growth.

For trees living in a continuously warm climate such as experienced in the American Southwest, the limiting factor on growth rate tends to be precipitation, arid years yielding very little growth and wet years yielding more growth. In a rainy environment, temperature might be the limiting factor. Thus, tree-ring widths indicate how the limiting climatic factor has changed over time. For instance, in a tree limited by precipitation, variations in ring width can be used to infer when droughts or particularly wet years occurred. This kind of analysis has yielded climatic records that go back several thousand years.

Tree rings are particularly useful climatic indicators because each band represents one year of growth. Radiocarbon dating can be used to determine the age of one of the bands, and from this age the total amount of time spanned by the tree can be found simply by counting the rings; thus, climatic events can be dated very precisely.

Marine Organisms and Oxygen Isotopes During glacial periods, $H_2{}^{16}O$ is preferentially extracted from the ocean

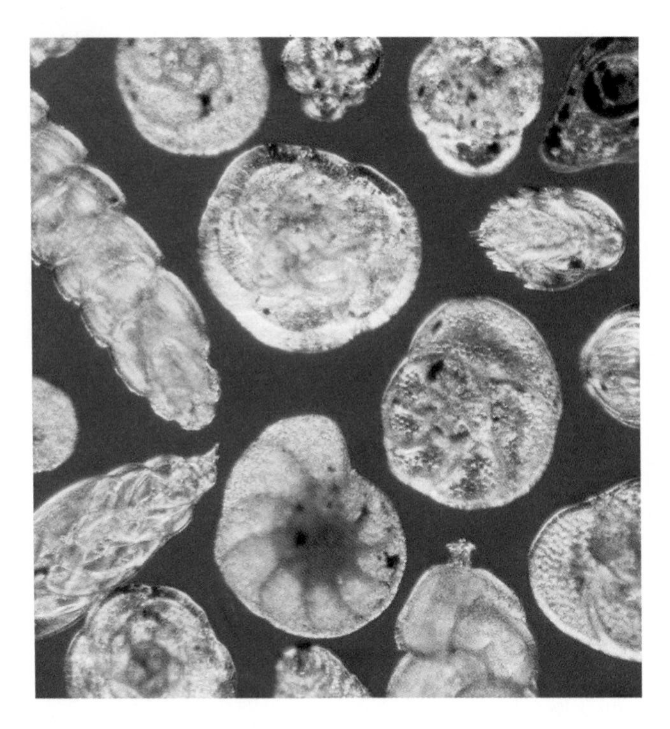

FIGURE 12-17 Geologists use the shells of foraminifera such as these to determine the oxygen isotopic composition of the ocean, from which they can infer global ice extent.

basins to form ice sheets. As a result, ocean water becomes heavier, i.e., more enriched in $H_2{}^{18}O$, during glacial periods. Tiny single-celled organisms living in the oceans known as **foraminifera** (Figure 12-17) secrete shells of calcium carbonate ($CaCO_3$) in equilibrium with the ocean water surrounding them. Thus, they contain a record of the past isotopic composition of the ocean, a record that can be used to determine when glaciations occurred on Earth. Using hollow drills, scientists have drilled through ocean sediments all over the world. The resulting cores usually contain a continuous record of sedimentation for thousands to hundreds of thousands of years.

Scientists have devised ways of measuring the abundance of ^{18}O relative to ^{16}O atoms in foraminifera. Higher proportions of ^{18}O suggest the existence of ice sheets, whereas lower proportions indicate little ice.

Isotopic composition of glacial ice and of the ocean vary as climate changes. During glaciations, the ocean is very enriched in ^{18}O, while glacial ice is very depleted in ^{18}O. During interglaciations, when there is little ice on the planet, the ocean is more depleted in ^{18}O, and glacial ice is more enriched in ^{18}O. The oxygen isotope record of climate change for the last 800,000 years, based on the shells of fossil foraminifera, is shown in Figure 12-18. The high proportion of ^{18}O at ~20,000 years ago corresponds to the last glacial maximum, when an ice sheet covered all of Canada and stretched south into parts of the United States.

- Indicators of environmental change can be broadly classified as either geologic or biologic.
- Arid environments may be recognized from the presence of sand dunes and mudcracks. Cross-bedding in dunes indicates the prevailing wind direction.
- Erosional and depositional landforms produced by glaciation are distinctive and allow geologists to infer the past extent of glacial ice. Glacial ice itself contains a wide variety of climatic information: Oxygen isotopes in the ice allow past air temperature to be determined, bubbles contain gases that record past atmospheric composition, and dust particles may record volcanic eruptions and wind strength.
- Lake shorelines in the arid western United States testify to previously wetter climates.
- Fossil plants and animals provide ecological information about temperature, amounts of precipitation, and lake salinity.
- Plants are used in three different ways in paleoenvironmental studies: The width of tree rings can indicate temperature or precipitation variations from year to year, and pollen and packrat middens record vegetational distribution at different times in the past.

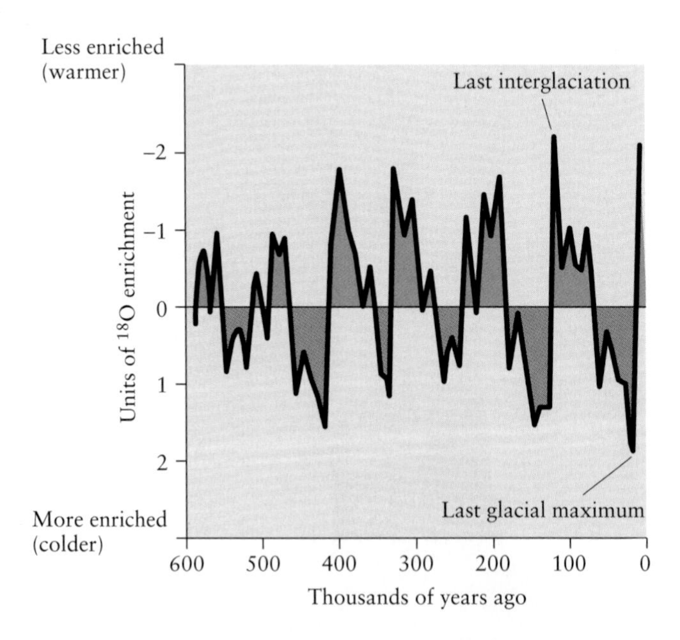

FIGURE 12-18 The marine oxygen isotopic record. Positive values of ^{18}O represent glaciations; negative values represent interglaciations.

Closing Thoughts

Earth's surface environments are in a constant state of flux because of the actions of many processes, some of which cause linear changes in climate and environment, others of which are more cyclic. These processes operate on a variety of time scales, some occurring so slowly that they are nearly unrecognizable on the time scale of a human life. As we go about our daily business, we do not notice that the continents beneath our feet are migrating to new positions, nor recognize the tiny changes in solar radiation intensity that accompany annual variations in Earth's orbit about the Sun. Nevertheless, the climate at any given point in time is dependent on these and many other factors, some fully understood by modern science, some probably yet to be discovered.

The continuous cycling of carbon between the atmosphere, hydrosphere, biosphere, and geosphere works to make Earth uniquely habitable among the planets in our solar system. However, this does not mean that all Earth environments are equally hospitable. Regional variations in the distribution of temperature and precipitation make some parts of Earth more livable than others, and climates have changed dramatically throughout geologic time and will continue to change.

The tools that we have at our disposal to assess these past changes are of varying kinds. Some give quantitative information about changes in temperature, precipitation, and vegetation patterns. Others are more qualitative, telling us only if climate was colder or warmer, wetter or drier. Still, any information we can gather about past climate and environment is critical to our understanding of how Earth works and what we might expect for the near future. As the human population grows, our ability to adjust to changes in climate becomes less certain. Modern societies cannot, as our early ancestors did, simply pack their bags and move to a new location if the current environment becomes inhospitable. We must be able to develop contingency plans to deal with climatic and environmental change in each community. The first step toward this goal is a full understanding of Earth's history of global environmental change.

Summary

- Were it not for the interaction of the hydrologic and carbon cycles with plate tectonics, Earth's surface temperature would probably be too warm or too cold to support liquid water and, therefore, life. Venus and Mars are each missing one of these three vital cycles and cannot support life.
- Climate is the by-product of interactions among the lithosphere, pedosphere, atmosphere, hydrosphere, biosphere, and solar radiation.
- Early Earth must have had a substantially larger concentration of greenhouse gas than at present to counteract the much lower luminosity of the young Sun. That this is true is evident from sedimentary rocks, which indicate that Earth had liquid water as early as 3.8 billion years ago.
- Processes that change climate operate on a variety of time scales. Volcanic eruptions can cool the planet for a matter of years. Variations in Earth's orbital parameters drive climate changes that occur over tens of thousands of years, and plate tectonics alters climate over time spans of millions to tens of millions of years. Climate at any point in Earth's history represents the aggregate of these different processes and time scales.
- Earth has been repeatedly glaciated in the last few million years because of cyclical variations in its orbital parameters that together produce periods of lowest insolation. Once ice began to grow, the positive feedback mechanism based on the reflectivity of ice probably enhanced that growth.
- Climate is recorded by a number of processes acting at Earth's surface, allowing us to reconstruct Earth's history of environmental change.
- Evidence of previously arid conditions are sand dunes, mudcracks, and salts deposited in saline lakes.
- Abandoned shorelines found high above valley floors in the western United States are evidence that climate was once much wetter in that region than it is today.
- Glaciers and ice sheets leave behind both erosional and depositional landforms. Some of these landforms indicate the past extent of ice, and some show direction of flow of the ice.
- Oxygen isotopes can be used to determine past atmospheric temperature, extent of ice on land, and relative lake level.
- Vegetational distribution, as determined from packrat middens and pollen grains, can give information about past rainfall amounts and temperatures.

Key Terms

eccentricity (p. 369)
precession of the equinoxes (p. 370)
Milankovitch cycles (p. 371)
cross-beds (p. 373)
mudcracks (p. 374)
moraines (p. 374)
striations (p. 374)
glacial erratics (p. 374)
till (p. 374)
destructional shorelines (p. 377)
constructional shorelines (p. 377)
middens (p. 378)
early wood (p. 378)
late wood (p. 378)
foraminifera (p. 380)

Review Questions

1. What cycles on Earth help keep its surface temperature within a livable range?
2. Which of these cycles is Mars missing? Venus?
3. How, and on what time scales, does plate tectonics influence climate change?
4. What role does the ability of a surface to absorb or reflect radiation play in setting global temperature?
5. How does oceanic circulation influence the climate of northern Europe?
6. What are Milankovitch cycles? What conditions are favorable to the growth of ice sheets in the northern hemisphere?
7. What are climatic feedbacks? How do they work?
8. What kinds of evidence can we examine to discover how climate has changed throughout Earth's history?
9. In what ways are oxygen isotopes used to unravel Earth's history of climate change?
10. How are plants used in climate-change studies?
11. What indicators supply information about the relative wetness of past environments?

Thought Questions

1. Will Earth's climate always remain suitable for life, or will our planet someday become like Venus or Mars?
2. Suppose that snow and ice were black rather than white. Would the positive feedback mechanism responsible for the formation of glacial ice continue to operate on Earth? Why or why not? What would happen to global temperature if Antarctica, the Arctic Ocean, and Greenland, presently covered by white snow and ice, were instead covered by black snow and ice?
3. Paleontologists have removed hundreds of bones from the La Brea tar pits, a naturally occurring sticky swamp of tar in Los Angeles. Many of these bones have been identified as belonging to mammoths and have been radiocarbon dated at 20,000 years. What does the presence of these bones imply about the past climate of Los Angeles?
4. Weathering of rocks is only one way in which carbon dioxide is removed from the atmosphere. Can you think of another? What sources besides volcanic eruptions supply CO_2 to the atmosphere?

Exercises

1. Pick any point on Earth. What conditions would be necessary to create the hottest possible summer at that position? The coldest possible winter? Use what you know about Earth's orbit and spin axis, reflectivity or absorption of energy, and temperature as a function of elevation to explain your answer.
2. Take a walk in a wooded area and note how many different species of plants you see. You don't have to be able to identify the plants to do this; just carry a sketchbook and draw a picture of the leaves each kind of plant has. Then, pick an area 10 feet by 10 feet and record how many of each different kind of plant you find there. Next, examine the leaf litter on the ground. Are all the plants in the area represented by leaves in the litter, or are leaves of some plants missing? Why might some plants not be represented in the litter? Are there leaves in the litter that come from plants outside your defined area? How did they get there? Are the relative proportions of leaves in the litter the same or different from the relative proportions of plants in the area? Why? If you were to try to infer what vegetation once flourished in an area based solely on a deposit of leaves, what difficulties might you have? These are the difficulties encountered by palynologists when they make environmental interpretations based on pollen grains.

Suggested Readings

Broecker, W. S., 1995. "Chaotic Climate." *Scientific American* (November): 62–68.

Frakes, L. A., 1979. *Climates Throughout Geologic Time.* New York: Elsevier.

Hodell, D. A., Curtis, J. H., and Brenner, M., 1995. "Possible Role of Climate in the Collapse of Classic Maya Civilization." *Nature* 375: 391–394.

Imbrie, J., and Imbrie, K. P., 1979. *Ice Ages.* Hillside, NJ: Enslow.

Kasting, J. F., Toon, O. B., and Pollack, J. B., 1988. "How Climate Evolved on the Terrestrial Planets." *Scientific American* (February): 90–97.

Ruddiman, W. F., and Kutzbach, J. E., 1991. "Plateau Uplift and Climatic Change." *Scientific American* (March): 66–75.

Stine, S., 1994. "Extreme and Persistent Drought in California and Patagonia During Mediaeval Time." *Nature* 369: 546–549.

CHAPTER 13

The Mediterranean Sea viewed through the Strait of Gibraltar from above the Atlantic Ocean. Evidence from sediment cores indicates that by 26 million years ago the Mediterranean had become land-locked and dried into a salt-encrusted desert. This occurred when growth of the Antarctic ice sheet caused sea level to fall beneath the level of the Strait of Gibraltar. Global warming subsequently caused glacial melting and re-filled the basin.

Tracing and Predicting Environmental Change

Robert Frost wrote, "Some say the world will end in fire, / Some say in ice." So far, the history of Earth is indeed one of fire and ice. From a fiery ball of molten rock, Earth cooled and differentiated into distinctive interior zones and a solid crust, liquid oceans, and gaseous atmosphere teeming with life. Over geologic time, continents grew, split apart, and migrated thousands of kilometers. Vast ice sheets waxed and waned. Occasionally, Earth suffered meteorite impacts so severe that they caused huge firestorms and dust clouds thick enough to block out the warming rays of the Sun. At least one meteorite impact is thought to have destroyed much of the life on the planet, making way for new species. One of Earth's new creatures, *Homo sapiens,* has recently outpaced some of these natural processes as an agent of environmental change.

An important indicator of environmental change is temperature. In some cases, fluctuations in global temperature have triggered a host of environmental changes. In other cases, Earth has cooled or warmed in response to evolving geological and biological systems. Whether causative or responsive, present hemispheric and global variations in temperature are linked to changes that occurred in the past and to changes that will occur in the future. In this chapter, we will:

- Place our modern environment within the context of Earth system processes and changes that occurred billions, millions, and thousands of years ago.
- Explore the relationship between environmental change and human evolution.
- Examine how modern human activities can affect the environment.

One of the most dramatic examples of the role of temperature in global environmental change was discovered in the geologic record of the Mediterranean Sea. The American research vessel *Glomar Challenger* set out in late summer of 1970 to drill into the Mediterranean floor and bring up long cores of the underlying material from which the formation of the basin could be analyzed.

A few years earlier, sound waves used to map the topography of the Mediterranean floor had revealed an unexpectedly hard layer about 100 to 200 m beneath the present floor. The nature of this layer remained a mystery until the cores recovered by the *Challenger* crew revealed the mineral **anhydrite** ($CaSO_4$). The scientists were stunned. Anhydrite typically forms when shallow water becomes hot and evaporates, increasing its saltiness until the mineral precipitates. How did it get to the seafloor?

The crew continued to drill in many locations. Time and time again they encountered the anhydrite layer. At one location, however, the retrieved core contained anhydrite interbedded with **stromatolites,** blue-green algae mats. Because algae are photosynthesizers and water filters sunlight, their presence indicates shallow waters. Such sequences of anhydrite and stromatolite deposits are found today in only one environment, an arid salt marsh typical of the coastal Middle East.

Shoreward of the anhydrite, the crew drilled into dolomite, a magnesium calcium carbonate mineral. Closer to the center of the basin, they hit halite (table salt). Neither mineral held any signs of life. Below the anhydrite, however, were muds containing fossilized marine organisms of Miocene age (26 million to 5 million years old) that were able to tolerate very high salt concentrations. The lowest sediments above the anhydrite layer contained only fossils of *plankton* (mostly microscopic animals and plants that float in the water column) and eventually graded upward into sediments containing fossils of both planktonic and seafloor-dwelling organisms.

All this evidence could point to no conclusion other than that the Mediterranean Sea had, during the Miocene epoch, dried out partially or completely in an event now known as the *Mediterranean salinity crisis.* In place of the sea stood a desert 6.9 million km^2 in area and 3000 m below sea level. For this to have happened, sea level must have dropped below the floor of the Strait of Gibraltar, a narrow and shallow passageway that is the only channel between the Mediterranean and the Atlantic. Such a drop in sea level in the Miocene is consistent with an episode of global cooling that tied up increasing amounts of water in the Antarctic ice sheet.

Because the evaporation rate in the area exceeds the rate of inflow from streams, the newly isolated Mediterranean slowly began to dry out. As the fossil record shows, organisms with lowest salinity tolerances were the first to die. Eventually, when the sea dried up sufficiently, all remaining organisms were killed off and the layers of salt deposited. The distribution of the salt layers also supports this hypothesis. When saline lakes evaporate to dryness, the least soluble minerals crystallize first, followed by salts with increasing solubilities. As a result, concentric "bathtub rings" of salts are deposited as lake levels recede. The Mediterranean shows just this pattern of salt deposits (Figure 13-1). Furthermore, the dried-up sea would explain how African horses, rodents, and antelopes could have migrated into Europe, and how hippopotami moved into the rivers of Cyprus from their ancestral home in the Nile.

At the beginning of the Pliocene, about 5 million years ago, Earth's temperature increased. The melting glacial ice raised the sea level, and water began to rush into the Mediterranean basin from the Atlantic. The waterfall over the Strait of Gibraltar would have been quite a sight to behold. Basing the estimate on the ages of sediments spanning the drying out and the subsequent refilling of the sea, scientists believe the flow was 40,000 km^3 per year, more than eight times the average flow of the Amazon today. As the Mediterranean basin refilled, plankton floated in with the water. Eventually, bottom-dwelling organisms crossed over.

The Mediterranean salinity crisis is but one example of the ways in which a change in Earth's temperature can trigger a host of climatic and environmental changes. Earth's history of environmental change has been pieced together over the last century from the work of countless geologists, astronomers, and biologists, and is continuously being revised by each succeeding generation of scientists. By understanding the past, we may gain insights into our future.

Earth Before Humans

Earth has existed over 4 billion years and undergone dramatic changes throughout its history. The conditions that surround us today are the result of those changes, and the future of Earth is governed by the same physical principles that governed its past. Foremost among the factors that have influenced our planet is temperature, and the theme of "fire and ice" recurs throughout Earth's history. Variations in global temperature have been controlled by numerous influences, each of which has been dominant at different times in the past. These influences include not only Earth's orbital parameters, as discussed in Chapter 12, but also Earth's internal temperature, the evolution of

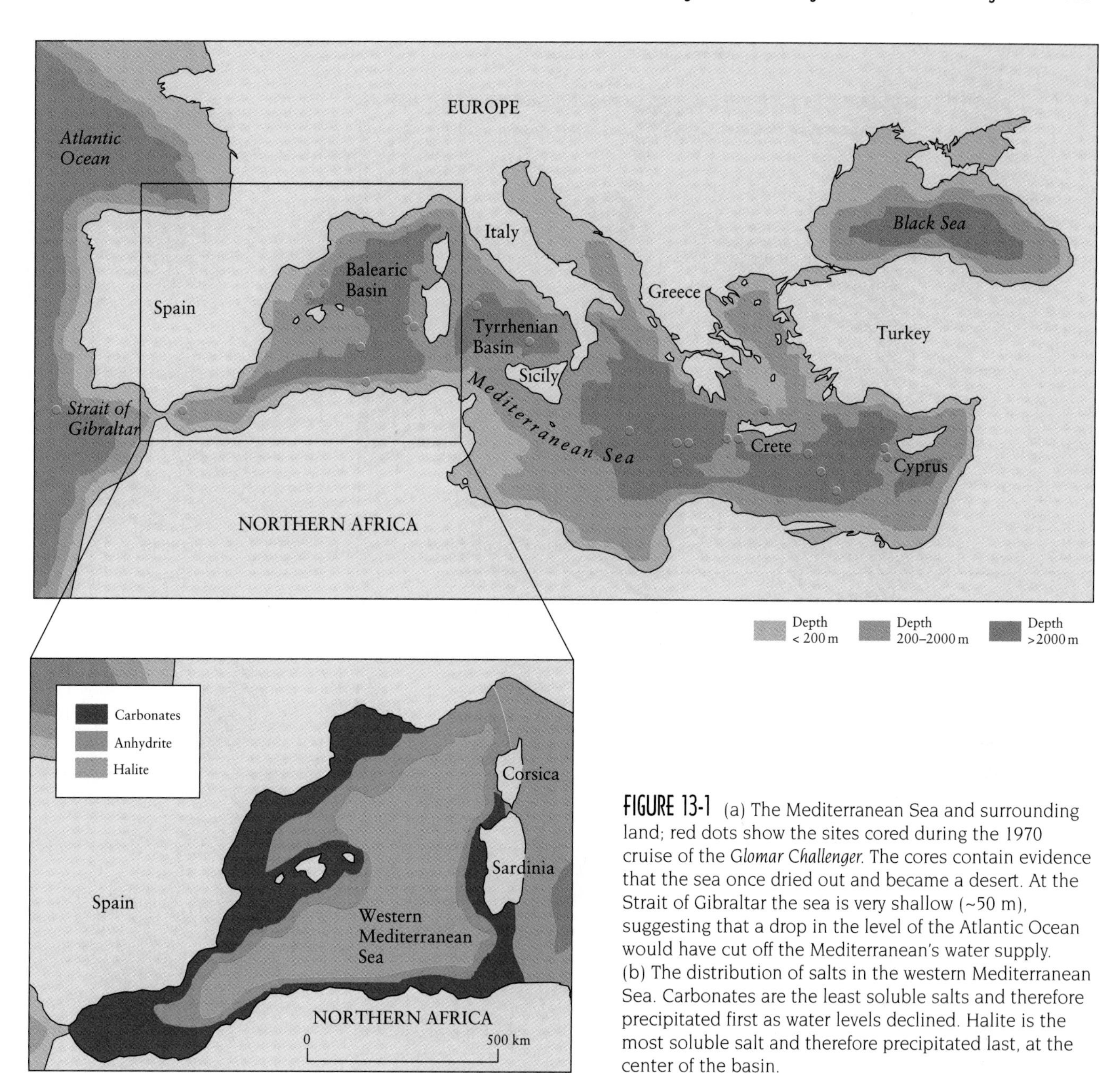

FIGURE 13-1 (a) The Mediterranean Sea and surrounding land; red dots show the sites cored during the 1970 cruise of the *Glomar Challenger.* The cores contain evidence that the sea once dried out and became a desert. At the Strait of Gibraltar the sea is very shallow (~50 m), suggesting that a drop in the level of the Atlantic Ocean would have cut off the Mediterranean's water supply. (b) The distribution of salts in the western Mediterranean Sea. Carbonates are the least soluble salts and therefore precipitated first as water levels declined. Halite is the most soluble salt and therefore precipitated last, at the center of the basin.

plants, the configuration of continents, and sudden catastrophic disturbances such as large-scale meteorite impacts. Figure 13-2 depicts how Earth's temperature changed in response to these factors from the formation of the planet to the present, and will be referred to throughout the discussion that follows.

Hothouse Earth

For its first few hundred million years, Earth probably was a "magma ocean," a boiling ball of molten rock. Occasionally the surface may have cooled enough to solidify in places, but those fragile bits of crust were easily remelted. The high heat flux came from the kinetic energy of Earth's accretion, the radioactive decay of elements, and from the gravitational energy associated with the differentiation of Earth into layers (see Chapter 2). This so-called Hadean eon lasted from about 4.6 billion to 3.8 billion years ago.

By the end of the Hadean eon, Earth's crust had begun to solidify and differentiate into dense, iron-rich basaltic rock that formed the ocean basins and lighter, more siliceous rock that rose buoyantly above the denser basaltic rock to become continental landmasses (see Chapter 4). As plate tectonics developed, heat and gases escaped from

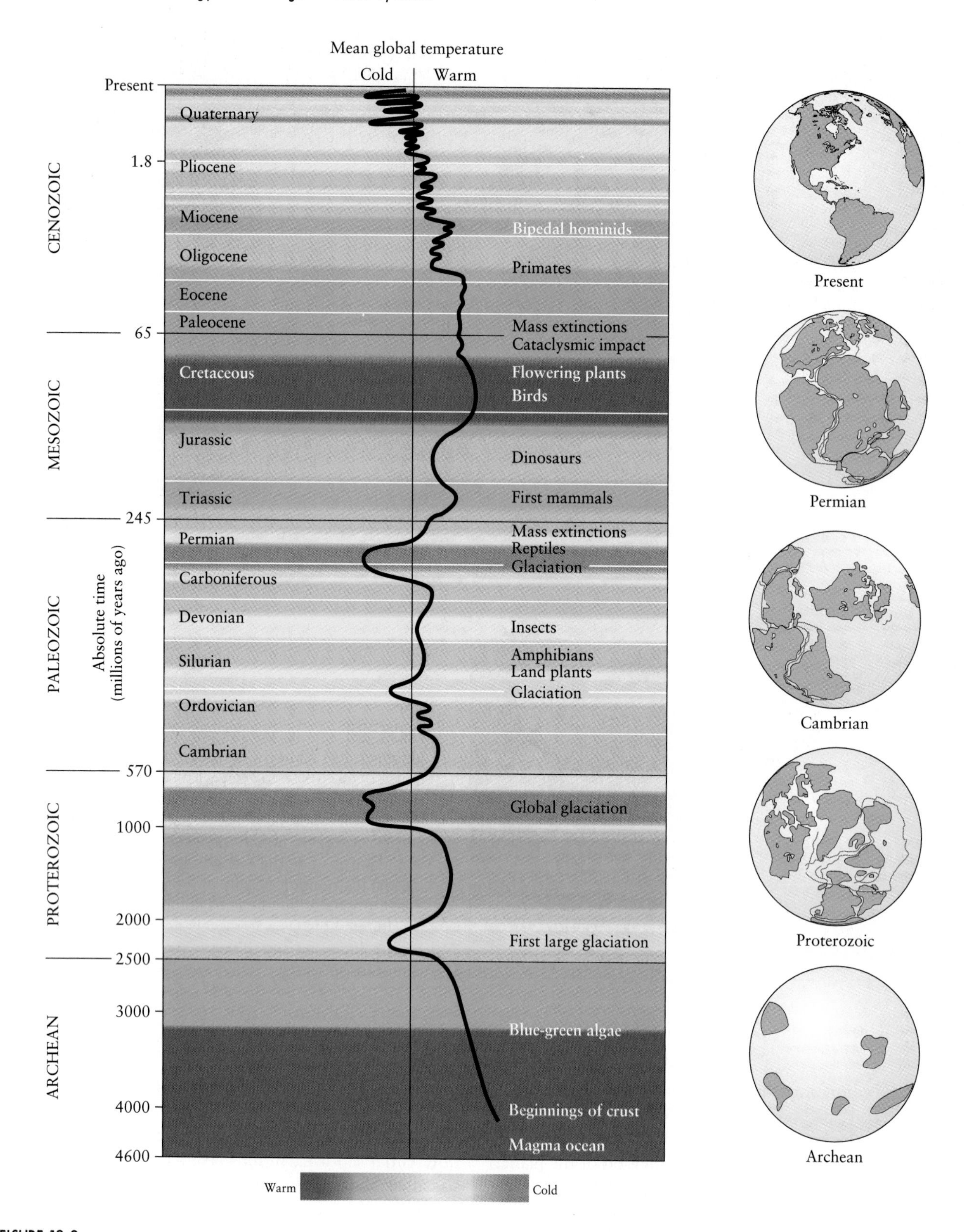

FIGURE 13-2 Early Earth was very warm because of high internal heat energy and an intense greenhouse effect. Single-celled marine plants arose by 3.5 billion years ago and probably caused a reduction in greenhouse gases that allowed glaciations to occur in the Proterozoic eon. The Paleozoic era was generally warm with the exception of two glaciations at 440 million and 305 million years ago, and the Mesozoic era was continuously warm, reaching a maximum temperature during the Cretaceous period. Since then, Earth has cooled and has experienced glacial-interglacial cycles over the last ~3 million years. Plate tectonics began to influence global climate about 570 million years ago.

(a)

(b)

FIGURE 13-3 (a) Modern cyanobacteria, or blue-green algae, flourish in warm, shallow coastal waters and sometimes cluster in oblong mats called stromatolites. (b) Proterozoic stromatolites became extinct before modern stromatolitic algae evolved, but some of the habits of the two microscopic marine plants are remarkably similar, so a similarity in temperature preferences is likely.

Earth's mantle. The gases included carbon dioxide, which made up a greater proportion of Earth's early atmosphere than it does today. They also included water vapor, which cooled and condensed into rain, filling the ocean basins. Rounded grains of sediment suggest the existence of a hydrosphere by this time. Between 3.8 billion and 2.5 billion years ago, a period known as the Archean eon, Earth's surface was dotted with tiny "microcontinents" separated by great expanses of shallow oceans (see Figure 13-2). By 3.5 billion years ago, the planet's surface was cool enough for single-celled, photosynthetic marine plants called cyanobacteria or blue-green algae to have developed in some locations.

As the planet cooled, radiation from the Sun was much less intense than it is today. If the powerful greenhouse effect of the abundant carbon dioxide in the atmosphere had not counterbalanced the low amounts of incoming solar energy, the planet would have frozen (see Box 12-1). Also contributing to surface warmth was the absence of large continental landmasses and the presence of vast oceans with very low reflectivities, which allowed more incoming solar energy to be absorbed than can be today.

Early Episodes of Global Cooling

Climatically, the Proterozoic eon (2.5 billion to 570 million years ago) was very different from the previous Archean eon. Although some ice existed on the planet from 2.7 billion years onward, not until 2.3 billion years ago did the first large glaciation develop on Earth (see Figure 13-2). This cooling might be attributable to the continued low luminosity of the early Sun combined with a weakened greenhouse effect on Earth as photosynthetic marine plants flourished and removed CO_2 from the atmosphere (see Chapter 10). Blue-green algae, clustered in mats called stromatolites, are much more abundant in Proterozoic than in Archean rocks (Figure 13-3).

From 2.3 billion to 900 million years ago, for reasons not yet understood, temperatures appear to have climbed substantially (see Figure 13-2). Large deposits of clay indicate significant chemical weathering, and hence warming, because the rate of reactions increases with increasing temperature. Furthermore, the clay in these deposits is kaolinite, a mineral found most frequently today in tropical regions.

Between 900 million and 600 million years ago came a second ice age, consisting of at least three glaciations. Paleomagnetic evidence (Box 13-1) indicates that the continents on which these glaciations occurred were at low latitudes, near the equator, where today temperatures are too warm to sustain large ice sheets. Decreased solar luminosity during the Proterozoic has been suggested as a possible cause for the glaciations, but again, it is likely that a weakening of the greenhouse effect was also necessary.

Climatic Impact of Plate Tectonics

With the advent of the Phanerozoic eon, which began 570 million years ago and in which we live today, plate tectonics processes began to dominate climatic changes. When a particular landmass moved to a new position on the globe, its climate and environment also changed. The science of paleomagnetism (see Box 13-1) has made it possible to determine whether an ancient continent was previously located at the equator, at a pole, or somewhere in between. Continents that bear evidence for glaciation

13-1 Geologist's Toolbox

Paleomagnetism

The centuries-old use of a compass for navigation is possible because Earth behaves like a giant magnet: It can attract metals such as iron. Magnetic attraction occurs along lines of force, and the area over which the force is exerted is called the magnetic field. Earth's magnetic field is believed to result from fluid flow within its molten iron outer core. This flow occurs because of heat convection within the core and because of Earth's rotation, and is thought to set in motion charged particles within the molten iron. A flow of charged particles creates an electric current, which in turn can generate a magnetic field. Like a bar magnet, Earth's magnetic field has a positive and a negative end, or a north and a south pole. Magnetic field lines emanate from the south pole, wrap around Earth, and reenter it at the north pole. Consequently, Earth's magnetic field points straight up at the south pole, straight down at the north pole, and lies parallel to Earth's surface at the equator.

Many rocks contain magnetic minerals such as magnetite (Fe_3O_4) and hematite (Fe_2O_3) that become magnetized by Earth's magnetic field and point toward its north magnetic pole. For instance, as a lava flow cools, the magnetic crystals it contains lock in the orientation of the field much the same way a compass needle becomes aligned with Earth's magnetic field. This means that if a lava flow were produced at Earth's north magnetic pole, the magnetization direction of the magnetite and hematite crystals within the lava would point straight down. If, on the other hand, the lava flow cooled at the equator, the magnetic crystals would be oriented parallel to Earth's surface. Once lava has cooled to a solid, the magnetization direction recorded in the

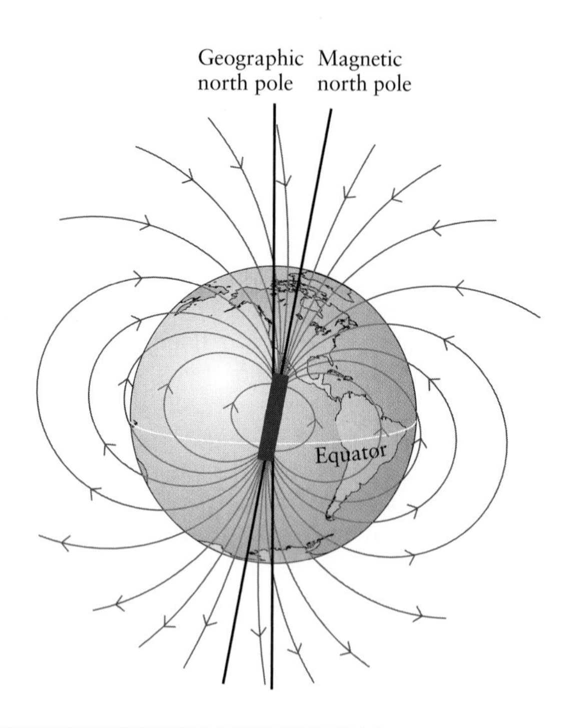

Earth's magnetic field is similar to that of a large bar magnet. Magnetic field lines come out of the south pole, wrap around Earth, and reenter it at the north pole.

in Phanerozoic rocks were commonly in a polar position at the time the glaciation occurred. Similarly, continents that have thick, ancient deposits of laterite soils (which are characteristic of the tropics) and fossilized tropical vegetation typically were located near the equator when those materials were deposited.

Furthermore, when the movement of continents redistributed most of the world's landmasses relative to the poles and the equator, the climate and environment of the entire globe changed. Continents scattered about the globe and surrounded by a warming ocean, as they were during the Archean eon, produced a climate that was moderate. Continents clumped together in a huge landmass near one of Earth's poles, as during the late Paleozoic era, produced cold terrestrial climates.

Paleozoic Climate Cycles The Paleozoic era, 570 million to 245 million years ago, was a time of great supercontinents, which periodically were inundated by shallow seas 100 to 300 m deep when rates of seafloor spreading increased (see Chapter 12). Paleomagnetic evidence tells us that most of the land in the Paleozoic era was located in the equatorial and low latitudes. Therefore, the terrestrial

crystals is locked in place. Similarly, magnetic minerals in sediments can align themselves with Earth's magnetic field before the sediments are turned into rock, and therefore the orientation of magnetic crystals in rocks indicates the direction of Earth's magnetic field at the time and place of the rock's formation.

Should plate movements relocate the rocks, the magnetic fields of the rocks will no longer match Earth's at their new location. Paleomagnetists, scientists who study Earth's magnetic field and its history, measure the magnetic field directions contained in rocks and compare them with Earth's at the same locations. These studies have led to the conclusion that many rocks have moved very long distances, in many cases thousands of kilometers, from the sites of their formation and so have confirmed the theory of plate tectonics. Furthermore, by measuring the differences in direction between the magnetic field in a rock and at Earth's surface, paleomagnetists can often pinpoint the latitude at which the rock was formed.

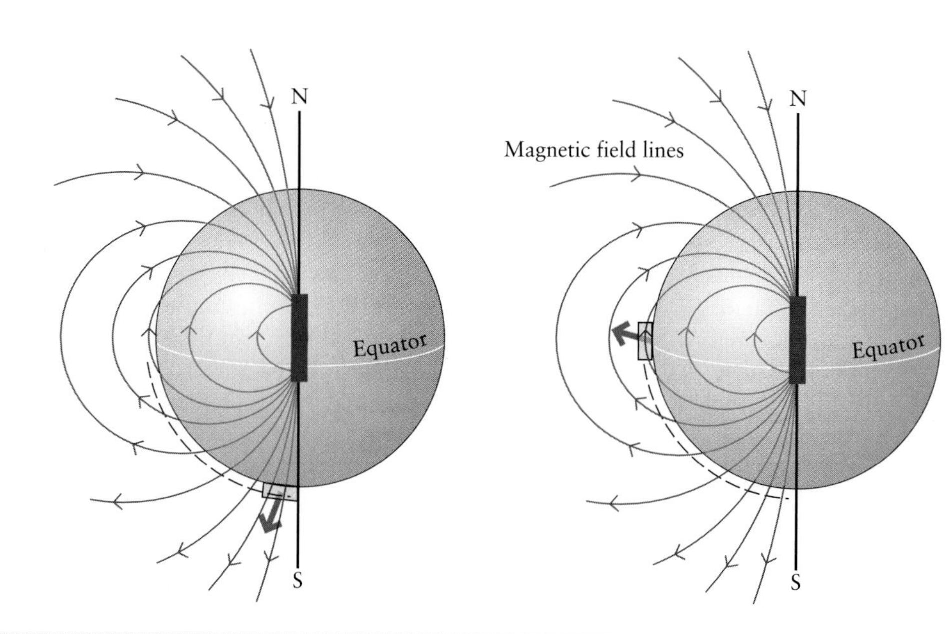

A volcanic rock formed near Earth's south magnetic pole travels to the equator with plate motion. In its new location, the direction of the rock's magnetic field, fixed when the rock cooled to a solid, no longer matches that of Earth.

climate was mild, except for two episodes of glaciation that began at approximately 440 million and 305 million years ago (see Figure 13-2) and probably stemmed from continental drift toward the poles. About 440 million years ago, for instance, what is now northern Africa became glaciated because it was situated near the south pole.

Directly after this glaciation, around 430 million years ago, land plants evolved in the Silurian period (Figure 13-4) and were followed by the first amphibians around 360 million years ago during the late Devonian period. Wide distribution of these early plants and amphibians suggests a warm climate, and indeed, much of the landmass continued to lie in the equatorial and low latitudes where it was bathed in sunlight. Warmth and wetness continued up to 286 million years ago, and swamps covered much of what would become North America and Europe. Some of the plant matter in these swamps ultimately was transformed into coal and gave the Carboniferous period its name (Figure 13-5).

With lower reflectivities than many rocks and sediments, plants would have increased the amount of incoming solar radiation being absorbed. At the same time,

FIGURE 13-4 Land plants evolved during the Silurian period, as shown in this artist's rendering.

FIGURE 13-5 Plants flourished in the swamps of the Carboniferous period, as seen in this reconstruction. Much of the coal we use today originates from this ancient vegetation.

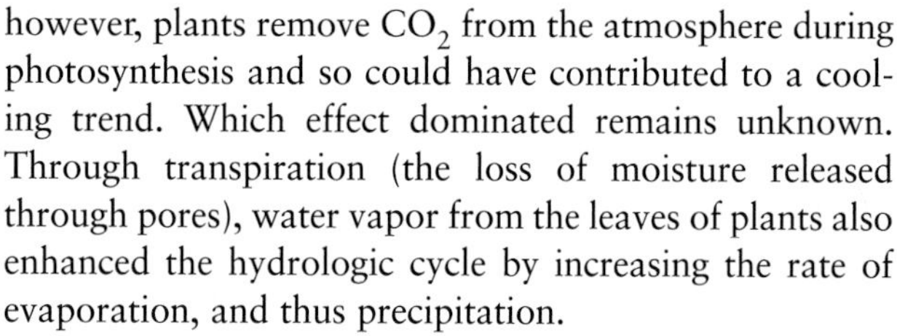

however, plants remove CO_2 from the atmosphere during photosynthesis and so could have contributed to a cooling trend. Which effect dominated remains unknown. Through transpiration (the loss of moisture released through pores), water vapor from the leaves of plants also enhanced the hydrologic cycle by increasing the rate of evaporation, and thus precipitation.

At the same time that warm swamps formed in the northern hemisphere, the southern hemisphere experienced a glaciation. About 305 million years ago, ice covered parts of **Gondwanaland,** the landmass that spread over much of the southern hemisphere, including the south pole (Figure 13-6). The glaciation lasted approximately 100 million years, but ice sheets grew and shrank repeatedly, causing sea level to fall and rise. These fluctuations in sea level are recorded in repeated cycles of sediments that are alternately marine (beach sands and offshore mud) and nonmarine (river channel and floodplain deposits). Called **cyclothems,** these sediment sequences contain much of the coal found in the United States today (Figure 13-7).

Pangaea: Mesozoic Supercontinent By the early Triassic period, 245 million years ago, all the continents on the globe had collided to form the supercontinent **Pangaea.** This supercontinent comprised the Gondwanan landmass and the landmasses that would eventually become North America, Asia, and Europe. As the landmasses collided,

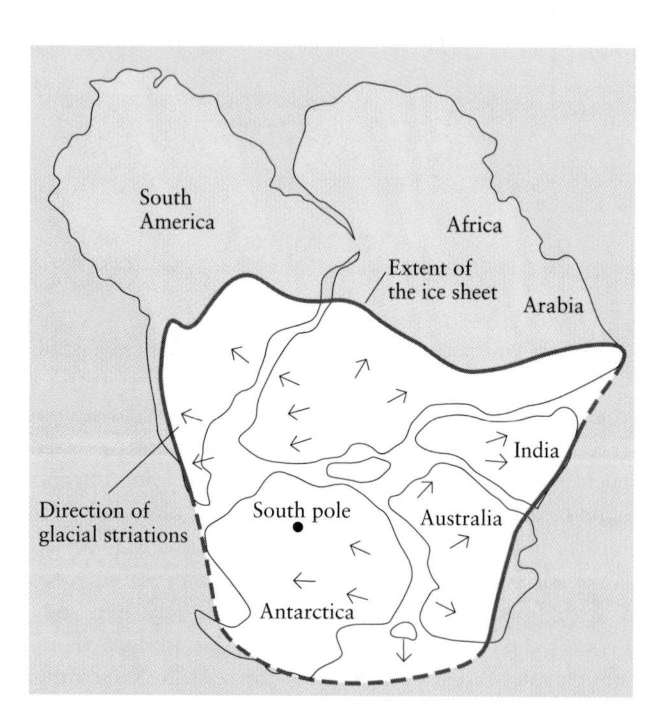

FIGURE 13-6 The supercontinent Gondwanaland experienced an extensive glaciation 305 million years ago; arrows show the direction in which the ice flowed. Alfred Wegener used the evidence of this glaciation to support his theory of continental drift. The current positions of the features resulting from the glaciation make it appear that ice flowed from the oceans onto land in many locations—a process that is physically impossible. Wegener, therefore, theorized that the continents on which this evidence was found had once been joined and covered by a large ice sheet.

FIGURE 13-7 A cyclothem sequence. The dark layers are the remains of coastal coal swamps.

inland areas became farther removed from the ocean and the interior of Pangaea experienced extreme continental climates, with differences of up to 50°C between summer and winter temperatures. Strong heating of the interior in summer caused the land to dry out. By the late Triassic and early Jurassic periods, evaporite deposits had formed where inland water bodies once stood and sand dunes accumulated where fertile soil might have developed. These ancient dunes, called **red beds** because of the reddish iron oxide that coats the sand grains, were part of a desert that rivaled the modern Sahara in size and now make up the rocks underlying many national parks in Utah, Arizona, New Mexico, and Colorado (Figure 13-8). The growth of Pangaea was accompanied by major changes in regional environments and life-forms. Mass extinctions occurred and other life-forms evolved. Dinosaurs came to prominence at this time.

In the middle Triassic, Pangaea began breaking up and the beginning of the Atlantic Ocean formed in the rift zone between the landmasses that became Europe and North America. The new ocean would eventually bring more maritime climates to the continental interiors. In the meantime, the warm trend continued during the Jurassic period, 208 million to 144 million years ago. Scientists have determined from oxygen isotopes in Jurassic fossils of squidlike organisms called belemnites that average temperatures at 75° south latitude in the seas just off Antarctica reached 14°C, or roughly 7°C higher than average temperatures today. Rifting continued until the Cretaceous period, 144 million to 65 million years ago, when North America finally pulled completely away from Europe and Africa.

Plate tectonics influences Earth environments not only by determining global geography but also through rates of tectonic processes. Recall from Chapter 12 that the Cretaceous period was marked by very high temperatures, attributed to an increase in seafloor-spreading rates and accompanying increases in atmospheric CO_2 concentration as volcanic gases issued from the spreading centers. Maximum warming occurred from 120 million to 90 million years ago (see Figure 13-2).

Meteorite Impacts and Environmental Change

The Cretaceous period and Mesozoic era ended with a cataclysmic event that brought about the extinction of dinosaurs and the ascendancy of mammals. The cause of the extinction (in which more than 50 percent of existing species perished) has been hotly debated for decades. One possibility is that the massive volcanism that created flood basalts in India at the time disturbed the environment enough to cause mass extinctions. However, the evidence now weighs in favor of the hypothesis that a huge meteorite or cluster of meteorites hit Earth 65 million years ago.

A buried impact crater 65 million years old has been identified just offshore Mexico's Yucatán peninsula. An impact powerful enough to gouge out such a crater would have vaporized the meteorite into a dust that would eventually settle in layers around the globe. Indeed, at sites throughout the world, a layer of clay that age has been

FIGURE 13-8 Triassic and Jurassic red beds near Zion National Park, Utah, are the remains of desert dunes.

(a)

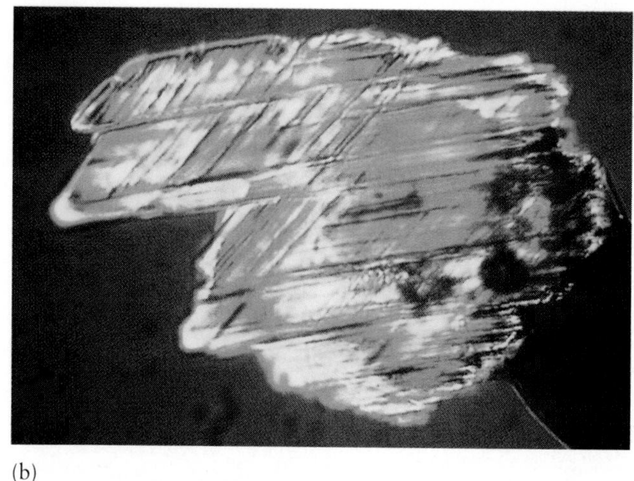
(b)

FIGURE 13-9 Evidence of a climate-changing meteorite that crashed to Earth near Mexico 65 million years ago has been found consistently in sediments of that age at sites throughout the world. (a) A clay layer formed 65 million years ago at Gubbio, Italy, contains 30 times as much iridium as the layers above and below it. Iridium at Earth's surface usually arrives in cosmic dust and meteorites. (b) An impact-metamorphosed quartz grain taken from iridium-rich sediments shows the internal fractures that can be produced only by the high pressures of meteorite impacts and nuclear explosions.

found containing unusually high quantities of **iridium,** an element common to certain types of meteorites but very rare in Earth materials [Figure 13-9 (a)]. Other metals such as rhodium, chromium, scandium, osmium, and titanium also are common constituents of meteorites and are present in unusually high concentrations in this sedimentary layer. North of the crater, in North America and southern Europe, some iridium-rich sites also contain impact-metamorphosed quartz grains, characterized by many internal parallel fractures that can be produced only by extremely high pressures and that are found only in connection with meteorite impacts and nuclear explosions [Figure 13-9(b)]. Large quantities of carbon, suggestive of global wildfires, are also present.

If such an impact did occur, its effects on the environment would have been catastrophic. The impact would have generated a massive fireball and winds in excess of 1200 km per hr. Much vegetation (as well as animal life) probably ignited, creating high concentrations of deadly carbon monoxide. The impact would have vaporized the meteor and the impact site, sending huge quantities of dust into the troposphere and stratosphere. The heat of the fireball would have enabled nitrogen, oxygen, and water vapor in the atmosphere to combine to form nitrogen oxides and nitric acid. Highly acidic rains would have caused the deaths of plant and aquatic animal species. Acid rain may also explain the evidence we have of enhanced chemical weathering at the time.

Land animals would have suffocated from the carbon monoxide and the noxious nitrogen oxides produced by the fireball. Also, the nitrogen oxides would have reacted with ozone in the stratosphere, thus removing Earth's protective shield against cell-damaging ultraviolet radiation. In the aftermath of the impact, the dust ejected into the atmosphere would have reflected much incoming solar radiation, causing both a decline in photosynthesis and abrupt global cooling. This cooling probably gave way to warming within a matter of years as the dust settled and the greenhouse effect of the increased carbon dioxide in the atmosphere from burning of the terrestrial biomass became dominant (Figure 13-10).

The Last Great Ice Age

Following the Cretaceous period, for reasons not yet understood, the Cenozoic era (65 million years ago to the present) remained quite warm for its first 27 million years or so. Sediments shed from Antarctica at this time were rich in kaolinite and smectite, clay minerals indicative of chemical weathering in a warm environment. Crocodiles ventured as far north as the Arctic Circle, and the high northern latitudes supported a temperate-zone forest of conifers and hardwood species. At present, these latitudes are covered by *tundra,* treeless plains bearing low grasses and shrubs characteristic of a very cold environment.

Why global climate remained so warm in the Eocene, long after accelerated seafloor spreading in the Cretaceous period had slowed and after the warming seas had receded, remains a mystery. In the Eocene, however, there began a sequence of one-way geographic and biologic changes that mark the onset of a prolonged trend in global cooling during the Cenozoic era (see Figure 13-2). After the Eocene, the rate of Cenozoic cooling changed from time to time, but the overall trend continued, favoring the rise of mammals, and driving Earth into the last great ice age.

Ocean Gateways and Plateau Uplift About 38 million years ago, Antarctica and Australia separated. Oxygen isotopic ratios in fossil foraminifera show an increase in the percentage of ^{18}O relative to ^{16}O in ocean water (see Chapter 12), suggesting that significant cooling began at this time. Sometime between 40 million and 30 million years ago, South America also split away from Antarctica, opening up the Drake Passage and allowing the establishment of the Antarctic Circumpolar Current, which isolated the continent from warm equatorial currents. Antarctic sea surface temperatures dropped by 10°C, leading to growth of the Antarctic ice sheet. Furthermore, the Transantarctic Mountains were uplifted at this time, further contributing to ice sheet growth. Ice-rafted debris around Antarctica first appears in marine sediments of this age.

During the Miocene, 26 million to 5 million years ago, the extent of the Antarctic ice sheet fluctuated widely. Initially it expanded, causing sea level to drop by 50 m and isolating the Mediterranean Sea from the Atlantic Ocean. The Mediterranean dried out and deposited a layer of salt 70 m thick on the floor of the basin. Time and time again sea level rose, refilled the Mediterranean, and then fell, isolating the Mediterranean and allowing it again to dry out. Eventually, a package of salts 2–3 km thick was deposited on the floor of the Mediterranean basin, perhaps 6 percent of the total dissolved salts once contained within the world ocean.

While the Antarctic ice sheet was expanding and contracting, the northern hemisphere appears to have remained consistently warm and ice free until about 3.2 million years ago. What finally caused the northern hemisphere to experience glaciations long after the Antarctic ice sheet formed? One hypothesis is that the uplift of the Tibetan and Colorado plateaus over the past 10 million to 5 million years may have led to increased chemical weathering that locked atmospheric carbon dioxide into marine sediments and thereby led to cooling and ice-sheet growth (see Box 6-1).

Biological Controls on Global Cooling Geologists have generally cited geographic modifications, tectonic forces, and weathering to explain persistent changes in atmospheric CO_2 concentration and subsequent climatic

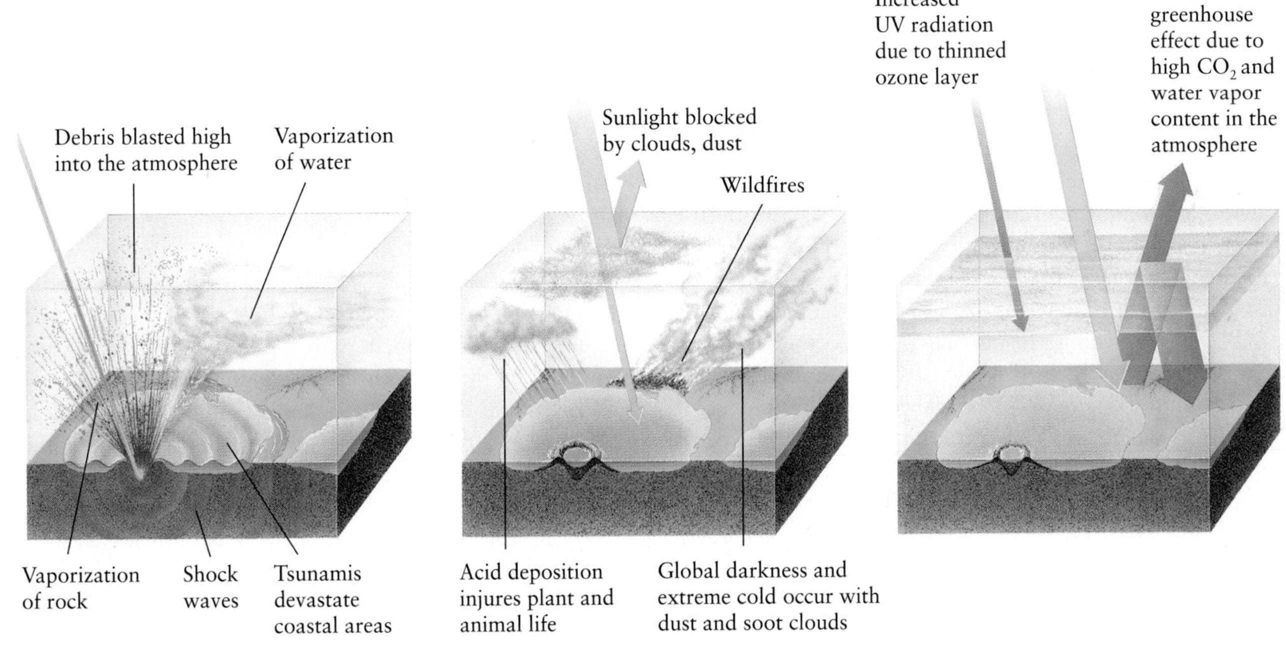

FIGURE 13-10 An impact great enough to have gouged out the crater discovered off Mexico would have had catastrophic effects. (a) Immediately on impact, the meteorite and a substantial amount of Earth material would have vaporized into the atmosphere. Vibrations from the impact would have touched off devastating earthquakes and tsunamis. (b) Residual heat from the impact would have started global fires, adding soot to the vaporized impact particles and further blocking out sunlight. The resulting global chill and acid rain would have wiped out many life-forms. (c) When the dust settled and sunlight could get through again, an enhanced greenhouse effect caused by residual carbon dioxide and water vapor would have caused a sudden warming, and destruction of ozone would have made the atmosphere increasingly transparent to cell-damaging ultraviolet radiation.

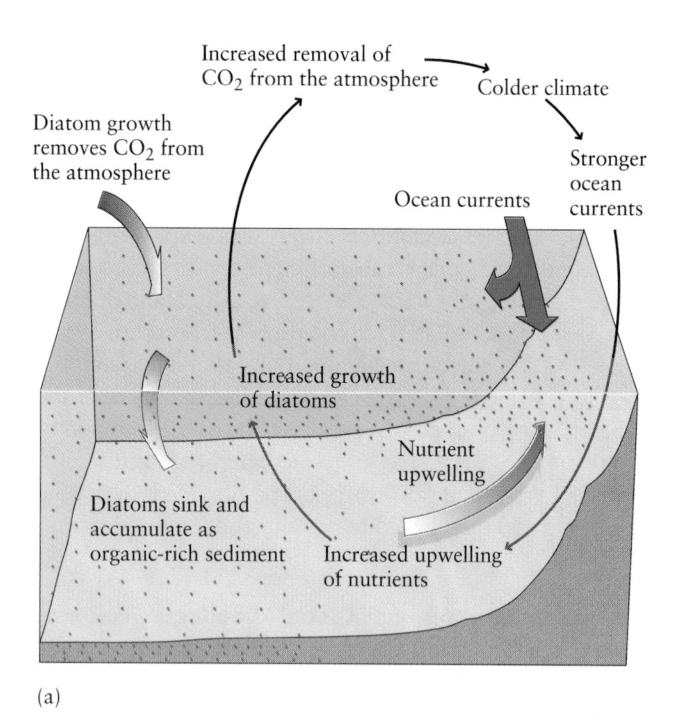

(a)

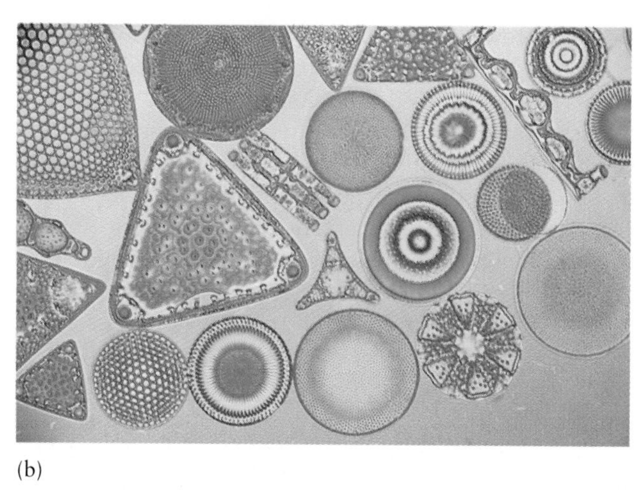

(b)

FIGURE 13-11 (a) The explosive growth of diatoms in the Cenozoic may have lowered the level of atmospheric CO_2 by photosynthesis and thus led to global cooling. (b) Diatoms are single-celled aquatic plants that secrete highly ornamented silica shells, as in these fossil diatoms from New Zealand.

changes. A controversial hypothesis, however, calls on the biosphere to account for Cenozoic cooling. This hypothesis points to the abundant growth of single-celled aquatic plants called diatoms around the rim of the Pacific Ocean, starting in the early Cenozoic and continuing into the Pleistocene. The hypothesis argues that the growing diatom population removed increasing amounts of CO_2 from the ocean and atmosphere and transferred it to the seafloor, where it exists as buried organic carbon. The growth of diatoms around the Pacific Rim is favored by the upwelling of nutrients, a process driven by ocean currents, which become stronger under colder global climates (Figure 13-11). Hence a feedback system is set up in which growth of diatoms lowers CO_2, promoting colder temperatures, leading to stronger ocean currents, more upwelling, and more diatom growth. The result could be a one-way cooling trend that may have led to the Pleistocene Ice Age. Thick sedimentary deposits rich in diatoms, such as the Monterey Formation (see the photograph that opens Chapter 3), provide some evidence for this hypothesis.

The gradual cooling that has occurred over the latter half of the Cenozoic, whatever its cause, has been accompanied by a gradual trend of increasing aridity. The evidence is that many jungles have been replaced by grasslands; increased amounts of surface particles have been blown about and deposited by winds; decreasing amounts of kaolinite, an indicator of wet conditions, have been deposited in marine sediments; and increasing amounts of charcoal have been found in marine sediments, suggesting greater frequency of fires.

Earth at the Last Glacial Maximum North Atlantic and North Pacific ocean sediments record the first presence of ice-rafted debris in the northern hemisphere 3.2 million years ago. Oxygen isotopes in fossil foraminifera are consistent with an ice sheet on Greenland that reached roughly one-quarter to one-half its current size by 2.4 million years ago. Today there are ice sheets only on Antarctica and Greenland, but from approximately 3 million years ago to the present, Earth's climate fluctuated between interglacial conditions, such as we experience today, and full glaciations in which North America and Asia were blanketed by thick ice sheets (see Figure 13-2).

What might Earth have looked like during one of these glaciations? If we were able to travel backward in time through the relatively stable current epoch, the Holocene, and into the last glacial maximum, which lasted from 22,000 to 14,000 years ago, we would find ourselves in a very unfamiliar environment. An ice sheet 18 million to 35 million km^3 in volume, known as the Laurentide ice sheet, extended as far south as present-day Illinois, Indiana, and Ohio and as far west as the Rocky Mountains. West of the Rockies, the Cordilleran ice sheet covered British Columbia and part of Alaska. Asia and Europe were covered by the massive Eurasian ice sheet. Smaller ice caps formed in the Rocky Mountains, the Sierra Nevada, the Andes, and the Himalayas. Ice and snow covered roughly 30 percent of Earth's land surface, compared with 10 percent today (Figure 13-12). The amount of water removed from the ocean basins to produce the glacial ice lowered sea level to between 120 and 160 m below its current level, exposing large areas of continental shelf.

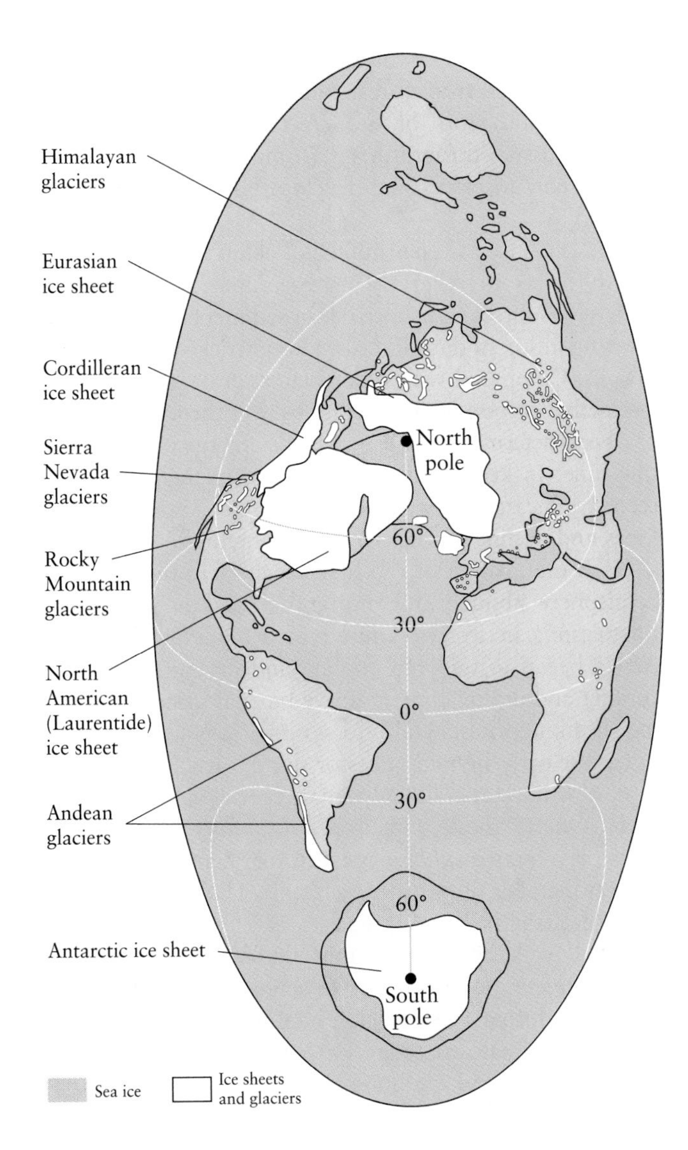

FIGURE 13-12 The extent of glaciers, ice sheets, and sea ice during the last glacial maximum.

Rivers emptying into the receding oceans etched the deep canyons found in the shelves today.

The positive feedback mechanism (see Chapter 2) whereby the highly reflective ice chills the atmosphere and the increasingly cold atmosphere encourages ice formation led to a net global cooling of 5° to 7°C. Local temperature drops may have ranged from 20°C for the atmosphere directly above the ice sheets to 2°–4°C at more tropical latitudes. Cold-loving spruce trees flourished as far south as the latitude of Alabama and Georgia until warming trends forced their northward migration to Canada starting about 12,000 years ago (see Chapter 12).

Climate-model experiments suggest that the Laurentide ice sheet deflected atmospheric currents and associated storms far to the south of their modern positions, adding significant amounts of moisture to what is now the southwestern United States. Such an influx of moisture would help to explain the development of lakes in this now-arid region. Modern-day remnants of these lakes include the Great Salt Lake in Utah, Pyramid Lake in Nevada, and Mono Lake in eastern California. Their shorelines suggest that these lakes were roughly 10 times larger during the last glacial maximum than at present. In contrast to the wet climate of the southwest, the area of the southeastern United States became more arid, as cold, dry winds blew southward off the ice sheet. In addition, cold, dry winds from the Eurasian ice sheet blew across northern Africa, causing lake levels to drop and deserts to expand.

Meltwater coming from the Laurentide ice sheet diluted the water of the North Atlantic, slowing or completely shutting off formation of North Atlantic Deep Water. With the sinking of cold water nearly halted, the entire thermohaline circulation of the ocean slowed or stopped, cutting off the supply of heat to Europe (see Chapter 10). France became covered in tundra, and Spain experienced a decline of nearly 15°C in the surface temperature of its coastal waters.

In the southern hemisphere, however, the Antarctic ice sheet expanded and sea level dropped, exposing more land on which to grow ice. The ice sheet, today about 24 million km^3 in volume, is thought to have grown to a maximum size of about 37 million km^3 during the last glacial maximum. Also in the southern hemisphere, ice sheets grew in Patagonia and New Zealand, and small glacial systems developed in Australia.

The ratios of oxygen isotopes in marine fossils indicate that the aggregate of warm intervals such as the present climate make up a mere 10 percent of the last few hundred thousand years (see Chapter 12). Most of the Pleistocene was a time of exceedingly cold and variable temperatures. What conditions caused the northern hemisphere to experience repeated episodes of glaciation? The periodicity with which these episodes occur suggests that they are related to the amount of solar energy reaching Earth's surface as a consequence of changes in its orbit, the so-called Milankovitch cycles (see Chapter 12).

- An early Earth too hot for crust formation gave way to an ice age during the Proterozoic eon, when the cooled surface experienced the combined effects of low-intensity solar radiation and a photosynthesis-related reduction in greenhouse gases.

- During the Paleozoic era, the continents repeatedly coalesced into giant supercontinents that were inundated by shallow seas and gave rise to coal-forming swamps. Climate was generally warm and wet but occasionally was interrupted by ice ages when continents migrated toward the south pole.

- Pangaea, the last supercontinent, split apart during the Mesozoic era. Accelerated seafloor spreading released much CO_2 into the atmosphere, resulting in an enhanced greenhouse effect and a nearly ice-free Earth in the Cretaceous.
- Global temperatures fell during the Cenozoic era, and an ice sheet formed on Antarctica after it separated from Australia and South America. Fluctuations in the size of this ice sheet, which began in the Miocene, led to repeated drying out and refilling of the Mediterranean Sea as the level of the Atlantic fell and rose. In the late Pliocene, ice began to form in the northern hemisphere, and glaciation occurred in response to Earth's orbital cycles.
- Warm periods, such as the one in which we presently live, represent only 10 percent of the geologic record of the last 500,000 years.
- During the last glacial maximum, about 20,000 years ago, ice covered 30 percent of the land surface, sea level dropped between 120 and 160 m, and average global temperature was 5° to 7°C cooler than today. A positive feedback mechanism developed between the highly reflective ice and the atmosphere, increasing the rate of ice formation.

Humans and Environmental Change

The trend toward global cooling in the Cenozoic and the strong and abrupt changes in environments that characterized the Pleistocene took place at a time when primate and human lines of evolution and culture were changing rapidly. The first primates (a biological classification that includes humans, apes, and monkeys; see Appendix 1) arose about 55 million years ago. Some anthropologists believe that subsequent climate changes led to the evolution of our human ancestors from knuckle-walking, apelike creatures to bipedal animals (Figure 13-13). More recent climate changes, while not causing evolutionary changes in humans, have led to cultural changes.

Aridification of Africa and Primate Evolution

Modern apes live primarily in forested regions and use their powerful arms to swing between tree branches. To walk on the ground, they generally bend toward and distribute their weight between their feet and their knuckles. Most of Africa initially was covered by forests, but about 7 million years ago, climatic changes brought about a gradual transformation of the dense forests to a mixture of forests and savannas, or grasslands. Upright walking requires much lower expenditures of energy than does knuckle-walking. Some scientists have proposed that the upright gait of humans evolved as an adaption to these new environmental conditions, enabling our human ancestors to travel from forest to forest across the savanna while the ancestors of modern apes remained in the forests. Fossilized footprints 3.75 million years old show that human ancestors had developed a fully bipedal gait by then.

The evidence for environmental change in Africa from moist forests to relatively dry savannas comes from a variety of sources. Offshore ocean sediments record an increasing influx of dust and siliceous particles derived from grasses. Compared with other plants, grasses also have a distinctive carbon isotopic composition through which it has been determined that grasslands emerged in Africa at this time. In keeping with the habitat change, the fossil record indicates an increase in grassland-dwelling antelopes and decline in forest-dwelling antelopes.

With the growth of the first ice sheet in the northern hemisphere about 3 million years ago and the subsequent fluctuations in its size caused by Milankovitch cycles, Africa began to undergo cooling and drying cycles. The cooling and drying climate drove bipedal primates out of the shrinking forests and into savanna grasslands. At this time, our early human ancestor disappeared and at least three new upright-walking primate species came into existence. One of these, named *Homo habilis* for the stone tools they created (*habilis* is Latin for "handy"), was the earliest member of our genus, *Homo*. The other two species eventually became extinct.

Fossil evidence suggests that the first individuals of *Homo sapiens*, our species, arose sometime about 125,000 years ago during the last large interglacial period, a time similar in climate to today or perhaps slightly warmer. These humans had the bone structure and cranial capacity of modern humans. The interglaciation lasted about 10,000 years, after which climate gradually became colder and ultimately reached the extreme conditions of the last glacial maximum. As glaciation progressed and sea level dropped, exposed land bridges connected the British Isles with the continent of Europe, the island of New Guinea with Australia, and North America with eastern Asia. It was at this time, anthropologists believe, that the first humans migrated into North America.

The Holocene Epoch and Human History

After the last glacial maximum, the ice sheets covering North America, Europe, and Asia began to retreat. By roughly 10,000 years ago, the amount of ice on Earth was about what it is today. Since then, Earth has undergone weaker and shorter episodes of warming and cooling. The maximum of interglacial warming apparently was reached about 6000 years ago, during a warm interval that lasted from 8000 to 4000 years ago. Rainfall decreased in western North America, and most of the lakes dried up. In northern Africa, however, rainfall increased and lakes were filled during the middle Holocene. These opposite responses in different regions of the world reflect

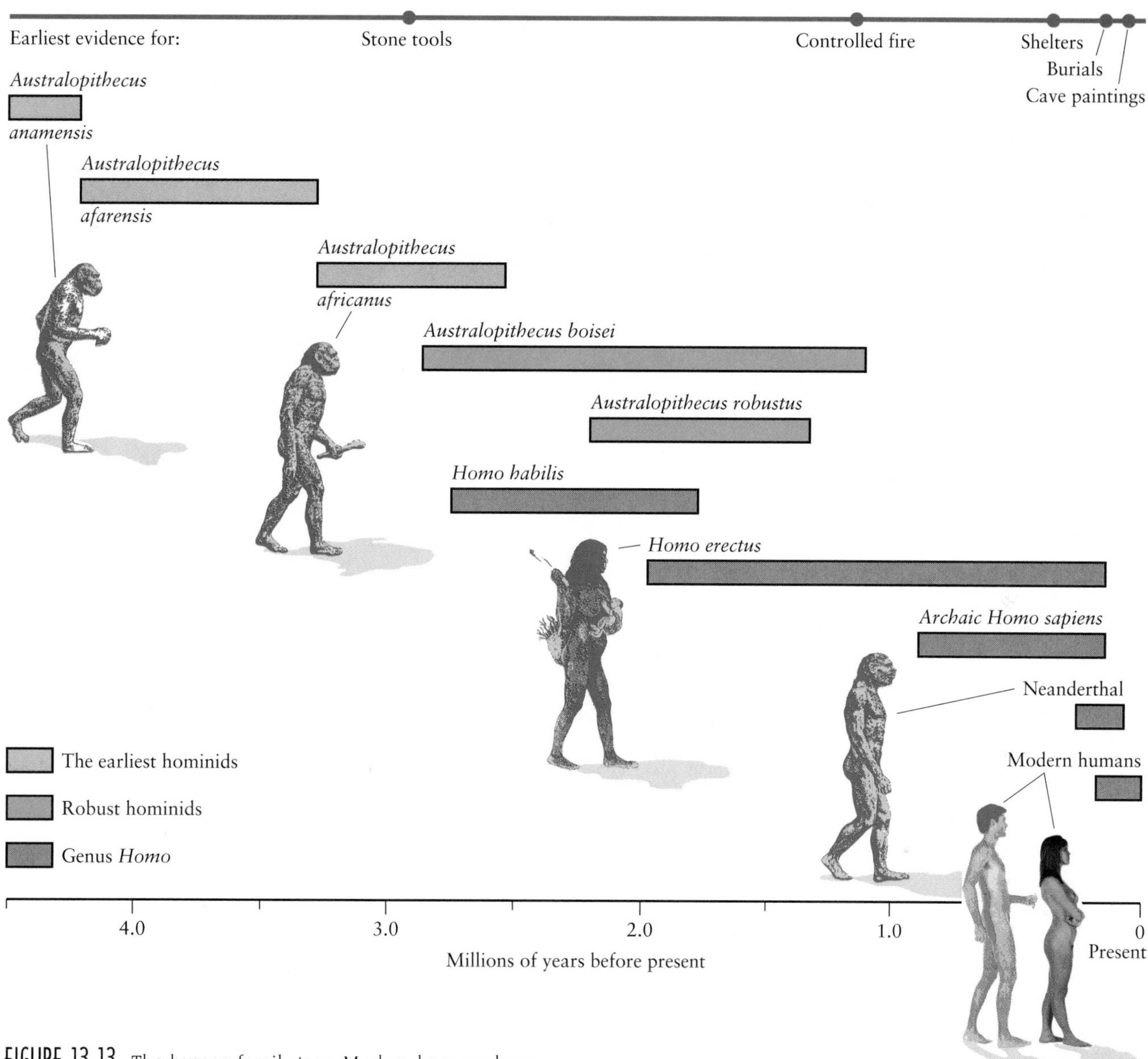

FIGURE 13-13 The human family tree. Modern humans have existed for about 125,000 years.

major changes in atmospheric circulation and in the supply of moisture reaching the interiors of continents. Such responses must have been of great consequence to developing human populations, requiring major adjustments to rapidly changing temperatures and amounts of precipitation.

The Medieval Warm Period

After the mid-Holocene warming in temperate latitudes, temperatures in Europe fell, but a warming followed that coincided with the rise of the Roman empire. The fall of Rome coincided with a return to cold conditions, which lasted from AD 500 to 1000. Beginning about 1100, the climate became warmer than it is at present in many parts of Europe. For 200 years, sea ice was greatly reduced and Vikings settled Greenland. Grapes grew in England, and glaciers retreated in Switzerland and Austria. Spruce forests in Canada migrated northward along with the cooler climes. This episode, known as the **Medieval Warm Period,** coincided with general prosperity of the human population. Similar warmings took place a little earlier, about 600 to 1000, in China and Antarctica.

The Little Ice Age

Following the Medieval Warm Period, glaciers advanced around the globe in what has come to be known as the **Little Ice Age.** This event, which lasted from the late 13th century to the late 19th century, actually consists of a pair of cold snaps (roughly the 1300s through the 1600s, and the 1800s) with an intervening warmer period. These cold snaps did not occur at the same time in all locations, but they did come to all parts of the globe. There are

(a)

(b)

FIGURE 13-14 The Argentière glacier in the French Alps. (a) Note the extent of the glacier in this 1850 etching, made during the Little Ice Age. (b) The glacier had shrunk markedly by the time this photograph was taken in 1966.

many sources of evidence for the Little Ice Age: historical records of temperature; maps, photographs, and landscape paintings showing advancing glaciers (Figure 13-14); observations of increased annual snow cover and freezing of lakes and rivers; and moraine and tree-ring records. One scientist even compiled statistical information on landscape paintings made before, during, and after the Little Ice Age and found that paintings created during the Little Ice Age contained the highest proportion of clouds and dark skies.

The Little Ice Age caused great social upheaval, particularly in Europe. The Vikings, who had settled Greenland during the Medieval Warm Period, found their marine trading routes choked with sea ice. Eventually their culture died out as trade and farming became impossible. Glaciers advanced across hunting trails used by the Inuit, the indigenous population of Greenland. Iceland suffered crop failures and famines as sea ice enveloped the country. Farmers in northern Norway abandoned their fields after successive years of failed harvests. In the Netherlands, canals vital for transporting goods between cities began to freeze in winter (see the illustration that opens Chapter 9). All these catastrophic changes resulted from average global temperature declines of a mere 1° to 2°C.

What caused the Little Ice Age? Because temperature drops occurred at different times in different locations, that question is very difficult to answer. However, ice cores indicate that elevated concentrations of dust and acidic droplets were deposited on the Greenland and Antarctic ice sheets during this time, so perhaps the Little Ice Age was caused by a period of frequent volcanism. If temperatures were lowered by ash eruptions sufficiently for new ice to form on the continents and in the oceans, the positive feedback mechanism between ice and atmosphere would have enhanced ice growth. Another possible cause of the Little Ice Age is a decline in output of solar energy. A decrease in solar radiation of only 1 percent would be sufficient to cause a 1° to 2°C drop in temperature on Earth, and during the 1600s the radiation given off by the Sun fell to a historically low value.

Whatever the cause of the Little Ice Age, the period illustrates how small changes in global temperature can have profound impacts on the environment and on human societies. Today we face similar climatic changes, some of them produced by the natural cycles Earth has been experiencing for millennia and others induced by human activities. Like the algae of the Proterozoic and the diatoms of the Miocene, we are altering atmospheric composition and thereby modifying the environment.

- The transition from knuckle-walking to bipedality in primates coincides with the aridification of Africa and the emergence of grasslands separating forests.
- Humans spread across the world by crossing land bridges exposed when sea level fell during the last glacial maximum.
- The Holocene represents the latest interglacial period in a cycle of repeating glaciations. The warmest part of this period occurred about 6000 years ago and caused the drying out of lakes that had formed during the last glacial maximum in what is now the western United States.
- Over the last few thousand years, temperatures have oscillated 1° to 2°C. Warm intervals, such as the Medieval Warm Period, brought prosperity. Cold intervals, such as the Little Ice Age, brought famines.

Humans As Agents of Environmental Change

In recent years, humans have surpassed natural processes as agents of environmental change. With bulldozers and backhoes, we cut into hillsides for construction or mining and greatly accelerate the natural process of hillslope evolution that would have occurred solely through running water and mass wasting. In our transportation systems and our production of electricity, we generate carbon dioxide, sulfur dioxide, nitrogen oxides, and other gases and particles that can alter atmospheric chemistries. By constructing dams for flood control and the generation of electricity, we change sediment transport rates in streams, leading to loss of wildlife habitat in some locations and growth of new habitat in others. Our farming practices bring about drastic environmental change: Growing rice in the arid central valley of California and cotton in parts of Arizona require that water be pumped from hundreds of miles away to irrigate wet-climate crops in the desert. Environment-damaging practices can have far-reaching, unintended consequences. Among those that we have experienced already or can foresee are global warming, metal poisoning, and the destructive aspects of deforestation.

Global Warming

Earth's surface is warmed by a group of atmospheric gases known as greenhouse gases, most of which—carbon dioxide, methane, water vapor, nitrous oxide, and ozone—have both natural and human-made sources. Chlorofluorocarbons (CFCs), however, are produced solely by humans, and are so damaging to the ozone layer that an international agreement was made to ban their production (see Chapter 9). CFC emissions have been significantly reduced, but CFC molecules may remain in the atmosphere well beyond the year 2050.

Since the Industrial Revolution, combustion of fossil fuels for electricity and transportation has led to great increases in atmospheric CO_2 concentrations (Figure 13-15). Every year 7 billion to 9 billion metric tons of CO_2 enter the atmosphere as a result of human activities. At the same time, deforestation is destroying one of the greatest sinks for atmospheric carbon—trees. Today CO_2 concentrations are 25 percent higher than in preindustrial days.

Methane concentrations have increased by 100 percent primarily because of expanding rice cultivation and cattle ranching. Rice cultivation requires swampy conditions, which release methane, while cows and other ruminants release methane in intestinal gas. Ozone is accumulating in the troposphere because the extreme heat generated in automobile engines splits oxygen molecules into highly reactive atoms (see Chapter 9). Nitrogen oxides result from the oxidation of fertilizers applied to farmlands and from industrial processes such as nylon production.

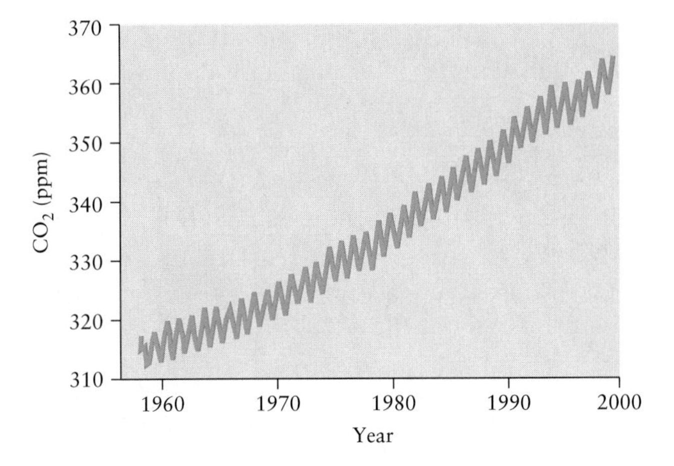

FIGURE 13-15 Since 1958, scientists have been monitoring atmospheric CO_2 concentrations at a volcanologic observatory on Mauna Loa in Hawaii, far from industrial and urban pollution. A steady increase in CO_2 concentration over the decades is due to industrial and agricultural activities and the use of motor vehicles. The oscillations in the curve reflect the annual "leafing out" of trees in the northern hemisphere.

While not nearly as abundant as carbon dioxide, these other gases are much more effective heat absorbers and cause greater temperature changes than the same volume of CO_2. Molecule for molecule, methane is 25 times as efficient and CFCs are 20,000 times as efficient as CO_2 at trapping heat. Although emissions of these other gases are low compared with CO_2 emissions, together they almost equal the warming effect of total CO_2 emissions (Figure 13-16).

In the last few decades, scientists have become increasingly aware that the human-induced increases in greenhouse gases have the potential to cause a significant warming of the planet. Greenhouse gases have residence times in the atmosphere ranging from 50 to 200 years, so even if human-induced emissions were stopped today, they would linger in the atmosphere for several generations. Long-term warming could have devastating consequences. First, it could change the distribution and amount of precipitation on the continents. Eventually, the polar ice caps could melt and cause a rise in sea level.

Depending on the climate model used to compute global temperature changes, a doubling of CO_2 may warm Earth's surface 2° to 5°C on average. Small shifts in average global temperature can cause significant environmental changes and social upheaval, as occurred during the Little Ice Age. Furthermore, this projected temperature increase may be an underestimate. If all the fossil fuels on Earth were used in the near future, atmospheric CO_2 could reach eight times the modern values and cause a 6° to 10°C rise in global temperature, making Earth as warm as it was during the Cretaceous, one of the warmest periods in the history of Earth. Finally, if greenhouse

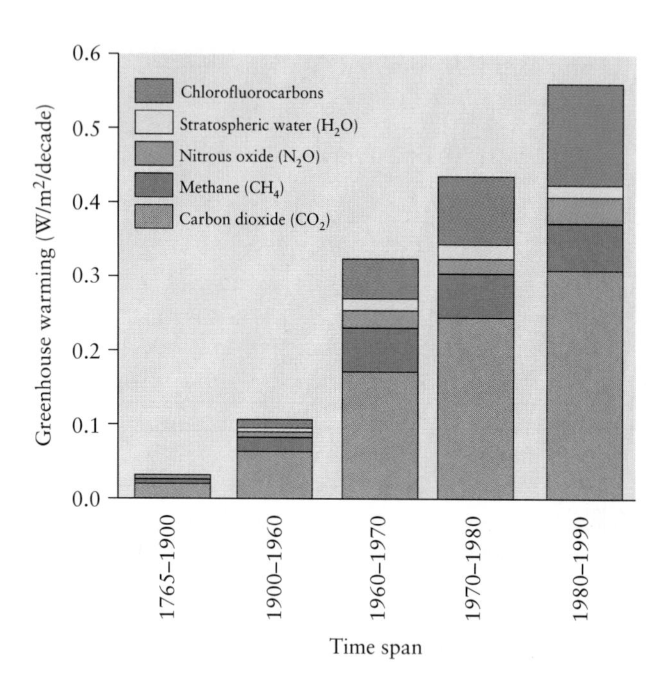

FIGURE 13-16 Concentrations of all greenhouse gases have risen since the advent of the Industrial Revolution. While carbon dioxide continues to produce the majority of greenhouse warming, the combined warming produced by the other gases is now nearly equivalent to that of carbon dioxide.

gases such as methane, nitrous oxide, and CFCs were allowed to increase, their concentrations would yield yet more warming.

Climate models exhibit varying degrees of sophistication and may differ in their mathematical treatment of certain Earth processes. These disparities result from gaps in our understanding of the physics underlying atmospheric processes. For instance, clouds may cause both a cooling and a heating effect on Earth. While clouds reflect some incoming solar radiation back to space, effectively cooling the planet, they also trap outgoing infrared radiation, keeping the surface warm. Suppose that Earth were blanketed with a thin, high-altitude cloud bank. The clouds could reflect enough incoming solar radiation to cool the planet's surface by 20°C or more. However, they could also trap enough infrared radiation to warm the surface by as much as 27°C. The difficulty of predicting the extent of cloud cover and the complex nature of the impact of clouds on temperature leads to uncertainty.

Consequences of Global Warming The consequences of an increase in global temperature are somewhat conjectural. Warmer oceans will release more moisture into the atmosphere and enhance the hydrologic cycle. Lands at high latitudes, such as Canada and Scandinavia, are expected to receive increased precipitation throughout the year. Lands at mid-latitudes, such as the United States and central Europe, would get increased precipitation only during the winter. In some areas, however, higher evaporation rates probably will also increase the frequency of summer droughts, causing problems for farmers; climate models indicate that parts of North America and Eurasia may be particularly hard hit (Figure 13-17).

Parts of the Antarctic ice sheet and mountain glaciers could melt, leading to higher sea levels and flooding of low-lying coastal areas. Island countries in the Pacific, Bangladesh, and the Netherlands, among other nations, are justifiably concerned given the predicted increase in sea level of between 5 and 40 cm by the year 2050. Furthermore, some models suggest a total rise of between

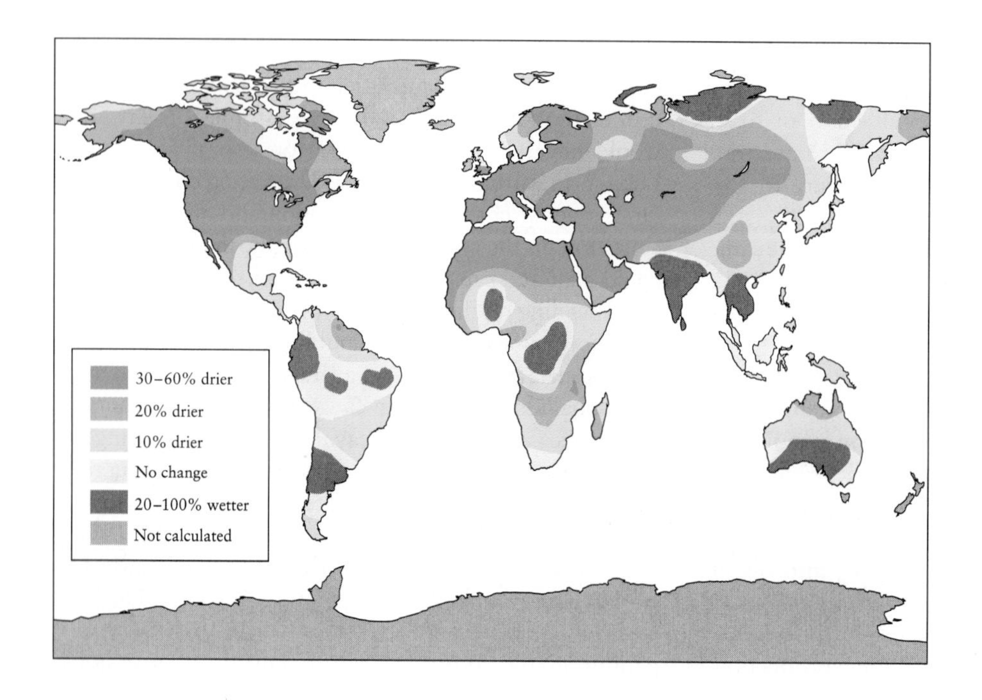

FIGURE 13-17 According to model predictions, most of the northern hemisphere will become 20 to 60 percent drier, with a doubling of pre-industrial CO_2 levels, and therefore experience a greater frequency of droughts, likely causing problems for agriculture.

5 and 35 m if all fossil fuels are burned and methane, ozone, and nitrous oxides continue to increase at their present rates. In the United States, the Atlantic and Gulf Coast states have cause for worry. The Environmental Protection Agency has estimated that Louisiana could lose up to 5000 square miles of wetlands and coasts if sea level were to rise 1 m. The higher sea-surface temperatures associated with global warming may lead to a greater frequency of hurricanes and tropical storms. Again, coastal regions will bear the brunt.

While global warming will most certainly be detrimental to some countries, those in cold climates may benefit. Russia and Canada, for example, stand to gain from a longer growing season, increasing perhaps up to 20 days. In addition, milder winters would reduce expenditures on heating oil, natural gas, and electricity.

Global Warming: Real or Imagined? What evidence do we have that global warming is occurring? Do two summers in a row with hotter than usual temperatures indicate a warming trend? What about locations that appear to be experiencing a cooling of summer temperatures while other locations are heating up? Partly because of such contradictory observations and partly because disagreements exist among scientists and between climate models, some argue that global warming is not a problem and thus requires no solution.

The danger in doing nothing, however, is that we risk losing an opportunity to slow or halt global warming before significant, and perhaps catastrophic, climatic changes occur. While Earth scientists may not agree about the future extent of global warming, enough experimental evidence exists for most to believe that it is now occurring. Measurements conducted over the last century show that Earth has warmed roughly 1°C in that time. Superimposed on this long-term warming trend are short periods of slight cooling, lasting a few years each and attributable to volcanic eruptions and changes in the intensity of solar radiation. With the benefit of more than a century of climatic data, we now know that mean global temperature is on the rise despite brief coolings.

Is the increase in temperature over the last century related to increased greenhouse gas concentrations from human activities, or has it resulted from some natural cause? Because short-term records are insufficient as a basis for conclusions about climate change, a better approach to answering this question is to combine historical data with much longer records of climatic change.

Data from tree rings and oxygen isotopes in ice cores suggest that the last 50 years have been warmer than at any time in the last few hundred to few thousand years. The coincidence of this warming trend with years of high greenhouse gas emissions does not conclusively prove a link, but it does provide suggestive evidence. Furthermore, ice-core data have shown that greenhouse gas concentrations and global temperature have varied together in the past 150,000 years, and there is no reason to think this relationship has changed in modern times. The composition of bubbles in ice cores has shown that the concentration of the atmospheric carbon dioxide fluctuated between about 180 and 280 ppm over the last 160,000 years up until the last two centuries (Figure 13-18). These fluctuations occurred gradually, over tens of thousands of years. In the last two centuries, however, concentrations of CO_2 have risen to between 350 and 355 ppm, a rate of increase unprecedented in recent Earth history. Whether ecosystems can tolerate this rapid change remains to be seen.

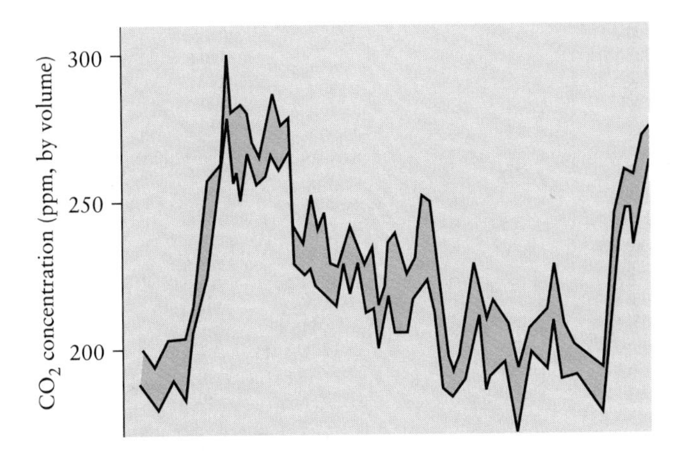

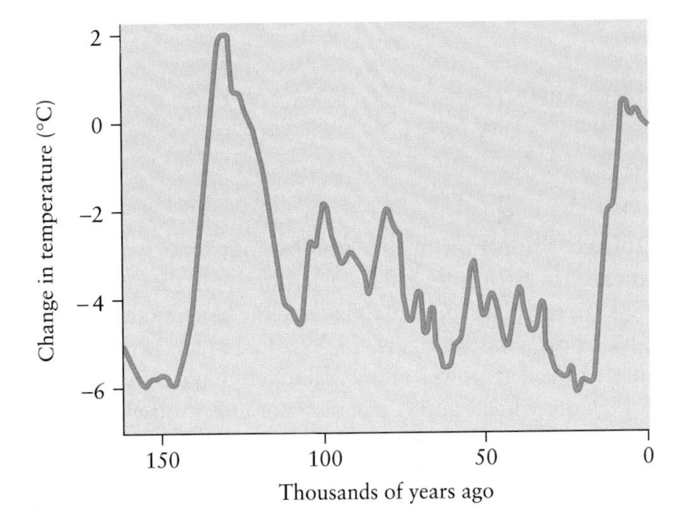

FIGURE 13-18 The concentration of atmospheric CO_2 over the last 160,000 years was obtained from bubbles in Antarctic glacial ice; global temperature was derived from the ratio of ^{18}O to ^{16}O in Antarctic glacial ice (see Chapter 12). Rises and falls in temperature correspond closely with rises and falls in CO_2 level.

Reducing Global Warming What, if anything, is to be done? Greenhouse gas emissions are largely the by-products of the activities of affluent societies dependent on fossil fuels for most of their industrial and transportation needs. However, greenhouse gas production is not

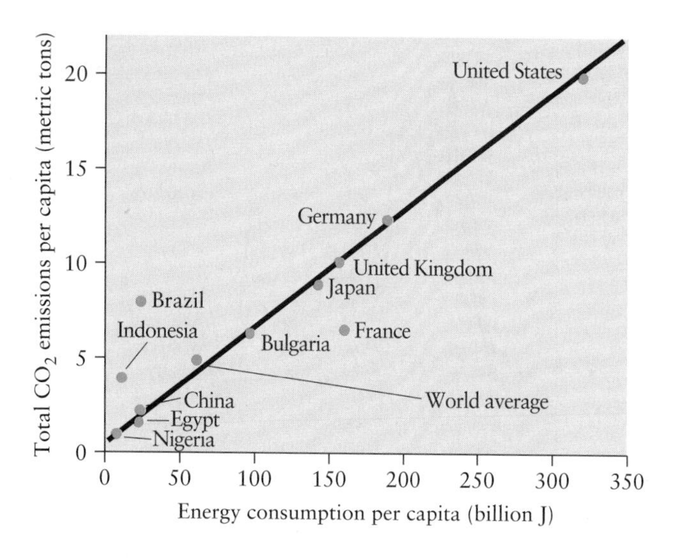

FIGURE 13-19 The relationship between per capita energy consumption and per capita production of carbon dioxide.

entirely a function of industrialization. France and Germany are both industrialized nations, yet they consume less energy per capita and produce far less CO_2 than does the United States (Figure 13-19). In France, widespread use of nuclear power allows relatively high consumption of energy while limiting emissions of greenhouse gases. In both France and Germany, the high cost of oil and other fossil fuels (between \$3 and \$5 per gallon of gasoline) limits energy consumption. Furthermore, distances between population centers in Europe tend to be much less than those in the United States, so less fuel is consumed in transportation. Bicycles are a common sight in European cities, and are even more common in China (Figure 13-20), particularly in the megacities of Beijing and Shanghai.

Many scientists and policymakers worry that economic development will replace all these bicycles with cars, but why shouldn't the rest of the world enjoy the same standard of living that Americans take for granted? Perhaps the most practical solution is for Americans voluntarily to increase their reliance on alternative energies such as solar and wind power, to improve the efficiency of appliances and motor vehicles dependent on fossil fuels, and to use more mass transit and modes of transportation such as walking and biking. These changes would enable us to make our supplies of fossil fuels last much longer, make us healthier from the added exercise, reduce concentrations of air pollutants such as ozone and nitrogen oxides which lead to ill health, and help curb global warming. Table 13-1 lists the energy savings and reductions in carbon emissions if individuals take some simple steps.

Global Metal Pollution

Human-induced pollution occurs not only in the atmosphere, but also in the pedosphere, hydrosphere, and biosphere, and often leads to serious impairment of human and animal health. Metals in the food supply can be especially toxic.

All life requires tiny quantities of various metals for normal functioning. Among the metals humans need are iron, copper, zinc, and selenium. Metals occur naturally in rocks and soils, from which they are extracted by plants.

FIGURE 13-20 Commuting by bicycle in Kunking, Yunnan Province, China. Each street contains wide lanes for bicycles (far left and far right) in addition to lanes for motor vehicles. Note the much larger number of bicycles than cars and the many buses.

TABLE 13-1 Ways to Save Energy and Reduce CO_2 Emissions

Action	Energy savings	Reduction in carbon emissions
Improving insulation of hot water heater	300 kWh/yr	55 kg (120 lbs)
Substituting 18-W compact fluorescent light for 75-W incandescent bulb (8 hrs/day)	170 kWh/yr	32 kg (400 lbs)
Carpooling 5 people instead of driving solo	3790 liters/yr (1000 gallons/yr)	2270 kg (5000 lbs)
Driving 10,000 miles in a car that gets 30 mpg (instead of 20 mpg)	630 liters (167 gallons/yr)	400 kg (880 lbs)
40 mpg	950 liters (250 gallons/yr)	600 kg (1320 lbs)
50 mpg	1140 liters (300 gallons/yr)	120 kg (265 lbs)
Regular car maintenance	190 liters (50 gallons/yr)	120 kg (265 lbs)
Planting trees to block the Sun	500–1500 kWh/yr	455 kg/yr (1000 lbs)

Source: Modified from F. Lyman et al., *The Greenhouse Trap: What We're Doing to the Atmosphere and How We Can Slow Global Warming* (Boston: Beacon Press), 1990, 132–133.

The metals are transferred to animals when they eat plants, and may then be transferred from prey to predator. In excessive concentrations, metals can cause deformity or death, and certain metals, such as lead and mercury, are inherently toxic regardless of concentration. In humans, lead and mercury can cause brain damage, irritability, hyperactivity or attention deficit disorder, and death.

Humans began contributing to global metal pollution 1.4 million years ago when they began to manipulate fire. Wood contains traces of a variety of metals, which become concentrated in the ash. At the time of the Roman empire, mining and smelting contributed large quantities of lead to the local atmosphere. Even then lead pollution was recognized as a health hazard to humans and domestic animals.

The invention of tall smokestacks in the 18th century allowed factory emissions filled with lead and other metal particulates to be carried high above the surface, where they were blown far away from their source. Indeed, particulate lead concentrations in Antarctic and Greenlandic ice jumped dramatically at this time and continued to rise as industrial activity became more widespread. Metal concentrations in the atmosphere peaked in the 1970s and then began to fall as countries enacted clean air legislation. Emissions of lead in particular have dropped dramatically since leaded gasoline was banned in North America and western Europe (Figure 13-21).

Some metals contaminate water rather than the atmosphere. Selenium, for example, is an element found in rocks and soils in much of the western United States. Soluble in water, selenium is readily picked up by irrigation runoff. Western farmers generally collect their irrigation runoff in evaporation ponds. Ducks, coots, geese, and other waterfowl nest there and drink the contaminated water. As a result, their offspring suffer horrific birth defects, if they survive long enough to hatch. Many chicks develop without eyes, with twisted beaks and legs, or with stunted, useless wings. At the Kesterson National Wildlife Refuge, adjacent to farm fields in the San Joaquin Valley of California, as many as 42 percent of the chick embryos of some

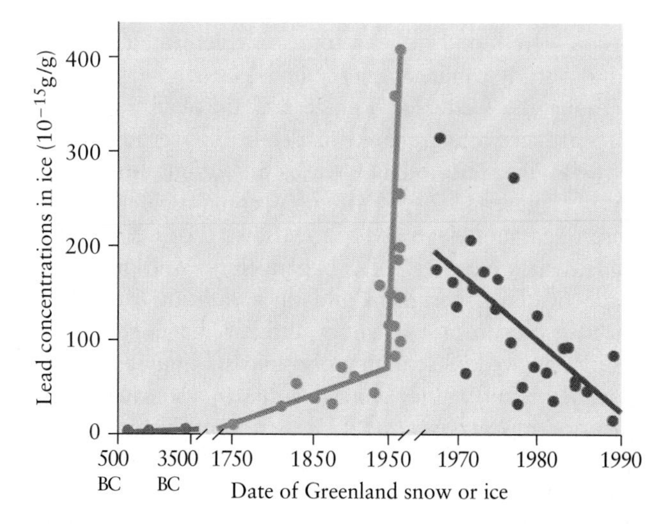

FIGURE 13-21 The concentration of lead in the Greenland ice sheet over time, measured in units of 10^{-15} gram of lead per gram of ice. The rapid increase in 1950 corresponds with the widespread use of leaded gasoline. A sharp decline in lead levels over the last three decades corresponds with the increasing substitution of unleaded fuels.

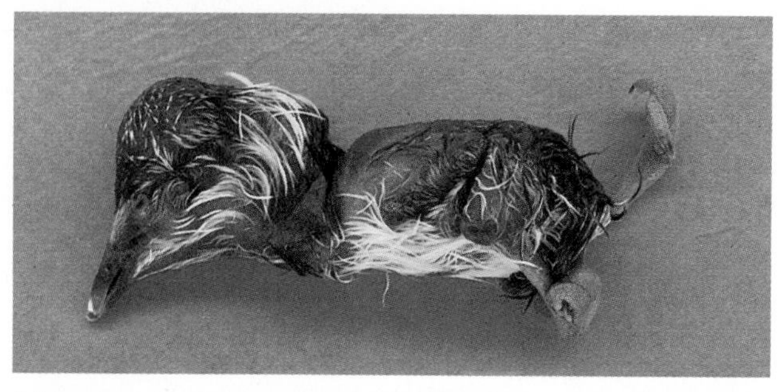

(a)

(b)

FIGURE 13-22 (a) Embryo of a killdeer deformed by selenium at Kesterson National Wildlife Refuge, California, in the 1980s. (b) Flocks of healthy waterfowl now flourish at Kesterson since agricultural runoff was stopped and selenium-contaminated ponds were filled.

species were found to be in some way deformed, according to a study conducted in the mid-1980s.

Today the Kesterson Refuge and the wildlife it supports are recovering, thanks to a shift in government policy away from use of the area as an agricultural runoff sump (Figure 13-22). Extremely contaminated ponds were filled, and clean water now flows through the wetlands, which have also been expanded by land purchases made by the government and duck hunting clubs. Reduced exposure to selenium directly benefits human health, as well. Selenium poisoning in humans causes liver, heart, and kidney damage, and can also cause infertility and miscarriages.

Another toxic metal that commonly contaminates aquatic systems is mercury. Beginning in the late 1950s, many Japanese were particularly hard hit by mercury poisoning when a corporation that manufactured fertilizers, plastics, and petroleum-derived chemicals began dumping waste water containing mercuric oxide into Minimata Bay in southern Japan. The mercuric oxide was readily ingested by fish, which in turn were eaten by humans. Thousands of people suffered loss of coordination, seizures, and other indications of brain damage, and by the late 1980s at least 730 people had died of acute mercury poisoning.

Although legislation such as the Clean Water and Clean Air acts in the United States has helped decrease the amount of metal pollution, it continues to be a serious problem throughout the world. Wildlife are particularly at risk. As habitat is destroyed by urbanization, industrialization, and farming, wild animals and birds are attracted to evaporation and industrial waste ponds. In wetlands, lead shot from hunting in the 1960s and 1970s is eaten by birds and is estimated to kill nearly 2.4 million of them annually. A rapid decline in the California condor population is directly attributable to lead poisoning caused by eating lead-contaminated prey.

Usually, the most effective way to restore metal-contaminated wetlands and water bodies is to leave them alone. Attempting to remove contaminated sediments only stirs them up and spreads them to other areas. Sometimes, as at the Kesterson Refuge, cleanup efforts involve burial of contaminated sediments beneath clean fill. This method can be successful as long as erosion does not re-expose the contaminants.

Global Land Transformation

Humans alter Earth systems not only by polluting air and water but also by modifying the land. Land transformation has two linked causes: changes in *land use* and changes in *land cover.* Land use includes cultivation, settlement, pasture, and recreation. Land cover includes surface vegetation, water, and pedosphere materials.

The earliest land-cover changes occurred when humans began to use fire. Fire was probably used to burn forests in order to create grasslands, which enabled prey animals to increase their populations, thereby ensuring a steady supply of meat for humans. The formation of grasslands also made hunting easier, as game animals could be spotted from a much greater distance. In addition, burning forests released nutrients to the soil, which initially may have increased soil fertility. The burning of forests is still practiced throughout the world, particularly to create and enhance rangeland for cattle (Figure 13-23).

Agriculture developed in the Middle East about 10,000 years ago, and it is this use of land that has caused the greatest transformation of Earth's surface. Changes in land cover from preagricultural times to the present are summarized in Table 13-2. Most of the 17.6×10^6 km^2 used for agriculture today have been converted from grasslands, woodlands, and temperate forests, each of which has experienced a decline of up to 20 percent. Tropical rain forests have declined 3.9 percent, but the rate at which they are being logged and burned today far outpaces the rate at which temperate forests were being destroyed several centuries ago.

In Europe, agriculture and logging accelerated during the Middle Ages, generally considered to be AD 1000–1400, and by 1400 most of the European forests had been

FIGURE 13-23 A satellite photograph taken over central South America shows smoke plumes rising from rain forests that are being burned to make way for farms and cattle ranches. Deforestation of tropical rain forests in South America and southeast Asia threatens to rival the destruction of temperate-zone forests in Europe since the Middle Ages and in North America since European settlement.

destroyed. Much of this change was caused by the need to feed a growing human population. Other factors included the need for wood to construct ships and for charcoal. The forests along the Adriatic coast, for example, were cut down to construct the Venetian shipping fleets. In the Americas, significant land transformation probably began with the Maya and greatly accelerated with the European colonization of North and South America. By 1900, many North American forests had been cut down, particularly in the eastern United States. Some of these forests have regrown as the main area of agricultural production has moved westward to the Midwest.

Changes in land cover often lead to changes in the rates and magnitudes of geologic processes. For example, the invention of the plow and the beginning of agriculture caused an acceleration in rates of soil erosion. Likewise, logging in the Pacific Northwest has resulted in a much greater frequency of landsliding there, and urbanization and the construction of impermeable surfaces such as concrete sidewalks and asphalt parking lots have resulted in the increased frequency and intensity of floods. Land-cover modifications can also cause changes in climate. A decrease in vegetative land cover increases the reflectivity of the land and can lead to cooling. Less vegetation also means less transpiration and a less active hydrologic cycle, interrupting the natural fixation of carbon by plants (Box 13-2). In many cases, land-use patterns were established before these risks were known. In other cases, humans ignored known risks out of foolishness, greed, or necessity.

Although only 1 percent of all land is used for human habitation, the proportion approaches 100 percent in heavily urbanized areas. Because people tend to cluster in temperate areas rather than in deserts, tundra, or glaciers, some of the most agriculturally and biologically productive lands are given over to urbanization. By 2000, urbanization probably will have claimed farmland capable of feeding 84 million people.

TABLE 13-2 Preagricultural and Current Area Devoted to Major Land-cover Types

Ecosystem	Preagricultural (millions of km^2)	Present (millions of km^2)	Change (millions of km^2)	% change
Forests				
Tropical rain forest	12.8	12.3	–0.5	–20.6
Other forest	34.0	27.0	–7.0	–3.9
Total	46.8	39.3	–7.5	–16.0
Woodland	9.7	7.9	–1.8	–18.6
Shrubland	16.2	14.8	–1.4	–8.6
Grassland	34.0	27.4	–6.6	–19.4
Tundra	7.4	7.4	0.0	0.0
Desert	15.9	17.6	0.3	–1.9
Cultivation	0.0	17.6	+17.6	—

Source: Modified from M. Williams, "Forests and Tree Cover," in *Changes in Land Use and Land Cover: A Global Perspective*, W. B. Meyer and B. L. Turner, eds. (Cambridge: Cambridge University Press, 1994), 104.

13-2 Case Study

Desertification in the Sahel

Between the Sahara and the equatorial rain forests of Africa lies a region 500 km wide known as the Sahel. This semiarid region extends nearly 5000 km from the Red Sea in the east to the Atlantic Ocean in the west. Within this zone live millions of people who survive through subsistence farming and livestock ranching. In the 1960s, overgrazing by livestock began to strip the soil of its protective cover of vegetation. Rates of soil erosion increased. In addition, the tropical sun dried out the exposed topsoil and baked it into a hard, impermeable surface. The hooves of animals, pressing down on the soil, compacted and hardened it further. Precipitation could no longer soak into the ground and instead began to run off the landscape, causing both further soil erosion and declining recharge of subsurface aquifers.

Native plant species could not exist in this environment. Soil productivity and then fertility decreased. As the land lost its vegetation, it became more reflective. Sunlight that the vegetation would have absorbed and converted into heat instead escaped to space, making the land cooler than it was before desertification. This set up a positive feedback process that has led to further desertification: The air above the land sinks as it is cooled from below (because of the decreased amount of solar radiation retained by the devegetated land). The cooled air is compressed by the weight of air masses that have moved in above it. This pressure warms the air as it subsides, increasing its capacity to hold moisture, which leads to high rates of evaporation of both standing water and soil moisture and to decreased precipitation.

During the late 1960s and again in the 1980s droughts struck the Sahel and led to widespread famine. Many earlier droughts, while bringing hard times, were not catastrophic; these were. Apparently the stress of desertification had made the environment incapable of coping with a fairly common climatic cycle.

Desertification has ravaged much of the Sahel, a semiarid belt of land situated between Africa's Sahara Desert to the north and equatorial rain forests to the south. Overgrazing of cattle and other livestock has led to a loss of vegetation and a consequent hardening of the soil. In addition, the higher reflectivity of the bare ground causes cooling and downward motion of atmospheric currents over the Sahel. The air warms as it sinks, resulting in less precipitation than would occur in an undesertified landscape.

Desertification in Africa is related to the transformation of cultures on that continent from nomadic to more sedentary agricultural societies, because agriculture

Moreover, urbanization concentrates sewage and garbage wastes and atmospheric pollutants such as ozone, sulfur dioxide, and nitrogen oxides, damaging plant, animal, and human health. These problems are particularly severe in developing countries such as India, Brazil, Mexico, and China, where a lack of environmental laws or the money to enforce them lets pollutants accumulate disastrously, and many people live in vast urban slums without benefit of sanitation and clean water.

The growth of urban areas over the last century has been rapid and promises to continue. In 1900, 14 percent of the human population lived in towns of 5000 or more. By 1991, that proportion had jumped to nearly 50 percent. In addition, the number of cities with 10 million or more inhabitants continues to grow. Many of these cities are in developing nations, which are experiencing high rates of population growth and unprecedented migrations of people from the countryside to cities in search of employment. Land-use problems associated with urbanization will continue to grow unless global population growth can be curbed, or developing nations can reach a level of affluence at which people will demand sewage treatment plants and strictly enforced environmental regulations. In 1997, the United Nations reported encourag-

exposes the same patch of ground to tilling and grazing indefinitely without time to recover. Furthermore, the decline in infant and animal mortality rates that followed widespread access to medical and veterinary care in the 1960s caused human and animal populations in the Sahel to explode. For this reason, even the most marginally productive land has now been forced into production to grow grain for humans or provide pasture for sheep and cattle. At present the droughts have abated, but much of the land has already experienced declines in productivity between 10 and 90 percent, and the droughts will return.

Desertification also is occurring in the western United States, Australia, Asia, and South America, and represents a huge threat to human well-being. To prevent further desertification and restore fertility require constructive agricultural practices. Overgrazing can be prevented by either decreasing the number of animals in an area or by allowing the land to lie fallow for some time to recover. The United Nations Environment Program (UNEP) has implemented a plan to combat desertification by encouraging replanting of native vegetation, limiting overgrazing, implementing better irrigation systems, and creating national insurance programs that would allow people to survive occasional droughts and crop failures without overtaxing remaining resources to the point of nonrenewability.

Extent of Desertifcation by Continent

	Used area (millions of km^2)	Desertified area (millions of km^2)	Desertified used area (%)	% desertified by overgrazing
World	80.56	19.43	24	35
Africa	17.99	4.94	27	49
North and Central America	10.44	1.58	15	24
South America	13.86	2.44	18	28
Asia	24.40	7.45	31	26
Europe	8.43	1.99	24	23
Oceania	5.44	1.03	19	80

Source: D. Graetz, in *Changes in Land Use and Land Cover: A Global Perspective,* W. B. Meyer and B. L. Turner, eds. (Cambridge: Cambridge University Press, 1994), 141.

ing progress toward this goal, with pronounced declines in Asian and Latin American population growth and a perceptible decline in African population growth. These results were attributed to education and an increasing willingness to recognize the rights of women.

- Earth's average temperature has risen nearly 1°C over the last century because of increasing levels of greenhouse gases in the atmosphere.
- Bubbles in ice cores contain a record of atmospheric CO_2 fluctuations over the last 160,000 years.

The rate of increase in CO_2 over the last two centuries is unprecedented.

- Human activities such as smelting, the use of leaded gasoline, and the discharge of industrial and agricultural waste into water bodies have led to toxic concentrations of metals in the environment, damaging the health of humans and animals.
- Humans alter the physical environment through changes in land use and land cover, principally by deforestation for agriculture.

Closing Thoughts

For billions of years, Earth's climate and geologic environments have changed as a result of migrations of continents, uplift of mountain ranges, variations in distance between Earth and Sun, and other factors. When environmental changes occur gradually, species can evolve adaptively or migrate to more hospitable environments. In contrast, rapid environmental changes can be devastating to the biosphere and lead to mass extinctions such as those linked to asteroid impacts between the Cretaceous and Tertiary periods.

Early *Homo sapiens,* when faced with rapid and extreme changes in the environment such as occurred during the last ice age, found ways to adapt and survive, in part because of their small numbers. Migration to more hospitable climates is far easier for a small population than for the billions of people on Earth today. This great human population explosion has taken place during the later Holocene, at a time when environmental conditions are stable and favorable for growth. However, climate and environments have rarely been as stable and favorable as those we have known.

Today, human actions are altering the composition of the atmosphere, increasing rates of soil erosion and mass wasting, causing desertification, and creating a host of other environmental changes. Moreover, these changes we are causing are occurring at unprecedented rates. Although glacial-interglacial transitions caused fluctuations in concentrations of greenhouse gases, those fluctuations occurred over thousands of years and the biosphere was able to keep up. Today, many ecosystems are stressed by the combined pressures of geologically induced climate change and human-induced factors such as overgrazing of domestic animals.

What does the future hold? The answer to that question depends in part on the choices that we make as a species. Environments on Earth could become inhospitable if population grows beyond Earth's carrying capacity, if overgrazing and plowing erode soils and turn fertile lands into barren deserts, and if we continue to increase our use of pollutant-producing internal combustion engines. The incidence of skin cancer could increase dramatically if the ozone layer is further decimated by CFC manufacture in defiance of international agreements, and many more cases of ill health, death, and destruction of ecosystems could occur if air and water pollution increase.

Alternatively, constructive choices may allow our planet of the future to be a pleasant and sustainable world. We may reduce fertility to rates at which population can stabilize at a sustainable level. We may rely on nondepletable energy sources such as solar and wind power. We may change our farming habits, growing various crops only in areas where they can be farmed sensibly, changing to drip irrigation systems in places where water is in short supply, and preventing overgrazing. Furthermore, we may choose to clean fully our industrial and sewage wastes before we allow the effluent to flow into streams, and we may allocate resources to cleaning up areas that are already contaminated.

Our human culture is at a crossroads. The United States, Canada, Germany, Japan, and other developed nations have instituted environmental laws to clean up air, water, and soil, and recycling programs are now found in many cities. It is harder to change individual behavior, however. Automobile use increases, people leave the lights and television on when no one is using them, some groups oppose population control. In less-developed countries, environmental concerns take a back seat to economic growth and improved standards of living. For these countries, the day-to-day issue of survival is more pressing than the question of causing damage to the environment. Ultimately the interests of all are best served if the developed countries insist on conserving their environment and help the less-developed countries do the same.

Summary

- Earth's history of climate change has been driven by gradual processes such as plate tectonics, the evolution of new species, and variations in orbital parameters, and by catastrophic events such as volcanic eruptions and asteroid impacts.
- The evolution of new plant species commonly led to global cooling as plants consumed CO_2 for photosynthesis.
- Paleomagnetism allows geologists to determine the past position of continents, and their findings support the theory of plate tectonics.
- Accelerated rates of seafloor spreading during the Cretaceous period increased the concentration of carbon dioxide in the atmosphere and led to global warming.

- Earth has cooled throughout much of the Cenozoic era. Explanations include increased chemical weathering in areas of relatively recent uplift that locked atmospheric CO_2 in sediments, and an increase in photosynthetic organisms that use CO_2.
- The Holocene and other warm interglacial periods constitute only 10 percent of the last half million years of Earth's history.
- During the last glacial maximum, vast ice sheets covered North America, Europe, and Asia, causing a decline in global sea level of 120 to 160 m.
- Temperature changes as small as 2°C brought either prosperity (in the Medieval Warm Period) or famine (in the Little Ice Age).
- Our earliest bipedal ancestor may have arisen because of the progressive aridification of Africa.
- Possible consequences of global warming include an increased frequency of hurricanes and droughts, retreat of glaciers and ice sheets leading to sea level rise, and a longer growing season for Russia, Canada, and other nations at high latitudes.
- Ice-core bubbles contain a record of atmospheric CO_2 fluctuations over the last 160,000 years. Carbon dioxide concentration varied between 180 and 280 ppm until the last two centuries, when it rose to 350 ppm. The rate of increase in CO_2 over the last two centuries is unprecedented.
- Many metals are necessary for normal metabolic functioning but are toxic in large doses. Human activities such as smelting, use of leaded gasoline, and discharge of industrial waste into water bodies have led to increased concentrations of these metals in the environment with attendant health problems for humans and animals.
- Urbanization concentrates wastes in small areas and destroys farmland.

Key Terms

anhydrite (p. 386)
stromatolite (p. 386)
Gondwanaland (p. 392)
cyclothem (p. 392)
Pangaea (p. 392)
red beds (p. 393)
iridium (p. 394)
Medieval Warm Period (p. 399)
Little Ice Age (p. 399)

Review Questions

1. What was the Mediterranean salinity crisis? Describe the factors that led to it.
2. What evidence did the *Glomar Challenger* scientists use to confirm the Mediterranean salinity crisis?
3. What factors led to the extreme warmth of early Earth?
4. What brought about the Proterozoic glaciations?
5. What is the relationship between geography of a continent and its climate?
6. How is paleomagnetism used to determine the former position of landmasses?
7. What evidence exists that an asteroid caused the extinctions at the end of the Cretaceous?
8. What ecological impacts would an asteroid impact have?
9. Give two hypotheses for the cause of Cenozoic cooling.
10. What is causing the cyclical glaciations that have characterized northern hemisphere climate for the last few million years?
11. How has climate change influenced human evolution and human societies?
12. What human activities have potential impacts on the climate system?
13. What evidence do we have that atmospheric CO_2 levels are increasing?
14. List some likely consequences of global warming. How will these consequences affect human societies?
15. Explain the causes and problems associated with global metal pollution.
16. What is the difference between change in land use and change in land cover?
17. How does urbanization affect the environment?

Thought Questions

1. Warm interglacial periods constitute only about 10 percent of the last several hundred thousand years. In general, these warm episodes last about 10,000 years, and since the Holocene began about 10,000 years ago, we should be coming to the end of the latest interglacial period. What do you think the future holds for climate? Consider what you know about recent global warming and about Milankovitch cycles.
2. Human cultures and civilizations have arisen at a time in Earth's history when climate has been relatively warm and stable, yet we know that Earth's climate has oscillated between warm and extremely cold periods. How might human civilizations respond to future climate change? What effect will population size have on society's ability to cope with future changes?
3. Throughout most of geologic time Earth has been warm. However, a few episodes of glaciation occurred early in the planet's history, and since the Eocene epoch, Earth has become progressively colder. What factors would have to come into play to reverse this cooling trend? On what time scales do they operate? How likely is it that Earth will return to a warmer temperature? Why?

Exercises

1. The volume of the Antarctic ice sheet is about 24 million km^3. If the entire ice sheet were to melt, how much would sea level rise around the globe? (Assume that no other ice sheets exist and that the oceans would continue to cover about 70 percent of the surface area of the globe.) The radius of the Earth is 6370 km and the surface area of a sphere is $A = 4\pi r^2$.
2. A gallon of gasoline produces roughly 10 kg of CO_2 when burned in a car's engine. Calculate how much CO_2 you create each year by driving. (You must be able to estimate how many miles you drive each year and your car's gas mileage under your driving conditions.) How many gallons of gas do you use each year? Now that you know your annual gas consumption, what is your annual CO_2 production rate?

Suggested Readings

Allman, W. F., and Wagner, B., 1992. "Climate and the Rise of Man." *U.S. News and World Report* 112: 60–67.

Frakes, L. A., 1979. *Climates Throughout Geologic Time.* New York: Elsevier Scientific Publishing.

Goudie, A., 1989. "The Changing Human Impact." In Friday, L., and Laskey, R. (eds.), *The Fragile Environment.* New York: Cambridge University Press.

Graedel, T. E., and Crutzen, P. J., 1995. *Atmosphere, Climate, and Change.* New York: W. H. Freeman and Company.

Grove, J. M., 1988. *The Little Ice Age.* New York: Methuen.

Hsü, K. J., 1983. *The Mediterranean Was a Desert.* Princeton, NJ: Princeton University Press.

Imbrie, J., and Imbrie, K. P., 1979. *Ice Ages.* Hillside, NJ: Enslow.

Keeling, C. D., and Bacastow, R. B., 1977. "Impact of Industrial Gases on Climate." In *Energy and Climate.* Washington, DC: National Academy of Sciences.

Raymo, M. E., and Ruddiman, W. F., 1992. "Tectonic Forcing of Late Cenozoic Climate." *Nature* 359: 117–122.

Schotterer, U., and Andermatt, P., 1992. *Climate—Our Future?* Minneapolis: University of Minnesota Press.

Appendix 1

Classification of Biological Organisms

To facilitate communication among scientists throughout the world, all living organisms are classified according to a system developed in the 18th century by Swedish botanist Carl von Linné, or Carolus Linnaeus, (1707–1778) and modified as biological knowledge increased. In the Linnaean classification, organisms are grouped into a hierarchy of sets and subsets, with species the narrowest of the subsets and the kingdom the overarching set.

A *species* is a population whose members are able to interbreed freely under natural conditions. Above species in the Linnaean hierarchy is the *genus* (plural *genera*), a group of species that are very similar and of more or less immediate common ancestry. Related genera make up *families,* related families make up *orders,* related orders make up *classes,* related classes make up *phyla,* and related phyla make up *kingdoms.* In general, the life-forms on Earth are divided into five kingdoms: animals, plants, fungi (molds, yeasts, and mushrooms), protists (amoebas, paramecia, and algae), and monerans (bacteria and related organisms).

Life-forms are generally referred to by genus and species names, which by convention are written in italics; the genus name begins with a capital letter. The genus and species designation for modern human beings is *Homo sapiens,* meaning "thinking man." Using ourselves as an example, we can trace our classification through increasing levels of detail.

Kingdom: Animals (organisms with many cells that live by taking in complex organic molecules)
Phylum: Chordates (animals with a notochord, a rudimentary spinal column)
Class: Mammals (animals that secrete milk to nourish their young)
Order: Primates (apes, monkeys, and related animals)
Family: Hominids (primates that walk upright)
Genus: *Homo* (large-brained hominids)
Species: *sapiens* (modern humans)

Appendix 2

Periodic Table of the Elements

Characteristics associated with different electron shell patterns are indicated on the table. Two special groups of rare elements—atomic numbers 58–71 and 90–103—have been omitted.

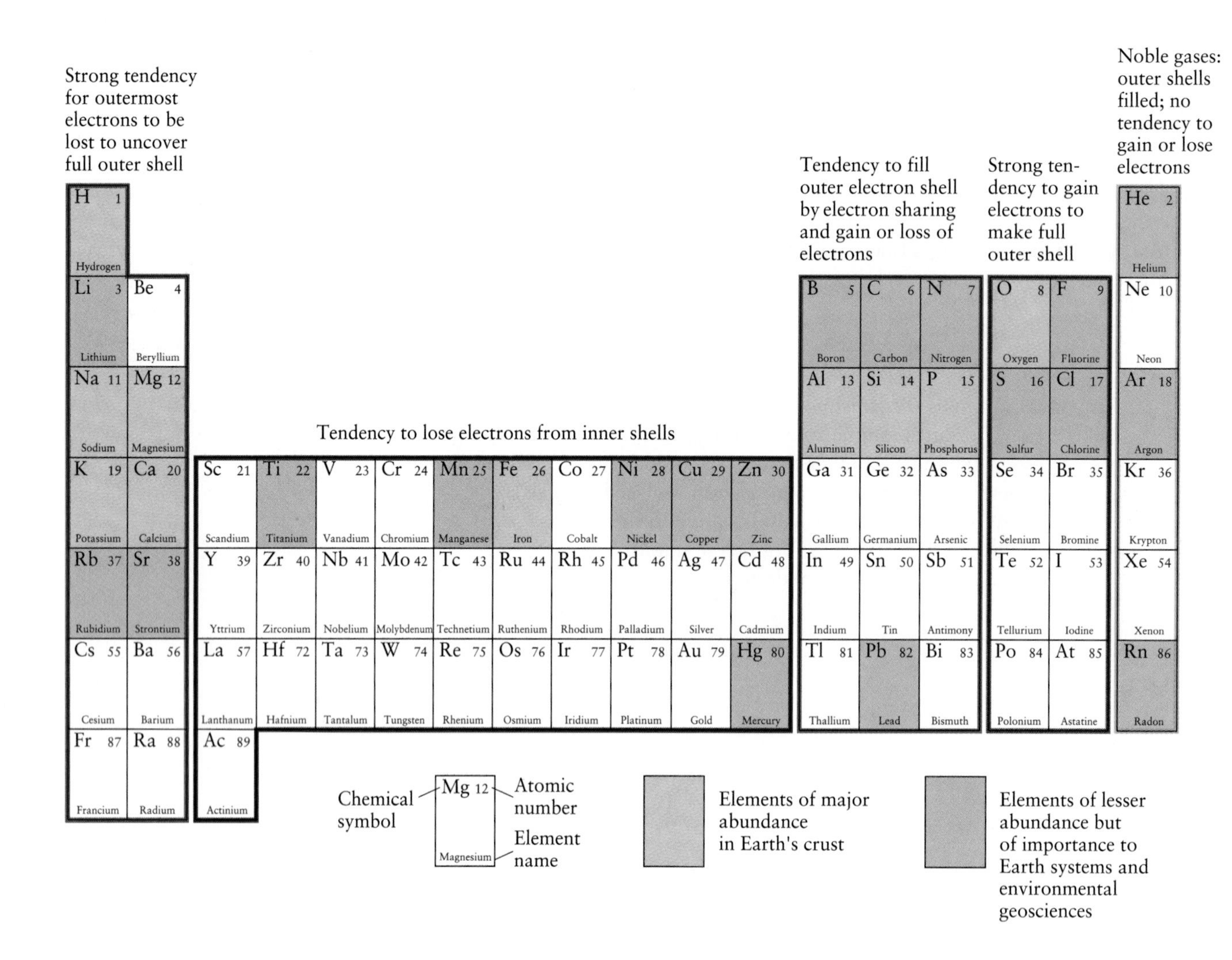

Appendix 3

Units and Conversions

In the English system of measurement, which is used by most Americans, length is measured in inches, feet, and miles, and mass is measured in units of ounces, pounds, and tons. In the metric system, which is used by many other countries and the scientific community worldwide, length is measured in centimeters, meters, and kilometers, and mass is measured in grams, kilograms, and metric tons. Both systems use the same units for time: seconds, minutes, hours, and years.

The metric system is based on powers of 10. A meter is equal to 10 decimeters, or 100 centimeters, or 1000 millimeters; 10 meters is called a dekameter, 100 meters a hectometer, 1000 meters a kilometer. Each of these numbers can be expressed as a power of 10: 10 is 10 to the first power, or 10^1, 100 is 10 to the second power, or 10^2 (10×10), and 1000 is 10 to the third power, or 10^3 ($10 \times 10 \times 10$). Numbers smaller than 1 are expressed as negative powers: one-hundredth of a meter, or 0.01 m, is the same as 10^{-2}, or $1 \div (10 \times 10)$. The table below lists some metric numbers with the most commonly used prefixes and their abbreviations.

In this book, we generally use the metric system, because it is the standard for scientific work. However, a few special cases warrant the use of different measures. Particularly in the United States, climatic and hydrologic data typically are recorded in English units. Rainfall, for example, commonly is given in inches per year, and stream discharge in cubic feet per second (see Chapter 7). Rather than convert standard data recorded by government agencies, we have retained the English system in these instances. In addition, navigators commonly measure ocean depth in fathoms, and governments used this unit in establishing offshore territorial rights (see Chapter 10).

Metric Numbers

10^{-9} = 0.000000001	nano (n)	10^2 = 100	hecto (h)
10^{-8} = 0.00000001		10^3 = 1,000	kilo (k)
10^{-7} = 0.0000001		10^4 = 10,000	
10^{-6} = 0.000001	micro (μ)	10^5 = 100,000	
10^{-5} = 0.00001		10^6 = 1,000,000	mega (M)
10^{-4} = 0.0001		10^7 = 10,000,000	
10^{-3} = 0.001	milli (m)	10^8 = 100,000,000	
10^{-2} = 0.01	centi (c)	10^9 = 1,000,000,000	giga (G)
10^{-1} = 0.1	deci (d)	10^{10} = 10,000,000,000	
10^0 = 1		10^{11} = 100,000,000,000	
10^1 = 10	deka (da)	10^{12} = 1,000,000,000,000	tera (T)

The charts below provide the factors needed to convert from one system of units to another in measuring length, area, volume, mass, pressure, energy, power, and temperature. Begin at the left: To convert a value expressed in inches to the equivalent in centimeters, for example, multiply by 2.5400.

Conversion Charts

from	to	multiply by
Length		
centimeters	inches	0.3937
fathoms (1 fathom = 6 feet)	meters	1.8288
feet	meters	0.3048
inches	centimeters	2.5400
kilometers	miles (statute)	0.6214
kilometers	feet	3281.5
meters	feet	3.2808
meters	yards	1.0936
meters	inches	39.37
miles (statute)	kilometers	1.6093
miles (nautical)	kilometers	1.8531
yards	meters	0.9144
Area		
acres (U.S.; 1 acre = 43,560 square feet)	hectares	0.4047
hectares (1 hectare = 10,000 square meters)	acres	2.471
square centimeters	square inches	0.1550
square inches	square centimeters	6.4516
square feet	square meters	0.0929
square meters	square feet	10.764
square meters	square yards	1.1960
square kilometers	square miles	0.3861
square miles	square kilometers	2.590
Volume		
acre feet	cubic meters	1234
barrels of petroleum (1 barrel = 42 U.S. gallons)	cubic meters	0.159
cubic centimeters	cubic inches	0.06102
cubic feet	cubic meters	0.02832
cubic inches	cubic centimeters	16.3871
cubic meters (1 cubic meter = 1000 liters)	cubic feet	35.314
cubic meters	cubic yards	1.3079
cubic yards	cubic meters	0.7646

gallons (U.S.)	liters	3.7853
liters (1000 cubic centimeters)	quarts (U.S.)	1.0567
liters	gallons	0.2642
quarts	liters	0.9463
Mass		
grams	ounces	0.03527
kilograms	pounds	2.20462
tonnes (metric tons) (1 metric ton = 1000 kilograms)	short tons	1.1023
ounces	grams	28.34952
pounds	kilograms	0.45359
tons (1 short ton = 2000 pounds)	kilograms	907.1848
tons (short)	tonnes (metric tons)	0.90718
Pressure		
bars (1 bar = 0.98692 atmosphere)	pascals	10^5
bars	newtons/square meter	10^5
kilograms/square centimeter	atmospheres	0.96784
kilograms/square centimeter	bars	0.98067
kilograms/square centimeter	pounds/square inch	14.2233
pascals (1 newton/square meter)	bars	10^{-5}
pounds/square inch	kilograms/square centimeter	0.70307
Energy (see Figure A-1)		
Btu (British thermal units)	ergs	1.054×10^{10}
Btu	joules	1.054×10^3
Btu	quads	10^{-15}
foot pounds	joules	1.356
ergs	calories (gram)	2.39006×10^{-8}
ergs	Btu	9.48451×10^{-11}
ergs	joules	10^{-7}
joules	ergs	10^7
joules	calories (gram)	0.2390
joules	Btu	9.484×10^{-4}
quads	Btu	10^{15}
quads	joules	1.05×10^{18}
Power		
ergs/second	watts	10^{-7}
horsepower (U.S.)	watts	7.4571×10^2
Btu/minute	watts	1.758×10^1
watts	ergs/second	10^7
watts	horsepower (U.S.)	0.001341
watts	Btu/minute	0.05688

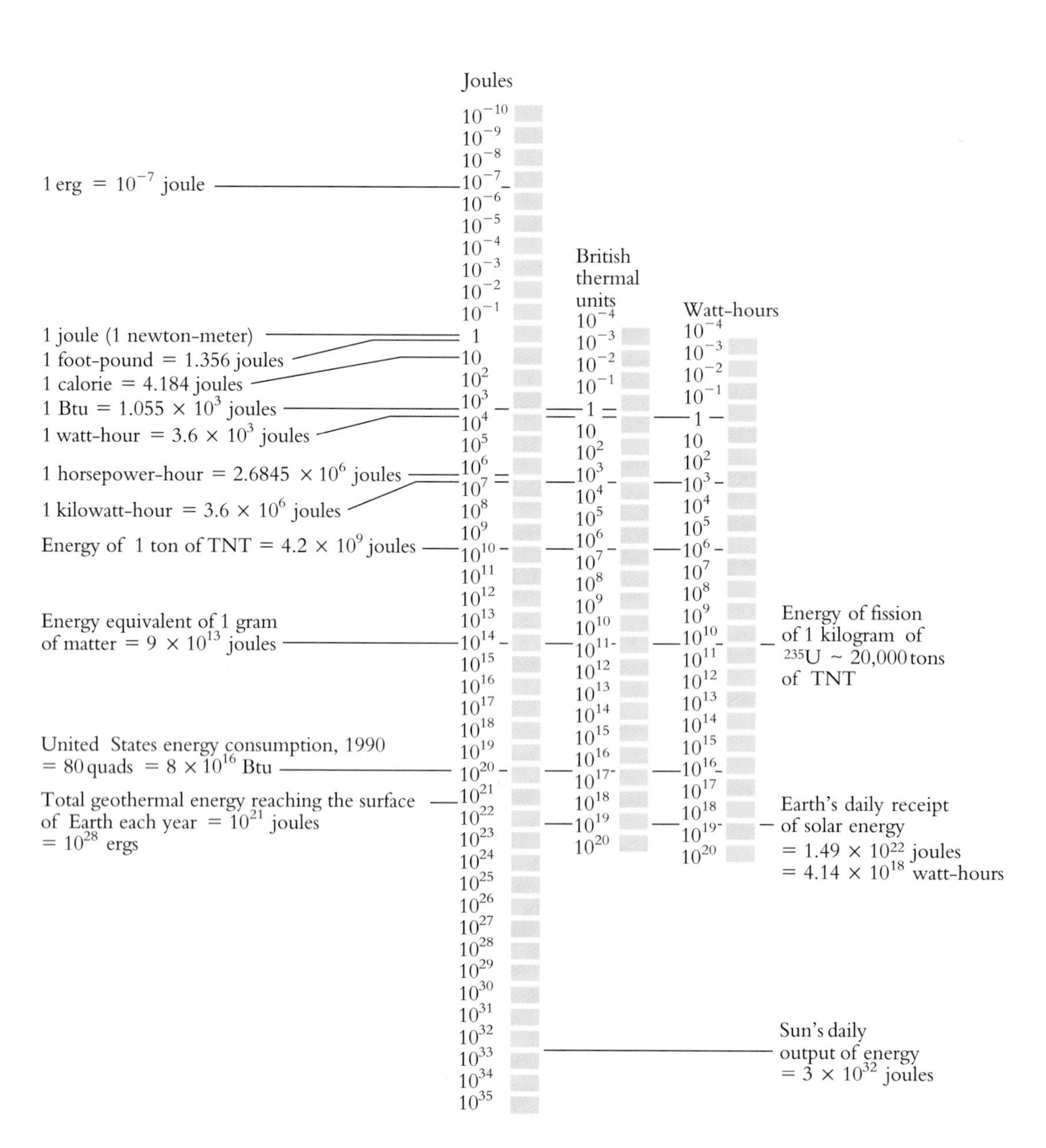

FIGURE A-1 Energy conversion from joules to selected, commonly used energy units.

Conversion Chart

from	to	compute
Temperature (see Figure A-2)		
Fahrenheit (°F)	Celsius (°C)	(°F – 32)/1.8
Celsius (°C)	Fahrenheit (°F)	(°C × 1.8) + 32

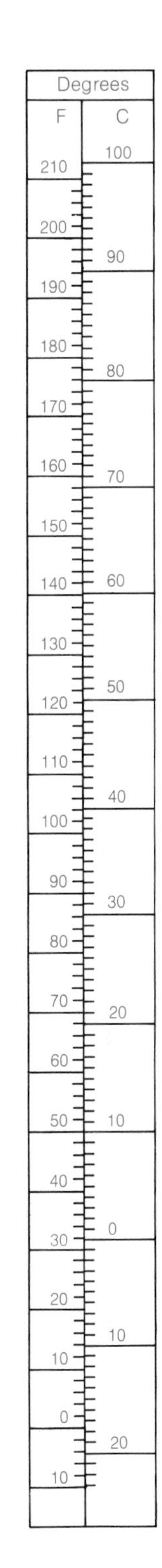

FIGURE A-2 Comparison of temperature equivalents of Fahrenheit and centigrade degrees.

Measuring Energy

Energy is measured in many different units; the most common are joules, calories, British thermal units, and watts.

One joule (J) is defined as 1 kg-m^2/sec^2. For a sense of what a joule is, consider how many joules of energy can be produced by burning 1 kg of carbon. Complete combustion releases 33 million J of energy, enough to drive an efficient subcompact car for at least 15 km.

Because global energy budgets and energy demands involve large amounts of energy, preferred units for these uses are the exajoule (10^{18} J), petajoule (10^{15} J), terajoule (10^{12} J), and gigajoule (10^9 J). The world's annual energy consumption is currently more than 300×10^{18} J (300 exajoules). In comparison, a human being needs 3×10^9 joules of energy each year, supplied by food, to breathe, move, and function. People generally use much more energy than this, however, because they also drive cars, fly in airplanes, and use goods manufactured with the aid of energy resources. In the United States, per capita energy consumption based on all types of activities is 320×10^9 joules per year.

The calorie is defined as the amount of energy necessary to raise the temperature of 1 g of liquid water 1° C. One calorie is equivalent to 4.184 joules. The energy content of our food is measured in Calories (upper case C), each of which is equivalent to 1000 calories or 4184 joules. A single candy bar contains about 400 Calories, or 400,000 calories, enough to raise the temperature of 400 kg of water by 1°C. The human body typically weighs 50–80 kg, and more than 70 percent of this weight is water. Normal body temperature is 37°C, so a great proportion of food intake is used to maintain the temperature of water at 37°C . To keep itself at a fairly constant temperature, the body burns an increased number of calories when the ambient temperature is lower than the body's temperature.

One of the most common units for measuring energy resources in the English-speaking world is the British thermal unit, or Btu. A Btu is the amount of energy required to raise the temperature of 1 pound of water 1°F from 39.2°F to 40.2°F. One barrel of oil, which contains 42 U.S. gallons, is equivalent to 5.8 million Btu. When discussing a large amount of an energy resource, such as that obtained from millions or billions of barrels of oil, the preferred unit of measure is the quad, which represents one quadrillion (10^{15}) Btu.

The amount of energy used over a given period of time is a measure of power, and a common unit for power is the watt. A watt is 1 joule per second, and a terawatt is 10^{12} watts. Earth's annual energy inflow is more than 173,000 terawatts. In comparison, energy requirements for all human activities in a single year are a mere 10 terawatts. It is evident, therefore, that although supplies of nonrenewable energy resources might become exhausted, the world will not run out of energy.

Appendix 4

Properties of Common Minerals

Mineral	Composition	Chemical classification	Hardness	Cleavage
Minerals with metallic luster (in general order of decreasing hardness)				
Pyrite	FeS_2	sulfide	6–6.5	none
Magnetite	Fe_3O_4	oxide	5.5–6.5	none
Hematite	Fe_2O_3	oxide	5.5–6.5	
Uraninite	UO_2 to U_3O_8	oxide	5–6	none
Limonite	$Fe_2O_3 \cdot H_2O$	oxide (hydrous)	1–5.5	none
Sphalerite	ZnS	sulfide	3.5–4	perfect in six directions
Chalcopyrite	$CuFeS_2$	sulfide	3.5–4	none
Copper	Cu	native element	2.5–3	none
Galena	PbS	sulfide	2.5	perfect in three directions at right angles
Gold	Au	native element	2.5	none
Graphite	C	native element	1	one perfect
Minerals with nonmetallic luster (in general order of decreasing hardness)				
Garnet	$Al_2(SiO_4)_3$ + other metallic elements	silicate	6.5–7.5	none
Quartz	SiO_2	silicate	7	none
Olivine	$(Fe,Mg)_2SiO_4$	silicate	6.5–7	none
Potassium feldspar (orthoclase)	$KAlSi_3O_8$	silicate	6–6.5	two perfect, at right angles
Plagioclase feldspar	$(Na,Ca)AlSi_3O_8$	silicate	6–6.5	two perfect, nearly at right angles

Common color(s)	Streak color	Other properties
pale brass yellow	greenish black, brownish black	cubic crystals common
black	black	strongly attracted to a magnet
silver to dark gray	reddish brown	reddish rust very common
black to dark brown	black to dark brown	dense; radioactive
silver to golden-brown	yellowish brown	yellowish rust very common
yellowish brown	brownish to light yellow	resinous luster; streak smells like rotten eggs
brass yellow	greenish black, brownish black	iridescent tarnish
copper, commonly stained green	reddish copper	can be cut with a knife; dense
silver gray	gray to black	cubic crystals common; dense
gold	golden yellow	malleable; can be flattened without breaking; dense
pencil-lead gray	black	forms slippery flakes; feels greasy
dark red to black	none	commonly 12-sided crystals; conchoidal fracture
clear, milky, purple, rose, smoky	none	conchoidal fracture; crystal faces common
green	none	conchoidal fracture
white to pink	none	prism-shaped crystals common
white to dark gray	none	fine, straight striations

(continued)

Properties of Common Minerals *(continued)*

Mineral	Composition	Chemical classification	Hardness	Cleavage
Minerals with nonmetallic luster (in general order of decreasing hardness) (continued)				
Amphibole group (hornblende is most common member)	complex Ca, Na, Mg, Fe, Al silicates	silicate	6	two, intersecting at 56° and 124°
Pyroxene group	$XYSi_2O_6$ (X,Y = Ca, Mg, and Fe)	silicate	6	two perfect, nearly at right angles
Fluorite	CaF_2	halide	4	perfect in four directions
Malachite	$CuCO_3 \cdot Cu(OH)_2$	carbonate	3.5–4	one perfect, but rarely seen
Azurite	$2\ CuCO_3 \cdot Cu(OH)_2$	carbonate	3.5–4	perfect in two directions
Dolomite	$CaMg(CO_3)_2$	carbonate	3.5–4	perfect in three directions at oblique angles
Serpentine	$H_4Mg_3Si_2O_9$	silicate	2.5–4	one perfect
Aragonite	$CaCO_3$	carbonate	3.5	poor in two directions
Calcite	$CaCO_3$	carbonate	3	perfect in three directions at oblique angles
Biotite	$K(Mg,Fe)_3AlSi_3O_{10}(OH)_2$	silicate	2.5–3	perfect in one direction
Muscovite	$KAl_2(AlSi_3O_{10})(OH)_2$	silicate	2–3	perfect in one direction
Bauxite	$Al_2O_3 \cdot 2\ H_2O$	oxide (hydrous)	2.5	none
Halite	NaCl	chloride	2.5	three perfect, at right angles
Sulfur	S	native element	1.5–2.5	very poor in two directions
Kaolinite	$Al_2Si_2O_5(OH)_4$	silicate	2–2.5	one perfect
Gypsum	$CaSO_4 \cdot 2\ H_2O$	sulfate	2	one perfect
Talc	$Mg_3Si_4O_{10}(OH)_2$	silicate	1–1.5	one perfect

Common color(s)	Streak color	Other properties
dark greeen to black or brown	white to pale green	long, six-sided crystals, irregular grains, and fibers
dark greeen to black	white to pale green	eight-sided stubby crystals and granular masses
highly variable	none or white	glassy; cubic crystals
green	green	copper ore mineral
blue	blue	copper ore mineral
clear, yellow, gray, or pink	none	glassy; crystals with rhomb-shaped faces
light to blackish green	white	splintered or layered appearance; greasy feel
clear or white	none or white	massive or slender, needlelike crystals; effervesces with dilute hydrochloric acid
clear, yellow, white	none or white	glassy; crystals with rhomb-shaped faces; effervesces with dilute hydrochloric acid
brown to black	black	transparent in thin sheets; flakes are elastic
clear to yellow-brown	white	transparent in thin sheets; flakes are elastic
white, tan, red, brown, black	tan to brownish red	pebblelike character; earthy appearance
clear to dark gray	none or white	glassy; cubic crystals; salty taste
yellow to red	pale yellow	conchoidal fracture; brittle
white, yellow, pink, reddish	white	chalky; soft, earthy masses
clear to yellow	white	glassy, transparent plates or satin-white rods
silvery to greenish white	white	layered appearance; greasy feel

Appendix 5

Recommended Journals and Web Sites

One way to keep abreast of important issues in science, technology, and the environment is by reading journals, either those written specifically for the interested and informed nonscientist or the more public-oriented of the professional publications. A list of eight such journals and magazines follows. These are well written, available in almost every college or university library, and highly regarded for their accuracy and timeliness. We have ranked them in order of accessibility, with the least technical placed first. The next time you go to the library, look for some of these and browse through them, reading articles that catch your attention. This is one of the best ways to stay informed about environmental issues, including desertification, earthquake hazards, climate change, acid rain, deforestation, oil spills, and coastal erosion. Another way is by using the Internet's World Wide Web.

The addresses and brief summaries of 20 Web sites that provide a wealth of geologic information appear below.

Journals

1. *Geotimes* is a monthly publication of the American Geological Institute, a nonprofit organization. Its audience is professional Earth scientists, educators, industry and government decision-makers, and scientifically literate members of the general public. It covers important events and news in the geosciences, follows political activities and technological advances, and reviews books, maps, and software. An especially interesting part of the magazine is a listing of current geologic phenomena, including events such as landslides, volcanic eruptions, and earthquakes.

2. *U. S. Water News,* a monthly newsmagazine copublished by U.S. Water News, Inc., and the Freshwater Foundation, has the most timely information available on issues such as bottled drinking water, flooding, and Superfund cleanup efforts by the U.S. Environmental Protection Agency. It contains global as well as national news about water supply, water quality, wetlands, climate and water, water rights, and water policy and legislation.

3. *Science News* is a weekly publication of Science Service, a nonprofit organization with the goal of increas- public understanding of science. Each issue con- to 10 short summaries of important events in t week, which might be the most recent data on acid rain in the United States or the use of radar to monitor movement of glaciers. Several longer articles appear on timely issues such as temperature measurements on Venus or estimates of rates of deforestation. To students, an especially interesting feature is the summary of recent research in archaeology, Earth sciences, environment, astronomy, biology, science and society, technology, and other fields.

4. *Earth: The Science of Our Planet,* a glossy magazine published bimonthly by Kalmbach Publishing, is a geoscientist's and outdoor explorer's dream. Each issue has full-color articles on such topics as tsunamis, earthquakes, caves, the ocean floor, the links between mountain building and climate change, and remaining supplies of oil in the world. The articles, many of them written by Earth scientists, are fascinating. The magazine does a superb job of covering complex scientific material in an exciting and easy to follow style. Special sections each month include reviews of books and software, maps, exploration gear, and other products. A calendar lists seminars, mineral and fossil shows, museum exhibits, and more.

5. *Oceanus: International Perspectives on Our Ocean Environment* is a colorful, authoritative journal published quarterly by Woods Hole Oceanographic Institute that features articles and reports on coral reefs, ocean floor mining, oil spills, ocean law, and many other topics. The writing style and scientific content are excellent, making it highly appealing to both the interested nonprofessional as well as professionals who want to follow important issues in ocean environments.

6. *Environment* is a journal published ten times a year by Heldref Publications, a division of the nonprofit Helen Dwight Reid Educational Foundation, in cooperation with the Scientists' Institute for Public Information. As one of the oldest and most respected environmental journals, it is a rich resource for term papers and projects, with many timely issues in science, environment, and technology explained and analyzed in well-written, thorough articles containing much factual information. Topics have included desertification, climate change, acid rain, ozone depletion, global warming, and antipollution technologies. It is a good source of reports on the outcome of environmental meetings, such as the United Nations Conference on Environ-

ment and Development, and it has a section of short reports on environmental news.

7. *GSA Today,* a monthly newsmagazine published by the Geological Society of America, is directed toward professional earth scientists and educators, but it has so many excellent articles on important issues that it is worth noting to nonscientists or future scientists. The feature article in each issue generally includes some fairly technical information, but the topic is always important and timely. Other sections include reports on public policy and debates on issues such as opening Alaskan public lands to oil drilling and exploration. The debate articles, written by experts in the field, are invaluable in presenting both sides of an issue.
8. *Scientific American* is a monthly newsmagazine that appeals to a wide audience of scientists and those interested in science. The magazine is highly regarded for its beautiful illustrations; its scientific artwork is some of the best in the publishing world. Each issue contains a short section of news and analysis followed by several fully illustrated feature articles on a wide range of topics in science, technology, and society, including computer technology, genetics, medicine, biology, chemistry, and mathematics. Recent articles in the geosciences have explored asbestos, sea level rise, global population, the nitrogen cycle, water on Mars, nuclear-waste disposal, and energy resources. Many of the articles are written by scientists at the forefront of their fields. Special issues, which are fine sources, are devoted to single topics, such as managing resources and energy for planet Earth.
9. *Discover,* a monthly newsmagazine with a broad audience, contains a news section that summarizes recent breakthroughs and articles on topics in chemistry, physics, medicine, linguistics, mathematics, biology, psychology, and Earth sciences.

Environmental Geology Web Site Sampler

UNITED STATES GEOLOGICAL SURVEY (USGS)
http://www.usgs.gov/
Contains a wealth of geologic information on hazards, resources, and current USGS projects. It can be used to follow current flooding and stream discharge at any of the USGS gauging stations, to access geologic maps, and much more.

EARTH AND ENVIRONMENTAL SCIENCE
http://information.er.usgs.gov/network/science/earth/index.html
Offers many links for climate, earthquakes, environment, hydrology, oceanography, and volcanoes. This list covers some things that the USGS does not handle, so between this list and the information from the USGS, almost every topic in geology is covered in detail. Great reference tool and fun to browse.

NATIONAL GEOPHYSICAL DATA CENTER
http://www.ngdc.noaa.gov/ngdc.html
Contains specific information on satellite data, glaciology, marine geology, geophysics, paleoclimatology, solar and terrestrial physics, and solid earth geophysics, and is best suited to answering a particular question or finding particular data.

COMET.NET EARTHQUAKE PAGE
http://www.comet.net/earthquake
News about recent earthquakes, eyewitness accounts, links to other sites, research centers and their projects, seismology departments. Slides, films and maps for sale. Site has numerous helpful links.

ASTEROID AND COMET IMPACT HAZARD
http://ccf.arc.nasa.gov/sst/
Spaceguard survey report, congressional testimony and statements, list of asteroids with near earth orbits and their orbital elements and future close passes. Lists of new books, articles and research. Latest news.

PIMA COMMUNITY COLLEGE CURRENT GEOEVENTS
http://www.azstarnet.com/~dshakel/linkspg1.html
List of current geological hazards, particularly volcanoes, with information and history of each "geoevent." Covers current eruptions well, but little on past eruptions.

GLOBE GALLERY
http://hum.amu.edu.pl/~zbzw/glob/glob1.htm
Many global images, some of which take a long time to load but are worth it if you are in need of specific information, such as ocean water temperatures or the changing configuration of the ozone hole.

UNITED NATIONS WATER RESOURCES
http://www.un.org/dpcsd/dsd/freshwat.htm
Comprehensive assessment of the freshwater resources of the world.

CARBON CYCLE AND MISSING CARBON
http://www.cru.uea.ac.uk/~markn/carbon/back grnd.htm
This site provides up-to-date information on carbon cycling in Earth systems.

GEOLOGIC SOCIETY OF AMERICA (GSA)
http://www.aescon.com/geosociety/index.htm
Information about GSA administration, policies, and membership; and the books and maps available through the GSA bookstore.

VOLCANO WORLD
http://volcano.und.nodak.edu/vw.html
General information about volcanoes, and parks and monuments of interest. Covers recently recorded eruptions.

JET PROPULSION LABORATORY (JPL)
http://www.jpl.nasa.gov
Current news about projects like NASA's Mission to Planet Earth, a long-term program that monitors Earth with satellite data. Information on current planned planetary exploration missions. Also good descriptions of past missions.

VIEWS OF THE SOLAR SYSTEM
http://bang.lanl.gov/solarsys
Educational tour of the solar system.

PRINCETON EARTH PHYSICS PROJECT
http://www.lasker.princeton.edu/pepp.shtml
Covers many activities related to earthquakes, and lists recent earthquake activity (date, time, location of major events). Links to other earthquake research institutes.

PLANET EARTH HOMEPAGE
http://godric.nosc.mil/planet_earth/information.html
Many links to Earth-related sites.

Environmental Geology HOME PAGE
http://www.whfreeman.com/environmental geology
Provides access to many additional Web sites for each chapter of *Environmental Geology.*

Glossary

Note: *Italicized* terms in the definitions also appear as separate entries.

Absolute age The numerical age of a *rock,* fossil, or geologic feature. Compare *Relative age.*

Absolute age dating Methods of estimating the *absolute age* of a *rock,* fossil, or geologic feature, most often by *radiometric dating.*

Abyssal plain A flat or nearly flat plain on the *deep-ocean* floor.

Acid mine drainage Acidic *runoff* from a mine or mine dump.

Acid rain Precipitation with a *pH* less than 5.0, usually caused by human activities that introduce significant amounts of acids into the local *hydrologic cycle.*

Active volcano A *volcano* that has erupted within recorded history.

Aerosols Extremely fine particles or droplets carried in atmospheric suspension.

Aftershock One of the smaller, residual shocks that frequently occur soon after the main *earthquake.*

Air mass A parcel of air of homogeneous temperature, humidity, and density that moves as an entity.

Alluvial fan A sloping deposit of *sediment* formed when a stream travels from a steep to a nearly level slope, or from a confined to an unconfined flow.

Alluvium *Sediment* deposited by a stream.

Alpha decay The type of *radioactive decay* in which an *atom* emits a helium *nucleus* (two *protons* and two *neutrons*), thereby reducing its *atomic number* by 2.

Andesite Fine-grained, volcanic (*extrusive igneous*) *rock* with a composition between *rhyolite* and *basalt;* usually found in *volcanic arcs*. Compare *Diorite.*

Angle of repose The steepest angle of slope that a *sediment* can maintain without becoming unstable.

Anhydrite A *mineral* ($CaSO_4$) composed of calcium and sulfate *ions* that forms when warm, salty surface water evaporates.

Anion An *ion* with a negative charge.

Anthracite The highest-grade, *metamorphic* form of *coal;* it is black, harder and shinier than the *sedimentary* forms, and contains more than 90 percent carbon.

Anticlinal trap A *petroleum trap* formed when layers of *rock* of varying *porosity* are folded into an arch and *petroleum* accumulates within the arch.

Aquifer A permeable body of *rock* or *sediment* that stores and transmits enough water to supply wells or springs.

Artesian flow The flow of *groundwater* in a *confined aquifer,* where the pressure is great enough to force the water in a well up past the level of the *confining layer.*

Artesian spring A natural flow of *groundwater* at Earth's surface that occurs when water under high pressure in a *confined aquifer* is able to reach the surface, for example along a *fault.*

Artesian well A well that taps water under high pressure in a *confined aquifer* and needs little or no pumping.

Asthenosphere The part of Earth that extends for about 250 km below the *lithosphere* and which deforms in a plastic manner.

Atmosphere The envelope of gases that surrounds Earth's surface.

Atom The smallest possible particle of a given *element* that retains all the physical and chemical properties of that *element;* typically composed of a *nucleus* containing *protons* and *neutrons,* and one or more *electron* shells.

Atomic number The number of *protons* in the *nucleus* of an *atom*; the number determines the *element* of that atom.

Barrier island A long, narrow island nearby and parallel to the shore, separated from the mainland by either open water or salt marsh.

Basalt Dark, fine-grained, *extrusive igneous rock* that is composed of mainly *silicate minerals* rich in iron and magnesium. Compare *Gabbro.*

Base level The level below which a stream cannot continue to erode a channel; usually sea level.

Baseflow The inflow from *groundwater* that determine the average water level in a stream channel or river.

Bathymetry The measurement of ocean depth, for the purpose of mapping seafloor topography.

Bauxite An ore of aluminum composed of a mixture of hydrous aluminum oxides and aluminum hydroxides.

Beach Loose *sediment,* usually *sand,* that covers the shore.

Beta decay The type of *radioactive decay* in which an *atom* releases an *electron* from the *nucleus,* thereby converting a *neutron* into a *proton* and increasing the *atomic number* by one.

Big Bang theory The scientific *theory* that all the matter and *energy* in the universe originated during an explosive cosmic event that took place about 10 billion to 20 billion years ago.

Biodiversity The multiplicity of life-forms on Earth.

Biological amplification The accumulation and concentration of chemical *compounds* at each succeeding level in the food chain; also referred to as biomagnification.

Biomass The quantity of living matter in a certain area; also, plant and animal matter and their wastes (e.g., dung), which can be used as a fuel source.

Bioremediation The use of microbes to clean up oil spills and contaminated *soils* and *aquifers.*

Biosphere The part of Earth's exterior shell that includes living and nonliving organic matter.

Bituminous coal A black *sedimentary rock* that is the intermediate grade of *coal* and contains 60 to 90 percent carbon.

Braided stream A high-*energy* stream carrying such a heavy load of *sediment* that the water forms many small interweaving channels.

Caldera A large crater (more than 1.5 km in diameter) resulting from the collapse of a *volcano* into the emptied *magma* chamber.

Carbon cycle The continuous flow, or cycling, of carbon from one Earth *system* to another.

Carbonation The *chemical weathering* process that occurs when carbonic acid dissolves *minerals* and *rocks,* particularly calcite and *limestone.*

arrying capacity The population of a species that can supported with food, water, and other necessities by en area of land.

nverter A *system* in the engine of a car that -producing waste into relatively benign

charged *ion.*

Channelization The process of replacing natural streambanks with artificial ditches, usually in order to straighten and deepen the natural channel and protect the surrounding land against floods.

Chemical weathering The dissolving or altering of *minerals* to other forms as a result of chemical reactions that take place in *rocks* exposed to water and the *atmosphere.* Compare *Physical weathering.*

Chlorofluorocarbons (CFCs) A group of organic *molecules* that are able to release chlorine *atoms,* which then destroy ozone molecules in the *stratosphere.*

Clastic sediments Particles deposited physically, by wind, running water, or ice.

Clay *Mineral* fragments smaller than 0.0039 mm, or a *sediment* composed of *clay minerals.*

Clay minerals Hydrous aluminum *silicate minerals* with a layered structure, such as kaolinite.

Cleavage The property by which minerals break at certain regular planes of weakness along the crystal lattice.

Climate The long-term atmospheric and surface conditions that characterize a particular region. Compare *Weather.*

Closed system A *system* that can exchange *energy* but not matter across its boundaries. Compare *Isolated system; Open system.*

Coal A *fossil fuel* encompassing three different grades of *rock* that originate as *peat* and contain mainly carbon *compounds* that burn easily; the lowest grade of coal is *lignite,* the intermediate grade is *bituminous coal,* and the highest grade is *anthracite.*

Compound A chemical substance consisting of two or more *elements* bonded in a specific way.

Concentration factor The ratio of the abundance of an *element* in a particular *ore deposit* to its average crustal abundance.

Conduction The mechanism of heat transfer by which thermally agitated *molecules* pass their *energy* to adjacent molecules.

Cone of depression The cone-shaped dip in the *water table* that forms around a well when the amount of water being removed from the well exceeds the flow of water into the well.

Confined aquifer An *aquifer* in which the water is held under pressure between *strata* that are impermeable or have very low *permeability.*

Confining layer A layer of *rock* or *sediment* that is impermeable, or has very low *permeability,* and impedes the flow of *groundwater.*

Conglomerate A *sedimentary rock* composed mainly of rounded pebbles, cobbles, and boulders.

Constructional shoreline A shoreline made up of tufa, *sedimentary rock* formed from freshwater deposits where springs empty into the margins of a saline lake.

Consumption overpopulation The use of *resources* at a very high rate by a small number of people.

Continental crust The buoyant, granitic *rock* that makes up the base of continents to a depth of about 35 to 50 km.

Continental drift The horizontal movements of the continents relative to each other across Earth's surface.

Continental margin The submerged area that spans the *continental shelf, continental slope,* and *continental rise.*

Continental rise A sediment-covered region of the seafloor that rises from the *abyssal plain* to meet the *continental slope.*

Continental shelf The submerged area of a continent that lies between the coast and the upper *continental slope.*

Continental slope The portion of the seafloor that slopes steeply away from the *continental shelf* and ends in the *continental rise.*

Convection The mechanism of heat distribution in a liquid or plastic solid that is heated from the bottom; hot material from the bottom rises and cooler surface material sinks, producing circulation.

Convergent plate boundary A boundary where lithospheric plates collide, resulting in either *subduction* or continental collision with crustal thickening.

Core Earth's center, composed mainly of iron and nickel and differentiated into two concentric regions: an outer layer kept liquid by extreme temperatures and pressures and an inner core that even higher pressures make solid.

Coriolis effect The force resulting from Earth's rotation about its axis that causes moving objects in the northern hemisphere to be deflected to the right of their direction of motion and moving objects in the southern hemisphere to be deflected to the left.

Covalent bond A bond between *atoms* that involves the sharing of *electrons.*

Creep The slow movement of *soil* and *rock* downhill under the influence of gravity.

Cross-beds The inclined planes of deposition in a *sediment* or *sedimentary rock* formed by the action of wind or water.

Crude oil The liquid form of *petroleum.*

Crust Earth's outermost solid layer, made up mostly of *basalt* in the *oceanic crust* and *granite* in the *continental crust.*

Crystallization The formation of crystals from a gas or liquid. See also *Precipitate.*

Cyclothems Repeated cycles of *sedimentary rock* layers that alternate between marine and nonmarine deposits and typically include *shale, limestone,* and *coal.*

Darcy's law A formula used to quantify the flow of fluids through a porous medium, such as an *aquifer;* according to this law, the rate of *groundwater* flow is proportional to the *hydraulic gradient* of the fluid and the *hydraulic conductivity* of the porous medium.

Daughter product An *atom* formed as the result of the *radioactive decay* of an *isotope* that is the *parent element.*

Debris flow The rapid downhill *mass movement* of a coarse mixture of *rock* and mud, caused by the force of gravity, in which the debris moves like a fluid.

Deep ocean The parts of oceans that reach depths of at least 1 km below the ocean surface.

Deep-sea trenches Long, deep depressions in the seafloor, found at *subduction* zones, where one tectonic plate is overriding another; associated with *volcanic arcs.*

Delta A deposit of *sediment,* usually triangular, formed in an ocean or lake at the mouth of a stream or river.

Depositional environment The environment in which *sediments* accumulate and often become *sedimentary rocks.*

Desertification The transformation of land to deserts by a number of processes, including loss of vegetation, reduced *soil* fertility, soil *erosion,* and the trampling of soil by animals.

Destructional shoreline A shoreline produced when waves strike the same place repeatedly, producing a wave-cut bench.

Diorite Coarse-grained *igneous intrusive rock*, the plutonic equivalent of *andesite.*

Discharge (stream) The rate of water flow through a stream, measured in volume per unit time.

Discharge (groundwater) The movement of *groundwater* to water bodies at the surface.

Divergent plate boundary A boundary between tv lithospheric plates that are moving apart, allow *magma* to rise up between them and form new *crust,* usually along a *mid-ocean ridge.*

Drainage basin An area of land that contributes water to a stream or lake.

Drainage divide The boundary between two *drainage basins* that have separate stream systems.

Drainage network The network of *tributaries,* large and small, belonging to a stream system.

Dynamic system A *system* in which *energy* is used to do work that causes the condition, or state, of the system to change with time.

Earth system science The scientific study of the interconnected whole-Earth systems, such as the *lithosphere, hydrosphere,* and *atmosphere,* the cycling of matter and *energy* through them, and the changes that occur in them with time.

Earthquake A sudden violent motion of Earth's crust as a result of movement along a *fault* or of volcanic activity.

Earthquake magnitude A measure of the *energy* released by an *earthquake.* See also *Moment magnitude; Richter magnitude.*

Eccentricity The degree to which a planet's orbit deviates from a circle.

Economic concentration factor The *concentration factor* necessary to make recovery of a *mineral* economically feasible.

Ecosystem An ecological unit comprising the resident organisms and the environment they inhabit.

Ekman transport The movement of water to the left of the wind direction in the southern hemisphere and to the right in the northern hemisphere as a result of the *Coriolis effect.*

El Niño A warm surface *ocean current* that appears every 3 to 7 years off the coasts of Ecuador and Peru, reducing the population of marine organisms.

Elastic rebound model The scientific *theory* that sudden movements of *rock* along a *fault* and the *earthquakes* associated with them result from the release of elastic *energy* along the fault.

ctron A negatively charged subatomic particle that its the *nucleus* of an *atom.*

pture A process of *radioactive decay* in which transforms into a *neutron* by capturing an the innermost orbit, or shell.

nce that consists only of *atoms* with *mber;* it cannot be separated chem- bstances.

Energy The ability to do work.

Energy efficiency The percentage of *energy* not lost as waste heat.

Environmental geology The study of the relationships between humans and their varied geological settings.

Epicenter The point on Earth's surface directly above the *hypocenter,* or focus, of an earthquake.

Epoch A subdivision of a geologic *period,* which can be chosen to correspond to a stratigraphic sequence.

Era A division of *geologic time* intermediate in length between a *period* and an *eon.*

Erosion The transport of *rock* and *sediment* over Earth's surface, largely by gravity, water, wind, and ice.

Estuary A partially enclosed body of water along the coast, where fresh and salt water mix.

Eutrophication The process that begins whereby an extreme influx of *nutrients* (often from sewage and agricultural *runoff*) into a body of water encourages unusually large algae blooms, which prevent light from penetrating the water and thereby inhibit subsurface *photosynthesis* and deplete the water of oxygen.

Evaporite A *sedimentary rock* that originates as a *precipitate* from water as it evaporates from a closed basin.

Evapotranspiration The return of water to the *atmosphere* through evaporation from Earth's surface and the escape of water vapor from plant leaves.

Exclusive economic zone The area extending 200 nautical miles (322 km) seaward from a country's shoreline, within which that country has exclusive mineral and fishing rights.

Expansive soils *Soils* that contain a substantial amount of *clay,* which causes them to expand when wet and shrink when dry.

Exponential growth The repeated multiplication of a quantity by a given exponent; this type of growth is approximated in populations, in which doublings of the population occur in such quick succession that the total number of organisms competing for *resources* increases rapidly.

Extrusive igneous rock A fine-grained *igneous rock* that has cooled from *lava* ejected onto Earth's surface.

Fault A fracture in *rocks* along which there has been some movement. See also *Earthquake.*

Fault trap A *petroleum trap* in which a *fault* has juxtaposed permeable and impermeable *rocks* so that *petroleum* accumulates in the permeable rocks.

Faunal and floral succession, principle of The assumption that fossil assemblages of species of plants and animals succeed each other over time; the basis for *relative age dating.*

Felsic A term describing a light-colored *igneous rock, lava,* or *magma* with high silica content and low iron and magnesium content.

Fission (nuclear) The breaking of the *nucleus* of an *atom* into other, lighter, *elements,* a process that releases *energy.*

Flood frequency The frequency with which floods of a particular magnitude tend to occur.

Floodplain A nearly planar landform composed of *sediment* that lies on either side of a stream and is commonly underwater during floods.

Flux The movement of material and *energy* from one *reservoir* to another.

Foliation A set of planes, produced by deformation, in a *metamorphic rock;* the planes can be either flat or wavy.

Foraminifera Surface and deep-sea protozoa with calcium carbonate shells.

Formation A set of *sedimentary rocks* with similar characteristics that covers a considerable area.

Fossil fuels Combustible deposits of altered organic material, such as *coal, crude oil,* and *natural gas.*

Front The boundary between two *air masses.*

Fusion (nuclear) The combination of two atomic nuclei to form one larger *nucleus,* a process that releases *energy.*

Gabbro The dark plutonic (*igneous intrusive*) equivalent of *basalt;* it contains calcium feldspar and pyroxene minerals.

Gaia hypothesis The *hypothesis* that Earth is a single, self-regulating *system* in which both living and nonliving materials are linked.

Geologic column A composite stratigraphic section that shows the subdivisions of all or part of *geologic time.*

Geologic time The length of time that began with Earth's formation and continues to the present.

Geologic time scale The subdivision of *geologic time* into *eras, periods,* and *epochs,* determined by means of *stratigraphy,* paleontology, and *radiometric dating.*

Geothermal energy *Energy* obtained from the heat contained in the volcanic regions of Earth's *crust.*

Glacial erratic *Sediment* transported by ice as much as hundreds, and sometimes thousands, of kilometers from its source region.

Glacier A mass of ice formed from compacted snow that moves downhill under gravity and lasts throughout the year.

Gondwanaland The supercontinent that existed in the southern hemisphere before the creation of *Pangaea.* Compare *Laurasia.*

Granite Coarse-grained, light-colored, *intrusive igneous rock* made of quartz, orthoclase, sodium-rich plagioclase, and mica; the plutonic equivalent of *rhyolite.*

Gravel Coarse *clastic sediment* made up of particles over 2 mm, such as cobbles and boulders.

Greenhouse effect The warming of Earth's surface by heat reradiated from carbon dioxide and other heat-absorbing gases in the *atmosphere.*

Groin A structure built perpendicular to a *beach* to catch *sand* transported alongshore.

Groundwater Underground water below the level at which all open spaces in *rock* and *sediment* are saturated. See also *Water table.*

Gully A small stream channel several to tens of meters deep, tens to hundreds of meters wide, and up to several kilometers long; typically formed by rapid incision into *sediments* associated with deforestation or the *erosion* of *soil.*

Hadley cell A circulation loop in Earth's *atmosphere* caused by the *Coriolis effect.*

Half-life The time needed for half the number of *atoms* of a given radioactive *isotope* to decay.

Heave A type of *mass movement* in which material on a hillslope is alternately raised and lowered in response to alternating expansion and contraction, typically caused by repeated wetting and drying or freezing and thawing.

Hot spot A spot on a lithospheric plate directly over a *mantle* plume, characterized by volcanic activity.

Humus Decayed organic matter found in a *soil.*

Hurricane A large storm with winds that blow at speeds greater than 120 km per hr.

Hydraulic conductivity The volume of *groundwater* that moves through a unit of area perpendicular to the direction of flow in a unit of time; also referred to as hydraulic *permeability.*

Hydraulic gradient The change in water pressure over horizontal distance in an *aquifer.*

Hydrocarbon An organic *compound* containing only hydrogen and carbon; hydrocarbons may be solid, liquid, or gas, and are major components of *coal, crude oil,* and *natural gas.*

Hydrologic budget The balance between rates of precipitation, *evapotranspiration,* surface *runoff,* and groundwater recharge.

Hydrologic cycle The path traveled by water as it moves from the ocean to the *atmosphere,* back to Earth's surface as precipitation, and returns to the sea through streams.

Hydrologic equation The statement that inflow of water into a *reservoir* or *system* is equal to outflow plus or minus changes in *stock;* for the surface water system, inflow is precipitation and outflow is *evapotranspiration* and *runoff.*

Hydrolysis A *chemical weathering* reaction between a *silicate mineral* and water, usually producing *clay minerals.*

Hydrosphere The shell of water around Earth's surface, made up of oceans, surface water, *groundwater,* water vapor, and *glaciers.*

Hydrothermal ore deposits *Minerals* deposited or altered by the circulation of hot *groundwater.*

Hydrothermal reservoir The store of *energy* from Earth's interior heat.

Hypocenter The point within Earth's *crust* where an *earthquake* occurs; also called the focus..

Hypothesis An explanation of observations based on physical principles.

Igneous ore deposits Deposits of *ore minerals* within *igneous rocks;* typically originate along *volcanic arcs* in zones of plate convergence.

Igneous rock *Rock* formed from cooling *magma* or *lava.*

Immiscible contaminants In *groundwater,* contaminants that are insoluble.

Infiltration capacity The ability of the ground to absorb water, usually expressed in terms of rate of infiltration.

Injection wells Wells through which water is pumped into the ground to maintain the pressure of the *reservoir* or to dispose of chemical wastes.

Inner core See *Core.*

Intermediate magma *Magma* that is intermediate in composition between *rhyolite* and *basalt,* forming *andesite* extrusively and *diorite* intrusively.

Intrusive igneous rock The new *rock* formed when *magma* that has intruded into the rock cools.

Ion An *atom* that has an electrical charge because it has gained or lost one or more *electrons.*

Ionic bond A bond between two *atoms* resulting from the attraction between *ions* with opposite charges.

Iridium An *element* common in certain types of meteorites but uncommon on Earth.

Island arc See *Volcanic arc.*

Isolated system A *system* that does not allow either *energy* or matter to cross its boundaries. Compare *Closed system; Open system.*

Isotope One of the possible forms of the same *element,* differing in the number of *neutrons* in the *nucleus* but having a constant number of *protons.*

Jetty A structure built out from the shore to prevent accumulation of *sediment* in a harbor.

Kinetic energy The *energy* of a moving body.

Land subsidence The slow sinking of an area of *crust* without major deformation.

Lateral migration (of streams) The process by which streams shift to one side over time because of their tendency to erode the outer banks of the bends in their channels and deposit *sediment* at the inner banks of the bends.

Laterite A *soil* rich in oxides, iron, and aluminum that is found in very humid, typically warm regions and forms as a result of extensive *chemical weathering* of feldspars and other *silicate minerals.*

Laurasia The northern supercontinent that existed before the creation of *Pangaea.* Compare *Gondwanaland.*

Lava *Magma* that has erupted onto Earth's surface.

Layered intrusive (ore deposit) A deposit containing *ore minerals* that forms when *metals* from a *magma* become segregated and layered in the magma chamber, commonly by sinking to the bottom of the chamber.

Leading-edge coast A coastal area that is the site of convergent or transform plate motion, which results in *volcanoes, earthquakes,* mountain building, and a steep coastline cut by high-*energy* waves. Compare *Trailing-edge coast.*

Levee A ridge of fine-grained *sediment* deposited along the banks of a river during floods.

Lignite A brown *sedimentary rock* that is the lowest grade of *coal,* softest and with the lowest carbon content (less than 60 percent). Compare *Anthracite; Bituminous coal.*

Limestone A *sedimentary rock* made mainly of calcium carbonate, that is either precipitated out of seawater or available from carbonate shell debris.

Lithification The process of converting *sediment* into *rock* by compaction, cementation, and *crystallization.*

Lithosphere Earth's outermost solid layer, about 100 km thick, made up of the *crust* and upper *mantle.* See also *Pedosphere.*

Little Ice Age An interval of cold *weather* that took place during the current interglacial stage and lasted approximately from the late 13th century to the late 19th century.

Mafic Describing *rocks* or *magmas* rich in iron and magnesium.

Mafic magma *Magma* rich in iron and magnesium that solidifies into *basalt* and *gabbro.*

Magma Molten *rock* below Earth's surface that can solidify to *igneous rock.*

Mantle The layer of Earth, composed of *silicate minerals* rich in iron and magnesium, that lies between the *core* and the *crust.*

Mass movement The movement of *rock* or *soil* downhill under the influence of gravity; can be either slow, as in *creep*, or very fast, as in landslides or *debris flows.*

Mass wasting The *erosion* of Earth materials that occurs when *soil* and *rock* move downslope as a result of gravity.

Meandering stream A stream characterized by a channel that winds freely on a broad floodplain in prominent, sinuous curves.

Medieval Warm Period A time of relatively warm conditions that existed in many places on Earth from about AD 1100 to AD 1300.

Meltdown The melting of the structure within a nuclear power plant that contains atomic reactor fuel; caused by uncontrolled *fission.*

Mercalli scale A method of rating the intensity of an *earthquake,* I to XII, based on the nature of the damage caused and the subjective perception of the event rather than on the actual amount of *energy.*

Mesosphere The rigid lower part of the *mantle* between the *asthenosphere* and the *core.*

Metal A class of substances, including copper, iron, and aluminum, that typically have a luster and are good conductors of electricity.

Metallic bond A *covalent bond* in which *electrons* are free to wander among the bonded metallic *ions.*

Metamorphic rock *Rock* that has been altered by heat or pressure so that its texture or composition has been transformed.

Midden An accumulation of twigs, leaves, and bark collected by packrats, used for carbon dating and reconstructions of past *climates.*

Mid-ocean ridge A long, underwater mountain chain found at a *divergent plate boundary*, where new ocean floor is being created.

Milankovitch cycles Variations in Earth's orbital parameters—*eccentricity*, tilt, and *precession*— that lead to varying receipts of solar radiation over time, and hence to changing climatic conditions on Earth.

Mineral A natural, inorganic solid with a definite chemical composition and crystal structure.

Mineral reserves Known recoverable *mineral* deposits that have not yet been removed.

Miscible contaminants In *groundwater,* contaminants capable of dissolving in water.

Modified Mercalli scale A version of the *Mercalli scale* used in North America.

Mohs scale of hardness A scale on which *minerals* are rated in respect to hardness from 1 to 10 by comparison with other minerals.

Molecule A group of two or more *atoms* bonded together; the smallest piece of a *compound* that retains its chemical characteristics.

Monsoons Winds created by the presence of a high plateau in the interior of a continent.

Moment magnitude A scale of *earthquake magnitude* based on the product of the area of slippage along a *fault,* the amount of slip, and *rock* rigidity. Compare *Richter magnitude.*

Moraine A deposit of poorly sorted glacial *till* left behind by a *glacier* as it retreats; varieties include lateral, medial, and terminal.

Mudcrack An indicator of dessication in fine-grained sedimentary deposits.

Natural disaster A sudden and destructive environmental change as a result of long-term geologic processes that appears to occur without notice.

Natural gas The gaseous form of *petroleum.*

Negative feedback The process by which a change in one direction leads to events that reverse the direction of change.

Net primary production The amount of organic matter produced in a given area over a given time interval less that part consumed by respiration.

Neutron An uncharged particle in the *nucleus* of an *atom.*

Nitrogen cycle The continuous flow, or cycling, of nitrogen from one Earth *system* to another.

Nonrenewable resources *Resources* such as *fossil fuels* that have a finite *stock* and are exhaustible.

Non–point-source pollutants Pollutants that occur in *runoff* from the land surface and cannot be traced to a specific source.

Nucleus The central portion of an *atom,* comprising nearly all its mass because it contains the *protons* and *neutrons.*

Nutrients *Compounds* needed to promote proper growth, such as, for plants, nitrates and phosphates.

Ocean currents Distinct masses of water that flow throughout the oceans and carry matter and *energy* from one part of Earth's surface to another.

Ocean trench A long depression on the ocean floor along the boundary between two lithospheric plates where one is being subducted beneath the other.

Oceanic crust The part of Earth's outermost solid layer that lies under the oceans; it is thin (~7 km) and mostly *basalt.*

Oil See *Crude oil.*

Oil shale *Shale* that contains substantial amounts of organic matter and can yield liquid or gaseous *hydrocarbons.*

Open-pit mining Mining done by excavating a large area of the surface, usually to obtain large quantities of low-grade ore.

Open system A *system* that allows both matter and *energy* to flow across its boundaries. Compare *Closed system; Isolated system.*

Ore deposit A *rock* containing *ore minerals* in sufficient concentrations and quantity to be mined economically.

Ore mineral A *mineral* that is commercially valuable and that occurs in concentrations large enough to be mined economically.

Original horizontality, principle of The assumption that all *sedimentary rocks* were deposited originally in horizontal layers.

Outer Core See *Core.*

Overbank deposition The deposition of fine *sediment* on the *floodplain* of a river during a flood.

Oxidation A process of *chemical weathering* in which an *element* combines with oxygen to form oxide or hydroxide minerals.

Paleoclimatology The study of ancient *climates* on Earth through fossil evidence.

Pangaea The single supercontinent that existed between 300 million and 200 million years ago and that comprised the progenitors of all the continents on Earth today.

Parent element An unstable *element* that through *radioactive decay* forms *daughter products.*

Peat A deposit of partially decayed plant remains that accumulated in a *wetland* environment and still contains substantial amounts of moisture; the basis for *coal* formation.

Pedosphere The layer of decomposed *rock* particles and organic matter covering Earth's surface.

People overpopulation A circumstance in which a large number of people have insufficient *resources* and therefore overuse them.

Period A unit of *geologic time* that subdivides an *era.*

Permeability The ability of a *rock* to allow the flow of *groundwater* through its pores. See also *Hydraulic conductivity.*

Perpetual resources *Resources* that are inexhaustible on the human time scale.

Petroleum A liquid or gaseous form of organic matter composed of *hydrocarbons, compounds* consisting of *atoms* of carbon and hydrogen. See also *Crude oil; Natural gas.*

Petroleum trap Any geologic structure that confines the flow of *petroleum.*

pH The "potential for hydrogen," a measure of the hydrogen *ion* concentration in a water-based solution; a pH of 7 is neutral, lower than 7 is acidic, and higher is alkaline.

Photic zone The uppermost layer of the ocean, where enough light penetrates to allow *photosynthesis.*

Photochemical smog Haze that forms when ultraviolet light strikes automobile exhaust emissions and transforms them into health-damaging oxidants. See also *Smog.*

Photosynthesis The metabolic process by which plants use chlorophyll and *solar energy* to convert water and carbon dioxide to larger organic molecules.

Photovoltaic cell A component of solar panels, made of a semiconductor material, that transforms *solar energy* into electrical *energy*.

Physical weathering The mechanical processes by which a *rock* is broken into smaller pieces. Compare *Chemical weathering*.

Placer An *ore deposit* formed when relatively dense *ore minerals*, such as gold, become mechanically segregated from other minerals, typically in streambeds along the quiet-water zones on the outsides of meander bends or in deep *pools* at the base of a waterfall.

Plate tectonics, theory of The scientific *theory* that describes the origin and interaction of lithospheric plates, explaining the occurrence of *continental drift, earthquakes, volcanoes,* and mountains.

Pleistocene An *epoch* of the Quaternary *period* between 1.6 million and ~10,000 years ago, a time of heavy glaciation.

Point-source pollutant A pollutant that is released from an identifiable source.

Pollution The contamination of a substance with another, undesirable, material.

Pollution overpopulation A circumstance in which a small or large number of people use technology that is grossly polluting.

Pool An area of deep water in a stream.

Population growth rate The rate of increase of a population, usually expressed as a percentage.

Porosity The ratio of pore volume in a *rock* to its total volume, expressed as a percentage.

Positive feedback The process by which a change in one direction promotes continued change in the same direction.

Positron decay The type of *radioactive decay* in which a *proton* turns into a *neutron* by emitting a positron.

Potential energy The *energy* of a body that results from its position within a *system*.

Potentially renewable resources *Resources* that can be depleted in the short term by rapid consumption but can be replaced in the long term by natural rates of replenishment.

Precession of the equinoxes The phenomenon in which the tilt of Earth's axis changes with time, resulting in a gradual shift of the onset of the different seasons (and hence the equinoxes) over a period of thousands of years.

Precipitate A *compound* that crystallizes and falls out of solution.

Primary wave (P wave) The faster subsurface *seismic wave* generated by an *earthquake*, traveling through solid *rock* as a compressional wave. Compare *Secondary wave (S wave)*.

Proton A positively charged particle in the *nucleus* of an *atom*.

Pycnocline A zone of ocean water in which density or temperature changes relatively rapidly with depth.

Pyroclastic Describing fragmented materials, such as boulders, cinders, and ash, that are ejected from a *volcano*.

Radioactive decay The emission of *energy* and/or subatomic particles from the nuclei of unstable *isotopes*; radioactive *elements* include radon and uranium.

Radiometric dating The dating of *rocks* by measuring the abundance of radioactive *isotopes* and their stable products.

Rainsplash A process of the *erosion* of *soil* in which raindrops hit exposed (unvegetated) soil during intense rainstorms and lift fine particles above the soil surface, causing some of the loosened soil to move downslope.

Recharge basins Basins that have been constructed to hold water in order to resupply *groundwater* by allowing the water to infiltrate over time.

Recurrence interval The average length of time separating repetitions of a geologic event such as a flood or *earthquake*.

Red beds Ancient dune deposits colored red by iron oxide stains on the grain surfaces.

Relative age The age of one geologic feature relative to another without reference to *absolute age*.

Relative age dating Methods of estimating the age of a *rock*, fossil, or geologic feature relative to that of some other object or feature.

Reserves Deposits of *minerals* or *fossil fuels* that are recoverable but not yet excavated. Compare *Resources*.

Reservoir A place of residence for a store of a particular material; e.g., the *pedosphere* in relation to carbon and nitrogen.

Residence time The average amount of time that an *element* stays in a *reservoir*.

Resources Deposits of *minerals, fossil fuels,* or other valuable material, discovered or undiscovered, that may be used at some time in the future. Compare *Reserves*.

Rhyolite Fine-grained, light-colored *extrusive igneous rock* consisting of quartz, orthoclase, sodium-rich plagioclase, and mica; the volcanic equivalent of *granite.*

Richter magnitude An approximation of the amount of *energy* released by an *earthquake,* found by taking the logarithm of the largest amplitude recorded on a seismogram and correcting for distance to the *epicenter.* Compare *Moment magnitude.*

Riffle An area of shallow water in a stream.

Rill A small stream channel several centimeters in width and depth that typically forms by rapid incision into *sediments* as a result of deforestation or the *erosion* of *soil.*

Rill erosion The carving of small channels up to 25 cm deep in *soil* by running water.

Rock Any naturally formed coherent, relatively hard, assemblage of *minerals.*

Rock cycle The geologic cycle governing the production, alteration, and destruction of *rocks* resulting from processes such as *volcanism, weathering, erosion,* and *lithification.*

Runoff The quantity of water, usually rainwater, that leaves an area in surface drainage and does not sink into the *soil.*

Saline intrusion The infiltration of salt water into the fresh-water supply of coastal regions.

Salinity The concentration of salt in seawater, expressed in parts per thousand.

Salt dome trap A *petroleum trap* in which salt forms large, impermeable, dome-shaped masses that trap *petroleum* along their walls.

Sand Small grains of *rock,* between 63 µm and 2 mm in diameter.

Sandstone A *sedimentary rock* made up of *sand*-sized particles, usually quartz, cemented together into solid *rock.*

Scarp A break in slope that results from movement along a *fault* and offsets physical and structural features, such as *rock* layers and roads.

Scientific method A method of investigation that involves the collection of data, the formation of hypotheses, and the testing of those hypotheses.

Seafloor spreading The process by which new seafloor is produced at *mid-ocean ridges,* causing neighboring lithospheric plates to move apart.

Seawall A structure made of *rock* or concrete built to protect land along the shore from waves.

Secondary wave (S wave) The slower subsurface *seismic wave* generated by an earthquake that arrives after the *primary wave (P wave);* it travels as a shearing wave but cannot travel far through water. Compare *Primary wave (P wave).*

Sediment Pieces of organic or inorganic material that are deposited by wind, water, or ice on Earth's surface.

Sedimentary rock *Rock* lithified from deposits of *sediment.*

Seismic wave Any of the elastic vibrations produced by *earthquakes,* such as *primary waves, secondary waves,* and *surface waves.*

Seismicity The degree, in both frequency and *earthquake magnitude,* to which *earthquakes* occur in a certain area.

Seismograph A graphic representation of the passage of *seismic waves* from an *earthquake.*

Seismometer An instrument that detects and records the passage of *seismic waves* from an *earthquake.*

Shale A layered *sedimentary rock* made of *silt* and *clay.*

Sheet wash erosion The removal of successive layers of *rock* or *soil* from a gentle slope by thin sheets of running water.

Shield volcano A gently sloping volcanic landform with a very broad base, formed by successive eruptions of basaltic *lava* that spreads out in thin sheets.

Silicate See *Silicate minerals.*

Silicate minerals Minerals based on the silicate *ion* (SiO_4); the most abundant mineral group in Earth's *crust.*

Silt Tiny particles, between 4 and 63 µm in diameter, intermediate in size between *clay* and *sand.*

Siltstone A *sedimentary rock* made up of *silt.*

Sinkhole A topographic feature caused by the collapse of underground *limestone* caverns.

Slide A type of *mass movement* in which relatively cohesive blocks of material move, or fail, along a well-defined plane.

Smog A complex mixture of ozone, nitrogen oxides, and *hydrocarbons* in the lower *atmosphere* that results from industrial activities and automobile exhausts.

Soil Loose material that accumulates on Earth's surface and is composed of *clay, sand,* and *humus.*

Soil horizons A unique layer of *soil* that differs from others because of its color, texture or composition.

Soil profile A section of *soil* through all its *soil horizons.*

Solar energy *Energy* emitted by the Sun.

Solar-thermal power plant A power plant that generates electricity by means of *solar energy* and heat *energy.*

Specific retention The ratio of *groundwater* retained by *surface tension* to total volume of water.

Specific yield The ratio of the volume of water that is drained from an *aquifer* by gravity to its total volume.

Static system A *system* in which no work is done and no change in state occurs.

Steady state A *system* in which inflows balance outflows.

Stock The content of a *reservoir* at any given time.

Stormflow Surface water *runoff* resulting from storms.

Strata Layers of *rock.*

Stratigraphy The study of the origin and significance of *strata* of *sedimentary rocks.*

Stratosphere The layer of the *atmosphere* above the *troposphere;* a major site of ozone production.

Stratovolcano A cone-shaped volcanic landform made up of alternating layers of viscous *lava* and *pyroclastic* material.

Strength The degree to which a material resists deformation.

Stress The amount of force acting on a body per unit area.

Striations Grooves found in *rocks* that have been ground against one another during transport by *glaciers.*

Stromatolites Matted masses of living or fossilized organisms, such as blue-green algae, that form in shallow water.

Subduction The sinking of one plate beneath another that occurs at *convergent plate boundaries.*

Subsidence See *Land subsidence.*

Superposition, principle of The assumption that a stratum that overlies another stratum is the younger of the two; the basis for *stratigraphy.*

Surface mixed layer The layer of the ocean, up to about 200 m deep, that is mixed by the wind.

Surface tension The attraction between *molecules* at the surface of a liquid.

Surface wave A *seismic wave,* slower than a *secondary wave,* that follows the surface of Earth.

System A group of interrelated and interacting objects and phenomena.

Systems thinking An approach whereby models of a *system* are constructed in order to test the possible impacts of changes on different parts of the system.

Tailings Waste *rock* discarded after mining.

Tar sand A sandy deposit containing tarry organic matter from which *hydrocarbons* can be extracted.

Tectonism The deformation of planetary *crust* by folding, faulting, volcanism, and other processes.

Theory A *hypothesis* that has survived repeated testing.

Thermal energy Heat-related *energy* produced by the motion of *atoms.*

Till A glacial deposit consisting of a mixture of particles ranging from *clay-* to boulder-size.

Tolerable soil loss The amount of *erosion* of *soil,* or rate of soil loss, below which that soil can sustain a high level of crop productivity indefinitely.

Tornado A violently rotating wind that forms a vertical, elongated vortex; common in the midwestern United States.

Trailing-edge coast A coastal area that is not the site of *subduction* or plate collision but is located in the interior of a lithospheric plate; characterized by gentle slopes and low-*energy* waves. Compare *Leading-edge coast.*

Transform fault A *fault* along which there is horizontal movement; found along boundaries between two plates.

Transform plate boundary A plate boundary where two lithospheric plates slide past each other.

Tributary A stream that contributes water to a larger stream.

Tropical storm A large, rotating storm over the ocean; less powerful than a *hurricane.*

Troposphere: The layer of the *atmosphere,* closest to Earth's surface, where *weather* occurs.

Trunk stream A major artery of a stream network.

Tsunami A large water wave caused by undersea *earthquakes* or volcanic eruptions.

Unconfined aquifer An *aquifer* that is not held under pressure by an impermeable layer.

Unconformity A plane between two layers of *rock* that represents a time when no deposition occurred and/or *erosion* removed some of the lower surface before deposition resumed.

Uniformitarianism, principle of The assumption that the processes occurring on Earth today are the same as those that operated in the past.

Universal soil loss equation A mathematical tool for predicting rates of the *erosion* of *soil* under a constellation of varying conditions, including variations in soil character, *climate,* land cover, and topography.

Viscosity A measure of a liquid's resistance to flow.

Volcanic ash Fine particles of *rock* less than 2 mm in diameter ejected from an erupting *volcano.*

Volcanic arc A chain of volcanic islands formed at a *subduction* zone by melted material from the subducted plate that rises upward through the overriding plate.

Volcano The eruption of *lava* and/or *pyroclastic* materials; the mountainous landform that accumulates at the site of such eruptions.

Water hardness The concentration of calcium and magnesium *ions* in water.

Water spreading The practice of spreading *runoff* water over a large land area to infiltrate permeable regions where the *water table* is not close to the surface.

Water table The two-dimensional surface beneath Earth's surface that divides the saturated zone, in which pore spaces in the *rock* are filled with *groundwater,* from the overlying unsaturated zone.

Weather Short-term (e.g., daily) fluctuations in temperature, wind speed, and precipitation in a particular region. Compare *Climate.*

Wetlands Areas characterized by extremely moist *soils;* swamps and marshes.

Illustration Credits

Cover photo, Kunio Owaki/The Stock Market. **Repeated opener photo of dunes,** B. Schmid/Photonica.

CHAPTER 1 Page 2 Rick Wessels, Arizona State University. **Fig. 1-1** Peter Francis, The Open University, Milton Keynes, UK. **Fig. 1-3** David Rind/NASA Goddard Institute for Space Studies. **Fig. 1-5** Peter W. Sloss, NOAA-NESDIS-NGDC. **Fig. 1-6** From W. Alvarez and F. Asaro, "What Caused the Mass Extinction? An Extraterrestrial Impact," *Scientific American*, October 1990, pp. 78–84. **Fig. 1-8** From data in Joel E. Cohen, *How Many People Can the Earth Support?* New York: W. W. Norton, 1995. **Fig. 1-9** Viviane Moos/The Stock Market. **Fig. 1-10 a** Jim Foster/The Stock Market; **b** J. Patrick Phelan/The Stock Market; **c** Lowell Georgia/Science Source/Photo Researchers. **Fig. 1-11** From E. L. McFarland et al., *Energy, Physics, and the Environment*, Winnipeg: Wuerz Publishing, 1994, p. 28, Fig. 2.1. **Fig. 1-12** From M. King Hubbert, "Exponential Growth as a Transient Phenomenon in Human History," 1976, in *Societal Issues, Scientific Viewpoints*, Margaret A. Strom (ed.) New York: American Institute of Physics, 1987, pp. 75–84. **Fig. 1-13 b** Corbis-Bettmann. **Fig. 1-14** V. Leloup/Gamma-Liaison. **Fig. 1-15** Jim Richardson, Denver, Colorado. **Box 1-1 (page 6) a** Runk/Schoenberger/Grant Heilman. **Box 1-2 (page 13)** Jeff Foott/Bruce Coleman.

CHAPTER 2 Page 28 Alberto Garcia/SABA. **Fig. 2-1** Robert M. Carey/NOAA/Science Photo Library/Photo Researchers. **Fig. 2-4** Data from USGS. **Fig. 2-5** Anglo-Australian Observatory 1991. **Fig. 2-6** F. Press and R. Siever, *Understanding Earth*, 2d ed. © 1998 by W. H. Freeman and Company. **Fig. 2-10** David L. Brown/The Stock Market. **Fig. 2-12** Johnson Space Center/NASA. **Fig. 2-15 a** and **b** Johnson Space Center/NASA. **Fig. 2-18 b** Peter W. Sloss, NOAA-NESDIS-NGDC. **Fig. 2-20** Pete Kopischke/Outback-Images, Steamboat Springs, Colorado. **Fig. 2-21** F. Press and R. Siever, *Understanding Earth*, 2d ed. © 1998 by W. H. Freeman and Company; data from E. K. Berner and R. A. Berner, *Global Environment*, Upper Saddle River, N.J.: Prentice Hall, 1996. **Fig. 2-22** NASA. **Fig. 2-23** Data from *World Water Balance and Water Resources of the Earth*, UNESCO, 1978. **Box 2-2 (page 40)** Chip Clark.

CHAPTER 3 Page 62 Jeff Marshall, Department of Geosciences, The Pennsylvania State University. **Fig. 3-1** U.S. Department of Energy, Yucca Mountain Site Characterization Project, North Las Vegas, Nevada. **Fig. 3-2** Modified from *Earth System Science: A Closer View*, NASA, 1988. **Fig. 3-3** Alessandro Montanari, Osservatorio Geologico di Coldigioco, Frontale di Apiro, Italia. **Fig. 3-6** AIP Emilio Segre Visual Archives, American Institute of Physics. **Fig. 3-9** M. Abbey/Photo Researchers. **Box 3-1 (page 67)** Arthur M. Greene/Bruce Coleman; **(page 68) left** Gary Rosenquist, Tacoma, Washington and Raincity, Shoreline, Washington; **right** After R. A. Bailey, P. R. Beauchemin, F. P. Kapinos, and D. W. Klick, USGS. **Box 3-2 (page 75)** After W. B. N. Berry, *Growth of a Prehistoric Time Scale* © 1968 by W. H. Freeman and Company. **Box 3-3 (page 78)** USGS; **(page 79) a** and **b** Landauer, Inc., Glenwood, Illinois. **Box 3-4 (page 83)** U.S. Department of Energy.

CHAPTER 4 Page 90 NASA/TSADO/Tom Stack & Associates. **Fig. 4-2** Art Wolfe, Seattle, Washington. **Fig. 4-4 a** Roberto de Gugliemo/Science Photo Library/Photo Researchers; **b** Geoff Tompkinson/Science Photo Library/Photo Researchers. **Fig. 4-5** Doug Sokell/Tom Stack & Associates. **Fig. 4-6 b** E. R. Degginger/Bruce Coleman. **Fig. 4-7** Paul Silverman/Fundamental Photographs. **Fig. 4-8** Chip Clark. **Fig. 4-9 a** Paul Silverman/Fundamental Photographs. **Fig. 4-10 a** Philip Hayson/Photo Researchers; **b** Alfred Pasieka/Science Photo Library/Photo Researchers. **Fig. 4-15 a** Dr. Eckart Pott/Bruce Coleman. **Fig. 4-16** Emil Muench/Photo Researchers. **Fig. 4-17** David Howell, USGS, Menlo Park, California. **Fig. 4-18 a** D. Merritts. **Fig. 4-19 a** R. E. Wallace/USGS. **Fig. 4-21** Basalt, Chip Clark; andesite, Joyce Photo/Photo Researchers; rhyolite and gabbro, Chip Clark; diorite and granite, Ward's Natural Science Establishment, Rochester, New York. **Fig. 4-22 a** Erich Schrempp/Photo Researchers; **b** A. de Wet; **c** Ben S. Kwiatkowski/Fundamentals Photographs. **Fig. 4-24 a** Thomas Kitchin/Tom Stack & Associates; **b** and **c** Jeff Marshall, Department of Geosciences, The Pennsylvania State University. **Fig. 4-25** Alan Kearney/Ellis Nature Photography. **Fig. 4-26 a** Brownie Harris/The Stock Market; **b** Scripps Institution of Oceangraphy, University of California, San Diego. **Box 4-1 (page 99) a, left** Don Thomson/Science Photo Library/Photo Researchers; **a, right** Dr. Jeremy Burgess/Science Photo Library/Photo Researchers; **b** David M. Doody/Tom Stack & Associates. **Box 4-3 (page 114) a** World Ocean Floor based on bathymetric studies by Bruce C. Heezen and Marie

Tharp. Painting by Heinrich C. Berann. Copyright © Marie Tharp, 1977.

CHAPTER 5 **Page 122** Freeport-McMoRan Copper and Gold, New Orleans. **Fig. 5-1 a** E. R. Degginger/ Photo Researchers; **b** Richard Hutchings/Photo Researchers. **Fig. 5-2** Data from U.S. Bureau of Mines, 1992. **Fig. 5-3** Freeport-McMoRan Copper and Gold, New Orleans. **Fig. 5-5** Dudley Foster/ Woods Hole Oceanographic Institution. **Fig. 5-6** Spence Titley. **Fig. 5-7 b** Wolfgang Bayer/Bruce Coleman. **Fig. 5-8** Spence Titley. **Fig. 5-11** A. Keith/USGS. **Fig. 5-12** The Natural History Museum, London. **Fig. 5-14 a** François Gohier/Photo Researchers; **b** Krafft/I & V/HOA-QUI/ Gamma-Liaison. **Fig. 5-15 a** Guido Cozzi/Bruce Coleman. **Fig. 5-16 b** Masao Hayashi/Dunq/Photo Researchers. **Fig. 5-17 a** T. L. Wright/USGS. **Fig. 5-18** After R. A. Bailey, P. R. Beauchemin, F. P. Kapinos, and D. W. Klick, USGS. **Fig. 5-19** T. L. Wright/USGS. **Fig. 5-20** Krafft/Photo Researchers. **Fig. 5-21** After T. Wright, T. Pierson, Living with Volcanoes, USGS Circular 1073, 1992. **Fig. 5-22** Herman Kokojan/Black Star. **Fig. 5-23** Bruce A. Bolt. **Fig. 5-25** Chuck O'Rear/Westlight. **Fig. 5-28** Mark Downey/Gamma-Liaison. **Fig. 5-29** Modified from USGS Circular 1079, 1992. **Box 5-1 (page 132)** Tony DiFronzo, Butte, Montana. **Box 5-2 (page 136) left** and **right** ASARCO Inc., New York. **Box 5-3 (page 143) right** D. J. Roddy/ USGS. **Box 5-4 (page 153)** R. Hyndman, "Giant Earthquakes of the Pacific Northwest," *Scientific American*, December 1995, pp. 68–75; **(page 154) left** Brian F. Atwater/USGS; **right** R. C. Buckman/USGS.

CHAPTER 6 **Page 158** Dr. Nigel Smith/Animals/ Animals, Earth Scenes. **Fig. 6-1 a** Library of Congress; **b** Corbis-Bettmann. **Fig. 6-4** Peter Kresan, Tucson, Arizona. **Fig. 6-5 a** D. Merritts; **b** Paul Bierman, Geology Department, University of Vermont. **Fig. 6-6** Arthur M. Greene/Bruce Coleman. **Fig. 6-7 left** Betty Derig/Photo Researchers. **Fig. 6-9** USGS/NMD EROS Data Center. **Fig. 6-10** Lowell Georgia/Photo Researchers. **Fig. 6-14** All photos, Loyal A. Quandt, National Soil Survey Center, NRCS, USDA. **Fig. 6-15** Lancaster Country Soil Conservation Service. **Fig. 6-16** Errol D. Sehnke/Bruce Coleman. **Fig. 6-17** Modified from C. Runnels, "Environmental Degradation in Ancient Greece," *Scientific American*, March 1995. **Fig. 6-18** Heidi Fassnacht, Oregon State University, Corvallis, Oregon. **Fig. 6-19** Culver Pictures. **Fig. 6-20 a** Larry Mayer/Gamma-Liaison; **b** Francois Gohier/ Photo Researchers. **Fig. 6-21** AP/Wide World Photos. **Fig. 6-23 (page 186)**, debris slide, Director of Information Services, Hong Kong; debris slump, John S. Shelton, La Jolla, California; rockslide, Joy Spurr/Bruce Coleman. **Fig. 6-23 (page 187)** debris flow, Michael Collier, Flagstaff, Arizona; creep, Robert S. Semeniok/The Stock Market. **Fig. 6-24 a** David R. Montgomery, University of Washington, Seattle. **Box 6-1 (page 166)** Eric Fielding, Jet Propulsion Laboratory, California Institute of Technology, Pasadena. **Box 6-2 (page 170)** J. Longstaffe, University Western Ontario. **Box 6-2 (page 171)** James Wilson/Woodfin Camp & Associates.

CHAPTER 7 **Page 192** Michael Collier, Flagstaff, Arizona. **Fig. 7-1 a** Johnson Space Center/NASA; **b** David Turnley, Detroit Free Press/Black Star. **Fig. 7-2** AP/Wide World Photos/Eric Gay. **Fig. 7-4** U.S. Department of Agriculture. **Fig. 7-5** Comstock. **Fig. 7-6** G. P. Thelin and R. J. Pike, USGS, 1991. **Fig. 7-7** D. Merritts. **Fig. 7-9** John Hyde/Bruce Coleman. **Fig. 7-10 a** and **b** EOSAT, Earth Observation Satellite Company, Lanham, Maryland. **Fig. 7-11** D. Merritts. **Fig. 7-14 b** Mike Andrews/Earth Scenes. **Fig. 7-16** Adapted from H. N. Fisk, *Geological Investigation of the Alluvial Valley of the Lower Mississippi River*, Mississippi River Commission, 1944. **Fig. 7-17** Compiled from sources including C. R. Kolb and J. R. Van Lopik, in *Deltas and Their Geological Framework*, M. L. Shirley (ed.), Houston Geological Society, 1966. **Fig. 7-18** AP/Wide World Photos/James A. Finley. **Fig. 7-19 a** and **b** Kansas State University Photographic Services. **Fig. 7-20** Modified from USGS Circular 1120-E, *Effects of Reservoirs on Flood Discharges in the Kansas and the Missouri River Basins*. **Fig. 7-21** Federal Emergency Management Agency, National Flood Insurance Program. **Fig. 7-23** South Florida Water Management District, West Palm Beach. **Fig. 7-24** Jeff Marshall, Department of Geosciences, The Pennsylvania State University. **Fig. 7-25** Adapted From R. W. Tiner, Jr., U.S. Fish and Wildlife Service. **Fig. 7-26** Modified from USGS Water Supply Paper 2425. **Fig. 7-27** Kul Bhatia/Photo Researchers. **Fig. 7-28 a** and **b** Nashua River Watershed Association, Nashua, Massachusetts. **Box 7-1 (page 202) top,** Peter Kresan, Tucson, Arizona; **bottom,** From C. W. Stockton and G. J. Jacoby, Lake Powell Research Project Bulletin Number 18, "Long-Term Surface Water Supply and Streamflow Trends in the Upper Colorado River Basin," University of California, Los Angeles, 1976. **Box 7-2 (page 213)** *Floods and People: A Geological Perspective*, Moss et al., Reproduced with permission of the publisher, the Geological Society of America, Boulder, Colorado USA. Copyright © 1978 Geological Society of America. **Box 7-3 (page 221) left** Modified from T. Dunne and L. Leopold, *Water in Environmental Planning* © 1979 by W. H. Freeman and Company; **right** Hubbard Brook Experimental Forest, West Thornton, New Hampshire.

CHAPTER 8 **Page 232** Richard Packwood, Oxford Scientific Films/Earth Scenes. **Fig. 8-1** Doris LeBlanc/ USGS. **Fig. 8-3** From Rasmussen, W. C. and G. E. Andreasen, Hydrologic budget of the Beaverdam Creek Basin, Maryland, U.S. Geological Survey Water-Supply Paper 1472106, 1959. **Fig. 8-4 a** Science Photo Library/Photo Researchers; **b** From Robert A. Robertson, *J. of Sedimentary Pretrology*, Vol. 37, pp. 355–64, 1967, in Robert R. Compton, *Interpreting the Earth,* Harcourt Brace Jovanovich; **c** Michael Collier, Flagstaff, Arizona. **Fig. 8-6** Paul Bierman, Geology Department, University of Vermont. **Fig. 8-9** Modified from F. Press and R. Siever, *Understanding Earth*, 2d ed., © 1998 by W. H. Freeman and Company. **Fig. 8-11** Laura Zito/ Photo Researchers. **Fig. 8-12** Derek Bayes, Aspect Picture Library/The Stock Market. **Fig. 8-13** Gary D. Kraus, National Research Council, Washington, D.C., photo by Robert Forvolden, University of Waterloo. **Fig. 8-17** Kevin Svitana, Columbus, Ohio. **Fig. 8-18 a** D. Merritts. **Fig. 8-22** Dan Oleski, Franklin and Marshall College, Lancaster, Pennsylvania. **Box 8-2 (page 247)** Charles O'Rear/Westlight. **Box 8-3 (page 255)** Modified from *National Water Summary*, USGS, 1986.

CHAPTER 9 **Page 262** Kunsthistorisches Museum, Vienna, Erich Lessing/Art Resource. **Fig. 9-1** Steve Berman/Gamma-Liaison. **Table 9-2** Adapted from Elson, Derek M., *Atmospheric Pollution: A Global Problem* 2d ed., © 1992 by Blackwell. **Fig. 9-2** Johnson Space Center/ NASA. **Fig. 9-4** Modified from T. E. Graedel and P. J. Crutzen, *Atmosphere, Climate, and Change,* © 1995, 1997 by Lucent Technologies. Used with permission of W. H. Freeman and Company. **Fig. 9-8** From J. Brown et al., *Ocean Circulation*, New York: Pergamon Press, 1991. **Fig. 9-16** Data from U.S. National Weather Service. **Fig. 9-17** Modified from *Meterology,* 4th ed., by Moran/Morgan, © 1994. Reprinted by permission of Prentice-Hall, Inc., Upper Saddle River, N.J. **Fig. 9-20 right** Keith Kent/Science Photo Library/Photo Researchers. **Fig. 9-21** Goddard Laboratory for Atmospheres/NASA. **Fig. 9-22** Wesley Bocxe/Photo Researchers. **Fig. 9-23** Tom McHugh/Photo Researchers. **Fig. 9-24** Martin Bond/Science Photo Library/Photo Researchers. **Fig. 9-25 a** and **b** Jean-Paul Davis, Pasadena. **Fig. 9-27** All photos, NASA/GSFC, Ozone Processing Team, Goddard Space Flight Center, data from United Nations Environment Program. **Box 9-1 (page 266)** R. P. Wayne, *Chemistry of Atmospheres*, 2d ed., Oxford, U.K.: Clarendon Press, 1991. By permission of Oxford University Press. **Box 9-3 (page 278) left** NOAA Central Library; **right** John Freeman/Gamma-Liaison.

CHAPTER 10 **Page 290** Robb Kendrick/Aurora, Bridgton, Maine. **Fig. 10-1** Greenpeace/Morgan, Greenpeace, Washington, D.C. **Fig. 10-3** Image created by David A. Hastings. Image generated solely from digital topographic data available from the National Geophysical Data Center, NOAA, Boulder, Colorado. **Fig. 10-4** Lincoln F. Pratson, University of Colorado and William F. Haxby, Columbia University "Panoramas of the Seafloor," *Scientific American,* June 1997, pp. 82–83. **Fig. 10-6** and **Fig. 10-7** Loren McIntyre, Arlington, Virginia. **Fig. 10-8** Landsat/NASA. **Fig. 10-10** D. P. Wilson/Science Source/Photo Researchers. **Fig. 10-11** Dr. Gene Feldman/NASA-GSFC/Science Photo Library/ Photo Researchers. **Fig. 10-20** Bruce Brander/Photo Researchers. **Fig. 10-21** Netherlands Ministry of Transport and Public Works. **Fig. 10-23** Robert Perron, Branford, Connecticut. **Fig. 10-24** Jack Dermid/ Photo Researchers. **Fig. 10-25** Lowell Georgia/Photo Researchers. **Fig. 10-26** The Center for Coastal, Energy, and Environmental Resources, Louisiana State University. **Fig. 10-27** Modified from R. A. Davis, Jr., *The Evolving Coast*, © 1994 by Scientific American Library. Used with permission of W. H. Freeman and Company. **Fig. 10-28** Frans Lanting/Photo Researchers. **Fig. 10-29** Adapted from U.S. Office of Technology Assessment, *Wastes in Marine Environments*, OTA-0-334. **Box 10-1 (page 296)** Al Duester, Woods Hole Oceanographic Insitution; **(page 297)** L. Praxton and W. Haxby, "Panoramas of the Seafloor," *Scientific American* June 1997, p. 86. **Box 10-2 (page 311) a** and **b** Photri, Alexandria, Virginia. **Box 10-3 (page 319)** Susan M. Glascock, Seaford, Virginia.

CHAPTER 11 **Page 326** Ned Haines/Photo Researchers. **Fig. 11-1 a** Data from U.S. Energy Information Agency; **b** D. Cahoon, NASA Langley Research Center; **c** A. de Wet. **Fig. 11-3** Frank Spooner/ Gamma-Liaison. **Fig. 11-5** Adapted from C. D. Masters, D. H. Root, and E. D. Attanasi, USGS, and U.S. Department of Energy. **Fig. 11-6** Ken Graham/Tony Stone Images. **Fig. 11-7** Vanessa Vick/Photo Researchers. **Fig. 11-8** Bob Gomel/The Stock Market. **Fig. 11-9 b** W. H. Freeman photo by Ken Karp. **Fig. 11-10** S. J. Gould, *The Book of Life*, Ebury Press. **Fig. 11-11** John S. Shelton, La Jolla, California. **Fig. 11-12 a** Steven Hilty/ Bruce Coleman. **Fig. 11-13** Peter Davey/Bruce Coleman. **Fig. 11-16** From Atmospheric Release Advisory Capability. **Fig. 11-17** Tom Martin/The Stock Market. **Fig. 11-18** *McGraw Hill Encyclopedia of Energy,* 2nd ed.,

© 1981 by McGraw Hill, Inc., New York. **Fig. 11-20** Robert Millman, Rocky Mountain Institute. **Fig. 11-21** Data from U.S. Department of Energy. **Fig. 11-22** David L. Brown/The Stock Market. **Fig. 11-23** Forrest Anderson/Gamma-Liaison. **Fig. 11-24** Steve McCurry/Magnum. **Fig. 11-26** A. de Wet, data from NASA. **Box 11-1 (page 335)** Modified from F. Press and R. Siever, *Understanding Earth*, 2d ed., © 1998 by W. H. Freeman and Company, photo, Halliburton Geophysical Services; **(page 336) top** Johnson Space Center/NASA; **bottom,** Stratamodel, Inc. from *Geotimes*, July 1993. **Box 11-2 (page 343) a** and **b** D. Merritts. **Box 11-3 (page 353)** Jacques Halber/Gamma-Liaison.

CHAPTER 12 **Page 362** Manufactured by Holiday Film Corp, photo by Cheryl Haselhorst, *The Columbian.* **Fig. 12-1** Scott Stine, Ph.D., Berkeley, California. **Fig. 12-2** A.S.P./Science Source/Photo Researchers. **Fig. 12-7 a** A. Berger, "Milankovich Theory and Climate," *Reviews of Geophysics* 26, 624–657, 1988; **b** Modified from T. J. Crowley and G. R. North, *Paleoclimatology*, New York: Oxford University Press, 1991, redrawn from W. Koeppen and A. Wegener, *Die Klimate der Geologischen Vorzeit*, Berlin: Gebrueder Borntraeger, 1924. **Fig. 12-8 b** Paul Steel/The Stock Market. **Fig. 12-11** Carr Clifton, Taylorsville, California. **Fig. 12-12** John and Ann Mahan, Gaylord, Michigan. **Fig. 12-13** C. U. Hammer et al., "Greenland Ice Sheet Evidence of Post-Glacial Volcanism and Its Volcanic Impact," *Nature* 288 (1980), pp. 230–255. **Fig. 12-14 a** and **b** K. Menking. **Fig. 12-15** T. Webb et al., "Vegetation, Lake Levels, and Climate in Eastern North America for the Past 18,000 Years," *Global Climates Since the Last Glacial Maximum*, H. E. Wright, Jr. et al. (eds.), Minneapolis: University of Minnesota Press, 1993. **Fig. 12-16** Adapted from R. S. Bradley, *Quaternary Paleoclimatology: Methods of Paleoclimatic Reconstruction*, New York: Chapman and Hall, 1984. **Fig. 12-17** Alfred Pasieka/Science Photo Library/Photo Researchers. **Fig. 12-18** Data from J. Imbrie et al., "The Orbital Theory of Pleistocene Climate," *Milankovitch and Climate, Part I*, Hingham, Mass.: D. Riedel, 1984, pp. 269–305. **Box 12-1 (page 366)** Adapted from J. F. Kasting et al., "How Climate Evolved on the Terrestrial Planets," *Scientific American* 258(1988), pp. 90–97. **Box 12-2 (page 376) a** R. J. Delmas, Laboratorie de Glaciologie et Gophysique de L'environment, Centre National de la Recherchf Scienifique; **b** Greenland Ice Sheet Project 2, Climate Change Research Center, Institute for the Study of Earth, Oceans and Space, University of New Hampshire, photo by Michael Morrison.

CHAPTER 13 **Page 384** NASA. **Page 385** From Edward Connery Lathem, *The Poetry of Robert Frost*, "Fire and Ice," Henry Holt and Company Publishers, New York, 1979. **Fig. 13-1** Modified from K. J. Hsu, *The Mediterranean Was a Desert,* © 1983 by Princeton University Press. Reprinted by permission of Princeton University Press. **Fig. 13-3 a** and **b** P. F. Hoffman. **Fig. 13-4** Chase Studio, Inc. **Fig. 13-5** The Field Museum, Chicago, Neg # GE0 85637C. **Fig. 13-6** From data in C. R. Scotese and J. Golonka, *Paleogeographic Atlas*, PALEOMAP Project, 1992; and in W. Sullivan, *Continents in Motion: The New Earth Debate*, New York: McGraw-Hill, 1974. **Fig. 13-7** Jerome Wyckoff, Ringwood, New Jersey. **Fig. 13-8** Clara Calhoun/Bruce Coleman. **Fig. 13-9 a** Prof. W. Alvarez/Science Photo Library/Photo Researchers; **b** Glen A. Izett, USGS. **Fig. 13-11 b** M. I. Walker/Science Source/Photo Researchers. **Fig. 13-12** Modified from W. S. Broecker an G. H. Denton, "What Drives Glacial Cycles?" *Scientific American*, January 1990, pp. 43–50. **Fig. 13-14 a** and **b** From E. L. Ladurie, *Times of Feast, Times of Famine: A History of Climate Since the Year 1000*, 1971, Doubleday, New York, in J. Imbrie and K. P. Imbrie, *Ice Ages*, 1979, Harvard Univeristy Press and Enslow Publishers, Springfield, New Jersey. **Fig. 13-15** Data after May 1974 from National Oceanic and Atmospheric Administration, Climate Monitoring and Diagnostics Laboratory, Carbon Cycle Group; data before May 1974 from Charles D. Keeling, Scripps Institution of Oceanography. **Fig. 13-16** Modified from J. T. Houghton et al. (eds.), *Climate Change: The IPCC Scientific Assessment,* Cambridge University Press, 1990. **Fig. 13-17** A. Revkin, *Global Warming: Understanding the Forecast*, Abbeville Press 1992, p. 123. Courtesy, Abbeville Press. **Fig. 13-18** Adapted from C. Iorius et al., "The Ice-Core Record: Climate Sensitivity and Future Greenhouse Warming," *Nature,* September 13, 1990, 347:141. **Fig. 13-19** Data from World Resources Institute, *World Resources 1994–95*, New York: Oxford University Press, 1994. **Fig. 13-20** Alain Evrard/Photo Researchers. **Fig. 13-21** C. F. Boutron et al., "Decrease in Anthropogenic Lead, Cadmium and Zinc in Greenland Snows Since the Late 1960s," *Nature* 353: 153–156. **Fig. 13-22 a** U.S. Fish and Wildlife Service/Samuel Woo; **b** U.S. Fish and Wildlife Service/G. R. Zahm. **Fig. 13-23** Johnson Space Center/NASA. **Box 13-2 (page 408)** Victor Englebert, Cali, Columbia, S.A.

Index

Page numbers in **boldface** refer to definitions; page numbers in *italics* refer to illustrations and tables. The Appendixes and Glossary are not covered by this index.